LUNG MACROPHAGES AND DENDRITIC CELLS IN HEALTH AND DISEASE

LUNG BIOLOGY IN HEALTH AND DISEASE

Executive Editor

Claude Lenfant
Director, National Heart, Lung and Blood Institute
National Institutes of Health
Bethesda, Maryland

1. Immunologic and Infectious Reactions in the Lung, *edited by Charles H. Kirkpatrick and Herbert Y. Reynolds*
2. The Biochemical Basis of Pulmonary Function, *edited by Ronald G. Crystal*
3. Bioengineering Aspects of the Lung, *edited by John B. West*
4. Metabolic Functions of the Lung, *edited by Y. S. Bakhle and John R. Vane*
5. Respiratory Defense Mechanisms (in two parts), *edited by Joseph D. Brain, Donald F. Proctor, and Lynne M. Reid*
6. Development of the Lung, *edited by W. Alan Hodson*
7. Lung Water and Solute Exchange, *edited by Norman C. Staub*
8. Extrapulmonary Manifestations of Respiratory Disease, *edited by Eugene Debs Robin*
9. Chronic Obstructive Pulmonary Disease, *edited by Thomas L. Petty*
10. Pathogenesis and Therapy of Lung Cancer, *edited by Curtis C. Harris*
11. Genetic Determinants of Pulmonary Disease, *edited by Stephen D. Litwin*
12. The Lung in the Transition Between Health and Disease, *edited by Peter T. Macklem and Solbert Permutt*
13. Evolution of Respiratory Processes: A Comparative Approach, *edited by Stephen C. Wood and Claude Lenfant*
14. Pulmonary Vascular Diseases, *edited by Kenneth M. Moser*
15. Physiology and Pharmacology of the Airways, *edited by Jay A. Nadel*
16. Diagnostic Techniques in Pulmonary Disease (in two parts), *edited by Marvin A. Sackner*
17. Regulation of Breathing (in two parts), *edited by Thomas F. Hornbein*
18. Occupational Lung Diseases: Research Approaches and Methods, *edited by Hans Weill and Margaret Turner-Warwick*
19. Immunopharmacology of the Lung, *edited by Harold H. Newball*
20. Sarcoidosis and Other Granulomatous Diseases of the Lung, *edited by Barry L. Fanburg*

21. Sleep and Breathing, *edited by Nicholas A. Saunders and Colin E. Sullivan*
22. *Pneumocystis carinii* Pneumonia: Pathogenesis, Diagnosis, and Treatment, *edited by Lowell S. Young*
23. Pulmonary Nuclear Medicine: Techniques in Diagnosis of Lung Disease, *edited by Harold L. Atkins*
24. Acute Respiratory Failure, *edited by Warren M. Zapol and Konrad J. Falke*
25. Gas Mixing and Distribution in the Lung, *edited by Ludwig A. Engel and Manuel Paiva*
26. High-Frequency Ventilation in Intensive Care and During Surgery, *edited by Graziano Carlon and William S. Howland*
27. Pulmonary Development: Transition from Intrauterine to Extrauterine Life, *edited by George H. Nelson*
28. Chronic Obstructive Pulmonary Disease: Second Edition, Revised and Expanded, *edited by Thomas L. Petty*
29. The Thorax (in two parts), *edited by Charis Roussos and Peter T. Macklem*
30. The Pleura in Health and Disease, *edited by Jacques Chrétien, Jean Bignon, and Albert Hirsch*
31. Drug Therapy for Asthma: Research and Clinical Practice, *edited by John W. Jenne and Shirley Murphy*
32. Pulmonary Endothelium in Health and Disease, *edited by Una S. Ryan*
33. The Airways: Neural Control in Health and Disease, *edited by Michael A. Kaliner and Peter J. Barnes*
34. Pathophysiology and Treatment of Inhalation Injuries, *edited by Jacob Loke*
35. Respiratory Function of the Upper Airway, *edited by Oommen P. Mathew and Giuseppe Sant'Ambrogio*
36. Chronic Obstructive Pulmonary Disease: A Behavioral Perspective, *edited by A. John McSweeny and Igor Grant*
37. Biology of Lung Cancer: Diagnosis and Treatment, *edited by Steven T. Rosen, James L. Mulshine, Frank Cuttitta, and Paul G. Abrams*
38. Pulmonary Vascular Physiology and Pathophysiology, *edited by E. Kenneth Weir and John T. Reeves*
39. Comparative Pulmonary Physiology: Current Concepts, *edited by Stephen C. Wood*
40. Respiratory Physiology: An Analytical Approach, *edited by H. K. Chang and Manuel Paiva*
41. Lung Cell Biology, *edited by Donald Massaro*
42. Heart–Lung Interactions in Health and Disease, *edited by Steven M. Scharf and Sharon S. Cassidy*
43. Clinical Epidemiology of Chronic Obstructive Pulmonary Disease, *edited by Michael J. Hensley and Nicholas A. Saunders*
44. Surgical Pathology of Lung Neoplasms, *edited by Alberto M. Marchevsky*

45. The Lung in Rheumatic Diseases, *edited by Grant W. Cannon and Guy A. Zimmerman*
46. Diagnostic Imaging of the Lung, *edited by Charles E. Putman*
47. Models of Lung Disease: Microscopy and Structural Methods, *edited by Joan Gil*
48. Electron Microscopy of the Lung, *edited by Dean E. Schraufnagel*
49. Asthma: Its Pathology and Treatment, *edited by Michael A. Kaliner, Peter J. Barnes, and Carl G. A. Persson*
50. Acute Respiratory Failure: Second Edition, *edited by Warren M. Zapol and Francois Lemaire*
51. Lung Disease in the Tropics, *edited by Om P. Sharma*
52. Exercise: Pulmonary Physiology and Pathophysiology, *edited by Brian J. Whipp and Karlman Wasserman*
53. Developmental Neurobiology of Breathing, *edited by Gabriel G. Haddad and Jay P. Farber*
54. Mediators of Pulmonary Inflammation, *edited by Michael A. Bray and Wayne H. Anderson*
55. The Airway Epithelium, *edited by Stephen G. Farmer and Douglas Hay*
56. Physiological Adaptations in Vertebrates: Respiration, Circulation, and Metabolism, *edited by Stephen C. Wood, Roy E. Weber, Alan R. Hargens, and Ronald W. Millard*
57. The Bronchial Circulation, *edited by John Butler*
58. Lung Cancer Differentiation: Implications for Diagnosis and Treatment, *edited by Samuel D. Bernal and Paul J. Hesketh*
59. Pulmonary Complications of Systemic Disease, *edited by John F. Murray*
60. Lung Vascular Injury: Molecular and Cellular Response, *edited by Ar nold Johnson and Thomas J. Ferro*
61. Cytokines of the Lung, *edited by Jason Kelley*
62. The Mast Cell in Health and Disease, *edited by Michael A. Kaliner and Dean D. Metcalfe*
63. Pulmonary Disease in the Elderly Patient, *edited by Donald A. Mahler*
64. Cystic Fibrosis, *edited by Pamela B. Davis*
65. Signal Transduction in Lung Cells, *edited by Jerome S. Brody, David M. Center, and Vsevolod A. Tkachuk*
66. Tuberculosis: A Comprehensive International Approach, *edited by Lee B. Reichman and Earl S. Hershfield*
67. Pharmacology of the Respiratory Tract: Experimental and Clinical Research, *edited by K. Fan Chung and Peter J. Barnes*
68. Prevention of Respiratory Diseases, *edited by Albert Hirsch, Marcel Goldberg, Jean-Pierre Martin, and Roland Masse*
69. *Pneumocystis carinii* Pneumonia: Second Edition, Revised and Expanded, *edited by Peter D. Walzer*
70. Fluid and Solute Transport in the Airspaces of the Lungs, *edited by Richard M. Effros and H. K. Chang*

71. Sleep and Breathing: Second Edition, Revised and Expanded, *edited by Nicholas A. Saunders and Colin E. Sullivan*
72. Airway Secretion: Physiological Bases for the Control of Mucous Hypersecretion, *edited by Tamotsu Takishima and Sanae Shimura*
73. Sarcoidosis and Other Granulomatous Disorders, *edited by D. Geraint James*
74. Epidemiology of Lung Cancer, *edited by Jonathan M. Samet*
75. Pulmonary Embolism, *edited by Mario Morpurgo*
76. Sports and Exercise Medicine, *edited by Stephen C. Wood and Robert C. Roach*
77. Endotoxin and the Lungs, *edited by Kenneth L. Brigham*
78. The Mesothelial Cell and Mesothelioma, *edited by Marie-Claude Jaurand and Jean Bignon*
79. Regulation of Breathing: Second Edition, Revised and Expanded, *edited by Jerome A. Dempsey and Allan I. Pack*
80. Pulmonary Fibrosis, *edited by Sem Hin Phan and Roger S. Thrall*
81. Long-Term Oxygen Therapy: Scientific Basis and Clinical Application, *edited by Walter J. O'Donohue, Jr.*
82. Ventral Brainstem Mechanisms and Control of Respiration and Blood Pressure, *edited by C. Ovid Trouth, Richard M. Millis, Heidrun F. Kiwull-Schöne, and Marianne E. Schläfke*
83. A History of Breathing Physiology, *edited by Donald F. Proctor*
84. Surfactant Therapy for Lung Disease, *edited by Bengt Robertson and H. William Taeusch*
85. The Thorax: Second Edition, Revised and Expanded (in three parts), *edited by Charis Roussos*
86. Severe Asthma: Pathogenesis and Clinical Management, *edited by Stanley J. Szefler and Donald Y. M. Leung*
87. *Mycobacterium avium*–Complex Infection: Progress in Research and Treatment, *edited by Joyce A. Korvick and Constance A. Benson*
88. Alpha 1–Antitrypsin Deficiency: Biology • Pathogenesis • Clinical Manifestations • Therapy, *edited by Ronald G. Crystal*
89. Adhesion Molecules and the Lung, *edited by Peter A. Ward and Joseph C. Fantone*
90. Respiratory Sensation, *edited by Lewis Adams and Abraham Guz*
91. Pulmonary Rehabilitation, *edited by Alfred P. Fishman*
92. Acute Respiratory Failure in Chronic Obstructive Pulmonary Disease, *edited by Jean-Philippe Derenne, William A. Whitelaw, and Thomas Similowski*
93. Environmental Impact on the Airways: From Injury to Repair, *edited by Jacques Chrétien and Daniel Dusser*
94. Inhalation Aerosols: Physical and Biological Basis for Therapy, *edited by Anthony J. Hickey*
95. Tissue Oxygen Deprivation: From Molecular to Integrated Function, *edited by Gabriel G. Haddad and George Lister*
96. The Genetics of Asthma, *edited by Stephen B. Liggett and Deborah A. Meyers*

97. Inhaled Glucocorticoids in Asthma: Mechanisms and Clinical Actions, *edited by Robert P. Schleimer, William W. Busse, and Paul M. O'Byrne*
98. Nitric Oxide and the Lung, *edited by Warren M. Zapol and Kenneth D. Bloch*
99. Primary Pulmonary Hypertension, *edited by Lewis J. Rubin and Stuart Rich*
100. Lung Growth and Development, *edited by John A. McDonald*
101. Parasitic Lung Diseases, *edited by Adel A. F. Mahmoud*
102. Lung Macrophages and Dendritic Cells in Health and Disease, *edited by Mary F. Lipscomb and Stephen W. Russell*
103. Pulmonary and Cardiac Imaging, *edited by Caroline Chiles and Charles E. Putman*

ADDITIONAL VOLUMES IN PREPARATION

Inhalation Delivery of Therapeutic Peptides and Proteins, *edited by Lex A. Adjei and Pramod K. Gupta*

Treatment of the Hospitalized Cystic Fibrosis Patient, *edited by David M. Orenstein and Robert C. Stern*

Dyspnea, *edited by Donald A. Mahler*

$Beta_2$-Agonists in Asthma Treatment, *edited by Romain Pauwels and Paul M. O'Byrne*

Gene Therapy for Diseases of the Lung, *edited by Kenneth L. Brigham*

Oxygen, Gene Expression, and Cellular Function, *edited by Donald J. Massaro and Linda Clerch*

Asthma and Immunological Diseases in Pregnancy and Early Infancy, *edited by Michael Schatz, Robert S. Zeiger, and Henry Claman*

Self-Management of Asthma, *edited by Harry Kotses and Andrew Harver*

Asthma in the Elderly, *edited by Robert A. Barbee and John W. Bloom*

LUNG MACROPHAGES AND DENDRITIC CELLS IN HEALTH AND DISEASE

Edited by

Mary F. Lipscomb
University of New Mexico
School of Medicine
Albuquerque, New Mexico

Stephen W. Russell
University of Kansas
Medical Center
Kansas City, Kansas

MARCEL DEKKER, INC. NEW YORK • BASEL • HONG KONG

Library of Congress Cataloging-in-Publication Data

Lung macrophages and dendritic cells in health and disease / edited by
Mary F. Lipscomb, Stephen W. Russell.
p. cm. — (Lung biology in health and disease ; v. 102)
Includes bibliographical references and indexes.
ISBN 0-8247-9817-1 (hardcover : alk. paper)
1. Lungs—Diseases—Immunological aspects. 2. Lungs—Immunology.
3. Macrophages. 4. Dendritic cells. I. Lipscomb, Mary F.
III. Russell, S. W. (Stephen W.) III. Series.
[DNLM: 1. Lung Diseases—immunology. 2. Macrophages, Alveolar—immunology. 3. Dendritic cells—immunology. W1 LU62 v.102 1997 /
WF 600 L96345 1997]
RC776.I45L86 1997
616.2'4079—dc21
DNLM/DLC
for Library of Congress

96-53510
CIP

The publisher offers discounts on this book when ordered in bulk quantities. For more information, write to Special Sales/Professional Marketing at the address below.

This book is printed on acid-free paper.

MARCEL DEKKER, INC.
270 Madison Avenue, New York, New York 10016

Current printing (last digit):
10 9 8 7 6 5 4 3 2 1

PRINTED IN THE UNITED STATES OF AMERICA

INTRODUCTION

Although it took centuries for immunity—natural or acquired—to receive its name, there is clear evidence that physicians have known since the oldest time that some infections stimulate lifelong immunity to their recurrence.

The modern concepts of immunity rest on the shoulders of such giants as Louis Pasteur, Robert Koch, Paul Ehrlich, and Elie Metchnikoff. Yet, it would appear that the recognition of biological incompatibilities had its origin in the discovery of the circulation by William Harvey in the early 1600s. As reported by Lois N. Magner, Harvey's followers became "engaged in attempts to transfuse and infuse many curious substances into animal and human veins" (1). It did not take long to go from this to blood transfusions. Jean Baptiste Dennis (1620–1704), fascinated by the potential of this technique, decided to use it in the treatment of mental patients. This was so far out, even in those days, that very adverse publicity following the death of his patients made him abandon this treatment. However, the notion of incompatibility was introduced. History tells us the rest of the story, which culminates with our understanding of molecular and cellular immunity.

Because of its very function as the interface between air and blood, the lung is "attacked" from outside and inside. The mucosal surfaces of the airways and the capillary bed are so large that they are prime targets of inhaled and blood-borne particulates and incompatible material. Fortunately, nature has provided ramparts against these attackers, including the pivotal macrophages and dendritic cells. Much has been written about macrophages, and while less has been said about dendritic cells, we know today that they play a role as important as, and certainly complementary to, that of macrophages.

The table of contents of this volume reveals a comprehensive treatise on mechanisms of defense in the lung in health and disease. Undoubtedly, both clinical investigators and basic researchers will be stimulated and challenged by it. The editors called on long-established experts from the United States and from other countries to contribute the latest information in this important area.

As the executive editor of this series of monographs, I am indebted to Drs. Lipscomb and Russell and to the contributors for the opportunity to include this volume in the series. The journey of our discovery of immunity and defense mechanisms has been long, and the road has been tortuous, but, fortunately, contributions like this volume pave the way to their understanding and at the same time provide greater benefits to patients.

Claude Lenfant, M.D.
Bethesda, Maryland

Reference

1. Magner LN. A History of the Life Sciences. 2d ed. New York: Marcel Dekker, 1994.

PREFACE

One of the earliest volumes in this series on Lung Biology in Health and Disease was on mechanisms of defense in the lung. Enormous progress has been made in understanding natural and acquired immunity and the role of lung macrophages, dendritic cells, lymphocytes, and other leukocytes in carrying out the important function of protecting the lung against infectious agents and other foreign materials, as well as repair of the damage that might occur following these processes. This volume was developed to review in depth the roles of two essential lung cells in host defense and immunity, inflammation, and repair.

Macrophages are among the most readily recovered cells from the lung, and many studies have established their pivotal role in lung homeostasis and disease. They exist in the alveoli and airways, in the connective tissue, and even within the pulmonary vessels of some species. Their capacity to secrete a large array of molecules makes them one of the most important cells in lung homeostasis. Dendritic cells, which are much less readily recovered, partly because they are far less numerous than pulmonary macrophages, have recently been studied intensively because of their crucial role in immune regulation. These cells, like macrophages, exist in airspaces and connective tissue, but their unique location within the airway epithelium allows them to sample aeroantigens and initiate primary immune responses. Dendritic cells play a similar role in the skin and in the gastrointestinal tract.

This collection of selected topics by established researchers in the field of lung biology draws together current knowledge about these two important residents of the respiratory tract, one type well known and populous and the other still at the threshold of establishing its importance in normal and abnormal lung function. Emphasis is on the latest information, with attention to molecular biological mechanisms.

The book is divided into five parts. The first focuses on general properties of predominantly nonpulmonary macrophages and dendritic cells, how their functions differ depending on organ and tissue site, how they compare functionally and phenotypically when isolated from experimental animals versus humans, and methods of isolation and study. The next two parts focus on lung macrophages and dendritic cells in the lung, describing their sites, phenotypes, and in vitro functions. Part four includes discussions of the role of macrophages in host defenses against pulmonary infections caused by viruses, fungi, and bacteria, including mechanisms of recruiting circulating monocytes into infected lungs. Lastly, the importance of macrophages and dendritic cells in selected lung noninfectious diseases, including asthma, pneumoconioses, transplantation, and neoplasms, will be covered.

Each chapter begins with a broad perspective of the defined topic. Important details are described where consensus seems to exist, and key experiments that have validated the prevailing view are included. The authors of most chapters have indicated where controversies exist and have identified important issues for future research.

Biologists and clinicians interested in host defenses, immunity, and inflammatory processes in the lung should find this volume a handy resource for comprehensive information regarding two cell types that are essentially involved in these processes.

Mary F. Lipscomb
Stephen W. Russell

CONTRIBUTORS

Joseph D. Brain, S.D. in Hyg. Director of Physiology Program, Department of Environmental Health, Harvard School of Public Health, Boston, Massachusetts

Ellen Broug Department of Cell Biology and Immunology, Medical Faculty, Free University, Amsterdam, The Netherlands

Stephen W. Chensue, M.D., Ph.D. Associate Professor, Department of Pathology, University of Michigan Medical School, Ann Arbor, Michigan

Paul J. Christensen, M.D. Assistant Professor, Department of Internal Medicine–Pulmonary, University of Michigan, Ann Arbor, Michigan

Aalzen de Haan, Ph.D. Research Division, Department of Cardiopulmonary Surgery, University Hospital Groningen, Groningen, The Netherlands

Isaiah J. Fidler, D.V.M., Ph.D. Professor and Chairman, Department of Cell Biology, The University of Texas M. D. Anderson Cancer Center, Houston, Texas

Siamon Gordon Professor, Sir William Dunn School of Pathology, University of Oxford, Oxford, England

Allan J. Hance, M.D. Director of Research, Faculté de Médecine Xavier Bichat, Institut National de la Santé et de la Recherche Médicale, Paris, France

Thomas S. Harrison, M.D. Evans Memorial Department of Clinical Research and Department of Medicine, Boston University Medical Center Hospital, Boston, Massachusetts

Gary B. Huffnagle, Ph.D. Research Faculty, Division of Pulmonary Medicine, University of Michigan, Ann Arbor, Michigan

Derralynn Hughes, M.D. Sir William Dunn School of Pathology, University of Oxford, Oxford, England

Kayo Inaba, Ph.D. Associate Professor, Department of Zoology, Graduate School of Science, Kyoto University, Kyoto, Japan

Jerald J. Killion, Ph.D. Associate Professor, Department of Cell Biology, The University of Texas M. D. Anderson Cancer Center, Houston, Texas

Eugenie S. Kleinerman, M.D. Professor, Departments of Cell Biology and Pediatrics, The University of Texas M. D. Anderson Cancer Center, Houston, Texas

Lester Kobzik, M.D. Assistant Professor of Pathology, Department of Pathology, Brigham and Women's Hospital, Harvard School of Public Health, Boston, Massachusetts

Henry Koziel, M.D. Instructor, Harvard Medical School, Division of Pulmonary and Critical Care Medicine, Deaconess Hospital, Boston, Massachusetts

Georg Kraal, Ph.D. Department of Cell Biology and Immunology, Free University, Amsterdam, The Netherlands

Richard L. Kradin, M.D. Associate Professor of Pathology, Harvard Medical School, and Director, Pulmonary Immunology and Molecular Biology, Pulmonary and Critical Care Unit, Massachusetts General Hospital, Boston, Massachusetts

Steven L. Kunkel, Ph.D. Professor, Department of Pathology, University of Michigan Medical School, Ann Arbor, Michigan

R. Clark Lantz, Ph.D. Associate Professor, Department of Cell Biology and Anatomy, University of Arizona College of Medicine, Tucson, Arizona

Richard J. Lemen, M.D. Professor of Pediatrics, Director, Lung Injury Laboratory, Department of Pediatrics, University of Arizona College of Medicine, Tucson, Arizona

Stuart M. Levitz, M.D. Associate Professor of Medicine and Microbiology, Department of Medicine, Boston University School of Medicine, Boston, Massachusetts

Mary F. Lipscomb, M.D. Professor and Chair, Department of Pathology, University of New Mexico School of Medicine, Albuquerque, New Mexico

Nicholas W. Lukacs, M.D. Department of Pathology, University of Michigan Medical School, Ann Arbor, Michigan

C. Richard Lyons, M.D. Assistant Professor, Department of Internal Medicine, University of New Mexico School of Medicine, Albuquerque, New Mexico

Barbara J. Masten, Ph.D., MT(ASCP), CLS(NCA) Research Associate, Department of Pathology, University of New Mexico School of Medicine, Albuquerque, New Mexico

Ramon M. Molina, D.V.M., M.P.H., Sc.D. Research Associate, Physiology Program, Environmental Health, Harvard School of Public Health, Boston, Massachusetts

David C. Morrison, M.D. Kansas Masonic Distinguished Professor, Department of Microbiology, Molecular Genetics and Immunology, and The Kansas Cancer Institute, University of Kansas Medical Center, Kansas City, Kansas

William J. Murphy, Ph.D. Research Assistant Professor, Department of Pathology and Laboratory Medicine, and The Kansas Cancer Institute, University of Kansas Medical Center, Kansas City, Kansas

Laurent P. Nicod, M.D. Department of Medicine, Respiratory Division, University Hospital of Geneva, Geneva, Switzerland

Jek H. A. Persoons, M.D., Ph.D. Department of Cell Biology and Immunology, Free University, Amsterdam, The Netherlands

Marc Peters-Golden, M.D. Professor of Internal Medicine, Division of Pulmonary and Critical Care Medicine, Department of Internal Medicine, University of Michigan, Ann Arbor, Michigan

Jochum Prop, M.D., Ph.D. Research Division, Department of Cardiopulmonary Surgery, University Hospital Groningen, Groningen, The Netherlands

William N. Rom, M.D., M.P.H. Professor and Chief, Division of Pulmonary and Critical Care Medicine, New York University Medical Center, New York, New York

Stephen W. Russell, D.V.M., Ph.D. Wilkinson Distinguished Professor of Cancer Research, Department of Pathology and Laboratory Medicine, and The Kansas Cancer Institute, University of Kansas Medical Center, Kansas City, Kansas

Larry S. Schlesinger, M.D. Associate Professor, Department of Internal Medicine, University of Iowa, Iowa City, Iowa

Eveline E. Schneeberger, M.D. Professor of Pathology, Department of Pathology, Massachusetts General Hospital, Boston, Massachusetts

Judd E. Shellito, M.D. Professor of Medicine, Section of Pulmonary and Critical Care Medicine, Louisiana State University Medical Center, New Orleans, Louisiana

Michael P. Sherman, M.D. Professor of Pediatrics, Section of Neonatology, Baylor College of Medicine, Houston, Texas

Paul Soler, Ph.D. Institut National de la Santé et de la Recherche Médicale, Paris, France

Ralph M. Steinman, M.D. Professor and Senior Physician, Laboratory of Cellular Physiology and Immunology, Rockefeller University, New York, New York

Robert M. Strieter, M.D. Professor, Department of Internal Medicine, University of Michigan Medical School, Ann Arbor, Michigan

Takafumi Suda, M.D., Ph.D. Department of Pathology, Electron Microscope Unit, Massachusetts General Hospital, Boston, Massachusetts

Abdellatif Tazi, M.D., Ph.D. Professor of Medicine, Institut National de la Santé et de la Recherche Médicale, Paris, France

Theo Thepen, M.D. Department of Dermatology, Utrecht University, Utrecht, The Netherlands

Galen B. Toews, M.D. Chief, Division of Pulmonary and Critical Care Medicine, Department of Internal Medicine, and Professor of Internal Medicine, University of Michigan, Ann Arbor, Michigan

Homer L. Twigg III, M.D. Associate Professor of Medicine, Division of Pulmonary and Critical Care, Indiana University Medical Center, Indianapolis, Indiana

Tadashi Uyama, M.D., Ph.D. Associate Professor, Second Department of Surgery, School of Medicine, University of Tokushima, Tokushima City, Japan

J. Freek van Iwaarden, Ph.D. Department of Cell Biology and Immunology, Free University, Amsterdam, The Netherlands

Angeline E. Warner, D.V.M., Sc.D. Assistant Professor, Department of Environmental Health, Harvard School of Public Health, Boston, Massachusetts

Jonathan C. Weissler, M.D. Maes M. Collins Professor of Biomedical Research, Department of Internal Medicine, University of Texas Southwestern Medical Center, Dallas, Texas

Lewis J. Wesselius, M.D. Veterans Affairs Medical Center, Kansas City, Missouri, and University of Kansas Medical Center, Kansas City, Kansas

David S. Wilkes, M.D. Assistant Professor of Medicine, Department of Pulmonary and Critical Care Medicine, and Department of Medicine, Indiana University Medical Center, Indianapolis, Indiana

Mark L. Witten, Ph.D. Research Associate Professor, Department of Pediatrics, University of Arizona College of Medicine, Tucson, Arizona

CONTENTS

Part One

MACROPHAGES AND DENDRITIC CELLS

1

Macrophages and Their Origins

Heterogeneity in Relation to Tissue Microenvironment

SIAMON GORDON and DERRALYNN HUGHES

Sir William Dunn School of Pathology
University of Oxford
Oxford, England

I. Introduction

Macrophages are specialized hemopoietic cells distributed throughout different tissues of the body where they play a central role in homeostasis, host defense, and the response to foreign materials, including particulates. Macrophages contribute to many disease processes by their trophic and cytotoxic activities. Cell production, differentiation, and recruitment are controlled by a range of specific mediators, including growth factors and chemokines. Their localization within different organs depends on various adhesion molecules and interactions with endothelium. Macrophages express common as well as site-specific properties within different tissue microenvironments, including various marker antigens and receptors for soluble, "matrix"- and cell-associated ligands. Mature macrophages display proficient phagocytic and endocytic activities and complex intracellular signalling pathways and membrane traffic linked to their adaptive biosynthetic and secretory responses. The macrophages of the body thus constitute a mobile, dispersed system of cells that are able to react to local injury and modify the properties of all cells in their immediate vicinity while communicating with other key regulatory systems of the intact host.

In this chapter, we review some of the general properties of macrophages, emphasizing their mobilization and specialization in responding to different stimuli. The lung is a special instance, but the principles of macrophage adaptation are comparable to those operating in other organs. We summarize the key aspects of macrophage life history that hold true for all anatomical sites and discuss briefly particular examples of macrophage properties in the lung, referring to other chapters as appropriate. Much of our knowledge of macrophage biology is based on the study of experimental animals, especially the mouse, and genetic, infectious, immunological, and other models of disease. We believe that in the main, similar features apply to humans, but will point out known species differences. We prefer to draw on information derived from studies on primary cells as close as possible to conditions that apply in vivo to approximate the expression of functions within the living organism and avoid some of the artefacts of cell culture. We refer to recent general reviews for further information (1–9), and we cite only selected examples of key findings to illustrate principles rather than provide exhaustive details.

II. Mononuclear Phagocyte System: General Aspects

Although ultimately all macrophages derive from precursor cells in hemopoietic organs, they gain access to diverse tissues via blood and lymph (Fig. 1), and, depending on a particular regional environment such as lung, they can be capable of considerable local self-renewal. Macrophages share many features with other myeloid as well as lymphoid cells, but they also display selected characteristics of connective tissue and endothelial cells, which share a common microenvironment. Blood monocytes and tissue macrophages are unique among hemopoietic cells in their ability to enter all tissues of the normal adult and developing organism and to migrate extensively within different compartments of the body. Populations of such "resident" macrophages are found constitutively in the absence of an overt inflammatory stimulus and are present in major lymphoid as well as nonlymphoid organs, as detailed below. These macrophages are relatively long-lived and retain considerable mRNA and protein-synthesizing activity even as they become refractory to growth factor stimulation of DNA synthesis. Resident macrophages display a varied profile of plasma membrane receptors (Fig. 2; Table 1), and they are able to endocytose and phagocytose a wide range of soluble and particulate ligands. Their local secretory responses of cytokines and other mediators help to initiate early changes in tissues that signify local inflammation. Altered adhesion ligands on local endothelium and changes in vascular permeability, combined with the effects on circulating leukocytes, including monocytes, contribute to local recruitment of various leukocytes. Monocytes then enter tissues in increased numbers ("induced recruitment") and as they migrate, phagocytose dying cells, and encounter microbial or host-derived molecules, they differentiate into "elic-

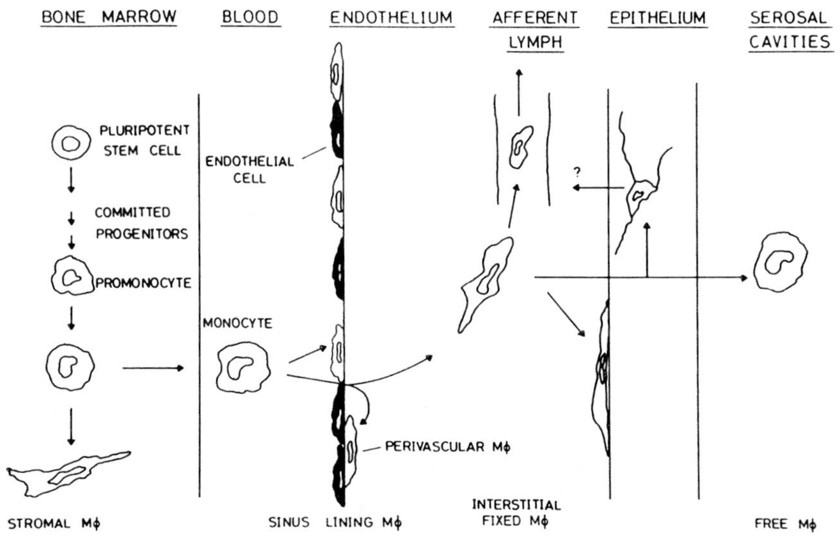

Figure 1 Differentiation and distribution of macrophages (Mϕ) within various tissue compartments. Blood monocytes derive from bone marrow in the adult host and enter tissues as sinus-lining or extravascular mature cells. Interstitial or intraepithelial adherent macrophages and serosal macrophages can enter lymph. Stromal macrophages in bone marrow may derive from circulating monocytes. The control of constitutive adhesion and migration within tissues is precise but not understood.

ited" or "immunologically activated" macrophages. We use the term *activated* to characterize a distinct set of properties, including enhanced antigen-presenting or pathogen-destroying functions, which are induced in macrophages by interactions with antigen-stimulated T lymphocytes. This concept is described in more detail below. Inflammatory and immunological stimulation has a profound effect on many macrophage properties; these include complex changes in gene expression irrespective of whether terminally differentiated resident cells have become "re-activated" or whether recently recruited monocytes, now present locally in larger numbers, have become fully activated to express higher levels of proinflammatory, cytocidal, and other properties. Enhanced turnover of macrophages accompanies such processes, although little is known about the relative importance of apoptosis compared with cell death by other mechanisms.

III. Ontogeny

Early mature macrophages can be detected in the yolk sac of the developing embryo by the expression of characteristic plasma membrane antigen or receptor

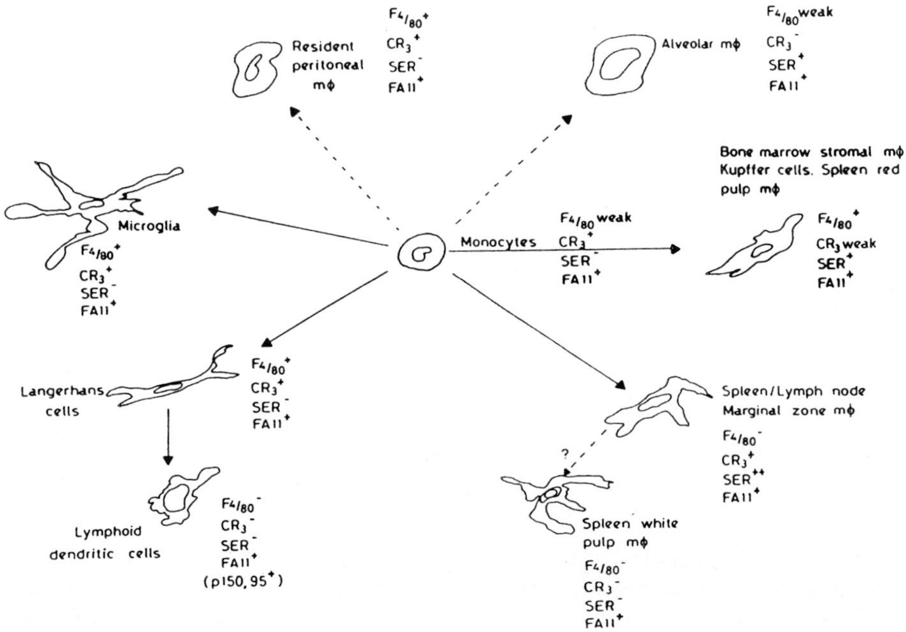

Figure 2 Phenotypic microheterogeneity of resident macrophage (Mφ) in different tissues. Murine differentiation antigens are listed in Table 1. See text for further details. Abbreviations: CR_3, type 3 complement receptor; SER, sheep erythrocyte receptor (sialoadhesin); FA 11, macrosialin, mouse CD68. Resident peritoneal and alveolar macrophages are relatively autonomous, although ultimately derived from blood monocytes.

markers such as the F4/80 molecule (see Table 1) or Fc receptors, which are described further below. Progenitor cells prior to this are still poorly defined. Macrophages are a major component of fetal hemopoiesis, being associated with erythropoiesis in fetal liver, spleen, and bone marrow before and after birth, as well as with myelopoiesis and lymphopoiesis, which are delayed relative to erythroid development (23). Macrophages in fetal liver, for example, represent a prominent part of the hemopoietic microenvironment and express specialized adhesion receptors involved in nonphagocytic binding of developing blood cells, which are often seen as clusters of erythroblasts on the surface of stromal macrophages (24). This property persists on specialized macrophages in other hemopoietic organs later in development and in the adult and is likely to reflect a poorly understood "trophic" role by these macrophages in hemopoietic cell differentiation. In addition, such macrophages are intensely phagocytic, ingesting discarded erythrocyte nuclei and probably other dying hemopoietic cells. In thymus, phagocytic macrophages clear apoptotic thymocytes (25), the majority population of cells that never differentiate fully for export to peripheral lymphoid organs.

Table 1 Selected Membrane Antigens that Define Macrophage Heterogeneity in the Mouse

Ag	mAb	Size (kDa)	Structure	Ligand	Expression	Regulation	Function	Ref.
F4/80	F4/80	160	Type I extracellular 7 EGF and seven-transmembrane domain (homologous to secretin family)	?	Plasma membrane Many mature resident, elicited and activated macrophages	↓T-cell areas	?	10,11
CR3	5C6	150/90	β_2-integrin	iC3b, ICAM-1, etc.	Plasma membrane and intracellular myelomonocytic	↓Many tissues	Recruitment, adhesion, and phagocytosis	12,13
Sialo-adhesin	SER-4 3D6	185	Type I, lectin 17 lg domains	Sialyl α2–3 gal β1-3 Gal NAc-glycoconjugates	Plasma membrane subpopulations macrophage lymphohemopoietic and elsewhere	↑Serum factor ↓IL-4	Nonphagocytic cell interactions	14–16
Macrosialin (mouse CD68)	FA11	~85	Type I, heavily glycosylated lgp family with macrophage-specific mucin	Oxidized LDL ? other	Endosomal Pan-macrophage and dendritic cells	Glycoforms modulated by inflammation and receptor-mediated phagocytosis	?	17,18
Scavenger receptor (A I, II)	2F8	~90 × 3	Type II trimer collagenous domain	Range poly-anions, e.g., modified proteins, LPS	Intracellular and plasma membrane Subpopulations of macrophages and hepatic endothelium	↑M-CSF ↓TGF-β	Endocytosis Phagocytosis Adhesion LPS clearance	19–21

mAb for human homologues only available for CR3 and CD68. For details of other antigens on macrophages (mouse, rat, human cells), see ref. 22.

Apart from their role in hemopoiesis, macrophages appear widely from mid-gestation in mesenchyme and all developing organs, including lung. They are prominent at sites of tissue remodeling, as in programmed cell death in limb buds, where they phagocytose local apoptotic cells (26). They enter the developing nervous system, where phagocytosis of naturally dying neurons has also been observed (27). It is likely that macrophages contribute to cell growth and differentiation, as well as their clearance, but such functions have not been directly demonstrated. There are examples of selected deficiencies of macrophage populations, for example, in the osteopetrotic (op/op) natural mutant mouse that lacks macrophage colony-stimulating factor (M-CSF) (28), a major macrophage growth factor to be discussed below. These mice lack some, but not all tissue macrophages, yet organs with markedly reduced macrophages do not display overt developmental deficiencies other than in bone, where resorption by osteoclasts, a macrophage-related cell, is impaired. More severe deficiencies of macrophage, together with B lymphocytes, can be observed in experimental gene knockout mice that lack the transcription factor Pu-1 (29). Complex mechanisms, including severe anemia, may contribute to lethality in these animals, and macrophages could play an additional indirect role in hemopoietic failure, presumably more critical than simple deficiency in scavenging functions.

IV. Production and Differentiation

Macrophages share a common $CD34^+$ stem cell progenitor with other hemopoietic cells; more committed precursors are shared with polymorphonuclear leukocytes (PMNs). Bipotential precursors for macrophages and PMNs can be readily identified by in vitro assays in semisolid media supplemented with specific colony-stimulating factors (CSFs), including granulocyte-macrophage CSF (GM-CSF), M-CSF, and granulocyte CSF (G-CSF). CFU-S, colony-forming units in spleen, are able to reconstitute macrophages and other hemopoietic lineages in irradiated animals. The cytokines/receptors required for early stages of differentiation include SCF (stem-cell factor), c-*kit* ligand, and interleukin-1 (IL-1); IL-3 supports multiple lineages, including macrophages, whereas GM-CSF and M-CSF act mainly on relatively late stages of differentiation to support cell survival at low levels of growth factor and macrophage growth at higher levels. These responses are mediated by specific plasma membrane receptors for various cytokines, with several of them being encoded by a group of linked genes on the human chromosome 5q region. Target cell responses vary with differentiation, which is accompanied by a progressive decrease in proliferation, and the induction of complex patterns of expression of genes associated with maturation. IL-4 and IL-13 are relatively poor growth factors, but they continue to act, weakly in comparison with the CSFs, on mature stages. Tumor necrosis factor-α (TNF-α) and other surface and secretory molecules of stromal cells also contribute to differentiation of hemopoietic cells, including macrophages. Several of these molecules (GM-CSF, IL-4, TNF-α) play a role in the modulation of macrophages to dendritic cell phenotype in vitro, which is described below.

The branch points between macrophages and other myeloid and lymphoid cells are not defined. Recent progress has been made in defining DNA binding transcription factors, such as Pu-1 (macrophages and B lymphocytes) and proteins of the NF-κB family (macrophages, dendritic cells, and lymphoid cells) (9). Plasma membrane receptors, for example, for M-CSF (c-*fms*) signal through tyrosine kinase activation, and important cytoplasmic phosphatases (PT-Pase 1) play a role in signal transduction. Genetic, functionally null mutants of M-CSF (op/op) and of the PTPase locus (moth-eaten) have been very useful in defining components of this pathway. These and other knockout mice, for example, of the GM-CSF gene, have provided insights into partial macrophage dependence on specific growth factors within the intact animal and have yielded unexpected observations; for example, GM-CSF knockout mice develop an alveolar proteinosis syndrome (30–32) to be discussed further below.

Macrophage differentiation also involves the selective inhibition of cell growth and suppression of gene expression. For example, deficiencies of transforming growth factor β (TGF β) (9) or inactivation of such genes by gene translocation result in excessive proliferation and infiltration of tissues by myeloid cells, including macrophages. Macrophages themselves play a role in inhibiting stem cell differentiation via macrophage inflammatory protein-1α (MIP-1α), also known as a chemokine that mobilizes monocytes and PMNs from bone marrow into the circulation. In spite of the above examples, it is fair to say that we have very limited understanding of the interplay between intrinsic and extrinsic factors in macrophage production and differentiation.

V. Blood Monocytes and Differentiation of Resident Tissue Macrophages

Monocytes circulate for 1–2 days before entering tissues either as sinus-lining macrophages, for example, Kupffer cells in liver, or at perivascular, interstitial, or intraepithelial locations in lymphoid and nonlymphoid organs (Fig. 1). Substantial numbers of macrophages are normally present in gut lamina propria, spleen, lymph nodes, lungs (alveolar macrophages), skin, and brain (microglia) as well as in serosal cavities, including pleural and peritoneal macrophages. These resident macrophages differ in phenotype, as illustrated in Figure 2, even within different regions of the same organs. Moreover, more specialized populations of cells that are closely related to macrophages and can be considered as distinct sublineages of mononuclear phagocytes are present in bone (osteoclasts) and lymphoid organs/

complex epithelia (dendritic and Langerhans' cells). Mononuclear precursors for osteoclasts also circulate in blood, and dendritic cells can be mobilized from peripheral organs into the bloodstream. These observations raise several questions which cannot yet be answered; how heterogeneous are circulating mononuclear cells and what determines their entry into tissues; are such cells already predetermined, for example, to give rise to Kupffer cells, microglia, osteoclasts, or alveolar macrophages; what are the local, microenvironmental factors in lung, spleen (red pulp, marginal zone, white pulp), brain, and epidermis (Langerhans' cells), for example, that yield the observed phenotypic pattern of gene expression?

When monocytes are recruited to local sites of inflammation or immune responses, as discussed below, they utilize distinct adhesion molecules and ligands to enter tissues; however, within a few days in their new (and also altered microenvironment), they begin to resemble macrophages ordinarily resident in those sites, which is presumably evidence of adaptation to organ-specific regional influences. It is, therefore, perhaps not surprising that it becomes increasingly difficult to distinguish recently recruited macrophages from locally reactivated preexisting cells.

The properties of blood monocytes are considered further below. They provide a readily available population in humans, whereas the mouse can provide less accessible tissue macrophages and models of inflammatory and immune recruitment. The properties of osteoclasts and dendritic cells (DCs) should be considered further. Osteoclasts are multinucleated, specialized bone-resorbing cells, reminiscent of macrophages in their catabolic properties, but with novel responses to hormones (e.g., parathormone). M-CSF is essential for their differentiation and maturation, as it is for selected macrophage subpopulations. Dendritic cells derive from circulating mononuclear precursors and mature in epithelia (Langerhans' cells in epidermis or in, e.g., trachea, cervix), where they express some markers of Mϕ, e.g., F4/80 antigen) but reduced phagocytic activity compared with many other macrophages (see Chapter 3). Following encounter with foreign antigens, Langerhans' cells become poorly adherent and migrate as veiled cells to draining lymph nodes via afferent lymph. These lymphoid dendritic cells localize in T-lymphocyte–dependent areas where they become highly specialized, potent antigen-presenting cells (APCs) for naive T lymphocytes. They express high levels of major histocompatibility (MHC) class II antigens constitutively, as well as other accessory molecules required for T-lymphocyte activation. Although phagocytic activity becomes reduced, the differentiating dendritic cell can express high levels of macropinocytosis, perhaps related to its APC function. Cytokines such as GM-CSF are critical for DC differentiation from bone marrow and blood mononuclear cells and survival in vitro but not necessarily in vivo, as shown in knockout mice (31). Human blood monocytes can be modulated in bulk by cultivation in GM-CSF, IL-4, and TNF-α to display many "DC"-like characteristics, which is consistent with the hypothesis that they are closely related cell types.

Little is known about the disposal of effete macrophages and DCs. The lymph nodes and spleen are likely graveyards, where they are cannibalized by local macrophages. Virtually no macrophages enter efferent lymph or the thoracic duct unless the lymph node filter is removed by surgery. Macrophages elsewhere are phagocytosed by surviving or younger macrophages. Although macrophages play a special role in recognition and uptake of all types of dead cells, little is known about programmed cell death in macrophages themselves as opposed to PMNs. Macrophages express fas and *fas*-ligand under certain circumstances, which may contribute to their turnover, especially after immune stimulation.

VI. Recruitment: Role of Chemokines and Adhesion Molecules (see Chapter 15)

The role of chemokines favoring myeloid and mononuclear cell recruitment to inflammatory or immune foci has been clarified by the discovery of a number of CC and CXC polypeptides with overlapping or distinct activities in vitro and to some extent in vivo. Species differences in the spectrum of cells responding to a particular chemokine have been noted. Progress in the cloning and characterization of receptors for various chemokines has been rapid. These receptors are members of seven transmembrane (serpentine) receptor families which are known to interact with G-proteins in signal transduction. Chemokines and receptors known to be involved in recruitment of monocytes, alone or with other cells, include Rantes, monocyte chemotactic protein (MCP-1) and MIP-1. Macrophages themselves, resident and recruited, are a major source of such chemokines on stimulation, for example, by lipopolysaccharide (LPS) or phagocytosis, but activated T lymphocytes and endothelial and epithelial cells are all able to produce chemokines. Their effects on leukocytes include shape changes, enhanced mobility, directed migration, and exocytosis/degranulation, depending on the cell target. Gradients of soluble chemokine can act as diffusible cues for cell recruitment, but increasing interest has been drawn to matrix-bound molecules, especially glycosaminoglycans in cell migration, but other actions, for example, as regulators of hemopoiesis, have also been postulated. The recruitment of different white blood cells to infectious, immunological, or metabolic stimuli can broadly be explained in terms of chemokines and receptor-specificity, but there is considerable overlap and complexities that are not yet fully understood.

One of the actions of chemokines, better documented for PMNs than monocytes, is to upregulate expression and/or function of adhesion receptors and counterligands. Other cytokines, such as TNF-α, act similarly on leukocytes as well as on local endothelium and other cell types, to modulate cell-cell interactions, adhesion, and migration. Now well established are the roles of the selectins (L-, E-, P-) as initial, carbohydrate-dependent adhesion molecules; of the

β_2-integrins, as more stable mediators of adhesion of leukocytes to inflamed endothelium; and of CD31, an immunoglobulin superfamily molecule involved in homotypic interactions that underlie leukocyte diapedesis. Our knowledge of the roles of these molecules in endothelial transmigration is based on the development of specific monoclonal antibodies (mAbs) that block their actions in vitro as well as in vivo, the characterization of rare human inherited deficiencies, such as the leukocyte adhesion–deficiency syndrome (LAD)-1, and the production of mice with disrupted genes (e.g., for P-selectin) (9). However, none of the above molecules account fully for monocyte-specific adhesion, in the absence of PMNs, to atherosclerotic lesions, for example, or for the constitutive and T-lymphocyte–directed recruitment of monocytes together with $CD4^+$ T lymphocytes in delayed-type hypersensitivity reactions. Additional candidate receptors and ligands include CD14, CD44, other (β_1- and β_3-) integrins, and the macrophage scavenger receptor, to be discussed further below. Once monocytes have passed through basement membrane, other plasma membrane molecules, poorly characterized in relation to their possible adhesion function, may be implicated in tissue macrophage-specific localization.

What are the effects of diapedesis and tissue migration on monocyte differentiation and what effects do monocytes and transmigrated macrophages exert on the vessel wall and its role in permeability control and platelet adhesion? Almost nothing is known about these issues, but since adhesion regulates many macrophage functions in vitro, we should expect major effects on gene expression and biosynthesis as well as on regulation of plasma membrane and cytoskeletal functions per se.

VII. Phenotypic Heterogeneity in Vivo and in Vitro: General Aspects

We have already commented on the role of different tissue microenvironments in generating diverse phenotypes of macrophages and related cells and the importance of inflammatory and other factors in mediating enhanced recruitment. Microbial products such as LPS and other cell wall constituents (e.g., of mycobacteria, streptococci, and fungi) can all act on monocytes and macrophages to promote their uptake into the vacuolar system, but these substances also profoundly influence signaling pathways, gene expression, biosynthesis, secretory activities, and macrophage behavior. Similarly, macrophages are acted on by diverse ligands in plasma, matrix, and cell surface or secretory products of other cells, especially activated T lymphocytes, as well as by macrophages themselves. These include cytokines, lipid mediators, peptides, modified amino acids, and metabolites. In turn, macrophages interact with neighboring cells to modulate their responses. Before dealing with some of these macrophage activities, it is necessary to outline some of the general principles regarding modulation of the macrophage phenotype. Our concepts derive from studies in vivo, from macrophages isolated from different tissue compartments, or macrophages generated in cell culture, which can provide a better-defined, closed system for analysis.

A. Resident Macrophages Display Microheterogeneity in Situ

Morphological appearance and extensive use of plasma membrane (F4/80) and endosomal antigen markers (macrosialin, the murine homologue of human CD68) as well as of other markers present on macrophage subpopulations (sialoadhesin) reveal consistent differences among mature macrophages in lymphohemopoietic and other organs (10) (see Fig. 2). Brain and spleen are good examples: In brain, we are able to distinguish macrophages in the parenchyma (neuropil), at the interface between blood and the cerebrospinal space (choroid plexus, blood brain barrier), in adventitial structures (meninges), and at perivascular sites; in spleen, it is possible to distinguish macrophages by distinctive marker expression in the marginal and marginal metallophil zone (specialized to capture pathogens and complex carbohydrate and other antigens) from macrophages in the red and white pulp. The latter region contains abundant lymphocyte subpopulations, which influence macrophage phenotype as a result of cellular and humoral immune responses. Circulating blood cells, including T and B lymphocytes and dendritic cells, also enter spleen and migrate within its complex structure. We are only beginning to define the different types of macrophages in these complex microenvironments and have little idea as to their mechanisms of localization, migration, and differentiation in situ.

B. Elicited and Immunologically Activated Macrophages Differ from Resident Macrophages but Can Also Be Distinguished (Fig. 3)

Recruited macrophages, whether elicited or immunologically activated, share properties irrespective of where they localize, but they also adopt features of local macrophages, as pointed out above. Many resident macrophages are deficient in their killing capacity, perhaps desensitized to protect their local environment, and host defenses and antimicrobial/cytotoxic activities may therefore depend on newly recruited monocytes. Even phagocytic capacity, the hallmark of the macrophages, can be markedly downregulated as tissue macrophages differentiate locally (e.g., in epidermis). The type 3 complement receptor (CR_3), which is well expressed on monocytes and has been implicated in myelomonocytic recruitment as well as phagocytosis, is often downregulated on tissue macrophages, including alveolar macrophages. Expression of MHC class II antigens, important in immune cell interactions, is upregulated on activated macrophages; other plasma membrane phagocytic and endocytic receptors, such as the mannosyl receptor (MR),

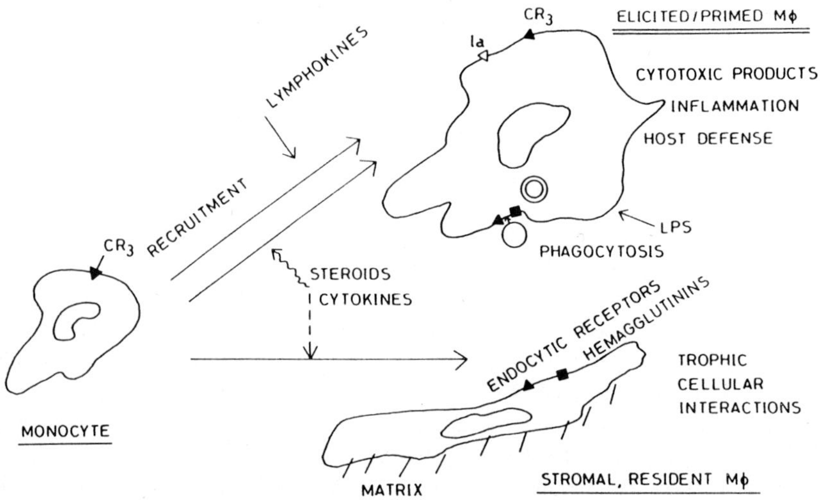

Figure 3 Elicited and immunologically primed macrophages (Mϕ) differ from resident macrophages. Monocyte recruitment is enhanced and yields macrophages with proinflammatory and cytotoxic properties. Resident macrophages, for example, in bone marrow, lack inflammatory functions but participate in trophic interactions, for example, with developing hemopoietic cells, as well as performing endocytosis.

are differentially expressed depending on the cytokine environment and may be involved in recognition and processing of microbial antigens as part of natural immunity.

C. Macrophage "Activation" Requires at Least Two Stages: Priming and Triggering

Unlike PMNs, with preformed granule products, macrophages require a distinct stage, priming by microbial stimuli or interferon-γ (INF-γ) before reacting to a further plasma membrane trigger to generate proinflammatory cytokines or cytotoxic products. INF-γ derived from natural killer (NK) and activated T lymphocytes is the major macrophage-priming cytokine (34), responsible for MHC class II induction, induction of inducible nitric oxide synthase, and enhanced cytotoxicity and antimicrobial activity as well as proinflammatory cytokine (TNF-α, IL-1β, IL-6) production. IL-12, produced by macrophages themselves, for example, after microbial uptake, enhances IFN-γ production. A new IL-1–like IFN-γ–inducing factor (IGIF), with activity broadly similar to that of IL-12, with which it can synergize, has recently been reported (35). Priming, a qualitative change in the cell involving altered gene expression and poorly understood changes in signal-transducing capacity, can be inhibited by other cytokines, such as TFG-β or IL-10, both of which can be produced by macrophages as well as by glucocorticosteroids. The advantage of localizing macrophage cytotoxicity should be apparent.

D. Macrophage Activation/Deactivation Covers a Spectrum of Phenotypic States (Fig. 4)

In addition to the polar opposites of macrophage activation mediated by IFN-γ and IL-10, other Th_2-type cytokines (IL-4 and IL-13, which are virtually indistinguishable) promote an alternative activation phenotype in which macrophages express enhanced endocytosis by MR, and induced MHC class II antigen, promoting humoral rather than cellular immunity (9). Other cytokines such as M-CSF and GM-CSF can be considered to be nonactivating with regard to these characteristics.

In real life, it may be impossible to separate these distinctive phenotypes. Different subpopulations of T helper (Th) lymphocytes can coexist in a disease state, and different cytokines may be produced simultaneously by different cell

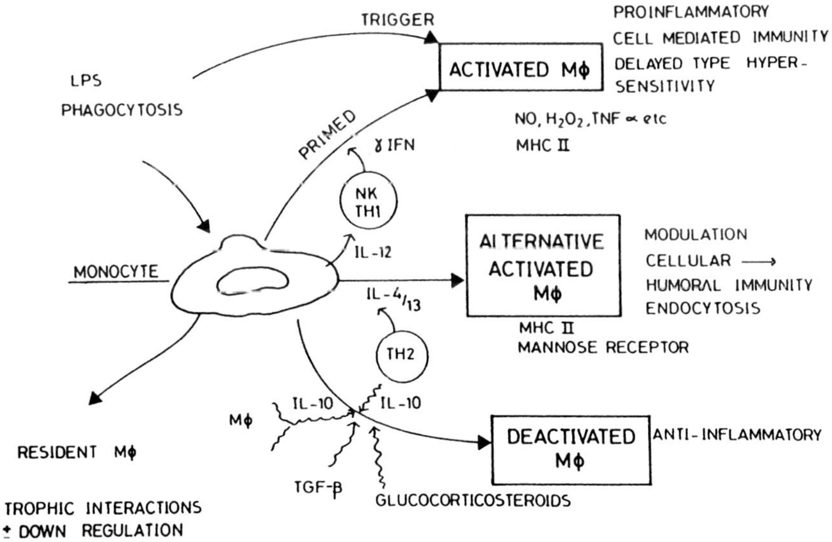

Figure 4 Modulation of macrophages (Mϕ) activation by microbial, phagocytic, and cytokine signals from T helper lymphocytes, macrophages themselves, and other cell types. Macrophage activation covers a spectrum of surface and secretory properties which can be classified as activated, alternatively activated, or deactivated, and which contributes to macrophage functions in cell-mediated as well as humoral immunity.

populations or even by the same cells at different times. Taken in combination with other modulators of macrophage function, such as proteolytic enzymes and plasma protein–derived polypeptides, the end result will be a macrophage population with marked heterogeneity.

VIII. Cell Biology of the Macrophage

Given the above perspectives, we shall detail some of the molecules and cellular processes that relate to the functions of macrophages.

A. Plasma Membrane Antigens and Receptors

Tables 1 and 2 list some of the membrane antigens that have been used as markers of macrophage distribution and of their functional heterogeneity. A few deserve special comment.

Mouse F4/80

Mouse F4/80 is highly restricted to mature macrophages in T-cell–independent regions of lymphoid organs. The molecule has an unusual chimeric structure, consisting of multiple extracellular Egf domains and a seven-transmembrane domain homologous to peptide hormone receptors of the secretin family. Its ligand(s) and function(s) are unknown. A human homologue has been identified (36), but tissue expression in humans has not yet been defined.

CR3 (CD11b, CD18)

This leukocyte β_2-integrin plays a dual role in myelomonocytic cell recruitment and phagocytosis. Monoclonal antibody directed against the mouse α chain, such as 5C6 (rat antimouse), block both functions, whereas M1/70 blocks only the latter (12). The mAb has been used to inhibit myelomonocytic recruitment in a range of inflammatory and immunological models in vivo (13). CR3 ligation by mAb can stimulate DNA synthesis and apoptosis in terminally differentiated macrophages (microglia) (37).

Scavenger Receptor (Type AI, II; SR-A)

This trimeric type II transmembrane molecule contains a collagen-helical domain involved in the binding of a range of polyanionic ligands, including LPS and modified proteins (38). It mediates adhesion in vitro (19), involving as yet unidentified serum protein ligands, endocytosis of ligands such as acetylated low-density lipoprotein, and phagocytosis of foreign particulates, including carbon and, in part, of apoptotic thymocytes (25). Because of its widespread tissue distribution

(20), including alveolar macrophages, it is thought to play a role in innate immunity. It is selectively upregulated by M-CSF (21). Macrophages and other cells also express other "scavenger receptors."

Fc Receptors

Macrophage express distinct Fc receptors (FcRs) for different Ig subclasses involved in antibody-dependent phagocytosis and the uptake of immune complexes. Receptor ligation can trigger the release of macrophage secretory products or downmodulate immune responses of the host, as shown in recent knockout mice experiments (39).

CD14

Mainly expressed by monocytes and modulated on mature macrophages, this GPI-anchored molecule is a receptor of LPS binding protein in plasma. Ligation results in the release of cytokines directly or via other membrane molecules.

Mannosyl Receptor (MR)

This lectin-like glycoprotein contains multiple C-type lectin domains involved in glycoconjugate binding, endocytosis, and nonopsonic phagocytosis, for example, of yeasts and *Pneumocystis carinii* (40). It is upregulated by IL-4/13 and downregulated, although functionally activated, by IFN-γ (9).

Sialoadhesin (Sheep Erythrocyte Receptor, SER)

Subpopulations of macrophages in lymphoid and other tissues express this nonphagocytic lectin-like receptor for sialylated glycoconjugate ligands on inflammatory PMNs (41), other mouse hemopoietic cells, and sheep erythrocytes. It has multiple Ig-like domains (15) and is regulated by cytokines (e.g., IL-4) (16) and a poorly defined plasma protein(s). Sialoadhesin is related to other sialic acid binding glycoproteins (e.g., CD22) on B lymphocytes (42) and is thought to play a role in cellular adhesion and perhaps trafficking.

Macrosialin (CD68)

This endosomal pan-macrophage glycoprotein is also expressed by dendritic cells (43). It is a member of the lysosomal glycoprotein (lgp) family, with a macrophage-specific extracellular mucin domain (17). Elicited/activated and phagocytically stimulated macrophages express distinctive glycoforms, possibly protective within the endosome (18).

In addition, macrophages display receptors for growth factors, cytokines (including chemokines), and other adhesion molecules, some of which have been

Table 2 Macrophage Plasma Membrane Receptors and Ligands

Receptor	Ligand(s)	Functions	Comment
FcR (several)	IgG, IgE cytotoxicity, release inflammatory mediators	Opsonic phagocytosis	Ig superfamily (IgSF)
CR_3	iC3b, ICAM-1 promiscuous	Opsonic phagocytosis nonopsonic pathogen entry. Migration Cell activation	β_2-integrin (also LFA-1, p150,95)
C1qR	C1q, collectins (mannose binding protein, surfactant protein-A)	?Opsonic	Ligands lectins with collagenous sequences
Macrophage scavenger R (SR-A)	Modified lipoproteins Selected polyanionic macromolecules	Adhesion and endocytosis	Expressed by selected tissue macrophages
AGE R	Advanced glycosylation endproducts	Endocytosis, cytokine release	Interacts with nonenzymatically glucosylated proteins, e.g., diabetes
VLA 1, 2, 4, 5, 6	Matrix/coagulation components, laminin, fibrogen	Adhesion/migration	Several β_1- and possibly β_3-integrins
VnR (CD51)	Vitronectin		α_v β_e
CD31	Endothelial cells, platelets	Diapedesis	Homophilic IgSF
CD44	Hyaluronic acid	Adhesion/migration	Hematopoietic variant
CD36	Thrombospondin	Adhesion	
L-selectin	Lectin: Sialyl-Lewis *x* antigen	Rolling on inflamed endothelium	Monocytes also express ligand for E- and P-selectins
Sialoadhesin	Lectin: Neu Acα2$\rightarrow$3 Galβ1$\rightarrow$3 Gal NAc on sialoproteins and gangliosides	Interactions hematopoietic cells	Stromal macrophages in bone marrow and lymphoid organs. Carbohydrate recognition and multiple IgGSF domains
Macrophage mannose receptor (MMR)	Lectin: Mannosyl glycoconjugates	Endocytosis, phagocytosis and secretion	Multiple C-type carbohydrate-recognition domains

Table 2 Continued

Receptor	Ligand(s)	Functions	Comment
Cytokine receptors:	M-CSF	Growth, differentiation of macrophages and osteoclasts, endocytosis	IgSF, macrophage-restricted c-*fms*
	GM-CSF	Growth, differentiation of macrophages and dendritic cells	Hematopoietin receptor superfamily (low- and high-affinity polypeptides)
	IFN-γ	Activation macrophages MHC class II ag↑, induction iNOS respiratory burst	Regulation by Th_1 cells. Two polypeptides
	IL-4	Alternate differentiation and activation, fusion	Regulation by Th_2 cells related to IL-13 receptor, which has similar effects
	TNF-α	Granuloma formation, antimicrobial activity	Two polypeptides (p75, p55)
	IL-10	Deactivation, anti-inflammatory	Regulation by Th_2 cells and stimulated macrophages
	MCP-1, Rantes	Recruitment monocytes	Multiple trans-membrane spanning
α_2-macroglobulin	Clearance proteinases	Non–macrophage specific	
Urokinase		Localization, endocytosis	
Transferrin		Endocytosis	
CD14		Receptor for plasma LPS binding protein transduces LPS activation	Macrophage specific
Peptides (many)		Modulation inflammatory functions	
Drugs (many)			

Source: Modified from Fraser, I., Gordon, S. An overview of receptors of MPS cells. Blood Cell Biochem 1993; **5**:1–27.

listed above. Other ligands thought to bind to macrophages are listed elsewhere, including hormones and peptides (44).

B. Phagocytosis and Endocytosis: General Remarks (Fig. 5)

The macrophage is a highly proficient phagocyte sharing the ability to take up particles and infectious agents with other professional and facultative phagocytic cells; in addition, it displays high levels of endocytosis mediated by specific receptors or by fluid phase mechanisms. As a result of extensive membrane flow to the cell surface and to phagosomes and endosomes, the macrophage vacuolar system is highly complex and adapted to host defense, antigen recognition, and pathogen clearance. Although historically recognized as the hall mark of macrophage activity in homeostasis, and extensively studied in relation to the pathogenesis of many disease processes, the molecular basis and regulation of macrophage phagocytosis, endocytosis, and signaled responses, such as secretion of bioactive products, are still being clarified.

The distinction between phagocytic and endocytic uptake is mainly one of ligand particle size, with resultant requirements for movement and remodeling of membrane in bulk and interactions with the cytoskeleton. Energy requirements may also differ. However, involvement of particular plasma membrane receptors is a key determinant of cellular responses, and we also consider these activities

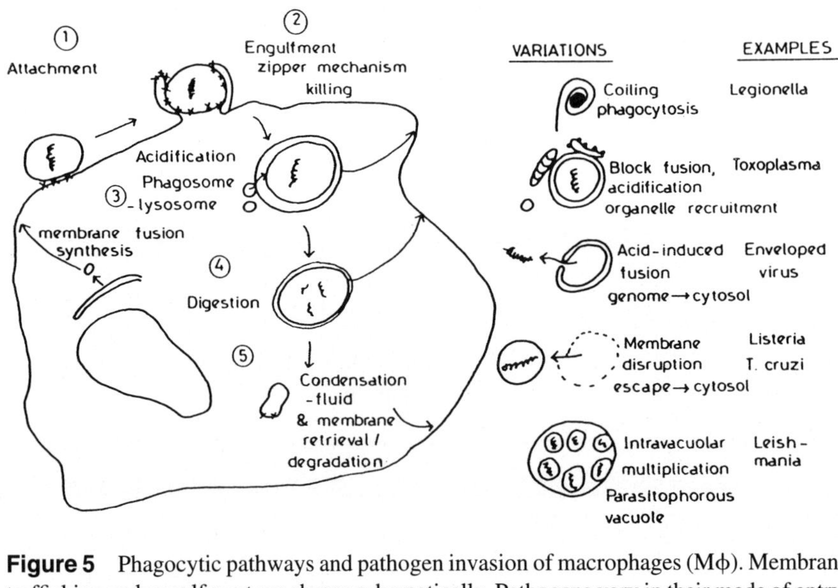

Figure 5 Phagocytic pathways and pathogen invasion of macrophages (Mϕ). Membrane trafficking and engulfment are shown schematically. Pathogens vary in their mode of entry, site of multiplication, and evasion of macrophage defenses.

as a part of a continuum. Additionally, the same molecules can contribute to adhesion to substrata and nonphagocytic extracellular interactions with other cell types. There is, therefore, considerable sorting of molecular and vesicular traffic within the different intracellular compartments and the plasma membrane. We cite several reviews which provide further information for the interested reader (7,45–47).

Endocytosis

We use the term *endocytosis* for nonphagocytic uptake. Receptor-mediated endocytosis is well understood as a result of studies with many cell types, including macrophages. Involvement of specific plasma membrane receptors, which may be cell specific or common, and which are often differentially expressed on different cells, such as alveolar macrophages, enhances the efficiency of ligand uptake compared with fluid phase mechanisms. There has been considerable recent interest in the phenomenon of macropinocytosis, in which solute or ligand uptake, for example, via mannose receptors, can also be considerably enhanced by cell surface trapping mechanisms mediated by plasma membrane ruffling (46). Macropinocytosis may be a particular feature of dendritic cells, where it contributes to antigen processing and presentation and is also relevant to invasion by pathogens, including viruses. Clathrin and other cytoplasmic coat molecules play a role in receptor internalization within vesicles; caveolin-dependent uptake has been implicated as an alternative process in the endocytosis of certain ligands, such as vitamin B_{12} binding proteins. This pathway has not been well defined in macrophages.

Extensive internalization of plasma membrane, as much as the total surface equivalent of the macrophages within 30 min in cell culture, poses major problems of fluid volume and membrane surface area regulation. There is sorting of membrane molecules, vesicle fusion and fission, and membrane retrieval by recycling as well as the addition of new molecules by synthesis and removal by degradation in endosomes and lysosomes. The intracellular fusion machinery is selective, has been reconstituted to some extent in cell-free systems, and involves interactions with cytoplasmic molecules of the G-protein family (48). Acidification and transmembrane ion pumps contribute to the modification of the intravacuolar milieu.

Within the alveolar space, there is active catabolism of pulmonary surfactant; collectins (collagenous lectin-like receptors) and scavenger receptors may be implicated in the recognition of surfactant proteins by alveolar macrophages and other specialized cells lining the alveolar space. The alveolar proteinosis syndrome that results from GM-CSF deficiency in the airway is not yet understood (30–33) but underlines the unexpected importance of specific cytokines acting on macrophages in this physiological process.

Phagocytosis

Macrophages are the major cells involved in the recognition and uptake of apoptotic cells as well as of cells dying by other mechanisms. The type A scavenger receptors account for part of this activity; other "scavenger receptors" such as CD36 and integrins (vitronectin receptors) have been implicated in the removal of senescent neutrophils and other cells (49). Foreign particulates, for example, carbon particles, may be recognized directly, but opsonization with plasma- and tissue-derived opsonins (complement and fibronectin, thrombospondin) contribute to its uptake and to the phagocytosis of a range of inert and infectious particles. Receptors involved in the recognition of biologically inert particles such as latex are not well characterized; once the alternative pathway of complement has been activated, a property of negatively charged or polysaccharide surfaces, the type 3 complement receptor (CR_3), especially, becomes a major mediator of enhanced uptake. There can also be recognition of particulate structures by other, so-called nonopsonic receptors of macrophages; for example, for mannosylated (MR) and polyanionic (SR) ligands. Both of the latter macrophage-restricted receptors are highly expressed on resident alveolar macrophages, which is unlike CR_3, which is present at low levels; however, newly recruited monocytes express CR_3. The ability of these different receptors to signal macrophage responses varies and is regulated precisely, as discussed further below. They may well coexist and cooperate in phagocytosis and play a major role in the invasion of macrophages by microorganisms, including intracellular pathogens such as *Mycobacterium tuberculosis* (see Chapter 17). Clearance of antibody-coated particles is mainly mediated by various Fc receptors, although complement activation can also result in CR involvement.

The main features of phagocytosis have been well described—it is useful to distinguish attachment from ingestion, phagosome formation, and subsequent interactions with lysosomes. Many of the membrane processes alluded to above are again involved: receptor-ligand interactions, internalization, and sorting and fusion as well as the modification of the intraphagosomal contents by acidification and intracellular digestive enzymes and interactions of phagosomes with actin and other constituents of the cytoskeleton. Progressive maturation of the phagosome is a dynamic process involving the retention or loss of plasma membrane molecules and the addition and removal of membrane and membrane-associated cytosolic molecules. A particle can be degraded or persist within the phagosome even after lysosomal fusion (residual bodies); conversely, pathogenic organisms can influence the maturation process and persist at different stages of phagosome development or escape from the vacuole by membrane lysis (e.g., *Listeria monocytogenes*) or by acid-dependent fusion of viral envelopes with endosomal membranes (see Fig. 5).

Phagocytic Mechanisms

Studies in the 1970s by Silverstein and his colleagues gave rise to the widely applicable "zipper" hypothesis of phagocytic ingestion (50). (For a recent evaluation, see ref. 46.) In the case of Fc and CR-mediated uptake of test particles, for example, opsonized erythrocytes, receptor-ligand interactions guide pseudopod extensions of macrophages from local contact to circumferential envelopment followed by actin polymerization and engulfment. Morphological studies are compatible with this hypothesis. It was also evident early on that there are alternative mechanisms that apply to particular agents, such as *Legionella pneumophila* (51). Horwitz and others showed that unopsonized living *Legionella* organisms enter macrophages by a coiling mechanism, in which microbial outer wall components induce wrapping of plasma membrane around the organism reminiscent of the myelin sheath. Opsonization by specific antibodies or killing of the *Legionella* organism brings about a zipper-type pathway. Once ingested, the live organism also maintained a niche for itself inside the specialized vacuole, inhibiting acidification and fusion with lysosomes and attracting strips of endoplasmic reticulum and mitochondria. These striking intracellular morphological features are not unique to this pathogen, but had been observed earlier in studies of *Toxoplasma gondii* (52).

Studies in the past decade of the invasion of macrophages and other cells by a range of pathogens, such as *Salmonella*, have also extended our understanding of phagocytic mechanisms. Contact with an organism can induce extensive membrane ruffling and macropinocytosis, resulting in the capture of organisms in spacious phagosomes rather than the tight apposition of membranes of host cell and pathogen that are evident in the zipper mechanism. Even other forms of triggered uptake, excluded by earlier studies, may become operative in particular cell-target interactions.

One critical determinant, apart from the organism-ligands, is the specific repertoire of macrophage receptors involved. Pathogens can also invade by recruiting distinct membranes from lysosomes. The exclusion of plasma membrane molecules can be subtle, and access of newly synthesized MHC class II molecules is thought to be critical to antigen presentation to naive and primed T lymphocytes. Escape into the cytosol would favor MHC class I association and antigen presentation to cytotoxic T lymphocytes. *Mycoplasma tuberculosis* persists within macrophage phagosomes by interfering with fusion and acidification via H^+ ATPase transfer and delaying vacuolar maturation (47). Entry of *M. tuberculosis* is, in part, attributable to CR_3 and MR has also been implicated, as discussed in Chapter 17. The absence of signaling of cytotoxic product release via CR_3 may account for its frequent use by pathogens to gain entry into macrophages.

MARCKS (myristoylated alanine-rich C-kinase substrate) is a cytoplasmic molecule which, on phagocytic activation, associates rapidly with the nascent phagosome (53). Its association with the actin cytoskeleton and membrane is regulated by Ca^{2+}-dependent phosphorylation and myristoylation, which is one example of a possible motor mechanism for ingestion.

The endosomal transmembrane glycoprotein, CD68 (macrosialin), is a relatively abundant, heavily sialylated molecule that is well expressed by all macrophage populations and at a lower level by dendritic cells (43). As noted above, the molecule is a member of the lamp family of lysosomal glycoproteins found in many cell types but bears a macrophage-restricted mucin-like extracellular domain (17). Complex changes in carbohydrates (O- as well as N-linked and other structures) are induced in macrophages by inflammatory recruitment, immune activation, and receptor-mediated phagocytosis (18). Its function is still unclear, but it may be protective in this acid, hydrolase-rich environment. It has recently been reported that oxidized lipoproteins interact directly with macrosialin (54); other possible ligands remain unknown.

Regulation of Phagocytosis

Expression of individual phagocytic receptor molecules varies during macrophage differentiation, and in response to extrinsic signals, especially the cytokines present in the microenvironment. For example, M-CSF upregulates SR-A, and the Th_2 cytokines IL-4 and IL-13 upregulate MR expression and endocytic functions. The control of ingestion is of considerable significance in host defense and inflammation. Noninternalizable particles, such as asbestos fibers, which provide a ligand for SR-A as well as a complement-activating surface, are too large to permit engulfment and are, in addition, nondegradable, so that persistent recruitment and activation of macrophages are set in train. The metabolic requirements for engulfment are not well understood; selective uptake of *Pseudomonas aeruginosa*, an important lung pathogen, is totally dependent on the presence of extracellular glucose (55). Other particles do not show this requirement in which glucose transporters may be involved. Earlier studies with 2-deoxyglucose showed an analogous inhibition of phagocytosis, but not of binding, and were thought to be independent of intracellular adenosine triphosphate (ATP) levels. In cell cultures models, adhesion of mature macrophages to extracellular matrix was shown to play a major role in the phagocytic capacity of CR_3, and this was thought to depend on ligation of β_1-integrins (e.g., fibronectin receptors) on the basal surface. The physiological state of the macrophage (recruitment, immune activation) also regulates the CR_3 receptor phagocytic activity. Finally, surface contact can be sufficient to trigger secretory and cytotoxic activities and intracellular signaling in the absence of ingestion, which may serve to limit the signal. Microbial products and surface molecules can interfere with any or multiple stages of the phagocytic process or induce alternative pathways to evade cellular defenses. Such mechanisms include the regulation of complement activation or of antibody opsonization (e.g., cleavage or binding of Fc fragments), protease release, interference with intracellular signaling (pertussis toxin), and expression of complex macromolecules (mucins, mycolic acids) which act extra- or intracellularly.

C. Biosynthetic and Secretory Responses (see Chapters 7 and 8)

Following engagement of various receptors (e.g., for cytokines, steroid hormones, peptides), a complex pattern of macrophage genes can be differentially expressed.

NF-κB is one of the better-understood regulators of macrophage gene expression, and reference has been made earlier to other DNA binding proteins that regulate transcription in macrophages. Other, more proximal signaling pathways involving MAP kinase, for example, and *ras*-like molecules are likely to be important in the differential responses of macrophages, but they have not been as well studied as in neutrophils. The adaptive response of macrophages includes the regulation at all levels of protein synthesis and modification as well as in the turnover of intracellular molecules by cytoplasmic and nuclear degradation. Secretory responses (surface expression and controlled release) have been intensively studied since it became evident that macrophages are major secretory cells of the immune system. Macrophages are able to produce a wide range of macromolecules and metabolites that play a major role in inflammation, immunomodulation, repair, host defense, and tissue injury, as is shown in Table 3. Cytokines, such as TNF-α, play a proximal role in complex networks of cellular interactions, and their generation and activity provide targets for therapy. Apart from their potential cytotoxicity (reactive oxygen and nitrogen intermediates, TNF-α, *fas* ligand), other macrophage products serve trophic functions, enhancing cell proliferation (e.g., growth factors), angiogenesis, and repair. Fibrogenic products of macrophages may contribute to scarring in lung, whereas neutral proteinases such as elastase, collagenase, and plasminogen activator, via plasmin formation, can foster connective tissue catabolism. Inhibitors, such as α_1-antiproteinase, are often coexpressed by macrophages, resembling the autoregulatory production of cytokines, such as INF-α/β, IL-10, and IL-12. Factors that influence macrophage secretory activity include phagocytosis, lysosomal storage of poorly degradable bioactive materials, and persistent immunologically specific and nonspecific irritants. These induce local physiological and pathological changes to cell surface and matrix-bound ligands. If excessive or prolonged, macrophage products can enter the circulation (IL-6) or induce vascular permeability (TNF-α), plasma cascade activation, acute phase, and other systemic responses of the host. Many of the pathological changes associated with chronic inflammatory disease can be ascribed to macrophage activation.

IX. Relevance to Lung and Conclusions

The above discussion has been of a general nature and has only referred in passing to lung relevance, because this will be dealt with elsewhere in this volume. Compared with extrapulmonary sites, the lung microenvironment is clearly highly specialized, and macrophages can be expected to adapt to factors, such as high oxygen tension, surfactant, mechanical-ventilatory stress, and an air-blood interface of an extraordinarily large surface area. The other cell types (bronchial epithelium, type II alveolar, and other cell types) can influence local macrophages directly and in turn be influenced by them. The airway is a major portal of entry of

Table 3 Macrophage Derived Secretory Products

Product	Function	Comments
ENZYMES		
Lysozyme	Antimicrobial	Constitutive in vitro, inducible in vivo; myelomonocytic and Paneth's cells
Urokinase	Plasminogen activation and fibrinolysis, neutral proteinase activation	Inducible by macrophage activation, and phagocytosis, cytokines.
Collagenase	Connective tissue catabolism	Inducible
Elastase		
Angiotensin-converting enzyme	Pressor	Steroid inducible
Acid hydrolases (many)	Lysosomal digestion	Mainly intracellular
CYTOKINES		
IL-1 (β) TNF-α	Multiple local and systemic host defense functions, e.g., endothelium, leukocytes, connective tissue	Endogenous pyrogen, tissue injury, weight loss, Superinducible
IFN-α/β	Antiviral, immune modulation	
IL-6	Acute phase response	
IL-10	Inhibitor proinflammatory cytokines and APC function	Also product Th_2 cells
IL-12	Stimulates IFN-γ production by NK and Th cells	
FGF	Fibroblast growth	
TGF-β	Inhibitor activation of macrophages and other targets	
GM-CSF	Granulocyte, macrophage and dendritic cell growth and differentiation	
IL-8	Granulocyte chemoattractant	Chemokine family
MIP-1α/β	Chemoattractant, hematopoietic regulator	Chemokine family
MCP-1	Monocyte recruitment	Chemokine family
Rantes	Recruitment of monocytes and Th cells	Chemokine family

Table 3 (Continued)

Product	Function	Comments
COMPLEMENT PROTEINS		
Most components classic and alternative pathway	Local opsonization and complement activation	Other sources, e.g., liver
Coagulation factors		
Several, including tissue factor	Local initiation and regulation clotting	Other major sources
Adhesion, matrix molecules		
Fibronectin, thrombospondin, proteoglycan	Localization, migration. Modulation of cellular interactions and phagocytosis	Other major sources
Transport proteins		
Transferrin, B12 binding protein, apolipoprotein E,	Transport Fe, vitamin, lipid	Other sources
α_2-macroglobulin	Proteinase inhibitor	
BIOACTIVE LIPIDS		
Cyclooxygenase, lipoxygenase products of arachidonate	Mediators of inflammation, e.g., effects of leukocytes, small vessels	Macrophage has high-level arachidonate in membrances, therefore potent source
PAF	Platelet activation	
REACTIVE OXYGEN INTERMEDIATES		
Superoxide anion, hydrogen peroxide, singlet oxygen, hydroxyl radicals	Killing and stasis of microorganisms and cells by activated macrophages	Tissue injury, modulated by radical scavengers
REACTIVE NITROGEN INTERMEDIATES		
Nitric oxide, nitrites, nitrates	Killing microbial, parasitic, and cellular targets by IFN-γ–activated macrophages	iNOS inducible, compared with low levels of NOS constitutive in other tissues. *Note*: NO not readily detectable in human monocyte/macrophage culture systems
DEFENSINS		
Polypeptides	Antibacterial	Produced only by some macrophages as well as PMN's and Paneth's cells

Source: Adapted from ref. 6, pp. 336–345.

foreign irritants, particles, antigens, and infectious agents, all of which can interact with resident alveolar macrophages. In addition, there are interstitial and pleural resident macrophages as well as the potential to recruit and activate macrophages by immune and other pathways. The macrophages of the lung play a major role in infection (e.g., *Mycobacterium*, *Pseudomonas*, *Legionella*, human immunodeficiency virus [HIV]); in chronic inflammation to particulates and pulmonary fibrosis (silicosis, asbestosis); in the response to cigarette smoke and air pollutants; in allergies and asthma; and in the pathogenesis of emphysema. New genetic mouse models of cytokine deficiency point to a role in surfactant metabolism.

The alveolar macrophage has been well studied in humans and the rat, but because of low yield, less extensively in the mouse. These cells are a semiautonomous population of macrophages, with a distinctive morphology and phenotype; heterogeneity defined in regard to resident and recruited macrophages in other tissues applies also to the lung, but further work is needed to examine the unique properties of the resident macrophages within the lung microenvironment.

Acknowledgments

Work in the authors' laboratory supported by grants from the Medical Research Council (UK) and the Arthritis Research Council. We thank Christine Holt for help in preparing the manuscript.

References

1. Van Furth R, ed. Mononuclear Phagocytes. Biology of Monocytes and Macrophages. Dordrect, The Netherlands: Kluwer, 1992. (Also see earlier volumes of Leiden Conferences.)
2. Lewis C, McGee J O'D, eds. The Macrophage. Oxford, UK: IRL Press, 1992.
3. Russell S, Gordon S, eds. Macrophage Biology and Activation. Current Topics in Microbiology and Immunology, Vol. 181. Berlin: Springer Verlag, 1992.
4. Zembala M, Asherson GL, eds. Human Monocytes. London: Academic Press, 1989.
5. Horton MA, ed. Blood Cell Biochemistry 5. Macrophages and Related Cells. New York: Plenum Press, 1993.
6. Gordon S. The mononuclear phagocyte system and host defense. In: McGee JO'D, Wright NA, Isaacson PG, eds. Oxford Textbook of Pathology. Vol. 1. Oxford, UK: Oxford University Press, 1992: 336–345.
7. Gordon S. Monocytes/phagocytes. In: McGee JO'D, Wright NA, Isaacson PG, eds. Oxford Textbook of Pathology. Vol. 1. Oxford, UK: Oxford University Press, 1992: 236–258.
8. Gordon S. Mononuclear phagocyte system and tissue homeostasis. In: Oxford Textbook of Medicine. Oxford, UK: Oxford University Press, 1995: 84–95.
9. Gordon S, Clarke S, Greaves D, Doyle A. Molecular immunobiology of macrophages: recent progress. Curr Opin Immunol 1995; 7:24–33.

10. Gordon S, Lawson L, Rabinowitz S, Crocker PR, Morris L, Perry VH. Antigen markers of macrophage differentiation in murine tissues. In: Russell S, Gordon S, eds. Current Topics in Microbiology and Immunology. Macrophage Biology and Activation. Vol. 181. Berlin: Springer-Verlag, 1992: 1–37.
11. McKnight AJ, MacFarlane AJ, Dri P, Turley L, Willis AC, Gordon S. Molecular cloning of F4/80, a murine macrophage-restricted cell-surface glycoprotein with homology to the G-protein linked transmembrane 7 receptor family. J Biol Chem 1996; 271:486–489.
12. Rosen H, Gordon S. Monoclonal antibody to the murine type 3 complement receptor inhibits adhesion of myelomonocytic cells in vitro and inflammatory cell recruitment in vivo. J Exp Med 1987; 166:1685–1701.
13. Rosen H, Gordon S. Adhesion molecules and myelomonocytic cell-endothelial interactions. Br J Exp Pathol 1989; 70:385–394.
14. Crocker PR, Gordon S. Mouse macrophage haemagglutinin (sheep erythrocyte receptor) with specificity for sialylated glycoconjugates characterised by a monoclonal antibody. J Exp Med 1989; 169:1333–1346.
15. Crocker PR, Mucklow S, Bouckson V, McWilliam A, Willis AC, Gordon S, Milon G, Kelm S, Bradfield P. Sialoadhesin, a macrophage-specific adhesion molecule for haemopoietic cells with 17 immunoglobulin-like domains. EMBO J 1994; 13:4490–4503.
16. McWilliam A, Tree P, Gordon S. Interleukin-4 regulates induction of sialoadhesin, the macrophage sialic acid specific receptor. Proc Natl Acad Sci USA 1992; 89:10522–10526.
17. Holness CL, da Silva RP, Fawcett J, Gordon S, Simmons DL. Macrosialin, a mouse macrophage restricted glycoprotein, is a member of the lamp/lgp family. J Biol Chem 1993; 268:9661–9666.
18. da Silva RP, Platt N, de Villiers WJS, Gordon S. Membrane molecules and macrophage endocytosis. Scavenger receptor and macrosialin as markers of plasma membrane and vacuolar functions. Biochem Soc Trans 1996; 24:220–224.
19. Fraser IP, Hughes DA, Gordon S. Divalent cation-independent macrophage adhesion inhibited by monoclonal antibody to murine scavenger receptor. Nature 1993; 364:343–346.
20. Hughes DA, Fraser IP, Gordon S. Murine macrophage scavenger receptor: in vivo expression and function as receptor for macrophage adhesion in lymphoid and non-lymphoid organs. Eur J Immunol 1995; 25:466–473.
21. de Villiers WJS, Fraser IP, Hughes DA, Doyle AG, Gordon S. M-CSF selectively enhances macrophage scavenger receptor expression and function. J Exp Med 1994; 180:705–709.
22. Gordon S. (sec. ed.). In: Weir DM, Herzenberg LA, Herzenberg LA, eds. Handbook of Experimental Immunology. 5th ed. Blackwell, 1996. In press.
23. Morris L, Graham CF, Gordon S. Macrophages in haemopoietic and other tissues of the developing mouse detected by monoclonal antibody F4/80. Development 1991; 112:517–526.
24. Morris L, Crocker PR, Fraser IP, Hill M, Gordon S. Expression of a divalent cation-dependent erythroblast adhesion receptor by stromal macrophages from murine bone marrow. J Cell Sci 1991; 99:141–147.

25. Platt N, Suzuki H, Kurihara Y, Kodama T, Gordon S. A role for the macrophage scavenger receptor in the elimination of apoptotic thymocytes. 1996; Proc Nat Acad Sci USA. In press.
26. Hopkinson-Woolley J, Hughes D, Gordon S, Martin P. Macrophage recruitment during limb development and wound healing in the embryonic and foetal mouse. J Cell Sci 1994; 107:1159–1167.
27. Perry VH, Gordon S. Macrophages and the nervous system. Int Rev Cytol 1991; 125: 203–244.
28. Witmer-Pack MD, Hughes D, Schuler G, Lawson L, McWilliam A, Inaba K, Steinman R, Gordon S. Identification of macrophages and dendritic cells in the osteopetrotic [op/op] mouse. J Cell Sci 1993; 104:1021–1029.
29. Scott EW, Simon MC, Anastasi J, Singh H. Requirement of transcription factor PU.1 in the development of multiple hematopoietic lineages. Science 1994; 265:1573–1577.
30. Lieschke GJ, Stanley E, Grail D, Hodgson G, Sinickas V, Gall JAM, Sinclair RA, Dunn AR. Mice lacking both macrophage- and granulocyte-macrophage colony-stimulating factor have macrophages and coexistent osteopetrosis and severe lung disease. Blood 1994; 84:27–35.
31. Dranoff G, Crawford AD, Sadelain M, Ream B, Rashid A, Bronson RT, Dickersin GR, Bachurski CJ, Mark EL, Whitsett JA, Mulligan RC. Involvement of granulocyte-macrophage colony-stimulating factor in pulmonary homeostasis. Science 1994; 264: 713–716.
32. Huffman AJ, Hull WM, Dranoff G, Mulligan RC, Whitsett JA. Pulmonary epithelial cell expression of GM-CSF corrects the alveolar proteinosis in GM-CSF–deficient mice. J Clin Invest 1996; 97:649–655.
33. Hallmann M. Lack of GM-CSF as a cause of pulmonary alveolar proteinosis (editorial). J Clin Invest 1996; 97:589–590.
34. Dalton DK, Pitts-Meek S, Keshav S, Figari IS, Bradley A, Stewart TA. Multiple defects of immune cell function in mice with disrupted interferon-γ genes. Science 1993; 259:1739–1742.
35. Okamura H, Tsutsui H, Komatsu T, Yutsudo M, Hakura A, Tanimoto T, Torigoe K, Okura T, Nukada Y, Hattori K, Akita K, Namba M, Tanabe F, Konishi K, Fukuda S, Kurimoto M. Cloning of a new cytokine that induces IFN-γ production by T cells. Nature 1995; 378:88–91.
36. Baud V, Chissoe SL, Viegas-Péquignot E, Diriong S, N'Guyen VC, Roe BA, Lipinski M. EMRI, an unusual member in the family of hormone receptors with seven transmembrane segments. Genomics 1995; 26:334–344.
37. Reid DM, Perry VH, Andersson P-B, Gordon S. Mitosis and apoptosis of microglia in vivo induced by an anti-CR3 monoclonal antibody which crossed the blood-brain barrier. Neuroscience 1994; 56:529–533.
38. Krieger M. Structures and functions of multiligand lipoprotein receptors: macrophage scavenger receptors and LDL receptor-related protein (LRP). Annu Rev Biochem 1994; 63:601–637.
39. Ravetch JV. Fc receptors: rubor redux. Cell 1994; 78:553–560.
40. Pontow SE, Kery V, Stahl PD. Mannose receptor. Int Rev Cytol 1992; 137B:221–244.
41. Crocker PR, Freeman S, Gordon S, Kelm S. Sialoadhesin binds preferentially to cells of the granulocytic lineage. J Clin Invest 1995; 95:636–643.

42. Kelm S, Pelz A, Schauer R, Filbra MT, Tang S, De Bellard M-E, Schnaar RL, Mahoney JA, Hartnell A, Bradfield P, Crocker PR. Sialoadhesin, MAG and CD22 define a new family of sialic acid-dependent adhesion molecules of the immunoglobulin superfamily. Curr Biol 4:965–972.
43. Rabinowitz S, Gordon S. Macrosialin, a macrophage-restricted membrane sialoprotein differentially glycosylated in response to inflammatory stimuli. J Exp Med 1991; 174:827–836.
44. Fraser I, Gordon S. An overview of receptors of MPS cells. In: Horton M, ed. Blood Cell Biochemistry 5: Macrophages and Related Cells. New York: Plenum Press, 1993:1–27.
45. Rabinovitch M. Professional and non-professional phagocytes: an introduction. Trends Cell Biol 1995; 5:85–87.
46. Swanson JA, Baer SC. Phagocytosis by zippers and triggers. Trends Cell Biol 1995; 5:89–92.
47. Russell DG. Mycobacterium and Leishmania: stowaways in the endosomal network. Trends Cell Biol 1995; 5:125–128.
48. Beron W, Alvarez-Dominguez C, Mayorga L, Stahl PD. Membrane trafficking along the phagocytic pathway. Trends Cell Biol 1995; 5:100–104.
49. Ren Y, Silverstein RL, Allen J, Savill J. CD36 gene transfer confers capacity for phagocytosis of cells undergoing apoptosis. J Exp Med 1995; 181:1857–1862.
50. Silverstein SC, Steinman RM, Cohn ZA. Endocytosis. Annu Rev Biochem 1977; 46: 669–722.
51. Horwitz MA. Formation of a novel phagosome by the Legionnaires' disease bacterium (Legionella pneumophila) in human monocytes. J Exp Med 1983; 158:1319–1331.
52. Jones TC, Hirsch JG. The interaction between Toxoplasma gondii and mammalian cells II. The absence of lysosomal fusion with phagocytic vacuoles containing living parasites. J Exp Med 1972; 136:1173–1194.
53. Allen HL-A, Aderem A. A role for MARCKS, the α isozyme of protein kinase C and Myosin I in zymosan phagocytosis by macrophages. J Exp Med 1995; 182:829–840.
54. Ramprasad MP, Fischer W, Witztum JL, Sambrano GR, Quehenberger O, Steinberg D. The 94- to 97-kDa mouse macrophage membrane protein that recognizes oxidized low density lipoprotein and phosphatidylserine-rich liposomes is identical to macrosialin, the mouse homologue of human CD68. Proc Nat Acad Sci USA 1995; 9580–9584.
55. Speert DP, Gordon S. Phagocytosis of unopsonized Pseudomonas aeruginosa by murine macrophages is a two-step process requiring glucose. J Clin Invest 1992; 90: 1085–1092.

2

Macrophage Activation

Concepts and Principles Applicable to the Lung

LEWIS J. WESSELIUS

Veterans Affairs Medical Center
Kansas City, Missouri
and University of Kansas Medical Center
Kansas City, Kansas

WILLIAM J. MURPHY, DAVID C. MORRISON, and STEPHEN W. RUSSELL

The Kansas Cancer Institute
University of Kansas Medical Center
Kansas City, Kansas

I. Introduction

There are many variables that influence the interactions that macrophages in the lung have with their environment. As a consequence, there is no single model of how pulmonary macrophages accomplish the many things that they do, which range from defense of the lung and the orchestration of tissue repair and remodeling to, in some instances, contributing to the pathogenesis of certain pulmonary diseases and to lung injury. One unifying suggestion is that often offered as to how these functions are accomplished, however, is that of macrophage "activation."

The purpose of this chapter is to summarize newer concepts that pertain to activation of macrophages in general and pulmonary macrophages in particular. Both the positive and negative consequence of activation will be considered in the context of various conditions and diseases that affect the lung. Because there is not sufficient space here to review the field comprehensively, it has been necessary to exclude many important works; however, the majority of these have either been considered in complementary chapters within this volume or in previously published reviews and books on macrophages (1–4) and macrophage activation (3,5–7), to which the reader's attention is drawn.

II. Macrophage Activation: Dependency on Gene Regulation

A. The Starting Point: Pulmonary Macrophages

The lung potentially contains two generic kinds of macrophage populations, one that is resident and another that is mobilized under conditions of inflammation. This combination ensures that an effective macrophage-mediated component can be integrated with other defensive mechanisms when the lung is threatened.

Resident Macrophages

Resident macrophages can be found within the interstitium, lining the alveoli, within the airways (both in the lumen and within the lining epithelium) and, in some species, perhaps including human beings (8), adherent to the endothelial surfaces of pulmonary blood vessels (9,10). The origins of these populations have been the subject of past controversy. At present, however, the general consensus is that there are likely to be two origins for resident pulmonary macrophages: (1) differentiation from blood monocytes that enter the lung from the circulation and (2) replication of precursors within the lung. With respect to the former source, the first unequivocal demonstration that alveolar macrophages derive from cells that circulate in the blood was provided by Ungar and Wilson in 1935 (11). They labeled elicited peritoneal macrophages with carbon and showed that, subsequent to their intravenous injection, many of the labeled cells could be found lining the alveolar spaces of the lungs. Their controls consisted of guinea pigs injected intravenously with carbon, an approach that produced very few labeled cells in the alveoli. Later, in the 1960s, Bowden and Adamson (12) proposed a model that has a number of attractive features. They hypothesized that resident pulmonary macrophages represent a differentiating continuum of mononuclear phagocytes originating from blood monocytes. These were proposed to enter the lung's interstitial tissue where they would differentiate into macrophages. These interstitial macrophages were then thought to translocate into the alveoli (13), and from there, according to Lehnert and his colleagues (14), would migrate cephalad into airways of increasing size. Whether or not this migratory pathway actually exists has not been proved conclusively. However, there is now no question that the great majority of resident pulmonary macrophages originate from the bone marrow, with a turnover time that has been estimated to range from approximately 27 days in mice (15) to 81 days in human beings (16). With respect to the second source of resident pulmonary macrophages, Sorokin et al. (17) concluded, on the basis of morphological evidence, that there is a replicating population of macrophage-like cells in the interstitium of fetal lung before bone marrow and circulating monocytes have developed. They suggested that these primitive cells could give rise to at least a portion of the resident pulmonary macrophage population. Taken together with the fact that pulmonary macrophages have low (<1%), but detectable, DNA synthetic activity (18) and are capable of division (19,20), it is probable that local replication of macrophages helps to sustain the resident pulmonary population. Given these probable mechanisms of development, coupled with the influence that environmental factors have on the phenotypic character of resident macrophages, it is not surprising that they are a heterogeneous population, both physically and functionally. For example, studies of density-separated rat alveolar macrophages showed differences in expression of Fc receptors, class II major histocompatibility antigen expression, and the capacity to become activated to kill tumor cells (21,22). Additional evidence of heterogeneity in lung macrophages includes functional differences that are reported to exist between rat alveolar and interstitial macrophages, including their capacity to produce nitric oxide (NO) when activated with the combination of interferon-γ (IFN-γ) and bacterial lipopolysaccharide (LPS) (23).

Inflammatory Macrophages

Resident pulmonary macrophages are immediately available to contend with microbial and other kinds of particulate material that enter the lung, either via the blood or after inhalation. Inflammatory macrophages, on the other hand, are mobilized as a result of changes that occur during the development of an inflammatory response. Early on in a host's inflammatory reaction, platelets aggregate and release their mediators. Local vessels are permeabilized by these and other locally produced mediators with the result that plasma proteins leak into the affected area where they are cleaved by proteinases and converted into active mediators. Endothelial cells and other cell types in the area also begin to produce proinflammatory mediators. One of the systemic effects of a developing inflammatory reaction is the increased production of bone marrow–derived monocytes (24,25), which, in the face of an inflammatory drain, would appear to remain in the circulation for a shorter than normal period of time. The latter phenomenon is directly attributable to the production of chemotactic agents that attract inflammatory cells from the blood. Their ability to do so is, in part, due to modulation of adhesion molecules on the surfaces of monocytes as well as on endothelial cells (26,27). Recruited monocytes then differentiate on site into macrophages. A replicative component also helps in a relatively minor way to augment the inflammatory build-up of macrophages (19,20,28).

B. General Properties of Activated Macrophages

Several different investigators have recently reviewed the complex interrelations that reciprocally exist between activated T lymphocytes and macrophages (29–33), how the principal macrophage-activating stimuli mediate their effects (5,34–

39), the receptors that are involved (1,4,40–44), and the products of macrophages that have been activated in various ways (1,4,45). We will not, therefore, review these same subjects here. Instead, our intention is to extend Paulnock's recent outstanding review of inducible genes and related transcription factors that are the bases for many activation phenomena (46); underscore the importance of the autocrine/paracrine feedback of macrophage-produced products as a means of regulating many kinds of activation; call attention to the fact that population-based responses to an activating stimulus can be heterogeneous as well as homogeneous; and apply what is known about macrophage activation to better understand the positive and negative effects that can accompany this phenomenon in the lung.

Localization of the Activation Process

Both from the standpoint of an efficient host defense and the prevention of negative consequences that can attend body-wide activation, a key need is to restrict the inflammatory activation of monocytes-macrophages to sites of need. The first of the related regulatory mechanisms is associated with emigration of circulating monocytes into the tissues where they first become adherent to endothelial cells and then to the extracellular matrix. This change in their physical status, adherence, provides one of the first "release points" for aspects of the activation process, because attachment to a substratum is required for expression of certain inducible genes by macrophages. The response can be modified further depending on the nature of the extracellular matrix that is encountered (47–50). Developmental regulation of the availability of specific transcription factors, e.g., NF–IL-6 (51) and Stat 1 (Stat is an acronym for signal transducer and activator of transcription) (52,53), likely contributes to some of these functional changes as monocytes mature into macrophages. An additional level of developmental control over macrophage activation has been observed by McDowell et al. (53a). These investigators demonstrated that Stat 1, a transcription factor that is activated in response to stimulation by IFN-γ (see below), was detectable in cells of the mouse macrophage line WEHI-3. Yet this cell line was not responsive to added IFN-γ. Thus, the activation of Stat 1, itself, could be developmentally regulated, perhaps through the availability (or lack thereof) of additional components needed in the activation process. Finally, some genes (54–56) are induced or augmented in their expression by serum components that are available only at sites of inflammation due to localized intravascular and extravascular clotting. *Collectively, these and other mechanisms help to ensure that macrophages develop activated functions only at sites where they are needed: that is, locally rather than on a systemic basis.* This is particularly important with respect to the secretion of cytokines, because the body-wide production of cytokines by macrophages can contribute to the pathogenesis of the systemic inflammatory response syndrome, which can lead to secondary multiple organ dysfunction and death (57). Tight regulation is also important at the local level, because activated macrophages are powerful secretory cells (1,40,58,59) that, when uncontrolled, can elaborate large quantities of highly noxious products. These may injure other cells, suppress aspects of the immune response, digest stroma, and inappropriately stimulate growth, to name but a few of the consequences that can negatively affect the lung.

Modulation of Macrophage Functions to Meet a Specific Need

One of the key functional characteristics of mononuclear phagocytes is that, depending on their stage of differentiation and microenvironmental factors that they encounter at a site of inflammation, they can be tailormade, that is, "activated," to have whatever functions are needed to deal most effectively with a given kind of inciting stimulus. *Implicit in this statement is the fact that there is no one kind of activated macrophage, nor is there a single process that leads to an activated state.* Furthermore, most, if not all, stimuli have the effect of activating more than a single function; more often than not by modulating the expression of the target macrophage's genes. Thus, the activated macrophage must be regarded as having the potential for great diversity, a point that has recently been emphasized by Adams and Hamilton (60). Accordingly, to convey the nature of an activated population as accurately as possible, one must specify the stimulus that was used, as well as the function for which the macrophages of a specific species was activated. A complete characterization of an activated population would, for example, be the activation of mouse macrophages for the production of NO by the combination of IFN-γ and LPS. Because the NO-producing system has so many properties in common with activation for other functions, and because it is one of the best understood at the molecular level, it will now be used as a model to introduce how gene expression allows macrophages to acquire new functions; that is, to become activated.

IFN-γ/LPS-Induced Production of NO as a Model of Gene-Based Activation

In many instances, the phenotype that an activated macrophage evolves within a developing lesion is often due to the expression of inducible genes that are under the influence of locally available proinflammatory stimuli. The extensive array of host-derived components that is concentrated in the inflammatory milieu serves to amplify the initial inflammatory or immune reaction until it becomes adequate to eliminate the inciting stimulus. At that time, without continued stimulation from the target, recruitment of inflammatory macrophages subsides, the local production of activating stimuli and augmenting factors wanes, and negative control mechanisms become dominant. These typical characteristics are all found in the NO production system.

Role of NO Synthase in Producing NO

Nitric oxide is a small reactive nitrogen intermediate that has broad and enormously important effects (61). Among these are the capacity to regulate vascular tone and neuronal transmission and to kill both microorganisms and tumor cells that are susceptible to its effects. The production of NO is dependent on the catalytic activity of an enzyme, NO synthase (NOS), that is available in three different isoforms (62). The vascular and neuronal production of NO is catalyzed by constitutively expressed endothelial and neuronal isoforms of NOS, which are responsible for the low-level production of NO at the sites implied by their names. The production of large quantities of NO, by contrast, is dependent on an inducible isoform of the enzyme, iNOS. Many cell types are capable of expressing this gene; however, macrophages are one of the most important (63). Maximal production by mouse macrophages of both iNOS and, therefore, NO is obtained when they are first primed by one of the interferons and then triggered for function by LPS (64). This observation provides a mechanistic explanation for the phenomena of "priming" and "triggering" that were associated in the late 1970s with the stepwise activation of mouse macrophages for tumor cell killing (65–67). This model has since proved to be equally applicable to the killing of NO-sensitive microorganisms (68). However, it must be emphasized that, although the production of NO is necessary to kill susceptible microbes and tumor cells, it alone is not sufficient to do so (69).

Role of Autocrine/Paracrine Feedback in LPS-Mediated Induction of iNOS/NO

Autocrine/paracrine feedback is an important regulatory mechanism in the production of NO by LPS. In fact, without such feedback, LPS alone would have little or no ability to induce the production of NO by mouse macrophages. Evidence of this fact includes the observation that NO production and killing are abrogated if the secretion of IFN-α/β by LPS-stimulated macrophages is either eliminated or neutralized (70,71). Thus, autocrine/paracrine feedback of one of the type I interferons is needed for LPS to induce iNOS gene expression. This activation mechanism is undoubtedly of importance during the early stages of an inflammatory response before immunologically produced IFN-γ is available to provide an augmenting stimulus. Given these facts, it is possible to understand mechanistically why only relatively high amounts of LPS will effect the substantive production of NO: Only those concentrations will induce the secretion of IFN-γ/β. Tumor necrosis factor-α (TNF-α), which is also an LPS-induced product of macrophages, similarly feeds back to help enhance the production of NO (72–74). Although these examples are both dependent on the initial stimulation by LPS, autocrine/paracrine feedback as a regulatory mechanism should not be restricted to this stimulus; it is of much greater importance. For example, IFN-γ upregulates the production of IFN-α/β by macrophages (72,75), which then can feed back to activate the same cells for the production of the complement component C1q (76). These examples, as well as those of negative regulation that will be described later, serve to underscore the importance that autocrine/paracrine feedback of a range of products has in regulating various kinds of macrophage activation.

Transcriptional and Posttranscriptional Mechanisms of Regulation of iNOS/NO Production

Much of the phenomenology of the activation of mouse macrophages for NO production can now be explained at the molecular level. The first major step was cloning of the cDNA for iNOS (77,78). Its availability allowed the mechanisms that underlie priming and triggering by IFN-γ and LPS, respectively, to be identified. The primary regulatory pathway is transcriptional. For example, little or no constitutive transcription of the mouse iNOS gene can be detected in unstimulated cells; LPS increases the rate of transcription to a modest extent; and IFN-γ markedly increases LPS-induced transcription, causing both earlier onset and a much higher maximal level of gene transcription (78,79). Although transcriptional regulation appears to be most important in the production of iNOS, there may be posttranscriptional regulation as well. For example, Weisz et al. (80) found that treatment of RAW 264.7 macrophages with either IFN-γ, LPS, or LPS/IFN-γ comparably increased the rate of iNOS gene transcription, but accumulated iNOS mRNA was more stable (i.e., had a longer half-life) and, therefore, was more abundant in LPS-stimulated cells (with or without IFN-γ). The majority of these findings were not made earlier by Lorsbach and his colleagues (79). The differences may be explainable on the basis of the cell lines that were employed, because the latter workers used RAW 264.7γNO(−) macrophages, a variant that does not produce NO in response to IFN-γ as cells of the parent RAW 264.7 cell line do.

Activation of Transcription Factors

The next key advance was cloning of the promoter that regulates the expression of the iNOS gene (81,82). Subsequent analysis showed that the 5′-flanking DNA immediately upstream from the coding region contains a TATA box and two more distal regulatory regions. One of these was LPS responsive (region I), whereas the other (region II) was responsive to both LPS and IFN-γ. Low-level, basal induction of transcription by LPS could be obtained with region I alone; however, region II was required to obtain the maximal production of iNOS under the combined influence of the two stimuli. Given this latter result, it was not surprising when Alley et al. (83) subsequently showed that region II had the characteristics of a classic enhancer element. At least five response elements are now known to be functionally active in the two regions—an octamer element in region I (84,85), two juxtaposed interferon-stimulated response elements (ISREs) (86,87), and INF-γ (gamma)–activated site (GAS) (88) in region II; as well as the κB site in each of region I (89,90) and region II (89). Alone, this information is not particularly helpful in understanding the process of macrophage activation by IFN-γ/LPS for the expression of the mouse iNOS gene. However, when it is appreciated that IFN-γ/LPS activate specific transcription factors that bind to these regulatory elements, it becomes apparent that *acquisition of an activated phenotype is, in many instances, attributable to the regulation of gene expression through activation of specific transcription factors.*

Low levels of LPS (in the nanogram range), which induce neither IFN-β nor NO production, cause the efficient activation of transcription factor NF-κB (89). Thus, although NF-κB is necessary for induction of the iNOS gene, it alone is not sufficient. The activation of NF-κB is accomplished by the induction of a kinase/protease pathway that leads to the phosphorylation and destruction of a protein, IκB, that binds members of the NF-κB family of proteins, thereby rendering them inactive (91). Once free of IκB, NF-κB proteins are able to translocate to the nucleus, where they become available to help regulate the expression of genes bearing κB elements in their promoters and that are induced primarily by NF-κB alone (e.g., TNF-α). The LPS-mediated activation of NF-κB is additionally dependent on the availability of reactive oxygen intermediates (ROI) (92). Silica dust, which is also an activator of NO production by rodent macrophages (93), like LPS, has the capacity to induce the production of reactive oxygen intermediates (94,95) and to activate NF-κB (96). Whether or not these inductive properties of silica can be linked is unclear, however, because, in the latter study, the antioxidant N-acetylcysteine had no effect on the silica-induced activation of NF-κB even though it completely inhibited the activation of that same transcription factor by LPS. In the activation of NF-κB by LPS (97) and other stimuli (98), the most important species of reactive oxygen appears to be the hydroxyl radical. The Fenton reaction, which is key to the generation of the hydroxyl radical, is dependent on the availability of iron (99,100). It is not surprising, therefore, that iron concentration affects the expression of the mouse iNOS gene, although in a way that might be unexpected: increasing iron suppressed the expression of the iNOS gene in a study by Weiss et al. (101). This may be related to the fact that cellular responses to oxidants are often biphasic. Therefore, if raising the concentration of iron increased the level of ROI beyond a certain point, the paradoxical effect of lowering gene expression could be the result.

The expression of another inducible enzyme in macrophages, prostaglandin G/H synthase-2 (PGHS-2; also called cyclooxygenase-2), has also been discovered to depend on ROI (102,103). This observation, coupled with what is known about the activation of the iNOS gene, has stimulated *increasing interest in reactive oxygen intermediates as important contributors to the biology of macrophage activation for a broad range of functions.*

IFN-γ activates the transcription factors Stat 1α and IRF-1. IRF-1 is synthesized de novo after macrophages are treated with IFN-γ, a process that requires the activation of Stat 1α (104). Stat 1α is one of a family of latent transcription factors that are tyrosine-phosphorylated by members of the Janus kinase (JAK) family. These latter enzymes are associated with the IFN-γ receptor (105). Once it becomes phosphorylated, Stat 1α forms a homodimer that translocates to the nucleus where it can associate with the GAS elements that are found in the promoters of the iNOS (88) and IRF-1 (106) genes. Thus, the effect of Stat 1α on the iNOS gene is both direct and indirect. The latter indirect effect is through the IRF-1 gene's product, which, once synthesized, translocates to the nucleus to interact with the iNOS gene (86,87). As mentioned earlier, the availability of Stat 1α and its ability to be activated is developmentally regulated in the mononuclear phagocyte lineage (52–53a). This observation explains, in part, why immature members of the mononuclear phagocyte lineage are not as responsive to IFN-γ as fully mature macrophages are.

IFN-α/β, like IFN-γ, can also cause the phosphorylation of Stat 1α (107) and, as a consequence, IRF-1 synthesis (108). However, compared with IFN-γ, it takes 500–1000-times more antiviral units (and also substantially more on a molar basis) of either IFN-α or INF-β to achieve the same stimulatory effect when mouse macrophages are activated in the presence of LPS for tumor cell killing (109). This is likely due to the fact that IFN-α/β causes the formation of an additional transcription factor complex, interferon-stimulated gene factor-3 (ISGF-3), to be produced, which contains phosphorylated Stat 1α as one of its components. Thus, in cells stimulated with IFN-α/β, activated Stat 1α is apportioned between ISGF-3, which the IRF-1 promoter cannot bind, and Stat 1α homodimers, which it can bind. The result of this more complex distribution of Stat 1α is most likely the lower availability of Stat 1α homodimers for *trans*-activation of the IRF-1 gene and, as a consequence, the lower production of IRF-1. Decreased cellular levels of Stat 1α and, as a consequence, IRF-1, could explain why macrophages stimulated with high concentrations of LPS (and autocrine/paracrine IFN-α/β) produce low levels of NO as compared with those stimulated with LPS+IFN-γ.

In addition to the κB, GAS, and ISRE binding transcription factors described above, octomer binding transcription factors are also involved in the transcriptional regulation of the iNOS gene (84). In vivo footprinting analysis reveals the protection of a guanosine residue on the bottom strand of the octamer sequence after macrophages are activated with either LPS or LPS/IFN-γ. The exact identity of the binding protein is uncertain, because mature macrophages contain both Oct-1 and Oct-2, and the detection of these binding proteins, as assayed in vitro by gel-shift analysis, is unaffected by treatment of cells with LPS, IFN-γ, or combinations thereof. Analysis of developing monocytes-macrophages, however, demonstrates a decrease in Oct-1 in maturing macrophages and a concomitant increase in Oct-2 (85,110), implicating the latter in iNOS gene induction.

The activation of two different sets of transcription factors by LPS and IFN-γ explains, at the molecular level, the synergistic activity between these two different activators when they are combined to stimulate rodent macrophages (Fig. 1). The activation of one set of factors or the other (e.g., by treatment of cells with either low stimulatory concentrations of LPS or, in some strains of mice, IFN-γ alone [111]) leads to little or no iNOS gene transcription. However, the activation of both sets of transcription factors, even by the use of suboptimal concentrations of LPS + IFN-γ, leads to the strong activation of the iNOS gene. In other words, *optimal rates of transcription can only be achieved when a full complement of transcription factors is activated in any given cell.* Thus, transcription of the iNOS gene appears to be an all or none phenomenon, which is most likely the cause of the heterogeneous response that populations of mouse macrophages make when they are exposed to activators for iNOS expression (see next section).

Augmentation by the IFNs of LPS-mediated production of iNOS/NO, as well as of other initiating stimuli as different from LPS as asbestos fibers (112), is typical of how a number of different proinflammatory macrophage genes are activated (113). Specific examples include TNF-α (114–117) and PGHS-2 (118,119). The enzyme produced by the latter gene is rate limiting in the production of large amounts of prostanoids by macrophages. Thus, *an important regulatory strategy that provides important checks and balances against the indiscriminate or excessive production of proinflammatory mediators by activated macrophages is the requirement that at least two stimuli be present in order to obtain the maximal expression of many different inducible genes.* Additional control, which contributes to the localization of the production of the mediators, stems from the fact that, as in the examples given above, one stimulus must be produced by the host (IFN) and the other comes from the invading microorganism (LPS). Without both stimuli, the large-scale local production of these potentially dangerous gene products is not possible.

Thus, *the functional diversity of the macrophage lineage is a reflection of several independent influences: the genetic character of the individual, the stage of development of the subject mononuclear phagocyte, the microenvironment in which the cell finds itself, and the types of substrata and combinations of mediators it encounters there.* These influences primarily determine which transcription factors will be available, which of them can be activated in response to a given stimulus, and how they will combine with one another qualitatively and quantitatively to cause expression or amplification of a given function.

Population Heterogeneity of Gene Expression in Response to IFN-γ/LPS

When a given function is induced by an activating stimulus, it is usually not possible to know whether all cells within the population or just a few develop the sought-after function. Similarly, when functional activity is augmented, for example, by increasing the concentration of an activating stimulus, it is not clear

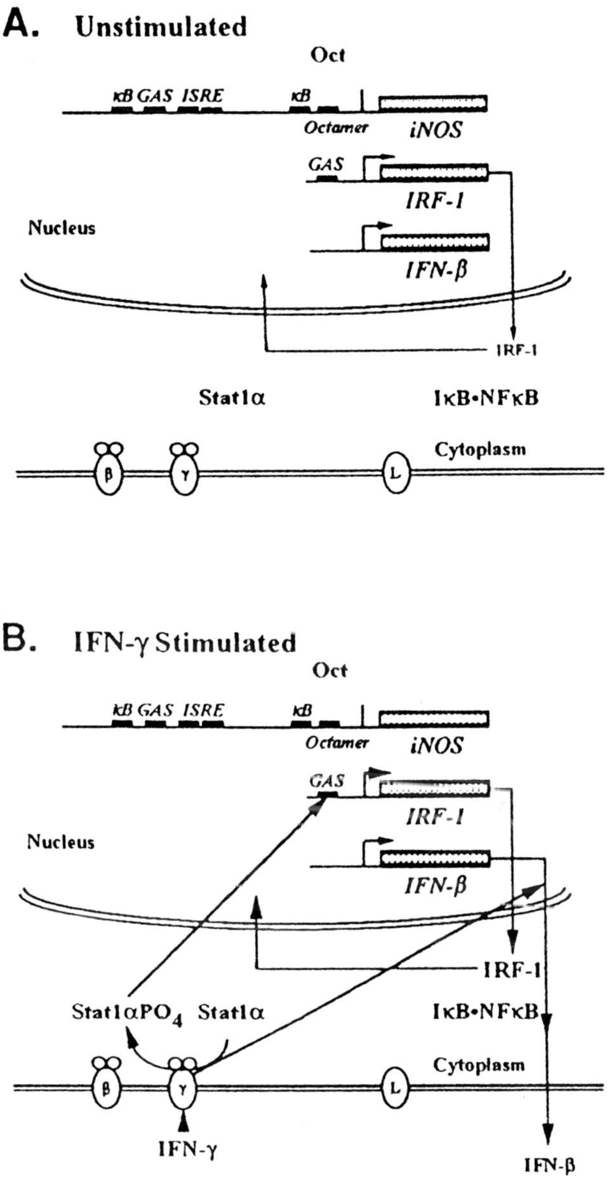

Figure 1 (legend on p. 45)

C. LPS Stimulated

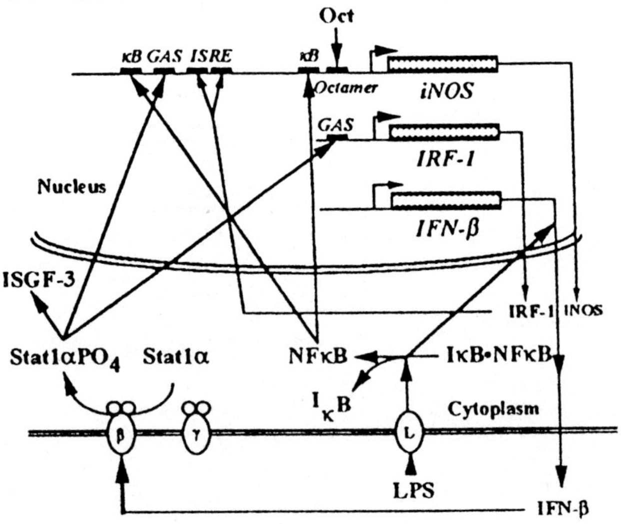

D. IFN-γ+LPS Stimulated

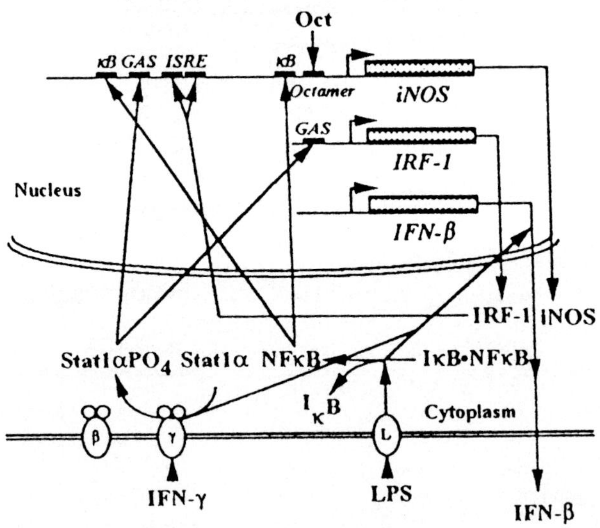

Figure 1 (continued)

Figure 1 Proposed transcription factor interactions that lead to different levels of mouse iNOS gene expression. At the bottom of each of the four panels, there are three ovals, which are meant to depict receptors in the plasma membrane of a mouse macrophage (β, IFN-βR, γ, IFN-γR; and L, LPSR). The two small spheres on the cytoplasmic side of each of the IFN receptors represent Janus kinases (Jak). The curved double line in the approximate middle of each panel represents the nuclear membrane. Above it are cartoons of the *iNOS* (top), *IRF-1* (middle), and *IFN*-β (bottom) genes and their associated promoters. Response elements in the promoters of the *iNOS* (two κB, one GAS, two closely associated ISREs, and one octomer) and *IRF-1* (only the GAS is shown) genes are depicted as raised segments. Gene products of the *IRF-1* and *IFN*-β genes are depicted in roman letters, with arrows indicating where these proteins have their effect(s). The cytoplasmic complex IκB·NF-κB is depicted in panels C and D to be dissociating, with IκB being degraded (letters no longer on the same line). (A) Unstimulated cells. Without stimulation, the iNOS gene is not transcribed. Both the IRF-1 and IFN-β genes are constitutively transcribed at low levels, but there is no detectable effect on the iNOS promoter. The reasons are quantitative. For example, although the IRF-1 gene is transcribed, only low levels of IRF-1 mRNA and protein are produced, and none of the other requisite factors (e.g., NF-κB) is present. In the case of the IFN-β gene, transcription results in neither mRNA nor protein under these conditions. (B) Stimulation with IFN-γ. Engagement of the receptor for IFN-γ leads to activation of the cytoplasmic transcription factor, Stat1α, with the result that homodimers of its phosphorylated form translocate to the nucleus, where it binds to the *IRF-1* GAS. This results in increased expression of the gene and, therefore, increased production of IRF-1 protein. It is hypothesized that Stat 1αPO_4 does not bind to the GAS of the *iNOS* promoter, because it must be part of a larger complex to do so. By mechanisms that are yet unknown, engagement of the IFN-γ receptor also stabilizes IFN-β mRNA, making low-level production of IFN-β possible. In some strains of mice, e.g., C57BL/6, treatment with IFN-γ alone can lead to iNOS gene expression, presumably because other transcription factors needed to combine with Stat1αPO_4 are constitutively produced and, therefore, available to form the hypothesized complex that *trans*-activates the gene. (C) Stimulation with high doses of LPS. This form of stimulation leads to the activation of the octamer and NF-κB transcription factors, the latter because IκB is phosphorylated and destroyed. LPS also stabilizes IFN-β mRNA, leading to the production of IFN-β that feeds back on the producer cell to activate Stat1α but with a different result than when the IFN-γ receptor is engaged, i.e., the production of the transcription factor complex ISGF-3, in addition to Stat1αPO_4 homodimers. The relatively low levels of Stat1αPO_4 homodimers that are available to bind under these conditions to the GAS of the IRF-1 gene are enough to allow moderate expression of the iNOS gene even though all of the other transcription factors needed collectively to form the hypothesized *trans*-activating complex are present. (D) Stimulation with LPs+*IFN*-γ. Maximal expression of the mouse iNOS gene is obtained under these conditions, because it leads to simultaneous activation of NF-κB and the production of high concentrations of Stat1αPO_4 homodimers. The latter induce increased synthesis of IRF-1. The availability, under these conditions, of optimal levels of all transcription factors leads to optimal formation of the putative *trans*-activating complex and, therefore, maximal expression of the *iNOS* gene. IFN-β synthesis is also induced under these circumstances, but the temporal appearance of IFN-β after that of IFN-γ exposure is unlikely to change the levels of available Stat1α.

whether all of the cells within the population fractionally increase their functional activity or, as an alternative, only a few maximally activated cells are initially producers, and this number is increased as the concentration of an activator goes up. In fact, both possibilities have been shown to exist depending on which gene is considered. For example, the LPS-induced production of IFN-β is homogeneous throughout a population of stimulated macrophages (120). The expression of the iNOS gene, on the other hand, occurs heterogeneously. By the latter statement, it is meant that, when mouse macrophages are stimulated for production of iNOS with LPS alone, only a few cells become producers. In this system, IFN-γ mediates its augmentative effect by increasing the number of cells in the population that express the iNOS gene as the concentration of IFN-γ increases (121). Such heterogeneity of gene expression within the population is most likely attributable to differences in the levels of activated transcription factors in individual cells. This hypothesis is consistent with the view that, for a gene to be expressed, a threshold level of activated transcription factors must be exceeded (122–125).

Activation of Human Macrophages for Expression of the iNOS Gene

There are distinct functional differences between the macrophages of different species; that is, *it is dangerous to assume that because the macrophages of one species respond to activating stimuli in a certain way that the response of another species' macrophages will be the same.* There is, perhaps, no better example of this than the difference between human and rodent macrophages with respect to the production of nitric oxide.

When stimulated with IFN-γ/LPS, rodent cells become efficient producers of NO because of their ability to become activated for the expression of the iNOS gene. The ease with which these kinds of macrophages produce NO makes the iNOS/NO system one of the primary antimicrobial defense mechanisms in these species. Human macrophages, by contrast, are very poor producers of iNOS and NO when stimulated similarly (126–128). It is emphasized that this finding does not mean that human macrophages cannot express iNOS and produce NO. Rather they require different stimuli than the cells of rats or mice in order to become productive (129,130). Thus, under the right conditions, human macrophages can produce NO (131–136) to include those in the lung (137). In the latter setting, iNOS was localized by immunological means to human alveolar macrophages, especially those obtained from sites of inflammation. However, the extent to which human iNOS was expressed was generally less than that found in similar investigations of rat and mouse cells. As a probable result of this quantitative difference, the iNOS/NO system appears not to have the same degree of antimicrobial importance in human beings that it has in rodents (138). Instead, the capacity that activators, such as IFN-γ, have to reduce important bacterial nutrients, such as tryptophan (139) and, to a greater extent, intracellular iron (140–144), appears to be of greater importance.

The reason that the combination of IFN-γ/LPS is ineffective in stimulating human macrophages has only recently become clear: Although the human iNOS promoter has a similar region II enhancer element than the mouse promoter has, in the human, there are inactivating point mutations in each of its κB, ISRE, and GAS response elements (145). Because of these changes, the human iNOS gene can make a basal response to LPS through region I, but it cannot make the augmented response that is attributable in the mouse gene to stimulation with IFN-γ and LPS through the region II enhancer.

C. Reprogramming of Macrophages for Activated Functions

Three of the hallmark characteristics that distinguish the macrophage from other inflammatory cell types include (1) the ability of these cells to respond to a variety of stimuli of diverse microbial origin, (2) the unusually low threshold concentrations of these stimuli that elicit macrophage responses in vitro; and (3) the broad spectrum of different effector molecules and other mediators of inflammation that can be secreted by these cells in response to stimulation by microbial products. Given these properties and the ubiquitous distribution of macrophages in many different tissues, it might appear that macrophages have been designed by nature to have an unselective, "shotgun" approach to eliminating the threat of an invading microbe. The available evidence would suggest, however, that this concept is not correct. Rather, the macrophage's response appears to be carefully orchestrated, both qualitatively and quantitatively, by prevailing conditions in its microenvironment. Reprogramming of functions is another example of such regulation.

As described above, macrophages readily become activated for some functions by concentrations of microbial products, such as LPS, in the nanogram/milliliter range. Such a response is augmented by IFN-γ to include making it possible for macrophages to recognize and respond to even lower concentrations of LPS; for example, in the picogram/milliliter range (146,147). It is clear, therefore, that macrophages have very sensitive means for sensing and responding to microbial antigens. A growing body of evidence suggests that initiation of such responses occurs, at least in part, through a single membrane receptor, the phosphatidylinositol-linked CD14 glycoprotein (148). Evidence also exists for pathways of activation through other surface molecules (149).

In addition to conventional responses that result in the activation of a variety of functions, there is growing appreciation that macrophages can be profoundly influenced by concentrations of microbial products that are well below the minimal threshold concentration required to elicit detectable activation. For example, whereas concentrations of LPS in the nanogram/milliliter range are required for macrophage activation, LPS alone in the 10–100 pg/ml range can influence macrophages, such that their subsequent interactions with other microbial stimuli (including LPS) are markedly altered. In addition, there exist some macrophage responses that are not generally elicited by LPS, yet pretreatment with LPS can markedly influence the magnitude of the response to secondary challenge with stimuli to which the macrophage does respond. In general, such secondary responses are either potentiated or suppressed. When a potentiated response is observed, such effects have been referred to in the literature as priming events. By way of example, Pabst and Johnston (150) reported that mouse peritoneal macrophages did not respond to LPS by the production of superoxide anion. However, preincubation with LPS followed by secondary challenge with either phorbol myristate acetate (PMA) or opsonized zymosan caused macrophages to increase their level of superoxide anion production substantially compared with PMA or zymosan alone. As another example, heat shock has the apparent effect of amplifying the ability of rat macrophages to produce NO in response to LPS alone (151). Alternatively, it has been shown by many investigators that the preincubation of macrophages with LPS and subsequent challenge with LPS will suppress the secondary response as assessed by TNF-α production. This latter phenomenon has been termed macrophage desensitization or tolerance. These examples serve to illustrate the fact that *preexposure of macrophages to potential activating stimuli will influence the profile of responsiveness upon subsequent challenge.*

Considerable insight into this process of priming/desensitization has derived from the studies of Zhang and Morrison (152), who investigated the effects of a wide range of low substimulatory doses of LPS on subsequent macrophage responsiveness, with the latter being assessed by LPS-initiated production of TNF-α and nitric oxide. The results of those studies indicated a biphasic response profile. At very low LPS pretreatment doses, subsequent LPS challenge resulted in markedly enhanced TNF-α secretion (priming effect) and concomitant reductions in the ability of the pretreated cells to secrete nitric oxide (desensitization effect). Increasing the pretreatment concentration of LPS while still remaining below the threshold concentration for activation reversed this phenomenon. The secretion of nitric oxide in response to LPS was enhanced (priming effect), whereas TNF-α secretion was suppressed (desensitization effect). There were similar observations when other microbial stimuli, such as zymosan or heat-killed *Staphylococcus aureus*, were employed as the challenge stimulus. In addition, secondary macrophage responses, as assessed by interleukin-6 (IL-6) production appeared to parallel those observed for TNF-α secretion, thus suggesting a more global degree of macrophage reorientation (153). In view of the biphasic nature of the responsiveness and the dependence of the observed phenomenon on both the pretreatment concentration and the nature of the subsequent response being reassessed, it has been proposed that the more general term *macrophage reprogramming* be employed rather than priming/desensitization.

There are a number of potential mechanisms by way of which macrophages may be reprogrammed. It is unlikely, however, that alterations in the level of receptor expression would serve as a major regulatory pathway unless more than one receptor signaling pathway were to contribute to activation. Although this is possible, the available evidence would suggest that, at least at the level of CD14 expression, little detectable difference is observed between normal monocytes and monocytes made refractory (tolerized) to stimulation by prior LPS exposure (154).

An alternative mechanism would involve intracellular events in which LPS pretreatment initiates changes in second-messenger enzyme systems or regulatory transcription factors. In this regard, there is evidence that treatment of macrophages with LPS, so as to render them refractory to subsequent challenge with LPS for TNF-α production (but not IL-1 synthesis), suppresses NF-κB activation but not the activation of protein kinase C (154). Because homodimers of the p50 component of the NF-κB transcription factor family have binding potential but lack the ability to *trans*-activate (155–158), the formation of these complexes can be inhibitory (159). Of relevance here is the fact that the pretreatment of macrophages with low concentrations of LPS, which downregulate subsequent secondary challenge with high concentrations of LPS, have been recently shown to induce the overexpression of p50 homodimers, with a concomitant underexpression of *trans*-activating p50/p65 heterodimers (160). The homodimeric complexes induced by low-dose LPS could render the iNOS gene refractory to subsequent activation by LPS either by blocking sites on the iNOS promoter/enhancer necessary for efficient activation or by diverting p50 away from the formation of *trans*-activating heterodimeric complexes. Neither of these mechanisms, however, would adequately explain why TNF-α responses (which are strongly dependent on *trans*-activation of the cognate gene by NF-κB) are enhanced, whereas iNOS levels are concomitantly suppressed (152).

A more likely explanation would be the autocrine/paracrine feedback regulation of macrophage reprogramming. For example, as mentioned earlier, to produce NO in response to LPS, mouse macrophages require autocrine/paracrine stimulation by IFN-α/β. Pretreatment of mouse macrophages with exogenously added IFN-β has been shown to decrease the level to which these cells can subsequently be activated by LPS to produce NO and NO-dependent killing of tumor cells either with or without IFN-γ (161,162). The mechanism of effect has recently been shown to be transcriptional downregulation of the iNOS gene (163). The effect of IFN-β, as noted above, may be to activate both ISGF-3 and Stat 1α homodimer formation, thus lowering the availability of the latter (and, therefore, of IRF-1) for iNOS gene induction.

Collectively, these points provide support for the concept that reprogramming events can prepare macrophages for selective responses on the basis of changes that they sense in their external environment. Although the principal stimulus used to date for reprogramming has been subthreshold activating concentrations of LPS, it is likely that other changes, such as in oxidation state or serum factors, may also cause reprogramming. Thus, although additional studies are needed to establish the exact mechanism by which macrophage functions are modulated, the phenomenon of reprogramming needs to be recognized as another means by which this pluripotential inflammatory cell mediates host defense.

D. Markers of the Activated Phenotype

Cellular Level

Because of the heterogeneous nature of macrophages activated in different ways for different functions, the most specific marker for a given kind of activated macrophage is the function or characteristic that is of interest. Alternatively, other kinds of markers that are *associated* with the activated function can be used. For example, stages of activation of mouse macrophages for tumor cell killing have been identified either by using monoclonal antibodies that bind to newly expressed antigens (164–166) or by recognizing newly expressed proteins in two-dimensional gels (167,168). Ideally, an associative marker must be one that is absent from unstimulated macrophages, present on activated macrophages, and absent from all other cell types even after they have been exposed to the macrophage-activating stimulus under study. In practical terms, these criteria have been exceptionally difficult to meet, often necessitating some degree of compromise or alternative strategy. As a case in point, the protein doublet p71/73 has been described as a marker that distinguishes LPS-stimulated mouse macrophages (167,168). This pair of proteins has recently been identified as two isoforms of the inducible enzyme PGHS-2 (169), an enzyme that is also produced by human alveolar macrophages that have been exposed to LPS (170). As mentioned earlier, PGHS-2 confers on macrophages and other cell types the capacity to increase dramatically their production of prostanoids. Because the expression of the related gene can be induced in other cell types by LPS (171–173), it cannot be regarded as a specific marker for LPS-mediated activation of macrophages. To make it useful, therefore, one would either have to purify macrophages from a mixed population (e.g., obtained by bronchoalveolar lavage), use macrophage cell lines, or, if mixed populations were to be analyzed directly, use the activation marker in conjunction with another characteristic that is specific for macrophages. For example, one of the most widely used means of identifying mouse macrophages is the detection of the glycoprotein that binds monoclonal antibody F4/80 (174). Until recently, it was thought to be absolutely specific for cells of the mononuclear phagocyte lineage. Now, however, there is a recent report that, under certain conditions, eosinophils can also bind F4/80 antibody (175). Similarly, CD14, another marker that is often described as "macrophage specific," is weakly expressed on granulocytes (176). *These observations indicate the need to use care when designing and interpreting marker studies, because "specificity" for mononuclear phagocytes may be more a reflection either of how few other cell types than mononuclear phagocytes have been examined or the sensitivity of the examining methods rather than absolute specificity.*

Once a useful marker and the conditions for its use have been chosen, it is next necessary to be certain that it is retained in/on macrophages that have the activated phenotype. This latter consideration has limited the utility of some markers, such as F4/80 and CD14, which either can increase (177), decrease (178), or disappear (179) on/from activated macrophages.

Although the use of two markers (one that identifies the macrophage and the other that signals acquisition of an activated function) on/in the same cell is conceptually feasible, it is technically problematic for a range of reasons. For example, in tissue sections, there are resolution problems with frozen material and potential loss of marker antigenicity/activity when the tissue has been fixed and embedded for processing. Even when working with cells in suspension, as soon as they are separated from the activating stimulus, unless they have been fixed in some way, activation markers can be lost rapidly during the time that is required to process the cells for examination.

The alternative to using individual markers at the cellular level is to use a panel of characteristics that has been associated with the activation phenomenon of interest. For example, upregulation of major histocompatibility (MHC) class II antigen is often used as a marker of cells that have been exposed to IFN γ (180) Certainty that macrophages have been exposed to the relevant activating stimulus may be increased if a panel of markers is used. One such panel is enhanced expression of MHC class II antigen, intercellular adhesion molecule-1 (ICAM-1), and Fc receptors (181). Another is the production of TNF, IL-1, and PGE_2 (182). A panel of lectin binding properties may also hold promise as a means of identifying activated macrophages and, perhaps, even subsets thereof (183).

Clinical Markers of Activation

For diagnostic, monitoring, and prognostic purposes, it is sometimes important to ascertain the extent to which macrophages in the body have been stimulated. The most convenient, least invasive approach to doing so is to quantify a product of macrophage activation in one of the body fluids, such as blood. The two compounds most often measured in human beings are tryptophan and neopterin, which decrease and increase respectively, when macrophages become activated (184). Tryptophan is catabolized at an increased rate by macrophages that have been exposed to IFN-γ, because this cytokine induces the expression of the gene that encodes for indoleamine-2,3-dioxygenase (IDO), which catalyzes the degradation of tryptophan. IDO is yet another example of a gene that is induced synergistically when cells are exposed to both IFN-γ and LPS (184). As an example of the clinical use of this marker enzyme, patients with various kinds of inflammatory interstitial lung disease had lowered levels of serum tryptophan and higher levels of

IDO activity in cells obtained by bronchoalveolar lavage (BAL) when compared to normal volunteers. In addition, the BAL cells of patients with inflammatory lung disease contain detectable levels of mRNA specific for IDO, whereas those of normals contained none (185). Similar serum findings have been made with patients infected by human immunodeficiency virus-1 (HIV-1) (186). The other commonly used marker is neopterin (179,186,187). Neopterin is a stable byproduct of the biosynthesis of tetrahydrobiopterin from guanosine triphosphate (GTP). The first and rate-limiting enzyme in the pathway is GTP cyclohydrolase I, which is inducible by several cytokines that are associated with macrophage activation (188). They include the interferons (184,189), with which LPS synergizes (184), and TNF-α (190). Although tryptophan and neopterin are of definite use as markers, and are strongly suggestive that macrophages have been stimulated, they are not as macrophage specific as some (191) have suggested. For example, IDO can be induced in a variety of cell types, including fibroblasts (192), and GTP cyclohydrolase I can be found newly expressed in pulmonary artery smooth muscle cells (193), activated T lymphocytes (194), and probably endothelial cells (195).

E. Terminal Pathways: Resolution or Chronic Response

As mentioned earlier, in the ideal situation, the outcome of an inflammatory process is the elimination of the stimulus that provoked the reaction. Once it is removed, the balance between regulatory influences that are positive and negative should shift toward the latter, thereby preventing lung injury due to the excessive production of proinflammatory cytokines and other mediators (196). Thus, with respect to the downregulation of activation, there are two main general mechanisms that account for subsidence. One of these mechanisms is passive, that is, progressive reduction in availability of upregulatory stimuli that are needed to maintain the activated state. The second mechanism is dependent on active inhibition, which is related to the negative influence of agents that are either found locally, as a consequence of the inflammatory process, or systemically, as a result of stress or therapy.

One mechanism of inhibiting macrophage activation for killing of bacteria is unique to the lung: ingestion of surfactant. The underlying mechanism(s) has not been worked out fully; however, the end result, at least in part, is the reduced production of reactive oxygen species (197). Another general mechanism of the inhibition of activation is dependent on mediators that have a direct or indirect negative regulatory effect on a wide range of activated functions. These mediators are produced at a site of inflammation concomitantly with mediators that are proinflammatory, thereby controlling the extent to which an inflammatory response can develop. Some of the most important mediators with *direct* negative effects on various macrophage functions are TGF-β (198–202), IL-4 (198, 203), IL-6 (204), and IL-10 (198,199,205–207) as well as macrophage-deactivating factor (208). *Macrophages, themselves, can be important producers of some of these (e.g., TGF-β [209]); again underscoring how important autocrine/paracrine feedback is in regulating various activation pathways.* Although the effect of an inhibitory cytokine on the macrophage may be direct, the outcome may depend on interference with a function of activated macrophages that affects another cell type. In other words, the actual effect on activation is an *indirect* one. For example, IL-4 and IL-10 inhibit the ability that inflammatory macrophages have to become activated for the production of IL-12 (200). IL-12 is a potent stimulator of the production of IFN-γ by natural killer (NK) cells (210) as well as the antigen-induced proliferation of the Th_1 subset of T lymphocytes (211), which are also producers of IFN-γ. Thus, reduction in the production of IL-12 interferes with both the T-lymphocyte–dependent and T-lymphocyte–independent production of one of the most important macrophage activators. A similar indirect mechanism of effect is associated with PGE_2, which can inhibit the production of IFN-γ by activated T lymphocytes (212). PGE_2 can also have a more direct effect, for example, on the expression of the mouse iNOS gene (213). Another example of indirect influence is the negative effect that IL-10 has on the production of TNF-α, which can have a costimulatory effect with IFN-γ in the induction of microbicidal activity (214). As would be expected, either the direct or indirect inhibition of the production of chemotactic and activating (for microbicidal activity) cytokines could have negative consequences in diseases like pneumonia (215). On the other hand, the production of inhibitory cytokines (e.g., IL-4 and IL-10) can result in an important protective effect under certain conditions, such as immune complex–mediated injury of the lung (196).

The inhibition of macrophage activation for a variety of functions can also be seen with the glucocorticoids, which are either produced as a result of inflammatory stress or are administered for therapeutic reasons. For example, they can inhibit the induction of the iNOS gene and, therefore, NO production (216,217); the expression of the PGHS-2 gene and, therefore, the production of prostanoids (49,119); the transcription of the early-response gene JE/macrophage chemotactic protein-1 (218) as well as the gene for macrophage inflammatory protein-1α (219); the production of IL-8 mRNA by human alveolar macrophages (220); and the synthesis by human alveolar macrophages of a number of metalloproteinases (221). At least some of these inhibitory effects may be the result of increased IκB gene expression (222,223). The result is reduction in availability of NF-κB transcription factors that are needed to express activation-related genes that have κB response elements in their promoters, such as TNF-α and iNOS. The importance of negatively regulating the inflammatory process is probably reflected in the fact that, among the proinflammatory augmenting effects that IFN-γ has, it concomitantly increases the number of glucocorticoid receptors that are expressed by mouse macrophages. The likely effect is to increase the sensitivity of the stimulated cells to negative regulation by glucocorticoids (224).

If the inciting stimulus either cannot be eliminated or there is an irreversible change in the phenotype of stromal cells (such that they constitutively produce cytokines that will upregulate adhesion molecules on inflammatory and endothelial cells as well as attract monocytes and lymphocytes), the nature of the mononuclear phagocyte response can be shifted to produce granulomatous inflammation (225–228). The hallmarks of this type of inflammation are so-called epithelioid cells and multinucleated giant cells. Both of these special cell types are derivatives of tissue macrophages. The former is a morphological and functional variant, whereas the latter is the result of the cytokine-induced fusion of individual macrophages to form a syncytium (229,230). There are two general kinds of granulomatous inflammation, immune and nonimmune. The latter kind, as the name implies, is most often associated with particles that do not have the capacity to evoke an immune response. The former type is the one that is associated with resistant microorganisms and other kinds of persistent antigens. The immune-type or hypersensitivity-type granuloma is the more active of the two; however, the cells in each type have the capacity to produce a range of mediators (231). Granulomas are highly cellular and can have the negative consequence in the lung of replacing tissue that is needed for oxygen exchange. This occurs through the development of space-occupying lesions, dissolution of tissue, the production of pulmonary fibrosis, or combinations of these. Because the cellular component in immune granulomas is constantly being replenished with monocytes and lymphocytes recruited from the blood, there is an ongoing process of activation. One would assume that the specialized forms of the mononuclear phagocyte lineage, that is, epitheloid cells and multinucleated giant cells, would be ultracapable with respect to functions that are needed to eliminate resistant pathogens. In fact, that may not be the case; if anything, many of their functions, such as phagocytosis, are moderately to greatly diminished. Several diseases of unknown cause, such as sarcoidosis, manifest themselves as pulmonary granulomatous inflammation. The pathogenesis of these diseases is one of the great unsolved mysteries of pulmonary pathophysiology.

III. Macrophage-Related Consequences of Invasion of the Lung by Microorganisms

Recent, detailed reviews of how macrophages interact with microorganisms (bacteria, viruses, protozoa, fungi, and macroparasites) are readily available to the reader. These include the chapters in this volume as well as those in recent books edited by Lewis and McGee (3) and by Zwilling and Eisenstein (7). Emphasis here will be placed, therefore, on the consequences to the lung of such interactions, many of which lead to the activation of related macrophages.

A. General Considerations

The intent of invading microorganisms is colonization of the host. As noted at the beginning of this chapter, the resident macrophage is one of the front-line defenders against colonization of the lung either by inhaled or blood-borne microorganisms. Some of these are sufficiently large, for example, large protozoa, fungal mycelia, and macroparasites, such as worms, that the principal defensive mechanisms are ones that the macrophage deploys extracellularly. This is also the case for virus-infected cells other than the macrophage. However, if the macrophage, itself, is the target for infection by a virus, then interactions that are most pertinent become those that affect subsequent functions of the infected cell. The most prominent examples of viruses that have this latter kind of effect is members of the lentivirus family (232). If, on the other hand, the invader is particulate in nature, such as a bacterium or small protozoan, phagocytosis and intracellular killing/stasis become the interactions of interest. The point of importance here is that *different kinds of invading microorganisms are defended against differently by marshaling a different spectrum of macrophage-mediated responses. As might be expected, therefore, the consequences for the lung, both positive and negative, potentially differ with each kind of invader.*

B. Relation to Activation

Innate (natural) immunity, that is, that which requires no prior experience, is characteristically involved in the first interactions that macrophages have with an invading microorganism. For example, engulfment by means that are not dependent on antibody, such as interaction with either the mannose (234,235) or β-glucan (234) receptors, might be involved depending on which kinds of molecules are present on the microorganism's surface. As another possibility, the microorganism might activate the alternative pathway of the complement system, resulting in opsonization of the particle with complement components and phagocytosis through the related receptor(s) (236). Ingestion of particles opsonized with complement component C3b by alveolar macrophages has been reported to increase the capacity that these cells have to release chemotactic factor(s) for neutrophils (237). The local synthesis of relevant complement components can also be increased when alveolar macrophages encounter a bacterial pathogen (238). However, the ability of these opsonic proteins to reach the airways in pneumonic lungs has been questioned by Coorod and Yoneda (239); their conclusion stemmed from the observation that *Streptococcus pneumoniae* organisms were opsonized in bronchoalveolar lavage fluid to a lesser extent than they were by incubating the bacteria in serum. Phagocytosis of bacteria as well as nonbacterial pathogens also appears to be enhanced in the lung by surfactant protein A through its ability to aggregate and opsonize the materials (240). The capacity of surfactant A to act as an opsonin may be due to its structural homology with complement component C1q (241).

Interactions with microbes can also lead to the nonimmune production of IFN-γ by NK cells (210). The mechanism is induction of secretion of IL-12, which then stimulates NK cells to produce IFN-γ.

Macrophages also contribute importantly to the development of active immunity against microorganisms by producing secretory products that stimulate lymphocytes (e.g., IL-1) (29,242). As active humoral immunity becomes established, there is an increasing availability of antibodies, some of which accentuate phagocytosis through Fc receptor–mediated mechanisms, whereas others make macrophages competent to effect antibody-dependent cell-mediated cytotoxicity. Alveolar macrophages express Fc receptors for IgA (243); IgE (low-affinity receptor; also called CD23), with greater expression in asthmatics (244); and IgG, including classes I, II, and III (245). In addition, larger quantities of the IFNs, especially IFN-γ, become available. These can enhance a variety of antimicrobial mechanisms, including (1) the production of reactive oxygen intermediates in association with the respiratory burst (246); (2) the generation of reactive nitrogen intermediates in rat and human pulmonary inflammatory cells by inducible nitric oxide synthase (137); and (3) the limitation of intracellular nutrients, such as tryptophan (184) and iron (140).

Regardless of the activation status of a macrophage, if a bacterium is phagocytosed, it will initially be contained within a phagosome, which subsequently fuses with one or more lysosomes. Escape from the phagolysosomal system into the cytoplasm is one way that a bacterium can evade a macrophage's intracellular killing mechanisms. Unstimulated macrophages cannot, for example, prevent such an escape by *Listeria monocytogenes*, whereas macrophages that have been activated with IFN-γ can (247,248). Recent evidence provided by Tanaka et al. (69) shows that this functional alteration of mouse macrophages activated for killing of bacteria is apparently dependent, in part, on the availability of the transcription factor, NF–IL-6 (also called C/EBPβ). This was shown by disrupting the related gene. For example, the IFN-γ/LPS–treated macrophages from (-/-) mice were fully able to produce nitric oxide, but they were unable to mediate significant levels of *Listeria* killing in vitro, because they allowed the organisms to escape from the phagolysosomal system into the cytoplasm. A concomitant reduction in the production of reactive oxygen species by macrophages (but not by neutrophils) seemed not to be of significance, because scavengers of ROI failed to affect killing by wild-type macrophages. Mice with the disrupted gene were also more susceptible to infection in vivo by *Listeria* and *Salmonella typhimurium*. Thus, activation of mouse macrophages for the killing of these microorganisms is apparently dependent on transcriptional regulation of one or more critical functions that are characteristic of activated, but not unstimulated, macrophages.

Inflammatory mononuclear phagocytes, that is, those that are recruited to a site of invasion, are by definition stimulated when they encounter the target microorganism. The initial source of stimulation is the chemotactic factor(s) that has attracted circulating monocytes. However, recruited and locally based mononuclear phagocytes may not be fully activated for functions that are relevant to elimination of an invading microorganism until stimuli in the microenvironment are encountered. There are four major sources of such stimuli: (1) the microorganism itself; (2) the responding macrophages, through autocrine/paracrine feedback; (3) secreted products of other cell types in the immediate area; and (4) plasma proteins. With respect to the first of these activating stimuli, cell wall components of certain bacteria can provide a potent activating signal. The best known of these is endotoxin/lipopolysaccharide. In the lung, endotoxin that is presented via the vascular system seems to have greater activating effects on interstitial, compared with alveolar, macrophages (249). However, endotoxin is not the only cell wall component to have stimulatory activity. For example, peptidoglycans (250,251) lipoarabinomannan, lipomannan, and phosphoinositolmannoside (252) have also been documented to cause activation for various functions. Interaction of the macrophage with these can occur either during engulfment of a microorganism or after such components have been released by intracellular killing/digestion (253,254). With respect to the latter category, the results of in vitro studies would suggest that endotoxin secreted by macrophages that have previously phagocytosed gram-negative microbes has been modified in the polysaccharide part of the molecule (i.e., the O antigen). However, the biologically active lipid A component is not modified by macrophages. Therefore, such molecules retain both their immunostimulatory and endotoxic properties.

The second major source of a macrophage-activating stimulus is the macrophages themselves. This is an especially important source, because it allows amplification of functions that are induced by bacterial components without the need for lymphocyte-derived stimulation. For example, endotoxin in sufficiently high concentration will cause human alveolar macrophages to release IFN-α/β (255), which in turn can provide the priming signal that is needed to augment and amplify the macrophage's capacity to kill (70,71,256) as well as to secrete a wide range of biologically active substances. Similarly, granulocyte-macrophage colony-stimulating factor (GM-CSF) can be a potent stimulator of macrophage activation for anticryptococcal activity, especially when it is combined with the augmenting effects of IFN-γ (257).

With respect to the third source of activators, cell types other than the macrophage, the most important source of activating stimuli, such as IFN-γ, is the T lymphocyte (30–33), especially after active immunity has been established. However, other inflammatory cell types, such as the neutrophil (258,259), eosinophil (260–262), and mast cell (260,263) can be important contributors, sometimes in a negative sense (264). Stromal cells, such as fibroblasts (258,265,266) and endothelial cells (258,267), also have to be considered as potential sources of activating stimuli.

Finally, as a fourth source of activators, there are numerous plasma proteins that are extravasated at sites of microbial invasion that can be important contributors to the process. Some of these, such as select components of the Hageman factor–related systems (kinens, intrinsic clotting, fibrinolysis and complement) (268), can be cleaved by locally available proteinases to become macrophage activators. For example, complement components C5b-9 cause human monocytes to release reactive oxygen intermediates, prostanoids, and IL-1 (269). Other serum proteins are immediately functional, such as lipopolysaccharide-binding protein (LBP) (270). The latter protein contributes importantly to the efficiency with which LPS/endotoxin can activate macrophages for various functions.

C. Positive and Negative Consequences of Activation Against Microbes

Given the fact that microbial invasion of the lung has the potential to elicit the local production of a wide range of activating stimuli from a number of different sources, the relatively large number of macrophages that normally reside in this organ, and the many more that can be recruited into it on short notice, it is not surprising that there can be both positive and negative consequences to the lung that are substantive. First among the positive, of course, is clearance of the offending microorganism. Enhanced phagocytosis contributes to this end, if the organism is of a size and composition to lend itself to engulfment, and some bacterial components, such as endotoxin, have the capacity to cause such enhancement (249). The fact that clearance is often the result of enhanced macrophage function, often without clinical evidence of infection, is readily apparent when the capacity to produce the usual activators of pulmonary macrophages is compromised, as is the case in acquired immunodeficiency syndrome (AIDS). The virus causes the loss of $CD4^+$ T lymphocytes, thereby reducing the availability of IFN-γ, one of the pulmonary macrophage's most important activators. The result is often colonization of the lung by opportunistic microorganisms. The realization that activating stimuli are locally deficient in such circumstances has led to various replacement strategies (271,272), which have met with modest experimental (180,273) and clinical (274) successes. Most of these approaches have utilized aerosolization as the means of delivering recombinant-activating stimuli. However, in the face of microbial invasion, the efficacy of delivery by this route is dependent on a number of variables; for example, dose, frequency of administration, and the depth to which an aerosol can penetrate into the lung through airways that are obstructed by exudate. Attention is being given to overcoming such obstacles in clinical trials that are currently ongoing.

Another positive consequence of activation can be enhancement of other host defensive mechanisms that are active in the eradication process. For example, activated macrophages are a rich source of mediators, such as prostaglandin E_2, that increase vascular permeability and, therefore, the availability of plasma proteins, such as antibody and complement components locally. Still another important positive consequence of pulmonary macrophage activation is the accelerated mobilization and removal of inflammatory debris that accumulates in alveoli and airways that have been invaded by a microorganism (275). As much as those who have affection for the macrophage may hate to admit it, "garbage collection" is one of this cell type's most important functions, and activated macrophages are better at this task than quiescent ones are.

Negative consequences are sustained in the lung when microbe-related activation is either inappropriate, excessive in either extent or duration, or suppressed. As an example of the former, infection of pulmonary macrophages with HIV-1 can inappropriately activate them to produce cytokines and other secretory products prior to the onset of full-blown AIDS (276). The full meaning of this is not yet understood; however, this phenomenon can promote the development of pneumonitis and interstitial fibrosis in this disease (277). Activation is also inappropriate when an intracellular microorganism uses it to induce the host cell to produce inhibitory cytokines, thereby suppressing the host's antimicrobial mechanisms. For example, the production of transforming growth factor-β (TGF-β) by macrophages infected with *Toxoplasma gondii* has been demonstrated (278), and respiratory syncytial virus can cause infected macrophages to produce IL-10 (279). In the case of the former, the negative consequences on the lung can be functional, structural, or both. With respect to function, a variety of mediators produced by macrophages has the potential to cause bronchial hyperreactivity and bronchoconstriction (280–283). This is particularly of concern in asthmatics (284,285). The secretion by activated alveolar macrophages of cytokines, such as IL-8 (286), leukotriene B_4 (287), and GRO-α (288), can also contribute to the recruitment of neutrophils into the lung. With relatively low numbers of bacteria, the result is usually clearance of the invaders with minimal damage to the lung; that is, a positive outcome. However, when large numbers of bacteria are involved, resulting in the recruitment of large numbers of neutrophils, extensive lung injury coupled with pulmonary edema can be the result (289). A macrophage-produced cytokine, IL-1, will cause neutrophil-dependent acute oxidative lung injury when it is introduced into the lung via the airways (290). The extent of injury is exacerbated when another secretory product of activated macrophages, TNF-α, is combined with the IL-1 (291), which may be of relevance in the pathogenesis of the lung injury that occurs in adult respiratory distress syndrome. Evidence for this is the fact that alveolar macrophages recovered from affected patients are activated for the increased production of both cytokines (292,293). Given the role that transcription factors play in many kinds of activation, it would be expected that, when excessive production of cytokines is seen in the lung, regulatory *trans-*

activating proteins like NF-κB would be activated. That is precisely the finding, for example, in bronchoalveolar lavage cells within 2.5 hr of the systemic administration of endotoxin (294). Another example of excessive activation is the overly exuberant production of proteinases by activated macrophages, which can result in the digestion of connective tissue in the vicinity of an invading microorganism (295). The result is that parenchymal tissue is irrevocably lost and either a cavity or scar tissue develops in its place. Should activated macrophages not be able to eliminate a microorganism, they may continue to produce cytokines for prolonged periods. Some of these, for example, IL-1 (296), TGF-β (297), and platelet-derived growth factor (298,299), can foster fibroplasia and the development of pulmonary fibrosis. When the full range of mediators that can induce lung injury is appreciated (300), it is apparent why activation for the killing/stasis of microorganisms can be a two-edged sword. Finally, given that activation of macrophages is an important means of eradicating microorganisms that have invaded the lung, premature suppression of the process must be seen as a potential negative factor. In fact, some of the most successful and, therefore, resistant pathogens are ones that have found one or more means to subvert macrophage activation. The induced secretion of inhibitory cytokines has already been mentioned. Decreased expression of costimulatory molecules that allow efficient interaction between infected macrophages and T lymphocytes is an additional means of suppressing activation in mycobacterial infections (301). If activation is suppressed, a macrophage may fail to kill an engulfed pathogen. Under such circumstances, macrophages can actually contribute to the spread of a microbial disease, such as tuberculosis, because of their motility; the macrophage becomes a carrier that allows the microorganism to gain access first to draining regional lymph nodes and then to spread systemically.

IV. Macrophage-Related Consequences Associated with Pulmonary Neoplasia

A. General Considerations

There are several recent, in-depth reviews of tumor-associated macrophages to which the reader's attention is drawn (302–304). Interest in macrophage activation as it relates to tumors stems from two observations made during the 1970s. The first of these was that macrophages could be activated to kill tumor cells (305–308). The second was that many, if not all, kinds of tumors contain appreciable numbers of macrophages (309), apparently because a gradient of chemotactic factors is produced that attracts monocytes into the tumors (310,311), and also because, in some instances, there is local replication of mononuclear phagocytes in the tumor (303,312). These original observations have been corroborated many times in many different laboratories. Quite predictably, interest was initially strong in harnessing the therapeutic potential that macrophages were thought to have; what better vehicle than a motile cell type that could be recruited to a specific location, where it could be activated to kill tumor cells? In addition, activated macrophages were known to distinguish between normal and neoplastic cells when mediating their cytotoxic effects (308,313,314), thereby minimizing the danger to normal, innocent bystander cells. It was hoped that here was finally the counter to the oncologist's greatest nemesis: the residual or metastatic tumor cells that escape surgical/chemotherapeutic/radiological debulking. Although activated macrophages have provided some successes in this area, as will be discussed later, they have not, unfortunately, proved to be the panacea for which everyone had hoped. In fact, as will be seen, activated tumor-associated macrophages can be a decided negative factor in some instances.

B. Relation to Activation

Tumor cell killing by macrophages is dependent on their becoming activated for one or more functions. Recently, activation by IFN-γ/LPS of mouse macrophages for the killing in vitro of P815 mastocytoma cells was shown to be dependent on the availability of the transcription factor NF–IL-6 (69). As mentioned earlier, these macrophages were also compromised with respect to their ability to kill phagocytosed bacteria. The fact that the macrophages with the disrupted NF–IL-6 gene had a marked reduction in their capacity to kill tumor cells, in spite of their activation to produce NO, strongly suggests that there either is a tumoricidal mechanism that is independent of the need to secrete NO or that the secretion of NO alone is insufficient to mediate tumor cell killing. In the latter case, some other step that is essential to NO-mediated killing would have to be compromised by the loss of transcriptional regulation that is mediated through NF–IL-6.

The activation of macrophages for tumor cell killing, including human alveolar macrophages (314,315), is an unassailable reality in vitro; however, this phenotype has never been directly demonstrated within a neoplasm. If we assume, however, that such a phenomenon does exist in vivo, there are important factors that need to be considered. For example, smoking suppresses the capacity that human alveolar macrophages have to become activated for tumor cell killing (314) and to synthesize or secrete a variety of cytokines (316,317). Presumably, such alterations could contribute to the increased incidence of lung cancer that is associated with smoking. The physical relations that exist between tumor cells and host inflammatory cells also need to be considered. Previous sections of this chapter make the point that maximal activation is often dependent on the availability of amplifying cytokines, such as IFN-γ, that are secreted in the immediate microenvironment of macrophages by T lymphocytes. In addition, it is clear from abundant independent observations that for an activated macrophage to kill a tumor cell by any one (or a combination) of several mechanisms (e.g., secretion of reactive oxygen and/or nitrogen intermediates, TNF-γ, and antibody-dependent cell-mediated cytotoxicity, to name some of the principal means that are currently known), it must be in contact with its target for a relatively prolonged period (312). Thus, any mechanism that restricts the entry of lymphocytes and/or macrophages into tumors would effectively prevent the destruction of the malignant cells, a fact that is often overlooked when therapeutic strategies involving macrophages are devised. In the former case, the local availability of activating cytokines would be reduced and, in the latter instance, all-important contact between activated macrophages and the target tumor cells would not be possible. The phenomenon of restricted entry of inflammatory cells into tumors was originally demonstrated in 1962 through the collaborations of a surgeon and a pathologist, who performed a very simple, but telling, experiment (318). They threaded pieces of silk suture material through experimentally induced tumors several days before removing them for histological examination. At the exit and entry points, that is, outside of the tumors, there was the expected inflammatory reaction to the silk. However, within the tumors, no inflammatory reaction was detectable. A parallel observation made more recently, using experimental mouse tumors that either regress spontaneously or grow progressively to kill the host (319), speaks to the same conclusion. The regressing tumors were replete throughout with a uniformly distributed cellular inflammatory infiltrate. In the progressor tumors, by contrast, the cellular infiltrate was relegated to the periphery of the neoplasm. In both cases, tumor-associated inflammatory cells consisted largely of T lymphocytes and macrophages (320). Mitotic figures, indicative of active growth of the tumor cells, were abundant in areas that lacked the cellular inflammatory infiltrate and absent from those where it was present (319). Thus, rejection of neoplastic cells by a cellular inflammatory response is apparently possible *providing* that the tumor and inflammatory cells are in close proximity.

Although macrophage activation for tumor cell killing has not been directly demonstrated in vivo, there are certainly indirect suggestions that such a phenomenon exists. For example, as noted above, in regressing experimental tumors that evoke a strong cellular immune response, there are large numbers of T lymphocytes and possibly NK cells that have the capacity to secrete IFN-γ. Regression of at least one such tumor is dependent on IFN-γ (321). As a presumed consequence, these tumors contain large numbers of macrophages activated for tumor cell killing (322). By contrast, progressively growing tumors of the same type contain few or no macrophages that are similarly activated (323,324). The concentration of IFN-γ in tumors of the latter kind is much lower than in the smaller, regressing neoplasms (321) either because there is a finite amount of this critical cytokine that can be produced under conditions of maximal stimulation or because its production is suppressed in some way.

C. Positive and Negative Consequences of Activation Against Neoplastic Cells

The effective activation of macrophages for tumor cell killing is the positive outcome for which everyone is working. Activated macrophages clearly have the potential to combat tumor cells in the lung; studies of animals with metastatic disease suggest that macrophages remain functional and capable of becoming activated until late in the metastatic process when tumors are quite large (325). One of the most promising means of achieving the therapeutic goal of eradicating residual and/or metastatic tumor cells is through the use of liposomes, a subject that is covered in depth in chapter 25. In this approach, an activator or mixture of activators that will maximally stimulate macrophages for tumor cell killing, and minimally stimulate them to develop functions that might be deleterious to the host, is loaded into liposomes that have been engineered to have a particular affinity for macrophages. Because the systemic activation of macrophages can lead to overproduction of cytokines and disastrous consequences, side effects might be expected; however, these have been minimal, presumably because activation is selective; that is, overstimulation resulting in the production of unwanted functions is avoided. Aerosolization of activating agents, such as IFN-γ, is another therapeutic approach that has shown some promise in prolonging the survival of mice bearing metastatic Lewis lung carcinoma (326).

With respect to negative consequences, the activation for functions other than tumor cell killing is most relevant and may have a role in the general biology of neoplasms. For example, during the earliest stages of tumor development and, in some cases, even after they have become established, the proliferation of tumor cells may be dependent on the local availability of growth factors (327); macrophages can be activated to provide such stimuli (328). As other examples, in order for either a primary or metastatic tumor cell to become established, it has to accomplish several things that are key to its ability to survive and proliferate into a fully developed neoplasm. First and foremost, if the nascent tumor is to grow beyond a very small size, that is, that which can be supported entirely by diffusion of nutrients and waste products, it must establish its own blood supply. Second, it must orchestrate the production of an extracellular matrix so that its cells will have a substratum to which they can adhere. Third, if components of the neoplastic cells can be recognized by the host's surveillance system as "foreign," the tumor has to subvert or avoid such defensive systems in some way. Each of these key steps in the development of a tumor can be fostered by macrophages that are inappropriately activated within the environment of a tumor. Beginning with the first need, macrophages can be activated to produce potent angiogenic factors (329). These have great importance in wound healing, in which macrophages play a critical role. However, activation for a wound-healing response within the environment of a tumor is potentially disastrous, a point that has been emphasized by

Dvorak (330) and Whalen (331). With respect to the production of extracellular matrix, macrophages can secrete growth factors for mesenchymal and endothelial cells (332) as well as cytokines, such as IL-1, that will stimulate fibroblasts to proliferate and lay down collagen (333). Again, the analogy with wound healing is appropriate. Finally, in addition to the restricted entry of inflammatory cells, there are other mechanisms that can interfere with a host's ability to mount an effective response against a developing tumor. For example, tumors—even those that have been handled to ensure that artifacts are minimized—contain prostaglandins (334). Such mediators can negatively affect both cytokine secretion by lymphocytes (335–337) and the activation of macrophages for tumor cell killing (338, 339). Cytokines, such as TGF-β (340), IL-4 (341,342), IL-10 (342,343), and IL-13 (342), can interfere with the capacity that macrophages have to become activated for both/either the secretion of proinflammatory cytokines and/or tumor cell killing. In some instances, such suppressors are produced by tumor cells (344, 345). In other cases, the inhibitory effect may be related to autocrine/paracrine feedback (343). These are only a few of the many mechanisms that can potentially be involved in subverting the host's macrophage response against tumors. Some of these may additionally activate functions of tumor-infiltrating macrophages that would foster the growth rather than the destruction of tumor cells.

Macrophages may also have a part in fostering the metastasis of tumor cells, many kinds of which reach and multifocally colonize the lung. This is potentially one of the most negative of all consequences of macrophage activation, because it is usually the metastatic tumor rather than the primary one that kills the patient. One possible mechanism that may contribute to the development of this negative outcome is the inappropriate activation of macrophages for secretion of proteinases, such as urokinase (346). These have the capacity to degrade stromal elements, thereby increasing the likelihood that tumor cells will invade adjacent tissue as well as blood and lymphatic vessels.

V. Conclusions and Future Directions

It is hoped that the reader can better appreciate that, depending on the kind, extent, and duration of pulmonary macrophage activation, the outcome for the lung can either be positive or negative. Given this fact, the importance of selective stimulation, that is, that which fosters activation that leads to a positive instead of negative outcome cannot be overemphasized. Such perspective is also important in developing understanding of how various irritants, such as particulates, mediate their effects on the lung. For example, understanding how different kinds of particulates stimulate macrophages, whether directly or indirectly (e.g., after becoming opsonized and engulfed), is key to understanding the pathogenesis of the pulmonary lesions that result. Because particulates are often not cleared, chronic stimulation of macrophages compounds the problem, often causing excessive macrophage responses, both temporally and quantitatively. Understanding the principles of how macrophages become activated for various functions is also potentially important as a means of obtaining meaningful clues that will help narrow the search for the cause of a variety of diseases of unknown origin, such as sarcoidosis and idiopathic pulmonary fibrosis. Finally, conceptual understanding of macrophage activation in all its manifestations, especially that which pertains to selective expression of inducible genes and to actuation of related posttranscriptional mechanisms, is key to the development of therapeutic strategies that will maximize the positive consequences of pulmonary macrophage activation.

Acknowledgments

The writing of this chapter was supported, in part, by a program project grant from the National Cancer Institute (PO1 CA 54474) and the Wilkinson Endowment for Cancer Research. The authors would like to thank Mari Lynn Lovelace for her assistance in preparing the manuscript.

References

1. Fels AOS, Cohn Z. The alveolar macrophage. J Appl Physiol 1986; 60:353–369.
2. Lehnert BE. Pulmonary and thoracic macrophage subpopulations and clearance of particles from the lung. Environ Health Perspect 1992; 97:17–46.
3. Lewis DE, McGee JO'D. The Macrophage. Oxford, UK: Oxford University Press, 1992.
4. Lohmann-Matthew M-L, Steinmüller C, Franke-Ullmann G. Pulmonary macrophages. Eur Respir J 1994; 1678–1689.
5. Adams DO, Hanna MG Jr. Macrophage Activation. New York: Plenum Press, 1984.
6. Russell SW, Gordon, S. Macrophage Biology and Activation. Berlin: Springer-Verlag, 1992.
7. Zwilling BS, Eisenstein TK. Macrophage-Pathogen Interactions. New York: Marcel Dekker, 1994.
8. Dehring DJ, Wismar BL. Intravascular macrophages in pulmonary capillaries of humans. Am Rev Respir Dis 1989; 139:1027–1029.
9. Chitko-McKown CG, Blecha F. Pulmonary intravascular macrophages: a review of immune properties and functions. Ann Rech Vet 1992; 23:201–214.
10. Warner AE, Brain JD. The cell biology and pathogenic role of pulmonary intravascular macrophages. Am J Physiol 1990; 258:L1–L12.
11. Ungar J Jr, Wilson GR. Monocytes as a source of alveolar phagocytes. Am J Pathol 1935; 11:681–691.

12. Bowden DH, Adamson IY, Grantham WG, Wyatt JP. Origin of the lung macrophage. Evidence derived from radiation injury. Arch Pathol 1969; 88:540–546.
13. Bowden DH, Adamson IYR. The pulmonary interstitial cell as immediate precursor of the alveolar macrophage. Am J Pathol 1972; 68:521–536.
14. Lehnert BE, Valdez YE, Sebring RJ, Lehnert NM, Saunders GC, Steinkamp JA. Airway intra-luminal macrophages: evidence of origin and comparisons to alveolar macrophages. Am J Respir Cell Mol Biol 1990; 3:377–391.
15. van Oud Alblas AB, van Furth R. Origin kinetics and characteristics of pulmonary macrophages in the normal steady state. J Exp Med 1979; 149:1504–1518.
16. Thomas ED, Ramberg RE, Sale GE, Sparkes RS, Golde DW. Direct evidence for a bone marrow origin of the alveolar macrophage in man. Science 1976; 192:1016–1018.
17. Sorokin SP, Hoyt RF Jr, Grant MM. Development of macrophages in the lungs of fetal rabbits, rats and hamsters. Anat Rec 1984; 208:103–121.
18. Golde DW, Byers LA. Proliferative capacity of human alveolar macrophages. Nature 1974; 247:373–375.
19. Pinkett MO, Cowdrey CR, Nowell PC. Mixed hematopoietic and pulmonary origin of 'alveolar macrophages' as demonstrated by chromosome markers. Am J Pathol 1966; 48:859–867.
20. Tarling JD, Lin H-S, Hsu S. Self-renewal of pulmonary alveolar macrophages: evidence from radiation chimera studies. J Leukoc Biol 1987; 42:443–446.
21. Chandler DB, Fuller WC, Jackson RM, Fulmer JC. Studies of membrane receptors and phagocytosis in subpopulations of rat alveolar macrophages. Am Rev Respir Dis 1986; 133:461–467.
22. Shellito J, Kaltreider HB. Heterogeneity of immunologic function among subfractions of normal rat alveolar macrophages. Am Rev Respir Dis 1985; 131:678–683.
23. Prokkorova S, Lavnikova N, Laskin DL. Functional characterization of interstitial macrophages and subpopulations of alveolar macrophages from rat lung. J Leukoc Biol 1994; 55:141–146.
24. van Furth R, Diesselhoff-den Dulk MMC, Mattie H. Quantitative study on the production and kinetics of mononuclear phagocytes during an acute inflammatory reaction. J Exp Med 1973; 138:1314–1330.
25. van Oud Alblas AB, van der Linden-Schrever B, van Furth R. Origin and kinetics of pulmonary macrophages during an inflammatory reaction induced by intra-alveolar administration of aerosolized heat-killed BCG. Am Rev Respir Dis 1983; 128: 276–281.
26. Prieto J, Eklund A, Patarroyo M. Regulated expression of integrins and other adhesion molecules during differentiation of monocytes into macrophages. Cell Immunol 1994; 156:191–211.
27. Vanhee D, Delneste Y, Lassalle P, Gosset P, Joseph M, Tonnel AB. Modulation of endothelial cell adhesion molecule expression in a situation of chronic inflammatory stimulation. Cell Immunol 1994; 155:446–456.
28. Velo GP, Spector WG. The origin and turnover of alveolar macrophages in experimental pneumonia. J Pathol 1973; 109:7–19.
29. Campbell PA. T cell involvement in resistance to facultative intracellular pathogens of the lung. Chest 1993; 103:113S–115S.

30. Doherty TM. T-cell regulation of macrophage function. Curr Opin Immunol 1995; 7:400–404.
31. Nacy CA, Meltzer MS. T-cell–mediated activation of macrophages. Curr Opin Immunol 1991; 3:330–335.
32. Paulnock DM. Macrophage activation by T cells. Curr Opin Immunol 1992; 4: 344–349.
33. Stout RD. Macrophage activation by T cells: cognate and non-cognate signals. Curr Opin Immunol 1993; 5:398–403.
34. Chen T-Y, Lei M-G, Suzuki T, Morrison DC. Lipopolysaccharide receptors and signal transduction pathways in mononuclear phagocytes. Curr Top Microbiol Immunol 1992; 181:169–188.
35. David M. Transcription factors in interferon signaling. Pharmacol Ther 1995; 65:149–161.
36. Hamilton TA, Ohmori Y, Tebo JM, Narumi S, Tannenbaum CS. Transmembrane and intracellular signaling events in lipopolysaccharide-stimulated macrophages. Immunol Ser 1994; 60:83–97.
37. Manthey CL, Vogel SN. Interactions of lipopolysaccharide with macrophages. Immunol Ser 1994; 6:63–81.
38. Shuai K. Interferon-activated signal transduction to the nucleus. Curr Opin Cell Biol 1994; 6:253–259.
39. Ulevitch RJ, Tobias PS. Recognition of endotoxin by cells leading to transmembrane signaling. Curr Opin Immunol 1994; 6:125–130.
40. Auger MJ, Ross JA. The biology of the macrophage. In: Lewis CE, McGee JO'D, eds. The Macrophage. Oxford, UK: Oxford University Press, 1992:1–74.
41. Farrar MA, Schreiber RD. The molecular and cell biology of interferon-γ and its receptor. Ann Rev Immunol 1993; 11:571–611.
42. Kielian TL, Blecha F. CD14 and other recognition molecules for lipopolysaccharide: a review. Immunopharmacology 1995; 29:187–205.
43. Suzuki T. Signal transduction mechanisms through Fcγ receptors on the mouse macrophage surface. FASEB J 1991; 5:187–193.
44. Tartaglia LA, Goeddel DV. Two TNF receptors. Immunol Today 1992; 13:151–153.
45. Kelley J. Cytokines of the lung. Am Rev Respir Dis 1990; 141:765–788.
46. Paulnock DM. The molecular biology of macrophage activation. Immunol Ser 1994; 60:47–62.
47. Juliano RL, Haskill S. Signal transduction from the extracellular matrix. J Cell Biol 1993; 120:577–585.
48. Standiford TJ, Kunkel SL, Kasahara K, Milia MJ, Rolfe MW, Strieter RM. Interleukin-8 gene expression from human alveolar macrophages: the role of adherence. Am J Respir Cell Mol Biol 1991; 5:579–585.
49. Tordjman C, Coge F, Andre N, Rique H, Spedding M, Bonnet J. Characterisation of cyclooxygenase 1 and 2 expression in mouse resident peritoneal macrophages in vitro; interactions of nonsteroidal anti-inflammatory drugs with COX2. Biochim Biophys Acta 1995; 1256:249–256.
50. Xing Z, Jordana M, Gauldie J. IL-1β and IL-6 gene expression in alveolar macrophages: modulation by extracellular matrices. Am J Physiol 1992; 262:L600–L605.
51. Natsuka S, Akira S, Nishio Y, Hashimoto S, Sugita T, Isshiki H, Kishimoto T.

Macrophage differentiation-specific expression of NF-IL6, a transcription factor for interleukin-6. Blood 1992; 79:460–466.

52. Eilers A, Seegert D, Schindler C, Baccarini M, Decker T. The response of gamma interferon activation factor is under developmental control in cells of the macrophage lineage. Mol Cell Biol 1993; 13:3245–3254.
53. Eilers A, Decker T. Activity of Stat family transcription factors is developmentally controlled in cells of the macrophage lineage. Immunobiology 1995; 193:328–333.

53a. McDowell MA, Lucas DM, Nicolet CM, Paulnock DM. Differential utilization of IFN-γ–responsive elements in two maturationally distinct macrophage cell lines. J Immunol 1995; 155:4933–4938.

54. DeWitt DL, Meade EA. Serum and glucocorticoid regulation of gene transcription and expression of the prostaglandin H synthase-1 and prostaglandin H synthase-2 isozymes. Arch Biochem Biophys 1993; 306:94–102.
55. Gilbert RS, Herschman HR. Transforming growth factor beta differentially modulates the inducible nitric oxide synthase gene in distinct cell types. Biochem Biophys Res Commun 1993; 195:380–384.
56. Orlandi M, Bartolini G, Belletti B, Spisni E, Tomasi V. Thromboxane A_2 synthase activity in platelet free human monocytes. Biochim Biophys Acta 1994; 1215:285–290.
57. Cipolle MD, Pasquale MD, Cerra FB. Secondary organ dysfunction. From clinical perspectives to molecular mediators. Crit Care Clin 1993; 9:261–298.
58. Nathan CF. Secretory products of macrophages. J Clin Invest 1987; 79:319–326.
59. Takemura R, Werb Z. Secretory products of macrophages and their physiological functions. Am J Physiol 1984; 246:C1–C9.
60. Adams DO, Hamilton TA. Molecular basis of macrophage activation: diversity and its origins. Immunol Ser 1994; 60:75–114.
61. Kröncke KD, Fehsel K, Kolb-Bachofen V. Inducible nitric oxide synthase and its product nitric oxide, a small molecule with complex biological activities. Biol Chem Hoppe Seyler 1995; 376:327–343.
62. Förstermann U, Gath I, Schwarz P, Closs EI, Kleinert H. Isoforms of nitric oxide synthase. Properties, cellular distribution and expressional control. Biochem Pharmacol 1995; 50:1321–1332.
63. Nathan C. Nitric oxide as a secretory product of mammalian cells. FASEB J 1992; 6:3051–3064.
64. Lorsbach RB, Russell SW. A specific sequence of stimulation is required to induce synthesis of the antimicrobial molecule nitric oxide by mouse macrophages. Infect Immun 1992; 60:2133–2135.
65. Hibbs JB Jr, Taintor RR, Chapman HA Jr, Weinberg JB. Macrophage tumor killing: influence of the local environment. Science 1977; 197:279–282.
66. Ruco LP, Meltzer MS. Macrophage activation for tumor cytotoxicity: development of macrophage cytotoxic activity requires completion of a sequence of short-lived intermediary reactions. J Immunol 1978; 121:2035–2042.
67. Russell SW, Doe WF, McIntosh AT. Functional characterization of a stable, noncytolytic stage of macrophage activation in tumors. J Exp Med 1977; 146:1511–1520.
68. Green SJ, Nacy CA, Meltzer MS. Cytokine-induced synthesis of nitrogen oxides in macrophages: a protective host response to *Leishmania* and other intracellular pathogens. J Leukoc Biol 1991; 50:93–103.

69. Tanaka T, Akira S, Yoshida K, Umemoto M, Yoneda Y, Shirafuji N, Fujiwara H, Suematsu S, Yoshida N, Kishimoto T. Targeted disruption of the NF-IL6 gene discloses its essential role in bacteria killing and tumor cytotoxicity by macrophages. Cell 1995; 80:353–361.
70. Fujihara M, Ito N, Pace JL, Watanabe Y, Russell SW, Suzuki T. Role of endogenous interferon-β in lipopolysaccharide-triggered activation of the inducible nitric-oxide synthase gene in a mouse macrophage cell line, J774. J Biol Chem 1994; 269: 12773–12778.
71. Zhang X, Alley EW, Russell SW, Morrison DC. Necessity and sufficiency of beta interferon for nitric oxide production in mouse peritoneal macrophages. Infect Immun 1994; 62:33–40.
72. Zhou A, Chen Z, Rummage JA, Jiang H, Kolosov M, Kolosova I, Stewart CA, Leu RW. Exogenous IFN-γ induces endogenous synthesis of IFN-α and -β by murine macrophages for induction of nitric oxide synthase. J Interferon Cytokine Res 1995; 15:897–904.
73. Jiang H, Stewart CA, Leu RW. Tumor-derived factor synergizes with IFN-γ and LPS, IL-2 or TNF-α to promote macrophage synthesis of TNF-α and TNF receptors for autocrine induction of nitric oxide synthase and enhanced nitric oxide-mediated tumor cytotoxicity. Immunobiology 1995; 192:321–342.
74. Nacy CA, Meierovics AI, Belosevic M, Green SJ. Tumor necrosis factor-alpha: central regulatory cytokine in the induction of macrophage antimicrobial activities. Pathobiology 1991; 59;182–184.
75. Hughes TK, Baron S. A large component of the antiviral activity of mouse interferon gamma may be due to its induction of interferon alpha. J Bil Regul Homeost Agents 1987; 1:29–32.
76. Kolosov M, Kolosova I, Leu RW. Stimulation of macrophage synthesis of complement C1q by interferon-γ mediated by endogenous interferon-α/β. J Interferon Cytokine Res 1996; 16:245–249.
77. Lowenstein CJ, Glatt CS, Bredt DS, Snyder SH. Cloned and expressed macrophage nitric oxide synthase contrasts with the brain enzyme. Proc Natl Acad Sci USA 1992; 89:6711–6715.
78. Xie QW, Cho HJ, Calaycay J, Mumford RA, Swiderek KM, Lee TD, Ding A, Troso T, Nathan C. Cloning and characterization of inducible nitric oxide synthase from mouse macrophages. Science 1992; 256:225–228.
79. Lorsbach RB, Murphy MJ, Lowenstein CJ, Snyder SH, Russell SW. Expression of the nitric oxide synthase gene in mouse macrophages activated for tumor cell killing. Molecular basis for the synergy between interferon-γ and lipopolysaccharide. J Biol Chem 1993; 268:1908–1913.
80. Weisz A, Oguchi S, Cicatiello L, Esumi H. Dual mechanism for the control of inducible-type NO synthase gene expression in macrophages during activation by interferon-γ and bacterial lipopolysaccharide. Transcriptional and post-transcriptional regulation. J Biol Chem 1994; 269:8324–8333.
81. Lowenstein CJ, Alley EW, Raval P, Snowman AM, Snyder SH, Russell SW, Murphy WJ. Macrophage nitric oxide synthase gene. Two upstream regions mediate induction by interferon-γ and lipopolysaccharide. Proc Natl Acad Sci USA 1993; 89: 6711–6715.

82. Xie QW, Whisnant R, Nathan C. Promoter of the mouse gene encoding calcium-independent nitric oxide synthase confers inducibility by interferon γ and bacterial lipopolysaccharide. J Exp Med 1993; 177:1779–1784.
83. Alley EW, Murphy WJ, Russell SW. A classical enhancer element responsive to both lipopolysaccharide and interferon-γ augments induction of the iNOS gene in mouse macrophages. Gene 1995; 158:247–251.
84. Murphy WJ, Parmely TJ, Russell SW. Expression of the mouse iNOS gene is absolutely dependent upon binding of its promoter-linked octamer response element. Submitted.
85. Sawada T, Murphy WJ, Pluznik DH, Falk LA. IL-6 induction of protein-DNA complexes via a novel regulatory region of the iNOS gene promoter: role of octamer binding proteins. J Immunol. Submitted.
86. Kamijo R, Harada H, Matsuyama T, Bosleand M, Gerecitano J, Shapiro D, Le J, Koh SI, Kimura T, Green SJ, Mak TW, Taniguchi T, Vilĉek. Requirement for transcription factor IRF-1 in NO synthase induction in macrophages. Science 1994; 263: 1612–1615.
87. Martin E, Nathan C, Xie QW. Role of interferon regulatory factor 1 in induction of nitric oxide synthase. J Exp Med 1994; 180:977–984.
88. Gao J, Russell SW, Morrison DC, Parmely TJ, Murphy WJ. Stat 1α binding to an interferon-γ activated site (GAS) enhances mouse iNOS gene expression in response to interferon-γ and LPS. Submitted.
89. Murphy WJ, Muroi M, Zhang X, Suzuki T, Russell SW. Both basal and enhancer κB elements are required for full induction of the mouse inducible nitric oxide synthase gene. Submitted.
90. Xie QW, Kashiwabara Y, Nathan C. Role of transcription factor NF-κB/Rel in induction of nitric oxide synthase. J Biol Chem 194; 269:4705–4708.
91. Ghosh S, Baltimore D. Activation *in vitro* of NF-κB by phosphorylation of its inhibitor IκB. Nature 1990; 344:678–682.
92. Müller JM, Ziegler-Heitbrock HWL, Baeuerle PA. Nuclear factor kappa B, a mediator of lipopolysaccharide effects. Immunobiology 1993; 187:233–256.
93. Blackford JA Jr, Antonini JM, Castranova V, Dey RD. Intratracheal instillation of silica up-regulates inducible nitric oxide synthase gene expression and increases nitric oxide production in alveolar macrophages and neutrophils. Am J Respir Cell Mol Biol 1994; 11:426–431.
94. Gossart S, Cambon C, Orfila C, Séguélas M-H, Lepert J-C, Rami J, Carré P, Pipy B. Reactive oxygen intermediates as regulators of TNF-α production in rat lung inflammation induced by silica. J Immunol 1996; 156:1540–1548.
95. Nyberg P, Klockars M. Interferon-γ and immunoglobulin enhanced mineral dust-induced production of reactive oxygen metabolites by human macrophages. Clin Immunol Immunopathol 1991; 60:128–136.
96. Chen F, Kuhn DC, Sun S-C, Gaydos LJ, Demers LM. Dependence and reversal of nitric oxide production on NF-κB in silica and lipopolysaccharide-induced macrophages. Biochem Biophys Res Commun 1995; 214:839–846.
97. Schreck R, Rieber P, Baeuerle P. Reactive oxygen intermediates as apparently widely used messengers in the activation of NF-κB transcription factor and HIV-1. EMBO J 1991; 10:2247–2258.

98. Ye J, Zhang X, Young HA, Mao Y, Shi X. Chromium (VI)-induced nuclear factor-κB activation in intact cells via free radical reactions. Carcinogenesis 1995; 16:2401–2405.
99. Halliwell B, Gutteridge JM. Oxygen free radicals and iron in relation to biology and medicine: some problems and concepts. Arch Biochem Biophys 1986; 246: 501–514.
100. Olakanmi O, McGowan SE, Hayek MB, Britigan BE. Iron sequestration by macrophages decreases the potential for extracellular hydroxyl radical formation. J Clin Invest 1993; 91:889–899.
101. Weiss G, Werner-Felmayer G, Werner ER, Grünewald K, Wachter H, Hentze MW. Iron regulates nitric oxide synthase activity by controlling nuclear transcription. J Exp Med 1994; 180:969–976.
102. Feng L, Xia Y, Garcia GE, Hwang D, Wilson CB. Involvement of reactive oxygen intermediates in cyclooxygenase-2 expression induced by interleukin-2, tumor necrosis factor-α, and lipopolysaccharide. J Clin Invest 1995; 95:1669–1675.
103. Hempel SL, Monick MM, He B, Yano T, Hunninghake GW. Synthesis of prostaglandin H synthase-2 by human alveolar macrophages in response to lipopolysaccharide is inhibited by decreased oxidant tone. J Biol Chem 1994; 269:32979–32984.
104. Sims SH, Cha Y, Romine MF, Gao P-Q, Gottlieb K, Deisseroth AB. A novel interferon-inducible domain: structural and functional analysis of the human interferon regulatory factor I gene promoter. Mol Cell Biol 1993; 13:690–702.
105. Ihle JN. Cytokine receptor signaling. Nature 1995; 377:591–594.
106. Rein T, Müller M, Zorbas H. *In vivo* footprinting of the IRF-1 promoter: inducible occupation of a GAS element next to a persistent structural alteration of the DNA. Nucleic Acids Res 1994; 22:3033–3037.
107. David M, Petricoin E III, Benjamin C, Pine R, Weber MJ, Larner AC. Requirement for MAP kinase (ERK2) activity in interferon α- and interferon β-stimulated gene expression through STAT proteins. Science 1995; 269:1721–1723.
108. Pine R, Canova A, Schindler C. Tyrosine phosphorylated p91 binds to a single element in the ISGF2/IRF-1 promoter to mediate induction by IFN-α and IFN-γ, and is likely to autoregulate the p91 gene. EMBO J 1994; 13:158–167.
109. Pace JL, Russell SW, LeBlanc PA, Murasko DM. Comparative effects of various classes of mouse interferons on macrophage activation for tumor cell killing. J Immunol 1985; 134:977–981.
110. Mantovani R, Malgaretti N, Giglioni B, Comi P, Cappellini N, Nicolis S, Ottolenghi S. A protein factor binding to an octamer motif in the γ-globin promoter disappears upon induction of differentiation and hemoglobin synthesis in K562 cells. Nucleic Acids Res 1987; 15:9349–9364.
111. Pace JL, Varesio L, Russell SW, Blasi E. The strain of mouse and assay conditions dictate whether MuIFN-γ primes or activates macrophages for tumor cell killing. J Leukoc Biol 1985; 37:475–479.
112. Thomas F, Ando T, Verma K, Kagan E. Asbestos-induced nitric oxide production: synergistic effect with interferon-γ. Ann NY Acad Sci 1994; 725:207–212.
113. Farber JM. A collection of mRNA species that are inducible in the RAW 264.7 mouse macrophage cell line by gamma interferon and other agents. Mol Cell Biol 1992; 12:1535–1545.

114. Collart MA, Baeuerle P, Vassalli P. Regulation of tumor necrosis factor alpha transcription in macrophages: involvement of four κB-like motifs and of constitutive and inducible forms of NF κB. Mol Cell Biol 1990; 10:1498–1506.
115. Drouet C, Shakhov AN, Jongeneel CV. Enhancers and transcription factors controlling the inducibility of the tumor necrosis factor-α promoter in primary macrophages. J Immunol 1991; 147:1694–1700.
116. Shakhov AN, Collart MA, Vassalli P, Nedospasov SA, Jongeneel CV. KB-type enhancers are involved in lipopolysaccharide-mediated transcriptional activation of the tumor necrosis factor α gene in primary macrophages. J Exp Med 1990; 171: 35–47.
117. Yu SF, Koerner TJ, Adams DO. Gene regulation in macrophage activation: differential regulation of genes encoding for tumor necrosis factor, interleukin-1, JE, and KC by interferon γ and lipopolysaccharide. J Leukoc Biol 1990; 48:412–419.
118. Arias-Negrete S, Keller K, Chadee K. Proinflammatory cytokines regulate cyclooxygenase-2 mRNA expression in human macrophages. Biochem Biophys Res Commun 1995; 208:582–589.
119. Riese J, Hoff T, Nordhoff A, DeWitt DL, Resch K, Kaever V. Transient expression of prostaglandin endoperoxide synthase-2 during mouse macrophage activation. J Leukoc Biol 1994; 55:476–482.
120. Alley EW, Zhang X, Russell SW. Contribution of clonally distinct subpopulations to heterogeneous production of inducible nitric oxide synthase by LPS-stimulated mouse macrophages. J Endotox Res 1994; 1:235–242.
121. Pace JL, Lowenstein CJ, Phillips TA, Chen LC, Morrison DC, Hunt JS, Russell SW. Population dynamics of inducible nitric oxide synthase production by LPS- and LPS/IFNγ-stimulated mouse macrophages. J Endotox Res 1994; 1:227–233.
122. Fiering S, Northrop JP, Nolan GP, Mattila PS, Crabtree GR, Herzenberg LA. Single cell assay of a transcription factor reveals a threshold in transcription activated signals emanating from the T-cell antigen receptor. Genes Dev 1990; 4:1823–1834.
123. Ko MSH, Nakauchi H, Takahashi N. The dose dependence of glucocorticoid-inducible gene expression results from changes in the number of transcriptionally active templates. EMBO J 1990; 9:2835–2842.
124. Laybourn PJ, Kadonaga JT. Threshold phenomena and long-distance activation of transcription by RNA polymerase II. Science 1992; 257:1682–1685.
125. Ross IL, Browne CM, Hume DA. Transcription of individual genes in eukaryotic cells occurs randomly and infrequently. Immunol Cell Biol 1994; 72:177–185.
126. Cameron ML, Granger DL, Weinberg BJ, Kozumbo WJ, Koren HS. Human alveolar and peritoneal macrophages mediate fungistasis independently of L-arginine oxidation to nitrite or nitrate. Am Rev Respir Dis 1990; 142:1313–1319.
127. Schneemann M, Schoedon G, Hofer S, Blau N, Guerrero L, Schaffner A. Nitric oxide synthase is not a constituent of the antimicrobial armature of human mononuclear phagocytes. J Infect Dis 1993; 167:1358–1363.
128. Weinberg JB, Misukonis MA, Shami PJ, Mason SN, Sauls DL, Dittman WA, Wood ER, Smith GK, McDonald B, Fachaus KE, Haney AF, Granger DL. Human mononuclear phagocyte inducible nitric oxide synthase (iNOS): analysis of iNOS mRNA, iNOS protein, biopterin, and nitric oxide production by blood monocytes and peritoneal macrophages. Blood 1995; 86:1184–1195.

129. Nussler AK, Di SM, Billiar TR, Hoffman RA, Geller DA, Selby R, Madariaga J, Simmons RL. Stimulation of the nitric oxide synthase pathway in human hepatocytes by cytokines and endotoxin. J Exp Med 1992; 176:261–264.
130. Sherman PA, Laubach VE, Reep BR, Wood ER. Purification and cDNA sequence of an inducible nitric oxide synthase from a human tumor cell line. Biochemistry 1993; 32:11600–11605.
131. Denis M. Tumor necrosis factor and granulocyte macrophage-colony stimulating factor stimulate human macrophages to restrict growth of virulent *Mycobacterium avium* and to kill avirulent *M. avium*: killing effector mechanism depends on the generation of reactive nitrogen intermediates. J Leukoc Biol 1991; 49:380–387.
132. Muñoz-Fernández MA, Fernández MA, Fresno M. Activation of human macrophages for the killing of intracellular *Trypanosoma cruzi* by TNF-α and IFN-γ through a nitric oxide-dependent mechanism. Immunol Lett 1992; 33:35–40.
133. Gyan B, Troye-Blomberg M, Perlmann P, Bjorkman A. Human monocytes cultured with and without interferon-γ inhibit *Plasmodium falciparum* parasite growth *in vitro* via secretion of reactive nitrogen intermediates. Parasite Immunol 1994; 16:371–375.
134. Tufano MA, Rossano F, Catalanotti P, Liguori G, Marinelli A, Baroni A, Marinelli P. Properties of *Yersinia enterocolitica* porins: interference with biological functions of phagocytes, nitric oxide production and selective cytokine release. Res Microbiol 1994; 145:297–307.
135. Zembala M, Siedlar M, Marcinkiewicz J, Pryjma J. Human monocytes are stimulated for nitric oxide release *in vitro* by some tumor cells but not by cytokines and lipopolysaccharide. Eur J Immunol 1994; 24:435–439.
136. de Maria R, Cifone MG, Trotta R, Rippo MR, Festuccia C, Santoni A, Testi R. Triggering of human monocyte activation through CD69, a member of the natural killer cell gene complex family of signal transducing receptors. J Exp Med 1994; 180:1999–2004.
137. Kobzik L, Bredt DS, Lowenstein CJ, Drazen J, Gaston B, Sugarbaker D, Stamler JS. Nitric oxide synthase in human and rat lung: immunocytochemical and histochemical localization. Am J Respir Cell Mol Biol 1993; 9:371–377.
138. Padgett EL, Pruett SB. Evaluation of nitrite production by human monocyte-derived macrophages. Biochem Biophys Res Commun 1992; 186:775–781.
139. Murray HW, Szuro-Sudol A, Wellner D, Oca MJ, Granger AM, Libby DM, Rothermel CD, Ruin BY. Role of tryptophan degradation in respiratory burst-independent antimicrobial activity of gamma interferon-stimulated human macrophages. Infect Immun 1989; 57:845–849.
140. Barnewall RE, Rikihisa Y. Abrogation of gamma interferon-induced inhibition of *Erlichia chaffeensis* infection in human monocytes with iron transferring. Infect Immun 1994; 62:4804–4810.
141. Byrd TF, Horwitz MA. Interferon gamma-activated human monocytes downregulate transferring receptors and inhibit the intracellular multiplication of *Legionella pneumophila* by limiting the availability of iron. J Clin Invest 1989; 83:1457–1465.
142. Gebran SJ, Newton C, Yamamoto Y, Widen R, Klein TW, Friedman H. Macrophage permissiveness for *Legionell pneumophila* growth modulated by iron. Infect Immun 1994; 62:564–568.

143. Murray HW, Granger AM, Teitelbaum RF. Gamma interferon-activated human macrophages and *Toxaplasma gondii*, *Chlamydia psittici* and *Leishmani donovani*: antimicrobial role of limiting intracellular iron. Infect Immun 1991; 59:4684–4686.
144. Weinberg ED. Iron depletion: a defense against intracellular infection and neoplasia. Life Sci 1992; 50:1289–1297.
145. Zhang X, Laubach VE, Alley EW, Edwards KA, Sherman PA, Russell SW, Murphy WJ. Transcriptional basis for hyporesponsiveness of the human inducible nitric oxide synthase gene to lipopolysaccharide/interferon-γ. J Leukoc Biol 1996; 59: 575–585.
146. Pace JL, Russell SW. Activation of mouse macrophages for tumor cell killing. I. Quantitative analysis of interactions between lymphokine and lipopolysacchride. J Immunol 1981; 126:1863–1867.
147. Pace JL, Russell SW, Torres BA, Johnson HM, Gray PW. Recombinant mouse γ interferon induces the priming step in macrophage activation for tumor cell killing. J Immunol 1983; 130:2011–2013.
148. Ulevitch RJ and Tobias PS. Receptor-dependent mechanisms of cell stimulation by bacterial endotoxin. Annu Rev Immunol 1995; 13:437–457.
149. Morrison DC, Lei M-G, Kirikae T, Chen T-Y. Endotoxin receptors on mammalian cells. Immunobiology 1993; 187:212–226.
150. Pabst MJ, Johnston RB Jr. Increased production of superoxide anion by macrophages exposed in vitro to muramyl dipeptide or lipopolysaccharide. J Exp Med 1980; 151:101–114.
151. Bernard C, Merval R, Esposito B, Tedgui A. Elevated temperature accelerates and amplifies the induction of nitric oxide synthesis in rat macrophages. Eur J Pharmacol 1994; 270:115–118.
152. Zhang X, Morrison DC. Lipopolysaccharide-induced selective priming effects on tumor necrosis factor α and nitric oxide production in mouse peritoneal macrophages. J Exp Med 1993; 177:511–516.
153. Hirohashi N, Morrison DC. Low-dose lipopolysaccharide (LPS) pretreatment of mouse macrophages modulates LPS-dependent interleukin-6 production in vitro. Infect Immun 1996; 64:1011–1015.
154. Takasuka N, Matsuura K, Yamamoto S, Akagawa KS. Suppression of TNF-α mRNA expression in LPS-primed macrophages occurs at the level of nuclear factor-κB activation, but not at the level of protein kinase C or CD14 expression. J Immunol 1995; 154:4803–4812.
155. Ballard DW, Dixon EP, Peffer NJ, Bogerd H, Doerre S, Stein B, Greene WC. The 65-kDa subunit of human NF-κB functions as a potent transcriptional activator and a target for v-Rel–mediated repression. Proc Natl Acad Sci USA 1992; 89:1875–1879.
156. Mercurio F, Didonato J, Rosette C, Karin M. Molecular cloning and characterization of a novel Rel/NF-κB family member displaying structural and functional homology to NF-κB p50/p105. DNA Cell Biol 1992; 11:523–537.
157. Schmid RM, Perkins ND, Duckett CS, Andrews PC, Nabel GJ. Cloning of an NF-κB subunit which stimulates HIV transcription in synergy with p65. Nature 1991; 352: 733–736.
158. Schmitz ML, Baeuerle PA. The p65 subunit is responsible for the strong transcription activating potential of NF-κB. EMBO J 1991; 10:3805–3817.

159. Franzoso G, Bours V, Azarenko V, Park S, Tomita-Yamaguchi M, Kanno T, Brown K, Siebenlist U. The oncoprotein Bcl-3 can facilitate NF-κB-mediated transactivation by removing inhibiting p50 homodimers from select κB sites. EMBO J 1993; 12:3893–3901.
160. Ziegler-Heitbrock HWL, Wedel A, Schraut W, Ströbel M, Wendelgass P, Sternsdorf T, Bäuerle PA, Haas JG, Riethmüller G. Tolerance to lipopolysaccharide involves mobilization of nuclear factor κB with predominance of p50 homodimers. J Biol Chem 1994; 269:17001–17004.
161. Deguchi M, Sakuta H, Uno K, Inaba K, Muramatsu S. Exogenous and endogenous type I interferons inhibit interferon-γ-induced nitric oxide production and nitric oxide synthase expression in murine peritoneal macrophages. J Interferon Cytokine Res 1995; 15:977–984.
162. Pace JL, MacKay RJ, Hayes MP. Suppressive effect of interferon-β on development of tumoricidal activity in mouse macrophages. J Leukoc Biol 1987; 41:257–263.
163. Pace JL, Lorsbach RB, Raval P, Russell SW. Pretreatment of mouse macrophages with IFN-β interferes with augmented transcription of the inducible nitric oxide synthase gene. Submitted.
164. Koestler TP, Rieman D, Muirhead K, Greig RG, Poste G. Identification and characterization of a monoclonal antibody to an antigen expressed on activated macrophages. Proc Natl Acad Sci USA 1984; 81:4505–4509.
165. Paulnock DM, Lambert LE. Identification and characterization of monoclonal antibodies specific for macrophages at intermediate stages in the tumoricidal activation pathway. J Immunol 1990; 144:765–773.
166. Taniyama T, Tokunaga T. Monoclonal antibodies directed against mouse macrophages in different stages of activation for tumor cytotoxicity. J Immunol 1983; 131: 1032–1037.
167. MacKay RJ, Russell SW. Protein changes associated with stages of activation of mouse macrophages for tumor cell killing. J Immunol 1986; 137:1392–1398.
168. MacKay RJ, Russell SW. Protein phenotypes of mouse macrophages activated in vivo for tumor cell killing. J Leukoc Biol 1987; 42:213–221.
169. Phillips TA, Kujubu DA, MacKay RJ, Herschman HR, Russell SW, Pace JL. The mouse macrophage activation-associated marker protein, p71/73, is an inducible prostaglandin endoperoxide synthase (cyclooxygenase). J Leukoc Biol 1993; 53:411–419.
170. Hempel SL, Monick MM, Hunninhake GW. Lipopolysaccharide induces prostaglandin H synthase-2 protein and mRNA in human alveolar macrophages and blood monocytes. J Clin Invest 1994; 93:391–396.
171. Gilbert RS, Reddy ST, Kujubu DA, Xie W, Luner S, Herschman HR. Transforming growth factor β_1 augments mitogen-induced prostaglandin synthesis and expression of the TIS10/prostaglandin synthase 2 gene both in Swiss 3T3 cells and in murine embryo fibroblasts. J Clin Pathol 1994; 159:67–75.
172. Gilbert RS, Reddy ST, Targan S, Herschman HR. TGF-β_1 augments expression of the TIS10/prostaglandin synthase-2 gene in intestinal epithelial cells. Cell Mol Biol Res 1994; 40:653–660.
173. Kawata R, Reddy ST, Wolner B, Herschman HR. Prostaglandin synthase 1 and prostaglandin synthase 2 both participate in activation-induced prostaglandin D_2 production in mass cells. J Immunol 1995; 155:818–825.

174. Starkey PM, Turley L, Gordon S. The mouse macrophage-specific glycoprotein defined by monoclonal antibody F4/80: characterization, biosynthesis and demonstration of a rat analogue. Immunology 1987; 60:117–122.
175. McGarry MP, Stewart CC. Murine eosinophil granulocytes bind the murine macrophage-monocyte specific monoclonal antibody F4/80. J Leukoc Biol 1991; 50:471–478.
176. Goyert SM, Ferrero EM, Seremetis SV, Winchester RJ, Silver J, Mattison AC. Biochemistry and expression of myelomonocytic antigens. J Immunol 1986; 137: 3909–3914.
177. Yang Z, Carter CD, Miller MS, Bochsler PN. CD14 and tissue factor expression by bacterial lipopolysaccharide-stimulated bovine alveolar macrophages in vitro. Infect Immun 1995; 63:51–56.
178. Menson EN, Wilson RA. Lung-phase immunity to *Schistosoma mansoni*: definition of alveolar macrophage phenotypes after vaccination and challenge of mice. Parasite Immunol 1990; 12:353–366.
179. Homolka J, Lorenz J, Zuchold HD, Muller-Quernheim J. Evaluation of soluble CD14 and neopterin as serum parameters of the inflammatory activity of pulmonary sarcoidosis. Cell Invest 1992; 70:909–916.
180. Skerret SJ, Martin TR. Intratracheal interferon-γ augments pulmonary defenses in experimental legionellosis. Am J Respir Crit Care Med 1994; 149:50–58.
181. Kawakami K, Kohno S, Morikawa N, Kadota J, Saito A, Hara K. Activation of macrophages and expansion of specific T lymphocytes in the lungs of mice intratracheally inoculated with *Cryptococcus neoformans*. Clin Exp Immunol 1994; 96: 230–237.
182. Pueringer RJ, Schwartz DA, Dayton CS, Gilbert SR, Hunninghake GW. The relationship between alveolar macrophage TNF, IL-1 and PGE_2 release, alveolitis, and disease severity in sarcoidosis. Chest 1993; 103:832–838.
183. Meyer KC, Powers C, Rosenthal N, Auerbach R. Alveolar macrophage surface carbohydrate expression is altered in interstitial lung disease as determined by lectin-binding profiles. Am Rev Respir Dis 1993; 148:1325–1334.
184. Meyer KC, Cornwell R, Carlin JM, Powers C, Irizarry A, Byrne GI, Border EC. Effects of interferons β or γ on neopterin biosynthesis and tryptophan degradation by human alveolar macrophages *in vitro*: synergy with lipopolysaccharide. Am J Respir Cell Mol Biol 1992; 6:639–646.
185. Meyer KC, Arend RA, Kalayoglu MV, Rosenthal NS, Byrne GI, Brown RR. Tryptophan metabolism in chronic inflammatory lung disease. J Lab Clin Med 1995; 126:530–540.
186. Fuchs D, Möller AA, Reibnegger G, Werner ER, Werner-Felmayer G, Dierich MP, Wachter H. Increased endogenous interferon-gamma and neopterin correlate with increased degradation of tryptophan in human immunodeficiency virus type 1 infection. Immunol Lett 1991; 28:207–212.
187. Baganha MF, Mota-Pinto A, Pêgo MA, Marques MAT, Santos Rosa MA, Robalo Cordeiro AJA. Neopterin in tuberculous and neoplastic pleural fluids. Lung 1992; 170:155–161.
188. Werner ER, Werner-Felmayer G, Wachter H. Tetrahydrobiopterin and cytokines. Proc Soc Exp Biol Med 1993; 203:1–12.

189. Werner ER, Werner-Felmayer G, Fuchs D, Hausen A, Reibnegger G, Yim JJ, Pfleiderer W, Wachter H. Tetrahydrobiopterin biosynthetic activities in human macrophages, fibroblasts, THP-1, and T24 cells. GTP-cyclohydrolase I is stimulated by interferon-gamma, and 6-pyruvoyl tetrahydropterin synthase and sepiapterin reductase are constitutively present. J Biol Chem 1990; 265:3189–3192.
190. Werner ER, Werner-Felmayer G, Fuchs D, Hausen A, Reibnegger G, Yim JJ, Wachter H. Impact of tumour necrosis factor-α and interferon-γ on tetrahydrobiopterin synthesis in murine fibroblasts and macrophages. Biochem J 1991; 280: 709–714.
191. Lissoni P, Rovelli F, Barni S, Ardizzoia A, Pittalis S, Tisi E, Perlangeli V, Tancini G. Modulation of macrophage response to interleukin-2 immunotherapy by interleukin-3 in cancer patients. Biol Chem Hoppe Seyler 1992; 373:1217–1222.
192. Dai W, Gupta SL. Regulation of indoleamine 2,3-dioxygenase gene expression in human fibroblasts by interferon-γ. Upstream control region discriminates between interferon-γ and interferon-α. J Biol Chem 1990; 265:19871–19877.
193. Nakayama DK, Geller DA, Di Silvio M, Bloomgarden G, Davies P, Pitt BR, Hatakeyama K, Kagamiyama H, Simmons RL, Billiar TR. Tetrahydrobiopterin synthesis and inducible nitric oxide production in pulmonary artery smooth muscle. Am J Physiol 1994; 266:L455–L460.
194. Ziegler I, Schott K, Lübbert M, Herrmann F, Schwuléra U, Bacher A. Control of tetrahydrobiopterin synthesis in T lymphocytes by synergistic action of interferon-γ and interleukin-2. J Biol Chem 1990; 265:17026–17030.
195. Andert SE, Griesmacher A, Zuckermann A, Müller MM. Neopterin release from human endothelial cells is triggered by interferon-gamma. Clin Exp Immunol 1992; 88:555–558.
196. Mulligan MS, Jones ML, Vaporciyan AA, Howard MC, Ward PA. Protective effects of IL-4 and IL-10 against immune complex-induced lung injury. J Immunol 1993; 151:5666–5674.
197. Nibbering PH, van den Barselaar MTH, van de Gevel JS, Leijh PCJ, van Furth R. Deficient intracellular killing of bacteria by murine alveolar macrophages. Am J Respir Cell Mol Biol 1989; 1:417–422.
198. Bogdan C, Nathan C. Modulation of macrophage function by transforming growth factor-β and interleukin-10. Ann NY Acad Sci 1993; 685:713–739.
199. Bogdan C, Paik J, Vodovotz Y, Nathan C. Contrasting mechanisms for suppression of macrophage cytokine release by transforming growth factor-β and interleukin-10. J Biol Chem 1992; 267:23301–23308.
200. Chensue SW, Ruth JH, Warmington K, Lincoln P, Kunkel SL. In vivo regulation of macrophage IL-12 production during type 1 and type 2 cytokine-mediated granuloma formation. J Immunol 1995; 155:3546–3551.
201. Corradin SB, Buchmuller-Rouiller Y, Smith J, Suardet L, and Mauël J. Transforming growth factor β1 regulation of macrophage activation depends on the triggering stimulus. J Leukoc Biol 1993; 54:423–429.
202. Vodovotz Y, Bogdan C, Paik J, Xie Q-W, Nathan C. Mechanisms of suppression of macrophage nitric oxide release by transforming growth factor β. J Exp Med 1993; 178:605–613.
203. Lehn M, Weiser WY, Engelhorn S, Gillis S, Remold HG. IL-4 inhibits H_2O_2

production and antileishmanial capacity of human cultured monocytes by IFN-γ. J Immunol 1989; 143:3020–3024.
204. Aderka D, Le J, Vilček J. IL-6 inhibits lipopolysaccharide-induced tumor necrosis factor production in cultured human monocytes, U937 cells, and in mice. J Immunol 1989; 143:3517–3523.
205. Berkman N, John M, Roesems G, Jose PJ, Barnes PJ, Chung KF. Inhibition of macrophage inflammatory protein-1α expression by IL-10. Differential sensitivities in human blood monocytes and alveolar macrophages. J Immunol 1995; 155:4412–4418.
206. de Waal Malefyt RW, Abrams J, Bennett B, Figdor CG, de Vries JE. Interleukin 10 (IL-10) inhibits cytokine synthesis by human monocytes: an autoregulatory role of IL-10 produced by monocytes. J Exp Med 1991; 174:1209–1220.
207. Cunha FQ, Moncada S, Liew FY. Interleukin-10 (IL-10) inhibits the induction of nitric oxide synthase by interferon-γ in murine macrophages. Biochem Biophys Res Commun 1992; 182:1155–1159.
208. Srimal S, Nathan C. Purification of macrophage deactivating factor. J Exp Med 1990; 171:1347–1361.
209. Nelson BJ, Danielpour D, Rossio JL, Turpin J, Nacy CA. Interleukin-2 suppresses activated macrophage intracellular killing activity by inducing macrophages to secrete TGF-β. J Leukoc Biol 1994; 55:81–90.
210. Gazzinelli RT, Hieny S, Wynn TA, Wolf S, Sher A. Interleukin 12 is required for the T-lymphocyte–independent induction of interferon gamma by an intracellular parasite and induces resistance in T-cell-deficient hosts. Proc Natl Acad Sci USA 1993; 90:5879–5880.
211. Kennedy MK, Picha KS, Shanebeck KD, Anderson DM, Grabstein KH. Interleukin-12 regulates the proliferation of Th1, but not Th2 or Th0, clones. Eur J Immunol 1994; 24:2271–2278.
212. Betz M, Fox BS. Prostaglandin E_2 inhibits production of the Th1 lymphokines but not of Th2 lymphokines. J Immunol 1991; 146:108–113.
213. Marotta P, Sautebin L, Di Rosa M. Modulation of the induction of nitric oxide synthase by eicosanoids in the murine macrophage cell line J774. Br J Pharmacol 1992; 107:640–641.
214. Oswald IP, Wynn TA, Sher A, James SL. Interleukin 10 inhibits macrophage microbicidal activity by blocking the endogenous production of tumor necrosis factor α required as a costimulatory factor for interferon γ-induced activation. Proc Natl Acad Sci USA 1992; 89:8676–8680.
215. Greenberger MJ, Strieter RM, Kunkel SL, Danforth JM, Goodman RE, Standiford TJ. Neutralization of IL-10 increases survival in a murine model of *Klebsiella* pneumonia. J Immunol 1995; 155:722–729.
216. Di Rosa M, Radomski M, Carnuccio R, Moncado S. Glucocorticoids inhibit the induction of nitric oxide synthase in macrophages. Biochem Biophys Res Commun 1990; 172:1246–1252.
217. Kunz D, Walker G, Eberhardt W, Pfeilschifter J. Molecular mechanisms of dexamethasone inhibition of nitric oxide synthase expression in interleukin 1β-stimulated mesangial cells: evidence for the involvement of transcriptional and post-transcriptional regulation. Proc Natl Acad Sci USA 1996; 93:255–259.

218. Kawahara RS, Deng Z-W, Deuel TF. Glucocorticoids inhibit the transcriptional induction of *JE*, a platelet-derived growth factor-inducible gene. J Biol Chem 1991; 266:13261–13266.
219. Berkman N, Jose PJ, Williams TJ, Schall TJ, Barnes PJ, Chung KF. Corticosteroid inhibition of macrophage inflammatory protein-1α in human monocytes and alveolar macrophages. Am J Physiol 1995; 269:L443–L452.
220. Standiford TJ, Kunkel SL, Rolfe MW, Evanoff HL, Allen RM, Strieter RM. Regulation of human alveolar macrophage- and blood monocyte-derived interleukin-8 by prostaglandin E_2 and dexamethasone. Am J Respir Cell Mol Biol 1992; 6: 75–81.
221. Shapiro SD, Campbell EJ, Kobayashi DK, Welgus HG. Dexamethasone selectively modulates basal and lipopolysaccharide-induced metalloproteinase and tissue inhibitor of metalloproteinase production by human alveolar macrophages. J Immunol 1991; 146:2724–2729.
222. Auphan N, DiDonato JA, Rosette C, Helmberg A, Karin M. Immunosuppression by glucocorticoids: inhibition of NF-κB activity through induction of IκB synthesis. Science 1995; 270:286–290.
223. Scheinman RI, Cogswell PC, Lofquist AK, Baldwin AS Jr. Role of transcriptional activation of IκBα in mediation of immunosuppression by glucocorticoids. Science 1995; 270:283–286.
224. Salkowski CA, Vogel SN. IFN-γ mediates increased glucocorticoid receptor expression in murine macrophages. J Immunol 1992; 148:2770–2777.
225. Adams DO, Hamilton TA. The activated macrophage and granulamatous inflammation. Curr Top Pathol 1989; 79:151–167.
226. Epstein WL, Fukuyama K. Mechanisms of granulomatous inflammation. Immunol Ser 1989; 46:687–721.
227. Kunkel SL, Strieter RM, Lukacs N, Chensue SW. Initiation and maintenance of the granulomatous response. Chest 1993; 103(suppl):135S–137S.
228. Warren KS. A functional classification of granulomatous inflammation. Ann NY Acad Sci 1976; 278:7–18.
229. McInnis A, Rennick DM. Interleukin-4 induces cultured monocytes/macrophages to form giant multinucleated cells. J Exp Med 1988; 167:598–611.
230. Weinberg JB, Hobbs MM, Misukonis MA. Recombinant human γ-interferon induces human monocyte polykaryon formation. Proc Natl Acad Sci USA 1984; 81: 4554–4557.
231. Chensue SW, Kunkel SL, Higashi GI, Ward PA, Boros DL. Production of superoxide anion, prostaglandins, and hydroxyeicosatetraenoic acids by macrophage from hypersensitivity-type (*Schistosoma mansoni* egg) and foreign body-type granulomas. Infect Immun 1983; 42:1116–1125.
232. Gendelman HE, Morahan PS. Macrophages in viral infections. In: Lewis CE, McGee JO'D, eds. The Macrophage. Oxford, UK: Oxford University Press, 1992: 157–213.
233. Warner AE, Molina RM, Brain JD. Uptake of bloodborne bacteria by pulmonary intravascular macrophages and consequent inflammatory responses in sheep. Am Rev Respir Dis 1987; 136:683–690.
234. Cross CE, Bancroft GJ. Ingestion of acapsular *Cryptococcus neoformans* occurs via

mannose and β-glucan receptors, resulting in cytokine production and increased phagocytosis of the encapsulated form. Infect Immun 1995; 63:2604–2611.
235. Stahl PD. The macrophage mannose receptor: current status. Am J Respir Cell Mol Biol 1990; 2:317–318.
236. Wessels MR, Butko P, Ma M, Warren HB, Lage AL, Carroll MC. Studies of group B streptococcal infection in mice deficient in complement component C3 or C4 demonstrate an essential role for complement in both innate and acquired immunity. Proc Natl Acad Sci USA 1995; 92:11490–11494.
237. Gadek JE, Hunninghake GW, Zimmerman RL, Crystal RG. Regulation of the release of alveolar macrophage-derived chemotactic factor. Am Rev Respir Dis 1980; 121: 723–733.
238. Alpert SE, Pennington JE, Colten HR. Synthesis of complement by guinea pig bronchoalveolar macrophages. Effect of acute and chronic infection with *Pseudomonas aeruginosa*. Am Rev Respir Dis 1984; 129:66–71.
239. Coonrod JD, Yoneda K. Complement and opsonins in alveolar secretions and serum of rats with pneumonia due to *Streptococcus pneumoniae*. Rev Infect Dis 1981; 3: 319–322.
240. McNeely TB, Coonrod JD. Aggregation and opsonization of type A but not type B *Hemophilus influenzae* by surfactant protein A. Am J Respir Cell Mol Biol 1994; 11: 114–122.
241. Tenner AJ, Robinson SL, Borchelt J, Wright JR. Human pulmonary surfactant protein (SP-A), a protein structurally homologous to C1q, can enhance FcR- and CR1-mediated phagocytosis. J Biol Chem 1989; 264:13923–13928.
242. Dinarello CA. Interleukin-1. Ann NY Acad Sci 1988; 546:122–132.
243. Gauldie J, Richards C, Lamontagne L. Fc receptors for IgA and other immunoglobulins on resident and activated alveolar macrophages. Mol Immunol 1983; 20:1029–1037.
244. Williams J, Johnson S, Masculi JJ, Smith H, Rosenwasser LJ, Borish L. Regulation of low affinity IgE receptor (CD23) expression on mononuclear phagocytes in normal and asthmatic subjects. J Immunol 1992; 149:2823–2829.
245. Kindt GC, Van de Winkle JGJ, Moore SA, Anderson CL. Identification and structural characterization of Fcγ-receptors on pulmonary alveolar macrophages. Am J Physiol 1991; 260:L403–L411.
246. Fels AOS, Nathan CF, Cohn ZA. Hydrogen peroxide release by alveolar macrophages from sarcoid patients and by alveolar macrophages from normals after exposure to recombinant interferons alpha, β and γ and 1,25-dihydroxyvitamin D3. J Clin Invest 1987; 80:381–386.
247. Portnoy DA, Schreiber RD, Connelly P, Tilney LG. γ interferon limits access of *Listeria monocytogenes* to the macrophage cytoplasm. J Exp Med 1989; 170:2141–2147.
248. Tilney LG, Portnoy DA. Actin filaments and growth movement, and spread of the intracellular parasite, *Listeria monocytogenes*. J Cell Biol 1989; 109:1597–1608.
249. Wizemann TM, Laskin DL. Enhanced phagocytosis, chemotaxis, and production of reactive oxygen intermediates by interstitial lung macrophages following acute endotoxemia. Am J Respir Cell Mol Biol 1994; 11:358–365.
250. Gupta D, Jin YP, Dziarski R. Peptidoglycan induces transcription and secretion of

TNF-α and activation of lyn, extracellular signal-regulated kinase, and rsk signal transduction proteins in mouse macrophages. J Immunol 1995; 155:2620–2630.
251. Mathison JC, Tobias PS, Wolfson E, Ulevitch RJ. Plasma lipopolysaccharide (LPs)–binding protein. A key component in macrophage recognition of gram-negative LPS. J Immunol 1992; 149:200–206.
252. Zhang Y, Broser M, Cohen H, Bodkin M, Law K, Reibman J, Rom WN. Enhanced interleukin-8 release and gene expression in macrophages after exposure to *Mycobacterium tuberculosis* and its components. J Clin Invest 1995; 95:586–592.
253. Duncan RL, Morrison DC. The fate of *E. coli* lipopolysaccharide after the uptake of *E. coli* by murine macrophages *in vitro*. J Immunol 1984; 132:1416–1424.
254. Duncan RL Jr, Hoffman J, Tesh VL, Morrison DC. Immunologic activity of lipopolysaccharides released from macrophages after the uptake of intact *E. coli in vitro*. J Immunol 1986; 136:2924–2929.
255. Nugent KM, Glazier J, Monick MM, Hunninghake GW. Stimulated human alveolar macrophages secrete interferon. Am Rev Respir Dis 1985; 131:714–718.
256. Riches DWH, Underwood GA. Expression of interferon-β during the triggering phase of macrophage cytocidal activation. J Biol Chem 1991; 266:24785–24792.
257. Chen GH, Curtis JL, Mody CH, Christensen PJ, Armstrong LR, Toews GB. Effect of granulocyte-macrophage colony-stimulating factor on rat alveolar macrophage anticriptococcal activity in vitro. J Immunol 1994; 152:724–734.
258. Strieter RM, Koch AE, Antony VB, Fick RB Jr, Standiford TJ, Kunkel SL. The immunopathology of chemotactic cytokines: the role of interleukin-8 and monocyte chemoattractant protein-1. J Lab Clin Med 1994; 123:183–197.
259. Xing Z, Jordana M, Kirpalani H, Driscoll KE, Schall TJ, Gauldie J. Cytokine expression by neutrophils and macrophages *in vivo*: endotoxin induces tumor necrosis factor-α, macrophage inflammatory protein-2, interleukin 1β, and interleukin-6 bu not RANTES or transforming growth factor-β_1 mRNA expression in acute lung inflammation. Am J Respir Cell Mol Biol 10; 1994:148–153.
260. Chavis C, Godard P, Michel FB, Crastes de Paulet A, Damon M. Sulfidopeptide leukotrienes contribute to human alveolar macrophage activation in asthma. Prostaglandins Leukot Essent Fatty Acids 1991; 42:95–100.
261. Moqbel R. Eosinophils, cytokines and allergic inflammation. Ann NY Acad Sci 1994; 725:223–233.
262. Spessotto P, Dri P, Bulla R, Zabucchi G, Patriarca P. Human eosinophil peroxidase enhances tumor necrosis factor and hydrogen peroxide release by human monocyte-derived macrophages. Eur J Immunol 1995; 25:1366–1373.
263. Dackiw AP, Nathens AB, Ribeiro MB, Cheung PY, Marshall JC, Rotstein OD. Mast cell modulation of macrophage procoagulant activity and TNF production. J Surg Res 1995; 59:1–5.
264. Dileepan KN, Lorsbach RB, Stechschulte DJ. Mast cell granules inhibit macrophage-mediated lysis of mastocytoma cells (P815) and nitric oxide production. J Leukoc Biol 1993; 53:446–453.
265. Okai Y. Heterogeneous macrophage-activating factors from extracellular matrix of human embryo fibroblasts. Immunol Lett 1988; 17:145–149.
266. Zöller M, Douvdevani A, Segal S, Apte RN. Interleukin-1 produced by tumorigenic fibroblasts influences tumor rejection. Int J Cancer 1992; 50:443–449.

267. Marfaing-Koka A, Devergne O, Gorgone G, Portier A, Schall TJ, Galanaud P, Emilie D. Regulation of the production of the RANTES chemokine by endothelial cells. Synergistic induction by IFN-γ plus TNF-α and inhibition by IL-4 and IL-13. J Immunol 1995; 154:1870–1878.
268. Wachtfogel YT, DeLa Cadena RA, Colman RW. Structural biology, cellular interactions and pathophysiology of the contact system. Thromb Res 1993; 72:1–21.
269. Hänsch GM, Seitz M, Betz M. Effect of the late complement components C5b-9 on human monocytes: release of prostanoids, oxygen radicals and of a factor inducing cell proliferation. Int Arch Allergy Appl Immunol 1987; 82:317–320.
270. Tobias PS, Mathison J, Mintz D, Lee JD, Kravchenko V, Kato K, Pugin J, Ulevitch RJ. Participation of lipopolysaccharide-binding protein in lipopolysaccharide-dependent macrophage activation. Am J Respir Cell Mol Biol 1992; 7:239–245.
271. Jaffe HA, Buhl R, Mastrangeli A, Holroyd KJ, Saltini C, Czerski D, Jaffe HS, Kramer S, Sherwin S, Crystal RG. Organ specific cytokine therapy. Local activation of mononuclear phagocytes by delivery of an aerosol of recombinant interferon-γ to the human lung. J Clin Invest 1991; 88:297–302.
272. Murray HW. Interferon-γ therapy in AIDS for mononuclear phagocyte activation. Biotherapy 1990; 2:149–158.
273. Beck JM, Liggitt HD, Brunette EN, Fuchs HJ, Shellito JE, Debs RJ. Reduction in intensity of *Pneumocystis carinii* pneumonia in mice by aerosol administration of gamma interferon. Infect Immun 1991; 59:3859–3862.
274. Chatte G, Panteix G, Perrin-Fayolle M, Pacheco Y. Aerosolized interferon gamma for *Mycobacterium avium*–complex lung disease. Am J Respir Crit Care Med 1995; 152:1094–1096.
275. Cox G, Crossley J, Xing Z. Macrophage engulfment of apoptotic neutrophils contributes to the resolution of acute pulmonary inflammation *in vivo*. Am J Respir Cell Mol Biol 1995; 12:232–237.
276. Buhl R, Jaffe HA, Holroyd KJ, Borok Z, Roum JH, Mastrangeli A, Wells FB, Kirby M, Saltini C, Crystal RG. Activation of alveolar macrophages in asymptomatic HIV-infected individuals. J Immunol 1993; 150:1019–1028.
277. Suffredini AF, Ognibene FP, Lack EE, Simmons JT, Brenner M, Gill VJ, Lane HC, Fauci AS, Parrillo JE, Masur H, Shelhamer JH. Nonspecific interstitial pneumonitis: a common cause of pulmonary disease in the acquired immunodeficiency syndrome. Ann Intern Med 1987; 107:7–13.
278. Bermudez LE. Production of transforming growth factor-β by *Mycobacterium avium* infected human macrophages is associated with unresponsiveness to interferon-γ. J Immunol 1993; 150:1838–1845.
279. Panuska JR, Merolla R, Rebert NA, Hoffmann SP, Tsivitse P, Cirino NM, Silverman RH, Rankin JA. Respiratory syncytial virus induces interleukin-10 by human alveolar macrophages. J Clin Invest 1995; 96:2445–2453.
280. Kips JC, Tavernier J, Pauwels RA. Tumor necrosis factor causes bronchial hyperresponsiveness in rats. Am Rev Respir Dis 1992; 145:332–336.
281. Tsukagoshi H, Sakamoto T, Xu W, Barnes PJ, Chung KF. Effect of interleukin-1β on airway hyperresponsiveness and inflammation in sensitized and nonsensitized Brown-Norway rats. J Allergy Clin Immunol 1994; 93:464–469.

282. Silbaugh SA, Stengel PW, Williams GD, Herron DK, Gallagher P, Baker SR. Effects of leukotriene B4 inhalation. Am Rev Respir Dis 1987; 136:930–934.

283. Wardlaw AJ, Chung KF, Moqbel R, MacDonald AJ, Hartnell A, McCusker M, Collins JV, Barnes PJ, Kay AB. Effects of inhaled PAF in humans on circulating and bronchoalveolar lavage fluid neutrophils. Am Rev Respir Dis 1990; 141:386–392.

284. Borish L, Mascali JJ, Dishuch J, Beam WR, Martin RJ, Rosenwasser LJ. Detection of alveolar macrophage–derived IL-1β in asthma. J Immunol 1992; 149:3078–3082.

285. Cembrzynska-Nowak M, Szklarz E, Inglot AC, Teodorczyk-Injeyan JA. Elevated release of tumor necrosis factor-α and interferon-γ by bronchoalveolar leukocytes from patients with bronchial asthma. Am Rev Respir Dis 1993; 147:291–295.

286. Strieter RM, Chensue SW, Basha MA, Standiford TJ, Lynch JP III, Baggiolini M, Kunkel SL. Human alveolar macrophage gene expression of interleukin-8 by tumor necrosis factor-α, lipopolysaccharide, and interleukin-1β. Am J Respir Cell Mol Biol 1990; 2:321–326.

287. Martin TR, Pistorese BP, Chi EY, Goodman RB, Matthay MA. Effects of leukotriene B_4 in the human lung. Recruitment of neutrophils into the alveolar spaces without a change in protein permeability. J Clin Invest 1989; 84:1609–1619.

288. Rogivue C, Car BD, Allmann II, Swahlen RD, Walz A. Bovine melanoma growth stimulatory activity: a new monocyte-macrophage-derived cytokine of the IL-8 family. Partial structure, function, and expression in acute pulmonary inflammation. Lab Invest 1995; 72:689–695.

289. Terashima T, Kanazawa M, Sayama K, Urano T, Sakamaki F, Nakamura H, Waki Y, Soejima K, Tasaka S, Ishizaka A. Neutrophil-induced lung protection and injury are dependent on the amount of *Pseudomonas aeruginosa* administered via airways in guinea pigs. Am J Respir Crit Care Med 1995; 152:2150–2156.

290. Guidot DM, Stevens EE, Repine MJ, Lucca-Broco, Repine JE. Intratracheal but not intravascular interleukin-1 causes acute edematous injury in isolated neutrophil-perfused rat lungs through an oxygen radical–mediated mechanism. J Lab Clin Med 1994; 123:605–609.

291. Wesselius LJ, Smirnov IM, O'Brien-Ladner AR, Nelson ME. Synergism of intratracheally administered tumor necrosis factor with interleukin-1 in the induction of lung edema in rats. J Lab Clin Med 1995; 125:618–625.

292. Jacobs RF, Tabor DR, Burks AW, Campbell GD. Elevated interleukin-1 release by human alveolar macrophages during the adult respiratory distress syndrome. Am Rev Respir Dis 1989; 140:1686–1692.

293. Nhieu JTV, Misset B, Lebarhy F, Carlet J, Bernaudin J-F. Expression of tumor necrosis factor-α gene in alveolar macrophages from patients with adult respiratory distress syndrome. Am Rev Respir Dis 1993; 147:1585–1589.

294. Blackwell TS, Holden EP, Blackwell TR, DeLarco JE, Christman JW. Cytokine-induced neutrophil chemoattractant mediates neutrophilic alveolitis in rats: association with nuclear factor κB activation. Am J Respir Cell Mol Biol 1994; 11: 464–472.

295. Shapiro SD. Elastolytic metalloproteinases produced by human mononuclear phagocytes. Potential roles in destructive lung disease. Am J Respir Crit Care Med 1994; 150:S160–S164.

296. Postlethwaite AE, Raghow R, Stricklin GP, Poppleton H, Seyer JM, Kang AH. Modulation of fibroblast functions by interleukin 1: increased steady-state accumulation of type I procollagen messenger RNAs and stimulation of other functions but not chemotaxis by human recombinant interleukin 1α and β. J Cell Biol 1988; 106:311–318.
297. Khalil N, Greenberg AH. The role of TGF-β in pulmonary fibrosis. Ciba Found Symp 1991; 157:194–207.
298. Bonner JC, Osornio-Vargas AR, Badgett A, Brody AR. Differential proliferation of rat lung fibroblasts induced by the platelet-derived growth factor-AA, -AB, and -BB isoforms secreted by rat alveolar macrophages. Am J Respir Cell Mol Biol 1991; 5: 539–547.
299. Shaw RJ, Benedict SH, Clark RAF, King TE Jr. Pathogenesis of pulmonary fibrosis in interstitial lung disease. Alveolar macrophage PDGF(B) gene activation and up-regulation by interferon gamma. Am Rev Respir Dis 1991; 143:167–173.
300. Laskin DL, Pendino KJ. Macrophages and inflammatory mediators in tissue injury. Annu Rev Pharmacol Toxicol 1995; 35:655–677.
301. Saha B, Das G, Vohra H, Ganguly NK, Mishra GC. Macrophage-T cell interaction in experimental mycobacterial infection. Selective regulation of co-stimulatory molecules on *Mycobacterium*-infected macrophages and its implication in the suppression of cell-mediated immune response. Eur J Immunol 1994; 24:2618–2624.
302. Bonta IL, Ben-Efraim S. Involvement of inflammatory mediators in macrophage antitumor activity. J Leukoc Biol 1993; 54:613–626.
303. Mantovani A, Bopttazzi B, Colotta F, Sozzani S, Ruco L. The origin and function of tumor-associated macrophages. Immunol Today 1992; 13:265–270.
304. Rees RC, Parry H. Macrophages in tumour immunity. In: Lewis CE, McGee JO'D, eds. The Macrophage. Oxford, UK: Oxford University Press, 1992:315–357.
305. Alexander P, Evans R. Endotoxin and double stranded RNA render macrophages cytotoxic. Nature New Biol 1971; 232:76–78.
306. Hibbs JB Jr, Lambert LH Jr, Remington JS. Resistance to murine tumors conferred by chronic infection with intracellular protozoa, *Toxoplasma gondii* and *Besnoitia jellisoni*. J Infect Dis 1971; 124:587–592.
307. Hibbs JB Jr, Lambert LH Jr, Remington JS. Possible role of macrophage mediated nonspecific cytotoxicity in tumour resistance. Nature New Biol 1972; 235:48–50.
308. Piessens WF, Churchill WH Jr, David JR. Macrophages activated in vitro with lymphocyte mediators kill neoplastic but not normal cells. J Immunol 1975; 114: 293–299.
309. Evans R. Macrophages in syngeneic animal tumours. Transplantation 1972; 14:468–473.
310. Graves DT, Jiang YL, Williamson MJ, Valente AJ. Identification of monocyte chemotactic activity produced by malignant cells. Science 1989; 245:1490–1493.
311. Martinet N, Beck G, Bernard V, Plenat F, Vaillant P, Schooneman F, Vignaud J-M, Martinet Y. Mechanism for the recruitment of macrophages to cancer site. In vivo concentration gradient of monocyte chemotactic activity. Cancer 1992; 70:854–860.
312. Steward CC, Beetham KL. Cytocidal activity and proliferative ability of macrophages infiltrating the EMT6 tumor. Int J Cancer 1978; 22:152–159.
313. Hibbs JB Jr. Discrimination between neoplastic and non-neoplastic cells in vitro by activated macrophages. J Natl Cancer Inst 1974; 53:1487–1492.

314. Thomassen MJ, Wiedemann HP, Barna BP, Farmer M, Ahmad M. Induction in *in vitro* tumoricidal activity in alveolar macrophages and monocytes from patients with lung cancer. Cancer Res 1988; 48:3949–3953.
315. Sone S, Moriguchi S, Shimizu E, Ogushi F, Tsubura E. *In vitro* generation of tumoricidal properties in human alveolar macrophages following interaction with endotoxin. Cancer Res 1982; 42:2227–2231.
316. Brown GP, Iwamoto GK, Monick MM, Hunninghake GW. Cigarette smoking decreases interleukin 1 release by human alveolar macrophages. Am J Physiol 1989; 256:C260–C264.
317. McCrea KA, Ensor JE, Nall K, Bleecker ER, Hasday JD. Altered cytokine regulation in the lungs of cigarette smokers. Am J Respir Crit Care Med 1994; 150:696–703.
318. Mahoney MJ, Leighton J. The inflammatory response to a foreign body with transplantable tumors. Cancer Res 1962; 22:334–338.
319. Russell SW, Gillespie GY. Nature, function and distribution of inflammatory cells in regressing and progressing Moloney sarcomas. J Reticuloendothel Soc 1977; 22: 159–168.
320. Russell SW, Gillespie GY, Hansen CB, Cochrane CG. Inflammatory cells in solid murine neoplasms. II. Cell types found throughout the course of Moloney sarcoma regression or progression. Int J Cancer 1976; 18:331–338.
321. Jarpe MA, Hayes MP, Russell JK, Johnson HM, Russell SW. Causal association of interferon-γ with tumor regression. J Interferon Res 1989; 9:239–244.
322. Russell SW, Gillespie GY, McIntosh AT. Inflammatory cells in solid murine neoplasms. III. Cytotoxicity mediated *in vitro* by macrophages recovered from disaggregated regressing Moloney sarcomas. J Immunol 1977; 118:1574–1579.
323. Russell SW, McIntosh AT. Macrophages isolated from regressing Moloney sarcomas are more cytotoxic than those recovered from progressing sarcomas. Nature 1977; 268:69–71.
324. Taniyama T, Holden HT. Cytolytic activity of macrophages isolated from primary murine sarcoma virus (MSV)–induced tumors. Int J Cancer 1979; 24:151–160.
325. Duffie GP, Young MRI. Tumoricidal activity of alveolar and peritoneal macrophages of C57BL/6 mice bearing metastatic or nonmetastatic variants of Lewis lung carcinoma. J Leukoc Biol 1991; 49:8–14.
326. Kessler R, Dumont S, Bartholeyns J, Weitzenblum E, Poindron P. Antitumoral potential of aerosolized interferon-γ in mice bearing lung metastases. Am J Respir Cell Mol Biol 1994; 10:202–206.
327. Pusztai L, Lewis CE, Lorenzen J, McGee JO'D. Growth factors: regulation of normal and neoplastic growth. J Pathol 1993; 169:191–201.
328. Rappolee DA, Werb Z. Macrophage-derived growth factors. Curr Top Microbiol Immunol 1992; 181:87–140.
329. Lewis CE, Leek R, Harris A, McGee JO'D. Cytokine regulation of angiogenesis in breast cancer: the role of tumor-associated macrophages. J Leukoc Biol 1995; 57: 747–751.
330. Dvorak HF. Tumors: wounds that do not heal. Similarities between tumor stroma generation and wound healing. N Engl J Med 1986; 315:1650–1659.
331. Whalen GF. Solid tumours and wounds: transformed cells misunderstood as injured tissue? Lancet 1990; 336:1489–1492.

332. Vignaud J-M, Marie B, Klein N, Plénat F, Pech M, Borrelly J, Martinet N, Duprez A, Martinet Y. The role of platelet-derived growth factor production by tumor-associated macrophages in tumor stoma formation in lung cancer. Cancer Res 1994; 54:5455–5463.
333. Schmidt JA, Mizel SB, Cohen D, Green I. Interleukin-1, a potential regulator of fibroblast proliferation. J Immunol 1982; 128:2177–2182.
334. Wechsler SJ, Eurell TE, Russell SW. Prostaglandin E_2 content of regressing and progressing Moloney sarcomas. J Leukoc Biol 1986; 40:159–168.
335. Hilkens CMU, Vermeulen H, van Neerven RJJ, Snijdewint FGM, Wierenga EA, Kapsenberg ML. Differential modulation of T helper type 1 (Th1) and T helper type 2 (Th2) cytokine secretion by prostaglandin E_2 critically depends on interleukin-2. Eur J Immunol 1995; 25:59–53.
336. Marcinkiewicz J, Chain BM. Differential cytokine regulation by eicosanoids in T cells primed by contact sensitisation with TNP. Cell Immunol 1993; 149:303–314.
337. Watanabe S, Yssel H, Harada Y, Arai K. Effects of prostaglandin E_2 on Th0-type human T cell clones: modulation of functions of nuclear proteins involved in cytokine production. Infect Immunol 1994; 6:523–532.
338. Chu E, Casey LC, Harris JE, Braun DP. Suppression of the development of tumoricidal function in gamma interferon-treated human peripheral blood monocytes by lipopolysaccharide: the role of cyclooxygenase metabolites. J Clin Invest 1993; 13: 49–57.
339. Schultz RM, Pavlidis NA, Stylos WA, and Chirigos MA. Regulation of macrophage tumoricidal function: a role for prostaglandins of the E serires. Science 1978; 202: 320–321.
340. Haak-Frendscho M, Wynn TA, Czuprynski CJ, Paulnock D. Transforming growth factor-β1 inhibits activation of macrophage cell line RAW 264.7 for cell killing. Clin Exp Immunol 1990; 82:404–410.
341. Nishioka Y, Sone S, Orino E, Nii A, Ogura T. Down-regulation by interleukin 4 of activation of human macrophages to the tumoricidal state. Cancer Res 1991; 51: 5526–5531.
342. Yano S, Sone S, Nishioka Y, Mukaida N, Matsushima K, Ogura T. Differential effects of anti-inflammatory cytokines (IL-4, IL-10 and IL-13) on tumoricidal and chemotactic properties of human monocytes induced by monocyte chemotactic and activating factor. J Leukoc Biol 1995; 57:303–309.
343. Strassmann G, Patil-Koota V, Finkelman F, Fong M, Kambayashi T. Evidence for the involvement of interleukin 10 in the differential deactivation of murine peritoneal macrophages by prostaglandin E_2. J Exp Med 1994; 180:2365–2370.
344. Alleva DG, Burger CJ, Elgert KD. Tumor-induced regulation of suppressor macrophage nitric oxide and TNF-α production. Role of tumor-derived IL-10, TGF-β, and prostaglandin E_2. J Immunol 1994; 153:1674–1686.
345. Gastl GA, Abrams JS, Nanus DM, Oosterkamp R, Silver J, Liu F, Chen M, Albino AP, Bander NH. Interleukin-10 production by human carcinoma cell lines and its relationship to interleukin-6 expression. Int J Cancer 1993; 55:96–101.
346. Chambers SK, Wang Y, Gertz RE, Kacinski BM. Macrophage colony-stimulating factor mediates invasion of ovarian cancer cells through urokinase. Cancer Res 1995; 55:1578–1585.

3

Dendritic Cells

KAYO INABA

Graduate School of Science
Kyoto University
Kyoto Japan

RALPH M. STEINMAN

Laboratory of Cellular Physiology and Immunology
Rockefeller University
New York, New York

I. Introduction

In this chapter, some general features of the dendritic cell system of antigen-presenting (APCs) cells are introduced. Dendritic cells (DCs) from different lymphoid and nonlymphoid organs exhibit a similar constellation of functional properties that are not readily observed with other cells. First, DCs are potent, with only small numbers being required to stimulate helper and killer T-cell responses. Second, DCs activate quiescent T cells, including immunologically naive lymphocytes. Third, DCs that bear antigen are able to prime or sensitize T cells in vivo in mice and rats in the absence of any other adjuvant. Fourth, when antigens are administered in vivo and then APCs are isolated, DCs are often the main cell carrying that antigen in a form that is immunogenic for T cells. Similar principles have been used to identify DCs in many different tissues and species. In each instance, mature DCs are large nonadherent cells that extend motile, sheet-like processes, or "veils," in several different directions from the cell body. DCs express high levels of antigen-presenting major histocompatibility complex (MHC) products, both class I and II, as well as the non-MHC CD1a molecule. Also abundant are several accessory molecules that function in T-cell binding and costimulation such as CD40, intracellular adhesion molecule-1 (ICAM-1)/CD54, leukocyte function–association antigen-3 (LFA-3)/CD58, and B7-2/CD86. A small number of DC-restricted markers are known (DEC-205 in mouse, OX62 in rat, CD83 and p55 in human), but none of these antigens is exclusive to DCs. Cytokines, especially granulocyte-macrophage colony-stimulating factor (GM-CSF), interleukin-4 (IL-4), and tumor necrosis factor-α (TNF-α), are being identified that mediate the development of DCs from less mature progenitors. The latter include proliferating progenitors in the CD34+ fraction of marrow and blood and nonproliferating precursors in many sites such as blood, skin, and lung. The maturation of DCs may be a pivotal control point in the onset of T-cell–mediated immunity in the lung and other organs. Macrophages and some of their specific cytokine products like IL-10 can block this development. During maturation, DCs best internalize and process a variety of different antigens, including small amounts of particulates. A new area of DC biology is the direct analysis of the endocytic vacuolar system which is specialized for antigen presentation rather than scavenging. Two known characteristics are the presence of lectin-like receptors for antigen uptake and numerous MHC class II–rich endosomal vesicles. Another key mechanistic feature of DCs, which is illustrated in the lung, skin, and afferent lymphatic vessels, is the capacity of DCs to capture antigens in a peripheral tissue and then migrate to the draining lymph node to initiate immunity. Several chapters in this volume consider the properties of pulmonary DCs directly.

II. Special Features of Dendritic Cells

DCs exhibit four functional features that are not readily observed in other types of APCs. These features can be appreciated by preparing enriched populations of DCs either from the small numbers that are present in most tissues or from the relatively large numbers that can be generated from newly recognized progenitors (see Section IV).

A. Potency as APCs

Very small numbers of DCs stimulate strong T-dependent responses in standard tissue culture systems. The first two systems to be studied were the primary mixed leukocyte reaction (MLR)(1) and the primary T-dependent antibody response (2). Some sample data with the mouse MLR are included in Table 1 to illustrate the kinds of results one should observe with populations that are highly enriched in mature DCs. If one tests graded doses of allogeneic DCs against a constant number of 300,000 T cells (as is standard in assays for T-cell responses) and measures the MLR in terms of [^{3}H]-thymidine uptake on the fourth day, then a DC:T-cell ratio of 1:100 gives a strong MLR. In the human MLR, or in the mouse MLR run in larger culture wells, strong MLRs are noted at doses of 1:1000 (1,3–5). DCs are typically 100 times more potent that unfractionated populations of

Table 1 MLR-Stimulating Activity of Spleen DCs

DCs prepared in	Type of response	Serum for MLR	MRL [^{3}H]-thymidine uptake × 10^3 cpm) to graded doses DCs				
			1 × 10^4	3 × 10^3	1 × 10^3	3 × 10^2	None
Fetal calf serum	Allo	FCS	189.2	112.5	42.3	9.3	0.2
	Syn	FCS	11.2	6.9	2.3	0.5	
	Allo	NMS	179.5	108.3	39.4	6.7	
	Syn	NMS	7.3	2.8	0.5	0.2	
Normal mouse serum	Allo	FCS	188.3	120.6	41.9	8.3	
	Syn	FCS	10.6	4.8	0.9	0.3	
	Allo	NMS	169.3	93.2	32.1	2.6	
	Syn	NMS	3.2	0.6	0.2	0.2	

Illustrative data for stimulation of the mixed leukocyte reaction by mouse DCs. In this case, the DCs were derived from the spleens of C57BL/6 mice and used to stimulate purified T cells from BALB/C mice ("Allo") or syngeneic T cells ("Syn"). The DCs were cultured overnight as adherent cells in either 5% fetal calf serum (FCS) or 0.5% normal mouse serum (NMS), and then the microcultures were run in medium supplemented with either serum. The 300,000 T cells were placed in triplicate flat-bottomed tissue culture wells, and graded doses of DCs were added to a final volume of 0.2 ml. The medium is either RPMI-1640 with 5% FCS or Click's medium with 0.5% mouse serum, each supplemented with 50 μM 2-mercaptoethanol and antibiotics. The MLR is quantified by adding 1 μCi of [^{3}H]-thymidine to the cultures for 12 hr on day 4. Note that a strong MLR takes place even at a DC to T cell ratio of 1:100 in these flat-bottomed microcultures. Stronger MLRs are seen in larger culture vessels such as 24-wellplates and upright culture flasks.

leukocytes such as mouse spleen suspensions or human blood mononuclear cells. The DC is a strong MLR stimulator whether it is cultured in fetal calf serum (FCS) or homologous serum (see Table 1), but the use of FCS increases the syngeneic MLR in both mouse and human T cells. The T cells that are stimulated in the MLR lie in both the naive (CD45RA$^+$, CD45RO$^-$) and memory (CD45RA$^-$, CD45RO$^+$) subsets.

In the primary antibody response that develops in mixtures of mouse spleen B and T lymphocytes, a DC to lymphocyte ratio of 1:100 again suffices to produce an antibody response that is equal or better than unfractionated spleen (2,6). Selective elimination of the major population of DCs in mouse spleen with the 33D1 monclonal and rabbit complement markedly reduces the antibody response, and the response is restored with small numbers of DCs (2).

In the primary MLR and antibody responses, DCs initiate the response. Once the T cell has been activated, it responds vigorously to antigens presented by other APCs like B cells and macrophages (3,6).

Low doses of DCs also present protein antigens to MHC class II–restricted, CD4$^+$, helper T cells. These include bulk lymph node T cells that have been primed by injection of protein antigens into a footpad with complete Freund's adjuvant (7,8), clonal populations including T-T hybridomas (9), and T-cell receptor (TCR) transgenic T cells that express a receptor for specific protein antigens (10,11). Likewise, low doses of DCs mediate presentation of antigens to MHC class I–restricted, $CD8^+$ cytolytic T cells, as in the mixed leukocyte reaction (12,13) and the response to several viruses, particularly influenza (14,15). There are several examples in which DCs can stimulate $CD8^+$ killer T-cell responses in the apparent absence of $CD4^+$ helper T cells (12–14,16).

B. Capacity to Stimulate Quiescent, Including Naive, T Cells

The efficacy with which the primary MLR and T-dependent antibody responses were stimulated by DCs (see above) led to the conclusion that the immune system had specialized APCs for initiating T-cell–mediated immunity. More stringent evidence for the capacity of DCs to stimulate primary or immunologically naive T cells has been obtained. In the MLR, DCs are able in small numbers to stimulate the naive cells that predominate in neonatal cord blood (4). For responses to foreign proteins, TCR transgenic T cells have been used (10,11). These T cells should not be encountering their cognate antigen in vivo unless the transgenic TCR-α chain has been replaced by an endogenous TCR-α gene. The T cells can be further selected for a naive immunological status by depleting cells with low levels of MEL-14 or CD62L or high levels of CD44, which are characteristic markers of memory T cells.

C. Adjuvant Action in Vivo

When DCs bearing antigen are injected into rodents, strong T-cell–mediated immunity can be induced. This was shown first in transplantation models. Lechler and Batchelor were able to induce tolerance of parental strain rats to kidneys from an F1 strain. Tolerance could be broken by the injection of small numbers of F1 DCs (10^5 per rat), whereas much larger numbers of macrophages and lymphocytes were without effect (17). Faustman et al. induced the acceptance of MHC-incompatible mouse pancreatic islet grafts by treating the graft with anti-DC monoclonal antibody (mAb) and complement (18). Rejection could be induced by an injection of DCs. Iwai et al. made similar observations with grafts of thyroid tissue (19).

DCs can be charged with protein antigens ex vivo and then injected into mice or rats. Again priming of the animal takes place. In vivo priming was first done for $CD4^+$ helper–type responses. Inaba et al. injected DCs into the footpad using DCs derived from mouse spleen (8) and later from proliferating bone marrow progenitors (20). Havenith et al. used DCs from spleen and injected the cells into the rat trachea (21). Liu and MacPherson obtained DCs from rat mesenteric lymph and gave the antigen into the intestinal lumen beforehand (22).

Sornasse et al. injected mouse spleen DCs into the footpad and measured help for antibody formation (23). In each case, it was necessary to use live DCs. In contrast, the priming that could be induced with macrophages also took place when killed macrophages were used (21).

Importantly, in vivo–primed T cells are restricted to the MHC of the injected DCs. If parental strain DCs were used to prime T cells from F1 mice, the latter were restricted to the MHC of the injected DCs (8,22). In other words, the DCs were priming the T cells directly and were not simply introducing antigen for presentation by host APCs.

In more recent experiments, it has been shown that DCs prime $CD8^+$ killer cell responses in mice (24). These killer, or cytolytic, T cells are able to exert significant protection against the corresponding antigen-bearing tumor, and even in some cases to reduce the burden of preexisting tumors (25– 27).

D. Efficacy in Antigen Capture in Vivo

Another facet of in vivo function is that DCs are the main cells that capture injected antigens for presentation to T cells. Macrophages are well known to be the main cells that capture antigens for purposes of clearance and destruction. However, if one administers proteins in vivo and then identifies the APCs that present efficiently to antigen-specific T cells, then DCs are the principal cells with activity (22,28–30). It is still not clear if this finding means that the DCs are presenting more MHC-peptide complexes, or for longer periods of time relative to other APCs, or if the accessory adhesion and costimulator molecules of DCs are more abundant.

III. Identification of Mature Dendritic Cells

In all tissues that have been studied, DCs are a trace cell type only rising to levels of 1–3% in a few special cases such as the epidermis and the afferent lymph. Nevertheless, mature DCs are relatively easy to identify. Some important features are diagrammed in Figure 1. Sample fluorescence-activated cell sorter (FACS) data illustrating the surface composition of immature and mature DCs will be shown in the next section, and sample MLR data were shown in Table 1.

A. Morphology

DCs extend large lamellipodia, or "veils," in more than one direction from the cell body. The processes are motile and are continually extended and retracted, so that the DCs are constantly changing shape. The cytoplasm is rich in mitochondria, and the nucleus is irregular in shape. There are relatively few lysosomes, but there can be many MHC class II–rich endosomes (see below).

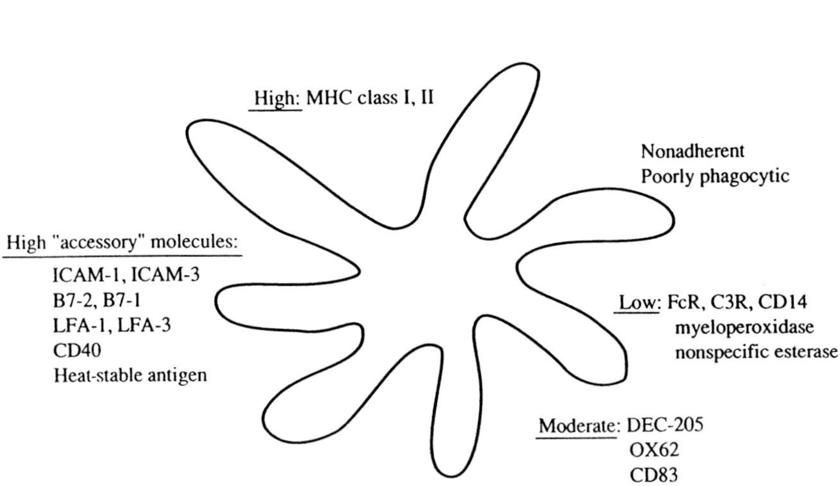

Figure 1 Diagram of the main features of mature DCs. (See text for details).

B. Features that Are Useful for Enrichment

Mature DCs do not adhere to glass or plastic, have a low buoyant density, and exhibit few or no Fc and C3 receptors. As such, DCs are often enriched by selecting for cells that fail to adhere to plastic or to bind immune complexes. Alternatively, one depletes cells expressing the critical marker of other lineages; for example, CD3, CD14, CD16, CD19, and 20. DCs are often enriched by floating on dense media such as dense bovine albumin and metrizamide.

C. High Levels of Antigen-Presenting MHC Products

Quantitative studies indicate $>10^6$ MHC class II locus products/cell (31) and a comparable number of superantigen-binding sites (32). The levels of class II on DCs are 10 or more times higher than other leukocytes. MHC class I products and the non-MHC–encoded presenting molecule CD1a are also abundant. Sample FACS data for DCs are shown and tabulated in Figure 2 and Table 2.

D. High Levels of T-Cells Adhesive and Costimulatory Molecules

Many membrane-bound molecular couples are known to mediate the APC–T-cell interaction, and the majority of these are expressed on DCs. Especially abundant are ICAM-1 (CD54), B7-2 (CD86), and CD40. LFA-3 (CD58), ICAM-3 (CD50), B7-1 (CD80), CD44, and several integrins (CD11a, CD11c, CD49d/CD29) are also abundant, with an exception being the DCs of skin which have little CD11a and CD11c but express more CD11b. All of these molecules are of course not DC

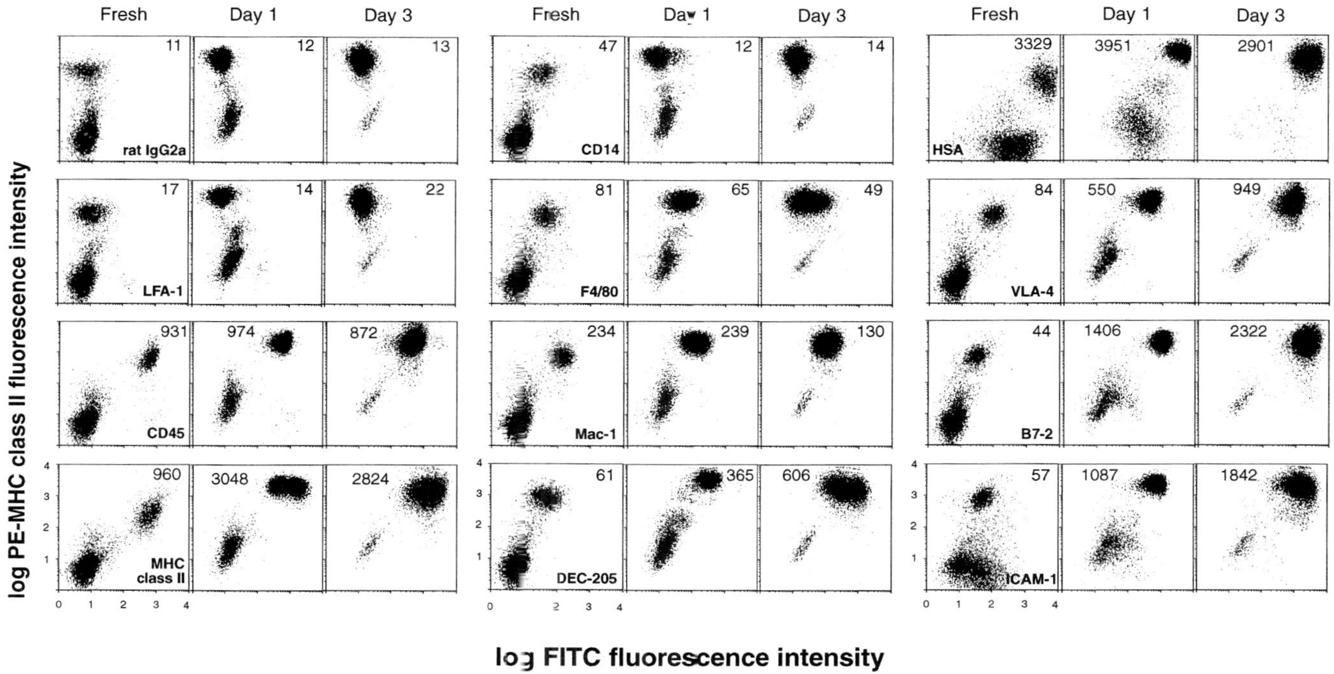

Figure 2 Sample FACS data showing on the x-axis staining of epidermal DCs with a panel of different mAbs. The mAbs are indicated in each panel, and the mean fluorescence indicies for the DC staining are shown in the panels and Table 2. The DCs in the epidermal suspension were selectively labeled with PE-conjugated anti–MHC class II. The DCs were studied in fresh epidermal suspensions or after 1 and 3 days of culture, as indicated along the top of the figure. During culture, DC maturation takes place, most likely because of GM-CSF secreted by the keratinocytes (63), and this allows the DCs to be enriched by flotation on dense media like metrizamide or bovine serum albumin as was done here. See text for details, but note the very high levels of adhesion and costimulatory molecules (heat-stable antigen [HSA], ICAM-1 [CD54], and B7-2 [CD86]).

Table 2 Phenotype of Immature and Mature Dendritic Cells in Cultured Mouse Skin

Surface antigen	Mean fluorescence intensity upon culture at		
	Day 0	Day 1	Day 3
Isotype control	11	12	13
CD45	931	975	572
MHC class II	960	3048	2824
CD14	47	12	14
F4/80	81	65	49
LFA-1, CD11a	17	14	22
Mac-1, CD11b	234	239	130
DEC-205	61	365	606
Heat-stable antigen	3329	3951	2901
ICAM-1, CD54	57	1087	1842
B7-2, CD86	44	1404	2322
VLA-4, CD49d	84	549	948

Illustrative data of the phenotype of immature and mature DCs using mouse epidermal suspensions that are analyzed fresh (day 0) or after culture (days 1 and 3). The data were tabulated from FACScan data of Figure 2.

specific, but what is special about DCs is the very high levels that can be expressed particularly of CD40, CD54, and CD86 (4,33–36) (see Fig. 2). Also, the molecules are rapidly upregulated when DCs are placed into culture (see Fig. 2 and Table 2). No stimulus like lipopolysaccharide (LPS) or anti-Ig needs to be applied for upregulation to take place, but underlying controls are not known (33).

Bhardwaj et al. did quantitative studies with superantigen to assess the efficacy of a known amount of MHC-associated ligand on DCs relative to other APCs (32). They found that at a given level of bound superantigen, human blood DCs were at least 30 times more effective than B cells at stimulating DNA synthesis in syngeneic blood T cells. Blood monocytes obtained in parallel from the same donor bound too little superantigen to conduct extensive studies of this cell type.

E. Macrophage and Lymphocyte Markers

CD14 is either absent or expressed at very low levels on DCs. CD13 and CD33, on the other hand, are more readily detected. CD16 and CD64 Fc receptors are absent, but CD32 is expressed by DCs prior to terminal maturation. Complement

(C3) receptors are weak (CD11b) or undetectable (CD21, CD35). Macrophages express a lysosome-associated membrane protein (LAMP)-like glycoprotein, called CD68 or macrosialin, within endocytic vacuoles (37). CD68 typically is present in small amounts in the perinuclear region of DCs.

DCs lack CD3 and TCRs but can express other "T-cell" markers. The most prevalent is CD4, but the levels can vary from undetectable to as high as that seen on T cells (38). Certain populations of DCs express other T-cell markers such as CD2 on some DCs in sheep afferent lymph (39), CD5 on some DCs in human blood (40), and CD8 (the αα homodimer) on some DCs in mouse lymphoid organs, including thymus (41), but many populations of DCs fail to express any of these. The CD8 subset of mouse DCs that has been identified by Shortman and colleagues is of some interest. The CD8 has been shown to be made by the DCs rather than being acquired from other cells (41). Recent data indicate that this subset expresses fas-ligand and can delete fas-expressing, responding T cells (42). Mature DCs always express the CD25 IL-2 receptor, but they have yet to be shown to be responsive to IL-2. Mature DCs are CD45R0 strong and CD45RA weak and thus would copurify with "memory" T cells (43). B-cell markers such as surface Ig, B220, and CD19/20/21 are typically absent.

F. DC-Restricted Markers

Although it is straightforward to identify and purify DCs according to the above criteria, much more needs to be learned in terms of DC-restricted markers. A number of interesting molecules are known to be more abundant on DCs relative to other leukocytes, but none is fully characterized functionally and none is entirely DC specific.

In mice, the 33D1 mAb was prepared against spleen DCs. It remains an excellent mAb for depleting many DCs from spleen using rabbit complement (44), but its molecular nature is as yet unknown. Two mAbs, M342 (45) and 2A1 (46), stain antigens that are abundant in endocytic vacuoles of DCs but not macrophages. These antigens also are found in the granules of B cells, but the corresponding molecules have not been identified. The integrin CD11c is expressed at relatively high levels on DCs in many sites but not in skin. Lower amounts of CD11c can be expressed by some mouse macrophages, but high levels of CD11c seem to identify mouse DCs in tissue sections and many lymphoid suspensions (47,48). DEC-205, originally called NLDC-145, has recently been cloned (49). It is expressed at high levels on several subsets of DCs, including cultured epidermal Langerhans cells and the DCs of the T-cell areas of lymphoid organs (50,51). As will be discussed below, DEC-205 enhances antigen uptake and presentation, but it is also found at high levels in several epithelia and on one macrophage population, the alveolar macrophage.

In rats, OX62 identifies a candidate integrin that is shared by DCs and γδ T cells (52). In humans, S100 is a cytoplasmic calcium-binding protein (so named because it is soluble in 100% ammonium sulfate) that is abundant in DCs but not other leukocytes (53). However, S100 is found in other cells such as neuroendocrine cells. Two new mAbs have been secured that react with mature DCs but not other leukocytes in human blood. CD83 is an Ig-like molecule of unknown function that was retrieved from an Epstein-Barr virus (EBV) library and can be found on some B cells in tissue sections (54). p55 is a presumptive actin-bundling protein that is found in some other tissues, particularly neurons and glia (55).

IV. Development of Dendritic Cells

The development of DCs is proving to be a complex topic in the sense that there may be several routes to the typical mature DC that is outlined above, and there are stages in their maturation process that differ significantly in function. Here we emphasize recent findings on the development of DCs from proliferating and nonproliferating precursors, but this field is moving rapidly.

A. Proliferating Progenitors—Suspension Cultures

Proliferating aggregates of immature DCs can be induced from precursors in mouse blood and bone marrow (46,56), rat marrow (57), human marrow (58), and cord blood (59,60). In humans, the progenitor carries the CD34 antigen. Different cytokines have been used, although GM-CSF is always one of them. In humans, TNF-α is also added (58,59), and we have recently confirmed that TNF-α enhances DC maturation in mouse bone marrow cultures. Some laboratories find it helpful to add IL-4 to the GM-CSF. In many of these culture systems, granulocytes and macrophages are also developing, so it becomes necessary to skew the development toward DCs and to figure out how to separate the DCs from the other myeloid progeny. The latter can be achieved in mouse marrow using the method of Inaba et al. in which most granulocytes and macrophages are removed by relatively simple adherence methods (46).

B. Proliferating Progenitors—Colony Cultures

Young et al. have studied $CD34^+$ progenitors from human marrow and have found that these populations actually contain two types of colony-forming units (CFUs) (5). One type gives rise to a mixture of myeloid cells, with only a trace DC constituent, much like the standard GM-CSF colonies derived from mouse marrow in Metcalf-Moore cultures (61). The other type, which require the addition of GM-CSF and TNF in combination, consists almost entirely of DCs. Therefore, there seem to be committed DC progenitors. DC commitment may also occur at a later stage. For example, Szabolcs et al. recently identified in human marrow cultures a post-CFU, biopotential intermediate that can develop along either the DC or macrophage pathway (62).

C. Maturation of DCs from Nonproliferating Immature Cells

The concept of DC maturation arose initially in studies of epidermal DCs or Langerhans cells (LCs) (31). LCs are clearly differentiated along a DC pathway in that the cells are nonadherent, are poorly phagocytic, especially for antibody-coated particles, and do not respond to macrophage colony-stimulating factor (M-CSF). However, it became apparent that LCs had to be cultured in the presence of cytokines, especially GM-CSF, to acquire the typical features of DCs (63,64). After such a "maturation" period in culture, the LCs are large and dendritic in shape, have very high levels of antigen-presenting and accessory molecules, especially B7-2 or CD86, and are powerfully immunostimulatory for resting T cells. A similar maturation phenomenon has also been detailed with the DCs that are found in human blood (38).

Importantly, the immature LC appears to be the most efficient at capturing proteins (9,65) and in some cases particles (66). Likewise, immature blood DCs best present soluble proteins (67).

Some FACS data illustrating the phenotype of fresh and cultured murine LCs are shown in Figure 2 and Table 2. The LCs are distinguished from keratinocytes and cutaneous T cells by selective labeling with a PE-conjugated, anti–MHC II mAb. After 1 and 3 days of culture, the maturing LCs acquire a low buoyant density and have been enriched by flotation on dense albumin columns; hence the large enrichment of class II–rich DCs in the cultured populations in Figure 2. Note the absence (CD14) or low levels (F4/80) of macrophage markers, the presence of DEC-205, and the very high levels of MHC classes I and II and several accessory molecules (ICAM-1, B7-2). Maturation includes the upregulation of all of these molecules that contribute to effective T-cell stimulation.

DCs in many tissues are immature in the sense that their full repertoire of properties only becomes apparent after a day or more in culture. In each case, however, the immature DC population is still a trace cell type. Recently, it has become evident that much larger numbers of DC progenitors are present in human blood (4,34). Maturation into DCs requires a more prolonged culture in a mixture of cytokines, including GM-CSF plus IL-4.

D. DC Development as a Control Point in the Induction of Immunity

Given the powerful stimulatory properties of DCs, the development of mature DCs is a control point in the induction of T-cell–mediated immunity (9,63). At the same time that important adhesion and costimulatory molecules are being upregulated, the capacity to capture antigens may be downregulated. Many nonlymphoid organs contain immature DCs that can capture antigens and begin to mature. In addition, there can be inputs of immature DCs from the bone marrow and circulation as described above. All of these pathways can be cytokine driven.

V. Mechanisms Underlying the Function of Dendritic Cells

As yet there is no single "magic bullet" to explain their efficacy of DCs as APCs. Instead, these cells exhibit many specializations, each of which likely contributes to their efficacy. We like to consider these specializations in three groups: mechanisms for antigen handling, accessory functions for T-cell stimulation and properties in situ.

A. Antigen Uptake, Processing, and Presentation

It had always been thought that antigen presentation by immunologically nonspecific cells like DCs would involve bulk mechanisms of antigen uptake followed by proteolysis in lysosomes. However, DCs are proving to be more specialized. Receptors that enhance the efficiency of antigen uptake and presentation are being found, as are distinct, intracellular vacuoles for antigen presentation. In other words, these cells are not only efficient APCs after antigens have been captured, but DCs also may be more efficient at capturing and presenting small amounts of external antigens. The latter property might be particularly important in areas of DC function that are not emphasized here; that is, the ability to mediate clonal deletion or negative selection during thymopoiesis (68,69). Efficient capture and processing mechanisms may be critical in self-tolerance, allowing thymic DCs to delete self-reactive T cells when only small amounts of self-protein are available. This is illustrated by the C5 component of complement which is efficiently presented by DCs and leads to negative selection (70,71).

Two multilectin-type receptors for adsorptive uptake have been found on DCs. These molecules have 8 or 10 C-type lectin, or carbohydrate-recognition, domains. Other perhaps more familiar C-type lectins in the immune system, for example, selectins, collectins, and natural killer (NK) receptors, have only a single C-type lectin domain per polypeptide chain. All C-type lectins are likely to recognize glycans, but the multilectins in addition mediate adsorptive endocytosis. One is the macrophage mannose receptor, which has eight contiguous, membrane-external, C-type lectin domains. This receptor binds mannosylated proteins and is abundant on alveolar macrophages as well as some DCs; for example, those that are induced from human blood progenitors using GM-CSF and IL-4 (72). A second multilectin receptor is a molecule called DEC-205. It has 10 contiguous C-type lectin domains and therefore is a "DECAlectin." The ligands for DEC-205 have not been identified, but antibodies to DEC-205 are internalized in coated pits, delivered to multivesicular endosomes, and presented 100 times more efficiently than nonspecific Ig to Ig-specific T cells (49). In other words, DEC-205 on a DC behaves much like surface Ig on a B cell.

Intracellular vacuoles that are rich in MHC class II products (and termed MIIVs or CIIVs) are abundant, especially in immature DCs. These include DCs from human blood (67,72), human skin (73), and mouse spleen (74). By either electron or light (confocal) microscopy, the MIIVs label intensely with gold-labeled anti–MHC II antibodies. In ongoing experiments, we are finding some MIIVs to be separable from standard lysosomes. In retrospect, this is not surprising, since lysosomes are sites for extensive proteolysis, degrading proteins down to the level of amino acids and showing little recycling to the cell surface. In contrast, presentation requires that intact peptides be retrieved for binding to MHC II products and the complexes access the cell surface.

B. T-Cell Binding and Activation

A striking feature of antigen and superantigen presentation by DCs in vitro is that most of the DCs attach for many days to the responding T cells in discrete multicellular aggregates. These aggregates represent a microenvironment in which antigen or superantigen-reactive T cells are efficiently bound and activated.

As mentioned above, many T cell–binding and costimulatory molecules are expressed on DCs. At a given level of MHC-associated ligand, DCs are at least 30 times more stimulatory than other APCs (32). In tissue culture, each molecular couple seems to contribute to function, and no one accessory molecule is known to play a preeminent role. Addition of a blocking mAb to any one of these molecular couples typically blocks the response by 50–70%, and combinations of two or more mAbs reduce T-cell function 90% or more (75). For example, the B7 system is considered to be the major costimulator of T cells, but the blocking of B7 in vitro (33, 36) does not fully inhibit DC-mediated responses, especially at higher doses of DCs. Likewise, T-cell priming in vivo takes place in mice that chronically express a CTLA-4Ig transgene (76). CD40 on DCs likely plays important roles in maintaining the viability of DCs that are in contact with CD40L on activated T cells and in inducing DC cytokine secretion (35).

The production of soluble lymphocyte-activating factors by DCs has been difficult to detect. IL-1β mRNA is present, but the production of IL-1 biological activity has not been shown. However, Enk and Katz have presented evidence that IL-1β is essential for the induction of contact sensitivity in the skin, and this quite likely is due to DC-derived IL-1 (77). More recently, attention has been given to IL-12, which can be a major product of DCs. Anti–IL-12 can interfere with the capacity of DCs to induce interferon-γ (IFN-γ) production by Th0- and Th1-type cells (78). Koch et al. have now documented very high levels of T-cell–dependent IL-12 release by DCs (79).

C. Maturation and Migration of DCs in Vivo

When DCs are tagged and reinfused into rodents, they home to the T-cell areas of the draining lymphoid tissues (80). Macrophages migrate to other sites; for example, the red pulp and marginal zone of spleen (81). Several recent studies describe the migration of pulmonary DCs from the airway to the draining mediastinal lymph nodes (82-84).

During DC migration, maturation also can take place. This has been studied in most detail in transplanted skin, where the migrating DCs upregulate expression of MHC II and B7 (85,86). A similar maturation takes place when epidermal DCs are placed in culture; that is, both antigen-presenting and accessory molecules are markedly upregulated (33,87–89). The biosynthesis of MHC products occurs early during the culture of LCs, and it is coordinated in time with the capacity of these cells to process exogenous proteins (7).

Therefore, the DC system seems to be designed to allow immunogens to be captured and presented at sites of antigen deposition. Then the DCs migrate and mature to stimulate the appropriate clones of reactive T cells, often in draining lymphoid organs.

VI. Some Issues of Dendritic Cell Biology in the Lung

There are many topics that are ripe for study in the lung. High on the list is the trigger for immunogenicity by lung DCs. The lung is constantly exposed to inhaled proteins. It is considered unlikely and undesirable that DCs should induce an immune response to all of these, since the lung could become chronically inflamed. A similar thesis applies to the gut where DCs are found in the lamina propria and are constantly trafficking in the mesenteric lymph. The prevailing view is that resident pulmonary DCs are immature and that alveolar macrophages block DC maturation (90). Alveolar macrophages can act as potent suppressor of DC function, perhaps by the release of nitric oxide (91), or perhaps by the release of cytokines like IL-10 which are known to inhibit DC development in other systems (92). An immunogenic stimulus is one that must override this suppression and must induce the release of DC maturation factors; for example, GM-CSF. The enhanced immunity that is observed following macrophage depletion in the lung (93) is consistent with this point of view.

The distinctions between immunity and allergic type responses is of great clinical impact in the lung. Whereas many chapters in this volume emphasize the similarities between pulmonary DCs and DCs elsewhere in the body, it remains possible that pulmonary DCs will have distinct features that can lead to the induction of allergy relative to Th1-type immunity. Lung DCs might be able to stimulate regulatory T-cell populations such as class I–restricted, $CD8^+$ T cells and CD1-restricted T cells. Possibly the lung environment causes the DCs to synthesize distinct products that are pertinent to asthma and allergy. There is evidence that if IL-4 were present during a response that is stimulated by DCs, the T cell would be skewed to a Th2 phenotype (94).

VII. Conclusions

The pursuit of pulmonary DCs should also help identify new probes for identifying these cells and their products. At the present time, the DCs in the airway epithelium and alveolar septa are visualized as elongate or stellate cells that are relatively rich in MHC II. OX62 is helpful as long as OX62-bearing T cells are ruled out. DEC-205 is not a useful DC marker in mouse lung, since the lung is one known organ where this molecule is abundant on macrophages. CD83 and p55 need to be assessed in human lung tissue. The recent development of methods for generating large numbers of DCs from immature progenitors should facilitate the future identification of DC-restricted products, particularly by the many new methods that are available for molecular cloning.

Acknowledgment

This work was supported by grants from the NIH (AI13013), the Yamanouchi Fund, and the Ministry of Education, Science, and Culture of Japan.

References

1. Steinman RM, Witmer MD. Lymphoid dendritic cells are potent stimulators of the primary mixed leukocyte reaction in mice. Proc Natl Acad Sci USA 1978; 75:5132–5136.
2. Inaba K, Steinman RM, Van Voorhis WC, Muramatsu S. Dendritic cells are critical accessory cells for thymus-dependent antibody responses in mouse and man. Proc Natl Acad Sci USA 1983; 80:6041–6045.
3. Inaba K, Steinman RM. Resting and sensitized T lymphocytes exhibit distinct stimulatory [antigen-presenting cell] requirements for growth and lymphokine release. J Exp Med 1984; 160:1717–1735.
4. Sallusto F, Lanzavecchia A. Efficient presentation of soluble antigen by cultured human dendritic cells is maintained by granulocyte/macrophage colony-stimulating factor plus interleukin 4 and downregulated by tumor necrosis factor α. J Exp Med 1994; 179:1109–1118.
5. Young JW, Szabolcs P, Moore MAS. Identification of dendritic cell colony-forming units among normal CD34$^+$ bone marrow progenitors that are expanded by c-kit-ligand and yield pure dendritic cell colonies in the presence of granulocyte/macrophage colony-stimulating factor and tumor necrosis factor alpha. J Exp Med 1995; 182: 1111–1120.

6. Inaba K, Steinman RM. Protein-specific helper T lymphocyte formation initiated by dendritic cells. Science 1985; 229:475–479.
7. Puré E, Inaba K, Crowley MT, Tardelli L, Witmer-Pack MD, Ruberti G, et al. Antigen processing by epidermal Langerhans cells correlates with the level of biosynthesis of major histocompatibility complex class II molecules and expression of invariant chain. J Exp Med 1990; 172:1459–1469.
8. Inaba K, Metlay JP, Crowley MT, Steinman RM. Dendritic cells pulsed with protein antigens in vitro can prime antigen-specific, MHC-restricted T cells in situ. J Exp Med 1990; 172:631–640.
9. Romani N, Koide S, Crowley M, Witmer-Pack M, Livingstone AM, Fathman CG, et al. Presentation of exogenous protein antigens by dendritic cells to T cell clones: intact protein is presented best by immature, epidermal Langerhans cells. J Exp Med 1989; 169:1169–1178.
10. Le Gros G, Ben-Sasson SZ, Seder R, Finkelman FD, Paul WE. Generation of interleukin 4 [IL-4]–producing cells in vivo and in vitro: IL-2 and IL-4 are required for in vitro generation of IL-4 producing cells. J Exp Med 1990; 172:921–929.
11. Croft M, Duncan DD, Swain SL. Response of naive antigen-specific CD4+ T cells in vitro: characteristics and antigen-presenting cell requirements. J Exp Med 1992; 176: 1431–1437.
12. Inaba K, Young JW, Steinman RM. Direct activation of CD 8^+ cytotoxic T lymphocytes by dendritic cells. J Exp Med 1987; 166:182–194.
13. Young JW, Steinman RM. Dendritic cells stimulate primary human cytolytic lymphocyte responses in the absence of $CD4^+$ helper T cells. J Exp Med 1990; 171:1315–1332.
14. Bhardwaj N, Bender A, Gonzalez N, Bui LK, Garrett MC, Steinman RM. Influenza virus-infected dendritic cells stimulate strong proliferative and cytolytic responses from human CD8+ T cells. J Clin Invest 1994; 94:797–807.
15. Bender A, Bui LK, Feldman MAV, Larsson M, Bhardwaj N. Inactivated influenza virus, when presented on dendritic cells, elicits human $CD8^+$ cytolytic T cell responses. J Exp Med 1995; 182:1663–1671.
16. Boog CJP, Boes J, Melief CJM. Stimulation with dendritic cells decreases or obviates the CD4+ helper cell requirement in cytotoxic T lymphocyte responses. Eur J Immunol 1988; 18:219–223.
17. Lechler RI, Batchelor JR. Restoration of immunogenicity to passenger cell-depleted kidney allografts by the addition of donor strain dendritic cells. J Exp Med 1982; 155: 31–41.
18. Faustman DL, Steinman RM, Gebel HM, Hauptfeld V, Davie JM, Lacy PE. Prevention of rejection of murine islet allografts by pretreatment with anti-dendritic cell antibody. Proc Natl Acad Sci USA 1984; 81:3864–3868.
19. Iwai H, Kuma S-I, Inaba MM, Good RA, Yamashita T, Kumazawa T, et al. Acceptance of murine thyroid allografts by pretreatment of anti-Ia antibody or anti-dendritic cell antibody in vitro. Transplantation 1989; 47:45–49.
20. Inaba K, Inaba M, Naito M, Steinman RM. Dendritic cell progenitors phagocytose particulates, including Bacillus Calmette-Guerin organisms, and sensitize mice to mycobacterial antigens in vivo. J Exp Med 1993; 178:479–488.
21. Havenith CEG, Breedijk AJ, Betjes MGH, Calame W, Beelen RHJ, Hoefsmit ECM. T

cell priming *in situ* by intratracheally instilled antigen-pulsed dendritic cells. Am J Respir Cell Mol Biol 1993; 8:319–324.
22. Liu LM, MacPherson GG. Antigen acquisition by dendritic cells: Intestinal dendritic cells acquire antigen administered orally and can prime naive T cells in vivo. J Exp Med 1993; 177:1299–1307.
23. Sornasse T, Flamand V, DeBecker G, Bazin H, Tielemans F, Thielemans K, et al. Antigen-pulsed dendritic cells can efficiently induce an antibody response in vivo. J Exp Med 1992; 175:15–21.
24. Porgador A, Gilboa E. Bone marrow-generated dendritic cells pulsed with a class I-restricted peptide are potent inducers of cytotoxic T lymphocytes. J Exp Med 1995; 182:255–260.
25. Zitvogel L, Mayordomo JI, Tjandrawan T, DeLeo AB, Clarke MR, Lotze MT, et al. Therapy of murine tumors with tumor peptide pulsed dendritic cells: Dependence on T-cells, B7 costimulation, and Th1-associated cytokines. J Exp Med 1996; 183: 87–97.
26. Celluzzi CM, Mayordomo JI, Storkus WJ, Lotze MT, Falo LD. Jr. Peptide-pulsed dendritic cells induced antigen-specific, CTL-mediated protective tumor immunity. J Exp Med 1996; 183:283–287.
27. Paplia P, Chiodoni C, Rodolfo M, Colombo MP. Murine dendritic cells loaded *in vitro* with soluble protein prime CTL against tumor antigen *in vivo*. J Exp Med 1996; 183: 317–322.
28. Crowley M, Inaba K, Steinman RM. Dendritic cells are the principal cells in mouse spleen bearing immunogenic fragments of foreign proteins. J Exp Med 1990; 172: 383–386.
29. Bujdoso R, Hopkins J, Dutia BM, Young P, McConnell I. Characterization of sheep afferent lymph dendritic cells and their role in antigen carriage. J Exp Med 1989; 170: 1285–1302.
30. Holt PG, Schon Hegrad MA, Oliver J. MHC class II antigen bearing dendritic cells in pulmonary tissues of the rat. Regulation of antigen presentation activity by endogenous macrophage populations. J Exp Med 1987; 167:262–274.
31. Schuler G, Steinman RM. Murine epidermal Langerhans cells mature into potent immunostimulatory dendritic cells in vitro. J Exp Med 1985; 161:526–546.
32. Bhardwaj N, Young JW, Nisanian AJ, Baggers J, Steinman RM. Small amounts of superantigen, when presented on dendritic cells, are sufficient to initiate T cell responses. J Exp Med 1993; 178:633–642.
33. Inaba K, Witmer-Pack M, Inaba M, Hathcock KS, Sakuta H, Azuma M, et al. The tissue distribution of the B7-2 costimulator in mice: abundant expression on dendritic cells in situ and during maturation in vitro. J Exp Med 1994; 180:1849–1860.
34. Romani N, Gruner S, Brang D, Kampgen E, Lenz A, Trockenbacher B, et al. Proliferating dendritic cell progenitors in human blood. J Exp Med 1994; 180:83–93.
35. Caux C, Massacrier C, Vanbervliet B, Dubois B, Van Kooten C, Durand I, et al. Activation of human dendritic cells through CD40 cross-linking. J Exp Med 1994; 180:1263–1272.
36. Caux C, Vanbervliet B, Massacrier C, Azuma M, Okumura K, Lanier LL, et al. B70/B7-2 is identical to CD86 and is the major functional ligand for CD28 expressed on human dendritic cells. J Exp Med 1994; 180:1841–1847.

37. Rabinowitz SS, Gordon S. Macrosialin, a macrophage-restricted membrane sialoprotein differentially glycosylated in response to inflammatory stimuli. J Exp Med 1991; 174:827–836.
38. O'Doherty U, Steinman RM, Peng M, Cameron PU, Gezelter S, Kopeloff I, et al. Dendritic cells freshly isolated from human blood express CD4 and mature into typical immunostimulatory dendritic cells after culture in monocyte-conditioned medium. J Exp Med 1993; 178:1067–1078.
39. Mackay CR, Hein WR, Brown MH, Matzinger P. Unusual expression of CD2 in sheep: implications for T cell interactions. Eur J Immunol 1989; 18:1681–1688.
40. Wood GC, Freudenthal PS. CD5 monoclonal antibodies react with human peripheral blood dendritic cells. Am J Pathol 1992; 141:789–795.
41. Vremec D, Zorbas M, Scollay R, Saunders DJ, Ardavin CF, Wu L, et al. The surface phenotype of dendritic cells purified from mouse thymus and spleen: investigation of the CD8 expression by a subpopulation of dendritic cells. J Exp Med 1992; 176:47–58.
42. Suss G, Shortman K. A subclass of dendritic cells kills CD4 T cells via Fas/Fas-ligand induced apoptosis. J Exp Med 1996; 183:1789–1796.
43. Freundenthal PS, Steinman RM. The distinct surface of human blood dendritic cells, as observed after an improved isolation method. Proc Natl Acad Sci USA 1990; 87: 7698–7702.
44. Steinman RM, Gutchinov B, Witmer MD, Nussenzweig MC. Dendritic cells are the principal stimulators of the primary mixed leukocyte reaction in mice. J Exp Med 1983; 157:613–627.
45. Agger R, Witmer-Pack M, Romani N, Stossel H, Swiggard WJ, Metlay JP, et al. Two populations of splenic dendritic cells detected with M342, a new monoclonal to an intracellular antigen of interdigitating dendritic cells and some B lymphocytes. J Leukoc Biol 1992; 52:34–42.
46. Inaba K, Inaba M, Romani N, Aya H, Deguchi M, Ikehara S, et al. Generation of large numbers of dendritic cells from mouse bone marrow cultures supplemented with granulocyte-macrophage colony stimulating factor. J Exp Med 1992; 176:1693–1702.
47. Metlay JP, Witmer-Pack MD, Agger R, Crowley MT, Lawless D, Steinman RM. The distinct leukocyte integrins of mouse spleen dendritic cells as identified with new hamster monoclonal antibodies. J Exp Med 1990; 171:1753–1771.
48. Witmer-Pack MD, Crowley MT, Inaba K, Steinman RM. Macrophages, but not dendritic cells, accumulate colloidal carbon following administration in situ. J Cell Sci 1993; 105:965–973.
49. Jiang W, Swiggard WJ, Heufler C, Peng M, Mirza A, Steinman RM, et al. The receptor DEC-205 expressed by dendritic cells and thymic epithelial cells is involved in antigen processing. Nature 1995; 375:151–155.
50. Kraal G, Breel M, Janse M, Bruin G. Langerhans cells, veiled cells, and interdigitating cells in the mouse recognized by a monoclonal antibody. J Exp Med 1986; 163: 981–997.
51. Witmer-Pack MD, Swiggard WJ, Mirza A, Steinman RM. Tissue distribution of the DEC-205 protein that is detected by the monoclonal antibody NLDC-145. II. Expression in situ in lymphoid and nonlymphoid tissues. Cell Immunol 1995; 163:157–162.

52. Brenan M, Puklavec M. The MRC OX-62 antigen: A useful marker in the purification of rat veiled cells with the biochemical properties of an integrin. J Exp Med 1992; 175:1457–1465.
53. Takahashi K, Isobe T, Ohtsuki Y, Sonobe H, Takeda I, Akagi T. Immunohistochemical localization and distribution of S-100 proteins in the human lymphoreticular system. Am J Pathol 1984; 116:497–503.
54. Zhou L-J, Tedder TF. Human blood dendritic cells selectively express CD83, a member of the immunoglobulin superfamily. J Immunol 1995; 154:3821–3835.
55. Mosialos G, Birkenbach M, Ayehunie S, Matsumura F, Pinkus GS, Kieff E, et al. Circulating human dendritic cells differentially express high levels of a 55-Kd actin bundling protein. Am J Pathol 1996; 148:593–600.
56. Inaba K, Steinman RM, Witmer-Pack M, Aya H, Inaba M, Sudo T, et al. Identification of proliferating dendritic cell precursors in mouse blood. J Exp Med 1992; 175:1157–1167.
57. Chen-Woan M, Delaney CP, Fournier V, Wakizaka Y, Murase N, Fang J, et al. A new protocol for the propagation of dendritic cells from rat bone marrow using recombinant GM-CSF, and their quantification using the mAb OX-62. J Immunol Methods 1995; 178:157–171.
58. Szabolcs P, Moore MAS, Young JW. Expansion of immunostimulatory dendritic cells among the myeloid progeny of human $CD34^+$ bone marrow precursors cultured with c kit-ligand, granulocyte-macrophage colony-stimulating factor, and TNFα. J Immunol 1995; 154:5851–5861.
59. Caux C, Dezutter-Dambuyant C, Schmitt D, Banchereau J. GM-CSF and TNF-alpha cooperate in the generation of dendritic Langerhans cells. Nature 1992; 360:258–261.
60. Santiago-Schwartz F, Belilos E, Diamond B, Carsons SE. TNF in combination with GM-CSF enhances the differentiation of neonatal cord blood stem cells into dendritic cells and macrophages. J Leukoc Biol 1992; 52:274–281.
61. Inaba K, Inaba M, Deguchi M, Hagi K, Yasumizu R, Muramatsu S, et al. Granulocytes, macrophages, and dendritic cells arise from a common major histocompatibility complex class II–negative progenitor in mouse bone marrow. Proc Natl Acad Sci USA 1993; 90:3038–3042.
62. Szabolcs P, Avigan D, Gezelter S, Ciocon DH, Moore MAS, Steinman RM, et al. Dendritic cells and macrophages can mature independently from a human bone marrow-derived, post-CFU intermediate. Blood 1996; 87:4520–4530.
63. Witmer-Pack MD, Olivier W, Valinsky J, Schuler G, Steinman RM. Granulocyte/macrophage colony-stimulating factor is essential for the viability and function of cultured murine epidermal Langerhans cells. J Exp Med 1987; 166:1484–1498.
64. Heufler C, Köch F, Schuler G. Granulocyte-macrophage colony-stimulating factor and interleukin-1 mediate the maturation of murine epidermal Langerhans cell into potent immunostimulatory dendritic cells. J Exp Med 1987; 167:700–705.
65. Koch F, Trockenbacher B, Kampgen E, Grauer O, Stössel H, Livingstone AM, et al. Antigen processing in populations of mature murine dendritic cells is caused by subsets of incompletely mature cells. J Immunol 1995; 155:93–100.
66. Reis e Sousa C, Stahl PD, Austyn JM. Phagocytosis of antigens by Langerhans cells in vitro. J Exp Med 1993; 178:509–519.
67. Nijman HW, Kleijmeer MJ, Ossevoort MA, Oorschot VMJ, Vierboom MPM, van de

Keur M, et al. Antigen capture and MHC class II compartments of freshly isolated and cultured human blood dendritic cells. J Exp Med 1995; 182:163–174.
68. Matzinger P, Guerder S. Does T-cell tolerance require a dedicated antigen-presenting cell? Nature 1989; 338:74–76.
69. Stockinger B, Lin Rh, Grant C. Presentation of a circulating self-protein [C5] to MHC class II restricted T cells. Inter Rev Immunol 1993; 10:357–364.
70. Tanaka Y, Mamalaki C, Stockinger B, Kioussis D. In vitro negative selection of alpha beta T cell receptor transgenic thymocytes by conditionally immortalized thymic cortical epithelial cell lines and dendritic cells. Eur J Immunol 1993; 23:2614–2621.
71. Zal T, Volkmann A, Stockinger B. Mechanisms of tolerance induction in major histocompability complex class II-restricted T cells specific for a blood-borne self-antigen. J Exp Med 1994; 180:2089–2099.
72. Sallusto F, Lanzavecchia A. Dendritic cells use macropinocytosis and the mannose receptor to concentrate antigen in the MHC class II compartment. Downregulation by cytokines and bacterial products. J Exp Med 1995; 182:389–400.
73. Kleijmeer MJ, Oorschot VMJ, Geuze HJ. Human resident Langerhans cells display a lysosomal compartment enriched in MHC class II. J Invest Dermatol 1994; 103: 516–523.
74. Kleijmeer MJ, Ossevoort MA, Van Veen CJH, Van Hellemond JJ, Neefjes JJ, Kast WM, et al. MHC class II compartments and the kinetics of antigen presentation in activated mouse spleen dendritic cells. J Immunol 1995; 154:5715–5724.
75. Young JW, Koulova L, Soergel SA, Clark EA, Steinman RM, Dupont B. The B7/BB1 antigen provides one of several costimulatory signals for the activation of CD4+ T lymphocytes by human blood dendritic cells in vitro. J Clin Invest 1992; 90:229–237.
76. Ronchese F, Hausmann B, Hubele S, Lane P. Mice transgenic for a soluble form of murine CTLA-4 show enhance expansion of antigen-specific CD4+ T cells and defective antibody production in vivo. J Exp Med 1994; 179:809–817.
77. Enk AH, Angeloni VL, Udey SI. An essential role for Langerhans cell-derived IL-1 beta in the initiation of primary immune responses in skin. J Immunol 1993; 150: 3698–3704.
78. Macatonia SE, Hosken NA, Litton M, Vieira P, Hsieh C, Culpepper JA, et al. Dendritic cells produce IL-12 and direct the development of Th1 cells from naive CD4+ T cells. J Immunol 1995; 154:5071–5079.
79. Heutler C, Koch F, Stanzl U, Topar G, Wysocka M, Trinchieri G, et al. Interleukin-12 is produced by dendritic cells and mediates development as well as interferon-γ production by T helper 1 cells. Eur J Immunol 1996; 26:659–668.
80. Austyn JM, Kupiec-Weglinski JW, Hankins DF, Morris PJ. Migration patterns of dendritic cells in the mouse. Homing to T cell-dependent areas of spleen, and binding within marginal zone. J Exp Med 1988; 167:646–651.
81. Rosen H, Gordon S. Adoptive transfer of fluorescence-labeled cells shows that resident peritoneal macrophages are able to migrate into specialized lymphoid organs and inflammatory sites in the mouse. Eur J Immunol 1990; 20:1251–1258.
82. Xia W, Pinto CE, Kradin RL. The antigen-presenting activities of Ia+ dendritic cells shift dynamically from lung to lymph node after an airway challenge with soluble antigen. J Exp Med 1995; 181:1275–1283.
83. Havenith CEG, van Miert PPMC, Breedijk AJ, Beelen RHJ, Hoefsmit ECM. Migra-

tion of dendritic cells into the draining lymph nodes of the lung after intratracheal instillation. Am J Respir Cell Mol Biol 1993; 9:484–488.

84. Hamilton-Easton A, Eichelberger M. Virus-specific antigen presentation by different subsets of cells from lung and mediastrimal lymph node tissue of influenza-infected mice. J Virol 1995; 69:6359–6366.
85. Larsen CP, Steinman RM, Witmer-Pack M, Hankins DF, Morris PJ, Austyn JM. Migration and maturation of Langerhans cells in skin transplants and explants. J Exp Med 1990; 172:1483–1493.
86. Larsen CP, Ritchie SC, Hendrix R, Linsley PS, Hathcock RJ, Lowry RP, et al. Regulation of immunostimulatory function and costimulatory molecule [B7-1 and B7-2] expression on murine dendritic cells. J Immunol 1994; 152:5208–5219.
87. Larsen CP, Ritchie SC, Pearson TC, Linsley PS, Lowry RP. Functional expression of the costimulatory molecule, B7/BB1, on murine dendritic cell populations. J Exp Med 1992; 176:1215–1220.
88. Romani N, Lenz A, Glassel H, Stossel H, Stanzl U, Majdic O, et al. Cultured human Langerhans cells resemble lymphoid dendritic cells in phenotype and function. J Invest Dermatol 1989; 93:600–609.
89. Teunissen MBM, Wormmeester J, Krieg SR, Peters PJ, Vogels IMC, Kapsenberg ML, et al. Human epidermal Langerhans cells undergo profound morphologic and phenotypical changes during in vitro culture. J Invest Dermatol 1990; 94:166–173.
90. Bilyk N, Holt PG. Inhibition of the immunosuppressive activity of resident pulmonary alveolar macrophages by granulocyte/macrophage colony-stimulating factor. J Exp Med 1993; 177:1773–1777.
91. Holt PG, Oliver J, Bilyk N, McMenamin C, McMenamin PG, Kraal G, et al. Downregulation of the antigen presenting cell function[s] of pulmonary dendritic cells in vivo by resident alveolar macrophages. J Exp Med 1993; 177:397–407.
92. Peguet-Navarro J, Moulon C, Caux C, Dalbiez-Gauthier C, Banchereau J, Schmitt D. Interleukin 10 inhibits the primary allogeneic T cell response to human epidermal Langerhans cells. Eur J Immunol 1994; 24:884–891.
93. Thepen T, Van Rooijen N, Kraal G. Alveolar macrophage elimination in vivo is associated with an increase in pulmonary immune response in mice. J Exp Med 1989; 170:499–509.
94. Seder RA, Paul WE, Davis MM, Fazekas de St. Groth B. The presence of interleukin 4 during in vitro priming determines the lymphokine-producing potential of CD4+ T cells from T cell receptor transgenic mice. J Exp Med 1992; 176:1091–1098.

Part Two

LUNG MACROPHAGES

4

Methods to Isolate and Study Lung Macrophages

LESTER KOBZIK

Brigham and Women's Hospital
Harvard School of Public Health
Boston, Massachusetts

I. Introduction

The lung contains macrophages in anatomically distinct locations which serve to define them. These are the alveolar, airway, and interstitial macrophages. In some species, pulmonary intravascular macrophages exist and are discussed in Chapter 5. In this chapter, I describe and critique the common methods used to obtain these cells, and discuss general aspects of the methodologies available for their in vitro study. I will also describe in detail the utility of one of these methodologies, flow cytometry, for studies of lung macrophages.

II. Isolation of Alveolar Macrophages

A. Bronchoalveolar Lavage Methods

Bronchoalveolar lavage (BAL) is the method of choice to obtain the most commonly studied lung macrophage, the alveolar macrophage (AM). The general outline of this procedure is beautifully simple. After filling all or part of the lungs with fluid for a minute or less, the liquid is withdrawn. This is repeated several times and the lavage fluid is pooled. The recovered fluid contains both cells, predominantly AMs in normal lungs, and components of the alveolar lining fluid (e.g., surfactant, proteins). After centrifugation and resuspension in biological buffer or media, cell counts and differentials provide quantitation of yield and purity. In most subjects, the population obtained is ≥95% AMs, suitable in general for *in vitro* experimentation without additional purification (see below). BAL can be performed in normal or ill subjects for experimental or diagnostic purposes. This chapter discusses the aspects of BAL used to obtain normal AMs. The interested reader will find discussion of the utility of this procedure in studying lung injury and disease models elsewhere in this volume and in specific reviews (1,2).

In small laboratory animals, the procedure is usually performed (after euthanasia) by cannulating the trachea and repeated lavages of the whole lungs in situ. For humans (and other larger animals), in vivo bronchoscopic lavage of lung segments distal to a third- or fourth-generation bronchus is used. Methods for in vivo lavage of small rodents under halothane anesthesia have also been reported (3,4). Additional procedural details for small animal or human BAL can be found in other sources (1,2).

Optimizing Lavage Yield

Typical results of BAL for several species are compiled in Table 1. It is worth noting that the yields reported for common experimental animals vary somewhat.

Table 1 BAL Yields of AMs: A survey

Species	No. of AMS in BAL × 10^6	Reference	Average
Mouse	0.36	21	
	0.53	78	
	1.36	20	0.75
Rat	3.8	3	
	6.0	79	4.9
Hamster	4.2	16	
	5.8	78	5.0
Rabbit	43	3	
	49	80	46
Human, nonsmoker, 100 ml lavage	7.3	81	
Human, nonsmoker, 150 ml lavage	12.8	82	
Human, nonsmoker, "expected value," 240 ml lavage	18.1	83	
Human smoker, "expected value," 240 ml lavage	59.5	83	

These yields do provide general targets against which one can judge one's own results. A number of procedural variables have been systematically studied and found to improve cell yield. These include the use of warm lavage fluid, using divalent cation-free fluid, multiple cycles of lavage, addition of a chelator such as EDTA, and massage of the chest (in laboratory animals). These results are summarized in Table 2. Addition of lidocaine to lavage fluid may improve yield, but this is presently avoided because of concerns raised by observations that lidocaine has deleterious effects on AM function in vitro (5). In one study of the effect of pH on yields, pH 5.5 versus 7.0 saline provided identical results in a small sample of normal humans undergoing bronchoscopic lavage (6). Nevertheless, most animal investigators use a buffered salt solution (e.g., phosphate-buffered saline). I recommend using a buffered, balanced salt solution (e.g., Hanks' BSS with HEPES 10 mM without Ca^{2+} or Mg^{2+}, pH 7.4). Although yields and initial viability are comparable with saline, PBS and HBSS, AM function and viability is improved in the presence of glucose (HBSS). This is especially useful for short-term "culture" (cells in suspension on ice) required for many same-day functional or biochemical analyses (unpublished observations).

Studies of human BAL have addressed in detail issues of variability among the normal population and within a given individual. There is indeed substantial variability in cell yield and (albeit less so) differential composition within samples of normal individuals even if the group is restricted to nonsmokers. Indeed, one extensive study of 111 never-smoking individuals concluded that BAL cellularity measurements do not conform to a normal distribution and that nonparametric statistics should be applied to comparisons of normal subjects and patients with lung disease (7). Ettensohn et al. also found intersubject variability, but they also compared repeated BAL results in 16 individuals performed at minimum intervals of 6 weeks. They found relatively frequent, isolated abnormal elevations of inflammatory cells in this normal population, notably in individuals without subjective respiratory complaints or pulmonary function test changes. The major practical consequence is not dire. Experimentalists using BAL cells from "normal" volunteers, are obliged to record the yield and cell differential to allow

Table 2 Variables Which Improve BAL Yield

Factor	Yield improved with	Reference
Divalent cations	Absence of Ca^{+2}, Mg^{+2}	45
Chelation of Ca^{+2}, Mg^{+2}	Addition of EDTA (0.6 mM)	20,45
Temperature	37°C > 25°C	6
Number of lavages	Multiple ≫ single	84
Massage	Improves yield during in situ lavage of animals	1

correlation with subsequent results, and they must consider if and what level of increased inflammatory cells will preclude further experimental use of the cells or data obtained therewith. Similar quality control should be practiced when using BAL cells from animals presumed to be normal.

How Well Does BAL Sample?

Two important questions arise when considering the use of AMs obtained by BAL: (1) what proportion of total AMs are recovered and (2) if less than all of the available AMs, is this sample representative or skewed in some way? The first question is easier to answer. The total number of AMs can be established by morphometric techniques and provide a "gold standard" against which yields of whole lung BAL can be measured (Table 3). One difficulty arises in that reported BAL yields vary among different investigators, even in the same species/strain of laboratory animal. Similarly, morphometric estimates vary in some cases, as seen in Table 3. Nevertheless, a comparison of average BAL yields against total AMs available in a number of species shows a yield of approximately 25–50% in most laboratory animals. It is worth noting that the multiple lavages in a typical BAL procedure can be divided into early versus late lavages and the cells obtained therein compared functionally. Lehnert and Morrow found these two populations essentially identical in viability, size, adherence, and Fc-mediated phagocytosis (8). Hence, the practice of pooling cells from multiple lavages is justified.

How do BAL AMs compare with the 50–75% of total AMs left behind? Surprisingly little definitive information is available to address this question directly. One study found similar lysozyme content and cell size in lavaged normal AMs and those studied in situ (9). Morphometric features of lavaged AMs (e.g., volume fractions of mitochondria, lysosomes) are similar quantitatively to those

Table 3 Comparison of BAL AM Yield to Morphometric Estimates of Total AMs in the Lungs

Species	No. of AMs by morphometry[a]	Reference	BAL AMs[a,b]	BAL % efficiency
Mouse[c]	Average ~ 2.3	14,85	0.75	29
Rat[d]	Average ~ 2.1	85,86	4.9	23
Hamster	7.2 ± 0.8	16	4.2	58
Rabbit	142 ± 30	85	46	32

[a]Cell numbers × 10^6.
[b]BAL AM yields from sources detailed in Table 1.
[c]Morphometry of CBA (14) and unspecified mouse strain (85).
[d]Morphometry of Fischer 344 rats.

measured in situ (10,11). One more direct approach is to remove the lavageable AMs and then homogenize and further extract the residual AMs for functional analysis. Studies of such a "postlavage in situ" population found pulmonary macrophages identical to BAL AMs by ultrastructural morphology. However, these cells manifest a greater metabolic response to phagocytosis and greater tumor cell cytotoxicity (10,12). However, the important control of subjecting the BAL AM population to the same physical stresses (e.g., homogenization, filtration) was not included, precluding precise comparison of the two cell populations. A similar hint of differences in the lavageable versus residual AM population is found in a comparison of BAL AMs with those obtained by mincing and rinsing human lung tissue (normal, away from tumor resection). Both populations were similar in the expression of HLA-DR antigen and in the production of interleukin-1 (IL-1). The minced lung AMs did stimulate a mixed lymphocyte reaction more than their BAL counterparts (13).

Exposure of animals to aerosolized particles has been followed by analysis of particle uptake by AMs found in early versus late lavage fractions (14). Talbot et al. found that AMs from in the first two lavages had a greater number of particles/cell, but that AMs within subsequent lavages from 3–30(!) showed no difference in the particle content. The basis for the greater uptake found in the early lavage (which may preferentially sample airway contents; see also below) remains unclear. Possibilities include a true greater phagocytic rate versus a greater opportunity because of increased deposition or because of admixture during lavage with a high number of free, unbound particles, as noted by these workers. This study also obtained a yield that was approximately 50% of the morphometric prediction for total lung AMs, and it pointed out that characteristics, including particle uptake, of the residual, unlavaged AMs remain unknown. Indeed, analysis of particle uptake in the residual postlavage AMs by fluorescence or electron microscopy would be an interesting approach to this problem. The residual AMs left after BAL have been studied after enzymatic tissue digestion during isolation of interstitial macrophages (see below). These "contaminating" AMs generally retain phagocytic and other features of their lavaged counterparts (15–17), but a precise comparison has not been performed and would arguably be clouded by the enzymatic treatment used to obtain these cells.

This discussion is perhaps best concluded in the context of two observations. First, macrophages (like other leukocytes) increase adhesiveness to other cells and matrix when activated (see also Chapter 2). Second, AMs are likely to encounter some low-level activating stimulus during their life span within a normal individual (e.g., inhaled dusts, fibers, oxidants). It is perhaps not surprising then that a spectrum of adhesiveness (interpretable as mild activation) is seen within AMs, and that some AMs are more resistant to lavage than others. Nevertheless, the lack of definitive proof that BAL AMs are fully representative of lung AMs should not preclude (and has not) their experimental use. It is reassuring that numerous observations on BAL AM responses in vitro are predictive of what is observed in vivo (e.g., phagocytosis, bacterial killing, cytokine production). These considerations, however, reinforce the importance of verifying in vitro findings in vivo whenever possible.

B. Digestion Methods

One alternate method to obtain AMs for experimental study combines physical/mechanical disruption of lung tissue with frequent rinses to loosen and collect the cells. Using this approach, a number of investigators have found quite satisfactory yields and purity of AMs (18). As detailed below, one laboratory uses mechanical disruption and rinsing of unlavaged rat lungs to obtain (and remove) AMs in high yield and purity prior to additional steps to isolate interstitial macrophages (17). This approach may be especially useful for investigators with access to surgically removed human lung tissue. These are generally from smokers undergoing resection of a tumor and may be subject to postobstructive pneumonitis or other abnormalities. Histological evaluation of a portion of such samples at some point in the experimental design is important, since samples that appear macroscopically normal or uninvolved by the tumor may contain significant inflammation or tumor infiltration. Nevertheless, by studying grossly normal tissue away from the tumor, a sample comparable to that obtained by BAL of a smoker subject can be obtained.

III. Isolation of Lung Interstitial Macrophages

Interstitial macrophages (IMs) constitute the second major subtype of lung macrophage. Morphometric studies find that the number of IMs is approximately one half to two thirds that of AMs in normal lungs (16,19). IMs most likely comprise a resident, self-renewing population and monocytes in transition en route to the alveolar space (16,20,21). IMs have been isolated for in vitro analysis using a variety of methods, and numerous differences between IMs and AMs have been observed. These include both morphological (size, intracellular organelles [22]) and functional (e.g., proliferative capacity [16] characteristics.

In general, IMs have been isolated by (1) exhaustive lavage of lungs to reduce AMs, (2) mechanical/enzymatic disruption of lung tissue to produce a single cell suspension containing IMs (and other cells), and (3) a purification step to identify and concentrate the IMs for further study. Each laboratory in this field seemingly follows protocols which are slightly to substantially different variations on these themes, precluding exact comparison. Most studies report numerical yield but few establish efficiency by comparison of yield to morphometric estimates of IM numbers in the animal being studied. Nevertheless, a number of broadly useful guidelines emerge from the literature.

Exhaustive lavage to remove AMs is never completely successful (~25–50%; see Table 3). Hence, all IM isolation procedures must deal with a substantial number of contaminating AMs in any postlavage sample. Enzymatic disruption of minced lung tissue greatly improves the yield of cells, including IMs. The method of mild digestion of lung tissue slices by collagenase first reported by Holt et al. (23) has been successfully used by other investigators (16,17). Collagenase digestion should not cause proteolytic damage to cell surface structures, as is possible with enzymes used in earlier studies, such as trypsin or pronase, although some residual trypsin–like activity can be present even in prescreened batches of collagenase (16,23).

An obvious, but difficult, requirement for IM isolation is a perfect marker of "IM"-ness. The ideal marker would be highly specific and sensitive, easy to use, and one whose application would not alter the IM phenotype so as to possibly affect subsequent studies. Unfortunately, no such perfect marker exists. A number of features have been used to identify and isolate IMs within the heterogeneous population of cells obtained by enzymatic disruption of lung tissue. These include cell size and bouyancy in density gradients (24), antigenic features of IMs versus AMs (16), phagocytosis of red blood cells, and adherence to tissue culture plastic or antibody-coated surfaces (15). Phagocytosis and adherence serve primarily to identify cells as macrophages rather than epithelial or other lymphoid cells. Of comparable importance are the methods used to identify residual, contaminating AMs. These cells can usually be distinguished by their larger size, strong labeling by nonspecific esterase staining, and/or antigenic features.

One highly successful methodology has been reported by Laskin and coworkers (17,25). These workers used additional rinsing and vortexing of intact and sieve-"disaggregated" lung tissue slices prior to enzymatic treatment. They found these steps reduced residual AM contamination to less than 1%, whereas preserving relatively high yields of IMs, which could then be purified by adherence. Although the macrophage response to adherence to plastic must be considered in subsequent experiments with such IM preparations, this protocol represents an important advance. Even better for the weary experimentalist, prelavage of the lungs was found to be unnecessary, as such AMs are collected during the rinses of the mechanically disrupted lung tissue (17).

IV. Isolation of Airway Macrophages

The relatively small number of macrophages found living in the airways have (understandably) not been frequently studied. Rankin et al. obtained airway lavage cells from humans using a balloon occlusion technique to lavage only the left mainstem bronchus (26). They obtained ~5×10^5 macrophages (~30% of the total yield) which differed slightly from AMs ultrastructurally but were similar in expression of a panel of cell surface antigens identified by monoclonal antibodies. Lehnert et al. (27) dissected the trachea and major bronchi from normal rat lungs and used perfusion to lavage these airways. These workers obtained an average of 3×10^4 airway macrophages per lung, representing 17% of the total cells obtained, the remainder being primarily exfoliated epithelial cells. These airway macrophages resembled AMs morphologically, but had somewhat diminished enzymatic and phagocytic capacity. The investigators concluded that airway macrophages are most likely functionally less active or effete AMs translocating out of the lung. Selective lavage of airways in humans (28) or rodents (29) yields greater numbers of cells in disease states or models.

V. Lung Macrophage Cell Lines

A number of alveolar macrophage cell lines have been developed and used experimentally (Table 4). These cell lines have arisen from spontaneous transformation (30) or by intentional application of immortalizing retroviruses (31). These cell lines share the advantage of providing large numbers of cells grown under defined conditions and purity. The major concern with the use of such cell lines is how representative they are of normal AMs. The best-characterized and most used of these cell lines is the rat NR8383 cell line developed by Helmke (30). These cells resemble normal AMs in many respects, including morphology, sensitivity to endotoxin, phagocytic function, and expression of cytokines. Like AMs, they show substantial heterogeneity while growing in culture and include a range of cell size and adherent and nonadherent subpopulations (generally 50% each). Murine cell lines have been described including the MH-S (32) and C11 cell lines (31,33).

VI. In Vivo Elimination of Alveolar Macrophages

For many experimental questions regarding the role of AMs in the lung, it would be most useful to study animals whose lungs are depleted of AMs. Although no

Table 4 Lung Macrophage Cell Lines

Name	Species	Origin	Reference
Nr8383	Rat	Normal rat AM, spontaneous transformant	30
MH-S	Mouse	SV-40 transformation of normal AMs	32
C3HAMSV40	Mouse	S4V-40 transformation of normal AMs	33
C11, etc.	Mouse	J2 retrovirus transformation of normal AMs	31
	Sheep	S4V-40 transformation of normal AMs	87

AM knockout mouse exists, some workers have reported success in chemically depleting AMs. The instillation of liposomes containing the toxin dichloromethylene diphosphate reduced the AM population by ~>70% (34,35). This methodology has been limited by the concomitant induction observed by some investigators of neutrophilic inflammation (36). Recently, Weiner-Kronish and coworkers have reported improvements to this method (37). By aerosolizing the liposomes, these workers were able to reduce the lavageable AMs by >95% without significant acute cellular inflammation. They observed substantially diminished chemokine production and neutrophil recruitment to the AM-depleted lungs of rats challenged with *Pseudomonas*.

VII. In Vitro Culture of Lung Macrophages

In vitro culture of lung macrophages uses techniques and concepts derived from experience with general cell culture as well as culture of monocytes and macrophages from other sites. The interested reader is directed to (references 38 and 39) as excellent starting points. This discussion focuses on a number of lung macrophage-specific issues.

A. Adherent or Suspension Cultures?

Lung macrophages can be cultured as adherent cells or in suspension. AMs adhere rapidly and firmly to tissue culture plastic. Suspension cultures can be achieved by using specially treated plastic to reduce its adhesiveness. These include commercially available proprietary plasticware (low binding plates, Costar Inc., Cambridge, MA), Teflon-coated bags or vessels (40) or coating with polyhydroxymethylacrylate (polyHEMA) (41,42). The latter is an easy and inexpensive method to render standard tissue culture plastic nonadherent for rodent and human AMs. The treatment lasts for 5–7 days, after which cracks in the polyHEMA gel diminish its efficacy. It should be noted that although AMs are prevented from sticking to the plastic under these conditions, they avidly adhere to each other, forming spheres of up to 50–100 cells floating in homotypic adhesion. These cells are readily dispersed during harvesting and resuspension. They resemble normal lavaged AMs in phagocytic function and in their lack of the basal expression of acute-response mRNAs (TNF, chemokines KC & MIP-2) (our unpublished observations).

Which culture method, adherent, or suspension more accurately simulates the normal AM milieu? Unfortunately for the lung macrophage biologist, the natural, normal state of the AM in situ seems to be somewhere in between these two experimentally convenient extremes. The AM observed in situ after fixation using vascular perfusion appears to be stretched out, as if adhering at least partially to subjacent structures (43). They also appear to be found preferentially at alveolar septal junctions (44), which implies a method to stay anchored there. The normal AM can be readily detached from its alveolar position by lavage with divalent cation-free buffers, and chelators of calcium such as EDTA further improve yields (45). These observations suggest a Ca^{2+}-dependent but rather weak basal adhesion by AMs (weak at least for those recovered by lavage) to components of the alveolar milieu. In contrast, AMs adherent to tissue culture plastic are much more firmly adherent. They are not readily detached by lavage-type buffers and usually require enzymatic treatment (trypsin) for release. Hence, adherence to plastic is clearly not representative of in situ conditions.

Similarly, AMs are not in free-floating suspension within the alveolus, and hence suspension cultures also cannot claim adequately to reproduce in situ conditions. Nevertheless, short-term suspension cultures are quite useful for the functional assay of AMs in the first few hours after lavage. Comparison of AMs in monolayer and suspension found greater phagocytosis of latex beads in suspension (46). Suspension cultures have the advantage of avoiding the initial transient expression of activation markers; for example, chemokines (47,48) seen on macrophage attachment to plastic. Both the cells and the surrounding medium are easily analyzed. Another approach is to allow AMs to attach, but simply to wait 1 day, at which point the AM basal expression of such markers is generally nil and comparable to that of freshly lavaged cells.

Two additional tissue culture approaches to study of lung macrophages merit mention. First, a number of investigators have successfully used organ culture methods to study portions of or entire rodent lungs. These include observations that macrophages will proliferate and differentiate within these lung organ cultures (49,50). Second, normal rat AMs in suspension cultures contain subpopulations capable of proliferating in response to macrophage active colony-stimulating factors (C. Farver and L. Kobzik, submitted). Since the progency cells obtained in this manner differ from mature AMs, this approach requires careful comparison with true normal lung macrophages to be informative.

VIII. Flow Cytometric Methods to Study Lung Macrophages

It is beyond the scope of any single chapter to discuss the galaxy of methodologies successfully applied to studies of the lung macrophage. The other chapters in this volume provide an excellent guide to their respective topics, within which specific methods can be found. The interested reader is also directed to excellent general compendia of methods for studies of macrophages (38,39), most which can be used for lung macrophages with little or no modification. I have also found portions of a handbook of neutrophil methods (51) and *Current Protocols in Immunology* (52) to be excellent sources for detailed protocols.

One specific methodology which has much to offers students of the lung macrophage is flow cytometry. Major advantages include speed (e.g., $\geq\sim10^3$ cells/

sec), high sensitivity, the ability to identify and analyze subpopulations within heterogeneous pools of cells, and the ability of some instruments to also sort live cells for further in vitro study. There are disadvantages, which include the expense of the instrumentation, the need for cells to be in suspension for analysis, and the inability to study dynamic events over time in a single cell. Two major uses of flow cytometry for study of lung macrophages are cell surface phenotyping and functional assays.

A. Immunophenotyping

The utility of quantitative analysis of cell surface antigens to study macrophage phenotype, differentiation, and (by inference) function is well established (53). For lung macrophage studies, a major methodological issue is the obstacle of high autofluorescence in certain AM populations. The level of autofluorescence varies among species, but is commonly a problem with studies of human AMs, especially those from smokers (due in large part to intracellular accumulation of smoke particulates and by-products) (54). In addition, glutaraldehyde fixation increases autofluorescence and should be avoided. A number of methods have been reported to overcome endogenous autofluorescence and allow immunofluorescent flow cytometric analysis of lung macrophages. In one method, AMs have been fixed with paraformaldehyde and then permeabilized with mild detergent treatment (0.1% n-octyl-B-d-glycopyranoside) to allow entry of a crystal violet solution. The crystal violent quenches autofluorescence, allowing enhanced detection of subsequently added anti–cell surface monoclonal antibodies (55). One drawback is the variability in crystal violet preparations, as discussed by these workers (55). Another approach is to use fluorochromes whose excitation and emission wavelengths overlap minimally with the major peak of autofluorescence in AMs which occurs in the blue-green range. For example, workers have used antibodies conjugated to allophycocyanin which excites and emits in longer wavelengths (red) (56). Newer fluorochromes and tandem fluorochrome reagents have expanded the possibilities for investigators to use more commonly available blue excitation sources (standard for fluorescein) for reporters in the red wavelengths (57,58). Having overcome these obstacles, the final barrier is availability of antibody reagents directed to the structure of interest. Two compendia catalogue commercially available antibodies (59,60).

B. Functional Assays

A number of critical lung macrophage functions can be easily and rapidly measured using flow cytometry. These include phagocytosis, regulation of intracellular pH, oxidative metabolism, and proliferation. Multiparameter analysis can be performed allowing correlation of one or more functions with other phenotypic features of the same cell (e.g., uptake of fluorescent latex beads vs respiratory burst [61]). Two applications used extensively in my own studies are discussed below: quantitation of AM phagocytosis and respiratory burst.

Phagocytosis

The AM is an avidly phagocytic cell (62,63) showing both a broader range of phagocytic targets and more rapid uptake than freshly isolated peritoneal macrophages or blood monocytes (61). These cells will rapidly bind and ingest fluorescently labeled targets (e.g., latex beads, bacteria, yeast). Both opsonin-dependent (Fc, complement) and opsonin-independent pathways exist (61,64). Latex beads offer the advantages of uniformity of size, composition, and shape. Distinct peaks of fluorescence can usually be resolved for cells with one to five beads, allowing precise calculation of uptake as particles/cell. More real-world phagocytic targets include micro-organisms and environmental particulates. Bacteria and yeast, generally studied as heat- or formalin-killed organisms, can be easily labeled with fluorescein or other fluorochromes. However, the amount of dye per organism varies considerably more than the precise levels found in latex beads. Hence, uptake data must generally be expressed as relative cell-associated fluorescence, and batch-to-batch variability in labeling must be considered in comparing different experiments. The uptake of environmental particulates (e.g., silica, titanium dioxide, diesel particulates) by AMs can be measured by flow cytometry, exploiting either increases in right-angle light scatter caused by cell-associated particles or by precoating the particulates with fluorescent-labeled phospholipids (65,66).

An important issue, at least hypothetically, is distinguishing particle binding from actual internalization. In practice, the AM is so avidly phagocytic that, in many cases, ≥85–90% of cell-associated particles will be found internalized (64). Depending on the experimental question, it is often possible to establish this parameter once, when initially characterizing an assay. Methods to identify "in versus out" include the use of dye quenching techniques, which reduce or eliminate the fluorescence of cell-attached particles, whereas not affecting fluorescence of internalized material. By analyzing samples with or without quenching, the proportion of signal in each compartment can be measured. Regrettably, these techniques have a number of disadvantages, including variability of dye solutions and a limited 'window of opportunity' for analysis before leakage of dye into cells or uptake by endocytosis occurs and obscures the analysis. Electron microscopy remains a gold standard assay but is relatively time consuming and laborious compared with the fluorescent assays for which it can provide validation. Another approach is to perform the phagocytosis assay, stop all cell activity by chilling the suspension to 4°C, and apply antibodies conjugated to a fluorochrome different from that of the particle. Attached, but not internalized, particles will be labeled with the second fluorescent reporter, and the proportion can be quantified in this manner (67).

Oxidative Metabolism

Lung macrophages can produce reactive oxygen intermediates (ROI) in response to phagocytic or other stimuli. Depending on the amount and site of production, these oxidants can contribute to intracellular signaling, participate in killing of ingested microorganisms, or cause injury to the cell itself or adjacent tissues (68,69). The indiscriminate production of oxidants by AMs interacting with otherwise innocuous environmental particles could have deleterious consequences for the lung, but this, thankfully, does not occur. Flow cytometric analysis of particle uptake and intracellular respiratory burst has enhanced our understanding of Am-particle interactions and determinants which lead to a respiratory burst or downregulation of Am responses (42,61,70,71).

The intracellular production of oxidants within lung macrophages can be measured using the fluorescent reporter molecules, dichlorofluorescin (DCFH), hydroethidine (HE), or dihydrorhodamine 123 (DHR). These probes share a common principle of action but differ in their final cellular fate. All three are nonfluorescent, reduced precursor molecules which are introduced into the cell during a preincubation or loading period. Once oxidized during basal metabolism or during a respiratory burst event within the cell, the reporter is converted to a fluorescent molecule whose concentration can be quantitated by flow cytometric (or fluorometric) techniques. For DCFH, the product is a dichlorofluoroscein (DCF), present diffusely in the cytoplasm, and having the spectral properties of fluorescein. Oxidation of HE (the sodium borhydride–reduced derivative of ethidium bromide) produces ethidium bromide, which intercalates with nuclear DNA and thereby enhances its red fluorescence (72). DHR is oxidized to rhodamine 123, a red fluorescent molecule accumulated in mitochondria (73). The relative specificity of these agents for different reactive oxidant species is unsettled. Initial characterization suggested that DCFH was specific for H_2O_2, whereas HE and DHR have been advanced as agents specific for O_2^- (72,74). In fact, DCFH can also be oxidized by nitric oxide (NO) or products of NO interaction with O_2^-(75) (see also below). HE shows sensitivity to both O_2^- and H_2O_2, as identified by the need for both superoxide dismutase and catalase to abrogate increased intracellular fluorescence in HE-loaded cells exposed to xanthine/xanthine oxidase as a source of ROI (70). Recent studies have established that DCFH can also be oxidized by nitric oxide or the product of NO and superoxide anion, peroxynitrite. Rodent AMs produce substantial amounts of NO after induction of type II (inducible) NO synthase with lipopolipaccharide (LPS). In such cells, a substantial l-arginine–dependent oxidation of intracellular DCFH (which is insensitive to mitochondrial inhibitors) occurs (L. Kobzik, manuscript in preparation). Hence, DCFH and perhaps the fluorescent agents used initially for detection of reactive oxygen intermediate production are not specific for ROI under some circumstances. Nevertheless, combination of these reporters with other functional or phenotypic analysis can provide important information.

In normal AMs, DCFH oxidation can be used to monitor the increased production of oxygen-derived intermediates, for example, O_2^- and H_2O_2, during the respiratory burst response to opsonized particles or microorganisms (61,70). The response is relatively rapid (easily detected in 10 to 30 min assays), is dose-dependent (increases with greater phagocytosis), and usually achieves a two to four fold increase in mean fluorescence over control cells at maximal stimulation. The relative fluorescence data provided by flow cytometry can be calibrated fluorometrically to allow expression of data in absolute terms of intracellular product produced (DCF). However, the precise stoichiometry of oxidant to DCF is unknown. In many cases, careful measurement of changes in mean relative fluorescence units observed using constant instrument conditions will suffice to compare experimental groups. Most of the AM respiratory burst detected with these agents is abrogated by treatment with mitochondrial inhibitors (e.g., antimycin), confirming older polarographic observations that most of the respiratory burst in AMs (in contrast to that of neutrophils) comprises mitrochondrial respiration (76).

Combining flow cytometric analysis of AM phagocytosis and oxidative metabolism has provided important information about the interaction of AMs with phagocytic particles. This is especially true for inert, unopsonized particles such as those inhaled environmental particulates which are a unique challenge to the AM's defense of the alveolar milieu. AM uptake of unopsonized latex beads results in little or no respiratory burst response. In fact, cells ingesting one or two beads shown a downregulation or inhibition of oxidative metabolism. In contrast, AM uptake of otherwise identical but antibody-opsonized beads causes a dose (particle number)–dependent respiratory burst. This allows AM clearance of inert particles without triggering a potential harmful respiratory burst and other pro-inflammatory responses. These observations with surrogates for environmental particles apply as well to AM interactions with "real-world" particulates with one significant exception. Multiparameter analysis of AM uptake of particulates like iron oxide and titanium oxide found no oxidant response to substantial uptake of these inert dusts, which is consistent with uptake via a nonactivating scavenger receptor pathway (71). However, AM uptake of pathogenic quartz was found to use the same pathway, yet cause substantial intracellular oxidant stress, which was detected by DCFH. This apparent contradiction can be reconciled by earlier observations that some environmental particles, for example, freshly "cracked" quartz, carry oxidant molecules on their surface (77). Hence, certain particles can cause AM activation and proinflammory responses despite initial binding and uptake via intrinsically nonactivating receptor systems. Flow cytometric analysis will undoubtedly contribute to further analysis of these processes.

Acknowledgment

Amy Imrich provided invaluable flow cytometry expertise to studies performed in the author's laboratory.

References

1. Brain JD, Beck BD. Bronchoalveolar lavage. In: Witschi HP, Brain JD, eds. Handbook of Experimental Pharmacology. Vol. 75. Berlin: Springer-Verlag, 1985: 203–226.
2. Reynolds HY. Bronchoalveolar lavage. Am Rev Respir Dis 1987; 135:250–263.
3. Mauderly J. Bronchopulmonary lavage of small laboratory animals. Lab Anim Sci 1977; 27:255–261.
4. Henderson RF. Use of bronchoalveolar lavage to detect lung damage. Environ Health Perspect 1984; 56:115–129.
5. Hoidal J, White J, Repine J. Influence of cationic local anesthetics on the metabolism and ultrastructure of human alveolar macrophages. J Lab Clin Med 1979; 93: 857–866.
6. Pingleton S, Harrison G, Stechschulte D, Wesselius L, Kerby G, Ruth W. Effect of location, pH, and temperature of instillate in bronchoalveolar lavage in normal volunteers. Am Rev Respir Dis 1983; 128:1035–1037.
7. Merchant R, Schwartz D, Helmers R, Dayton C, Hunninghake G. Bronchoalveolar lavage cellularity. The distribution in normal volunteers. Am Rev Respir Dis 1992; 146:448–453.
8. Lehnert B, Morrow P. Size, adherence, and phagocytic characteristics of alveolar macrophages harvested 'early' and 'later' during bronchoalveolar lavage. J Immunol Methods 1984; 73:329–335.
9. Kimura A, Goldstein E, Donovan R, Verwoerd N, Ploem J. Comparison of lavaged and intrapulmonary alveolar macrophages in respect to lysozyme content and size in the rat. Am Rev Respir Dis 1984; 129:149–154.
10. Drath DB, Karnovsky ML. Superoxide production by phagocytic leukocytes. J Exp Med 1975; 141:257–261.
11. Lum H, Tyler W, Hyde D, Plopper C. Morphometry of in situ and lavaged pulmonary alveolar macrophages from control and ozone-exposed rats. Exp Lung Res 1983; 5: 61–78.
12. Drath D. Enhanced superoxide release and tumoricidal activity by a postlavage, in situ pulmonary macrophage population in response to activation by Mycobacterium bovis BCG exposure. Infect Immun 1985; 49:72–75.
13. Weissler J, Lyons R, Lipscomb M, Toews G. Human pulmonary macrophages. Functional comparison of cells obtained from whole lung and by bronchoalveolar lavage. Am Rev Respir Dis 1986; 133:473–477.
14. Talbot R, Kellington J, Morgan A. Are macrophages recovered from mouse lung by bronchoalveolar lavage representative of the alveolar macrophage pool? Inhal Toxicology 1996; 8:201–212.

15. Dethloff LA, Lehnert BE. Pulmonary interstitial macrophages: isolation and flow cytometric comparisons with alveolar macrophages and blood monocytes. 1988; 43: 80–90.
16. Kobzik L, Godleski J, Barry B, Brain J. Isolation and antigenic identification of hamster lung interstitial macrophages. Am Rev Respir Dis 1988; 138:908–914.
17. Lavnikova N, Prokhorova S, Helyar L, Laskin DL. Isolation and partial characterization of subpopulations of alveolar macrophages, granulocytes, and highly enriched interstital macrophages from rat lung. Respir Cell Mol Bio 1993; 8:384–392.
18. Streck R, Jezewski H, Rodriguez M, Hurley E, Rich G, Braun K, Pauly J. A method for isolating human lung macrophages. J Immunol Methods 1994; 174:67–82.
19. Heppelston A, Young A. Population and ultrastructural changes in murine alveolar cells following $^{239}PuO_2$ inhalation. J Pathol 1985; 146:155–166.
20. Blusse van ould Alblas A, van Furth R. Origin, kinetics and characteristics of pulmonary macrophages in the normal steady state. J Exp Med 1979; 149:1504–1518.
21. Bowden D, Adamson I. Adaptive responses of the pulmonary macrophagic system to carbon I. Kinetic studies. Lab Invest 1978; 38:422–429.
22. Sebring RJ, Lehnert BE. Morphometric comparisons of rat alveolar macrophages, pulmonary interstitial macrophages, and blood monocytes. Exp Lung Res 1992; 18: 479–496.
23. Holt P, Degebrodt T, Venaille T, Krska K, Flexman J, Farrell H, Shellam G, Young P, Penhale J, Robertson T, Papadimitriou J. Preparation of interstitial lung cells by enzymatic digestion of tissue slices: preliminary characterization by morphology and performance in functional assays. Immunology 1985; 54:139–147.
24. Chandler D, Kennedy J, Fulmer J. Studies of membrane receptors, phagocytosis, and morphology of rat lung interstitial macrophages. Am Rev Respir Dis 1986; 134: 542–547.
25. Prokhorova S, Lavnikova N, Laskin DL. Functional characterization of interstitial macrophages and subpopulations of alveolar macrophages from rat lung. J Leukoc Biol 1994; 55:141–146.
26. Rankin J, Marcy T, Rochester C, Sussman J, Smith S, Buckley P, Lee D. Human airway macrophages. Am Rev Respir Dis 1992; 145:928–933.
27. Lehnert B, Valdez Y, Sebring R, Lehnert N, Saunders G, Steinkamp J. Airway intraluminal macrophages: evidence of origin and comparisons to alveolar macrophages. Am J Respir Cell Mol Biol 1990; 3:377–391.
28. Thompson A, Daughton D, Robbins R, Ghafouri M, Oehlerking M, Rennard S. Intraluminal airway inflammation in chronic bronchitis. Am Rev Respir Dis 1989; 140:1527–1537.
29. Long N, Skornik W, Kobzik L, Weller E, Shore S. Respiratory tract inflammation during the induction of chronic bronchitis in rats: role of C-fibers. submitted for publication 1996.
30. Helmke R, Boyd R, German V, Mangos J. From growth factor dependence to growth factor independence: the genesis of an alveolar macrophage cell line. In Vitro Cell & Dev Biol 1987; 23:567–574.
31. Palleroni A, Varesio L, Wright R, Brunda M. Tumoricidal alveolar macrophage and tumor infiltrating cell lines. Int J Cancer 1991; 49:296–302.
32. Mbawuike I, Herscowitz H. MH-S, a murine alveolar macrophage cell line: morpho-

logical, cytochemical and functional characteristics. J Leukoc Biol 1989; 46: 119–127.
33. Ryan L, Vermeulen M. Alveolar macrophages form C3H/HeJ mice show sensitivity to endotoxin. Am J Respir Cell Mol Biol 1995; 12:540–546.
34. Berg JT, Lee ST, Thepen T, Lee CY, Tsan MF. Depletion of alveolar macrophages by lipsome-encapsulated dichloromethylene diphosphonate. J Appl Physiol 1993; 74: 2812–2819.
35. Holt P, Oliver J, Bilyk N, McMenamin C, McMenamin P, Kraal G, Thepen T. Downregulation of the antigen presenting cell functions(s) of pulmonary dendritic cells in vivo by resident alveolar macrophages. J Exp Med 1993; 177:397–407.
36. Berg JT, White JE, Tsan MF. Response of alveolar macrophage-depleted rats to hyperoxia. Exp Lung Res 1995; 21:175–85.
37. Hashimoto S, Pittet J, Hong K, Folkesson H, Bagby G, Kobzik L, Frevert C, Wantanabe K, Tsurufuji S, Wiener-Kronish J. Depletion of alveolar macrophage macrophages abolishes neutrophil chemotaxis to airspace infections. Am J Physiol 1996. In press.
38. Adams D, Edelson P, Koren H. Methods for Studying Mononuclear Phagocytes. New York: Academic Press, 19981:1023.
39. Herscowitz H, Holden H, Bellanti J, Ghaffar A. Manual of Macrophage Methodology. In: Rose N, ed. Immunology Series. Vol. 13. New York: Dekker, 1981.
40. Helinski E, Bielat K, Ovak G, Pauly J. Long-term cultivation of functional human macrophages in teflon dishes with serum-free media. J. Leukoc Biol 1988; 44: 111–121.
41. Folkman J, Moscona A. Role of cell shape in growth control. Nature 1978; 273: 345–349.
42. Kobzik L. Huang S, Paulauskis JD, Godleski JJ. Particle opsonization and lung macrophage cytokine response. In vitro and in vivo analysis. J Immunol 1993; 151:2753–2759.
43. Brain J, Gehr P, Kavet R. Airway macrophages. The importance of the fixation method. Am Rev Respir Dis 1984; 129.
44. Parra S, Burnette R, Price H, Takaro T. Zonal distribution of alveolar macrophages, type II pneumonocytes, and alveolar septal connective tissue gaps in adult human lungs. Am Rev Respir Dis 1986; 133:908–912.
45. Brain J, Frank R. Alveolar macrophage adhesion: wash electrolyte compositions and free cell yield. J Appl Physiol 1973; 34:75–80.
46. Mueller H, Guilmette R, Muggenburg B. Uptake of inert particles by dog alveolar macrophages in vitro—a comparison of monolayer and suspension techniques. J Appl Toxicol 1989; 99:135–143.
47. Huang S, Paulauskis J, Godleski J, Kobzik L. Expression of macrophage inflammatory protein-2 and KC mRNA in pulmonary inflammation. Am J Pathol 1992; 141: 981–988.
48. Kashahara K, Streiter R, Chensue S, Standiford T, Kunkel S. Mononuclear cell adherence induces neutrophil chemotactic factor/Interleukin-8 gene expression. J Leukoc Biol 1991; 50:287–295.
49. Sorokin S, Hoyt R. Pure population of nonmonocyte derived macrophages arising in organ cultures of embryonic rat lungs. Anat Rec 1987; 217:35–52.

50. Sorokin S, Kobzik L, Hoyt R, Godleski J. Development of surface membrane characteristics of "premedullary" macrophages in organ cultures of embryonic rat and hamster lungs. J Histochem Cytochem 1989; 36:950–958.
51. Metcalf J, Gallin J, Nauseef W, Root R. Laboratory Manual of Neutrophil Function. New York: Raven Press, 1986.
52. Coligan J, Kruisbeek A, Margulies D, Shevach E, Strober W. Current Protocols in Immunology. In: Colco R, ed. New York: Wiley, 1994.
53. Gordon S, Lawson L, Rabinowitz S, Crocker PR, Morris L, Perry VH. Antigen markers of macrophage differentiation in murine tissues. Curr Top Microbiol Immunol 1992; 181:1–37.
54. Skold CM, Eklund A. Hallden G, Hed J. Autofluorescence in human alveolar macrophages from smokers: relation to cell surface markers and phagocytosis. Exp Lung Res 1989; 15:823–35.
55. Hallden G, Skold C, Eklund A, Forslid J, Hed J. Quenching of intracellular autofluorescence in alveolar macrophages permits analysis of fluorochrome labelled surface antigens by flow cytofluorometry. J Immunol Methods 1991; 142:207–214.
56. Fuchs H, McDowell J, Shellito J. Use of allophycocyanin allows quantitative description by flow cytometry of alveolar macrophage surface antigens present in low numbers of cells. Am Rev Respir Dis 1988; 138:1124–1128.
57. Viksman MY, Liu MC, Schleimer RP, Bochner BS. Application of a flow cytometric method using autofluorescence and a tandem fluorescent dye to analyze human alveolar macrophages surface markers. J Immunol Methods 1994; 172:17–24.
58. Lohmeyer J, Friedrich J, Rosseau S, Pralle H, Seeger W. Multiparameter flow cytometric analysis of inflammatory cells contained in bronchoalveolar lavage fluid. J Immunol Methods 1994; 172:59–70.
59. Linscott W. Linscott's Directory of Immunological and Biological Reagents. Santa Rosa, CA: 1996.
60. Weimer R. MSRS Catalog. Primary Antibodies.Birmingham, MI: 1993.
61. Kobzik L, Godleski J, Brain J. Selective down-regulation of alveolar macrophage oxidative response to opsonin-independent phagocytosis. J Immunol 1990; 144:4312–4319.
62. Brain JD. Macrophages in the respiratory tract. In: Fishman AP, ed. Handbook of Physiology. Vol. I. Bethesda, MD: American Physiological Society, 1985: 447–471.
63. Fels AO, Cohn Z, A. The alveolar macrophage. J Appl Physiol 1986; 60:353–369.
64. Parod R, Brain J. Immune opsonin-independent phagocytosis by pulmonary macrophages. J Immunol 1986; 136:2041–2047.
65. Stringer B, Imrich A, Kobzik L. Lung macrophage uptake of environmental particulates: a flow cytometric assay. Cytometry 1995; 20:23–32.
66. Stringer B, Imrich A, Kobzik L. Lung epithelial cell (A549) uptake of unopsonized environmental particulates. Exp Lung Res 1996; In press.
67. Parod R, Godleski J, Brain J. Inhibition of immune opsonin-independent phagocytosis by antibody to a pulmonary macrophage cell surface antigen. J Immunol 1986; 136:2048–2054.
68. Fantone JC, Ward PA. Role of oxygen-derived free radicals and metabolites in leukocyte-dependent inflammatory reactions. Am J Pathol 1982; 107:397–418.
69. Halliwell b, Gutteridge. Free Radicals in Biology and Medicine. Oxford: Clarendon Press, 1993.

70. Kobzik L, Godleski J, Brain J. Oxidative metabolism in the alveolar macrophage: analysis by flow cytometry. J Leukoc Biol 1990; 47:295–303.
71. Kobzik L. Lung macrophages uptake of unopsonized environmental particulates: role of scavenger-type receptors. J Immunol 1995; 155:367–376.
72. Rothe G, Valet G. Use of hydroethidine (HE) and 2,7-dichlorofluorescin (CDFH) for the flow-cyotmetric measurement of NADPH-oxidase and mitochondrial oxygen radical formation in phagocytes. Cytometry 1987; 1 (suppl):77.
73. Chen LB. Fluorescent labelling of mitrochondria. Methods Cell Biol 1989; 29:103–123.
74. Rothe G, Emmendorffer A, Oser A, Roesler J, Valet G. Flow cytometric measurement of the respiratory burst activity of phagocytes using dihydrorhodamine 123. J Immunol Methods 1991; 138:133–135.
75. Rao K, Padmanabhan J, Kilby D, Cohen H, Currie M, Weinberg J. Flow cytometric analysis of nitric oxide production in human neutrophils using dichlorofluorescein diacetate in the presence of a calmodulin inhibitor. J Leukoc Biol 1992; 51:496-500.
76. Rossi F, Bellavite P, Dobrina A, Dri P, Zabuchi G. Oxidative metabolism of mononuclear phagocytes. In: van Furth R, ed.Mononuclear Phagocytes. Part II. The Hague: Nijhoff, 1980:1187–1217.
77. Vallyathan V, Shi X, Dalal N, Irr W, Castranova V. Generation of free radicals from freshly fractured silica dust. Potential role in acute silica-induced lung injury. Am Rev Respir Dis 1988; 138:1213–1219.
78. Warheit D, Overby L, George G, Brody A. Pulmonary macrophages are attracted to inhaled particles through complement activation. Exp Lung Res 1988; 14:51–56.
79. Rinaldo J, Moore S, Lee R, Dauber J. Shift in subfractions of rat alveolar macrophages in vivo during endotoxin-induced alveolitis. J Leukoc Biol 1987; 42:230–238.
80. Henderson R, Mauderly J, Pickrell J, Hahn F, Muhle H, Rebar A. Comparative study of bronchoalveolar lavage fluid: effect of species, age, and method of lavage. Exp Lung Res 1987; 13:329–342.
81. Ettensohn D, Jankowski M, Duncan P, Lalor P. Bronchoalveolar lavage in the normal volunteer subject. Chest 1988; 94:275–280.
82. Reynolds H, Newball H. Analysis of proteins and respiratory cells obtained from human lungs by bronchial lavage. J Lab Clin Med 1974; 84:559-573.
83. Committee BCGS. Bronchoalveolar lavage constituents in healthy individuals, idiopathic pulmonary fibrosis, and selected comparison groups. Am Rev Respir Dis 1990; 141:S169–S202.
84. Brain J, Frank N. Recovery of free cells from rat lungs by repeated washings. J Appl Physiol 1968; 25:63–69.
85. Stone K, Mercer R, Gehr P, Stockstill B, Crapo J. Allometric relationships of cell numbers and size in the mammalian lung. Am J Respir Cell Mol Biol 1992; 6:235–243.
86. Adler K, Callahan L, Evans J. Cellular alterations in the alveolar wall in bleomycin induced pulmonary fibrosis in rats. Am Rev Respir Dis 1986; 133:1043–1048.
87. Gendelman H, Narayan O, Kennedy-Stoskopf S, Clements J, Pezeshpour G. Characterization of a transformed cell line of sheep alveolar macrophages. Lab Invest 1984; 51:547–55.

5

Pulmonary Intravascular Macrophages

JOSEPH D. BRAIN, RAMON M. MOLINA, and ANGELINE E. WARNER

Harvard School of Public Health
Boston, Massachusetts

I. Introduction

Not all pulmonary macrophages are alveolar macrophages (1). Macrophages exist throughout the lungs in addition to their conspicuous activity in alveolar spaces. As this book amply demonstrates, lung macrophages are present in airways, in connective tissues, in the pleural space, and finally, in pulmonary capillaries. This chapter introduces pulmonary intravascular macrophages (PIMs), an extensive population of mature phagocytic cells adherent to the pulmonary capillary endothelium in selected species. These macrophages are not prevalent in lungs of commonly studied laboratory animals, such as rodents, and thus have only been recently appreciated. However, their potential role in host defense and the pathogenesis of lung injury has attracted interest, since a number of studies have demonstrated pulmonary localization of circulating particles, microbes, and endotoxin by PIMs. This chapter introduces the reader to this interesting lung cell and tries to answer key questions about its structure, function, and role in the pathogenesis of acute lung injury, both in animal species and possibly in the human.

The unlikely event of "discovery" of an undescribed macrophage as late as the 1970s occurred in the case of the pulmonary intravascular macrophage in large part because of the species in which these cells are prevalent and the need for ultrastructural resolution to study them clearly. The animal species that have intravascular macrophages in their lungs are not those routinely studied, that is, rodents and rabbits, and it was not until chance observation of intravascular macrophages in the lungs of much larger species was confirmed by uptake studies that their presence was conclusively shown. The question of intravascular macrophages in human lungs is still a matter of debate, as will be discussed below. Light microscopic study of pulmonary capillaries does not provide adequate resolution to see macrophage cytoplasm closely adherent to endothelial cells or to determine whether mononuclear cell nuclei within capillaries are circulating cells or adherent to the capillary wall. Thus, the first observation of PIMs per se was in 1974 (2), and their phagocytic function was not studied until 1986 (3,4). Since then we have learned a great deal about the function of these unusual cells and their potential for participation in host defense against circulating pathogens and in contribution to inflammatory lung injury.

II. What Are Pulmonary Intravascular Macrophages?

PIMs are large (20–80 μm), mature macrophages located adjacent to the pulmonary capillary endothelium in selected animal species. They have been found only in pulmonary capillaries, never in larger vessels. They are a distinct population from alveolar macrophages, which reside in the air space, and currently there is no evidence supporting macrophage traffic between the capillary lumen and air space. PIMs have an indented nucleus characteristic of mononuclear phagocytes, pseudopodia that extend along the endothelial cell surface, and numerous phagosomes containing ingested material (Fig. 1). Two additional characteristics have emerged as useful criteria for definitive identification of PIMs. The first is the unusual adhesion complexes formed between the intravascular macrophage and the underlying endothelial cells. Although formation of junctional complexes between cells of epithelial and mesenchymal origin is extremely unusual, numerous such adhesion complexes can be found between an intravascular macrophage and the underlying capillary endothelial cell, and adhesion complexes have been found in all species with PIMs (5). There is increased electron density on the cytoplasmic side of the plasma membrane in both cells and an intercellular space of 12–15 nm (Fig. 2) (6). When compared with the classically described junctional complexes between epithelial cells, these PIM-endothelial complexes most closely resemble zonula adherens (7). A second distinguishing feature for PIMs is a surface coat or glycocalyx. Using normal preservation techniques, the surface coat is seen at areas of plasma membrane infolding, giving the characteristics micropinocytosis vermiformis appearance (6). Micropinocytosis vermiformis structures appear tubular with a characteristic central electron-dense lamella. The tubular profiles are continuous with the plasma membrane and the central lamella

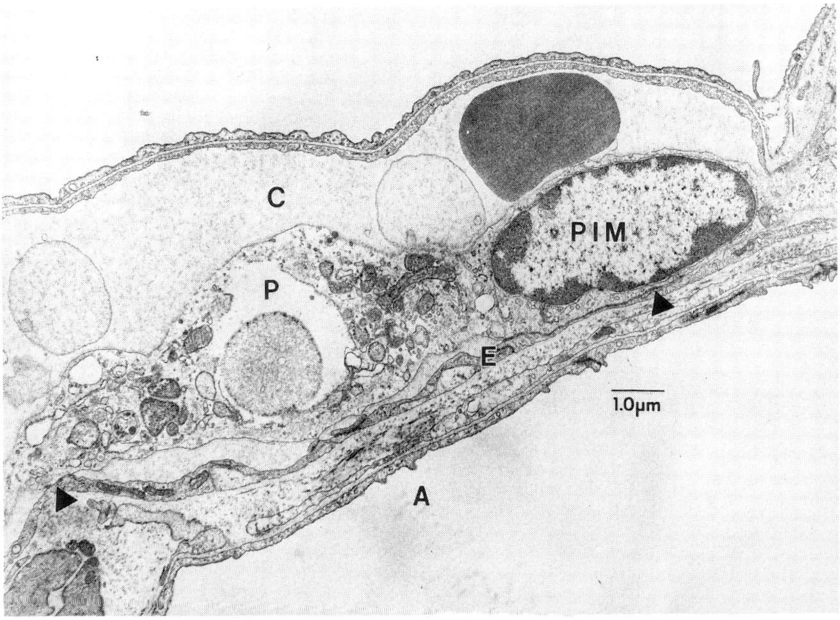

Figure 1 Pulmonary intravascular macrophage in sheep lung. A large intravascular macrophage (PIM) is shown within a pulmonary capillary (C). The macrophage contains a large phagosome (P) with ingested debris and is attached to the underlying capillary endothelial cell (E) at two adhesion complexes (arrowheads). A, alveolar space.

with the extracellular glycocalyx. When tannic acid is added to the fixation protocol, the surface coat appearance is enhanced, but it is removed by lipolytic lipase digestion, suggesting a lipoprotein composition (8). Phagosomes containing ingested debris, including partially digested erythrocytes, neutrophils, and platelets, are also characteristic of PIMs in all species where they are found (see Fig. 1).

Morphometric studies have provided some fascinating data on PIMs. First, they occupy a substantial portion of the pulmonary capillary lumen; 15.3% in sheep (6) and 25% in swine (9). In the latter species, they cover up to 16% of the capillary endothelial surface. They are located mainly on the thick side of the air-blood barrier, however, and seem positioned for minimal impediment of gas exchange (10). In mature sheep, PIMs present 16 m^2 of membrane surface available for phagocytosis of circulating particles (6). Estimates of cell density are 8×10^3 PIMs/mm^3 lung parenchyma in sheep and $14 \times 10^3/mm^3$ in pigs (10).

PIMs are apparently not prevalent at birth in pigs but rapidly colonize pulmonary capillaries after birth. By 3 days of life, monocytic cells with mitotic

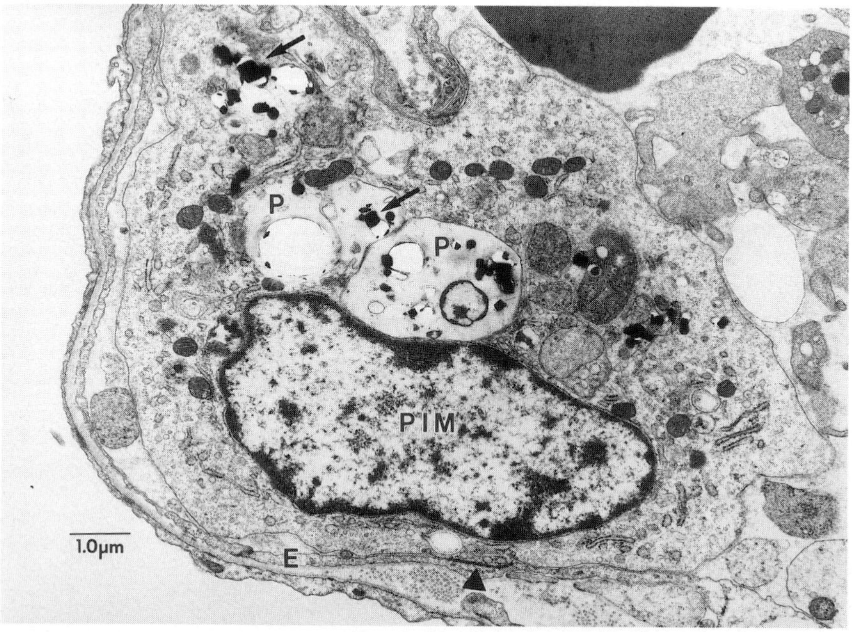

Figure 2. Pulmonary intravascular macrophage with ingested iron oxide particles in cat lung. A large macrophage (PIM) is shown with several phagosomes (P) containing ingested electron iron oxide particles (arrows). A prominent adhesion complex between the PIM and underlying endothelial cell (E) is shown (arrowhead).

figures are seen attached to the endothelium, and by 7–14 days, large macrophages with pseudopodia, phagocytic vacuoles, and endothelial adhesion complexes are present (11). Similarly, newborn lambs appear to have few if any PIMs, but by 14 days, show functional evidence of an active PIM population (12).

III. Pulmonary Intravascular Macrophage

A. Phagocytosis of Inert Tracer Particles

In all species where they have been demonstrated, PIMs are active phagocytic cells; and given the large membrane surface facing the circulation, they are poised to function as an important component of the mononuclear phagocyte system (MPS). Functional studies have borne out this speculation. In contrast to rodents, where 80–90% of injected tracer particles are engulfed by hepatic macrophages, in sheep, 99% of 0.1 μm diameter ferrimagnetic iron oxide particles and 60% of smaller (30 nm) colloidal gold particles are retained by the lungs (6). Pulmonary retention is not due to embolism of particles but to phagocytic uptake by PIMs. This has been demonstrated ultrastructurally using iron oxide particles, which are clearly within phagocytic vacuoles (see Fig. 2) (6) or using fluorescent latex beads (4). Wheeldon and Hansen-Flaschen occluded the posterior vena cava just prior to death to rule out agonal migration of hepatic or splenic macrophages into pulmonary capillaries. Similar results have been found in calves (3), swine (13), and ponies (14). A tracer substance that has been widely used in the study of PIMS is monastral blue, a copper-containing pigment that can be visualized by both light and electron microscopy. Sheep injected intravenously with monastral blue develop pulmonary hypertension, suggesting that PIM ingestion may be followed by release of vasoactive mediators (15).

These tracer experiments clearly show that in selected species, PIMs function as part of the MPS, as might be expected given their strategic location within a capillary bed that receives the entire cardiac output. PIMs can function as a very effective filter for any particulate material carried by the venous circulation, including effete erythrocytes. Erythrophagocytosis has been seen in bovine, ovine, caprine, feline, and porcine lungs (Fig. 3) (10). Both swine and cattle with bacterial respiratory infections have PIMs laden with fibrin, platelets, and occasional inflammatory cell debris (16,17). In these models of respiratory disease, the infectious agents enter the host via the airways and may or may not disseminate across the air-blood barrier. The subsequent inflammatory response, however, generates debris carried by the circulation and cleared by PIMs. The role of PIMs in host defense against these bacterial agents in animals has not yet been studied in depth.

B. Localization of Pathogens and Endotoxin

The observations described above have all been restricted to PIM uptake of inert tracer particles or cellular debris. However, PIMs have an important potential role in the clearance of circulating infectious agents and endotoxin or lipopolysaccharide (LPS). It is well accepted that hepatic Kupffer cells function in the removal of circulating pathogens and LPS that have breached the epithelial barrier of the gastrointestinal tract and are of major importance in limiting the extent of bacteremia or endotoxemia. Classic studies have shown 60–95% removal of circulating bacteria by hepatic and splenic macrophages in mice and guinea pigs (18). When species with PIMs are studied, however, the lungs emerge as an important site of pathogen removal. For example, when rats (a species without PIMS) are injected with live *Pseudomonas aeruginosa*, over 90% of bacteria recovered using quantitative culture are localized in the liver and spleen. The same injection in sheep, however, results in 93% pulmonary localization of recovered bacteria (19). Bacteria localized within sheep lungs are within phagosomes of PIMs (Fig. 4), and uptake in sheep lungs is associated with generalized micro-

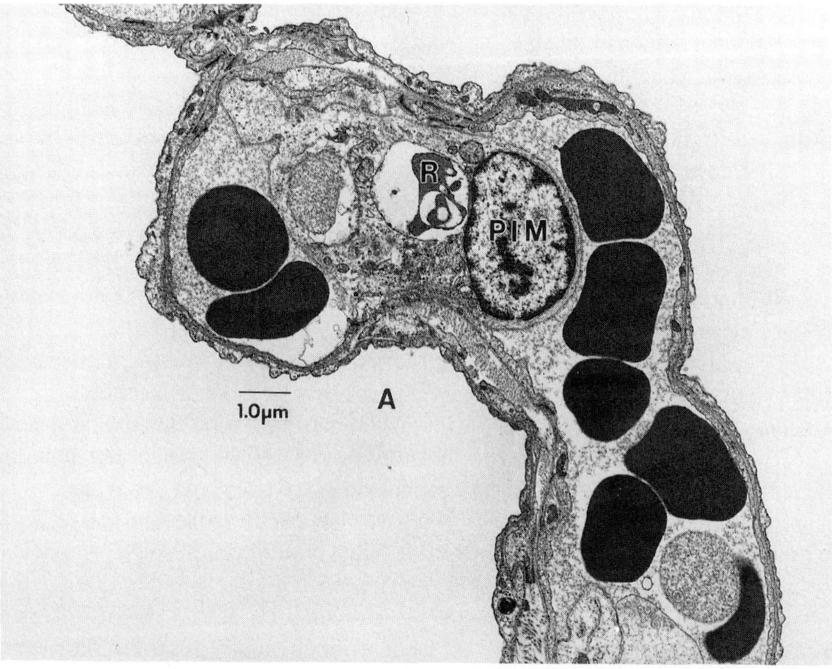

Figure 3. Sheep PIM demonstrating erythrophagocytosis. A large intravascular macrophage (PIM) is shown with a phagosome containing an ingested and partially degraded red blood cell (R). A, alveolar space.

vascular congestion, formation of fibrin microthrombi, and accumulation of neutrophils and platelets in pulmonary capillaries (Fig. 5). The injected *Pseudomonas* organisms are cleared before the development of neutropenia, confirming that neutrophils, although they accumulate in pulmonary capillaries, are not primarily involved in the localization of bacteria within the lungs. Similar studies in swine have shown clearance of 45–65% of injected *P. aeruginosa* and 75% of *Staphylococcus aureus* by the lungs (20). Pulmonary localization of the gram-negative bacteria in pigs is associated with systemic hypertension, arterial hypoxemia, and pulmonary edema. Thus, PIMs appear to be capable of the phagocytic clearance of circulating bacteria, and uptake of gram-negative bacteria is associated with local inflammatory changes.

The uptake pattern is similar when *Escherichia coli* endotoxin is injected intravenously in animal species with PIMs; preferential pulmonary sequestration of radiolabeled LPS is found in sheep, but the familiar pattern of predominant hepatic uptake is confirmed in rats (21). Once again, pulmonary localization of

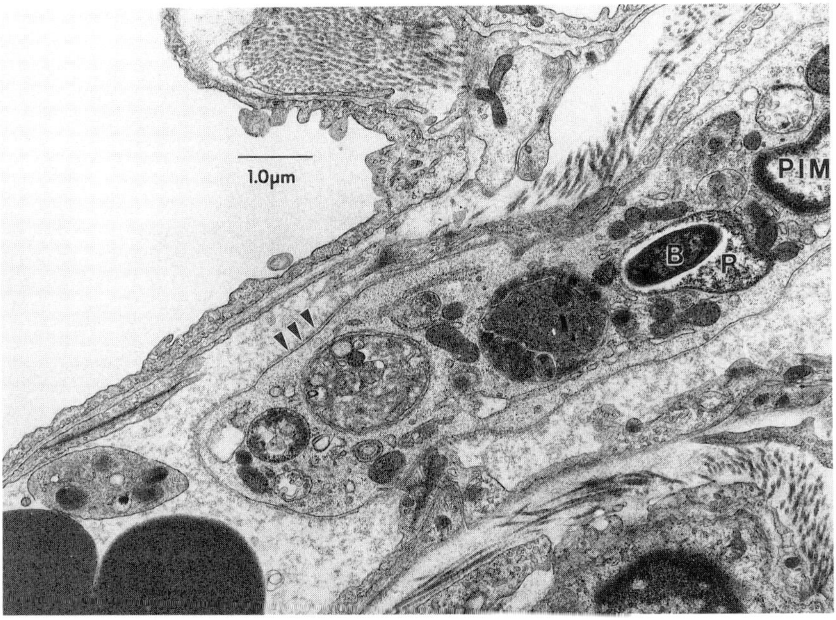

Figure 4. Sheep PIM with ingested *Pseudomonas aeruginosa*. A large intravascular macrophage (PIM) is shown with a phagosome (P) containing an ingested bacterial cell (B). The PIM surface coat is now more apparent (arrowheads), suggesting activation of the macrophage by ingestion of gram-negative bacteria.

LPS is associated with the development of local inflammatory changes and capillary endothelial injury (Fig. 6). Thus, PIM clearance of circulating endotoxin appears to take place in animal species, but although the host is protected from the effects of endotoxemia, the effects of the pulmonary localization of pathogens or endotoxin appear to be serious local inflammation, and PIMs are a likely source of the requisite mediators.

C. Source of Inflammatory Mediators

As noted above, PIM uptake of monastral blue initiates pulmonary hypertension in sheep (15). Injection of liposomes in sheep also results in transient pulmonary hemodynamic changes which can be prevented by the administration of indomethacin, implicating thromboxane as the mediator (22). In fact, large increases in the stable metabolite thromboxane B_4 can be found in arterial blood after stimulation. Newborn lambs provide a model of animals without PIMs that subsequently develop an active macrophage population. Injection studies using liposomes and

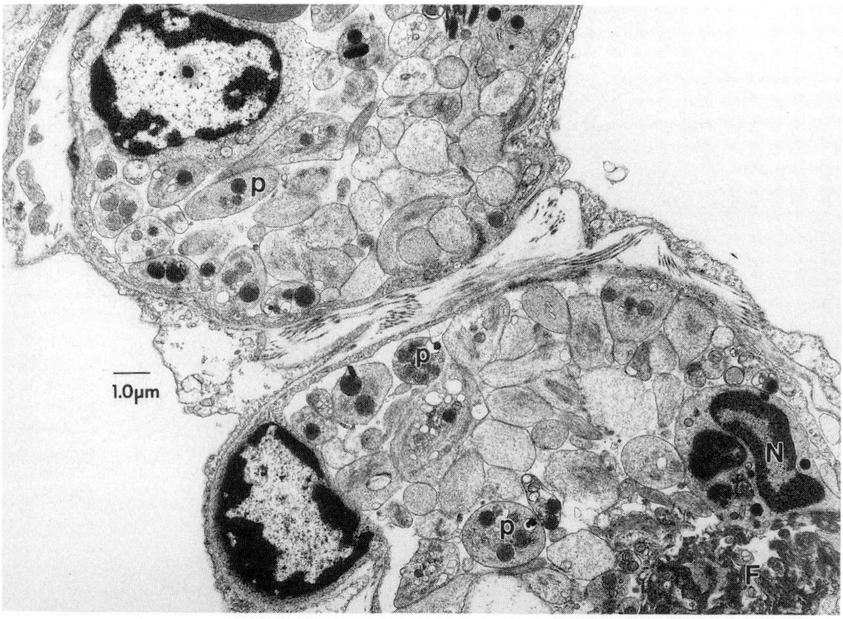

Figure 5 Sheep lung capillary following PIM localization of gram-negative bacteria in the lungs. Two pulmonary capillary loops are shown from sheep lungs 1 hr after injection of *Pseudomonas aeruginosa* that was cleared from the circulation by PIMs. Pulmonary capillaries are congested with neutrophils (N), fibrin aggregates (F), and platelet (p) microthrombi.

monastral blue at various points in the development of lambs show that the pulmonary production of thromboxane in response to injection of particles correlates temporally with the development of demonstrable pulmonary uptake of the particles (12). These two physiological studies support the hypothesis that the PIM uptake of certain particulates can lead to macrophage activation and the release of inflammatory mediators. Confirmation of this hypothesis has been particularly problematic, however, in part because of difficulty in isolation of PIMs for in vitro study, as will be discussed below.

IV. Species Distribution of Pulmonary Intravascular Macrophages

Early comparative studies showed species-specific retention patterns of intravenously injected particulates (23–26). It was shown that cats and sheep, but not

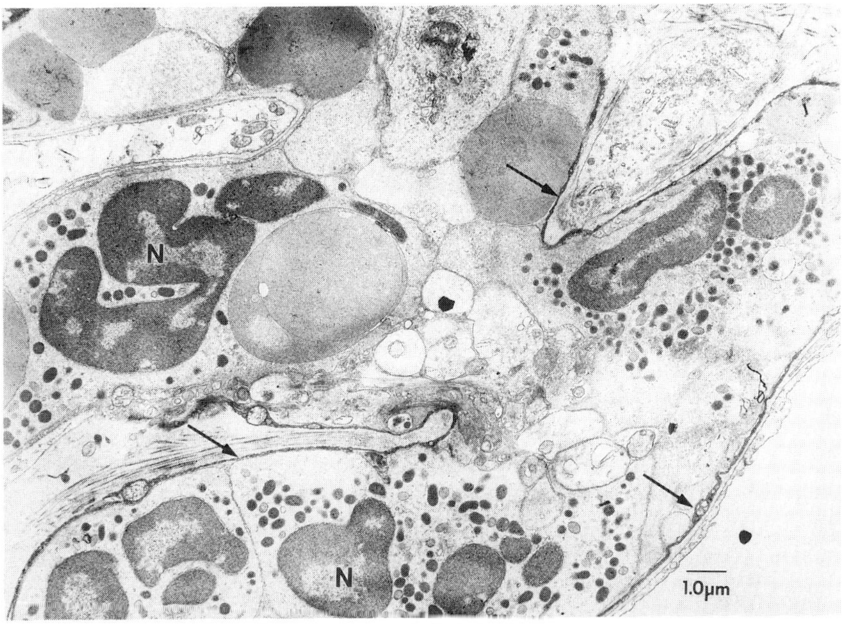

Figure 6 Sheep lung capillaries following pulmonary endotoxin localization. A pulmonary capillary is shown from a sheep lung 1 hr after injection of *Escherichia coli* endotoxin that was localized in the lungs. The capillary lumen is congested with neutrophils (N), and the endothelial cells has become electron dense (arrows), suggesting irreversible cell injury.

rabbits, had substantial lung retention of intravenously injected manganese dioxide particles in addition to typical hepatic and splenic uptake (24,25). Likewise, Hopkins and Parker (26) recovered hemolytic streptococci from the liver and spleen of cats and rabbits after intravenous injection, but significant numbers of colonies were also recovered from cat lungs. Using light microscopy, they found bacteria localized in large mononuclear cells. Because of the limited resolution of light microscopy, PIMs were interpreted as swollen capillary endothelial cells. Similar quantitative findings were reported by Drinker and Shaw (24) in cats injected with manganese dioxide. Lund et al. (25) gave a single intravenous injection of manganese dioxide to cats, rabbits, dogs, guinea pigs, chicken, rats, and turtles. Based on chemical measurement of the manganese content of different tissues, they found that the tracer particle was predominantly retained in the spleen and liver of all animals studied except for cats, which retained up to 50% of injected manganese in their lungs. Wislocki (23) intravenously injected carbon particles into sheep, cats, and rabbits. These particles were retained in the liver and spleen of all species. However, substantial amounts of carbon particles were also localized by large mononuclear phagocytes in the pulmonary capillaries of cats and sheep.

It was not until 50 years later that the presence of PIMs in calf lungs was conclusively shown using electron microscopy (2). Since then, it has been demonstrated that sheep, calves, horses, reindeer, goats, and pigs have significant pulmonary uptake of intravenously injected particles, bacteria, or endotoxin (3,6,9, 12,14,17,19,21,27,28). These studies revealed that distinct populations of resident PIMs are responsible for particle uptake. However, inevitably different investigators have used a variety of tracer particles and protocols to study the removal of particles from the circulation and to demonstrate the existence of PIMs.

Brain et al. (1,29,30) systematically surveyed 16 animal species for functional and morphological evidence of presence of PIMs. By using standard dose, particle, and protocol, they showed that sheep, calves, goats, horses, cats, and pigs localized 85% to almost 100% of injected iron oxide particles in their lungs. These particles were shown to be exclusively phagocytized by PIMs by electron microscopy. Minimal pulmonary uptake was observed in rats, dogs, baboons, monkeys, mice, hamsters, guinea pigs, hyraxes, chicken, and rabbits. Particles localized in the lungs of these species were found in monocytes or neutrophils. Thus, it appears that ruminants (sheep, calves, goats, reindeer), pigs, cats, and horses possess significant numbers of PIMs, whereas the common laboratory animals do not. The presence of PIMs in other species awaits further investigation.

V. The Role of Pulmonary Intravascular Macrophages in Lung Injury

A. Endotoxin-Induced Acute Lung Injury

The experimental studies in which sheep have been injected with gram-negative bacteria or LPS have demonstrated inflammatory changes in the pulmonary microvasculature after clearance by PIMs (19,21). Pathological changes in sheep lungs include microvascular congestion, neutrophil accumulation within pulmonary capillaries, the formation of platelet and fibrin microthrombi, and capillary endothelial cell injury. These findings suggest that localization of pathogens or endotoxin in the lungs via PIM phagocytosis may cause local pulmonary injury and also suggest that the potential role of cells of mononuclear lineage in acute inflammatory lung injury should be examined.

The most closely investigated example of acute inflammatory lung injury is the adult respiratory distress syndrome (ARDS), a frequent cause of respiratory failure seen in human patients that have experienced gram-negative sepsis, endotoxemia, major burn injury, or serious traumatic injuries (31). Typically, patients experiencing any one of these catastrophic events may go on to develop noncardiogenic pulmonary edema, decreased lung compliance, and hypoxemia, the classic signs of ARDS. Sepsis is the disorder most commonly associated with ARDS; 25–40% of the patients with the sepsis syndrome develop ARDS, and cases of ARDS with underlying sepsis have a less optimistic prognosis (31). Mortality from ARDS in this country has recently improved slightly, but it hovers at around 50% (32).

Both the clinical syndrome of ARDS and the pathological features represent a generalized response of pulmonary tissue to injury, and it is often impossible to distinguish the original insult by the pathological changes, which are termed acute lung injury. The final common pathway is severe injury to the alveolocapillary unit, with extravasation of intravascular fluid into the interstitium and alveolar space (33). This noncardiogenic pulmonary edema is subsequently followed by cellular necrosis, epithelial hyperplasia, and inflammation, termed diffuse alveolar damage. ARDS progresses histopathologically through three phases. An early exudative phase is characterized by capillary congestion, neutrophil recruitment, interstitial and alveolar edema, and the formation of hyaline membranes consisting of plasma proteins and cellular debris condensed along the alveolar surface. A proliferative phase with organization of the exudate and proliferation of fibroblasts and type 2 epithelial cells is finally followed by remodeling and collagen deposition in the fibrotic phase (33). Pulmonary vascular changes include transient vasoconstriction and the formation of platelet-fibrin microthrombi in capillaries and arterioles.

The pathogenesis of ARDS is now known to involve a combination of the effects of circulating endotoxin and complement activation, with endotoxin priming allowing amplification of complement-mediated responses (34). Neutrophils are known to play an important role in local pulmonary injury during ARDS through their capacity to release proteases, inflammatory arachidonic acid metabolites, platelet-activating factor, and toxic oxygen radicals. Recently, neutrophil-independent pathways in ARDS pathogenesis have been recognized, and macrophages have enjoyed renewed interest as potential causal cells. Macrophages have a similar armamentarium of inflammatory mediators as neutrophils plus inflammatory cytokines such as tumor necrosis factor, macrophage inflammatory proteins (MIPs) 1 and 2, and interleukin-8. Many of the hemodynamic changes in ARDS that were previously attributed to histamine and lipid mediators are now recognized as the result of macrophage-derived cytokines (34). Macrophages bind circulating endotoxin through membrane CD14, resulting in priming for the release of these mediators (35). Like neutrophils, they have the capacity for direct injury to contiguous cells. Thus, if present in the pulmonary microcirculation, macrophages are poised to play a pivotal role in the initiation of inflammatory lung injury in response to endotoxemia and in the amplification of the response through recruitment of circulating neutrophils.

The species distribution of PIMs supports the hypothesis that these macrophages play a role in endotoxin-mediated acute lung injury. They are abundant in those species that have emerged as our most productive animal models for human ARDS, swine and sheep. These species are of an appropriate size for instrumentation for physiological studies, but they are also very susceptible to endotoxin-induced acute lung injury. In fact, the endotoxin dose required experimentally to induce lung injury is much less in horses, cattle, swine, and sheep (species with abundant PIMs) than in dogs or rats (species without PIMs) (36).

B. Inactivation and Animal Models of ARDS

One experimental approach to confirm the potential role of PIMs in animal models of acute lung injury is macrophage inactivation through phagocytosis-inhibiting agents or agents that selectively destroy macrophages. Liposome-encapsulated dichloromethylene diphosphonate (clodronate) has been used transiently to deplete sheep of 80–90% of the normal PIM population (37,38). Such depletion is associated with reduced particle uptake and attenuated pulmonary vascular pressure response and lung lymph flow in response to endotoxin administration.

The detergent Tyloxapol (triton WR-1339) given intravenously is reported to inhibit PIM phagocytosis in sheep without macrophage destruction (39). Tyloxapol administration in sheep blocks the pulmonary artery pressure increase in response to endotoxin administration and blocks both the endotoxin-induced pulmonary hypertensive response and neutropenia in horses (40).

The rare earth salt gadolinium chloride has been reported to block hepatic macrophage phagocytosis and to cause macrophage destruction (41). Gadolinium chloride has been used in sheep to significantly reduce phagocytic uptake of magnetic iron oxide particles by PIMs (42) and to reduce pulmonary localization of injected gram-negative bacteria in sheep from 94 to 26% of the recovered dose (43). The pursuit of animal models of endotoxin-induced lung injury in which PIMs have been inactivated offers an exciting approach to understanding the role of these macrophages in the pathogenesis of lung injury.

C. Isolation for in Vitro Study

Confirmation of the hypothesis that PIMs can be activated to release inflammatory mediators requisite in the initiation of diffuse alveolar damage could best be approached through isolation of PIMs and in vitro study. However, these investigations have proved very difficult, owing to the unique location and endothelial adhesion of PIMs. There have been numerous attempts to flush them from the pulmonary capillaries of pigs using perfusion of the pulmonary circulation with saline, heparin solutions, chelating solutions, and collagenase-containing solutions (7,44). The mononuclear cells isolated by these procedures are smaller than alveolar macrophages and may not be mature PIMs. They do, however, release both cyclooxygenase and lipoxygenase arachidonic acid metabolites in response to calcium ionophore or LPS stimulation. A modified protocol with in situ perfusion of the pulmonary vasculature in swine (45) retrieved small mononuclear cells (mean 10.4 μm diameter) that resemble peripheral blood monocytes. Mononuclear cells isolated by perfusion from pig lungs have been used in a number of immune functional assays (46), but, once again, the isolated cells are smaller and morphologically less complex than PIMs seen in situ. The lack of a specific marker for PIMs has made such studies problematic and difficult to interpet.

A recent study by Rogers et al. (47) approached PIM isolation by proteolytic lung digestion using collagenase in a protocol similar to those used for Kupffer cell isolation from liver. In addition, the investigators prelabeled PIMs by allowing in vivo phagocytosis of injected magnetic iron oxide particles one day prior to the isolation procedure both in sheep and swine. One can argue that the phagocytic stimulus may alter the adhesion between PIMs and the pulmonary endothelium, but an objective criterion for PIM identification among cells of the digested lung (phagosomes containing visible iron oxide particles) was established. The lung digestion procedure significantly improved PIM yield when compared with perfusion isolation of similarly prelabeled cells. In addition, the magnetic iron oxide provided an objective comparison for PIM "content" of lung tissue from one animal before versus after collagenase perfusion. The iron oxide content of the lung tissue (on a per gram basis) was not significantly decreased by collagenase perfusion of the pulmonary vessels, and ultrastructural study of perfused lung showed retained PIMs with endothelial adhesion complexes intact. Thus, it would appear that proteolytic digestion of lung tissue is essential to release PIMs from the pulmonary capillaries. In addition, prior phagocytosis of an identifiable particle remains the best "marker" for isolated PIMs. In vitro studies on the macrophage population obtained after these perturbations show normal phagocytic ability and intracellular motility (48). However, in vitro studies on these isolated PIMs are open to the criticism that the cells have been substantially altered by the isolation procedure itself.

VI. Pulmonary Intravascular Macrophages in Human Lungs

A. Functional and Morphological Evidence

The question of whether intravascular macrophages are present in human pulmonary capillaries has raised some controversy. Liver scans performed with radiocolloids are a frequent diagnostic procedure and depend on mononuclear phagocytic uptake of the labeled colloid for liver imaging. The procedure is essentially equivalent to the experimental uptake studies described earlier which have been used to demonstrate the presence of PIMs in animal species. The vast majority of diagnostic liver scans performed using technetium sulfur colloid are not complicated by the pulmonary uptake of the radiocolloid, which would suggest that phagocytic uptake by macrophages in pulmonary capillaries is not a frequent occurrence in the human. There is at least one morphological study, however, that describes PIMs in human lung capillaries (49). The lung tissue obtained for this study was from patients undergoing lung biopsy, and thus there is some concern as to how representative the tissue was of normal human lung. A rigorous morphometric study of human lungs bears no report of macrophages or unusual mononuclear cells in pulmonary capillaries (50). Thus, the preponderance of information, both morphological and functional, is most consistent with absence of PIMs in the normal human lung.

B. Increased Pulmonary Uptake of Radiocolloids in Human Lungs

In horses and ponies, species with abundant PIMs, the administration of technetium sulfur colloid (TcSC) is followed by extensive pulmonary uptake of the radiocolloid and delineation of the lungs as well as liver during scintigraphy (51). It is highly likely that PIMs are responsible for TcSC uptake in equine lungs. Dogs do not normally show pulmonary uptake of TcSC, but there is a redistribution of radiocolloid away from the liver and a significant increase in relative pulmonary uptake after experimental bile duct ligation (52). These investigators speculate that the development of portosystemic shunts result in decreased hepatic blood flow, allowing colloid particles to bypass hepatic Kupffer cells and be extracted from the circulation by other compartments of the mononuclear phagocyte system; in this case in the lungs. They also note that the pulmonary uptake found suggests that fixed macrophages can become resident in the pulmonary vasculature. Thus, the pulmonary uptake of TcSC does appear to indicate the presence of phagocytic cells in the vasculature, either PIMs or some induced counterpart.

There are numerous examples in the literature suggesting that under some circumstances there is active phagocytic localization of circulating colloid particles in human lungs. Klingensmith (53) reported on 27 human patients who demonstrated a remarkable pulmonary uptake of TcSC. The majority had concurrent abnormalities in the hepatic and splenic scintigraphic images. Patients with marked lung uptake tended to have abnormal liver function tests. Keyes (54) also reported on human patients with an increased pulmonary uptake of radiocolloid during diagnostic liver scans. In these patients, improper preparation of the colloid which could have resulted in pulmonary embolization was ruled out. The majority of these patients had some type of liver disease, and many had neoplasia. In a more recent case report with enhanced TcSC pulmonary uptake, the patient had alcoholic liver disease and the investigators attribute the unusual uptake to increased pulmonary "reticuloendothelial" activity (55). Thus, there are numerous anecdotal reports of the pulmonary uptake of TcSC in human lungs. However, there has been no systematic investigation of the phenomenon.

C. Induction of PIMs

Although it appears that human lungs do not have PIMs under normal circumstances, the description of increased pulmonary localization of technetium sulfur colloid during liver-lung scans noted above suggests there may be clinical circumstances when circulating monocytes become adherent to pulmonary capillaries and differentiate into phagocytic macrophages. Experimental induction of these cells in animal species that do not normally possess PIMs provides a model for studying the induction of PIMs in human lungs and their role in lung diseases. Such models also offer a system to study mechanisms of leukocyte-endothelial cell interactions.

A single injection of endotoxin results in increased pulmonary uptake of tracer particles in rats (56) and rabbits (57), and marginated mononuclear cells are responsible for particle uptake. Chronic endotoxemia achieved by repeated injection of endotoxin to rats results in the induction of cells in pulmonary capillaries that appear to be mature macrophages with adhesion plaques characteristic of PIMs. This phenomenon has been studied further in rabbits (58). Monocytes removed during endotoxemia and radiolabeled before reinjecting into rabbits also marginated in the lungs, suggesting that prior endotoxin exposure of circulating monocytes is sufficient for margination to occur. These marginated monocytes also take up injected colloidal carbon. Induced macrophages in rats also localize injected gram-negative bacteria, suggesting the potential for increased risk for sepsis-induced lung injury and inflammation by prior endotoxemia (56).

It is now widely recognized that chronic liver disease may induce the development of PIM-like cells in animal species that do not normally have PIMs (59). Chronic liver cirrhosis has been induced in a rat model of chronic bile duct ligation, and 4 weeks after ligation, large macrophage-like cells were found in pulmonary capillaries (60). These cells are responsible for the increased pulmonary localization of injected particulates and resemble PIMs in appearance and function. Lungs of cirrhotic rats with induced PIMs have a fivefold increase in the uptake of radiolabeled endotoxin over sham controls. Moreover, their sensitivity to endotoxin-induced mortality and edema formation significantly increased over controls (60). The mechanism of PIM induction in this model is unclear but appears to be more complex than chronic endotoxemia model (59). The investigators ruled out migrating Kupffer cells as source of induced PIMs (60). Recent studies showed that these PIM-like cells likely originate from blood monocytes (61). It appears that there may be clinical situations where circulating monocytes become activated and marginate transiently or even adhere permanently in the lung capillaries. The induction of PIMs in humans has not been investigated. If this phenomenon occurs in humans, it may provide a potential mechanism for the susceptibility of patients with chronic liver diseases to development of ARDS.

VII. Conclusions

Pulmonary intravascular macrophages are phagocytic components of the MPS that reside permanently in the lung capillaries of sheep, calves, goats, pigs, horses, reindeer, and cats. They clear effete erythrocytes, inflammatory cells, platelets, fibrin, bacteria, and probably endotoxin from the circulation. When activated by phagocytic uptake of gram-negative bacteria or endotoxin, they are capable of inflammatory mediator release and thus can initiate or amplify acute pulmonary inflammation that began with the localization of pathogens in the lungs. PIMs are probably not prevalent in normal human pulmonary capillaries, but endotoxin or other types of lung or liver injury may promote adhesion of circulating mononuclear cells to the pulmonary endothelium and their differentiation into macrophages. Repeated stimulation may even promote differentiation of some adherent monocytes into resident macrophages. Such a change could put the lungs at an increased risk for the localization of pathogenic bacteria or endotoxin and consequent additional injury.

References

1. Brain JD, AE Warner, RM Molina, MM DeCamp. Pulmonary intravascular macrophages are an important part of the mononuclear phagocyte system in ruminants and cats (abstr). Am Rev Respir Dis 1988; 137:147.
2. Rybicka K, BD Daly, JJ Migliore, JC. Norman. Intravascular macrophages in normal calf lung. An electron microscopic study. Am J Anat 1974; 139:353–368.
3. Warner AE, JD Brain. Intravascular pulmonary macrophages: a novel cell removes particles from the blood. Am J Physiol 19 1986; 250:R728–R732.
4. Wheeldon EB, JH Hansen-Flaschen. Intravascular macrophages in the sheep lung. J Leukoc Res 1986; 40:657–661.
5. Warner AE, JD Brain. The cell biology and pathogenic role of pulmonary intravascular macrophages. Am J Physiol 1990; 258:L1–L12.
6. Warner AE, BE Barry, JD Brain. Pulmonary intravascular macrophages in sheep. Morphology and function of a novel constituent of the mononuclear phagocyte system. Lab Invest 1986; 55:276–288.
7. Morton D, TA Bertram. Isolation and preliminary characterization of the porcine pulmonary intravascular macrophages. J Leukoc Biol 1988; 43:403–410.
8. Atwal OS, KJ Minhas, BG Ferenczy, DS Jassal, D Milton, VG Mahadevappa. Morphology of pulmonary intravascular macrophages (PIMs) in ruminants: ultrastructural and cytochemical behavior of dense surface coat. Am J Anat 1989; 186: 285–299.
9. Winkler GC, Cheville NF. Postnatal colonization of porcine lung capillaries by intravascular macrophages: An ultrastructural, morphometric analysis. Microvasc Res 1987; 33:223–232.
10. Winkler GC. Pulmonary intravascular macrophages in domestic animal species: Review of structural and functional properties. Am J Anat 1988; 181:217–234.

11. Winkler GC, NF Cheville. Monocytic origin and postnatal mitosis of intravascular macrophages in the porcine lung. J Leukoc Biol 1985; 38:471–480.
12. Longworth KE, AM Westgate, MK Grady, JY Westcott, NC Staub. Development of pulmonary intravascular macrophage function in newborn lambs. J Appl Physiol 1992; 73:2608–2615.
13. Wismar BL, DJ Dehring, BD Lowery, CT Cloutier, LC Carey. Electron microscopic evidence of intravascular phagocytosis in the lungs of a porcine model of septic acute respiratory failure (abstr). Am Rev Respir Dis 1984; 129:A290.
14. Longworth KE, KA Jarvis, WS Tyler, EP Steffey, NC Staub. Pulmonary intravascular macrophages in horses and ponies. Am J Vet Res 1994; 55:382–388.
15. Albertine KH, NC Staub. Vascular tracers alter hemodynamics and airway pressure in anesthetized sheep. Microvasc Res 1986; 32:279–288.
16. Whiteley LO, SK Maheswaran, DJ Weiss, TR Ames. Immunohistochemical localization of Pasteurella haemolytica A1–derived endotoxin, leukotoxin, and capsular polysaccharide in experimental bovine Pasteurella pneumonia. Vet Pathol 1990; 27: 150–161.
17. Bertram TA. Intravascular macrophages in lungs of pigs infected with *Haemophilus pleuropneumonia*. Vet Pathol 1986; 23:681–691.
18. Benacerraf B, MM Sevestyen, A Schlossman. A quantitative study of the kinetics of blood clearance of P^{32}-labeled *Escherichia coli* and Staphylococcus by the reticuloendothelial system. J Exp Med 1959; 110:27–48.
19 Warner AE, RM Molina, JD Brain. Uptake of bloodborne bacteria by pulmonary intravascular macrophages and consequent inflammatory responses in sheep. Am Rev Respir Dis 1987; 136:683–690.
20. Dehring DJ, SH Crocker, BL Wismar, SM Steinberg, BD Lowery, CT Cloutier. Comparison of live bacteria infusions in a porcine model of acute respiratory failure. J Surg Res 1983; 34:151–158.
21. Warner AE, MM DeCamp, RM Molina, JD Brain. Pulmonary removal of circulating particles of circulating endotoxin results in acute lung injury in sheep. Lab Invest 1988; 59:219–230.
22. Miyamoto K, E Schultz, T Health, MD Mitchel, KH Albertine, NC Staub. Pulmonary intravascular macrophages and hemodynamic effects of liposomes in sheep. J Appl Physiol 1988; 64:1143–1152.
23. Wislocki GB. On the fate of carbon particles injected into the circulation with special reference to the lungs. Am J Anat 1924; 32:423–445.
24. Drinker CK, LA Shaw. Quantitative distribution of particulate material (manganese dioxide) administered intravenously to the cat. J Exp Med 1921; 33:77–100.
25. Lund CC, LA Shaw, CK Drinker. Quantitative distribution of particulate material (manganese dioxide) administered intravenously to the dog, rabbit, guinea pig, rat, chicken, and turtle. J Exp Med 1921; 33:231–238.
26. Hopkins JG, JT Parker. The effects of injections of hemolytic streptococci on susceptible and insusceptible animals. J Exp Med 1918; 27:1–26.
27. Staub NC, A Nicolysen, G Nicolysen. Pulmonary intravascular macrophages in reindeer (abstr). FASEB J 1992; 6:A1242.
28. Atwal OS, KA Saldanha. Erythrophagocytosis in alveolar capillaries of goat lung: ultrastructural properties of blood monocytes. Acta Anat 1985; 124:245–254.

29. Warner AE, RM Molina, JD Brain. Endotoxemia enhances pulmonary uptake of circulating particles in the baboon (abstr). Am Rev Respir Dis 1990; 141:A96.
30. Frevert CW, AE Warner, ET Adams, MW O'Callaghan. Pulmonary intravascular macrophages are an important part of the mononuclear phagocyte system in the horse (abstr). J Vet Intern Med 1991; 5:148.
31. Matthay MA. The adult respiratory distress syndrome. Clin Chest Med 1990; 11: 575–580.
32. Suchyta MR, TP Clemmer, CG Elliott, JF Orme, LK Weaver. The adult respiratory distress syndrome: a report of survival and modifying factors. Chest 1992; 101:1074–1079.
33. Tomashefski JF. Pulmonary pathology of the adult respiratory distress syndrome. Clin Chest Med 1990;11:593–619.
34. Rinaldo JE, JW Christman. Mechanisms and mediators of the adult respiratory distress syndrome. Clin Chest Med 1990; 11:621–632.
35. Watson RW, HP Redmond, D Bouchier-Hayes. Role of endotoxin in mononuclear phagocyte-mediated inflammatory responses. J Leukoc Biol 1994; 56:95–102.
36. Frevert CW, AE Warner. Respiratory distress resulting from acute lung injury in the veterinary patient. J Vet Intern Med 1992; 6:154–165.
37. Staub NC, Y. Sone. Effect of pulmonary intravascular macrophage depletion on the lung endotoxin response in anesthetized sheep (abstr). Am J Resp Crit Care Med 1994; 149:A582.
38. Staub NC, A. Nicolaysen, Y. Stone. Quantification of sheep lung intravascular macrophages after depletion using clodronate liposomes (abstr). Fed Assoc Soc Exp Biol J 1995; 9:A276.
39. Serikov V., NC Staub. The detergent triton WR-1339 blocks the pulmonary hemodynamic response to monastral blue particles in anesthetized sheep (abstr). Fed Assoc Soc Exp Bio J 1995; 9:A276.
40. Longworth KE, BL Smith, NC Couper, V Serikov, NC Staub, EP Steffey. Tyloxapol blocks early lung vascular and leukocyte responses to endotoxin in horses (abstr). Fed Assoc Soc Exp Biol J 1995; 9:A964.
41. Hardonk MJ, FW Dijkhuis, CE Hulstaert, J Koudstaal. Heterogeneity of rat liver and spleen macrophages in gadolinium chloride-induced elimination and repopulation. J Leukoc Biol 1992; 52:296–302.
42. Molina RM, AE Warner, JD Brain. Uptake of circulating particles by sheep pulmonary intravascular macrophages (PIMs) is significantly decreased by intravenous gadolinium chloride (abstr). Fed Assoc Soc Exp Biol J 1995; 9:A570.
43. Warner AE, RM Molina, MM DeCamp, JD Brain. Intravenous gadolinium chloride significantly decreases lung uptake of circulating bacteria by pulmonary intravascular macrophages (PIMs) in sheep (abstr). Am J Resp Crit Care Med 1995; 151:A764.
44. Bertram TA, LH Overby, AR Brody, TE Eling. Comparison of arachidonic acid metabolism by pulmonary intravascular and alveolar macrophages exposed to particulate and soluble stimuli. Lab Invest 1989; 61:457–466.
45. Fowler A A, PD Carey, CJ Walsh, CN Sessler, VR Mumaw, DE Bechard, SK Leeper-Woodford, BJ Fisher, CR Blocher, TK Byrne, HJ Sugerman. *In situ* pulmonary vascular perfusion for improved recovery of pulmonary intravascular macrophages. Microvasc Res 1991; 41:328–344.

46. Chitko-McKown CG, SK Chapes, RE Brown, RM Phillips, RD McKown, F Blecha. Porcine alveolar and pulmonary intravascular macrophages: comparison of immune functions. J Leukoc Biol 1991; 50:364–372.
47. Rogers RA, DR Tasat, AE Warner, JD Brain. Quantitative recovery of pulmonary intravascular macrophages from sheep lungs. J Leukoc Biol 1994; 56:692–701.
48. Warner AE, D Tasat, JD Brain. Sheep pulmonary intravascular macrophages isolated by lung digestion show normal phagocytic activity and intracellular motility (abstr). J Leukoc Biol 1992; (suppl. 3):15.
49. Dehring DJ, BL Wismar. Intravascular macrophages in pulmonary capillaries of human. Am Rev Respir Dis 1989; 139:1027–1029.
50. Zeltner TB, JH Caduff, J Gehr, J Pfenninger, PH Burri. The postnatal development and growth of the human lung. I. morphometry. Respir Physiol 1987; 67:247–267.
51. Theodorakis MC, AJ Bermudez, JP Manning, GD Koritz, CJ Hillidge. Liver scintigraphy in ponies. Am J Vet Res 1982; 43:1561–1565.
52. Koblik PD, WJ Hornof, Y Chi-Kwan, PE Fisher, J Komtebedde. Use of technetium-99m sulfur colloid to evaluate changes in retriculoendothelial function in dogs with experimentally induced chronic biliary cirrhosis and portosystemic shunting. Am J Vet Res 1995; 56:688–693.
53. Klingensmith SC, SL Yang, HN Wagner. Lung uptake of Tc-99m sulfur colloid in liver and spleen imaging. J Nucl Med 1978; 19:31–35.
54. Keyes JW, GA Wilson, JD Quinonest. An evaluation of lung uptake of colloid during liver imaging. J Nucl Med 1973; 14:687–691.
55. Stewart CA, G Hung, ME Siegel, JG McHutchison. Unusual pattern of lung uptake of technitium-99m sulfur colloid seen on the liver scan of a patient with pulmonary tuberculosis. Clin Nucl Med 1989; 14:271–274.
56. Warner AE, RM Molina, CF Bellows, JD Brain. Endotoxemia enhances pulmonary mononuclear cell uptake of circulating particles and pathogens in a species without pulmonary intravascular macrophages. Chest 1994; 105:50S-51S.
57. Cooper JA, BK McCandless, TJ Ferro, AB Malik. Pulmonary intravascular phagocytic activity in the rabbit and the sheep (abstr). Am Rev Respir Dis 1988; 137:398.
58. Ohgami M., C. M. Doerschuk, R. P. Gie, D. English, J. C. Hogg. Late effects of endotoxin on the accumulation and function of monocytes in rabbit lungs. Am Rev Respir Dis 1992; 146:190–195.
59. Chang SW, N Ohara. Pulmonary intravascular phagocytosis in liver disease. Clin Chest Med 1996; 17:137–150.
60. Chang SW, N Ohara. Chronic biliary obstruction induces pulmonary intravascular phagocytosis and endotoxin sensitivity in rats. J Clin Invest 1994; 94:2009–2019.
61. Chida M, PHS Sporn, X Wang, SW Chang. Perfusion isolation of pulmonary intravascular macrophages from rats with biliary cirrhosis (abstr). Am Rev Respir Dis 1993; 147:A307.

6

Lipid Mediator Synthesis by Lung Macrophages

MARC PETERS-GOLDEN

University of Michigan
Ann Arbor, Michigan

I. Introduction

The term *pulmonary macrophage* designates a group of mononuclear phagocytic cells located in various compartments of the lung, including the alveolar space (alveolar macrophages [AMs]), interstitial space, airway lumen, and, in certain species, intravascular space. All of these pulmonary macrophage populations ultimately originate from bone marrow–derived precursors which circulate as peripheral blood monocytes (PBMs). Since macrophages are the major resident inflammatory and immune effector cells of the lung, and since the alveoli and airways comprise the largest interface with the external environment of any organ, nowhere is the significance of macrophages more critical than in the lung. Here they are called on to deal with a myriad of inhaled particles, microbes, and antigens. The efferent limb of this task entails communication with other resident and circulating cells, which is accomplished largely by the secretion of a multitude of mediator molecules. Among the classes of mediators elaborated by pulmonary macrophages, one of the best-studied is a group of lipid metabolites derived primarily from membrane phospholipids. These include platelet-activating factor (PAF), the 20-carbon fatty acid arachidonic acid (AA), and a family of oxygenated derivatives of arachidonate, which are collectively known as eicosanoids.

Eicosanoids are synthesized from AA via several parallel series of enzyme-catalyzed reactions. The major subgroups of eicosanoids include the prostanoids (prostaglandins, prostacyclin [PGI_2], and thromboxane [Tx] A_2), derived from prostaglandin endoperoxide synthase (also known as cyclooxygenase, or COX); the leukotrienes (LTs), hydroxyeicosatetraenoic acids (HETEs), and lipoxins, derived from a group of lipoxygenases; and arachidonate epoxides, derived from the actions of cytochrome P450 enzymes. Eicosanoids are multifunctional mediators which modulate the functions of cells in both autocrine and paracrine fashions, and which have been shown to have important roles in both normal homeostasis and pathophysiological conditions. Although these substances have been under investigation since the 1960s, interest in eicosanoids and appreciation of their importance have increased substantially over the last several years. This is the result of new advances in (1) understanding the biochemistry, molecular biology, and cell biology of the enzymes involved in eicosanoid synthesis; (2) identifying many of the cellular receptors which mediate their actions in target cells; (3) recognizing their role in diverse biological and disease processes; (4) the development of improved reagents and analytical methodologies; and (5) the development of potent and specific pharmacological agents which hold promise for therapeutic applications in human disease.

The general purpose of this chapter is to review our current understanding of the biosynthesis and actions of PAF and eicosanoids, their role in pulmonary homeostasis and disease, and their interactions with other inflammatory mediator cascades. Particular emphasis will be placed on the expression, regulation, and peculiarities of these pathways in pulmonary macrophages, as well as the unique methodological and biological considerations which apply to these cells. Although macrophages participate in the processing of lipoproteins and pulmonary surfactant, these aspects of lipid metabolism will not be considered here.

II. Overview of Metabolic Pathways

Synthesis of both PAF and eicosanoids (Fig. 1) is initiated by a single common enzymatic step, cleavage of AA (or other unsaturated fatty acids) at the *sn*-2 position of membrane phospholipids (1). This is thought usually to result from the direct hydrolytic attack on the *sn*-2 acyl ester bond by the enzyme phospholipase A_2 (PLA_2). A low level of PLA_2-mediated hydrolysis occurs constitutively and continuously in all cells as a component of the process of phospholipid remodeling. Most of the fatty acids released by this process are rapidly reacylated back into lysophospholipids to regenerate intact phospholipid molecules. By contrast, generation of inflammatory mediators usually involves the *stimulated* hydrolysis of phospholipids in response to a cellular agonist. This response to agonist involves the activation of signal transduction pathways and generation of second

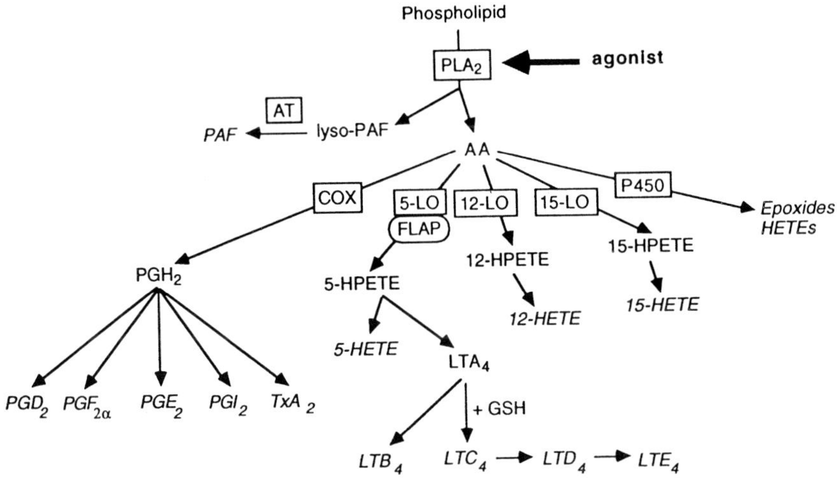

Figure 1 Pathways of lipid mediator formation. Enzymes are indicated by enclosure in rectangles, and terminal lipid mediator products are indicated by italics. AA, arachidonic acid; PAF, platelet-activating factor; PG, prostaglandin; HPETE, hydroperoxyeicosatetraenoic acid; HETE, hydroxyeicosatetraenoic acid; LT, leukotriene; GSH, reduced glutathione; PLA_2, phospholipase A_2; AT, acetyltransferase; COX, cyclooxygenase; LO, lipoxygenase; FLAP, 5-LO activating protein; P450, cytochrome P450.

messengers, which in turn activate PLA_2. It is this initial PLA_2 step which is often the rate-limiting step for lipid mediator formation. Instances in which generation of inflammatory mediators can occur in the absence of stimulated phospholipid hydrolysis have been described. These involve *inhibition* of enzymatic reacylation (2), or increases in the amounts or activities of distal enzymes without a change in available precursor (3).

Besides PLA_2 action, another factor which influences the availability of AA for conversion to eicosanoids is the relative proportion of AA among the unsaturated fatty acids esterified at the *sn*-2 position of phospholipids. Interestingly, although the mole percentage of AA within phospholipids is only ~3% in most cell types, it is 20–25% in macrophages, including AMs (4). The relative amounts of AA among *sn*-2 fatty acids are also influenced by dietary intake, and increased consumption of n-3 fatty acids which are enriched in fish oils, such as eicosapentaenoic and docosahexaenoic acids, results in decreased AA content within phospholipids and a corresponding decrease in eicosanoid synthesis (5). Lipids are known to undergo constant remodeling, with rapid movement of fatty acids among various lipid species being possible. The lipid species in which AA is found may also influence its availability for PLA_2-mediated hydrolysis (6).

The primary products of this hydrolytic reaction, lysophospholipid and AA, are the precursors for the synthesis of PAF and eicosanoids, respectively. PAF formation requires that the fatty acid at the *sn*-1 position of the parent phospholipid molecule be linked to the glycerol backbone by an ether rather than an acyl bond, and a subset of all phospholipid molecules exhibit such a linkage. Since ether-linked phospholipids are enriched in AA, PAF synthesis and AA release are thereby linked (7). The *sn*-1 ether-linked lysophospholipid (lyso-PAF) generated by PLA_2 is then acetylated by the enzyme acetyltransferase to generate PAF. Arachidonate released from the *sn*-2 position of any parent phospholipid is potentially available for bioconversion to eicosanoids via a variety of metabolic pathways, each generally referred to by the name of its initial oxygenase enzyme.

The two pathways about which the most information exists are the cyclooxygenase (COX) and the 5-lipooxygenase (5-LO) pathways (8). In both of these, the oxygenase catalyzes the synthesis of an unstable intermediate, which is further metabolized by other more distal enzymes. Thus, COX acts on AA to form the unstable endoperoxides prostaglandin (PG) G_2 followed by PGH_2, which can then be converted by a variety of isomerases to the individual bioactive prostanoids: PGE_2, $PGF_{2\alpha}$, TxA_2, PGI_2, and PGD_2. The COX enzyme has no specific cofactor requirements. It is important to note that COX is the pharmacological target for the nonsteroidal anti-inflammatory drugs; aspirin irreversibly inhibits COX by acetylating it, whereas the remainder of the agents in this class (such as indomethacin and ibuprofen) reversibly inhibit the enzyme. 5-LO is a Ca^{2+}-dependent and ATP-activated enzyme which acts on AA to form 5-hydroperoxyeicosatetraenoic acid (5-HPETE); this unstable compound can either be reduced to the hydroxy derivative (5-HETE), or it can be metabolized further by 5-LO to form another unstable intermediate, LTA_4 (9,10). Recently, a helper protein for 5-LO action, known as 5-LO–activating protein, or FLAP, has been identified. It is currently believed that FLAP is an AA-binding protein which "presents" endogenous AA liberated by PLA_2 to 5-LO. LTA_4 is then either hydrolyzed (by LTA_4 hydrolase) to LTB_4 or is conjugated with a glutathione molecule by LTC_4 synthase to form LTC_4. In some cell types, the amino acids of glutathione are sequentially removed by γ-glutamyl transpeptidase (to yield LTD_4) and then a dipeptidase (to yield LTE_4). LTs C_4, D_4, and E_4 are collectively known as the sulfidopeptide, or cysteinyl, LTs. Lipoxygenases which oxygenate AA at either carbons 12 (12-LO) or 15 (15-LO) generate 12- and 15-HETE, respectively. Sequential oxygenation by 5-LO followed by either 12- or 15-LO yields another group of products known as lipoxins (9). Various monooxygenases of the cytochrome P450 family can also act on AA to form epoxyeicosatrienoic acids (11). Recently, it has been discovered that AA can also be nonenzymatically oxygenated to form a group of compounds known as F_2-isoprostanes (12); although these appear to be excellent markers of oxidative stress, it is not yet known whether they possess any important biological actions.

Although virtually all cell types express one or more of these pathways, the specific pathways expressed, and therefore the spectrum of products synthesized, depends on the cell type. As a general rule, structural or parenchymal cell types (such as epithelial cells, endothelial cells, fibroblasts, and smooth muscle cells) metabolize AA primarily via the COX pathway, often with some modest capability for 12- or 15-LO metabolism. By contrast, bone marrow–derived cells tend to synthesize large amounts of 5-LO products, and some also form substantial amounts of COX, 12-LO, or 15-LO metabolites. A given cell type tends to produce specific COX and/or 5-LO products which are dictated by its complement of distal enzymes. Thus, the predominant COX product of endothelium is PGI_2, whereas that of epithelium is PGE_2; platelets make primarily TxA_2 and mast cells make PGD_2. The major 5-LO product of neutrophils (PMNs) is LTB_4, whereas that of eosinophils and mast cells is the cysteinyl LTs. Macrophages tend to synthesize a diverse profile of eicosanoids, including a somewhat balanced proportion of COX as well as 5-LO products (LTB_4, LTC_4, and 5-HETE) and lesser amounts of 12- and/or 15-HETE.

III. Biological Actions of Lipid Mediators

The biological effects of PAF and eicosanoids are assumed to be mediated by second messengers generated as a result of their binding to specific receptors on the plasma membrane of target cells. A consideration of the current knowledge about lipid mediator receptors is beyond the scope of this chapter, but it has been reviewed recently (13). This discussion of the biological actions of lipid mediators will emphasize those aspects about which the most is known or which are particularly relevant to the lung or to pulmonary macrophages.

The classically recognized actions of lipid mediators are summarized in Table 1. Certain lipid mediators possess actions which are unique for this class of substances. Examples of this include the effects of TxA_2 and PGI_2 to promote and inhibit, respectively, platelet aggregation, and the ability of PGE_2 to cause fever. However, many other actions are shared by a number of lipid mediators, and these common actions involve the modulation of such processes as microvascular permeability, smooth muscle tone, including that of vessels and bronchi, airway mucus secretion, and leukocyte recruitment and activation. Such redundancy is also characteristic of other inflammatory cascades such as cytokines.

It should be noted that many of these classically recognized actions of PAF and eicosanoids involve "macro"-level effects exerted on target tissues. Although these effects are undoubtedly of pathophysiological significance, a multitude of more subtle, intracellular or autocrine effects of lipid mediators have been recognized in recent years. These include the regulation of ion channels, of activities of protein kinases, and of transcription of growth-related and cytokine genes. The biological consequences of such actions are nothing less than the modulation of

Table 1 Major Biological Actions of Lipid Mediators

Lipid mediator	Platelet aggregation	Airway or vascular tone	Bronchial responsiveness	Vascular permeability	Leukocyte recruitment	Release of mediators	Fibroblast proliferation and collagen synthesis	Immune function
PGE_2		↓	↓	↑	↓	↓	↓	↓
PGI_2	↓	↓		↑	↓	↓	↓	
$PGF_{2\alpha}$		↑						
TxA_2	↑	↑						
PGD_2		↑	↑					
LTB_4			↑		↑	↑	↑	↑
5-HETE					↑	↑	↑	↑
$LTC_4/D_4/E_4$		↑	↑	↑		↑	↑	↑
PAF		↑	↑	↑	↑	↑		

such fundamental processes as cellular activation, proliferation, and differentiation (13).

Another feature of lipid mediators which is reminiscent of cytokines is the fact that the actions of certain of these molecules oppose or antagonize the actions of others. Thus, certain lipid mediators tend to promote the processes of inflammation, immune function, and fibrogenesis, whereas others suppress these processes. This yin-yang relationship is exemplified by 5-LO products and PGE_2, two classes of eicosanoids which are major products of macrophages.

LTB_4 is a potent chemotactic factor for PMNs and, to a lesser extent, eosinophils. The in vivo relevance of this action has been verified in mice whose 5-LO gene has been disrupted (14) and in the human lung following its bronchoscopic instillation (15). Although AMs can be induced to synthesize a number of potent chemotactic peptides (e.g., chemokines) in response to inflammatory stimuli, this response requires transcription and translation, and so the contribution of these molecules to leukocyte recruitment is a delayed one. In contrast, LTB_4 probably accounts for the majority of the neutrophil chemotactic activity elaborated by AMs immediately on stimulation (16). In addition to recruitment of leukocytes, LTB_4 promotes their adherence to endothelium and also stimulates them to release reactive oxygen intermediates, proteases, and other inflammatory mediators. Although less well delineated, the actions of 5-HETE appear to be similar to those of LTB_4. The cysteinyl LTs (LTC_4, D_4, and E_4) comprise the biological activity known for many years as "slow-reacting substance," reflecting their ability to evoke marked and prolonged contraction of smooth muscle in airways and vessels. They also increase airway responsiveness to bronchoconstrictors, increase microvascular permeability, and stimulate airway mucus production. In addition to these classic actions of 5-LO products, cysteinyl LTs have also been shown to promote, or play a permissive role in, the synthesis by macrophages of various cytokines and growth factors (17). LTs can also exert direct effects on fibroblasts and other mesenchymal cells; for example, they stimulate fibroblast chemotaxis, proliferation, and collagen synthesis (18). Although the fundamental intracellular mechanisms by which 5-LO products exert these proinflammatory, profibrogenic, and proproliferative effects remain to be clearly elucidated, their abilities to activate protein kinase C (PKC), increase intracellular Ca^{2+} concentrations, cause actin polymerization, and stimulate gene expression of the growth-related oncogenes *c-fos* and *c-jun* (19–21) may be pertinent.

PGE_2 has traditionally been viewed as a proinflammatory mediator, in large part based on the salutary effects of COX-inhibitory nonsteroidal agents. In fact, the application of PGE_2 does reproduce some of the cardinal manifestations of inflammation; namely, fever (reflecting its direct hypothalamic action), pain (reflecting its ability to sensitize pain fibers), and swelling (reflecting an increase in microvascular permeability). However, its effects on inflammatory and immune cells, mesenchymal cells and airways are in opposition to virtually all of the above actions of 5-LO products (22). Thus, PGE_2 inhibits leukocyte chemotaxis, adherence, and release of mediators (23), suppresses cellular and humoral immune functions (24), inhibits collagen synthesis as well as proliferation of mesenchymal cells (25), and has bronchoprotective effects in the airways (26). Many of these anti-inflammatory, immunosuppressant, and antiproliferative effects of PGE_2 seem to be mediated by the activation of adenyl cyclase, with resultant increases in cyclic AMP and activation of protein kinase A (27); inhibitory effects on the activation of mitogen-activated protein (MAP) kinase (28) may contribute to these actions.

The direct products of PLA_2 action, AA and lysophospholipids, are usually viewed solely as precursors for the bioactive metabolites eicosanoids and PAF, respectively. However, these substances are themselves increasingly recognized to possess inherent bioactivity independent of their subsequent metabolism. For example, AA promotes the release of Ca^{2+} from intracellular stores, activates PKC and MAP kinase, activates the NADPH oxidase for superoxide generation, plays a role in macrophage adherence and spreading, and stimulates heat shock protein expression (20,30). Interestingly, virtually all of these actions are shared by other unsaturated fatty acids. Lysophospholipids have long been known to possess potent detergent-like effects on cells. More recently, they too have been recognized to activate PKC (31).

Future advances in our understanding of the actions of lipid mediators are likely to revolve around three issues. The first will be the continued identification of the novel intracellular effects of precursors (e.g., AA) as well as products (e.g., eicosanoids). The second will be the characterization of specific receptors and delineation of their roles in mediating both the classic and novel effects of lipid mediators. The third is elucidation of the signaling mechanisms which link mediator receptors (or mediators themselves if they act independent of receptors) to cellular events. Fundamental investigations designed to address these issues coupled with emerging information derived from in vivo studies employing enzyme inhibitors, receptor antagonists, and transgenic animal models will enhance our understanding of the roles of lipid mediators in normal homeostasis and in disease.

IV. Key Enzymes Involved in Biosynthesis of Lipid Mediators

The last several years have seen dramatic advances in our state of knowledge about a number of the enzymes involved in lipid mediator synthesis. Many have been purified or molecularly cloned. For several of these enzymes, abundant information is now available regarding their synthetic regulation; their catalytic mechanisms, requirements, and modulation; and their intracellular compartmentalization. An unexpected degree of complexity in each of these arenas has emerged from these studies. This, in turn, has revealed many new questions which must be answered if the biological roles of each of these enzymes is to be fully understood. In this section, a synopsis of this information is presented for several key enzymes.

A. PLA_2

Among the numerous mammalian PLA_2 species which have been described, those that are the best characterized are (1) a ~14-kDa Ca^{2+}-dependent enzyme termed secretory PLA_2, or $sPLA_2$; (2) a Ca^{2+}-dependent 85-kDa enzyme termed cytosolic PLA_2 ($cPLA_2$); and (3) two Ca^{2+}-independent enzymes (reviewed in refs. 32 and 33). There are a number of examples in which a given cell type has been shown to contain multiple forms of PLA_2; $P388D_1$ macrophages (34) and AMs (35) are among these. This fact raises important questions about the roles of these various enzymes. It is likely that the particular biochemical characteristics of each of these forms of PLA_2 (such as Ca^{2+} optima and *sn*-2 fatty acid substrate preferences) provide clues as to their intracellular functions. Experimental strategies to determine the role of a given PLA_2 in stimulated lipid mediator synthesis have depended on pharmacological inhibitors with putative selectivity, transfected cells overexpressing a particular isoform, and antisense oligonucleotides to inhibit the expression of a particular isoform. Surprisingly, $sPLA_2$, $cPLA_2$, and Ca^{2+}-independent PLA_2 have all been implicated in stimulated eicosanoid synthesis as a result of studies employing these strategies. It is highly likely that the importance of a particular PLA_2 in stimulated lipid mediator synthesis will vary depending on the cell type and the agonist employed. It is also possible that distinct PLA_2s generate precursors for different lipid mediators or provide AA which serves as a precursor for eicosanoids.

At present, $cPLA_2$ is generally considered the best candidate for an intracellular enzyme which mediates agonist-stimulated phospholipid hydrolysis coupled to lipid mediator generation. This is based on its preference for substrates containing AA in the *sn*-2 position, its requirement for physiological levels of Ca^{2+}, and its phosphorylation-mediated activation by biologically relevant agonists. As its name implies, $cPLA_2$ is located predominantly in the cytosol of resting cells. On agonist stimulation, $cPLA_2$ undergoes a Ca^{2+}-dependent translocation to membranes, where it can gain access to its phospholipid substrates. It is now clear that the membrane to which $cPLA_2$ translocates is the *nuclear* membrane (36); the probable rationale for this unexpected finding will be apparent later in this section.

B. COX

The initial and rate-limiting enzyme in the metabolic pathway leading to prostanoid synthesis is COX. COX catalyzes two enzymatic reactions: a cyclooxygenase reaction, in which AA is converted to PGG_2, and a peroxidase reaction, in which PGG_2 is reduced to PGH_2 (the immediate precursor for prostanoids) (37). These reactions are catalyzed by distinct active sites in the enzyme, and it is the former which is inhibited by nonsteroidal drugs. A body of recent work (reviewed in ref. 38) indicates that there are two separate COX genes encoding antigenically distinct protein products. Although the two gene products are closely related at both the nucleic acid and amino acid levels, and both are enzymatically active, their expression, tissue distribution, intracellular localization, and regulation differ. COX-1, the originally described form of the enzyme, is constitutively expressed in a variety of tissues and its expression is usually (although exceptions exist) unaffected by exogenous agents. COX-2 is the more recently described isoenzyme; it is not generally expressed under normal conditions (although exceptions exist) but can be induced by diverse stimuli (including lipopolysaccharide [LPS]; phorbol esters; PAF; hormones; cytokines, such as IL-1; and mitogens, such as epidermal growth factor) and downregulated by glucocorticoids and certain anti-inflammatory and immunosuppressive cytokines (IL-4 and IL-10). In most instances, increased COX-2 protein involves transcriptional activation of the COX-2 gene; however, the mNRA encoding this protein is very unstable, and precedent also exists for post-transcriptional regulation.

It has been suggested that the two COX isoforms are localized to different intracellular locales, with COX-1 residing primarily in the endoplasmic reticulum and COX-2 primarily at the nuclear membrane (39). However, it is unknown whether this difference has functional implications. Given that $cPLA_2$ seems to translocate exclusively to the nuclear membrane on agonist stimulation, a metabolic enzyme which is also localized there would be ideally positioned to oxygenate the AA released at this site. If this is an important consideration in the coupling of liberated AA to PGH_2 synthesis, then COX-2 should hold an advantage over COX-1 in this regard. Data from cells transfected with the cDNA encoding either COX-1 or COX-2 are consistent with this possibility; in these experiments, both COX-1 and COX-2 transfectants were capable of converting exogenously supplied AA to PGs, but only COX-2 transfectants were able to synthesize PGs from endogenous AA, which was mobilized by agonist stimulation (40). Although it is attractive to envision that the two isoforms of COX might coexist to permit differential regulation of constitutive versus stimulated prostanoid synthesis, questions remain about the roles of COX-1 versus COX-2. For instance, the magnitude of increase in PG synthetic capacity from saturating concentrations of exogenous AA is typically far less than the magnitude of increase in COX-2 protein (41). In one study, both AMs and peritoneal macrophages (PMs) showed severalfold induction of COX-2 protein in response to LPS, yet increases in PG synthetic capacity from either exogenous or endogenous substrate were only ~10–35% (42).

It is thought that the well-known gastric and renal toxicity of nonsteroidal agents are the result of inhibition of PGs which serve homeostatic functions. The paradigm which is currently in favor holds that the anti-inflammatory effects of standard nonsteroidal agents, which inhibit both COX isoforms with essentially similar potencies, are mediated by inhibition of COX-2 at sites of inflammation, whereas their side effects are the result of inhibition of COX-1, which is constitutively expressed in the stomach and kidneys. It is therefore theoretically possible to obtain the beneficial effects of nonsteroidals without their side effects by selectively inhibiting COX-2 (43). Indeed, pharmacological agents with marked selectivity (up to 5000-fold) for inhibition of COX-2 have been synthesized, and initial studies in animal models support this hypothesis (44). However, an alternative perspective is also possible, in which induction of COX-2 at sites of inflammation can be viewed as representing a mechanism whereby the augmented production of PGE_2 could restore homeostasis by downregulating inflammation, immune function, and fibrogenesis. It is attractive to postulate such a feedback loop in the lung, since it is an organ in which PGE_2 has so many demonstrated beneficial effects, and since preservation of the lung's gas-exchange properties is of such paramount importance. In this regard, fibroblasts isolated from the lungs of patients with idiopathic pulmonary fibrosis have recently been shown to manifest a defect in induction of COX-2 with concomitantly reduced PGE_2 synthesis, as compared with normal pulmonary fibroblasts (45). In such an alternative scenario, inhibition of COX-2 might exacerbate the pathogenic process

C. 5-LO/FLAP

Unlike PLA_2 and COX, 5-LO (reviewed in ref. 10) appears to be a single gene product. The recombinant enzyme exhibits extensive sequence homology and similar catalytic mechanisms to other mammalian and even plant lipoxygenases. However, 5-LO is unique among lipoxygenases in its cofactor requirement for Ca^{2+} and its enhancement of activity by ATP. Increased levels of 5-LO protein in response to incubation with granulocyte-macrophage colony-stimulating factor have been demonstrated in PMNs, with evidence implicating both transcriptional and translational mechanisms (46). Expression of the helper protein FLAP can also be upregulated, such as by exogenous 1,25-dihydroxyvitamin D_3 (47).

An important advance in our understanding of the mechanism of 5-LO activation was provided by the demonstration that the enzyme undergoes a Ca^{2+}-dependent redistribution or translocation from its locale within a soluble intracellular compartment in resting cells to a membrane compartment following agonist activation (48). This process of translocation to membrane likely serves to bring 5-LO in proximity to its membrane-derived substrate. Recent studies have elucidated the intracellular localization of 5-LO and FLAP in both resting and activated cells, with surprising results. First of all, the membrane site to which 5-LO translocated had long been assumed to be the plasma membrane. However, experiments employing subcellular fractionation with immunoblot analysis as well as immunolocalization at both the light and electron microscopic levels have revealed in all cell types examined to date that it is instead the nuclear membrane (49–51). By these same techniques, the nuclear membrane was also shown to be the primary locale of FLAP irrespective of agonist stimulation. In stimulated cells, therefore, 5-LO and FLAP are colocalized at the nuclear membrane, along with $cPLA_2$. Thus, it would appear that, as might be true for prostanoid synthesis by COX-2, the nuclear membrane is also the site at which LT synthesis is initiated from endogenous AA. It is certainly possible, therefore, that some of the novel intracellular effects of eicosanoids reflect their synthesis at or near the nucleus. A second unexpected finding was that resting cells of different types contain 5-LO in different locales. Thus, PMNs, PMs, and PBMs all have a predominantly cytosolic localization (50,52). By contrast, AMs and mast cells contain both cytosolic and intranuclear pools of enzyme, both of which are enzymatically active in cell-free assays. Evidence also exists for agonist-induced translocation of both of these 5-LO pools (50,51).

What advantage to the cell is gained by compartmentalizing 5-LO to the nucleus? In view of the constant exposure of AMs to potential agonists in the inspired air and the potentially detrimental consequences for gas exchange of indiscriminate synthesis of proinflammatory LTs, it is teleologically attractive to speculate that AMs would evolve a means to limit activation of this enzyme. If the nuclear pool of enzyme had a higher threshold for activation, then nuclear localization might thereby enable a cell to be more "discriminating" and less "promiscuous" about the conditions necessary for LT synthesis. Such an adaptation might be similarly advantageous for mast cells, which reside adjacent to mucosal surfaces where they are exposed to large antigenic loads, but not to PMNs, which reside in a compartment (the vascular space) which is ordinarily inaccessible to potential agonists and are likely to encounter these only when they are recruited to tissue sites for the express purpose of mounting an inflammatory response. It is interesting to note in this regard that maximal LT synthesis requires 5- to 10-fold higher concentrations of Ca^{2+} ionophore in AMs than in PMNs (53). Further studies will be necessary to address this hypothesis. Other important issues which stem from these new findings about 5-LO compartmentalization and which require study include delineating the molecular mechanism by which differential compartmentalization is regulated and determining whether 5-LO or its products might have novel effects within the nucleus.

V. Methodological Considerations in Studying the Synthesis of Lipid Mediators in Macrophages

Generally, determining the effect of an experimental intervention on the pathways of lipid mediator synthesis most appropriately begins with measurements of the mediators themselves. In order to determine the precise enzymatic step which is affected, it is usually necessary to measure products derived from distinct metabolic pathways. Specific immunoassays as well as a variety of chromatographic or mass spectroscopic approaches are available for these analyses; although these will not be discussed in detail, it is essential that the investigator is aware of the conceptual and practical advantages and limitations of each. The sensitivity of chromatographic techniques can be greatly increased if the mediators being separated are radiolabeled, and this is usually accomplished by preincubating cells with tracer amounts of radiolabeled AA, which gets incorporated into cellular lipids; in response to an agonist which causes phospholipid hydrolysis, the prelabeled cells release and then metabolize both isotopic and unlabeled AA. The radiolabeled AA and products are separated, for example, by high performance liquid chromatography, and quantitated. The relative quantities of labeled products are then assumed to reflect the total mass of products released by the cell. However, there are a number of caveats about this assumption that must be taken into consideration, and the complexity of these phenomena require that data from radiolabeled cells be subjected to appropriate scrutiny (7). These concerns can be minimized by using prelabeling times long enough (16–24 hr) to allow isotope to equilibrate among the various lipid pools of the cell; however, it is usually advisable to confirm isotopic data with some independent measure of mediator mass, such as immunoassay.

Once an alteration in mediator synthesis and the enzymatic step likely to be responsible have been identified, other approaches are necessary to elucidate the precise mechanism involved. These usually include (1) an assay of catalytic activity for the specific enzyme in question, and (2) an assessment of steady-state levels of the protein. Analysis of mRNA can ascertain whether a change in a particular protein is due to a pretranslational mechanism, but this does not take the place of direct protein analysis. Comparisons between metabolic data in intact cells and these protein levels and activity assays can suggest whether the operative mechanism involves (1) a change in the steady-state level of a particular enzyme protein (2) a posttranslational modification (such as altered state of phosphorylation) which changes enzyme activity, or (3) some other mechanism, such as a change in either the amount of a cofactor (such as ATP or Ca^{2+}) or in the subcellular distribution of the protein.

Macrophages have traditionally been studied following their adherence to plastic culture dishes. However, the process of adherence itself has been shown to activate the release of reactive oxygen intermediates as well as AA and eicosanoids (54). Although the constitutive activation observed in freshly isolated and adhered cells subsides over a number of hours, an unavoidable consequence of a prolonged "cooling down" period is the possibility that by the time the cells are to be studied, they will have already lost certain of their in vivo characteristics. For example, the AM's capacity for LT synthesis declines over time in culture and has already done so to a modest extent by 24 hr (55), and this may reflect the loss of influence of some factor in the alveolar milieu (see below). One other alternative is to study macrophages cultured in Teflon containers, to the surface of which they adhere only loosely. It is clear from the foregoing discussion that there is no perfect way to study macrophages, and the most appropriate state of adherence and duration of culture must be chosen on the basis of the endpoints to be examined and on empiric determinations. Finally, if appropriate conclusions about pulmonary macrophages are to be reached from in vitro experiments, then the dangers of extrapolating from macrophage cell lines or from nonpulmonary macrophage populations must be understood. Subsequent sections will discuss in detail the differences in lipid mediator synthesis among macrophage populations.

One approach to examining macrophages which avoids most of the problems attendant to cell isolation and culture is to examine them in situ. Levels of expression of an enzyme protein may be assessed by immunohistochemical analysis (or in situ hybridization analysis of mRNA). If, on the other hand, increased product formation reflects translocation of a specific enzyme (such as $cPLA_2$ or 5-LO), immunohistochemical analysis may reveal a pattern of nuclear membrane staining, corresponding to the translocated or activated enzyme. Another advantage of these in situ approaches is that they have the potential to establish the particular cell type(s) in which changes in amounts or distribution of enzyme proteins occur.

VI. Changes in Lipid Mediator Synthesis During Macrophage Differentiation in the Lung

PBMs undergo a wide variety of morphological and functional changes as they differentiate into tissue macrophages. Some of these phenotypic changes may relate generically to the state of maturation but others are tissue specific. An example of the former is provided by the finding that in vitro differentiation of PBMs into macrophages is associated with a marked diminution in PAF synthetic capacity, which is attributed to a large increase in synthesis of the degradative enzyme PAF acetylhydrolase (56). Pulmonary AMs have been shown to differ not only from PBMs but also from other tissue macrophages in many respects, and lipid mediator generation is no exception. Differences in the intracellular localization of 5-LO have been discussed previously. Here, differences in eicosanoid profiles and the enzymatic mechanisms which account for them will be considered.

Both PBMs as well as most mature tissue macrophage populations (e.g., from peritoneum, pleura, liver, breast milk, and brain) metabolize endogenous AA to relatively greater quantities of COX products than 5-LO products (57–59). By contrast, AMs from humans as well as other species exhibit the opposite pattern (5-LO > COX) when stimulated with either particulate (zymosan) or soluble

(Ca^{2+} ionophore A23187) stimuli (58–60). With respect to their abundant capacity for LT synthesis, AMs appear to be unique among all macrophage populations. In view of the opposing biological actions of these two groups of eicosanoids, these results suggest that, at least in the "naive" state, AMs would tend to have less of a capacity to downregulate inflammatory and immune processes but a greater maximal capacity to elaborate proinflammatory molecules, including chemotactic factors. These contrasts in arachidonate metabolic patterns raise a number of questions regarding the biochemical and ontogenetic mechanisms responsible.

Two lines of evidence suggest that increases in 5-LO metabolism and decreases in COX metabolism which accompany macrophage differentiation in the lung represent primary and independent events during the differentiation of AMs. The first is that selective inhibition of either pathway in cultured macrophages fails to shunt AA down the uninhibited metabolic pathway but merely results in an accumulation of unmetabolized free AA (58). The second is that differences in the steady-state expression of proteins from both pathways have been described in AMs (see below).

Per milligram of cell protein, AMs contain severalfold more 5-LO protein (61,62) and 40-fold more FLAP (61) than do PBMs. It is likely that the difference in FLAP expression is the critical factor responsible for increased LT synthetic capacity in AMs (61). It is probable that the increases in 5-LO and FLAP proteins which accompany PBM maturation to AMs reflect the interaction of generic as well as tissue-specific signals. It has also been suggested that the increased 5-LO metabolic capacity of AMs may reflect a constitutive degree of PKC activation in these cells, but the possible mechanism(s) for this effect is not known (63). As discussed above, 5-LO is predominantly cytosolic in distribution in PBMs as well as PMs, but it is located in both nuclear and cytosolic compartments in AMs, however, it is not clear whether this difference contributes to the metabolic differences between the two.

The reduced basal COX metabolism and enzyme activity in rat AMs as compared with PMs is associated with (and is most likely due to) a ~50% lower expression of COX-1 protein in the former (42). Induction of COX-2 protein in response to LPS has been demonstrated in human as well as rodent AMs (42,64). However, although COX-2 induction in rat AMs was substantially less than that observed in autologous PMs (42), the magnitude of induction in human AMs appeared to exceed that in PBMs (64). Thus, it is not yet clear whether there exists any consistent difference between AMs and nonpulmonary macrophages with respect to inducible PG synthetic capacity and the role of specific COX isoforms therein. It should be noted that, unlike macrophages from many other sites, AMs synthesize little or no PGI_2 (detected by its stable breakdown product, 6-keto-$PGF_{1\alpha}$).

Is the altered AA metabolic pattern exhibited by AMs as compared with nonpulmonary macrophage populations a function of residence merely in the lung or in the alveolar space per se? To address this question, macrophages residing in the pulmonary interstitium (through which PBMs emigrate en route to the alveolus) were isolated from rat lungs and studied in parallel with AMs and PMs. On agonist stimulation, interstitial macrophages produced a ratio of 5-LO : COX products which most closely resembled that of the PMs (58). Thus, it is likely that most of the shift in eicosanoid profile occurs only once cells enter the alveolar space. Another piece of evidence consistent with the notion that high-level LT synthetic capacity represents a specific consequence of residence in the alveolar space is the fact that AM LT synthesis declines over several days in culture to a much greater extent than occurs in PMs or PBMs (55). Ruminant animal species contain intravascular macrophages which are distinct from PBMs, and these appear to synthesize an eicosanoid profile which is similar to that of AMs (65). The origin and differentiation of intravascular macrophages are not entirely clear.

The alveolar milieu differs from other tissue microenvironments in a number of respects. The presence of pulmonary surfactant, secreted by alveolar epithelial cells (AECs), is unique to this site. In addition, the oxygen tension of the alveolar space exceeds that of any other tissue. The alveolar milieu is further conditioned by the myraid of nonsurfactant products of AEC and of lymphocytes, many of which are potentially capable of modulating the expression or activity of eicosanoid synthetic enzymes. Although it is likely that the uniqueness of the alveolar milieu with respect to its effects on the expression of macrophage eicosanoid synthesis results from some combination of oxygen tension, surfactant, and the other mediators elaborated by AECs and lymphocytes, the relevant determinants await identification. One factor which is an attractive candidate in this regard is 1,25-dihydroxyvitamin D_3. Following incubation with this agent, PBMs exhibit a selective increase in 5-LO metabolic capacity along with an associated increase in the expression of FLAP, but not 5-LO (47), resulting in a phenotype which more closely resembles that of AMs. A physiological role for vitamin D in regulating FLAP was suggested by the demonstration in AMs obtained from vitamin D–deficient rats of reduced FLAP levels as well as reduced LT synthetic capacity (66). Thus, although the effects of 1,25-dihydroxyvitamin D_3 cannot explain the reductions in COX protein or metabolism observed in AMs, relatively high levels of this vitamin in the alveolar space could contribute to the increases observed in LT synthetic capacity by these cells.

VII. Species Differences in Alveolar Macrophage Eicosanoid Synthesis

The general profile of eicosanoids synthesized by stimulated human AMs (predominance of 5-LO over COX metabolites with intermediate amounts of 12-HETE) is, for the most part, representative of that which is observed for AMs from other species (including rat, mouse, rabbit, guinea pig, goat, and cow) (67,68).

Of the nonhuman species, the most information is available for rat AMs. These cells resemble human AMs in their profile of eicosanoids synthesized (58). In addition, rat AMs have high-level expression of 5-LO and FLAP, and they also manifest an intranuclear pool of 5-LO (51). Although mouse AMs conform to the general rule of more 5-LO products than COX products, they differ from human and rat AMs in an important way: LTC_4, rather than LTB_4, is their predominant 5-LO product (67). Considering the differences in biological actions between these two LTs, it would be anticipated that the 5-LO products of mouse AMs would have less leukocyte chemotactic activity but more smooth muscle contractile activity than would be the case for AMs from the human or rat; the indirect or intracellular consequences of this difference in LT profile must also be considered. With the growing reliance on transgenic murine models to infer biological roles of various gene products, it is essential that this fundamental difference of mouse AMs be kept in mind.

VIII. Lipid Mediator Synthesis by Macrophage Cell Lines

Because of the failure of primary isolates of macrophages, including AMs, to proliferate in culture, cells must be freshly isolated for each experiment. A number of continuous cell lines derived from macrophages of various species are available for investigative use, and proliferation of these cells in the presence of mitogens or serum offers a decided practical advantage. Unfortunately, no macrophage cell line studied to date is capable, in its unmodified state, of appreciable AA metabolism via the 5-LO pathway. However, when certain leukemic lines are induced to differentiate along either granulocytic or monocytic pathways, the capacity for LT synthesis is acquired. HL60 cells lack both 5-LO and FLAP. On differentiation along granulocytic or monocyte/macrophage pathways, these cells exhibit coordinate induction of both of these proteins and acquire the capacity for 5-LO metabolism (3). U937 cells and THP-1 cells, by contrast, were refractory to induction of LT synthetic capacity (69). Several cell lines derived from rat or murine AMs have been described; however, no data are available regarding the AA metabolite profile of any of these lines. It should be apparent from the foregoing discussion that macrophage cell lines are of limited utility as models of AA metabolism and its regulation in AMs. In addition, extrapolating data from cell lines to AMs must be done with caution.

IX. Interactions in Arachidonic Acid Metabolism between Alveolar Macrophages and Other Cell Types

Although studies with isolated cells have permitted the characterization of their individual metabolic repertoires, eicosanoid synthesis in any tissue in vivo reflects interactions among neighboring cells. One type of cell-cell interaction is known as transcellular eicosanoid synthesis. In this scenario, one cell type releases an intermediate in the AA metabolic pathway, which is taken up and processed to eicosanoids by the second cell type (70). It is clear from such studies that the products contributed by a given cell type in vivo may not be predictable based solely on its metabolic capability when studied in isolation.

Relatively few studies have examined transcellular eicosanoid synthetic interactions involving macrophages. PMs have been shown to synthesize PGs from either mast cell–derived AA (71) or platelet-derived endoperoxides (72), whereas human PBMs have been shown to donate both AA and LTA_4 to platelets (73). 5-LO products produced by PMN-AM cocultures were greater than the sum of products from each cell type alone (53); the particular interactions which were operative include AM donation of free AA to PMNs and PMN donation of LTA_4 to AMs.

Since AMs and AECs represent the two major cell types situated at the air-tissue interface of the lung, cocultures of AMs adhered on monolayers of AECs have been utilized as a model of the alveolar surface. In this model, stimulation of cocultures resulted in transcellular eicosanoid synthesis via the bidirectional transfer of free AA, with AMs synthesizing their exclusive 5-LO products from AEC-derived AA, and AECs synthesizing their exclusive PGs from AM-derived AA (74). These studies demonstrated that, as compared with AMs alone, the presence of AECs could either augment AM LT production (by providing additional AA substrate) or reduce it (by providing an additional metabolic "sink"). The ability of a given cell type to function as an AA acceptor in this system was determined by its capacity for oxygenation, whereas its ability to serve as a fatty acid donor depended on its ability to release an amount of AA which exceeded its own capacity for oxygenation. Another conclusion from these studies was that transcellular interactions could account for the apparent shunting to lipoxygenase products which has been documented in isolated organs or in vivo following treatment with nonsteroidal agents.

X. Agonist Specificity of Eicosanoid Synthesis

A partial list of agonists which have been shown to rapidly trigger AA release and metabolism in various cell types is found in Table 2. These "complete agonists" must be distinguished from certain other agents which can increase eicosanoid synthesis only after a lag period of several hours by increasing the expression of various enzymes in the pathway, but they cannot rapidly increase the availability of AA; an example of the latter is IL-1. Some of these complete agonists are physiologically or pathophysiologically relevant (e.g., hormones, microbes, antigens, and oxidants), whereas some are merely model agents which have experimental utility (e.g., Ca^{2+} ionophore or phorbol esters). Some effect lipid metabo-

Table 2 Representative Agonists Which Activate Macrophage Lipid Mediator Generation

Soluble "model" agonists	Microorganisms
Ca^{2+} ionophores	Parainfluenza virus
phorbol esters	*Pneumocystis carinii*
Receptor-mediated agonists	*Staphylococcus aureus*
thrombin	*Pseudomonas aeruginosa*
bradykinin	*Candida albicans*
interleukin-8	*Toxoplasma gondii*
complement	Phagocytic particles
fMLP	zymosan
HIV gp120	silica
endothelin	asbestos
immune complexes	carbonyl iron
leukotrienes	Oxidants
PAF	hyperoxia
	NO_2
	ozone
	t-butyl hydroperoxide
	H_2O_2

lism by direct actions on cellular components (e.g., ionophore, phorbol esters, oxidants, certain pharmaceuticals, or xenobiotics), whereas others do so as a consequence of receptor binding and signal transduction (hormones, complement, zymosan [a preparation of killed yeast cell walls], antigens). Agonists in the former group generally act on all cell types, whereas those in the latter group only act on cells which contain the appropriate receptor.

It is important to recognize that AA metabolism is modulated by participants in many other inflammatory cascades. These substances include cytokines, reactive oxygen metabolites, nitric oxide, and proteases. A particularly complex relationship exists between reactive oxygen metabolites and eicosanoids. Both are integral products of activated macrophages. Oxygen metabolites are produced during oxygenation of AA. They also modulate the activity of various enzymes in the pathways of lipid mediator synthesis, being capable of activating PLA_2 (75), inhibiting AA reacylation (2), and activating (at low concentrations) or inhibiting (at high concentrations) the oxygenase enzymes themselves (37). Finally, oxidant species may play an obligate role in the induction of COX-2 (76). Since eicosanoids, in turn, influence the release of many of these same other inflammatory mediators, it is clear that AA products occupy a central position in a multitude of cellular responses. It should also be noted that certain lipid mediators can themselves modulate AA release and the synthesis of other lipid mediators. Thus,

LTs can directly stimulate the formation of PGs, PAF can stimulate the formation of eicosanoids, and PGE_2 can inhibit the release of AA and the synthesis of LTB_4.

It was recognized more than a decade ago that different agonists could result in the production of different eicosanoids by a particular macrophage population (77), and this was subsequently found to be the case for AMs as well (78). Thus, ionophore and zymosan elicit the production of 5-LO and COX metabolites, whereas LPS and phorbol ester elicit primarily COX metabolites. The explanation for these differences remains to be determined. It has been suggested that these different agonists release AA from distinct intracellular pools, which are differentially coupled to one oxygenation pathway or the other. However, these "pools" of AA have been distinguished only on functional grounds, and the nature of such distinctions have not heretofore been defined. It is now possible to speculate though that different agonists might release different pools of AA by activating distinct isoforms of PLA_2. If different PLA_2s effect phospholipid hydrolysis at different membrane sites (e.g., $cPLA_2$ at the nuclear membrane and $sPLA_2$ at the plasma membrane), then the potential for the synthesis of particular eicosanoids could depend on the topographic proximity between enzymes effecting AA release and metabolism. This possibility remains to be evaluated.

Another possible explanation for the fact that some agonists fail to trigger the synthesis of LTs might be that they do not cause a sufficient increase in intracellular Ca^{2+}. It has been demonstrated that the Ca^{2+} threshold for LT synthesis in intact cells exceeds that for AA release (79), and this may reflect the higher Ca^{2+} requirement for 5-LO than for $cPLA_2$ activity in cell-free assays. In addition to Ca^{2+}, other second messengers generated in response to agonist stimulation could also influence whether LT synthesis occurs. For example, PKC activation could augment 5-LO metabolism and increases in cyclic AMP could inhibit it.

In addition to activating AA release, some agonists concomitantly alter the concentrations of small molecules which serve as substrates or cofactors for certain enzymes in the arachidonate cascade; this results in an altered profile of eicosanoids produced. For example, reduced glutathione is both a cofactor for PGE isomerase and a substrate for LTC_4 synthase. Various oxidants and xenobiotics trigger AA release in AMs but also decrease intracellular concentrations of reduced glutathione. As a result, there is a relative inability to synthesize PGE_2 and LTC_4 and a shunting of PGH_2 and LTA_4 to alternative eicosanoids (80). Hydrogen peroxide, by contrast, triggers AA release in AMs, but because it concomitantly depletes ATP, it elicits the selective synthesis of prostanoids (81).

XI. Role of Macrophage-Derived Eicosanoids in Lung Disease and Homeostasis

The determination that a given mediator participates in the pathogenesis of a disease process is based on the fulfillment of three criteria modified from Koch's postulates for microbial pathogenesis: (1) its administration to experimental subjects reproduces the pathophysiological features of the disease process; (2) it can be identified in appropriate body fluids (or it is synthesized by relevant cells) obtained from individuals with the disease; and (3) its antagonism or inhibition by pharmacological, immunological, or genetic strategies ameliorates the manifestations of the disease. It has been convincingly demonstrated that the intrapulmonary administration of lipid mediators to animals and to humans causes a variety of pathophysiological responses. However, substantial difficulties have complicated attempts to fulfill the last two criteria. For example, is the relevant body compartment being sampled? If so, might enzymatic or nonenzymatic metabolism, or even transport of these lipid-soluble substances between compartments (82), compromise the degree to which measured levels reflect the quantities of eicosanoids synthesized? These pitfalls have been only partially circumvented by the development of methods for analysis of urinary metabolites of eicosanoids. In addition, we have witnessed many examples in which ultimately erroneous conclusions about the role of eicosanoids were drawn from experiments utilizing drugs which, in retrospect, were inadequately specific or potent to justify such confidence. Agents meeting these standards have been available only recently. Together with the development of transgenic animal models, these are providing new insights into the clinical importance of eicosanoids. Abnormalities of eicosanoid production have been implicated in numerous disease processes involving the lungs, as listed in Table 3. These include both human diseases as well as animal models of disease involving airway inflammation, acute lung injury, interstitial fibrosis, and defects of host defense.

The yin-yang relationship between 5-LO products and certain prostanoids such as PGE_2 provides a useful framework for summarizing the roles of eicosanoids in lung homeostasis and pathophysiology. More information is available for asthma and models of airway inflammation than any other condition. As previ-

Table 3 Lung Diseases in Which Abnormalities of Lipid Mediator Production Are Implicated

Airway diseases	Vascular disorders
asthma	pulmonary hypertension
chronic bronchitis	pulmonary thromboembolism
cystic fibrosis	Host defense abnormalities
Acute lung injury	HIV infection
ARDS	protein-calorie malnutrition
oxygen toxicity	smoking
Interstitial lung disease	
idiopathic pulmonary fibrosis	
asbestosis	
sarcoidosis	

ously described, the actions of LTs promote airway inflammation, mucus hypersecretion, and bronchoconstriction, whereas those of PGE_2 have opposite effects. Increased levels of cysteinyl LTs (or their metabolites) have been identified in sputum, BAL fluid, plasma, and urine during spontaneous episodes of asthma as well as following challenge with known bronchoconstricting stimuli (83). Moreover, 5-LO enzyme inhibitors and LTD_4 receptor antagonists have proven to be surprisingly effective in the acute and chronic treatment of asthma (84). Conversely, a number of studies have demonstrated that aerosol administration of PGE_2 is bronchoprotective in asthmatics (26). A similar paradigm is likely to be operative in the pulmonary parenchyma, in which overproduction of LTs (85,86) and underproduction of PGE_2 (45,87) may be implicated in fibrotic lung diseases.

The ready availability of lavageable AMs has facilitated an examination of their capacity for eicosanoid generation in a number of animal models and human disease processes. There are theoretical reasons why AMs would and also would not be expected to contribute in a meaningful way to the LT overproduction which characterizes asthma. Arguing against a role for AMs is the fact the cysteinyl LTs are only minor products of these cells, at least under usual circumstances in vitro. On the other hand, AMs are located on the airway surface, produce substantially more LTC_4 than do PMNs, and share with mast cells the presence of plasma membrane IgE receptors. Director comparisons of the in vitro LT synthetic capacities of AMs from asthmatics versus normal subjects have, in fact, yielded conflicting results (88,89).

AMs from patients with both idiopathic pulmonary fibrosis and asbestosis have, however, been shown to constitutively elaborate increased amounts of LTs (86,90). Constitutive overproduction of LTB_4 has also been observed by AMs from normal subjects exposed to 50% oxygen for a mean of 44 hr, as compared with air-exposed subjects (91). However, the design and interpretation of such studies are fraught with difficulties related to the questions of whether the process of AM isolation has itself contributed to any alterations in eicosanoid synthesis which might be observed, whether the observed alterations are present constitutively or only following addition of an exogenous agonist, and whether the duration or conditions of ex vivo culture have been sufficient to render the results of in vitro metabolic studies unrepresentative of the metabolic activity of these cells in vivo. A fundamentally different approach to determining that the AM 5-LO enzyme is activated in a particular disease process, which avoids the above limitations, capitalizes on the findings from in vitro studies that the intracellular localization of the enzyme differs in resting and agonist-stimulated cells. Using immunohistochemical staining with an anti–5-LO antibody of sections of lung tissue from patients with idiopathic pulmonary fibrosis, it has recently been possible to identify an "activated" distribution of enzyme (i.e., staining concentrated at the nuclear membrane) in a proportion of stained AMs, whereas in histologically normal lung tissue, a "resting" distribution of 5-LO (i.e., diffuse intranuclear > cytoplasmic staining) in AMs is the norm (86). Further studies will be necessary to determine whether this approach will indeed be useful in other disease processes and in other cell types.

If PMN chemotactic factors derived from the AM 5-LO pathway are important for antimicrobial host defense, then a *deficient* capacity of AMs to produce LTs may also be clinically relevant. Studies of this nature have in fact demonstrated a relative defect in AM LT synthetic capacity in both animal models and human conditions which are known to be associated with increased risks of pneumonia. These include protein-calorie malnutrition (92) and vitamin D deficiency (66) in animals and cigarette smoking (93,94) as well as human immunodeficiency virus infection in humans (95).

In view of its anti-inflammatory, antifibrogenic, immunosuppressive, and bronchoprotective effects, a deficiency in PGE_2 synthesis might likewise be expected to have clinical implications. There is, in fact, evidence for deficient PGE_2 production by pulmonary fibroblasts (45) and by airway epithelial cells (96) in idiopathic pulmonary fibrosis and in an animal model of asthma, respectively. Although AMs may be less important a pulmonary source of this molecule than these other cell types, their capacity to produce PGE_2 has also been shown to be reduced as compared with cells from normal subjects in subjects with idiopathic pulmonary fibrosis (97), sarcoidosis (98), and a history of cigarette smoking (94). The decreased AM capacity for PGE_2 synthesis could conceivably contribute to the inflammation, enhanced immune function, or fibrosis which characterizes these conditions.

XII. Therapeutic Strategies to Alter the Synthesis or Actions of Lipid Mediators

Given the importance of lipid mediators in normal homeostasis and in the pathogenesis of a variety of disease processes, modulation of their tissue concentrations or actions might be exploited for therapeutic purposes. The generic strategies which can be considered are (1) administration of a potentially beneficial substance (or overexpression of the gene encoding it), (2) inhibition of the synthesis of a potentially deleterious substance, and (3) inhibition of the actions of a potentially deleterious substance. The best example of the former is inhalation of PGE_2 or of stable analogues thereof. Such a strategy has proven feasible as replacement therapy for the deficient endogenous production of PGE_2 (in patients with idiopathic pulmonary fibrosis) (87) and as purely supplemental treatment designed to achieve pharmacological effects (in asthma) (26). Another possible means to effect overproduction of beneficial prostanoids such as PGE_2 or PGI_2 is transfection of the cDNA encoding COX; for example, overexpression of COX-1 by liposome-mediated transduction in the pulmonary vascular endothelium of rabbits was shown to protect their lungs from LPS-induced pulmonary vasoconstriction and edema (99). The feasibility of applying a similar approach to humans with diseases of the airway, vasculature, or parenchyma remains to be established. Possibilities for inhibiting the synthesis or actions of potentially deleterious lipid mediators have received far more attention. The remainder of this section discusses potential strategies and pharmacological targets by which this might be accomplished, the current status of investigational agents which do so, and the effects of currently available pharmacological agents whose mechanisms of action might include the interruption of lipid mediator generation.

In considering possible strategies for inhibiting the synthesis or actions of lipid mediators, the relative merits of one which is highly selective for a single mediator versus one which inhibits more than one member of this group must be weighed. Agents which target a single mediator include specific receptor antagonists, such as those available for the LTD_4 receptor. Although it is theoretically possible to design agents which inhibit the synthesis of a single eicosanoid, such as TxA_2 or LTC_4, the development of such agents has been hampered by our relatively rudimentary understanding of these most distal enzymes.

The arachidonate oxygenases represent targets of intermediate selectivity. Inhibitors of 5-LO and its helper protein FLAP are currently in various stages of development, and these agents appear to show great promise in clinical trials of asthma. Since 5-LO–derived LTs, 5-HETE, and lipoxins all possess proinflammatory actions, an agent which inhibits all of these products might exhibit superior efficacy over one which inhibits only one of them. One theoretical concern is whether chronic 5-LO inhibition will be accompanied by impaired host defense capabilities. Nonsteroidal agents which inhibit both COX isoforms have, of course, been available for many years, and selective COX-2 inhibitors are under active development. The fact that COX metabolites include products which are potentially deleterious (e.g., TxA_2 and PGD_2) as well as those which are potentially salutory (e.g., PGE_2 and PGI_2) may explain some of the adverse effects of COX inhibitors, such as aspirin-induced asthma.

Finally, the least selective strategy would be one which interferes with the release of AA from membrane lipids. The attractiveness of this strategy derives from its potential to interfere with the synthesis of PAF as well as all eicosanoids. This could be accomplished either by the inhibition of PLA_2-mediated deacylation or by reducing the arachidonate content of lipids via dietary modifications of fatty acid intake. A major theoretical limitation to any strategy which reduces the levels of available free AA is that synthesis of PGE_2 and PGI_2 would be expected to be reduced along with potentially deleterious eicosanoids. Even this concern could be overcome, however, if it were determined that there was a particular species of PLA_2 which provided AA for 5-LO metabolism but not COX metabolism and which could be selectively targeted for inhibition.

The anti-inflammatory actions of glucocorticoids can be ascribed to their ability to downregulate both the production of a variety of mediators as well as certain cellular responses. One important effect of glucocorticoids is their ability to inhibit cellular AA release and eicosanoid synthesis. The inhibition of AA release appears to depend on new protein synthesis, and one glucocorticoid-induced protein which has been implicated in mediating this response is lipocortin-1 (although it is a matter of some controversy whether lipocortin is indeed responsible, and if so, whether it acts by inhibiting PLA_2 directly or by sequestering its phospholipid substrates). In vitro pretreatment of AMs with glucocorticoids has been demonstrated to inhibit AA release as well as the synthesis of prostanoids and LTs and also to induce lipocortin-1 (100). Moreover, the administration of prednisone to normal and asthmatic human subjects has been shown to inhibit the subsequent capacity of AMs to release eicosanoids when stimulated in vitro (101,102). Interestingly, eicosanoid synthesis by blood leukocytes from these same individuals was not inhibited (101,102), and this correlated with lipocortin-1 induction in AMs but not blood leukocytes. Taken together, these data strongly suggest that arachidonate release by AMs is an important target for the actions of glucocorticoids in vivo. In any case, like other agents which modulate the availability of AA, the fact that glucocorticoids ultimately reduce the synthesis of PGE_2 along with that of LTs and PAF may compromise their efficacy. In this regard, if COX-2 induction is viewed as a protective homeostatic response to inflammation, the ability of glucocorticoids to potently inhibit its induction could further limit the amounts of PGE_2 present at sites of inflammation. It remains to be determined whether these inhibitory effects of steroids on prostanoid synthesis might account for therapeutic failures in certain disease states.

There are a number of other pharmacological agents which are in current use for various indications and which have the potential to inhibit the generation of lipid mediators. These include nedocramil (88), β-adrenergic agonists and theophylline, colchicine (103), the oral gold compound auranofin (104), and the antifungal agent ketoconazole (105). The ultimate importance of inhibition of lipid mediator synthesis in the actions of these agents remains to be established.

XIII. Conclusions

A multitude of recent advances in the basic as well as clinical arenas have strengthened our appreciation of the importance of lipid mediators in lung physiology and pathophysiology. The likelihood is very high that in the near future some of these advances will be translated into new treatment options which will be applicable to a variety of lung diseases. It is equally likely that novel insights into the previously unsuspected cellular effects of lipid mediators and interactions with other mediator cascades will broaden our understanding of cellular function and pave the way for a subsequent generation of therapeutic advances.

References

1. Irvine R. How is the level of free arachidonic acid controlled in mammalian cells? Biochem J 1982; 204:3–16.
2. Sporn P, Marshall T, Peters-Golden M. Hydrogen peroxide increases the availability of arachidonic acid for oxidative metabolism by inhibiting acylation into phospholipids in the alveolar macrophage. Am J Respir Cell Mol Biol 1992; 7:307–316.
3. Kargman S, Rouzer CA. Studies on the regulation, biosynthesis, and activation of 5-lipoxygenase in differentiated HL60 cells. J Biol Chem 1989; 264:13313–13320.
4. Mason R, Stossel T, Vaughn M. Lipids of alveolar macrophages, polymorphonuclear leukocytes, and their phagocytic vesicles. J Clin Invest 1972; 51:2399–2407.
5. Lefkowith J, Jakschik B, Stahl P, Needleman P. Metabolic and functional alterations in macrophages induced by essential fatty acid deficiency. J Biol Chem 1987; 262: 6668–6675.
6. Triggiani M, Oriente A, Marone G. Differential roles for triglyceride and phospholipid pools of arachidonic acid in human lung macrophages. J Immunol 1994; 152: 1394–1403.
7. Chilton F, Connell T. 1-Ether-linked phosphoglycerides: major endogenous sources of arachidonate in the human neutrophil. J Biol Chem 1988; 263:5260–5265.
8. Needleman P, Turk J, Jakschik BA, Morrison AR, Lefkowith JB. Arachidonic acid metabolism. Annu Rev Biochem 1986; 55:69–102.
9. Samuelsson B, Dahlen S-E, Lindgren J, Rouzer C, Serhan C. Leukotrienes and lipoxins: structures, biosynthesis, and biological effects. Science 1987; 237:1171–1176.
10. Ford-Hutchinson A, Gresser M, Young R. 5-Lipoxygenase. Annu Rev Biochem 1994; 63:383–417.
11. Capdevila J, Falck J, Estabrook R. Cytochrome P450 and the arachidonate cascade. FASEB J 1992; 6:731–736.
12. Morrow J, Awad J, Boss H, Blair I, Roberts L 2d. Non-cyclooxygenase-derived prostanoids (F2-isoprostanes) are formed in situ on phospholipids. Proc Natl Acad Sci USA 1992; 89:10721–10725.
13. Goetzl E, An S, Smith W. Specificity of expression and effects of eicosanoid mediators in normal physiology and human diseases. FASEB J 1995; 9:1051–1058.
14. Chen X, Sheller J, Johnson E, Funk C. Role of leukotrienes revealed by targeted disruption of the 5-lipoxygenase gene. Nature 1994; 372:179–182.
15. Martin T, Pistorese B, Chi E, Goodman R, Matthay M. Effect of leukotriene B_4 in the human lung: recruitment of neutrophils into the alveolar spaces without a change in protein permeability. J Clin Invest 1989; 1989:1009–1019.
16. Martin T, Raugi G, Merritt T, Henderson W Jr. Relative contribution of leukotriene B_4 to the neutrophil chemotactic activity produced by the resident human alveolar macrophage. J Clin Invest 1987; 80:1114–1124.
17. Phan S, McGarry B, Loeffler K, Kunkel S. Regulation of macrophage-derived fibroblast growth factor release by arachidonate metabolites. J Leukoc Biol 1987; 42:106–113.
18. Phan S, McGarry B, Loeffler K, Kunkel S. Binding of leukotriene C_4 to rat lung

fibroblasts and stimulation of collagen synthesis in vitro. Biochemistry 1988; 27: 2846–2853.
19. Peppelenbosch M, Tertoolen L, Hage W, de Laat S. Epidermal growth factor-induced actin remodeling is regulated by 5-lipoxygenase and cyclooxygenase products. Cell 1993; 74:565–575.
20. Pace J, Hayman M, Galan J. Signal transduction and invasion of epithelial cells by S. typhimurium. Cell 1993; 72:505–514.
21. Stankova J, Rola-Pleszczynski M. Leukotriene B_4 stimulates *c-fos* and *c-jun* gene transcription and AP-1 binding activity in human monocytes. Biochem J 1992; 282: 625–629.
22. Goodwin J. Are prostaglandins proinflammatory, antiinflmmatory, both or neither? J Rheumatol 1991; 18(suppl 28):26–29.
23. Ham E, Soderman D, Zanetti M, Dougherty H, McCauley E, Kuehl F Jr. Inhibition by prostaglandins of leukotriene B4 release from activated neutrophils. Proc Natl Acad Sci USA 1983; 80:4349–4353.
24. Tripp C, Wyche A, Unanue E, Needleman P. The functional significance of the regulation of macrophage Ia expression by endogenous arachidonate metabolites in vitro. J Immunol 1986; 137;3915–3920.
25. Bitterman P, Wewers M, Rennard S, Adelberg S, Crystal R. Modulation of alveolar macrophage-driven fibroblast proliferation by alternative macrophage mediators. J Clin Invest 1986; 77:700–708.
26. Pavord I, Wong C, Williams J, Tattersfield A. Effect of inhaled PGE_2 on allergen-induced asthma. Am Rev Respir Dis 1993; 148:87–90.
27. Saltzman L, Moss J, Berg R, Hom B, Crystal R. Modulation of collagen production by fibroblasts: effects of chronic exposure to agonists that increase intracellular cyclic AMP. Biochem J 1982; 204:25–30.
28. Li X, Zarinetchi F, Schrier R, Nemenoff R. Inhibition of MAP kinase by prostaglandin E_2 and forskolin in rat renal mesangial cells. Am J Physiol 1995; 269:C986–C991.
29. Sumida C, Graber R, Nunez E. Role of fatty acids in signal transduction: Modulators and messengers. Prostaglandins Leukot Essent Fatty Acids 1993; 48:117–122.
30. Jurivich D, Sistonen L, Sarge K, Morimoto R. Arachidonate is a potent modulator of human heat shock gene transcription. Proc Natl Acad Sci USA 1994; 91:2280–2284.
31. Asaoka Y, Oka M, Yoshida K, Sasaki Y, Nishizuka Y. Role of lysophosphatidylcholine in T-lymphocyte activation: involvement of phospholipase A_2 in signal transduction through protein kinase C. Proc Natl Acad Sci USA 1992; 89:6447–6451.
32. Dennis E. Diversity of group types, regulation, and function of phospholipase A_2. J Biol Chem 1994; 269:13057–13060.
33. Mayer R, Marshall L. New insights on mammalian phospholipase A_2(s): Comparison of arachidonoyl-selective and -nonselective enzymes. FASEB J 1993; 7:339–348.
34. Ross M, Deems R, Jesaitis A, Dennis E, Ulevitch R. Phospholipase activities of the P388D1 macrophage-like cell line. Arch Biochem Biophys 1985; 238:247–258.
35. Hidi R, Vargaftig B, Touqui L. Increased synthesis and secretion of a 14-kDa phospholipase A_2 by guinea pig alveolar macrophages. J Immunol 1993; 151:5613–5623.

36. Glover S, Bayburt T, Jonas M, Chi E, Gelb M. Translocation of the 85-kDa phospholipase A_2 from cytosol to the nuclear envelope in rat basophilic leukemia cells stimulated with calcium inophore or IgE/antigen. J Biol Chem 1995; 270: 15359–15367.
37. Smith W, Marnett L. Prostaglandin endoperoxide synthase: structure and catalysis. Biochem Biophys Acta 1991; 1083:1–17.
38. DeWitt D, Smith W. Yes, but do they still get headaches? Cell 1995; 83:345–348.
39. Morita I, Schindler M, Regier M, Otto J, Hori T, DeWitt D, Smith W. Different intracellular locations for prostaglandin endoperoxide H synthase-1 and -2. J Biol Chem 1995; 270:10902–10908.
40. Reddy S, Herschman H. Ligand-induced prostaglandin synthesis requires expression of the TIS10/PGS-2 prostaglandin synthase gene in murine fibroblasts and macrophages. J Biol Chem 1994; 269:15473–15480.
41. DeWitt DL. Prostaglandin endoperoxide synthase: regulation of enzyme expression. Biochim Biophys Acta 1991; 1083:121–134.
42. Wilborn J, DeWitt D, Peters-Golden M. Expression and role of cyclooxygenase isoforms in alveolar and peritoneal macrophages. Am J Physiol 1995; 268:L294-L301.
43. Vane J. Towards a better aspirin. Nature 1994; 367:215–216.
44. Masferrer J, Zweifel B, Manning P, Hauser S, Leahy K, Smith W, Isakson P, Seibert K. Selective inhibition of inducible cyclooxygenase-2 in vivo is antiinflammatory and nonulcerogenic. Proc Natl Acad Sci USA 1994; 91:3228–3236.
45. Wilborn J, Crofford L, Burdick M, Kunkel S, Strieter R, Peters-Golden M. Fibroblasts isolated from patients with idiopathic pulmonary fibrosis have a diminished capacity to synthesize prostaglandin E_2 and to express cyclooxygenase-2. J Clin Invest 1995; 95:1861–1868.
46. Pouliot M, McDonald P, Khamzina L, Borgeat P, McColl S. Granulocyte-macrophage colony-stimulating factor enhances 5-lipoxygenase levels in human polymorphonuclear leukocytes. J Immunol 1994; 152:851–858.
47. Coffey MJ, Gyetko M, Peters-Golden M. 1,25-Dihydroxyvitamin D_3 upregulates 5-lipoxygenase metabolism and 5-lipoxygenase activating protein in peripheral blood monocytes as they differentiate into mature macrophages. J Lipid Mediators 1993; 6:43–51.
48. Rouzer CA, Kargman S. Translocation of 5-lipoxygenase to the membrane in human leukocytes challenged with ionophore A23187. J Biol Chem 1988; 263:10980–10988.
49. Woods J, Evans J, Ethier D, Scott S, Vickers P, Hearn L, Charleson S, Heibein J, Singer I. 5-Lipoxygenase and 5-lipoxygenase activating protein are localized in the nuclear envelope of activated human leukocytes. J Exp Med 1993; 178:1935–1946.
50. Woods J, Coffey M, Brock T, Singer I, Peters-Golden M. 5-Lipoxygenase is located in the euchromatin of the nucleus in resting human alveolar macrophages and translocates to the nuclear envelope upon cell activation. J Clin Invest 1995; 95: 2035–2040.
51. Brock TG, McNish RW, Peters-Golden M. Translocation and leukotriene synthetic capacity of nuclear 5-lipoxygenase in rat basophilic leukemia cells and alveolar macrophages. J Biol Chem 1995; 270:21652–21658.

52. Brock TG, Paine R, Peters-Golden M. Localization of 5-lipoxygenase to the nucleus of unstimulated rat basophilic leukemia cells. J Biol Chem 1994; 269:22059–22066.
53. Grimminger F, Sibelius U, Seeger W. Amplification of LTB_4 generation in AM-PMN cocultures: transcellular 5-lipoxygenase metabolism. Am J Physiol 1991; 261: L195–L203.
54. Kouzan S, Nolan R, Fournier T, Bignon J, Eling T, Brody A. Stimulation of arachidonic acid metabolism by adherence of alveolar macrophages to a plastic surface: modulation by fetal bovine serum. Am Rev Respir Dis 1988; 137:38–43.
55. Sporn P, Murphy T, Peters-Golden M. Complex effects of in vitro hyperoxia on alveolar macrophage arachidonic acid metabolism. Am J Respir Cell Mol Biol 1990; 2:81–90.
56. Elstad M, Stafforini D, McIntyre T, Prescott S, Zimmerman G. Platelet-activating factor acetylhydrolase increases during macrophage differentiation. J Biol Chem 1989; 264:8467–8470.
57. Vicenzi E, Biondi A, Bordignon C, Rambaldia A, Benedetta Donati M, Mantovani A. Human mononuclear phagocytes from different anatomical sites differ in their capacity to metabolize arachidonic acid. Clin Exp Immunol 1984; 57:385–392.
58. Peters-Golden M, McNish RW, Hyzy R, Shelly C, Toews GB. Alterations in the pattern of arachidonate metabolism accompany rat macrophage differentiation in the lung. J Immunol 1990; 144:263–270.
59. Balter M, Toews G, Peters-Golden M. Different patterns of arachidonate metabolism in autologous human blood monocytes and alveolar macrophages. J Immunol 1989; 142:602–608.
60. Bigby TD, Holtzman MJ. Enhanced 5-lipoxygenase activity in lung macrophages compared to monocytes from normal subjects. J Immunol 1987; 138:1546–1550.
61. Coffey MJ, Wilcoxen SE, Peters-Golden M. Increases in 5-lipoxygenase activating protein expression account for enhanced capacity for 5-lipoxygenase metabolism which accompanies differentiation of peripheral blood monocytes into alveolar macrophages. Am J Respir Cell Mol Biol 1994; 11:153–158.
62. Pueringer R, Bahns C, Hunninghake G. Alveolar macrophages have greater amounts of the enzyme 5-lipoxygenase than do monocytes. J Appl Physiol 1992; 73:781–786.
63. Peters-Golden M, McNish R, Sporn P, Balazovich K. Basal activation of protein kinase C in rat alveolar macrophages: implications for arachidonate metabolism. Am J Physiol 1991; 261:L462–L471.
64. Hempel S, Monick M, Hunninghake G. Lipopolysaccharide induces prostaglandin H synthase-2 protein and mRNA in human alveolar macrophages and blood monocytes. J Clin Invest 1994; 93:391–396.
65. Bertram T, Overby L, Danilowicz R, Eling T, Brody A. Pulmonary intravascular macrophages metabolize arachidonic acid in vitro: comparison with alveolar macrophages. Am Rev Respir Dis 1988; 138:936–944.
66. Coffey MJ, Wilcoxen SE, Phare SM, Simpson RU, Gyetko MR, Peters-Golden M. Reduced 5-lipoxygenase metabolism of arachidonic acid in macrophages from 1,25-dihydroxyvitamin D_3-deficient rats. Prostaglandins 1994; 48:313–329.
67. Rouzer C, Scott W, Hamill A, Cohn Z. Synthesis of leukotriene C and other arachidonic acid metabolites by mouse pulmonary macrophages. J Exp Med 1982; 155:720–733.

68. Taylor S, Liggitt H, Laegreid W, Silflow R, Leid R, Arachidonic acid metabolism in bovine alveolar macrophages: effect of calcium ionophore on lipoxygenase products. Inflammation 1986; 10:157.
69. Nolfo R, Rankin J. U937 and THP-1 cells do not release LTB_4, LTC_4, or LTD_4 in response to A23187. Prostaglandins 1990; 39:157–165.
70. Maclouf J, Fitzpatrick F, Murphy R. Transcellular biosynthesis of eicosanoids. Pharmacol Res 1989; 21:1–7.
71. Wei Y, Heghinian K, Bell R, Jakschik B. Contribution of macrophages to immediate hypersensitivity reaction. J Immunol 1986; 137:1993–2000.
72. Smith E, Wise W, Halushka P, Cook J. Macrophage eicosanoid formation is stimulated by platelet arachiodonic acid and prostaglandin endoperoxide transfer. Biochim Biophys Acta 1987; 923:355–361.
73. Bigby TD, Meslier N. Transcellular lipoxygenase metabolism between monocytes and platelets. J Immunol 1989; 143:1948–1954.
74. Peters-Golden M, Feyssa A. Transcellular eicosanoid synthesis in cocultures of alveolar epithelial cells and macrophages. Am J Physiol 1993; 264:L438–L447.
75. Chakraborti S, Gurtner G, Michael J. Oxidant-mediated activation of phospholipase A_2 in pulmonary endothelium. Am J Physiol 1989; 257:L430-L437.
76. Feng L, Xia Y, Garcia G, Hwang D, Wilson C. Involvement of reactive oxygen intermediates in cyclooxygenase-2 expression induced by interleukin-1, tumor necrosis factor-α, and lipopolysaccharide. J Clin Invest 1995; 95:1669–1675.
77. Humes J, Sadowski S, Galavage M, Goldenberg M, Subers E, Bonney R, Kuehl F Jr. Evidence for two sources of arachidonic acid for oxidative metabolism by mouse peritoneal macrophages. J Biol Chem 1982; 257:1591–1594.
78. Brown G, Monick M, Hunninghake G. Human alveolar macrophage arachidonic acid metabolism. Am J Physiol 1988; 254:C809–815.
79. Tripp CS, Mahoney M, Needleman P. Calcium ionophore enables soluble agonists to stimulate macrophage 5-lipoxygenase. J Biol Chem 1985; 260:5895–5898.
80. Peters-Golden M, Shelley C. Modulation of alveolar macrophage-derived-5-lipoxygenase products by the sulfhydryl reactant, N-ethylmaleimide. J Biol Chem 1987; 262:10594–10600.
81. Sporn PHS, Peters-Golden M. Hydrogen peroxide inhibits alveolar macrophage 5-lipoxygenase metabolism in association with depletion of ATP. J Biol Chem 1988; 263:14776–14783.
82. Westcott JY, McDonnell TJ, Voelkel NF. Alveolar transfer and metabolism of eicosanoids in the rat. Am Rev Respir Dis 1989; 139:80–87.
83. Taylor G, Black P, Turner N, Taylor I, Maltby N, Fuller R, Dollery C. Urinary leukotriene E_4 after antigen challenge and in acute asthma and allergic rhinitis. Lancet 1989; 1:584–588.
84. Israel E, Rubin P, Kemp J, Grossman J, Pierson W, Siegel S, Tinkelman D, Murray J, Busse W, Segal A, Fish J, Kaiser H, Ledford D, Wenzel S, Rosenthal R, Cohn J, Lanni C, Pearlman H, Karahalios P, Drazen J. The effect of inhibition of 5-lipoxygenase by zileuton in mild-to-moderate asthma. Ann Intern Med 1993; 119:1059–1066.
85. Ozaki O, Hayashi H, Tani K, Ogushi F, Yasuoka U, Ogura T. Neutrophil chemotactic

factor in the respiratory tract of patients with chronic airway diseases or idiopathic pulmonary fibrosis. Am Rev Respir Dis 1992; 145:85–91.
86. Wilborn J, Bailie M, Coffey M, Burdick M, Strieter R, Peters-Golden M. Constitutive activation of 5-lipoxygenase in the lungs of patients with idiopathic pulmonary fibrosis. J Clin Invest 1996; 97:1827–1836.
87. Borok Z, Gillissen A, Buhl R, Hoyt R, Hubbard R, Ozaki T, Rennard S, Crystal R. Augmentation of functional prostaglandin E levels on the respiratory epithelial surface by aerosol administration of prostaglandin E. Am Rev Respir Dis 1991; 144: 1080–1084.
88. Damon M, Chavis C, Daures J, Crastes de Paulet A, Michel F, Godard P. Increased generation of the arachidonic metabolites LTB_4 and 5-HETE by human alveolar macrophages in patients with asthma: effect of in vitro nedocramil sodium. Eur Respir J 1989; 2:202–209.
89. Balter M, Eschenbacher W, Peters-Golden M. Arachidonic acid metabolism in cultured alveolar macrophages from normal, atopic, and asthmatic subjects. Am Rev Respir Dis 1988; 138:1134.
90. Garcia J, Griffith D, Cohen A, Callahan K. Alveolar macrophages from patients with asbestos exposure release increased levels of leukotriene B_4. Am Rev Respir Dis 1989; 139:1494–1501.
91. Griffith D, Garcia J, James H, Callahan K, Iriana S, Holiday D. Hyperoxic exposure in humans: effects of 50 percent oxygen on alveolar macrophage leukotriene B_4 synthesis. Chest 1992; 101:392–397.
92. Skerrett S, Henderson W, Martin T. Alveolar macrophage function in rats with severe protein calorie malnutrition: arachidonic acid metabolism, cytokine release, and antimicrobial activity. J Immunol 1990; 144:1052–1061.
93. Laviolette M, Coulombe R, Picard S, Braquet P, Borgeat P. Decreased leukotriene B_4 synthesis in smokers' alveolar macrophages in vitro. J Clin Invest 1986; 77:54–60.
94. Balter M, Toews G, Peters-Golden M. Multiple defects in arachidonate metabolism in alveolar macrophages from young asymptomatic smokers. J Lab Clin Med 1989; 114:662–673.
95. Coffey M, Phare S, Kazanjian P, Peters-Golden M. 5-Lipoxygenase metabolism in alveolar macrophages from subjects infected with the human immunodeficiency virus. J Immunol 1996; 157:393–399.
96. Gray P, Derksen F, Broadstone R, Robinson N, Peters-Golden M. Decreased airway mucosal PGE_2 production during airway obstruction in an animal model of asthma. Am Rev Respir Dis 1992; 146:586–591.
97. Ozaki T, Moriguchi H, Nakamura Y, Kamei T, Yasuoka S, Ogura T. Regulatory effect of prostaglandin E_2 on fibronectin release from human macrophages. Am Rev Respir Dis 1990; 141:965–969.
98. Bachwich P, Lynch J III, Kunkel S. Arachidonic acid metabolism is altered in sarcoid alveolar macrophages. Clin Immunol Immunopathol 1987; 42:27–37.
99. Conary J, Parker R, Christman B, Faulks R, King G, Meryick G, Brigham K. Protection of rabbit lungs from endotoxin injury by in vivo hyperexpression of the prostaglandin G/H synthase gene. J Clin Invest 1994; 93:1834–1840.
100. Peters-Golden M, Thebert P. Inhibition by methylprednisolone of zymosan-induced

leukotriene synthesis in alveolar macrophages. Am Rev Respir Dis 1987; 135:1020–1026.

101. Sebaldt R, Sheller J, Oates J, Roberts LI, FitzGerald G. Inhibition of eicosanoid biosynthesis by glucocorticoids in humans. Proc Natl Acad Sci USA 1990; 87:6974–6978.
102. De Caterina R, Sicari R, Giannessi D, Paggiaro P, Paoletti P, Lazzerini G, Bernini W, Solito E, Parente L. Macrophage-specific eicosanoid synthesis inhibition and lipocortin-1 induction by glucocorticoids. J Appl Physiol 1993; 75:2368–2375.
103. Reibman J, Haines KA, Rich AM, Cristello P, Giedd KN, Weissman G. Colchicine inhibits inophore-induced formation of leukotriene B_4 by human neutrophils: the role of microtubules. J Immunol 1986; 136:1027–1032.
104. Peters-Golden M, Shelly C. Inhibition of alveolar macrophage 5-lipoxygenase metabolism by auranofin. Biochem Pharmacol 1989; 38:1589–1595.
105. Beetens J, Loots W, Somers Y, Coene M, De Clerck F. Ketoconazole inhibits the biosynthesis of leukotrienes in vitro and in vivo. Biochem Parmacol 1986; 35:883–891.

7

Macrophage-Derived Cytokines in Lung Inflammation

STEVEN L. KUNKEL, STEPHEN W. CHENSUE, NICHOLAS W. LUKACS, and ROBERT M. STRIETER

University of Michigan Medical School
Ann Arbor, Michigan

I. Introduction

The pathogenesis of pulmonary inflammation after exposure to an inciting agent is dependent on a complex interaction of cellular and humoral mediators. Although the precise mechanisms that orchestrate the initiation, maintenance, and resolution of the inflammatory response remain to be fully elucidated, it is evident that the etiology of a variety of lung diseases can be attributed to a programmed response involving a variety of macrophage-derived chemical mediators. These chemical signals include lipids, nucleotides, reactive nitrogen and oxygen intermediates, peptides and polypeptides. This latter group of inflammatory agents contains a large group of important mediators collectively identified as cytokines. Interestingly, mononuclear phagocytic cells, including monocytes, macrophages, and fixed tissue macrophages, are an important source of these key mediators during various phases of an immune response. Research into the mechanistic role of macrophage-derived cytokines in various diseases is currently being conducted in both clinical and basic research settings, as these polypeptide mediators are known to be key participants in a diverse array of diseases.

In the lung, inflammatory responses are usually identified as being stereotypic and can be classified as either an acute or chronic response depending on both the duration of the inflammation, the participating chemical signals, and the specific leukocyte subpopulations. For example, acute lung inflammation can be observed during the development of bacterial infections and usually involves the active participation of neutrophils and neutrophil-derived mediators. These peripheral blood leukocytes are recruited rapidly to the lungs and are involved in the initial stages of eliminating the infectious pathogen. The neutrophils do play a major role in bacterial clearance; however, the final resolution of acute inflammation includes the active participation of other leukocyte populations, such as mononuclear phagocytes. These cells are known to be instrumental in the reparative phase of the acute response. Although mononuclear phagocytes are important to the resolution phase of acute inflammation, this cell population is the sine qua non of chronic pulmonary inflammation. During chronic lung disease, the macrophage and its derivatives, epitheloid and giant cells play an important role via a number of activities including phagocytosis, antigen presentation and lymphocyte communication, and the secretion of cytokines.

II. Macrophage-Derived Early Response Cytokines

The importance of inflammatory mediators to the evolution of acute lung injury has been under intense investigation for may years. Although a number of serum-derived mediators have provided insight into the evolution of lung injury, research conducted during the past decade has truly underscored the importance of cytokines in orchestrating the inflammatory response. A number of studies have established the role of macrophage-derived cytokines as important communication links during the evolution of acute lung inflammation (1–5).

Although not structurally related, macrophage-derived interleukin-1 (IL-1) and tumor necrosis factor (TNF) possess similar functional activities during the evolution of an inflammatory response and have been termed early-response cytokines (6–9). This designation stems from the fact that both TNF and IL-1 are rapidly expressed at a site of inflammation and provide the foundation for establishing paracrine and endocrine communication links. These cytokines are produced primarily by monocytes, inflammatory macrophages, and tissue macrophages. On stimulation, these cells can generate high levels of IL-1 and/or TNF which possess a wide range of overlapping activities on most cells in the body. The importance of IL-1 and TNF as key cytokines in initiating inflammation is underscored by the fact that all nucleated cells appear to possess a functional receptor for IL-1 and TNF.

Early-response cytokines have been associated with a variety of cell-activation events, and circumstantial evidence suggests that these mediators are involved in the initiation of lung inflammation. These polypeptide mediators clearly function as communication links via their ability to establish cytokine networks, whereby one cytokine can bind to an appropriate receptor on a specific cell and induce the expression of an additional cytokine(s). This newly synthesized cytokine may then interact with other cells and cause the expression of yet another cytokine in a cascade-like manner. As shown in Figure 1, an exogenous challenge can induce the expression of IL-1 or TNF by alveolar or interstitial macrophages which can subsequently activate nonimmune lung cells, resulting in the production of additional cytokines. The importance of this cytokine network lies in the fact that (1) a variety of immune and nonimmune cells can fully participate in the evolution of the inflammatory response, (2) a relatively small group of mediators may initiate diverse inflammatory/immune responses, and (3) control of the evolving response in under the direction of many transcriptional and post-transcriptional regulatory steps. These aspects of cytokine networking are of interest in designing novel therapeutic strategies, as (1) immunoregulation may include suppressing the activation of structural cells of an organ, (2) suppression of polypeptide mediators which function early in the response may control the subsequent steps of inflammation, and (3) the regulation of protein synthesis at the transcriptional and posttranscriptional levels inherently allows for multiple levels to regulate the full expression and activity of specific cytokine cascades.

IL-1 and TNF have been associated with a variety of cell-activation events, and circumstantial evidence suggests that these mediators are involved in the initiation of specific human lung disease. For example, a number of experimental and clinical studies have shown that mononuclear cell–derived IL-1 and TNF appear to play a significant role in the pathogenesis of septic shock and the development of acute lung injury. Clinical investigations have assessed elevated TNF levels in serum samples from patients with meningococcal septicemia with associated acute lung injury (10). These studies demonstrated a significant cor-

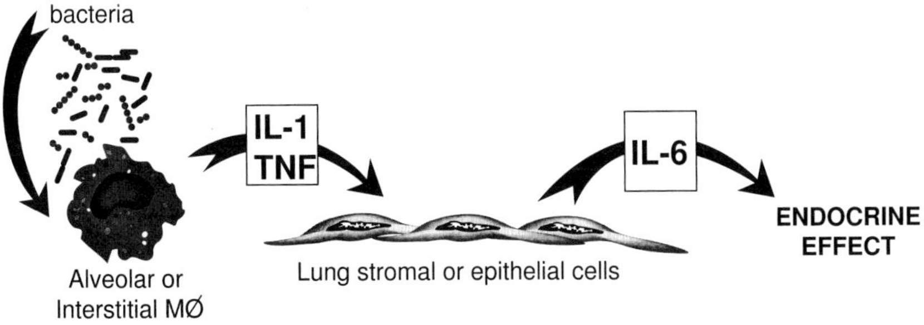

Figure 1 Cytokines are key polypeptide mediators in directing cell to cell communication via cytokine networks. In this scenario, the expression of one cytokine can set in motion a cascade of events culminating in the production of another cytokine with potent biological activities.

relation between serum TNF levels and mortality. Subsequent investigations have correlated TNF and IL-1 levels with mortality and clinically diagnosed patients with sepsis due to meningococcemia (11). In an additional prospective study, serum samples of patients with clinical correlates of septic shock possessed elevated levels of TNF (12). The magnitude of TNF was found to correlate with a higher incidence and severity of adult respiratory distress syndrome (ARDS), lung injury, and subsequent mortality. Interestingly, TNF levels were elevated with equal frequency in patients in shock with either gram-negative or gram-positive bacteriemia.

The role of macrophage-derived TNF in the pathogenesis of sepsis has been examined through the use of either established animal models of endotoxemia, cecal ligation and puncture, or by the direct administration of TNF (13–18). Interestingly, the systemic administration of TNF induces similar lung pathology and physiological alterations as compared with either the infusion of endotoxin or live gram-negative bacteria. These animals all demonstrated elevated body temperature and circulating levels of catecholamines, metabolic acidosis, consumptive coagulopathy, pulmonary dysfunction, alterations in circulating levels of leukocytes, and hypotension resulting in shock. The concomitant administration of both IL-1 and TNF resulted in a synergistic effect in mediating many of the above pathophysiological alterations. Investigations into the in vivo mechanism of TNF-induced injury demonstrated that protein synthesis was crucial to the host's ability to regulate the pathophysiology, suggesting that de novo protein synthesis is needed in vivo to protect against TNF-induced tissue injury and lethality.

The inhibition of endogenous TNF activity during gram-negative bacteria-induced sepsis has been shown significantly to attenuate the associated lung pathology and morality (17). The administration of neutralizing antibodies against TNF prior to challenging baboons with a lethal dose of *Escherichia coli* resulted in a significant reduction in morality. However, delaying antibody therapy until after *E. coli* challenge did not salvage the experimental animals, suggesting that the production of TNF was both early and set in motion a cascade of events which ultimately ended in the death of the challenged animal. One of the interesting pathological findings in these animals was the histological changes in the lungs, which included leukocyte infiltration and a vascular leak syndrome. Similar studies in rodents have demonstrated that macrophage-derived TNF is expressed very rapidly post–endotoxin challenge. After a lethal challenge of endotoxin, TNF levels peaked in vivo at 1 hr and then rapidly declined to undetectable levels (14). Human volunteers challenged with low concentrations of endotoxin also demonstrated a rapid rise and abrupt fall in the circulating levels of TNF (19). These studies support previous findings that the expression of TNF is under strict regulation in vivo.

III. Chemokines and Leukocyte Recruitment

The elicitation of specific leukocyte populations to an inflammatory lesion is one of the most fundamental of all host immune responses (20). For well over a century, the pathology of various infectious diseases has been characterized by the accumulation of leukocyte populations to a developing area of inflammation. Many of these early studies recognized that leukocytes could adhere to the vascular wall and move into the interstitium in response to unknown factors. One of the first investigations directed at understanding directed migration to a known stimulus was performed using dead microorganisms or the products of human leukocytes. These rather descriptive studies were important in providing the groundwork for a number of subsequent seminal investigations demonstrating that isolated N-formylmethionyl peptides from bacteria or synthetic N-formylmethionyl derivatives were potent inducers of leukocyte activation and chemotaxis (21).

The landmark studies identifying that both bacterial-derived peptides and proteins derived from the fifth component of complement were highly chemotactic for leukocytes provided some of the first information regarding the mechanism for the recruitment of leukocytes during infectious processes (22). However, these studies did not address the specificity associated with leukocyte recruitment during inflammation. The identification of other chemotactic factors such as leukotrine B_4, platelet-activating factor, and split products of fibronectin only complicated the issue of leukocyte movement, as these additional chemotactic agents also lack any degree of specificity for the recruitment of a given leukocyte population (23). An important aspect of all the initial studies identifying various leukocyte chemotactic factors was the redundancy in the number and types of factors which induce leukocyte chemotaxis. This fact alone stresses the importance of cell recruitment during an infectious/inflammatory process.

The fact that multiple chemotactic agents possess little in the way of specificity was at odds with the pathology of certain diseases which demonstrated that acute inflammatory reactions may be dominated by the presence of polymorphonuclear leukocytes, whereas many chronic immune reactions are characterized by the presence of specific mononuclear leukocyte populations. This conundrum has been recently addressed by investigations into related supergene families of chemotactic cytokines, which possess a relatively high degree of specificity for the elicitation of leukocyte populations (24–26). The discovery of these polypeptide mediators of inflammation has proven to be important, as they provide a potential mechanism for the selective recruitment of peripheral blood leukocytes to a specific site of inflammation.

Chemokines belong to two groups of related polypeptides, which are identified by the location of two of the four cysteine amino acids comprising their primary amino acid structure (24–26). Accumulating evidence supports the concept that members of these chemokine families have proinflammatory and reparative activities. In their monomeric form, chemokines are less than 10,000 Da and are basic heparin–binding proteins. One of the chemokine families displays a conserved amino acid motif characterized by the location of two amino-terminal cysteines separated by one nonconserved amino acid residue. This chemokine family is designated as CXC chemokines and appear to have specificity for the elicitation of mainly neutrophils. The CXC chemokines are all clustered on human chromosome 4 and possess approximately 20–55% homology in their primary structure. Interest in this area is exemplified by investigations which have identified a number of different CXC chemokines (Table 1), including IL-8, platelet factor-4, platelet basic protein, connective tissue–activating protein III, β-thromboglobulin, neutrophil-activating factor-2, growth-related oncogenes (GROs) (α, β, and γ), interferon-γ (INF-γ) inducible protein (IP-10), epithelial neutrophil–activating protein-78, monokine induced by INF-γ (MIG), and granulocyte chemotactic protein-2 (GCP-2).

Recent investigations have demonstrated that an important feature in the primary structure of CXC chemokines may account for the neutrophil chemotactic and activating properties of these mediators (Table 2). These studies identified three critical amino acid residues immediately preceding the first N-terminal cysteine residue, which are important in binding to a neutrophil receptor and activating neutrophils. These amino acids are Glu-Leu-Arg or the ELR motif, which is absent in certain members of the CXC chemokine family. In particular, IP-10, platelet factor-4, and MIG all lack the ELR motif and are not potent in neutrophil-activating assays. Interestingly, when the ELR motif was synthetically introduced into platelet factor-4 this polypeptide gained chemotactic activity. Therefore, certain members of the CXC supergene family may have different biological activities. Platelet factor-4 (a non-ELR containing CXC chemokine) was one of the first members of this family to be described. This factor was

Table 1 Representative Members of Chemotactic Cytokines Which Belong to Either the CXC Chemokine or CC Chemokine Supergene Family

CXC chemokines	CC chemokines
Interleukin-8	Monocyte chemoattractant protein
Growth-related oncogenes α, β, γ	RANTES
Neutrophili-activating factor-2	Macrophage inflammatory protein
β-thromboglobulin	Eotaxin
Platelet factor-4	
Interferon-inducible protein	
ENA-78	

Table 2 CXC Chemokines

ELR-CXC chemokines	Non-ELR CXC chemokines
IL-8	MIG
GCP-2	IP10
GRO-α	PF-4
GRO-β	
Gro-γ	
ENA-78	
NAP-2	

The ELR amino acid motif comprises the primary structure of specific CXC chemokine family members. The ELR-containing chemokines possess activity for the directed movement of neutrophils, whereas non-ELR–containing CXC chemokines are not potent inducers of neutrophil chemotaxis.

originally identified for its ability to bind heparin, leading to the inactivation of the anticoagulation function of heparin. All members of the chemokine supergene family appear to possess heparin binding domains in their primary amino acid structure. IL-8 was first identified as a monocyte-derived neutrophil chemotactic factor and has been the most studied CXC chemokine. Interestingly, it is produced by an array of cells, including primary cultures of monocytes, alveolar macrophages, neutrophils, keratinocytes, mesangial cells, epithelial cells, hepatocytes, fibroblasts, and endothelial cells (27–31). In addition, IL-8 is expressed by a number of neoplasms and transformed cell lines.

An additional related supergene family of polypeptide chemokines is the CC chemokine family. This group is defined by the juxaposition of the two N-terminal cysteines. The CC chemokine family includes macrophage inflammatory protein-1α and β; RANTES; monocyte chemoattractant protein-1, -2, and -3; I-309; and eotaxin (see Table 1). In general terms, the CC chemokine family has relative specificity for the elicitation of mononuclear cells. Certain members of the CC chemokine family also appear to be potent chemotactic factors for eosinophils. Like the CXC family, certain members of the CC members have been identified as products from a number of cellular sources, including lymphocytes, monocytes, neutrophils, and a variety of noninflammatory cells and tumor cell lines.

IV. Role of Macrophage-Derived Cytokines in the Recruitment of Leukocytes During Acute Lung Injury

As previously discussed, the sequestration of neutrophils is a characteristic associated with a variety of acute lung inflammatory reactions. However, in pathological processes associated with endotoxemia and sepsis, the lungs are often identified as the most susceptible target of all effected organs. The presence of neutrophils in the interstitium or alveolar space is due to the production of chemotactic factors which establish a gradient from the vasculature to the site of inflammation. Various peptides, polypeptides, and lipids have been identified as macrophage-derived chemotactic factors; however, none of these "historic chemotactic factors" have specificity for recruiting a particular leukocyte population. In addition, the pulmonary origin of chemotactic factors has historically been an enigma. Although substantial information has been generated regarding alveolar macrophages as a source of chemotactic factors, recent studies have identified other cellular constituents of the alveolar capillary wall as important sources of chemotactic factors. Both pulmonary fibroblasts and pulmonary type II–like epithelial cells can express chemokines in a stimulus-specific pattern (27,32). Although alveolar macrophages have the ability to generate a number of cytokines in response to LPS, the noninflammatory cells in the lung are generally unresponsive to this activating agent. The participation of these latter cells in an inflammatory event appears to be dependent on cytokine networking; a repeating scenario which is likely common to all organs.

The activation of this pathway in the lungs may be especially important in disease states characterized by bacteremia, endotoxemia, sepsis or septic shock, as the lungs are particularly vulnerable to these pathological states. Clearly, the alveolar and interstitial lung macrophages are important in generating early response signals (IL-1 and TNF) when challenged with lipopolysaccharide (LPS). These early-response cytokines can then act in either an autocrine or paracrine fashion to stimulate the production of more distal mediators of inflammation by the nonimmune cells found in the lung. To test this hypothesis, pulmonary fibroblasts were challenged with LPS, TNF, or IL-1 and the conditioned fibroblast media assessed for neutrophil chemotactic activity. Both IL-1 (IL-1β) and TNF caused a significant increase in the production of a neutrophil chemotactic factor by 4 hr postchallenge, which peaked by 8 hr and continued to be expressed for the next 20 hr. In these studies, IL-1β was a more potent agonist for the production of the neutrophil chemotactic factor than TNF. However LPS over a wide concentration range did not induce pulmonary fibroblasts to produce a neutrophil chemotactic factor over the 24-hr study period. These observations led to subsequent investigations to determine if the chemotactic activity was due, at least in part, to IL-8. Pulmonary fibroblasts were challenged with IL-1β (20 ng/ml), TNF (20 ng/ml), or LPS (1 μg/ml), and the expression of steady-state IL-8 mRNA was determined in a kinetic fashion. Both IL-1β and TNF were effective in causing the expression of IL-8 mRNA in a rapid and prolonged fashion, whereas LPS did not cause the expression of this chemokine gene. By 1 hr poststimulation, IL-1β and TNF induced detectable levels of IL-8 mRNA that persisted for 24 hr. The expression of steady-state IL-8 mRNA levels were also found to be elevated in a dose-dependent manner in response to either IL-1β or TNF but not LPS. Concentrations of IL-1β and TNF in the picogram/milliliter range could cause a significant increase in the expression of IL-8 mRNA from the pulmonary fibroblasts. These same findings were also observed at the translational level, as both IL-1β and TNF, over a wide dose-response range, could increase the production of antigenic IL-8. However, the production of antigenic IL-8 by LPS-stimulated pulmonary fibroblasts was not significantly elevated over control cells (Fig. 2).

In the above experiments, signal specificity for the generation of IL-8 by pulmonary fibroblasts was demonstrated. Both IL-1 and TNF stimulation of the pulmonary fibroblasts resulted in the expression of IL-8 mRNA, antigenic IL-8 protein, and biologically active IL-8, whereas LPS was not efficacious in this system. Since the resident alveolar macrophage is susceptible to LPS stimulation with the rapid generation of IL-1 and TNF, studies were designed to determine if alveolar macrophage–derived soluble mediators could influence the production of pulmonary fibroblast–derived IL-8. In this study, normal human alveolar macrophages were challenged with 1μg/ml of LPS for 24 hr, and the macrophage supernatant was collected. The supernatant was then overlaid on cultured fibroblasts in the presence or absence of neutralizing TNF or IL-1β antibody. The pulmonary fibroblast were placed back in culture for an additional 8 hr, and then total fibroblast RNA isolated (Fig. 3). The addition of LPS-stimulated alveolar macrophage–conditioned media to the cultured pulmonary fibroblasts resulted in significant steady-state expression of pulmonary fibroblasts–derived IL-8. However, the LPS-stimulated alveolar macrophage–conditioned media treated with either neutralizing antibody to IL-1β or TNF resulted in an approximate 40% reduction in steady-state IL-8 mRNA expression. Additional studies using a combination of neutralizating antibodies to both IL-1 and TNF resulted in over a 90% inhibition of fibroblast-derived IL-8 mRNA.

To further address the networking between alveolar macrophages and pulmonary fibroblasts, a coculture system using both normal human alveolar macrophages and normal lung fibroblasts was utilized. These cells were cultured together in the absence or presence of 1μg/ml LPS. At 1, 4, 8, and 24 hr after the addition of LPS, the cells were processed for immunohistochemical localization of antigenic IL-8 (Fig. 4). On the addition of LPS, the alveolar macrophages rapidly expressed antigentic IL-8, which was clearly observed by 4 hr. However, at the 4-hr time point, the expression of antigenic IL-8 by the lung fibroblasts in the coculture system was low. By 8 hr, significant levels of antigenic IL-8 could be detected by the fibroblasts. Finally, by 24 hr, most of the pulmonary fibroblasts were found to be expressing antigenic IL-8. At this time point, over 85% of the fibroblasts were positive for IL-8. These studies support the role of alveolar macrophages in aiding the granulocyte recruitment process to the lung interstitium and alveolar space via cytokine networking with lung resident cells. This resident lung macrophage can respond to a number of initiating agents and rapidly generate

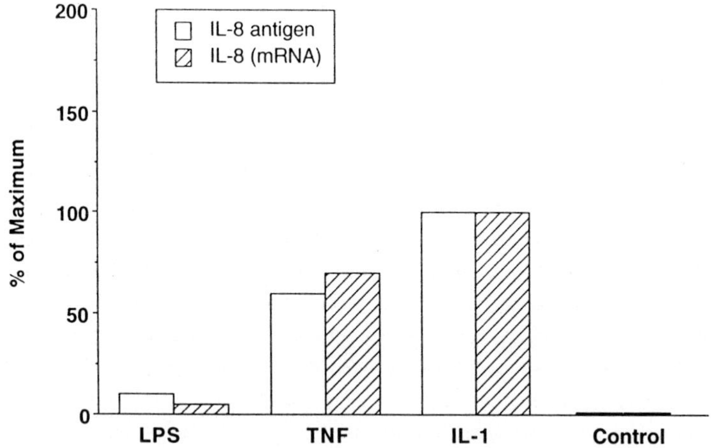

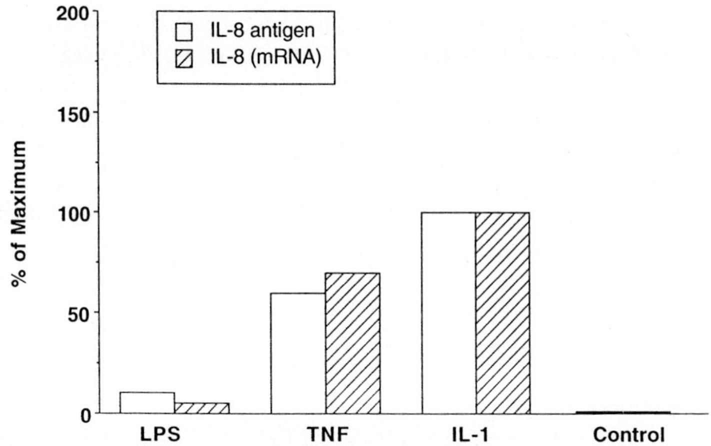

Figure 2 The expression of pulmonary fibroblast–derived IL-8 is dependent on early cytokines; however, direct LPS challenge did not elevate the levels of IL-8.

early-response cytokines. These mediators are important in orchestrating the inflammatory response by directly activating inflammatory cells, as well as serving in a paracrine role in networking with resident normal tissue. This latter activity elicits an effector response from the surrounding normal cells and augments the initiation and maintenance response of the inflammatory reaction.

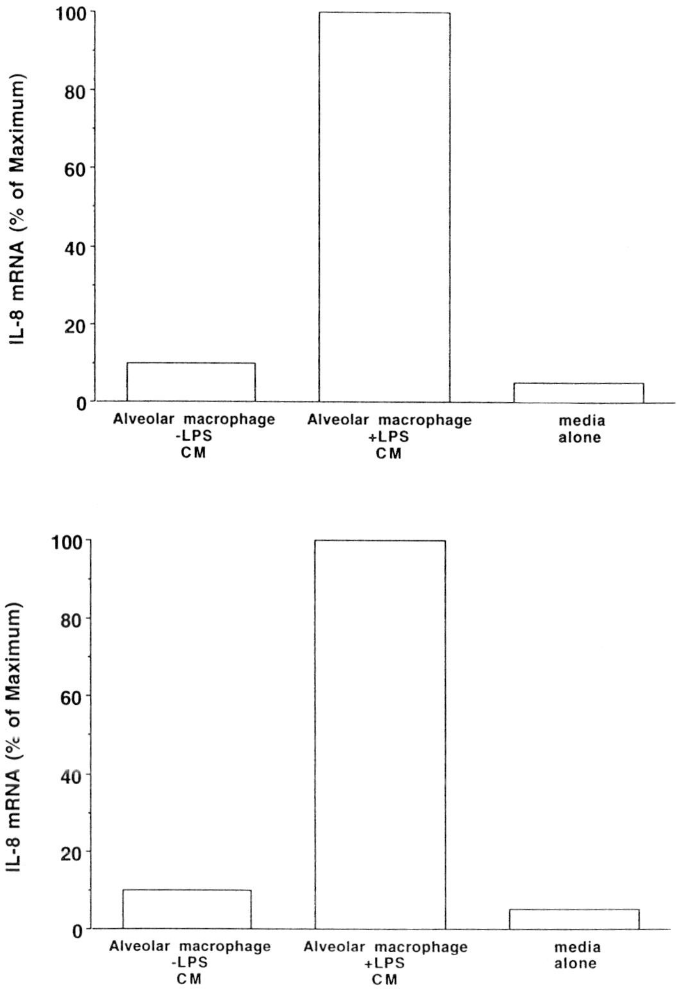

Figure 3 Conditioned media from alveolar macrophages challenged with LPS was effective in increasing steady-state levels of fibroblast-derived IL-8 mRNA. The expression pattern induced by the conditioned media was similar to IL-1–treated and TNF-treated fibroblasts.

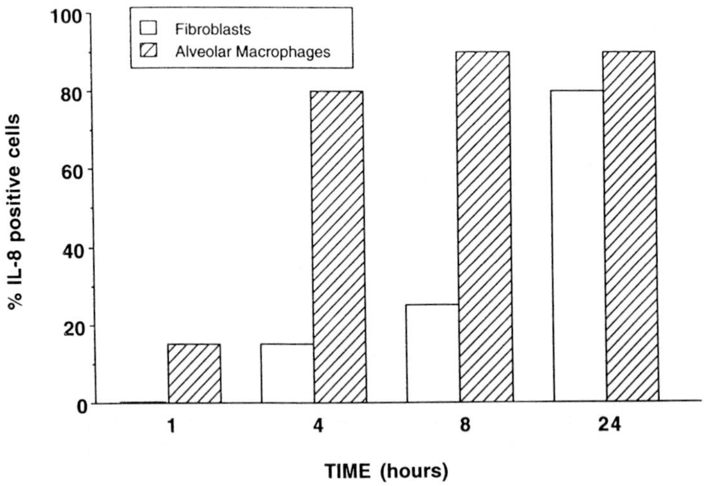

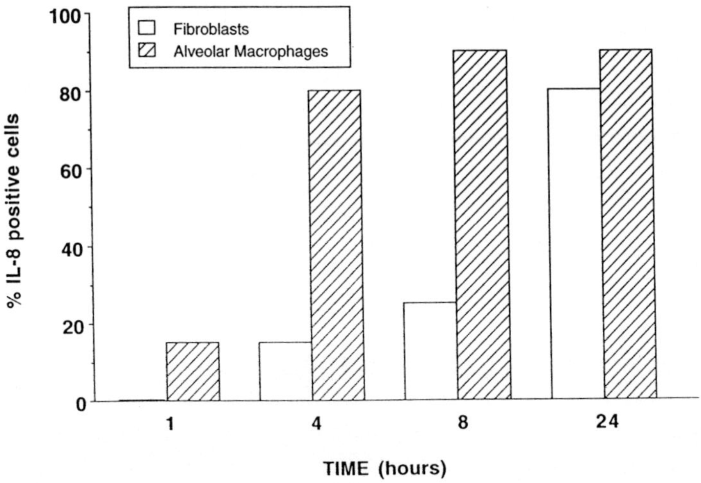

Figure 4 LPS stimulation of cocultured alveolar macrophages and fibroblasts resulted in the rapid expression of IL-8 antigen by alveolar macrophages, but a delay in the antigenic expression of fibroblasts. Other neutralization studies demonstrated that Il-1 and TNF produced by the alveolar macrophages was responsible for the fibroblast-derived Il-8.

V. Macrophage-Derived Cytokines in Chronic Lung Inflammation

Studies from our laboratory and others have begun to address the role of macrophage-derived IL-1 and TNF in the development of lung granulomas (33,34). For example, elevated levels of IL-1 have been measured in aqueous extracts of whole lungs containing delayed-type hypersensitivity (DTH) granulomas. This study demonstrated that the expression of IL-1 was prominent during the initiation phase of the DTH lung lesion. Subsequent studies have demonstrated that in vivo implantation of recombinant IL-1 impregnanted into a slow-release carrier (ethyl vinyl acetate disks) resulted in the rapid development of a macrophage-rich granulomatous lesion. Similar experiments showed that pulmonary embolization of IL-1 coupled to Sephadex beads could induce lung granulomas with cellular characteristics of a DTH response (33). Complementary studies have also supported the role of TNF in the maintenance of DTH granulomatous responses, as mycobacterial (BCG) granuloma development coincides with local lung macrophage TNF production (33). Passive immunization of experimental animals with neutralizing antibodies to TNF reduced both the size and the number of granuloma lesions. More importantly, the neutralization of TNF appeared to be useful as a therapeutic option, as fully developed BCG-induced granulomas regressed after antibody administration. Additional compelling evidence for the involvement of TNF in granuloma formation was demonstrated by the reconstitution of pulmonary *Schistosoma mansoni* egg–induced granulomas in SCID mice by exogenously administered TNF (35). Our laboratory has also addressed the potential role of macrophage-derived IL-1 and TNF in pulmonary granuloma development, as we have demonstrated that the production of both IL-1 and TNF was potentiated during the evolution of the DTH granulomas compared with foreign body–type responses.

By evaluating the expression of IL-1 and TNF during synchronously developing *S. mansoni* egg–induced lung granuloma, a defined pattern in the expression of IL-1 and TNF has emerged. Granuloma macrophages isolated from the initiation phase or early developing lesions (days 2–4) produced three- to fourfold more IL-1 than during the maintenance phase (days 8–16) of the granulomatous lesion. Thus, the biological activity of IL-1 appears to be essential during the early initiation phase of the DTH granuloma, but it diminishes during the maintenance phase. We have also demonstrated that levels of biologically active TNF were low during the initiation phase but increased significantly during the maintenance phase of the evolving granuloma (36). These findings suggest that IL-1 may be critical to the early recruitment and local cellular activation, whereas TNF contributes to the continued organization of the maturing granuloma. In contrast, the developing foreign body–type pulmonary granuloma produced lower levels of IL-1 and TNF as compared with DTH lesions. These observations and others likely reflect the degree in which mononuclear cells participate in the developing granuloma. In additional studies using this well-defined model of T-cell–mediated pulmonary granulomas, we have found that specific lymphocyte-derived mediators, IL-4 and INF-γ, can influence the development of the granulomatous response (37). The sequential production of IL-4 and INF-γ was determined and related to the temporal changes in granuloma macrophage production of IL-1, TNF, and superoxide anion production. High levels of IL-4 were produced early during granuloma formation (days 1–2) in sensitized animals, whereas IFN/γ production was found in greatest amounts on days 4–8 post initiation of the lesion. Maximal granuloma macrophage–derived superoxide anion release tended to parallel INF-γ production. Interestingly, passive immunization of experimental animals with antibody to murine IL-4 significantly reduced the size of the developing lesion, whereas neutralization of IFN-γ had little effect.

VI. Role of Cytokine Inhibitors in the Development of Macrophage-Dependent Lung Inflammation

The above studies demonstrate that a number of cytokines, including macrophage-derived IL-1 and TNF, may play a prominent role in the development of an immune response; however, the importance of a regulated cytokine response during inflammation is underscored by the identification of endogenous cytokine-inhibitory proteins. Some of these host-derived, cytokine-specific inhibitory proteins have been shown to block the biological activity of either IL-1 or TNF (38,39). The mechanism of these inhibitors appears to be via the interference with the formation of a functional cytokine-cell surface receptor complex, thus negating classic cell signaling events. There appears to be two separate means by which IL-1 and TNF inhibitors function (Fig. 5). One type of inhibitor is represented by the IL-1 receptor antagonist (IRAP)(38). This polypeptide is a true receptor-binding antagonist, which competes with IL-1 for the same receptor-binding site on the target cell without intrinsic agonist activity. The second type of inhibitor is a "shed" soluble TNF receptor (39). Macrophages have two distinct cell surface receptors for TNF designated p60 and p80, which are classified according to their molecular weights of 60 and 80 kDa, respectively (40). Recent evidence has shown that the murine TNF receptors can mediate distinct cellular responses. In addition, a dimeric soluble receptor construct, generated by fusing the TNF receptor with the Fc portion of an IgG molecule, has been generated and identified as a potential therapeutic agent (41). This construct possesses a higher affinity and is a much better inhibitor of TNF activity than the monomeric TNF receptor. Most soluble cytokine-binding proteins are usually the product of proteolytically cleaved membrane-bound receptors, but they may also represent alternatively spliced mRNA coding for soluble verses membrane cytokine receptors. These soluble, extracellular, ligand-binding segments of specific cytokine receptors

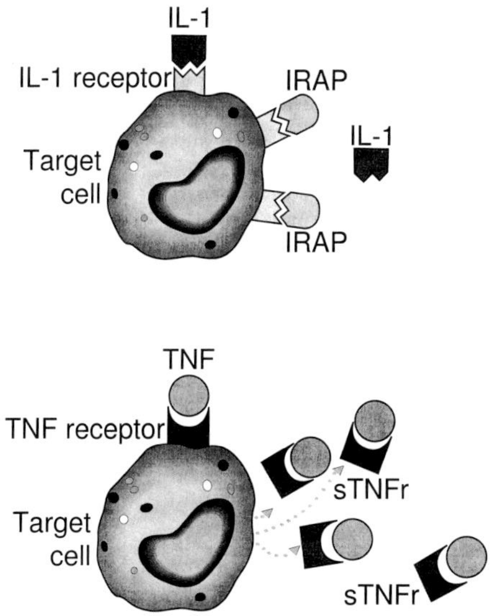

Figure 5 Endogenous inhibition of macrophage-derived IL-1 and TNF activity can be demonstrated by the interleukin-1 receptor antagonist (IRAP) and soluble TNF receptor (sTNFr), respectively. IRAP competes for the IL-1 molecule at the receptor level, whereas the sTNFr binds released TNF. In both cases, the biological activity of the specific cytokine is suppressed.

occur naturally and additional evidence supports their role as endogenous regulators of cytokine activity. For example, the importance of soluble cytokine receptors to the regulation of inflammation is highlighted by the interesting observation that various pox viruses encode polypeptides with both functional and structural homology to the extracellular sections of the TNF, IL-1 and IFN-γ receptor, suggesting that these polypeptides aid the virus in the subversion the host's immune response.

The potential therapeutic use of endogenous cytokine inhibitors to treat specific acute inflammatory responses are just beginning to be appreciated, as both in vivo and in vitro experimental studies have demonstrated the ability of recombinant IRAP and soluble dimeric TNF receptor construct to inhibit inflammatory responses (38,39). However, the majority of these studies have not investigated the ability of either IL-1 or TNF inhibitors to alter chronic immune or DTH responses. Recent experiments have identified the presence of lung macrophage–derived IRAP by immunolocalization in lesions from patients with pulmonary sarcoidoisis (42), but the biological significance of IRAP in this pulmonary granulomatous disease is unknown. The role of IRAP in either acute or chronic disease states is even more intriguing, as an intracellular form of IRAP has been discovered which is structurally identical to secreted IRAP except the former lacks an appropriate leader sequence. This cell-associated form of IRAP may serve as a means to accumulate high concentrations of IRAP locally on cell death or alternatively serve an important physiological role inside the viable cell.

The importance of endogenous inhibitors, which can specifically block the activity of either macrophage-derived IL-1 or TNF is further highlighted by the fact that adhesion molecules are rapidly expressed by endothelial cells exposured to these cytokines. This event is important, as it serves to target specific mononuclear cell populations to the developing area of chronic inflammation. Leukocyte localization is subsequently followed by migration from the lumen of the vessel to an area of inflammation. In the case of chronic DTH immune reactions, leukocyte recruitment may continue for a prolonged period of time depending on the continued expression of cytokines and adhesion molecules. The continued elicitation of lymphocytes and monocytes is important in DTH reactions and are likely dependent on the expression of endothelial cell–derived intracellular adhesion molecule-1 (ICAM-1) and vascular cell adhesion molecule-1 (VCAM-1). Since IL-1 and TNF may play a prominent role in the long-term mononuclear leukocyte recruitment and activation, via ICAM-1– and VCAM-1–dependent adhesion mechanisms, then the local balance of these cytokines and their specific inhibitors are important to the outcome of the inflammatory response. In addition, the biological significance and interdependence of IL-1, TNF, and adhesion molecules are further demonstrated during the evolving immune response, as these molecules play an important role in antigen recognition and T-cell sensitization. The above information supports the interrelated nature of many of these mediators during the evolution of a cell-mediated, DTH immune response.

In addition to IL-1 and TNF, a variety of cytokines have been linked with the evolution of pulmonary granulomas, including IL-2, IL-4, IL-6, IL-12, IFN-γ, growth factors, and chemokines. This list includes macrophage-derived mediators which possess early activation, chemotactic, growth and differentiation, and remodeling activities in the lung (Table 3). Data generated from a number of studies have provided circumstantial evidence regarding the role of particular cytokines in granulomatous lesions; however, many of these investigations neither provide mechanistic information regarding the contribution of macrophage-derived cytokines to the developing inflammation, nor do they fully support a defined role for a given cytokine during the various phases of the granulomatous response. However, there is a growing body of scientific evidence suggesting that the cytokine profile of the natural immune/inflammatory response likely determines the disease phenotype responsible for either resolution or progression to end–stage disease. Much of the supporting evidence is derived from studies demonstrating that macrophage-derived IL-12 and TNF and lymphocyte-derived IFN-γ have profound effects on the evolution of granulomatous inflammation induced by intracellular infectious agents, whereas lymphocyte-derived IL-4, IL-5, and IL-10 appear to be dominant cytokines expressed during granulomatous inflammation

Table 3 Sequential Expression of Macrophage-Derived Cytokines During the Evolution and Resolution of Chronic Inflammation Involves Cytokines with Various Activities

1. Early activation	3. Growth
tumor necrosis factor	Colony-stimulating factors
interleukin-1	Fibroblast growth factor
2. Chemotactic	4. Remodeling
monocyte chemoattractant protein-1	platelet-derived growth factor
RANTES	Transforming growth factor

caused by certain extracellular infectious agents or persistent antigens (43,44). These cytokine profiles are in line with the Th1- and Th-2 type expression bias, respectively. Furthermore, investigations have demonstrated that cytokine "cross-talk" occurs between the two types of inflammatory responses (44), resulting in cross-regulatory activity on the evolving immune reactions. Although a bias toward a specific cytokine profile may play an integral role in the development of chronic inflammation, the mechanisms which exert either a negative or positive influence on the type of immune response is not well understood.

VII. Conclusions

The role that macrophage-derived polypeptide mediators play in directing cell to cell communication are now known as being fundamental components of the immune/inflammatory response. Cytokine expression often occurs via an amplified cascade or network that involves cellular activation by an early response cytokine which triggers the expression of more distal cytokines. The generation of cytokine networks is necessary for both the pathogenesis and resolution of a variety of acute and chronic lung diseases, as these mediators initiate, maintain, and finally resolve the inflammatory response. The interest in the cytokine biology of macrophages is underscored by studies aimed at identifying their mechanistic role in mediating lung disease as well as developing therapeutic strategies to alter their effects. Future investigations are likely to provide the impetus for the development of novel cytokine-directed treatment for many inflammatory lung diseases.

Acknowledgment

This research was supported by NIH Grants 1PO50HL46487, HL31693, HL35276, and HL50057.

References

1. Ayala A, Perrin MM, Kisala JM, Chaudry IH. Polymicrobial sepsis selectively activates peritoneal but not alveolar macrophages to release inflammatory mediators. Circulatory Shock 1992; 36:191–199.
2. Hazuda DJ, Lee JC, Young PR. The kinetics of interleukin-1 secretion from activated monocytes. J Biol Chem 1988; 263:8473–8479.
3. Brigham, KL Meyrick B. Endotoxin and lung injury. Am Rev Respir Dis 1986; 133: 913–927.
4. Cybulsky, MI Chan, MKW, Movat HZ. Acute inflammation and microthrombosis induced by endotoxin, interleukin-1, and tumor necrosis factor, and their implication in gram-negative infections. Lab Invest 1988; 58:365–378.
5. Sibille Y, Reynolds HY. Macrophages and polymorphonuclear neutrophils in lung defense and injury Am Rev Respir Dis 1990 141:471–501.
6. Beutler B. The tumor necrosis factors: Cachectin and lymphotoxin. Hosp Pract 1990; 2:45–46.
7. Sherry B. Cerami A. Cachectin/tumor necrosis factor exert endocrine, paracrine, and autocrine control of the inflammatory response. J Cell Biol 1988; 107:1269–1277.
8. Larrick JW, Kunkel SL. The role of tumor necrosis factor and interleukin-1 in the immunoinflammatory response. Pharm Res 1988; 5:129–139.
9. Le J, Vilcek J: TNF and IL-1: Cytokines with multiple overlapping biological activities. Lab Invest 1987; 56:234–282.
10. Waage A, Halstensen A, Espevik T. Association between tumor necrosis factor in serum and fatal outcome in patients with meningococcal disease. Lancet 1987; 1: 355–357.
11. Girardin E, Grau GE, Dayer JM, Roux-Lombard P. Tumor necrosis factor and interleukin-1 in the serum of children with severe infectious purpura. N Engl J Med 1988; 319:397–400.
12. Marks JD, Marks CB, Luce JM, Montgomery AB, Turner J. Metz CA, Murray JF. Plasma tumor necrosis factor in patients with septic shock: Mortality rate, incidence of adult respiratory distress syndrome. Am Rev Respir Dis 1990; 141:94–97.
13. Hinshaw LB, Tekamp-Olson P, Chang ACK, Lee PA, Taylor FB, Murray CK, Peer GT, Emergon TE, Poassey RB, Juo GC. Survival of primates in LD100 septic shock following therapy with antibody to tumor necrosis factor (TNF alpha). Circulatory Shock 1990; 30:279–292.
14. Remick DG, Strieter RM, Lynch JP, Nguyen D, Eskandari M, Kunkel SL. In vivo dynamics of murine tumor necrosis factor gene expression: Kinetics of dexamethasone-induced suppression. Lab Invest 1989; 60:766–771.
15. Shalaby MR, Halgunset J, Haugen OA, Aarset H, Aarden L, Waage, A. Matsushima K, Kvithyll H. Boraschi D, Lamvil L, Espevik T: Cytokine associated tissue injury and lethality in mice: a comparative study. Clin Immunol Immunopathol 1991; 61: 69–82.
16. Standiford TJ, Strieter RM, Lukacs NW, Kunkel SL. Neutralization of IL-10 increases lethality in endotoxemia. J Immunol 1995; 155:2222–2229.
17. Tracey KJ, Fong Y, Hesse DG, Manogue KR, Lee AT, Kuo GC, Lowry SF, Cerami A.

Anti-cachectin/TNF monoclonal antibodies prevent septic shock during lethal bacteremia. Nature 1987; 330:662–664.
18. van der Poll T, Mardfhant A, Buurman WA, Berman L, Keogh CV, Lazarus DD, Nguyen L, Goldman M, Moldawer LL, Lowry SF: Endogenous IL-10 protects mice from death during septic peritonitis. J Immunol 1995; 155:5397–5401.
19. Michie HR, Mangue KR, Spriggs DR, Revhaug A, O'Wyer S, Dinarello CA, Cerami A, Wolff SM, Wilmore DW. Detection of circulating tumor necrosis factor after endotoxin administration. N Engl J Med 1988; 318:1481–1486.
20. Sibille Y, Reynolds HJ. Macrophages and polymorphonuclear neutrophils in lung defense and injury Am Rev Respir Dis 1990 141:471–501.
21. Shiffman E, Corcoran BA, Wahl, SM. N-formylmethionyl peptides as chemoattractants for leukocytes. Proc Natl Acad Sci USA 1975; 72:1059–1062
22. Ward PA Newman LJ. A neutrophil chemotactic factor from human C′5. J Immunol 1969; 102:93–99.
23. Ford-Hutchinson AW, Bray MA, Doig MV, Smith MJ. Leukotriene B4, a potent chemokinetic and aggregating substance released from polymorphonuclear leukocytes. Nature 1980; 286:262–265.
24. Baggiolini M, Walts A, Kunkel SL. NAP-1/IL-8, a novel cytokine that activates neutrophils. J Clin Invest 1989; 57:2837–2841.
25. Matsushima K Oppenheim JJ. Interleukin-8 and MCAF: Novel inflammatory cytokines inducible by IL-1 and TNF. Cytokine 1991; 1:2–13.
26. Leonard EJ, Yoshimura T. Human monocyte chemoattractant protein-1 (MCP-1) Immunol Today 1990; 11:97–100.
27. Rolfe MW, Kunkel SL, Standiford TJ, Orringer MB, Phan SH, Evanoff HL, Burdick MD, Strieter RM, Expression and regulation of human pulmonary fibroblast-derived monocyte chemotactic peptide-1. Am J Physiol 1992; 7:536–545.
28. Strieter, RM, Chensue, SW, Basha, MA, Standiford, TJ, Lynch, JP, Kunkel, SL. Human alveolar macrophage gene expression of interleukin-8 by TNF-a, LPS and IL-1b. Am J Respir Cell Mol Biol 1990; 2:321–326.
29. Strieter RM, Kunkel SL, Showell HJ, Remick DG, Phan SH, Ward PA, Marks RM. Endothelial cell gene expression of a neutrophil chemotactic factor by TNF, LPS, and IL-1. Science 1989; 243:1467–1469.
30. Strieter, RM, Phan, SH, Showell, HJ, Remick, DG, Lynch, JP, Genard, M, Raiford, C, Eskandari, M, Marks, RM Kunkel SL. Monokine-induced neutrophil chemotactic factor gene expression in human fibroblasts J Biol Chem 1989; 264:10621–10626.
31. Thornton, AJ, Strieter, RM, Lindley, I, Baggiolini, M, Kunkel, SL, Cytokine-induced gene expression of a neutrophil chemotactic factor/interleukin-8 by human hepatocytes. J Immunol 1990; 144:2609–2613.
32. Standiford TJ, Kunkel SL, Chensue SW, Lynch JP, Toews GB, Westwick J, Strieter RM. Interleukin-8 gene expression by a pulmonary epithelial cell line; a model for cytokine networks in the lung. J Biol Chem 1991; 266:9912–9918.
33. Chensue SW, Ottrness IG, Higashi GI, Forch C, Kunkel SL. Monokine production by hypersensitivity (Schistosoma mansoni egg) and foreign body (Sephadex bead) –type granuloma macrophages. Evidence for sequential production of IL-1 and TNF. J Immunol 1989; 142:1281–1286.
34. Sherry B, Cerami A. Cachectin/tumor necrosis factor exert endocrine, paracrine, and

autocrine control of the inflammatory response. J Cell Biol 1988; 107:1269–1277.
35. Amiri PR, Locksley M, Parslow TG, Sadick M, Rector E, Ritter D, McKerrow JH. Tumor necrosis factor restores granulomas and induces parasite egg laying in schistosome-infected SCID mice. Nature 1992; 356:604–607.
36. Cytokine responses during mycobacterial and schistosomal antigen-induced pulmonary granuloma formation. J Immunol 1994; 145:1105–1113.
37. Chensue SW, Terebuh PD, Warmington KS, Hershey SD, Evanoff HL, Kunkel SL, Higashi GI. Role of IL-4 and IFN-gamma in schistosoma mansoni egg-induced hypersensitivity granuloma formation. J Immunol 1992; 148:900–906.
38. Arend WP. Interleukin-1 receptor antagonist a new member of the interleukin-1 family. J Clin Invest 1991; 88:1445–1451.
39. Jansen J, van der Poll T, Levi M, ten Cate H, Galati J, ten Cate JW, van Deventer SJH. Inhibition of the release of soluble tumor necrosis factor receptors in experimental endotoxemia by an anti-tumor necrosis factor antibody. J Clin Invest 1995; 15:45–54.
40. Higuchi, M, Aggarwal, BB. Differential roles of two types of the TNF receptor in TNF-induced cytotoxicity, DNA fragmentation, and differentiation. J Immunol 1994; 152:4017–4025.
41. Peppel, K, Crawford, D, and Beutler, B. A tumor necrosis factor (TNF) receptor-IgG heavy chain chimeric protein as a bivalent antagonist of TNF activity. J Exp Med 1991; 174:1483.
42. Chensue SW, Warmington KS, Berger AE, Tracey DE. Immunohistochemical demonstration of interleukin-1 receptor antagonist (IRAP) and IL-1 in human lymphoid tissue and granulomas. Am J Pathol 1992; 140:269–274.
43. Chensue SW, Ruth JH, Warmington K, Lincoln P, Kunkel SL. In vivo regulation of macrophage IL-12 production during type 1 and type 2 cytokine-mediated granuloma formation. J Immunol 1995; 155:3546–3551.
44. Chensue SW, Warmington KS, Ruth J, Lincoln PM, Kunkel, SL. Crossregulatory role of gamma interferon, IL-4, and IL-10 in schistosome egg granuloma formation: In vivo regulation of Th activity and inflammation. Clin Exp Immunol 1994; 98:395–400.

8

The Role of Alveolar Macrophages in Pulmonary Immune Function

GEORG KRAAL, ELLEN BROUG, J. FREEK VAN IWAARDEN, and JEK H. A. PERSOONS

Free University
Amsterdam, The Netherlands

THEO THEPEN

Utrecht University
Utrecht, The Netherlands

I. Introduction

The respiratory system is the principal organ for gas exchange in air breathing animals. It can be subdivided in an upper and lower respiratory tract, in which the former has a predominant conductive function, whereas the latter is the actual respiratory, gas-exchanging part. For optimal gas exchange, lungs are provided with an extremely large surface area, vital for its function, but at the same time exposing the internal environment of the body to the external air.

Together with oxygen, antigenic material present in ambient air will enter the lungs and make contact with the respiratory epithelium. Considering the large surface area of the respiratory epithelium and the volume of air inspired on a daily base, it is remarkable that there is only a low prevalence of inflammation under normal conditions. This indicates that defense mechanisms against airborne antigens, among which potentially harmful pathogens are tightly controlled and regulated. For this, the lungs are equipped with an elaborate immune system comprising organized lymphoid structures, including lung-draining lymph nodes and mucosa-associated lymphoid tissues, and also a variety of immunocompetent cells dispersed throughout the tissues of the lungs and participating in its defense mechanisms.

In this chapter, the role of the alveolar macrophage in the regulation of the immunological responses in the lungs is described. Emphasis is on the suppressive regulatory role of alveolar macrophages on the local immune system. In addition, recent developments on the influence of neuroendocrine regulatory networks on alveolar macrophage function are described.

II. Alveolar Macrophages Suppress Pulmonary Immune Functions

Of all lymphoid and nonlymphoid cell types of the lung, the alveolar macrophage (AM) is probably best studied, not in the least because the cells are easily accessible in bronchoalveolar lavages (1–4). As the alveolar macrophage is situated at the interface of air and tissue in the alveoli and alveolar ducts, it is the first cell to encounter inhaled antigens and is therefore thought to play an important role in regulating local immune responses.

From many in vitro studies, it has become clear that AMs obtained from rodent lungs had a suppressive effect on the proliferation of antigen- or mitogen-stimulated T lymphocytes when compared with blood monocytes or macrophages from other tissues (3,5–7). This inability to provide an accessory function was also seen in the activation of B cells to produce antibodies. Differences in the activating function of AMs were observed between species, and even enhancing effects have been described. This became especially apparent when in the in vitro experiments different ratios of alveolar macrophages and T cells were used. At low AM/lymphocyte ratios some stimulatory activity can be seen, whereas at high AM/lymphocyte ratios, AMs are strongly suppressive. This has been explained by the complex secretory activity of the AMs, whereby at low AM concentrations, stimulating activity prevails, and at high AM/lymphocyte ratios, suppressive factors will predominate (8,9). Such a dilution can be attributed to factors produced by AMs themselves, but it can also be explained by contaminating dendritic cells in the AM population. The strong stimulatory activity of dendritic cells, which can be seen at even very low cell ratios, may become apparent when the suppressive activity of the AMs drops below a certain level.

Not only the presence of other cell types in populations of isolated AMs will influence results obtained in vitro studies, it is also clear that AMs themselves are quite heterogeneous. Especially under circumstances of inflammation, the influx of monocytes will lead to a more permissive antigen presentation capacity, whereas under steady-state suppression of antigen presentation will prevail.

Further evidence for heterogeneity of AMs based on phenotype was found in humans using the RFD1 and RFD7 anitbodies. Suppressive AMs were described to be RFD1^{+} RFD7^{+}, and it has been suggested that this specific, suppressive phenotype was induced by local factors in the lung microenvironment.

In summary, there is ample evidence that AMs function poorly as accessory cells and that during differentiation in the lung microenvironment they acquire this immunomodulatory function.

Further evidence for the downregulating effect of AMs on immune responses has come from series of experiments in which toxic liposomes were instilled into the lungs, by which method alveolar macrophages could selectively be removed.

III. Studies After in Vivo Elimination of Alveolar Macrophages

A. Effects on Pulmonary Immune Function

From a vast series of experiments, it has become clear that chlodronate containing liposomes is extremely efficient in killing macrophages in tissues such as spleen and liver, and at the same time giving minimal damage to other cell types. The method is based on the avid capacity of macrophages to ingest larger particles, after which the accumulation of the chlodronate inside the cell is leading to a rapid cell death through an until now undefined mechanism (12,13). When the method was applied to AMs a rapid and selective killing of AMs could be seen after the intratracheal instillation of chlodronate liposomes. A single administration of the liposomes led to a rapid depletion of the cells, so that within 24–48 hrs approximately 10% or less of the cells were still present (Fig. 1). The depletion did not lead to any apparent damage to other cells and no signs of inflammation were observed. The remaining cells in the alveoli were over 90% AMs. Furthermore, this condition lasted for at least a week, after which AMs slowly started to reappear, thereby creating an animal model with a useful timeframe to study the in vivo function of the AMs (14).

From the immunization studies that were performed with this animal model, the suppressive nature of the AMs was confirmed. Whereas in control animals antigen administration into the lungs leads to virtually no response, the same amount of antigen administered to AM-depleted animals gave rise to a drastic increase of the humoral immune response. This was seen, not only in the local lung-draining lymph nodes but also in the lung parenchyma, where, especially around larger blood vessels, accumulation of antibody-forming cells was found in the course of the immunization. Such cells could never be seen in nondepleted animals.

The initial experiments were performed with TNP-KLH, a T-cell–dependent protein antigen. When similar experiments were done using other types of antigens, an interesting phenomenon occurred. The enhancement of the humoral immune response after depletion of AM could only be observed with T-dependent antigens but not with polysaccharide antigens such as TNP-Ficoll or other T-independent type 2 antigens such as DNP-liposomes (15).

Responses against these type of antigens administered via the lungs were

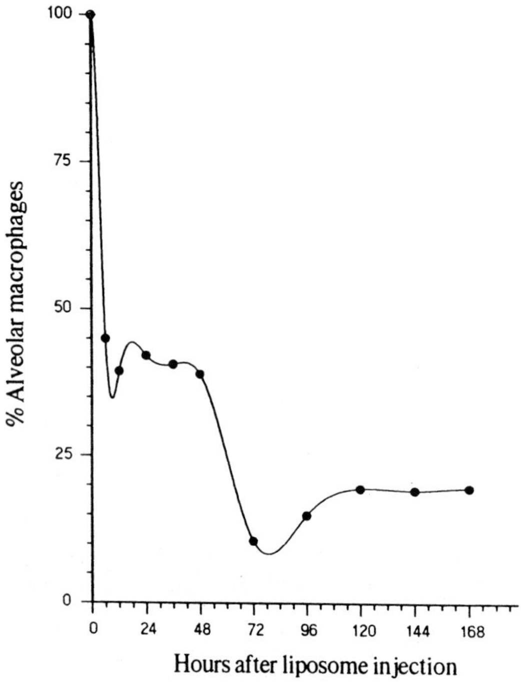

Figure 1 Number of cells in broncholavage fluid at different time points after intratracheal chlodronate liposome administration to rats. At all time points tested, >90% of the cells lavaged were AMs as tested by the expression of the ED1 macrophage-specific monoclonal antibody. (From ref. 24.)

predominantly systemic, albeit low. This argues against a mere scavenger function of AMs, so that elimination of the cells will lead to more free antigen but is in line with a suppressive nature, in particular on subsequent events leading to T-cell activation.

B. Alveolar Macrophages and T-Cell Function

T cells in the lungs can be found in small proportions in the bronchoalveolar fluid, but a larger population of T cells is found in the lung parenchyma, where they occur as single cells (16,17). The functional capacity, in particular the cloning efficiency of these interstitial T cells, is distinctly reduced compared with T cells in the blood, from which they are assumed to be derived. This indicates that the lung microenvironment has a downregulating effect on these cells. When the cloning efficiency of these T cells was assayed after depletion of the AMs it was clear that the proliferative capacity of the interstitial T cells was rapidly restored. The depletion had no effect on the number of T cells in the lung walls, nor did it change the recirculation patterns through the organ (18,19). These results clearly demonstrate that the alveolar macrophages are involved in the downregulation of T-cell functions in the lung, thereby preventing T-cell activation and inflammation.

This lymphocytostatic signal from the AMs to the T cell was further investigated using coculture systems where alveolar macrophages were incubated with various cytokines. Overnight pretreatment of resident alveolar macrophages with granulocyte-macrophage colony-stimulating factor (GM-CSF) abolished the downregulatory activity, which could be enhanced by the addition of tumor necrosis factor-α (TNF-α) (20). The overnight treatment seemed to be obligatory, and interestingly, the downregulating activity was reestablished after prolonged culture of the AMs. When the reactivity of the T cell was studied in more detail, it was found that in the presence of AMs, T cells downregulate CD3 expression, upregulate the interleukin-2 (IL-2) receptor, as well as the production of IL-2, but they can not respond to IL-2 (21). So although AMs allow for activation of the T cell by antigen, they selectively inhibit the proliferative capacity of the T cells. This was found in both human and rodent T cells.

C. Macrophage Depletion and the Response to Recall Antigens

As stated above, responses against inhaled antigens are extremely low under normal conditions; in fact, repeated inhalation with aeroallergens leads to a state of specific immunological tolerance (22,23). This tolerance induction takes place in the upper respiratory tract and has an important function in protecting the body against overreacting by shutting down immune responses, especially of the IgE type, against that antigen. This mucosal tolerance is in essence comparable to oral tolerance induction as seen in the gastrointestinal tract. Experiments to indicate a role for alveolar macrophages in this tolerance induction using the depletion technique made it clear that AMs are not involved in tolerance and that this intrinsic effect of the lungs was induced by other cell types (24). However, a completely different picture was observed when pulmonary immune responses were studied in rats that had previously been primed systemically and subsequently been challenged in the lungs with an antigen. In these experiments, the animals were primed with ovalbumin (OVA), using low doses capable of priming and inducing immunological memory but not eliciting an immune response. When such animals were now challenged, a low, insignificant immune response could be measured in the lungs. In contrast, when primed animals were first depleted of their AMs with liposomes and then challenged with an aerosol of OVA, a strong immune response was found. In particular, the IgE response was strongly increased, which was demonstrated by the presence of large numbers of anti-OVA IgE–secreting B cells in draining lymph nodes and also in the lung parenchyma. In addition to a strongly increased anti-OVA IgE response, many IgE-producing cells were also found with other specificity than against OVA. This bystander effect on IgE switching in the AM-depleted animals was especially apparent in the lung tissue. IgE-producing cells could never be found in tissues of the control animals. Further evidence for a disturbed regulation in the depleted animals could be inferred from the massive influx of T cells into the lungs in the course of the secondary, lung-induced immune response (25).

These results clearly show the importance of the downregulation exerted via the alveolar macrophages on the elicitation of a local immune response in the lung to recall antigens.

D. Alveolar Macrophage Depletion and Antigen Presentation

Although alveolar macrophages are not equipped to perform as antigen-presenting cells, professional antigen-presenting dendritic cells are abundantly present throughout the airways. They are characterized by high expression of major histocompatibility complex (MHC) class II and a high antigen-presenting capacity for naive and immune T cells (17,26). Only after careful removal of the alveolar macrophages from lung cell suspensions, could the population of lung dendritic cells exert its full antigen-presenting capacity in vitro (27).

In the lung, dendritic cells can be found in the proximity of alveolar macrophages, which is suggestive for a regulatory interaction of the two cell types.

Evidence for this came from experiments in which dendritic cells from the lungs are allowed to acquire increasing antigen-presenting capacity after overnight culture. This situation resembles closely the differentiation step of Langerhans cells from the skin (28). This in vitro maturation of lung dendritic cells could be abrogated by coculturing the dendritic cells with alveolar macrophages across a semipermeable membrane, which is suggestive for a role of macrophage-derived soluble factors. Interestingly, this downregulating effect could be blocked by adding monomethyl arginine, an inhibitor of nitric oxide production, to the cultures. Also a role for TNF-α in the suppression of dendritic cell activity was found (27). When dendritic cells were freshly isolated from animals that had been depleted from their alveolar macrophage previously, it was found that they had an antigen-presenting capacity that was comparable to dendritic cells isolated from normal animals but allowed to mature overnight. This indicated that, under normal conditions, alveolar macrophages inhibit optimal antigen presentation by dendritic cells, thereby preventing the initiation of immune responses and limiting the negative consequences of local inflammatory conditions.

IV. Regulatory Activity of Alveolar Macrophages Mediated by Secreted Factors

A. Mechanisms of Regulation

Although alveolar macrophages may have close contacts with cells in the parenchyma of the lungs, and this way can have a direct influence on the behavior of the cells by cell-cell communication, it is obvious from many of the data on the suppressive role of alveolar macrophages that in particular the secretory activity of the macrophages is extremely important. It has been documented that alveolar macrophages can secrete up to 100 different secretory products after stimulation by a variety of signals from other cell types or from themselves (29). Some of these products have opposite effects, and, depending on the stimulus, the production of these factors will be upregulated or downregulated.

In general, three types of regulatory mechanisms can be distinguished by which alveolar macrophage functions are controlled. First, most of the products from macrophages have either a short intrinsic half-life, such as is the case with nitric oxide, or are rapidly degraded by enzymatic breakdown, leading to only local influence or toxicity of the secreted products.

A second route of regulation involves the role of receptors on the macrophage. A good example of this type of regulation can be seen with the production of IL-1 and the IL-1 receptor antagonist (IL-1ra) in response to LPS (30). It is found that the IL-1ra is secreted later in time than the IL-1, which is suggestive for a downregulating role of this antagonist once immune activity has been initiated by IL-1. It can also be envisaged that downregulation of the receptors themselves can regulate the effects of macrophage secretory products (31). Intriguing in this respect is the finding that isolated alveolar macrophages can produce significant amounts of IL-1 receptor antagonist in the absence of any stimulus (32,33).

Direct inhibition of the production of activating signals is a third way to regulate the activity of macrophages. It has been shown that cytokines such as IL-4, IL-10, prostaglandin E_2, and transforming growth factor-β (TGB-β) can indirectly downregulate macrophage activity by suppression of the IFN γ production by proliferating Th1 helper cells (34). Also direct downregulating activity of these factors has been described (35). Interestingly, several of the factors that are known to have a regulatory effect on macrophages, such as IL-10, prostaglandin E_2 and TGF-β are also produced by macrophages themselves (36). This suggests that macrophages can regulate themselves either within the population, as a result of heterogeneity, or by autocrine mechanisms.

B. Autocrine Regulation of Cytokine Production

An example of such regulation of the production of one factor by the other in alveolar macrophages has recently been found by us for the interaction of IL-1 and nitric oxide. Culture of alveolar macrophages in the presence of LPS leads to a dose-dependent nitric oxide production, which can be blocked completely by the addition of the nitric oxide synthase inhibitor NMMA (37). No production was seen when the cells were not stimulated. When the production of IL-1β, IL-6, and TNF-α was measured after LPS stimulation, a biphasic dose-response relationship was found. Again, no production of any of these cytokines was found in the absence of a stimulus. Interestingly, when the nitric oxide production was inhibited by the addition of NMMA, a significant increase in IL-1β and IL-6 production was found on stimulation with high amounts of LPS, but no changes were found for the production of TNF-α (37). In fact, the production of IL-1β and nitric oxide by alveolar macrophages was clearly inversely related (Fig. 2). It was ruled out that nitric oxide itself was detrimental for the cytokines in the culture, because adding exogenous IL-1 or IL-6 to NO-containing cultures did not show any oxidizing effects on the cytokines. This indicates that the observed effects are based on regulatory events in the cell. How these regulatory pathways act is unclear. A major mechanism of the action of NO is to stimulate the soluble enzyme guanylate cyclase, thereby accumulating cGMP (37). Cyclic GMP itself has been shown to be an intracellular messenger for the induction of cytokine synthesis. However, other pathways cannot be excluded.

What is the physiological role of this autoregulation of cytokine production as observed in alveolar macrophages? In vitro, alveolar macrophages can directly affect many cellular functions through this L-arginine–nitric oxide pathway, such as T-cell cytokine production, T-cell proliferation, and antigen presentation by dendritic cells. But also, indirect effects via IL-1 can be important in the regulation of immune responses. It has been demonstrated that under asthmatic conditions the alveolar macrophages show reduced suppressive characteristics and increased levels of IL-1β (38,39). This suggests a role for dysregulation of factor production by alveolar macrophages in the pathogenesis of asthma. Therefore, the auto-

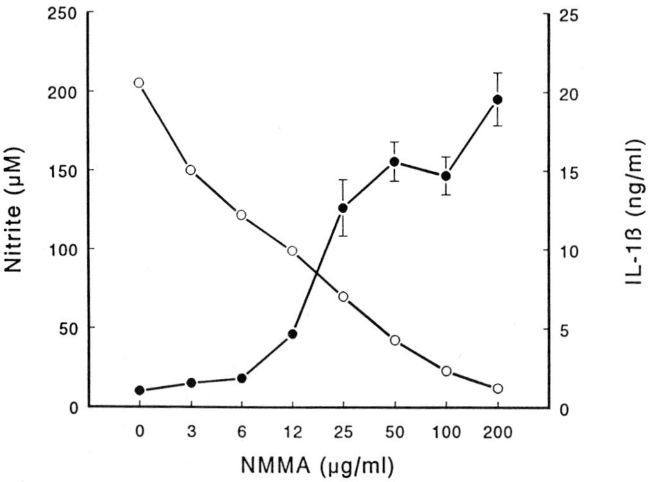

Figure 2 Effect of nitric oxide synthase inhibition on the LPS-induced IL-1β production by AMs. AMs were cultured with various concentrations of the nitric oxide synthase inhibitor NMMA in the presence of 100 µg/ml LPS. After 24 hr of culture, nitrite (open circles) and IL-1β (closed circles) concentrations were determined. Statistical analysis revealed that IL-1β and nitrite levels are inversely related. (From ref. 37.)

regulation of IL-1 and IL-6 by nitric oxide represents an important mechanism by which an initial inflammatory response can be regulated by nitric oxide levels.

V. Mechanisms Which Control Alveolar Macrophage Activity

The above results indicate that alveolar macrophages are capable of responding differentially to changing conditions in the microenvironment, so that with flexibility the most adequate response to occurring events can be given.

This directly implies that the macrophages, which are themselves so important in the maintenance of the lung homeostasis, are regulated in their function by factors in the lung.

Recently, we have started a series of experiments to unravel the regulatory mechanisms that determine the activity of alveolar macrophages in situ. In particular, we have focused on the role of the neuroendocrine system on the secretory activity of macrophages and the way this affects the immune reactivity in the lung.

A. Neuroendocrine Regulation

The relationship between the immune system and the neuroendocrine system is becoming more apparent. Immunocompetent cells express receptors for a great variety of hormones and neuropeptides (41–43), and binding of these hormones and neurotransmitters to receptors can result in changes in important immune functions such as antibody secretion, cytokine synthesis, and natural killer cell activation (44, 45). But activated lymphocytes can also produce hormones such as prolactin, thyrotropin, and growth hormone, demonstrating the bidirectional communication between the two systems (46).

Furthermore, various immunological compartments are highly innervated, and even close contacts between nerve endings and immune cells have been observed (46,47). This is especially the case in the lungs, by which the function of the organ can be controlled in many ways. The autonomic nervous system in the lung is primarily involved in controlling the air transport by regulating the airway caliber, the pulmonary blood flow, and the mucociliary clearance. In addition, the close contacts between nerve fibers and immune cells in the lungs is suggestive for regulation of the suppressive nature of the immune response in the lungs by neuroendocrine hormones. But, although emotional arousal can affect the pulmonary function by bronchoconstriction under asthmatic conditions, a direct contribution of stress to the onset or severity of asthma has not been proven.

B. Effects of Stress on Immune Reactivity in the Lung

Using an animal model for stress, the lung immune response and the secretory capacity of alveolar macrophages was investigated. To do this, rats were exposed to mild inescapable foot shock stress. This type of stressor is considered to be comparable to emotional stress. To test first whether immune responses in the lung could be affected by this stressor, rats were immunized intratracheally with TNP-KLH as a protein antigen. When serum titers of these animals were assayed for anti-TNP antibodies, a substantial increase of specific Ig was found. In particular, the levels of IgG, IgA, and IgE isotype antibodies against TNP were enhanced (49). Similar results, but with substantial higher absolute amounts of Ig, were seen when the animals had been primed systemically and were then stressed and challenged with the antigen. When the responses were analyzed at tissue level, the production of antibody seemed to be restricted to the lung-draining lymph nodes (49).

C. Effects of Stress on the Activity of Alveolar Macrophages

Having established a direct role for the stressor in upregulation of the local immune response in the lung, and in the light of the above-stated downregulating function of AMs, we wanted to see whether stress could have a direct effect on the activity of AMs. Therefore, animals were exposed to stress and immediately thereafter AMs were isolated from their lungs. The cells were cultured for 24 hr in the presence of varying doses of LPS, and their ability to produce IL-1β, TNF-α, nitric oxide, and IL-6 was tested.

AMs from nonstressed animals readily produced these different factors after stimulation with LPS. For the cytokines, a biphasic dose-response relationship was observed, but this was not the case for NO production. When the animals had been exposed to stress, the isolated AMs showed a marked increase of IL-1β and TNF-α secretion on LPS stimulation in vitro, but no changes were observed in the production of IL-6. In contrast, a strong decrease of NO production was observed after exposure to the stressor (Figs. 3 and 4).

Stress therefore results in rapid and apparent changes in the functional activity of alveolar macrophages. Furthermore, these changes seemed to be divergent for the different mediators that were assayed. This divergence could be related to differences within the population of AMs tested, possibly related to differences in LPS receptor expression or affinity (51,52). Alternatively, stress could affect the translation and transcription of cytokine genes.

How can stress affect the function of AMs? Various neuropeptides and receptors have been demonstrated at important parts of the lungs (47). In the bronchus, associated lymphoid tissue neuroimmune connections have been found (48). Furthermore, nerve endings in the alveolar walls have been demonstrated (53,54), indicating that at the level of the AM, a direct neuroimmune link may exist, which is in line with the presence of many neuropeptide receptors on macrophages (55,56). In fact, comparable effects as seen with our stress experiments on IL-1β secretion were found in experiments where AMs were incubated

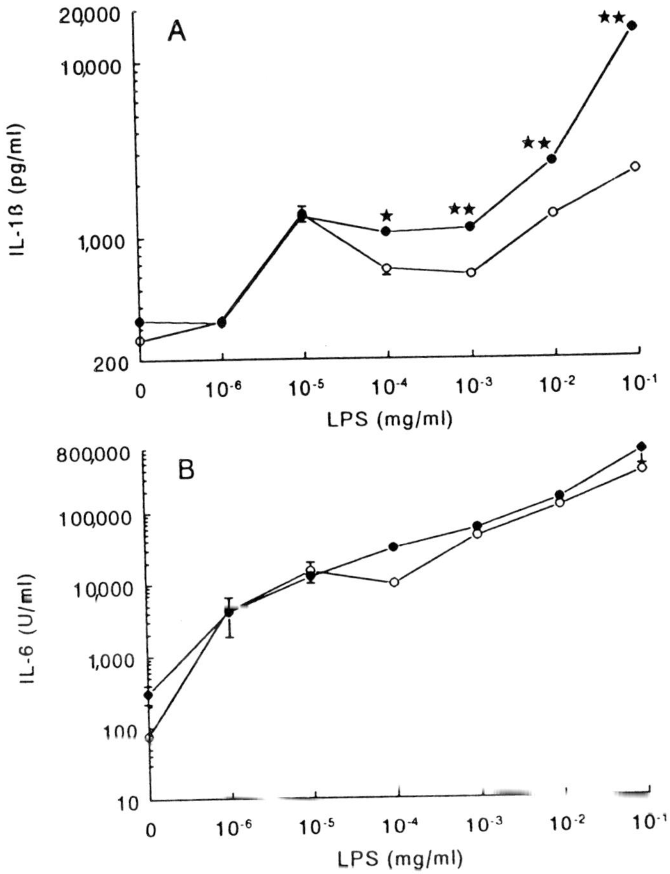

Figure 3 Effects of mild inescapable electric foot shocks on LPS-induced Il-1β and Il-6 secretion by AMs. AMs from stressed (closed circles) and nonstressed animals (open circles) were cultured for 24 hrs with indicated concentrations of LPS. Stress resulted in a significant enhancement of IL-1β production but had no effect on the production of IL-6 (From ref. 50.)

with neuropeptides such as bombesin and neurotensin (57,58). Therefore, it may very well be that neural pathways are involved in the observed effects of stress on AM function.

In addition to direct neuronal effects, a role for the classic stress hormones such as adrenocorticotropin (ACTH), corticosterone (CORT), and prolactin, as well as the catecholamines cannot be excluded. Although glucocorticoid response elements are functionally linked by cytokine-promotor regions in the genome

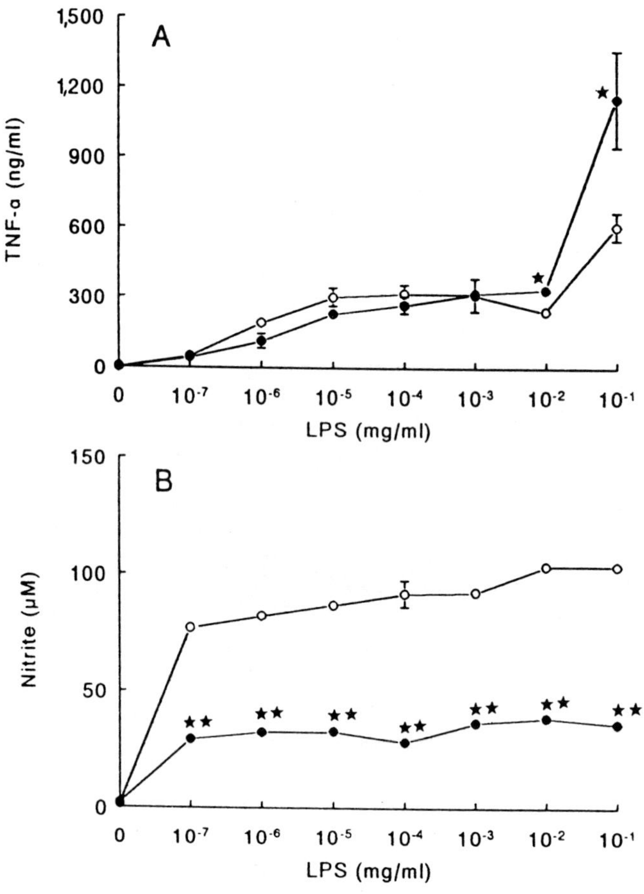

Figure 4 Effects of mild inescapable electric foot shocks on LPS-induced TNF-α and nitric oxide secretion by AMs. AMs from stressed (closed circles) and nonstressed animals (open circles) were cultured for 24 hrs with indicated concentrations of LPS. Stress resulted in a significant downregulation of nitric oxide secretion and resulted in enhanced TNF-α secretion with high doses of LPS stimulation. (From ref. 50.)

(59,60), it is unclear whether enhanced corticosteroid levels as seen after stress can be responsible for the observed enhancements of IL-1β and TNF-α production as seen in our experiments. Most data on the relationship between the production of cytokines and corticosteroids point to a downregulating, suppressive activity of CORT. However, the evidence for a role of corticosteroids in the positive activation of alveolar macrophages was recently obtained in experiments where isolated AMs were incubated with a corticosteriod analogue, dexamethasone (61). It turned out that dexamethasone incubation led to enhanced IL-1β production, provided that the stimulus was only given for a short time. Longer incubation with the same dexamethasone concentration led to reduction in cytokine production.

This clearly suggests that corticosteroids, especially under conditions that they are present in relative high concentrations but for only a short time, in fact the conditions that are seen after stress in vivo, can have a stimulating effect on the production of cytokines, thereby preparing the system to attack an impending inflammatory stimulus associated with a stressful event. This is in line with the observed enhancement of the humoral immune response in rats that had been subjected to stress.

Catecholamines, norepinephrine (noradrenaline) from sympathetic nerve endings, and epinephrine (adrenaline) from the adrenal medulla activate α- and β-adrenoceptors on various cell types in the respiratory tract (62). The presence of β-adrenoceptors on AMs has been demonstrated, and it is assumed that this cell population accounts for the majority of β-receptors in the deep lungs (63–66). The role of these receptors in the functional activity of AMs is not clear. The β-adrenoceptors are coupled to adenylate cyclase, and stimulation of the receptor induces profound increases of cAMP (64,65). However, stimulation of the receptor on AMs did not affect oxidative metabolism or the uptake of bacteria (64), but cytoplasmic motility was downregulated (67). Furthermore, from studies on peripheral macrophages, it has become clear that stimulation of the α-adrenoceptor leads to enhanced TNF-α production (68–70).

D. Stress and Lung Disease

Enhanced activation of the pulmonary immune functions may well be detrimental for the organism. Stress can perturb the well-defined downregulating role of alveolar macrophages by altering the capacity to produce nitric oxide, since this factor has been indicated as crucial for the immune-suppressive activity of AMs. Of interest is that IL-1 and TNF-α have both been implicated in the pathogenesis of asthmatic conditions (71–74), implying that stress-induced enhancement of the capacity of AMs to produce the cytokines can contribute to the facilitation of inflammatory processes in this disease.

Furthermore, nitric oxide also affects broncodilation by its mode of action on smooth muscle cells (75,76), so that reduced NO production by AMs after stress may contribute to bronchoconstriction as observed in asthmatics.

Also, the antimicrobial role of nitric oxide is important (77), and it can be imagined that stress exposure may interfere with the ability to deal with pathogenic microorganisms, perhaps contributing to the occurrence of recurrent pulmonary infections as seen in asthmatics.

VI. Conclusions

AMs are central cells in the maintenance of homeostasis in the lungs. In addition to their scavenger function, they regulate the activity of other important cell types, such as dendritic cells and T lymphocytes in the interstitial tissue of the lung. This is primarily mediated by the differential production of humoral factors. Whether these factors act directly on the target cells is unclear, although close contacts between the AMs and interstitial cells have been observed. Indirect action can also be envisaged, for which intermediate cells, such as alveolar wall cells, are likely candidates. In fact, preliminary data indicate that AMs can influence the secretory activity of alveolar type II cells (E. Broug, J. F. van Iwaarden, G. Kraal, in preparation). These type II cells constitute a minority of alveolar wall cells (approximately 15% [78]), but they are known for their active secretion of a variety of proinflammatory cytokines and chemokines. It is clear from the above that the activity of AMs themselves is also subject to control, which is executed by several mechanisms. The neuroendocrine system seems to play an important role in this regulation, either via stress hormones such as corticosteroids or via the autonomic nervous system. Again, these influences can be direct or indirect. In fact, we have recently shown that stress severely affects the activity of type II alveolar wall cells.

Although at this point still speculative, it may well be that AMs and alveolar type II cells can either enhance or control each other's activity.

Adding to the complexity of the system, it emphasizes the importance of the tight regulation of the immune functions in the lung, in which may cell types seem to play a role. Lung disease and asthmatic conditions can therefore result from disturbances of this regulatory system at many levels. Understanding the many aspects of regulation will hopefully lead to better strategies for treatment.

References

1. Lipscomb, MF, Lyons CR, Nunez, G, Ball EJ, Stastny P, Vial W, Lem V, Weissler J, Miller LM, Toews GB. Human alveolar macrophages: HLA-DR positive macrophages that are poor stimulators of a primary mixed leukocyte reaction. J Immunol 1986; 136:497–504.
2. Ettensohn DB, Roberts NJ. Human alveolar macrophage support lymphocyte responses to mitogens and antigens: analysis and comparison with autologous peripheral blood-derived monocytes and macrophages; Am Rev Respir Dis 1983; 128:516–522.
3. Shellito J, Kaltreider HB. Heterogeneity of immunological function among subfractions of normal rat alveolar macrophages Am Rev Respir Dis 1984; 129:747–753.
4. Lehnert BE, Valdez YE, Holland LM. Pulmonary macrophages: alveolar and interstitial populations. Exp Lung Res 1985; 9:177–190.
5. Holt PG. Inhibitor activity of unstimulated alveolar macrophages on T-lymphocyte blastogenic response. Am Rev Respir Dis 1978; 118:791-793.

6. McCombs CC, Michalski JP, Westerfield BT, Light RW. Human alveolar macrophages suppress the proliferative response to peripheral blood lymphocytes. Chest 1982; 82:266–271.
7. Toews GB, Vial WC, Dunn MM, Guzzeta P, Nunez G, Stastny P, Lipscomb MF. The accessory cell function of human alveolar macrophages in specific T cell proliferation. J Immunol 1984; 132:181–186.
8. Herscowitz HB. In defense of the lung: paradoxal role of the pulmonary alveolar macrophage. Ann Allergy 1985; 55:634–650.
9. Holt PG. Down-regulation of immune responses in the lower respiratory tract: the role of alveolar macrophages. Clin Exp Immunol 1986; 63:261–270.
10. Murphy MA, Herscowitz HB. Heterogeneity among alveolar macrophages in humoral and cell mediated immune responses: separation of functional subpopulations by density gradient centrifugation on percoll. J Leuk Biol 1984; 35:39–54.
11. Spiteri MA, Poulter LW. Characterization of immune inducer and suppressor macrophages from the normal human lung. Clin Exp Immunol 1991; 83:157–162.
12. Van Rooijen N, Sanders A. Liposome mediated depletion of macrophages: mechanism of action, preparation of lipsomes and applications. J Immunol Methods 1994; 174:83–93.
13. Van Rooijen N, Kors N, Kraal G. Macrophage subset repopulation in the spleen: differential kinetics after liposome mediated elimination. J Leukoc Biol 1989; 45: 97–104.
14. Thepen T, Van Rooijen N, Kraal G. Alveolar macrophage elimination in vivo is associated withban increase in pulmonary immune response in mice. J Exp Med 1989; 170:499–509.
15. Thepen T, Hoeben K, Breve J, Kraal G. Alveolar macrophages down-regulate local pulmonary immune response against intratracheally administered T cell deponent but not T cell independent antigens. Immunology 1992; 76:60–64.
16. Holt PG, Degebrodt A, O'Leary C, Krska K, Plozza T. T cell activation by antigen presenting cells from lung tissue digests: suppression by endogenous macrophages. Clin Exp Immunol 1985; 62:586–593.
17. Holt PG, Schon-Hegrad MA. Localisation of T cells, macrophages and dendritic cells in rat respiratory tract tissue: implications for immune function studies. Immunology 1987; 62:349–356.
18. Nelson S, Strickland D and Holt PG. Selective attrition of non-recirculating T cells during normal passage through the lung vascular bed. Immunology 1990; 69:476–481.
19. Strickland DH, Thepen T, Kees UR, Kraal G, Holt PG. Regulation of T cell function in lung tissue by pulmonary alveolar macrophages. Immunology 1993; 80:266–272.
20. Bylik N, Holt PG. Cytokine modulation of the immunosuppressive phenotype of pulmonary alveolar macrophage populations. Immunology 1995; 86:231–237.
21. Upham JW, Strickland DH, Bylik N, Robinson BWS, Holt PG. Alveolar macrophages from humans and rodents selectively inhibit T-cell proliferation but permit T-cell activation and cytokine secretion. Immunology 1995; 84:142–147.
22. Sedgwick JD, Holt PG. Induction of IgE isotype specific tolerance by passive antigenic stimulation of the respiratory mucosa. Immunology 1985; 50:625–630.
23. Holt PG, McMennamin C. Defence against allergic sensitization in the healthy lung: role of inhalation tolerance. Clin Exp Allergy 1989; 19:255–262.

24. Thepen T, McMennamin C, Oliver J, Kraal G, Holt PG. Regulation of immune response to inhaled antigen by alveolar macrophages: differential effects of in vivo alveolar macrophage elimination on the induction of tolerance vs. immunity. Eur J Immunol 1991; 21:2845–2850.
25. Thepen T, McMennamin C, Girn B, Kraal G, Holt PG. Regulation of IgE production in pre-sensitized animals: in vivo elimination of alveolar macrophages preferentially increases IgE responses to inhaled allergen. Clin Exp Allergy 1992; 22:1107–1114.
26. Holt PG, Schon-Hegrad MA, Oliver J, Holt BJ, McMennamin C. A contiguous network of dendritic antigen presenting cells within the respiratory epithelium. Int Arch Allergy Appl Immunol 1990; 91:155–159.
27. Holt PG, Oliver J, Bilyk N, McMenamin C, Kraal G, Thepen T. Downregulation of antigen presenting cell functions of pulmonary dendritic cells in vivo by resident alveolar macrophages. J Exp Med 1993; 177:397–407.
28. Schuler G, Steinman RM. Murine epidermal Langerhans cells mature into potent immunostimulatory dendritic cells in vitro. J Exp Med 1985; 161:526–546.
29. Sibille Y, Reynolds HY. Macrophages and polynorphonuclear neutrophils in lung defense and injury. Am Rev Respir Dis 1990; 141:471–501.
30. Arend WP, Smith MF, Janson RW, Joslin FG. IL-1 receptor antagonist and IL-16 production in human monocytes are regulated differently. J Immunol 1991; 147: 1526–1530.
31. Ding AH, Sanchez E, Srimal S, Nathan CF. Macrophages rapidly internalize their tumor necrosis factor receptors in response to bacterial lipopolysaccharide. J Biol Chem 1989; 264:3924–3929.
32. Moorse A, Strieter RM, Rolfe MW, Standiford TJ, Burdick MD, Kunkel SL. Expression and regulation of human alveolar macrophage-derived interleukin-1 receptor antagonist. Am J Respir Cell Mol Biol 1992; 6:569–575.
33. Galve-de Rochemonteix B, Nicod LP, Chicheportiche R, Lacraz S, Baumberger C, Dayer JM. Regulation of interleukin-1ra, interleukin-1 alpha, interleukin-1 beta production by human alveolar macrophages with phorbol myristate acetate, lipopolysaccharide and interleukin-4. Am J Resp Cell Mol Biol 1993; 8:160–168.
34. Hsu D-H, Moore KW, Spits H. Differential effects of IL-4 and IL-10 on IL-2 induced IFN-γ synthesis and lymphokine activated killer activity. Int Immunol 1992; 4:563–569.
35. Monick M, Glazier J, Hunninghake GW. Human alveolar macrophages suppress interleukin-1 (IL-1) activity via the secretion of prostaglandin E_2. Am Rev Respir Dis 1987; 135:72–77.
36. Fiorentino DF, Zlotnik A, Mosmann TR, Howard M, O'Garra A. IL-10 inhibits cytokine production by activated macrophages. J Immunol 1991; 147:3815–3822.
37. Persoons JHA, Schornagel K, Tilders FFH, de Vente J, Berkenbosch F, Kraal G. Alveolar macrophages autoregulate IL-1 and IL-6 production by endogenous nitric oxide. Am J Respir Cell Mol Biol 1996; 14:272–278.
38. Waldman SA, Murad F. Cyclic GMP synthesis and function. Pharmacol Rev 1987; 39:163–196.
39. Lee TH, Poston R, Godard P, Bousquet J. Macrophages and allergic asthma. Clin Exp Allergy 1991; 21:22–23.
40. Borish L, Mascali JJ, Dishuck J, Beam WR, Martin RJ, Rosenwasser LJ. Detection of alveolar macrophage-derived IL-1β in asthma. J Immunol 1992; 149:3078–3082.

41. Kay AB. Asthma and inflammation. J Allergy Clin Immunol 1991; 87:893–910.
42. Tonnell AB, Joseph M, Gosset P, Fournier E, Capron A. Stimulation of alveolar macrophages in asthmatic patients after local provocation tst. Lancet 1983; 1:1406–1408.
43. Herscowitz HB. In defense of the lung: paradoxical role of the pulmonary alveolar macrophages. Ann Allergy 1985; 55:634–648.
44. Martin M, Resch K. Interleukin-1: more than a mediator between leukocytes. TIPS 1988; 9:171–177.
45. Dinarello CA, Ikejima T, Warner SJC, Orencole SF, Lonneman G, Cannon JG, Libbey P. Interleukin-1 induces interleukin-1. J Immunol 1987; 139:1902–1910.
46. Mantovani A, Dejana E. Cytokines as communication signals between leukocytes and endothelial cells Immunol Today 1989; 11:370–375.
47. Barnes PJ, Baraniuk JM, Belvisi MG. Neuropeptides in the respiratory tract. Part I. Am Rev Respir Dis 1991; 144:1187–1198.
48. Nohr D, Weihe E. The neuroimmune link between in the bronchus associated lymphoid tissue (BALT) of cat and rat: peptides and neural markers. Brain Behav Immun 1991; 5:84–101.
49. Persoons JHA, Berkenbosch F, Schornagel K, Thepen T, Kraal G. Increased specific IgE production in lungs after induction of acute stress in rats. J Allergy Clin Immunol 1995; 95:765–770.
50. Persoons JHA, Schornagel K, Breve J, Berkenbosch F, Kraal G. Acute stress affects cytokines and NO production by alveolar macrophages differently. Am J Respir Crit Care Med 1995; 152:619–624.
51. Wright SD. Multiple receptors for endotoxin. Curr Opin Immunol 1991; 3:83–90.
52. Morrison DC, LeiM-G, Kirikae T, Chen T-Y. Endotoxin receptors on mammalian cells. Immunobiology 1993; 187:212–226.
53. Hertweck MS, Hung KS. Ultrastructural evidence for the innervation of pulmonary alveoli. Experientia 1980; 15:112–113.
54. Fox B, Bull TB, Guz A. Innervation of alveolar walls in the human lungs. An electron microscopical study. J Anat 1980; 131:683–692.
55. Peck R. Neuropeptides modulating macrophage function. Ann NY Acad Sci 1987; 496:264–270.
56. Hartung H. Activation of macrophages by neuropeptides. Brain Behav Immun 1988; 2:275–281.
57. Lemaire I. Bombesin-related peptides modulate interleukin-1 production by alveolar macrophages. Neuropeptides 1991; 20:217–223.
58. Lemaire I. Neurotensin enhances IL-1 production by alveolar macrophages. J Immunol 1988 140:2983–2988.
59. Beato, M. Gene regulation by steroid hormones. Cell 1989; 56:335–344.
60. Barnes PJ, Adcock I. Anti-inflammatory actions of steroids: molecular mechanisms. TiPS 1993; 14:436–441.
61. Broug E, Kraal G. Dose and time dependent activation of rat alveolar macrophages by glucocorticoids. Clin Exp Immunol 1996; 104:332–336.
62. Barnes PJ. Adrenergic receptors of normal and asthmatic airways. Eur J Respir Dis 1984; 65:72–79.
63. Nijkamp FP, Engles F, Henricks PAJ, Van Oosterhout AJM. Mechanisms of β-adren-

ergic receptor regulation in lungs and its implications for physiological responses. Physiol Rev 1992; 72:323–367.
64. Henricks PAJ, Van Esch B, Van Oosterhout AJM, Nijkamp FP. Specific and nonspecific effects of β-adrenoceptor agonists on guinea pig alveolar macrophage function. Eur J Pharmacol 1988; 152:321–330.
65. Hjemdahl P, Larsson K, Johansson MC, Zeterlund A, Eklund A. Beta-adrenoceptors on human alveolar macrophages isolated by elutriation. Br J Clin Pharmacol 1990; 30:673–682.
66. Ligett SB. Identification and characterization of homogeneous population of β-adrenergic receptors on human alveolar macrophages. Am Rev Respir Dis 1989; 139: 552–555.
67. Fukushima T, Sekizawa K, Jin Y, Jamaya M, Sasaki H, Takishima T. Effects of β-adrenergic receptor activation on alveolar macrophage cytoplasmic motility. Am J Physiol 1993; 265:L67–L72.
68. Spengler RN, Allen RM, Remick DG, Strieter RM, Kunkel SL. Stimulation of α-adrenergic receptor augments the production of macrophage-derived tumor necrosisi factor. J Immunol 1990; 145:1430–1434.
69. Hu X, Goldmuntz EA, Brosnan CF. The effect of norepinephrine on endotoxin-mediated macrophage activation. J Neuroimmunol 1991; 31:35–42.
70. Ignatowski TA, Spengler RN. Regulation of macrophage derived tumor necrosis factor production by modification of adrenergic receptor sensitivity. J Neuroimmunol 1995; 61:62–70.
71. Pujol J, Cosso B, Daures J, Clot J, Michel F, Godard P. Interleukin-1 release by alveolar macrophages in asthmatic patients and healthy subjects. Int Arch Allergy Appl Immunol 1990; 91:207–210.
72. Borish L, Mascali JJ, Dishuck J, Beam WR, Martin RJ, Rosenwasser LJ. Detection of alveolar macrophage-derived IL-1β in asthma. J Immunol 1992; 149:3078–3082.
73. Gosset P, Tsicopoulos A, Wallaert B, Vannimenus C, Joseph M, Tonnel AB, Capron A. Increased secretion of tumor necrosis factor α and interleukin 6 by alveolar macrophages consecutive to the development of the late asthma reaction. J Allergy Clin Immunol 1991; 88:561–571.
74. Cembrzynska-Nowak M, Szklarz E, Inglot AD, Teodorczyk-Injeyan JA. Elevated release of tumor necrosis factor α and interferon-γ by broncho-alveolar leukocytes from patients with bronchial asthma. Am Rev Respir Dis 1993; 147:291–295.
75. Ellis JL, Undem BJ. Inhibition by L-NG nitro-L-arginine of nonadrenergic-non-cholinergic-mediated relaxations of human isolated central and peripheral airways. Am Rev Respir Dis 1992; 146:1543–1547.
76. Jorens PG, Vermeire PA, Herman AG. L-arginine dependent nitric oxide synthase: a new metabolic pathway in the lungs and airways. Eur Respir J 1993; 6:258–266.
77. Nathan CF, Hibbs JB, Jr. Role of nitric oxide synthesis in macrophage antimicrobial activity. Curr Opin Immunol 1991; 3:65–70.
78. Crapo JB, Barry BE, Gehr P, Bachofen M, Weibel ER. Cell number and cell characteristics of the normal human lung. Am Rev Respir Dis 1982. 125:322–337.

Part Three

LUNG DENDRITIC CELLS

9

Methods to Isolate and Study Lung Dendritic Cells

BARBARA J. MASTEN and MARY F. LIPSCOMB

University of New Mexico School of Medicine
Albuquerque, New Mexico

I. Introduction

In Chapter 3, dendritic cells (DCs) are depicted as major antigen-presenting cells in the body, indicating the likelihood that these cells are central in the development of pulmonary immune responses. The purpose of this chapter is to describe techniques useful in identifying, isolating, and measuring the function of lung dendritic cells and highlighting many difficulties. Subsequent chapters in this section will describe aspects of lung DC ontogency, function, and migration.

Dendritic cells originate from major histocompatibility (MHC) class II Ia-negative progenitors in the bone marrow, circulate in the blood, and immigrate into all lymphoid organs and most nonlymphoid organs, including skin, heart, liver, kidney, gut and lung (1). In the respiratory tract, DCs develop into Ia-positive cells with the capacity to endocytose and process antigens (2–6). These cells may then migrate into secondary lymphoid tissues, that is, lung-associated lymph nodes, and mature into lymphoid DCs with a specialized costimulatory function capable of activating naive T cells and initiating a primary immune response (7–9).

II. Location of Lung Dendritic Cells

Lung DCs have been studied in a number of species, including mice (10–12), rats (13–21), guinea pigs (10,13), and humans (10,22–29). Like DCs from other sites, lung DCs constitutively express both class I and II MHC and fail to express pan–T cells, natural killer (NK) cells, most B cells, and many macrophage markers. They demonstrate long dendritic processes both in tissue sections and in cell suspensions. A number of investigators have exploited these attributes of DCs to examine their location in the lung using histological techniques. DCs form an interdigitating network in the airway epithelium of all species in whom they have been investigated (10,16,17,19,26,29), similar to the network described for Langerhans cells in the skin. DCs are localized within the bronchoalveolar lumen, airway epithelium, alveolar septa, and in connective tissue surrounding pulmonary veins and airway vessels. Evidence suggests that the density, distribution, and phenotype of DCs reflect the level of stimulation provided by inhalation of airborne irritant stimuli. For example, intraepithelial DCs are particularly dense in the trachea and gradually diminish in concentration as the airways branch, but they are increased at sites of inflammation (7,8,17). Furthermore, the number of DCs lavaged from the alveoli can be increased using bacillus Calmette-Guérin (BCG) as an inflammatory agent (30).

In the normal lung, DCs constitute less than 1% of all lung cells, and sequential steps are required to obtain them in highly enriched populations free of lung parenchymal cells and other leukocytes. Direct immunoselection is not currently possible, because there are no specific differentiation markers for lung DCs. However, a variety of cell surface markers in the mouse (Table 1), rat (Table 2), and human (Table 3) are useful in differentiating DCs from other cell types and characterizing the phenotype and function of DCs. These markers include products of MHC classes I and II, differentiation antigens, costimulatory molecules, adhesion molecules, and Fc and complement receptors. In every tissue and species in which they have been studied, DCs constitutively express MHC class I and II proteins, have a lower buoyant density than macrophages, are dendritic in shape, and are initially adherent, but then readily detach from plastic; that is, they are loosely adherent. These attributes were useful for splenic DC isolation (31) and are used to isolate DC-enriched lung cell populations from various physical locations in the lung; that is, the parenchyma, bronchoalveolar lumen, and airway epithelium. Several enrichment steps must be used to obtain a cell fraction highly enriched for DCs, because macrophages and DCs cannot be readily separated based entirely on surface antigens. The enrichment procedures are monitored by identifying cells with elongated cell processes, eccentric lobulated nuclei, delicate nucleoli, and a high density of surface MHC class II antigens. Assays evaluating DC function include the stimulation of periodate-treated lymphocytes (15), antigen-induced stimulation of immune T cells (7,13,14,18,20,32–35), and allo-

Table 1 Antibodies Used to Differentiate and Characterize Murine Lung DCs

Antigen recognized	Antibody	Reactivity with DCs in cell suspension	Reference
MHC class II (I-A/I-E)	M5/114.15.2	+	10,11,17
Thy1.2	30-H12	−	10,11
CD2	RM2-5	−	12
CD4	GK1.5	−	11
CD8	YTS169.4	−	11
CD11a (LFA-1 α)	2D7	+	12
CD11b (C3biR)	M1/70	+[a]	10,11
CD23 (FcεR)	B3B4	−	12
CD24 (HSA)	J11d	+[a]	11
CD25 (IL-2Rα)	7D4	+[a]	11
CD28	37.51	−	12
CD32/CD16 (FcγRII/III)	2.4G2	+[a]	10,11
CD40	3/23	+	12
CD40L	MR1	−	12
CD45R (B220)	RA3-6B2	−	12
CD45RA	14.8	−	11,12
CD54 (ICAM-1)	3E2	+	12
CD80 (B7-1)	1G10	+[a]	12
CD86 (B7-2)	GL1	+	12
CTLA4L	CTLA4-Ig	+	12
T200	M1/9	+	10
sIgM (B cell)		−	10
Unknown (macrophage)	F4/80	−	11
DEC-205	NLDC-145	+[a]	11
Unknown (lymphoid DC)	33D1	−	11
Unknown (T200)	T29/33	+	10

[a]Reacts with a subset of DCs.

geneic (11,12,18,21,22,24,28,36–39) and autologous (8,18,25) mixed lymphocyte reactions (MLRs).

III. Isolation of Dendritic Cells from Whole Lung

Conventional methods by which DCs are enriched from whole lung tissue require bronchoalveolar lavage, perfusion of the pulmonary vasculature, enzymatic digestion of minced or sliced lung tissue, and density fractionation (11,13,22). Bronchoalveolar lavage removes alveolar macrophages (AMs), which have been

Table 2 Antibodies Used to Differentiate and Characterize Rat Lung DCs

Antigen recognized	Antibody	Reactivity with DCs in cell suspension	Reference	Reactivity with lung DCs in situ	Reference
MHC-II	OX-6	+	6–8,18,20,21, 32,34,35,39	+	5,8,9,14,17, 19,52,53,54
MHC I	OX-27	+	6		
CD4	W3/25	±[a]	7,18,19,34	+[a]	17,19,54
CD5	OX-19	−	20		
CD8	OX-8	−	7,18,19,34	−	19
CD11a (LFA-1α)	WT.1	+[a]	19	−	19
CD18 (integrin β2)	WT.3			−	19
CD18 (integrin β2)	OX-42			+[a]	17,54
CD43 (leukosialin)	W3/13	±[a]	18,32	−	19
CD45R	OX-1	+	7,18,19,34	+	19
CD45RA	OX-33	−	7		
CD45RC	OX-22	−	7,18,34	−	19
CD54 (ICAM-1)	IA29	+[a]	18,19,34	−	19
IgM (B cell)	Polyclonal	−	7,18,19,32	−	19
Unknown (macrophage)	OX-41	+[a]	18,19,34	+[a]	17,19,54
Unknown (macrophage)	OX-43	±[a]	18,19,34	−	19
Unknown (macrophage)	RMA	±[a]	18,19,32	±[a]	19,34
Unknown (B cell)	His 14	−	20		
Unknown (MPS)[b]	ED1	±[a]	7,20	±[a]	17,52,54
Unknown (MPS)	ED2	−	7,20	−	6,8,17,52,54
Unknown (MPS)	ED3	−	20		
Unknown (MPS)	ED4	+[a]	20	−	52
Unknown (MPS)	ED7	+[a]	20		
Unknown (MPS)	ED8	+[a]	20	+[a]	17,
Unknown (MPS)	ED9	+[a]	20	+[a]	17,52
Unknown (T cell)	OX-52	−	20	−	54,
Unknown (T cell)	OX-19	−	20	−	54,
Unknown (pan DC, γδ T-cell subset)	OX-62	+	6	+[a]	5,6,53
Unknown (DC subset)	C11b	+[a]	19	+[a]	19
Unknown (κ light chain)	OX-12	−	7		
Unknown (B cell)	OX-33			−	9

[a]Reacts with a subset of DCs.
[b]Mononuclear phagocyte system.

Table 3 Antibodies Used to Differentiate and Characterize Human DCs

Antigen recognized	Antibody	Reactivity with DCs in cell suspension	Reference	Reactivity with lung DCs in situ	Reference
HLA-DR	OK1a	+	28,36	+	27,29
HLA-DR	L243	+	22,24,25,27		
HLA-DQ	Leu10			+[a]	29
HLA-DR/DQ	9.3F10	+	38		
CD1a	OKT6	±[a]	22,25,28,36	+[a]	27,29
CD2	OKT11	−	36		
CD3	Leu4	−	28	−	29
CD3	OKT3			−	10,29
CD4	OKT4	−	36		
CD4	Leu-3			−	10
CD5	Leu-1			−	10
CD8	OKT8	−	36		
CD8	Leu-2			−	10
CD11b (C3biR)	OKM1	−	22	−	10
CD14	My4	+[a]	28	+[a]	29
CD14	LeuM3	−	28	−	29
CD14	MO_2	−	24,25		
CD14	UCHM1			−	29
CD19	B4			−	29
CD19	Leu 12	−	36		
CD20	B1				10
CD22	Leu 14	−	22,25		
CD24	BA_1	−	24		
CD64 (FcγRI)	32.2	−	28		
CD68	KIM6			+[a]	29
CD68	KIM7	+[a]	28	+[a]	29
CD68	EBM11	+	38	+[a]	29
sIgA, IgG, IgM				−	10
T200	T29/33			+	10
Factor VIII		−	25		
Unknown (AM, DC, B cell)	RFD1	+[a]	27,28,36	+[a]	27,29
Unknown (DC and B cell)	L25	+[a]	27,28	+[a]	27,29
DRC1 (follicular DC)	R4/23	−[a]	25		
Unknown (AM)	PAM_1	−	25		
Unknown (monocyte)	M_3	−	36		
Unknown (macrophage)	RFD7	−	28	−	29
Unknown (macrophage)	RFD9	−	28	−	29

[a]Reacts with a subset of DCs.

shown in most species to be suppressive for DC-initiated T-cell proliferation (21,33,37,40). The pulmonary vasculature is perfused to remove peripheral blood cells. Collagenase is the enzyme of choice for extraction of viable interstitial lung cells. The use of proteolytic enzymes such as elastase and hyaluronidase may damage cell surface structures such as FcR (13). DNase is included in digestion mixtures to avoid cell losses owing to DNA-induced clumping. Cell types identified in the dissociated population include macrophages, T and B lymphocytes, NK cells, eosinophils, mast cells, neutrophils, DCs, and significant numbers of type II and ciliated epithelial cells (13). Pulmonary mononuclear cells are usually separated from the dissociated cells on a density gradient into high-density lung cells, primarily containing lymphocytes, and low-density DC-enriched lung cells, which are also highly enriched in interstitial macrophages (IMs).

Further enrichment of DCs can be obtained by subsequently including a short adherence step to separate nonadherent lymphocytes from adherent cells and an overnight culture to help separate transiently adherent cells (i.e., DCs) from the more firmly adherent cells (i.e., macrophages) (11,18,22). The disadvantage of using adherence as an enrichment step is that significant changes may occur in DC function and phenotype either owing to the time in culture or as a consequence of interactions with other cells from the pulmonary interstitium. For example, overnight culture downregulates the antigen-processing capacity of skin DCs (41,42) and increases the expression of accessory molecules (43) involved in stimulating T cells. The loosely adherent (LAd) lung cell population still contains a large number of contaminating IMs, which have been shown in some species to enhance the antigen-presenting function of DCs (12,21,35).

Effective procedures for isolating fairly pure DC populations require exploiting either the phagocytic or autofluorescent properties of lung IMs and/or the high Ia expression of DCs. Negative selection based on phagocytosis of silica particles (11,22), latex beads (11), or carbonyl iron particles (21) allows for the removal of phagocytic cells from lung cell preparations. The disadvantage of eliminating all phagocytic cells is that DCs, depending on their state of maturation, can be weakly phagocytic (12,44-46). For example, DCs which had phagocytized one latex bead were regularly seen in low-density LAd murine lung cells incubated with latex beads (12). Positive selection of cells bearing surface Ia antigens provides a means to enrich for lung DCs in the rat (18,21) and mouse (11,12), but it cannot be applied to human lung, because human macrophages also express Ia (22,47). As an alternative, by exploiting autofluorescence, fluorescent-activated cell sorting (FACS) can be used to remove interstitial macrophages (22). Lung cells are fractionated into brightly autofluorescent macrophages and relatively nonautofluorescent DC-enriched populations. Autofluorescence has been correlated to the number of lysosomes (48), which are very low in DCs and high in macrophages (49). The separation of low-autofluorescent DCs from high-autofluorescent IMs is also used for rat (20) and human (28) bronchoalveolar lavage (BAL) cells.

A combination of enrichment techniques is used in a variety of species to isolate fairly pure populations of DCs from enzyme-digested low–buoyant density lung cells since no one enrichment technique permits a pure population to be obtained. For the mouse, Pollard (50) and Masten et al. (12) described a multistep enrichment protocol in which Ia-positive, nonphagocytic, dendritic-shaped cells were isolated at 92–99% purity from LAd cells by dual laser FACS. LAd cells were incubated with FITC-labeled latex beads and stained with biotinylated anti-Ia monoclonal antibody and strepavidin allophyocyanin (AP) as the second fluorochrome. The forward/side scatter (FS/SSC) profile (Fig. 1a) of FITC latex bead–treated, Ia-stained LAd cells was used to set a gate to exclude debris, free latex beads, and small lymphocytes. To optimize sorting, this gate was refined by making AP-positive, FITC-bead negative cells the predominate population within the gate. We noted that the majority of AP-positive, FITC-bead-negative cells stained negative for CD45RA, a marker found at a high density on B cells. To exclude the majority of AP-positive, CD45RA-positive cells from the light scattering gate, in some experiments phycoerythrin (PE) -conjugated CD45RA monoclonal antibody was added to separate aliquot of bead-treated, Ia-stained cells, and the light scatter of CD45RA-positive cells was used to refine the FS/SSC gate. AP-positive, FITC-bead-negative cells falling within the FS/SSC gate were collected and represent Ia-positive, nonphagocytic DCs (Fig. 1b). These sorted cells displayed elongated cell processes by phase and were Ia-positive as determined by fluorescent microscopy, and they initiated strong allogeneic T-cell proliferation.

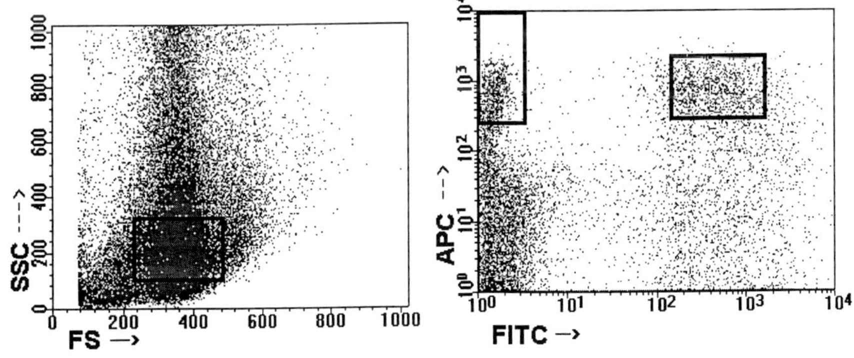

Figure 1 Dot plot representations used to set the gates for sorting DCs from FITC-bead–treated, Ia-stained LAd cells. (a) Light scatter plot (FS/SSC) of LAd cells and first sort gate (indicated with box). (b) Dot plot of FITC (latex bead positive) versus AP fluorescence (Ia positive) of gated FS/SSC LAd. The FITC-negative/AP-positive gate is used to sort Ia-positive, nonphagocytic DCs. The FITC-positive/AP-positive gate is used to sort Ia-positive, phagocytic DCs.

More recently, three-color sorting has been done in which PE-conjugated CD45R (B220) monoclonal antibody is used to exclude B cells more effectively.

Murine lung DCs isolated by this procedure express CD11a, CD54, and CD86; were phenotypically heterogeneous with respect to CD11b, CD32, CD24, CD25, CD80 and NLDC-145; and utilized CD11a, CD54, CD80, and CD86 in their role as initiators of primary T-cell responses (11,12,50). Furthermore, CD2 and CD40L on T cells were also clearly important in DC–T-cell interaction (12). Ia-positive dendritic-shaped cells which has phagocytized one to two beads were consistently seen in the LAd population (see Fig. 1b) and were able to stimulate T cells in an allogeneic MLR, although slightly less efficiently than the Ia-positive, nonphagocytic DC population. This multistep method yielded ~4.1×10^3 nonphagocytic DCs/mouse lung, representing ~0.3% of the original loosely adherent cell population and ~11% of the total stained LAd population.

Nicod et al (22) demonstrated that in humans a combination of adherence and phagocytosis could be used to enrich for nonphagocytic LAd cells with potent antigen-presenting function from low-density lung cells. LAd cells were incubated with silica particles and silica-laden phagocytic cells were separated from nonphagocytic cells over a Ficoll-Hypaque gradient. Enrichment for nonphagocytic cells with a dendritic morphology enhanced the MLR-stimulating capacity of LAd. The LAd cells were OKT6, CD22 negative, positive for HLA-DR, and heterogeneous for OKM1. Furthermore, a combination of adherence and autofluoresence (24) was used to purify nonautofluorescent, loosely adherent cells that were potent simulators in an MLR. The majority of low-autofluorescent cells was composed of nonspecific esterase, BA_1 (CD24), MO_2 (CD14)–negative cells which expressed HLA-DR.

In the rat, Rochester et al. (15) used a combination of adherence and ingestion of silica particles to obtain a population of DCs, ~76% pure, from low-density lung cells. These nonadherent, nonphagocytic low-density cells produced a dose-dependent increase in periodate-induced lymphocyte proliferation. Xia et al. (18) described a purification protocol for rat pulmonary DCs that utilized adherence and Ia expression to select loosely adherent Ia-positive DCs at a >90% purity. LAd cells were subjected to immunopanning with anti-rat OX-6 monoclonal antibody that reacts specifically with class II (la) MHC antigens. The purified Ia-positive lung cells displayed the morphological and functional features of dendritic accessory cells, including extended cell processes, the absence of nonspecific esterase-staining, rapid-clustering with T lymphocytes, and costimulation of T-cell mitogen responses. Immunophenotyping by cytofluorimetry and immunohistology showed that the purified DCs expressed Ia (OX-6) and CD45R and were heterogeneous for CD43, CD54, OX-41, rat macrophage activator (RMA), and FcR antigens. This method yielded ~3.5×10^5 DCs/rat, representing 2% of the low-density enzyme-digested cells and 9% of LAd cells. Armstrong et al. (21) showed that a fairly pure population of rat interstitial DCs could be isolated, without an adherence step, from low–boyant density lung cells by exploiting the phagocytic property of lung macrophages and the high Ia expression of DCs. Their method avoided subjecting lung DCs to prolonged incubations in the presence of other cells that might modulate DC function. Low-density pulmonary lung cells were allowed to ingest FITC-labeled latex beads to identify phagocytic cells and subsequently stained with murine OX-6 monoclonal antibody followed by a PE-conjugated secondary antibody prior to FACS. After initial and secondary sorting, the OX6 bright population was 70–90% and 88–90% pure, respectively. This method yielded 1×10^5 DCs/rat, representing ~0.02% of the original low-density pulmonary cell population. Purified Ia-positive, nonphagocytic cells were potent stimulators of an allogeneic MLR.

IV. Isolation of Dendritic Cells from the Bronchoalveolar Lumen

DCs in the bronchoalveolar lumen represent a trace subset of antigen-presenting cells in the lung. Bronchoalveolar lavage (BAL) of healthy human subjects yielded a cell population containing >90% AMs, ~4% lymphocytes, ~2% granulocytes, ~2% monocyte-like cells, and ~0.4% DCs (29), and similar percentages were observed in the rat (20) and mouse (11). AMs suppress DC-induced T-cell proliferation in a concentration-dependent manner. Therefore, enrichment procedures that clearly separate AMs from DCs must be used if DC phenotype and function are to be studied.

One effective method to isolate DCs from the alveolar lumen requires performing BAL to acquire bronchoalveolar cells and then sorting using flow cytometry to distinguish differences in autofluorescence between DCs and alveolar macrophages. Thus, BAL cells can be separated into brightly autofluorescent AMs and lymphocytes, granulocytes, and DCs, all exhibiting low autofluorescence. In contrast to whole lung preparations, density fractionation of BAL cells is not an effective enrichment procedure for DCs unless the fractionation is done on LAd cells (11,20). Prior to culture, BAL DCs and a large number of AMs have a similar density. However, during culture, DCs undergo morphological changes and become less dense than AMs.

In the mouse, 1±2% of BAL cells were Ia-positive with a dendritic shape (11). These cells were enriched in the LAd fraction which stimulated a strong allogeneic MLR. In the rat, Havenith et al. (20) sorted BAL from BCG-challenged lungs and found the low-autofluorescent fraction consisted of more than 95% Ia-positive cells, of which 75–85% showed acid phosphatase activity in the vicinity of the nucleus, indicating DCs. Acid phosphatase activity differentiated AMs, monocytes, and DCs; AMs exhibited acid phosphatase activity throughout the cytoplasm, monocytes had minimal acid phosphatase activity in the cytoplasm, and DCs were Ia-positive with acid phosphatase activity in a central spot near the nucleus. The remaining 15–25% Ia-positive cells demonstrated no acid phosphatase activity and were classified as monocytes. The low-autofluorescent fraction, enriched for DCs, stimulated glutamine-tyrosine–primed T cells in a dose-dependent manner. After overnight culture, most of the cells in the low-autofluorescent fraction changed to a lower density and developed morphologically into characteristic DCs with many processes. Immunophenotyping showed that the low-autofluorescent cells were heterogeneous for the mononuclear phagocytic system (MPS) markers ED1, ED4, ED7, ED8, and ED9 and did not express ED2, ED3, nor T- and B-cell markers.

In the human, van Haarst et al. (28) used a similar separation technique, that is, FACS based on differences in autofluorescence between DCs and AMs to study the immunocytological and functional properties of DCs isolated from human BAL. Freshly isolated low-autofluorescent cells formed clusters with T cells and vigorously stimulated naive T-cell proliferation in an allogeneic MLR. On overnight culture under nonadherent conditions, half of the low-autofluorescent cells developed cytoplasmic extensions characteristic of DCs, whereas the other half kept a monocyte-like appearance. The low-autofluorescent fraction was composed of acid phosphatase, RFD9, RFD7, LeuM3 (CD14), CD64-negative cells, which were strongly positive for HLA-DR, L25, RFD1, and CD86 and were heterogeneous for CD1a and My4 (CD14).

V. Isolation of Intraepithelial Dendritic Cells

Evidence in the rat suggest that intraepithelial DCs constitute a distinct subset of pulmonary DCs (19). Specifically to isolate DCs in the airway epithelium, Gong et al. (19) devised a method whereby the entire tracheobronchial tree is cleanly separated from the lung parenchyma before DCs were isolated from either compartment. A novel microdissection technique (51) combined with the use of dispase, a neutral protease that cleaves at the level of the basement membrane, enabled the exclusive isolation of DCs from airway epithelium. Like the isolation of DCs from whole lung, the bronchoalveolar lumen was lavaged and the pulmonary vasculature perfused prior to isolation. Effective procedures for isolating fairly pure, >90%, intraepithelial DCs from the dissected airways required multistep enrichment protocols similar to those used for whole lung DCs.

Gong et al. (19) found that the phenotype and function of low-density, nonadherent intraepithelial and parenchymal DCs in the rat were different. All of the intraepithelial DCs expressed moderate levels of Ia, >50% expressed FcγR, and less than half were intracellular adhesion molecule-1 (ICAM-1) positive. In contrast, all of the parenchymal DCs stained intensely with Ia, <5% expressed FcγR, and the majority were ICAM-1 positive. Intraepithelia DCs were more effective than parenchymal DCs in presenting soluble and particulate antigens to antigen-sensitized T cells. By contrast, parenchymal DCs were significantly more efficient in stimulating the proliferation of naive and resting T cells in an autologous MLR. These studies are discussed in detail in Chapter 10.

VI. Methods to Dissect Dendritic Cell Heterogeneity

A phenotypically diverse population of DCs exist in the lung. Physical location provides a means to divide lung DCs into subsets. However, DC populations isolated from the bronchoalveolar lumen, airway epithelium, and lung parenchyma are within themselves phenotypically heterogeneous. In the mouse, DCs from whole lung were phenotypically heterogeneous with respect to CD11b, CD24, CD25, CD32 (FcRγII), CD80, and NLDC-145 (DEC-205). However, all expressed high levels of Ia, CD54, and CD86 and lacked 33D1, a splenic DC marker (11,12). Variation in the expression of cytoplasmic and surface markers may reflect DCs at diverse maturational and/or functional states. Supporting this concept. Langerhans' cells freshly isolated from skin express FcRγII, which is downregulated as the cells mature in culture.

Expression of FcγRII has been used as a marker for identifying and studying FcγRII-positive and -negative subsets of lung DCs. Separation of FcγRII-positive and -negative subsets for study is achieved by immunopanning enriched populations of DCs on heat-aggregated IgG (34) or rosetting with IgG-coated erythrocytes followed by density gradient separation of rosetted and nonrosetted cells (11,19,22,25,35).

In the rat, approximately half of whole lung DCs expressed FcγRII (34). The FcγRII-positive and -negative subsets differ in morphological, phagocytic, and functional characteristics. Whereas FcγRII-positive DCs were round with few villous projections, FcγRII-negative cells were elongated and showed numerous cytoplasmic extensions. Only the FcγRII-positive subset was able to phagocytize latex beads and Ig-coated red blood cells and showed intracellular phagolysosomes by electron microscopy. The FcγRII-negative population was a more potent stimulator of naive T cells in an MLR and in response to lectins, but both populations were equally capable of presenting soluble and particulate antigens to immune T cells (34). Comparable percentages of the FcγRII-positive and -negative subsets expressed RMA and OX-41. The OX-43 antigen was only expressed by the FcγRII-positive subset. CD54 was expressed in a smaller percentage of FcγRII-positive cells.

In the mouse, DCs in the epithelium of the trachea are FcγRII-positive, and approximately half of whole lung DCs express FcγRII (10,11). Unlike rat DCs, no differences were found in the capability of the FcγRII-positive and -negative subsets to stimulate an allogeneic MLR (11). These studies were complicated by the large number of contaminating macrophages in the DC-enriched LAd population. In the human, enrichment for FcγRII-negative cells increased the capability of LAd cells enriched in DCs to stimulate both an allogeneic MLR (22) and antigen-induced T-cell proliferation (25).

Phagocytic capability and Ia expression have also been used to distinguish subsets of lung DCs. Masten et al. (12) found murine lung DCs could be divided into weakly (phagocytizing one or two beads) and nonphagocytic subsets. Weakly phagocytic DCs were able to stimulate T cells in an allogeneic MLR, although slightly less efficiently than the nonphagocytic DC population. MHC class II antigen is differentially expressed on lung DCs. Nelson et al. (6) found, in the rat, that DC precursors seeding into the respiratory tract tissues during fetal life were Ia negative or Ia low and subsequently developed high Ia expression in situ after birth. These findings suggest that mediators, owing to local inflammation in the lung and airways, regulate Ia expression on DCs.

VII. Conclusions

In the lung, DCs constitute only a small proportion of total cells and require multiple enrichment steps to both isolate and purify. In rodents, effective procedures for isolating fairly pure DC populations take advantage of the phagocytic properties of lung macrophages and the high MHC class II antigen expression of DCs. In humans, autofluorescence has been used to fractionate lung cells into brightly autofluorescent macrophages and relatively nonautofluorescent DC-enriched populations, because both macrophages and DCs are HLA-DR positive. The enrichment procedures are monitored by identifying cells with elongated cell processes, a high density of surface MHC class II antigens, and the capability to stimulate T-cell proliferation. All studies indicate that lung DCs are heterogeneous. Whether this heterogeneity reflects distinct functional subsets or disparate levels of maturation is unknown.

Acknowledgment

Studies done in the authors' laboratories were supported by research funds form NIH grant RO1 AI 21951.

References

1. Steinman RM. The dendritic cell system and its role in immunogenicity. Ann Rev Immunol 1991; 9:271–296.
2. McCarthy KM, Gong JL, Telford JR, Schneeberger EE. Ontogeny of la$^+$ accessory cells in fetal and newborn rat lung. Am J Respir Cell Mol Biol 1992; 6:349–356.

3. Simecka JW, Davis JK, Cassell GH. Distribution of Ia antigens and T lymphocyte subpopulations in rat lungs. Immunology 1986; 57:93–98.
4. van Rees EP, van der Ende MB, Sminia T. Ontogency of macrophage subpopulations and Ia-positive dendritic cells in pulmonary tissue of the rat. Cell Tissue Res 1991; 263:367–373.
5. Holt PG, Haining S, Nelson DJ, Sedgwick JD. Origin and steady-state turnover of class II MHC-bearing dendritic cells in the epithelium of the conducting airways. J Immunol 1994; 153:256–261.
6. Nelson DJ, McMenamin C, McWilliam AS, Brenan M, Holt PG. Development of the airway intraepithelial dendritic cell network in the rat from class II major histocompatibility (Ia)–negative precursors: differential regulation of Ia expression at different levels of the respiratory tract. J Exp Med 1994; 179:203–212.
7. Xia W, Pinto CE, Kradin RL. The antigen-presenting activities of Ia+ dendritic cells shift dynamically from lung to lymph node after an airway challenge with soluble antigen. J Exp Med 1995; 181:1275–1283.
8. McWilliam AS, Nelson D, Thomas JA, Holt PG. Rapid dendritic cell recruitment is a hallmark of the acute inflammatory response at mucosal surfaces. J Exp Med 1994 179:1331–1336.
9. Havenith CE, van Miert PP, Breedijk AJ, Beelen RH, Hoefsmit EC. Migration of dendritic cells into the draining lymph nodes of the lung after intratracheal instillation. Am J Respir Cell Mol Biol 1993; 9:484–488.
10. Sertl K, Takemura T, Tschachler E, Ferrans VJ, Kaliner MA, Shevach EM. Dendritic cells with antigen-presenting capability residue in airway epithelium, lung parenchyma, and visceral pleura. J Exp Med 1986; 163:436–451.
11. Pollard AM, Lipscomb MF. Characterization of murine lung dendritic cells: similarities to Langerhans cells and thymic dendritic cells. J Exp Med 1990; 172:159–167.
12. Masten BJ, Yates JL, Pollard-Koga AM, Lipscomb MF. Characterization of accessory molecules in murine lung dendritic cell function: Roles for CD80, CD86, CD54, and CD40L. Am J Respir Cell Mol Biol (submitted).
13. Holt PG, Degebrodt A, Venaille T, C OL, Krska K, Flexman J, Farrell H, Shellam G, Young P, Penhale J, et al. Preparation of interstitial lung cells by enzymatic digestion of tissue slices: preliminary characterization by morphology and performance in functional assays. Immunology 1985; 54:139–147.
14. Holt PG, Schon-Hegrad MA, Oliver J. MHC class II antigen-bearing dendritic cells in pulmonary tissues of the rat. Regulation of antigen presentation activity by endogenous macrophage populations. J Exp Med 1988; 167:262–274.
15. Rochester CL, Goodell EM, Stoltenborg JK, Bowers WE. Dendritic cells from rat lung are potent accessory cells Am Rev Respir Dis 1988; 138:121–128.
16. Holt PG, Schon-Hegrad MA, Oliver J, Holt BJ, McMenamin PG. A contiguous network of dendritic antigen-presenting cells within the respiratory epithelium. Int Arch Allergy Appl Immunol 1990; 91:155–159.
17. Schon-Hegrad MA, Oliver J, McMenamin PG, Holt PG. Studies on the density, distribution, and surface phenotype of intraepithelial class II major histocompatibility complex antigen (Ia)–bearing dendritic cells (DC) in the conducting airways. J Exp Med 1991; 173:1345–1356.
18. Xia WJ, Schneeberger EE, McCarthy K, Kradin RL. Accessory cells of the lung. II.

Ia+ pulmonary dendritic cells display cell surface antigen heterogeneity. Am J Respir Cell Mol Biol 1991; 5:276–283.
19. Gong JL, McCarthy KM, Telford J, Tamatani T, Miyasaka M, Schneeberger EE. Intraepithelial airway dendritic cells: a distinct subset of pulmonary dendritic cells obtained by microdissection. J Exp Med 1992; 175:797–807.
20. Havenith CE, Breedijk AJ, van Miert PP, Blijleven N, Calame W, Beelen RH, Hoefsmit EC. Separation of alveolar macrophages and dendritic cells via autofluorescence: phenotypical and functional characterization. J Leukoc Biol 1993; 53: 504–510.
21. Armstrong LR, Christensen PJ, Paine RR, Chen GH, McDonald RA, Lim TK, Toews GB. Regulation of the immunostimulatory activity of rat pulmonary interstitial dendritic cells by cell-cell interactions and cytokines. Am J Respir Cell Mol Biol 1994; 11:682–691.
22. Nicod LP, Lipscomb MF, Weissler JC, Lyons CR, Albertson J, Toews GB. Mononuclear cells in human lung parenchyma. Characterization of a potent accessory cell not obtained by bronchoalveolar lavage. Am Rev Respir Dis 1987; 136:818–823.
23. Soler P, Moreau A, Basset F, Hance AJ. Cigarette smoking-induced changes in the number and differentiated state of pulmonary dendritic cells/Langerhans cells. Am Rev Respir Dis 1989; 139:1112–1117.
24. Nicod LP, Lipscomb MF, Toews GB, Weissler JC. Separation of potent and poorly functional human lung accessory cells based on autofluorescence. J Leukoc Biol 1989; 45:458–465.
25. Nicod LP, Lipscomb MF, Weissler JC, Toews GB. Mononuclear cells from human lung parenchyma support antigen-induced T lymphocyte proliferation. J Leukoc Biol 1989; 45:336–344.
26. Holt PG, Schon-Hegrad MA, Phillips MJ, McMenamin PG. Ia-positive dendritic cells form a tightly meshed network within the human airway epithelium. Clin Exp Allergy 1989; 19:597–601.
27. vanHaarst JMW, de Wit HJ, Drexhage HA, Hoogsteden HC. Histology and Immunophenotyping of dendritic cells in the human lung. In: EWA Kamperdijk EWA, Nieuwenhuis P, Hoefsmit ECM, eds. Dendritic Cells in Fundamental and Clinical Immunology. New York: Plenum Press, 1993:577–580.
28. vanHaarst JM, Hoogsteden HC, de Wit HJ, Verhoeven GT, Havenith CE, Drexhage HA. Dendritic cells and their precursors isolated from human bronchoalveolar lavage: immunocytologic and functional properties. Am J Respir Cell Mol Biol 1994; 11: 344–350.
29. van Haarst JM, de Wit HJ, Drexhage HA, Hoogsteden HC. Distribution and immunophenotype of mononuclear phagocytes and dendritic cells in the human lung. Am J Respir Cell Mol Biol 1994; 10:487–492.
30. Havenith CE, Breedijk AJ, Hoefsmit EC. Effect of bacillus Calmette-Guérin inoculation on numbers of dendritic cells in bronchoalveolar lavages of rats. Immunobiology 1992; 184:336–347.
31. Steinman RM, Kaplan G, Witmer MD, Cohn ZA. Identification of a novel cell type in peripheral lymphoid organs of mice. V. Purification of spleen dendritic cells, new surface markers, and maintenance in vitro. J Exp Med 1979; 149:1–16.

32. Kradin RL, McCarthy KM, Xia WJ, Lazarus D, Schneeberger EE. Accessory cells of the lung. I Interferon-gamma increases Ia^+ dendritic cells in the lung without augmenting their accessory activities (comment). Am J Respir Cell Mol Biol 1991; 4:210–218.
33. Holt PG, Oliver J, Bilyk N, McMenamin C, McMenamin PG, Kraal, Thepen T. Downregulation of the antigen presenting cell function(s) of pulmonary dendritic cells in vivo by resident alveolar macrophages. J Exp Med 1993; 177:397–407.
34. Kradin RL, Xia W, McCarthy K, Schneeberger EE. $FcR^{\pm}$ subsets of Ia^+ pulmonary dendritic cells in the rat display differences in their abilities to provide accessory co-stimulation for naive ($OX\text{-}22^+$) and sensitized ($OX\text{-}22^-$) T cells. Am J Pathol 1993; 142:811–819.
35. Gong JL, McCarthy KM, Rogers RA, Schneeberger EE. Interstitial lung macrophages interact with dendritic cells to present antigenic peptides derived from particulate antigens to T cells. Immunology 1994; 81:343–351.
36. Nicod LP, Galve-de Rochemonteix B, Dayer JM. Dissociation between allogeneic T cell stimulation and interleukin-1 or tumor necrosis factor production by human lung dendritic cells. Am J Respir Cell Mol Biol 1990; 2:515–522.
37. Lipscomb MF, Pollard AM, Yates JL. A role for TGF-beta in the suppression by murine bronchoalveolar cells of lung dendritic cell initiated immune responses. Reg Immunol 1993; 5:151–157.
38. Havenith CE, van Haarst JM, Breedijk AJ, Betjes MG, Hoogsteden HC, Beelen RH, Hoefsmit EC. Enrichment and characterization of dendritic cells from human bronchoalveolar lavages. Clin Exp Immunol 1994; 96:339–343.
39. Christensen PJ, Armstring LR, Fak JJ, Chen G-H, McDonald RA, Toews GB. Regulation of rat pulmonary dendritic cell immunostimulatory activity by alveolar epithelial cell-derived granulocyte macrophage colony-stimulating factor. Am J Respir Cell Mol Biol 1995; 13:426–433.
40. Yarbrough WC, Jr., Wilkes DS, Weissler JC. Human alveolar macrophages inhibit receptor-mediated increases in intracellular calcium concentration in lymphocytes. Am J Respir Cell Mol Biol 1991; 5:411–415.
41. Pure E, Inaba K, Crowley MT, Tardelli L, Witmer-Pack MD, Ruberti G, Fathman G, Steinman RM. Antigen processing by epidermal Langerhans cells correlates with the level of biosynthesis of major histocompatibility complex class II molecules and expression of invariant chain. J Exp Med 1990; 172:1459–1469.
42. Girolomoni G, Simon JC, Bergstresser PR, Cruz PD Jr. Freshly isolated spleen dendritic cells and epidermal Langerhans cells undergo similar phenotypic and functional changes during short-term culture. J Immunol 1990; 145:2820–2826.
43. Inaba K, Witmer-Pack M, Inaba M, Hathcock KS, Sakuta H, Azuma M, Yagita H, Okumura K, Linsley PS, Ikehara S, et al. The tissue distribution of the B7-2 co-stimulator in mice: abundant expression on dendritic cells in situ and during maturation in vitro. J Exp Med 1994; 180:1849–1860.
44. Reis e Sousa C, Stahl PD, Austyn JM. Phagocytosis of antigens by Langerhans cells in vitro. J Exp Med 1993; 178:509–519.
45. Koch F, Trockenbacher B, Kampgen E, Grauer O, Stossel H, Livingstone AM, Schuler G, Romani N. Anitgen processing in populations of mature murine dendritic

cells is caused by subsets of incompletely matured cells. J Immunol 1995; 155: 93–100.
46. Steinman RM, Swanson J. The endocytic activity of dendritic cells (comment). J Exp Med 1995; 182:283–288.
47. Toews GB, Vial WC, Dunn MM, Guzzetta P, Nunez G, Stastny P, Lipscomb MF. The accessory cell function of human alveolar macrophages in specific T cell proliferation. J Immunol 1984; 132:181–186.
48. Kradin RL, McCarthy KM, Preffer FI, Schneeberger EE. Flow-cytometric and ultrastrucutural analysis of alveolar macrophage maturation. J Leukoc Biol 1986; 40: 407–417.
49. Beelen RHJ, Bos HJ, Kamperdijk EWA, Hoefsmit ECM. In: Zembala M, Asherson, GL, eds. Human Monocytes. London: Academic Press, 1989:7–26.
50. Pollard AM. Isolation and Characterization of Murine Lung Accessory Cells. Dallas: University of Texas Southwestern Medical School at Dallas, 1990.
51. Plopper CG, Chang AM, Pang A, Buckpitt AR. Use of microdissected airways to define metabolism and cytotoxicity in murine bronchiolar epithelium. Exp Lung Res 1991; 17:197–212.
52. Holt PG, Oliver J, McMenamin C, Schon-Hegrad MA. Studies on the surface phenotype and functions of dendritic cells in parenchymal lung tissue of the rat. Immunol 1992; 75:582–587.
53. Nelson DJ, McWilliam AS, Haining S, Holt PG. Modulation of airway intraepithelial dendritic cells following exposure to steroids. Am J Respir Crit Care Med 1995; 151:475–481.
54. Holt PG, Schon-Hegrad MA. Localization of T cells, macrophages and dendritic cells in rat respiratory tract tissue: implications for immune function studies. Immunol 1987;62:349–356.

10

Ontogeny and Heterogeneity of Lung Dendritic Cells

EVELINE E. SCHNEEBERGER and TAKAFUMI SUDA

Massachusetts General Hospital
Boston, Massachusetts

I. Introduction

Of all the mammalian mucosal surfaces located at the interface with the environment, that of the lung is one of the largest and most vulnerable. In the adult human, it comprises a surface area of more than 70 m^2 and, to enable efficient gas exchange, is <1 μm in thickness in its most attenuated portions (1). During the course of 24 hr, this mucosal surface is exposed to as much as 15,000 L of ambient air that may be contaminated to varying degrees with organic and inorganic pollutants. A number of physical and functional barriers serve to protect the lung from the deleterious effects of such contaminants (mucociliary clearance, phagocytosis by alveolar macrophages, epithelial tight junctions) (2–5). Should these barriers be penetrated, however, the immune system must then discriminate between innocuous and harmful antigens, as well as limit the extent of tissue damage that results from nonspecific and immune-mediated inflammation. This implies that there are mechanisms in place in the lung that control and limit cellular activation at both the T-cell and accessory cell level.

The newborn period in many species is a particularly critical period, since some of these immune mechanisms are not fully developed and the newborn is, therefore, particularly vulnerable to infections and to sensitization by a variety of inhaled antigens. The purpose of this chapter is to review the ontogeny of dendritic cells (DCs) in the lung and to summarize the results of studies that have examined the specialized phenotype and function of the DCs that populate four distinct compartments of the lung (lung vasculature, alveolar space, airway epithelium, lung parenchyma) and to relate these to the turnover of these important accessory cells.

II. Role of Accessory Cells in T-Cell–Mediated Immunity

A. Antigen Presentation

In order to respond to antigen, T cells require the cooperation of accessory cells whose function is to process and present antigen in association with either class I or class II major histocompatibility complex (MHC) glycoprotein. The reader is referred to several excellent, recent reviews on this topic (6,7), and only a brief outline, as it pertains to DCs, is included. In order to be recognized by T cells, antigen must first be endocytosed, partially degraded, bind to an MHC molecule, and be delivered to the surface of the accessory cell where it is presented to the T cell as an MHC/antigenic peptide complex. Endogenous antigens processed within the cells are presented by MHC class I molecules (8) and are recognized by CD8 T cells, whereas exogenous antigens bound to MHC class II molecules are recognized by CD4 T cells (9). To maintain a divergence between the MHC class I and II pathways within the cell, a further protein, the invariant chain, acts as a chaperone to the MHC class II molecule. Its function is to ensure the correct folding of the α/β chain of the MHC class II molecule, to inhibit binding of antigen prior to entry into the endosomal compartment, and to direct the post-Golgi targeting of the MHC class II molecule to an endosomal compartment (10–12).

Recent immunocytochemical observations made with freshly isolated and cultured human blood DCs indicate that colloidal gold–labeled bovine serum albumin is rapidly endocytosed and accumulates within 30 min in a well-developed MHC class II–enriched compartment (MIIC) (13). The uptake of exogenous antigen occurs more readily in freshly isolated than in cultured DCs; however, the latter express more surface Ia than do freshly isolated DCs. The MIIC is an acidic compartment that contains many of the components required for antigen processing/presentation, including invariant chain, MHC class II glycoprotein, and the recently described HLA-DM molecule that contributes to antigen presentation (14). This, together with the localization of exogenous antigen in the MIIC, indicates that DCs, particularly those that are relatively immature, play a key role in the processing and presentation of exogenous antigen.

B. Costimulatory Signals

In addition to presenting antigenic peptides in the context of MHC molecules, the accessory cell must also provide important adhesion and costimulatory signals in order to activate fully the T cell; presentation of antigen in the absence of these signals results in T-cell anergy (9). Many of these costimulatory and/or adhesion molecules as well as their corresponding receptors on T cells have been identified. They include leukocyte function–associated antigen-3 (LFA-3) and CD2 (15,16); intercellular adhesion molecule-1 (ICAM-1) and LFA-1 (17); B7-1, B7-2, B7-3, and CD28 and/or CTLA-4 (18,19); CD72 and CD5 (20); class I and CD8 (21); and, class II and CD4 (22). All these adhesion and costimulatory molecules are expressed on DCs at some stage of their maturation or state of activation (23,24).

C. Types of Accessory Cells

The requirements for activating naive T cells differs from that of memory T cells in that naive T cells have a T-cell receptor (TCR) of relatively low affinity and an absolute requirement for costimulatory signals. By contrast, memory T cells bind antigen with greater affinity, become saturated over a narrow range of antigen concentration, and are activated primarily via the TCR (9,25). It follows, therefore, that naive T cells require the help of highly specialized accessory cells, which has led to classifying these cells into two broad categories: (1) *Specialized accessory cells* that include DCs, Langerhans' cells, and B cells; and (2) *facultative accessory cells* that require cytokine stimulation for optimal accessory function and comprise a wide range of cell types. These include macrophages, astrocytes, microglial cells, endothelial and epithelial cells, and fibroblasts (26).

DCs in the lung, as in other nonlymphoid organs, are viewed as having three broad functions (23). (1) At the mucosal interface, they have a *sentinel role*, surveying for and capturing ambient antigens that penetrate beyond the physical and functional barriers of the air/tissue interface; (2) after ingesting antigen, DCs then adopt a *migratory role* carrying their antigenic load to local lymph nodes; and (3) during DC migration and/or within the lymph node itself, their phenotype undergoes a subtle change, so that in the environment of lymphoid tissues, they can assume their T-cell *activation* or *adjuvant role*.

In the adult lung, the number, phenotype, and function of DCs are governed by a balance between stimulatory (27–31) and inhibitory factors (29,30). By contrast, in the intrauterine environment, although potentially exposed to circulating maternal factors, the fetal lung is protected from inhaled antigens and/or environmental pollutants that lead to an inflammatory response. Under these conditions, it is possible to determine the time point in fetal and neonatal development at which DCs begin to populate the lung and to establish when these cells assume full accessory cell activity. In the following section, studies are reviewed in which the phenotype and function of DCs in fetal and neonatal lungs are compared with those populating the adult lung.

III. Ontogeny of Dendritic Cells in the Lung

The neonatal period is a critical time when the newborn is first exposed to an array of inhaled antigenic material, rendering the neonate particularly vulnerable to respiratory infections and to sensitization by inhaled antigens. An attempt to clarify the cellular basis for this apparently susceptibility to infection has prompted a number of studies concerning the ontogeny of DCs in the lung (32–35). Immunocytochemical staining of fetal and neonatal rat lungs for MHC class II antigen (Ia) revealed positively stained DCs in the mesenchyme of fetal lungs between 15 (33) and 16 days (32) of gestation. Interestingly, by 17 days, a rare Ia^+ DC was observed within the airway epithelium of these animals, a time when they are not exposed to inhaled antigens (33) (Fig. 1). Morphometric determinations indicated that the numerical density of Ia^+ DCs in the lung increased steadily during the last third of gestation and the first week of neonatal life, although at the end of this period, their number had not fully reached adult levels (Fig. 2a). During this time, the appearance of Ia^+ DCs in the fetal lung preceded that of lymphocytes (Fig. 2b). The accessory cell function of DCs isolated from fetal rats at days 20 and 21 of gestation was 40 and 60%, respectively, that of DCs isolated

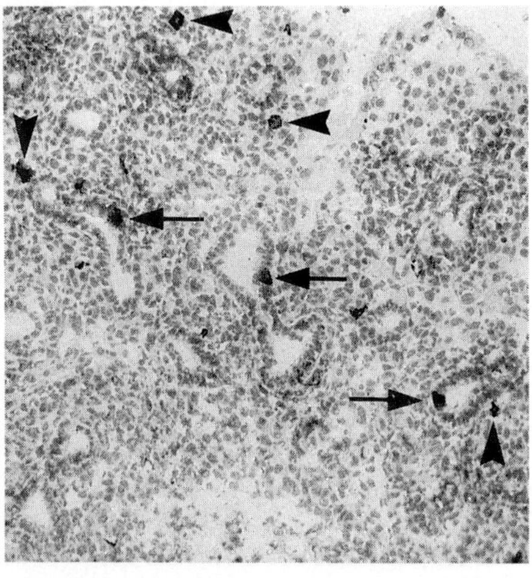

Figure 1 Frozen section of lung from an 18-day-old fetal rat immunostained with anti-Ia (OX6) monoclonal antibody. Ia^+ cells are present in the mesenchyme (arrowheads) and within the epithelium lining air spaces (arrows). ×94. (From ref. 33.)

by the same protocol from the lungs of adult rats, as determined by allogeneic mixed leukocyte reaction (MLR) assay (Fig. 3). However, in the immediate newborn period, this difference was no longer observed (33), suggesting that in this strain of rat and under the experimental conditions employed, a subpopulation of lung DCs is fully functional during the first week of life.

More recent studies on the ontogeny of DCs in the rat lung have yielded somewhat different results (34). In contrast to the earlier observations, significant numbers of Ia^+ DC were not detected until the second day after birth. However, utilizing a recently developed monoclonal antibody (OX62) that recognizes a subset of DCs as well as γ/δ T cells (36), a significant number of $OX62^+$ cells that were identified as DCs were in fact detected in the airway epithelium and lung interstitium of both fetal and neonatal rats. Although dual labeling confirmed that most of these cells were MHC class II negative, many expressed adult levels of MHC class I antigen. This interesting observation suggests that early in the postnatal period, DCs in the lung may have the capability to prime CD8 but not CD4 T cells.

A subsequent study of the functional capabilities of DCs harvested from the lungs of neonatal rats (35) indicated that both the Ia^-, $OX62^+$ as well as Ia^+ lung DCs from these animals appeared to be functionally immature. Both subpopulations of DCs responded poorly to granulocyte-macrophage colony-stimulating factor (GM-CSF) in vitro. In vivo, their response to inhaled bacterial antigen and to the parenteral administration of interferon-γ (IFN-γ) was attentuated relative to adult controls. Interestingly, endogenous macrophage-mediated suppression of DC activity was enhanced in the neonatal lung, suggesting that not only immaturity but also exposure to heightened suppressor activity contributed to the relatively ineffective accessory cell function of DCs in the neonatal lung. The purity of the DC preparations from neonatal lungs utilized in the in vitro assays ranged from 40 to 90%, and it is possible that contaminating macrophages may have contributed, in part, to the observed suppression of DC function.

A comparison of the results obtained by McCarthy et al. (33) with those of Nelson et al. (35) indicates a number of important and interesting differences. First, the method of DC isolation used by McCarthy et al. selected for Ia^+ DCs and included a number of adherence steps that effectively eliminated contaminating macrophages and/or monocytes. This is significant for the following reasons: (1) by using this protocol, the known inhibitory effect of lung macrophages was removed, yielding a highly purified population of Ia^+ DCs whose accessory cell function was then tested in an MLR assay; and (2) by selecting for Ia^+ DCs, however, the functional capabilities of only the most highly differentiated population of DCs was tested. Second, both genetic and environmental factors may have contributed to the much earlier detection of Ia^+ DCs in the lungs of fetal rats. The strains of rats employed in the studies of van Rees et al. (Wistar) (32) and McCarthy et al. (Lewis) (33) were different from those utilized in the Nelson

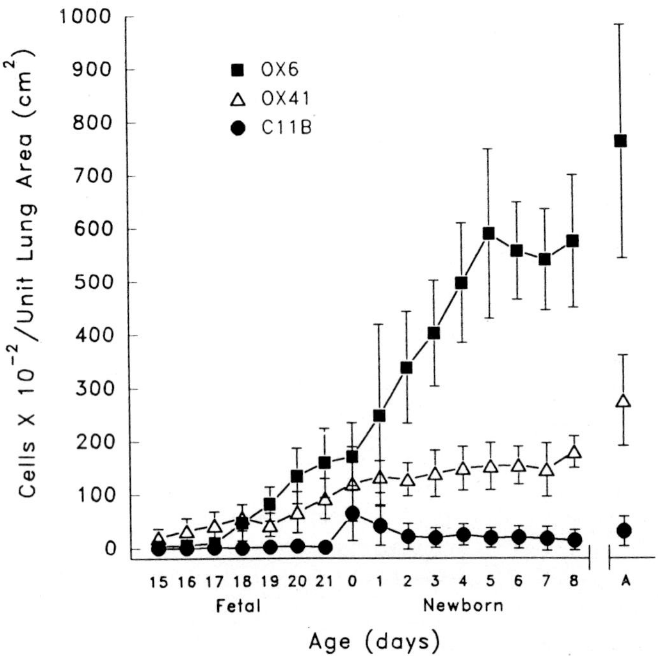

(a)

Figure 2 (a) Graph of the numerical density ± SEM of OX6+ (Ia+ DC), OX41+ (macrophages), and C11B+ (DC subclass) in the lungs of fetal (15–21 days of gestation), newborn (1–8 days old), and adult female rats. (b) Graph of the numerical density ± SEM of T cell (W3/13), MHC class II–restricted T-cell (W3/25), and MHC class I–restricted T-cell subsets (OX8) in the lungs of fetal, newborn, and adult female rates. (From ref. 33.)

studies (Wistar albino, brown Norway, Wistar Furth, PVG). In contrast to the studies of Nelson et al. (34,35), in which the animals were maintained in a dust-free environment, the rats utilized in the studies of van Rees et al. (32) and McCarthy et al. (33) were housed in a restricted-access environment but with conventional wood-derived bedding. This difference raises the interesting speculation that maternal factors, released into the circulation during the inhalation of environmental dust, may upregulate the expression of MHC class II by fetal lung DCs, which is a possibility that should be explored in future studies.

IV. Heterogeneity of Lung Dendritic Cells

Immune responses in the lung must be tightly controlled in order to maintain a balance between effective host defenses against incoming pathogens and the containment of tissue destruction involving the vitally important air-blood barrier

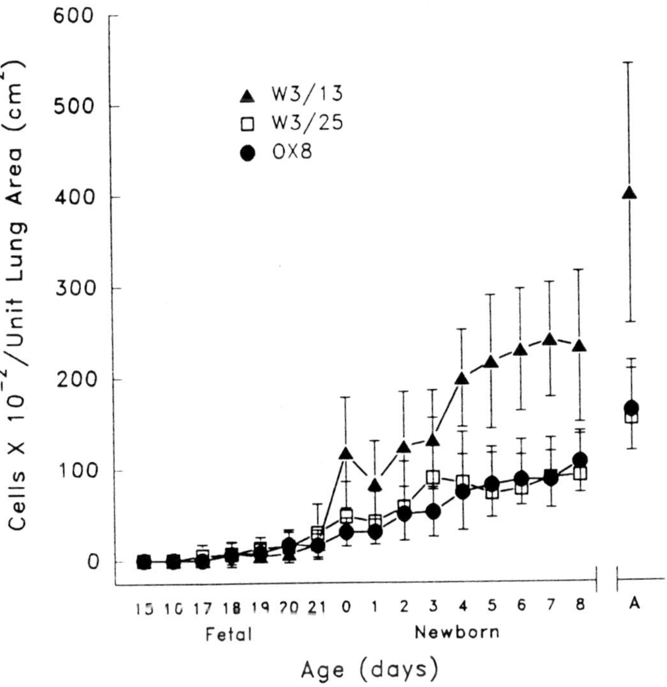

(b)

on which the organism depends for its gas exchange. The apparent phenotypic and functional heterogeneity of tissue DCs, in general and of those in the lung, in particular, is the result of an interplay of factors that include tissue-derived and inflammatory cell–derived cytokines, chemokines, soluble inhibitors, and hormones, as well as the normal turnover rate of DCs in any given tissue. In the lung, the phenotype and function of DCs in the various tissue compartments (vascular, alveolar, interstitial, airway epithelial) are beginning to be defined and a number of distinctive features are emerging.

A. Dendritic Cells in the Pulmonary Vascular Compartment

DCs are bone marrow–derived cells (37,38) that migrate via the bloodstream to periperal tissues (39). After endocytosing antigen (40,41) or reacting to locally released cytokines (42–44), they migrate via lymphatics to local lymph nodes or via the blood stream to the spleen (23). Together with granulocytes and macrophages, DCs originate from a common Ia^- progenitor in the bone marrow (37–39,45) from which they are released into the circulation where they represent <1% of circulating leukocytes. This trace population of circulating DCs in human blood appears to represent at least two subpopulations: the first comprises cells

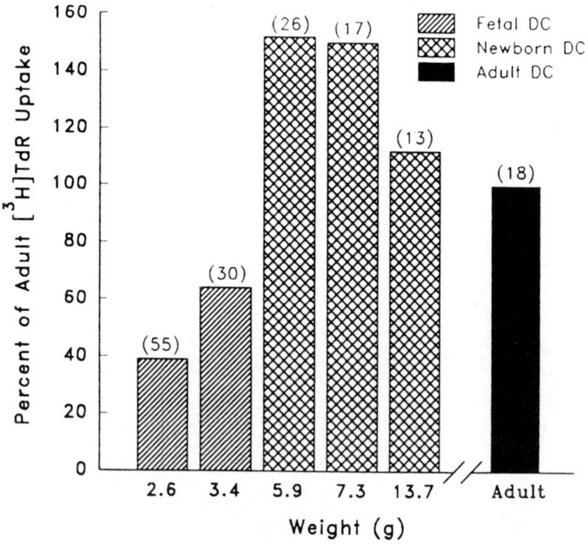

Figure 3 Stimulation of [^{3}H]thymidine incorporation of splenic T cells from adult Long Evans rats by fetal and newborn pulmonary DCs isolated from Lewis rats, expressed relative to that of DCs derived from simultaneously isolated adult pulmonary DCs. The numbers in parentheses indicate the number of animals required to obtain the graphed results. Background counts were <260 cpm, and the viability of the isolated DC was >90%. (From ref. 33.)

that express the myeloid marker $CD33^{bright}$, the adhesion molecules LFA-3, ICAM-1, and ICAM-2 and the β_2-integrin, CD11c; the DCs in this subpopulation are potent T-cell stimulators. A second, somewhat larger group, comprises small, $CD33^{dim}$, $CD11c^-$, functionally immature cells that require monocyte-derived cytokines to attain the mature DC phenotype (46). Both subgroups express MHC class II antigen, although expression is greater in the $CD11^+$ that in the $CD11^-$ group. Freshly isolated blood DCs of either subpopulation, however, do not express the B7 costimulatory molecules which are only upregulated after culture in vitro (47). These data are interpreted to indicate that DCs expressing the immature phenotype likely represent cells that have been recently released from the bone marrow, and are in transit to nonlymphoid tissues. By contrast, those DCs of mature phenotype are cells probably in transit from nonlymphoid tissues to the spleen after either having encountered antigen (40,41,48) or interacted with inflammatory cytokines, including tumor necrosis factor-α (TNF-α) (42–44). More recently, a third population of circulating DCs has been identified in human blood. These cells, although phenotypically mature, are poor T-cell stimulators, but they are the only DC population infectable with the human immunodeficiency virus (49).

The chemotactic and/or chemokine agents that induce DC precursors to leave the circulation and to populate the lung under nonperturbed, steady-state conditions remain to be identified. Early biodistribution studies using DCs harvested from the spleen have shown that very few, if any, of these mature DCs take up residence in the lung; instead they migrate to the spleen and liver (50). Similar results have been obtained recently by intravenously injecting bone marrow DCs cultured in vitro with GM-CSF (51). On the other hand, initiation of an inflammatory response by the intratracheal administration of heat-killed bacteria results in a rapid recruitment of Ia^+ DCs to the mucosal surface airways (31). Finally, the systemic administration of lipopolysaccharide (LPS) induces both an emigration of DCs from nonlymphoid tissues (52) as well as a recruitment of DC progenitors to the depopulated nonlymphoid tissues (53). These observations together with the biodistribution studies suggest that mature Ia^+ DCs do not recirculate once they reach the spleen or liver, instead it is the immature, Ia^- DC progenitors that emigrate from the blood to the lung interstitium. The route by which these cells are transported to the lung interstitium, however, is not established, and it is uncertain whether it is primarily via the bronchial and/or the pulmonary circulation.

That there are in fact a significant number of DC progenitors in circulation is evident when the leukocyte fraction of murine blood is cultured with GM-CSF. The addition of this growth factor to the medium enables approximately 100 DC clusters per 1×10^6 leukocytes to form (39). Under normal conditions, approximately two to three times as many leukocytes are sequestered in the pulmonary vascular compartment as are present in the systemic circulation (54,55). To determine whether this also pertains to circulating DCs and their precursors, the leukocyte fraction obtained by perfusing the pulmonary vascular compartment was compared with that harvested from the blood in the vena cava of the same rat. Under these conditions, there is an almost twofold enrichment of Ia^{dim} DC precursors in the blood obtained from the lung vascular compartment as compared with those obtained from the vena cava (56). When these leukocyte preparations are cultured for 3 days with GM-CSF, there is a 10- and 20-fold increase in the number of Ia^{bright} cells with the typical morphology of mature DCs (eccentric nucleus, a large amount of cytoplasm, and prominent cell processes) in the leukocyte cultures from the vena cava and lung, respectively (T. Suda et al., unpublished observations). The nature and possible role of adhesion molecules that may be involved in the apparent retention of DC precursors in the lung vasculature has not been established.

B. Dendritic Cells in the Lung

From the perspective of the immune response, the lung can be viewed as having two anatomically distinct components: (1) the *upper respiratory tract*, comprising the trachea and airways , and (2), the *lower respiratory tract*, including terminal bronchioles and alveoli (3). Lymphocytes are found in both locations where, depending on the species, they form aggregates and/or are distributed diffusely in the submucosa in the upper respiratory tract and are replenished from the systemic circulation. In the normal lower respiratory tract, interstitial lymphocytes are very sparse and, when present, are predominantly memory T cells (57). In terms of immunoglobulin production, a distinction can also be made between the two anatomical compartments of the lung. Although all major immuoglobins are represented in bronchial secretions, locally produced IgA predominates, whereas in bronchoalveolar lavage fluid from the lower respiratory tract, IgG derived from serum is the major immunoglobulin isotype (58). Mechanisms that limit or suppress IgE responses to inhaled allergens in the upper respiratory tract appear to be distinct and more highly developed than those in the alveolar portions of the lung (59,60). Analysis of the local lymph nodes suggest that the suppressor activity resides primarily in the superficial cervical lymph nodes that drain the oronasal cavity (59,60). Recently, this suppression of the IgE response has been shown to be mediated by $CD8^+$ TCR $\gamma\delta^+$ T cells which produce high levels of IFN-γ, a cytokine known to suppress Th2 cell proliferation (61). Although $\gamma\delta$ T-cell receptors can recognize antigen directly without antigen processing (62), in highly purified $\gamma\delta$ T cells from animals exposed in inhaled antigens, accessory cells, including DCs are required. The presence of nonpolymorphic class I molecules, which may present antigens to $\gamma\delta$ T cells has not been determined in DCs. However, CD1, which is a class I–like molecule, may serve as a restriction element for $\gamma\delta$ T cells, and this protein is abundant in DCs (23).

Dendritic Cells in Alveolar Spaces

Examination of lavage fluid from control animals and normal individuals indicate that there is a trace population of mature DCs present in the alveolar space compartment. Whether these cells continue to have a specific accessory cell function in this location or are destined to leave the lung via the mucocilary escalator is at present unclear. Recent studies, however, indicate that antigen-primed splenic DCs instilled into alveolar spaces not only ranslocate to local lymph nodes but also initiate an immune response (63). Antigen-primed DCs killed by freezing and thawing, on the other hand, are unable to prime T cells. By contrast, both viable and freeze/thaw–killed, antigen-pulsed alveolar macrophages (AMs) are capable of initiating an immune response, suggesting that antigenic peptides released by these cells may be endocytosed and/or absorbed by bystander DCs. Support for such a mechanism has recently been obtained in studies utilizing purified interstitial lung macrophages and DCs (29). The possibility cannot be excluded in the study by Havenith et al. (48) that the instilled DCs were able to gain access to the interstitium via traumatic breaks in the epithelium incurred during the instillation process. Whether antigen-loaded DCs in alveolar spaces are in fact capable of traversing normal, intact epithelium in order to migrate via the interstitium and lymphatics to local lymph nodes and the identity of the chemotactic signals that might be involved remains to be established.

In addition to a readily recognizable trace population of mature, Ia^+ DCs (0.4%), recent data indicate that human bronchoalveolar lavage fluid from normal individuals also contains a small number of monocyte-like cells (2%) (64,65). These two cell populations are distinguished by flow cytometry from AMs by their low autofluorescence. Interestingly, after overnight culture, the monocyte-like cells develop prominent cytoplasmic processes and adopt a phenotype characteristic of DCs. The latter, as well as the typical DCs harvested from alveolar spaces, are acid phosphatase and RFD9 (a macrophage marker) negative and strongly positive for HLA-DR, L25 (B cell, DC marker), RFD1 (a class II–associated epitope present on activated B cells, DCs, AMs). They function as typical DCs by forming clusters with T cells and inducing strong T-cell proliferation in an allogeneic MLR (64,65). These observations suggest that mature DCs and apparent DC precursors are normally present in small numbers in alveolar spaces, where their accessory cell function, however, appears to be downregulated by factors released by highly suppressive AMs (30,66,67). It remains to be determined whether these DCs are in transit to the mucocilliary escalator and, therefore, destined to be eliminated from the lung or whether, as in the rat model (48), they may migrate back across the lining epithelium in transit to the local lymph nodes.

Intraepithelial Airway Dendritic Cells

Before discussing the subset of DCs intercalated between the epithelial cells of airways, it is useful to review briefly the change in phenotype and function that DCs undergo as they mature from small, monocyte-like cells to the large, mature DCs. Early in differentiation, DCs share many characteristics usually associated with monocytes-macrophages. These include the presence of cytoplasmic compartments rich in acid phosphatase (68), the expression of FcγRII (69), and the ability to phagocytose particulates (70). As DCs assume their mature phenotype, however, these three attributes are lost. Although they appear to share a common progenitor in the bone marrow, their pathway of differentiation diverges; a process that is controlled, in part, by cytokines. In contrast to macrophages, DCs lack receptors to macrophage colony-stimulating factor (M-CSF) and instead have abundant receptors to GM-CSF and interleukin-1 (IL-1)(71). Freshly isolated Langerhans' cells from the epidermis are weak stimulators of resting T cells (72), and only after in vitro culture do they assume a mature phenotype that includes the expression of surface MHC class II, downregulation of invariant chain synthesis, and the ability to stimulate resting T cells (73).

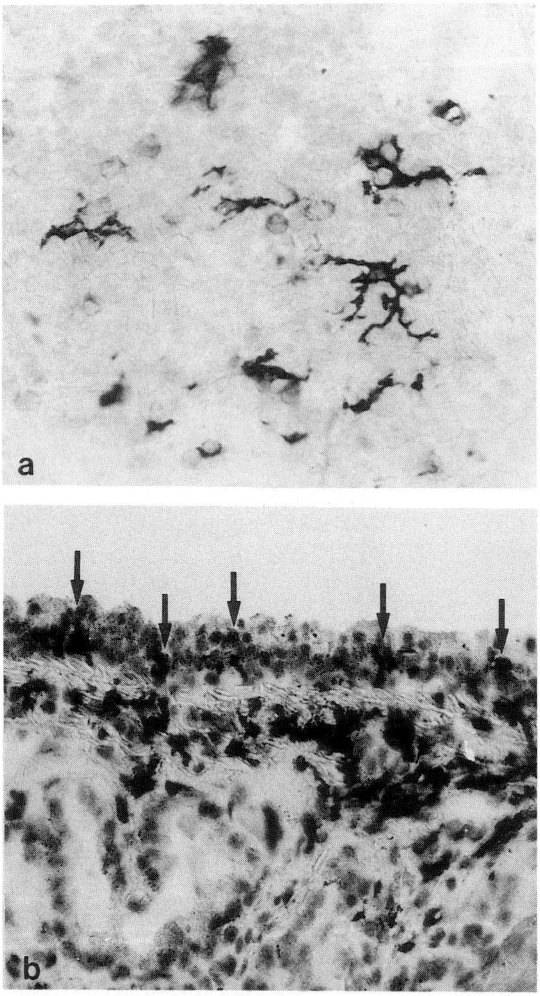

Figure 4 Frozen section of tracheal epithelium cut parallel (a) and perpendicular (b) to the plane of the epithelium and immunostained for Ia. Ia^+ airway DCs with arborizing cell processes are intercalated between epithelial cells (arrows). ×390. (From ref. 28.)

In view of the central role of DCs, in the capture, processing, and presentation of antigen and their close proximity to inhaled antigens, the question arises whether the phenotype and function of DCs observed within the airway epithelium (Fig. 4a,b) (74,75) resemble Langerhans' cells of the epidermis. To examine this question required that this subset of airway intraepithelial DCs be selectively harvested. To this end, Gong et al. (28) modified a microdissection method (76) that enables the dissection of almost the entire airway tree from the lung parenchyma (Fig. 5a). By adding dispase, an enzyme that specifically cleaves at the level of the basement membrane (Fig. 5b), to the agarose instilled into the dissected airway preparation, it became possible after incubation at 37°C to flush out the entire airway epithelium with its contained DCs, thereby facilitating their subsequent isolation. The fragments of lung tissue dissected from the airway tree provided the source of interstitial lung DCs with which the phenotype and function of airway DCs were compared. DCs isolated from airway epithelium (ADCs) (Fig. 6a) and the lung interstitium (LDCs) (Fig. 6b) are similar except that the nuclei of ADCs were not as lobulated as those from the lung interstitium, suggesting that they were less mature. After overnight culture, approximately 56% of airway DCs, but only 4% of interstitial lung DCs expressed FcR, as determined by IgG-coated erythrocyte rosetting. Since in vitro culture promotes DC maturation, it is possible that this is an underestimation of the proportion of airway DCs that express FcR. The "immature" phenotype of airway DCs was reflected in their greater ability to endocytose, process, and present soluble and particulate antigen to antigen-sensitized T cells relative to lung interstitial DCs. By contrast, lung interstitial DCs are more effective in stimulating resting T cells in an allogeneic MLR assay, an attribute unique to mature, fully differentiated DCs (Fig. 7).

Morphometric measurements of intraepithelial airway DCs immunostained for MHC class II antigen reveals the highest numerical density to be present in the trachea, with a progressive decrease in their number toward the more peripheral segments of the airway tree. Furthermore, the numerical density of Ia^+ DCs in the airway epithelium is augmented by treatment of the animals with parenteral IFN-γ (Fig. 8) (28). Recent evidence indicates that this is due to the upregulation of Ia expression by Ia^- DC precursors rather than an augmented migration of DC precursors to the lung (51,77). Interestingly, in the immediate neonatal period, Nelson and Holt observed that, in rats, intraepithelial DCs are first detected in the nares; several days later, MHC class II–expressing DCs begin to appear in the epithelium of the trachea and even later in the bronchial epithelium. It is worth noting that the prevalence of intraepithelial airway DCs is much lower in most strains of mice, and in fact in normal mice they are limited only to the upper regions of the trachea (T. Suda et al., unpublished observations). In summary, the functional studies of Gong et al. (28) indicate that the subset of intraepithelial airway DCs resemble epidermal Langerhans' cells in both their phenotype and function. Their demonstrated ability to endocytose, process, and present antigen indicates that they are well suited to act as sentinals at the air/epithelial interface of airways.

Dendritic Cells in Lung Parenchyma

In the absence immunological reagents that specifically immunolabel all DCs, investigators have relied on MHC class II expression and morphological criteria to

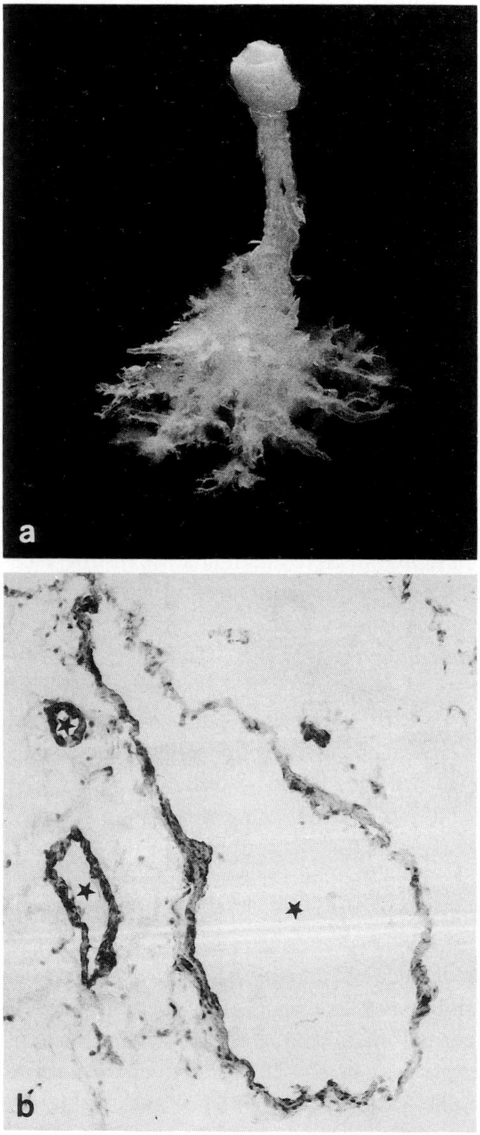

Figure 5 (a) Tracheobronchial tree dissected free of lung tissue before the isolation of airway DCs. ×2. (b) Small airway in the distal portion of the tracheobronchial tree (asterisks). Even small airways are denuded of their epithelium with the contained DCs following digestion with dispase. ×138. (From ref. 28.)

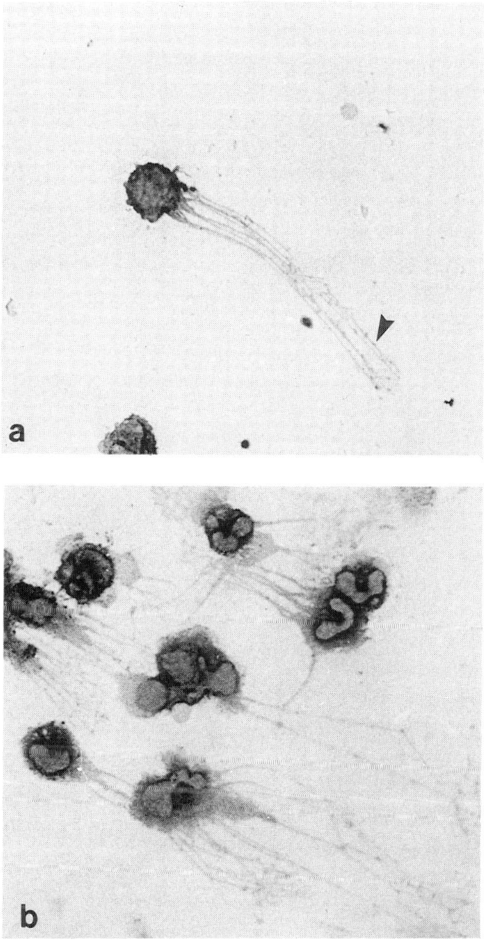

Figure 6 Cytocentrifuge preparations of isolated airway DCs (a) and lung DCs (b) immunostained for Ia. DCs isolated from both locations have long cytoplasmic processes (arrowhead), but those from the airway epithelium have nuclei that are not lobulated as are those from the lung interstitium, suggesting a less mature phenotype. ×538. (From ref. 28.)

identify DCs in situ in the lung. These include a relatively large cell size, distinctive cell processes, and a multilobulated eccentrically placed nucleus (27,75). Using these criteria, DCs in the lung parenchyma are localized to the connective tissue cuff around airways and major vessels and the connective tissue in alveolar walls where they are sparse in number. By electron microscopy, some of the DCs in the alveolar area of the lung are in fact in capillary lumens (27). Relying solely on MHC class II expression, however, does not identify Ia^- precursors. That these

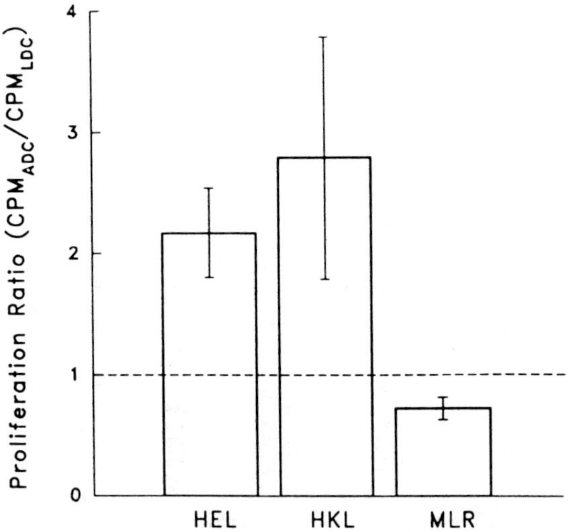

Figure 7 Graph relating the proliferation of T cells sensitized to hen egg lysozme (HEL) or heat-killed listeria (HKL), or allogeneic spenic T cells (MLR) induced by airway DCs to that induced by interstitial lung DCs. The bars that extend above the dotted line indicate the greater stimulatory capacity of airway DCs relative to lung DCs in antigen presentation. By contrast, lung DCs are superior to airway DCs in stimulating T cell proliferation in the one-way allogeneic MLR. The data are derived from the mean ± SD of a minimum of three separate isolations for each of the three functional assays. (From ref. 28.)

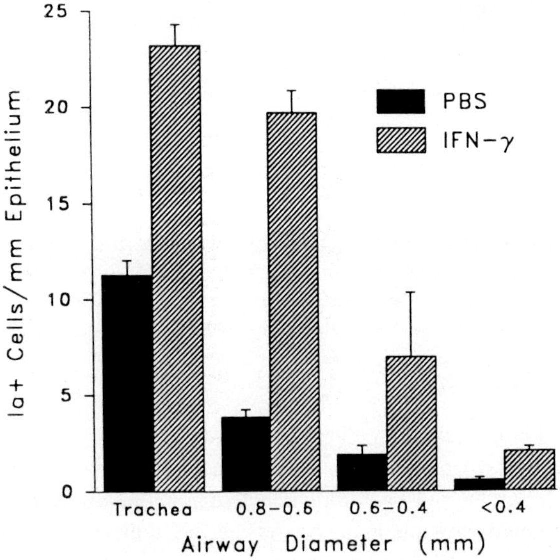

Figure 8 Histogram relating the number of Ia^+ airway DCs per millimeter of airway epithelium in control (phosphate-buffered saline, PBS, solid bars) and IFN-γ–injected (hatched bars) rats. (From ref. 28.)

may be present in the lung is suggested by the observation that in the fetal rat lung, OX62$^+$ DCs are present at a time in gestation when no Ia$^+$ DCs are observed (35). The distribution and localization of OX62$^+$ cells has not been systematically documented in the normal adult rat lung parenchyma. The importance of examining DC phenotype in situ is stressed, since DCs undergo significant phenotypic changes when they are subjected to isolation procedures and in vitro culture. For example, when immunostained in situ, intraepithelial airway DCs do not express ICAM-1 or LFA-1; however, after being subjected to isolation procedures and overnight culture, 43 and 78% of the isolated DCs expressed ICAM-1 and LFA-1, respectively (28).

Normal murine lungs immunostained for the important accessory molecules B7-1 and B7-2 reveal that a small subset of Ia$^+$ DCs express B7-2 and that these cells are localized predominantly in the connective tissue around airways and major vessels (Fig. 9a,b). This subset may represent mature DCs and/or DCs that have interacted with antigen and are in transit to local lymph nodes. By contrast, only a very rare DC in the lung parenchyma of such animals expresses B7-1.

Flow cytometry performed on DCs isolated from the murine lung indicates that the cells appear to be of mixed phenotype (78). Approximately half of the cells expressed FcRII, 61% expressed the C3bi receptor, and 81% had surface IL-2 receptors. Immunolabeling of the isolated lung DCs with monoclonal antibodies (mAbs) that recognize a number of murine DC subsets revealed that 45% of these cells stained with the NLDC-145 mAb (specific for interstitial DCs) and 67% were recognized by the J11d mAb (B cells and thymocytes). Lung DCs were not recognized by the 33D1 mAb (splenic DCs). Virtually none of the isolated lung DC reacted with the macrophage-specific mAb F4/80. These data, together with the results obtained from functional assays, suggested that lung interstitial DCs are similar to other tissue DCs, including epidermal Langerhans' cells and thymic DCs (78). In these lung DC preparations, however, no distinction was made between DCs derived from the lung parenchyma and those associated with the airways.

V. Turnover of Lung Dendritic Cells

Utilizing a radiation chimera model in which congenic rats expressed different allotypic variants of CD45, Holt and colleagues evaluated the turnover rate of DCs in the airway epithelium and lung parenchyma (79). The lung and airway DC population was depleted by x-irradiation or high-dose dexamethasone treatment. Based on the rate of DC depletion, an approximation was made of the half-life of the DC population in these two lung compartments and the results compared with those obtained for epidermal DCs from the same animals. Using this approach, the half-life of airway DCs was estimated to be about 2 days, whereas those in the epidermis of the same animals was approximately 15 days. The half-life of lung parenchymal DCs was intermediate between these two extremes and was esti-

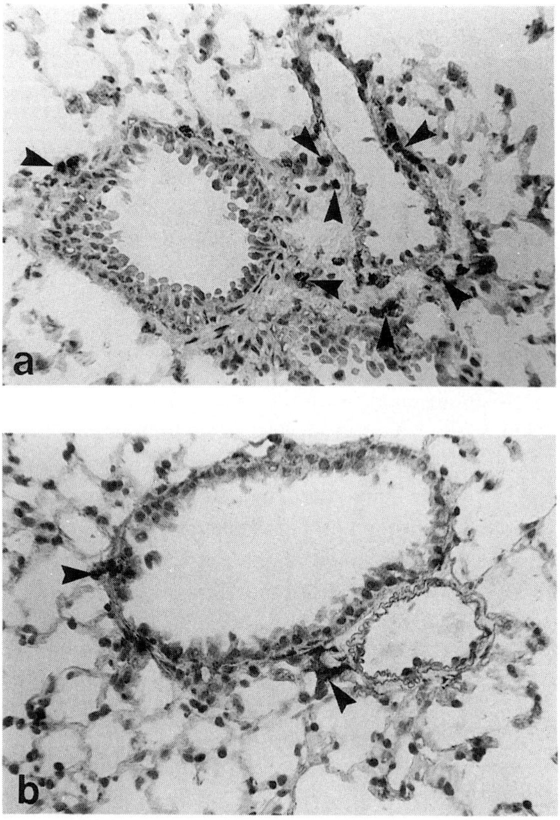

Figure 9 (a) Frozen section of mouse lung stained for Ia by immunoperoxidase techniques. Several Ia^+ DCs surround a small artery and adjacent intrapulmonary airway. (b) Frozen section from the same lung as in (a) stained for B7-2. Note the very sparse number of DCs that stain for B7-2.

mated to be about 6 days. After congenic bone marrow transplantation and a lag phase during which the transplanted marrow became established, the airway DC population was rapidly restored. The only other peripheral tissue with a comparably rapidly turning over DC population is the intestinal wall, indicating that DCs play a critical role in the surveillance of mucosal interfaces, including those of the lung. The identity of the chemokines/cytokines and endothelial integrins that are involved in the recruitment of DC to the various compartments of the lung, as part of their normal turnover or in response to an inflammatory focus, is an area requiring further investigation.

VI. Modulation in Situ of Lung Dendritic Cell Phenotye and Function

The apparent phenotypic and functional heterogeneity of DCs in lung and upper airways appears to be the result of a balance between locally secreted factors that either augment or depress DC accessory cell activity and that may regulate the rate of DC turnover in a given tissue compartment of the lung.

A. Epithelial Cell Factors

Both bronchial epithelial cells (80) and alveolar type II cells (81) synthesize and secrete GM-CSF, a factor that promotes both the cell division and differentiation of DCs in vitro (38,39). The interesting observation that DCs infiltrate those lung tumors that produce GM-CSF (82) supports the notion that the local secretion of this cytokine renders the microenvironment hospitable to DCs. Recent studies in mice carrying a null allele of the GM-CSF gene reveal that these animals do not have perturbations of hematopoiesis, but that they are susceptible to opportunistic infections, their alveolar macrophages are filled with surfactant, and prominent lymphocytic cuffs surround airways and major vessels (83,84). The distribution and accessory cell function of lung DCs was not examined in these animals. Although it appears that macrophage dysfunction contributed to the increased susceptibility to infection, the role of altered DC function in the perturbed immune state of these animals was not examined.

B. Macrophage Derived Factors

In order to contain the potentially damaging effects of uncontrolled immune responses in the lung requires, in addition to differentiation promoting factors, that there be counterbalancing inhibitory mechanisms in place. That AMs may have such a role was clearly demonstrated in mice whose AMs were eliminated by the intratracheal administration of liposomes containing dichloro-methylene-diphosphonate (Chlodronate) (85). Following immunization of these animals with TNP-KLH, there was both an increase in the number of DCs in local lymph nodes and in the lung, as well as a marked upregulation of antibody production. A subsequent study clearly showed that elimination of AMs enhanced local secondary antibody responses in the lung parenchyma but did not affect immune tolerance induction in the upper airways (86). This phenomenon was subsequently explained by the fact that suppressor cells that mediate tolerance are activated in lymph nodes draining the oropharynx and trachea and not by those draining the lung (61).

A number of factors released by alveolar macrophages may contribute to the suppression of accessory cell activity. Incubation of AMs with indomethacin and catalase partially restores the accessory cell activity of bronchoalveolar lavage cells, suggesting that prostaglandins and/or hydrogen peroxide might be involved

(87). Using antibodies to tumor growth factor-β (TGF-β), Lipscomb and colleagues provided evidence that this macrophage-derived factor contributes to the inhibitory activity of AMs (88). More recent data indicate that AM-derived NO plays a key role in the immune suppression of DCs (30). Coculture of lung DCs with AMs, separated by a semipermeable membrane, markedly suppressed DC function, and the inhibition was abrogated by the addition of NG-monmethylarginine, an NO synthase inhibitor. Interestingly, the addition of TNF-α, another macrophage-derived cytokine, synergizes with NO's inhibitory effect on DC function. Ultrastructural studies of perfusion-fixed rat lungs indicate that alveolar macrophages, by adhering to alveolar type I cells, are separated from the underlying interstitial lung DCs by only a 0.2-μm inch thickness layer of cytoplasm. In view of the proximity of AMs, the NO released from these cells can readily diffuse across this thin cytoplasmic barrier and downregulate DC phenotype and function in situ. Such a mechanism could explain the common observation that, once removed from the inhibitory environment created by released macrophage inhibitors, the accessory cell activity of freshly isolated lung DCs is upregulated after overnight culture. TNF-α, in addition to synergizing with NO to depress DC function, in other tissues, including the skin, it stimulates Langerhans' cells to migrate to local lymph nodes (42). It is possible, therefore, that in an inflammatory focus in the lung, TNF-α released from AMs could enhance lung DC migration to local lymph nodes where they would stimulate sensitized T cells to proliferate.

C. T-Cell–Derived Factors

One of the hallmarks of the Th1 subclass of Th cells is the secretion of IFN-γ. The administration of cytokine to mice (89) and rats (28) results in a marked upregulation of MHC class II expression on many cells, including epithelial cells (89,90), and an increase in the number of interstitial DCs. Biodistribution and immunolabeling studies utilizing cultured bone marrow DCs injected intravenously into IFN-γ treated mice have recently shown that this apparent increase in the number of Ia^+ cells in the lungs of these animals is not due to an increased migration of circulating Ia^+ DC to the lung. Rather, it appears that there is an induction of Ia expression either by recently arrived or resident Ia^+ DC precursors (77). Although IFN-γ has little effect on the accessory cell function of fully differentiated DCs (27), it appears to augment Ia expression on DC precursors and thus has an amplifying effect on the immune response.

VII. Conclusions

Dendritic cells are present in the lung early in gestation at a time when the lung is not exposed to inhaled antigens. However, indirect evidence suggests that maternal factors may upregulate Ia expression by DCs in the fetal lung, and this might explain the apparent discrepancy in the literature as to when MHC class II antigen first appears on these cells and the time point in fetal/neonatal development that they attain their full accessory cell potential. DCs are a heterogeneous population in all four tissue compartments of the lung (vascular, alveolar, airway epithelial, and lung interstitial). Their phenotype and function is governed by locally secreted factors. Under normal steady-state conditions, their immune function appears to be suppressed in situ by locally released alveolar macrophage factors that include such freely diffusible inhibitors as nitric oxide. TNF-α not only synergizes with NO but also induces DCs to migrate to local lymph nodes. The factors that induce DC precursors, recently released form the bone marrow, to migrate to peripheral tissues and the factors that govern their turnover and migration from the lung are topics requiring further study.

Acknowledgments

Supported by NIH grant HL36781. We would like to acknowledge the many valuable contributions made by Karin M. McCarthy, Jian Lin Gong, Joel R. Telford, Joanne M. McCormack, and Quynh Vu without whose help, hard work, and good humor none of this could have been accomplished.

References

1. Weibel ER. Design of airways and blood vessels considered as branching trees. In: Crystal RW, West JB, Barnes PJ, Cherniak, NS, Weibel ER, eds. The Lung: Scientific Foundations. New York: Raven Press, 1991:795–805.
2. Sleigh MA. Mucus propulsion. In: Crystal RW, West JB, Barnes PJ, Cherniak, NS, Weibel ER, eds. The Lung: Scientific Foundations. New York: Raven Press, 1991: 189–196.
3. Kaltreider HB. Normal Immune response. In: Crystal RW, West JB, Barnes PJ, Cherniak, NS, Weibel ER, eds. The Lung: Scientific Foundations. New York: Raven Press, 1991: 499–510.
4. Crystal RG. Alveolar macrophages. In: Crystal RW, West JB, Barnes PJ, Cherniak,NS, Weibel ER, eds. The Lung: Scientific Foundations, New York: Raven Press, 1991:527–538.
5. Schneeberger EE, Lynch RD. Structure, function and regulation of cellular tight junctions. Am J Physiol 1992; 262:L647–L661.
6. Germaine RN. The biochemistry and cell biology of antigen processing and presentation. Annu Rev Immunol 1993; 11:403–450.
7. Barber LD, Parham P. Peptide binding to major histocompatibility complex molecules. Annu Rev Cell Biol 1993; 9:163–206.
8. Townsend A, Bodmer H. Antigen recognition by class I-restricted T lymphocytes. Annu Rev Immunol 1989; 7:601–624.

9. Geppert TD, Davis LS, Gur H, Wacholtz MC. Lipsky PE. Accessory cell signals involved in T-cell activation. Immunol Rev 1990; 117:5–66.
10. Anderson MS, Miller J. Invariant chain can function as a chaperone protein for class II major histocompatibility complex molecules. Proc Natl Acad Sci USA 1992; 89: 2282–2286.
11. Roche PA, Cresswell P. Invariant chain association with HLA-DR molecules inhibits immunogenic peptide binding. Nature 1990; 345:615–618.
12. Baake O, Dobberstein B. MHC class II-associated invariant chain contains a sorting signal for endosomal compartments. Cell 1990; 63:707–716.
13. Nijman HW, Kleijmeer MJ, Ossevoort MA, Oorschot VMJ, Vierboom MPM, Van de Keur M, Kenemans P, Kast WM, Geuze HJ, Melief CJM. Antigen capture and major histocompatibility class II compartments of freshly isolated and cultured human blood dendritic cells. J Exp Med 1995; 182:163–174.
14. Morris P, Shaman J, Attaya M, Amaya M, Goodman S, Bergman C, Monaco JJ, Mellins E. An essential role for HLA-DM in antigen presentation by class II major histocompatibility molecules. Nature 1994; 368:551–554.
15. Springer TA, Dustin ML, Kishimoto TK, Marlin SD. The lymphocyte function associated LFA-1, CD2 and LFA-3 molecules: cell adhesion receptors of the immune system. Annu Rev Immunol 1987; 3:361–372.
16. Moingeon P, Chang HC, Syre PH, Clayton LK, Alcover A, Gardner P, Reinherz EL. The structural biology of CD2. Immunol Rev 1989; 111:111–114.
17. Springer TA. Adhesion molecules of the immune system. Nature 1990; 346:425–434.
18. Linsley PS, Brady W, Grosmaire L, Aruffo A, Damle NK, Ledbetter JA. Binding of the B cell activation antigen B7 to CD28 costimulates T cell proliferation and interleukin-2 mRNA accumulation. J Exp Med 1991; 173:721–730.
19. Linsley PS, Brady W, Urnes M, Grosmaire LS, Kamle NK, Ledbetter JA. CTLA-4 is a second receptor for the B cell activation antigen B7. J Exp Med 1991; 174: 561–569.
20. Van de Velde H, Von Hoegen I, Luo W, Parnes JR, Thielemans K. The B-cell surface protein CD72/Lyt-2 is the ligand for CD5. Nature 1991; 351:662–665.
21. Norment AM, Salter RD, Parham P, Engelhard VH, Littman D. Cell-cell adhesion mediated by CD8 and MHC class I molecules. Nature 1988; 336:79–81.
22. Doyle C, Strominger JL. Interaction between CD4 and class II MHC molecules mediates cell adhesion. Nature 1987; 330:256–259.
23. Steinman RM. The dendritic cell system and its role in immunogencity. Annu Rev Immunol 1991; 9:271–296.
24. Lenz A, Heine M, Schuler G, Romani N. Human and murine dermis contain dendritic cells. J Clin Invest 1993; 92:2587–2596.
25. Streilein JW, Grammer SF, Yoshikawa T, Demidem A, Vermeer M. Functional dichotomy between Langerhans cells that present antigen to naive and to memory/ effector T lymphocytes. Immunol Rev 1990; 117:159–183.
26. Klinkert WE. Lymphoid dendritic accessory cells of the rat. Immunol Rev 1990; 117: 103–119.
27. Kradin RL, McCarthy KM, Xia W, Lazarus D, Schneeberger EE. Accessory cells of the lung. I. Interferon-gamma increases Ia+ dendritic cells in the lung without augmenting their accessory activities. Am J Respir Cell Mol Biol 1991; 4:210–218.

28. Gong JL, McCarthy KM, Telford JR, Schneeberger EE. Intraepithelial airway dendritic cells: A distinct subset of pulmonary dendritic cells obtained by microdissection. J Exp Med 1992; 175:797–807.
29. Gong JL, McCarthy KM, Rogers RA, Schneeberger EE. Interstitial lung macrophages interact with dendritic cells to present antigen peptides derived from particulate antigens to T-cells. Immunology 1994; 81:343–351.
30. Holt PG, Oliver J, Bilyk N, McMenamin C, McMenamin PG, Kraal G, Thepen T. Downregulation of the antigen presenting cells function(s) of pulmonary dendritic cells in vivo by resident alveolar macrophages. J Exp Med 1993; 177:397–407.
31. McWilliams AS, Nelson D, Thomas JA, Holt PG. Rapid dendritic cell recruitment is a hallmark of the acute inflammatory response at mucosal surfaces. J Exp Med 1994: 179:1331–1336.
32. Van Rees EP, Van der Ende MB, Sminia T. Ontogeny of macrophage subpopulations and Ia-positive dendritic cells in pulmonary tissue of the rat. Cell Tissue Res 1991; 263:367–373.
33. McCarthy KM, Gong JL, Telford JR, Schneeberger EE. Ontogeny of Ia+ accessory cells in fetal and newborn rat lung. Am J Respir Cell Mol Biol 1992; 6:349–356.
34. Nelson DJ, McMenamin C, McWilliam AS, Brenan M, Holt PG. Development of the airway intraepithelial dendritic cell network in the rat from class II major histocompatibility (Ia)–negative precursors: Differential regulation of Ia expression at different levels of the respiratory tract. J Exp Med 1994; 179:203–212.
35. Nelson DJ, Holt PG. Defective regional immunity in the respiratory tract of neonates is attributable to hyporesponsiveness of local dendritic cells to activation signals. J Immunol 1995; 155:3517–3524.
36. Brenan M, Puklavec M. The MRC OX-62 antigen: A useful marker in the purification of rat veiled cells with the biochemical properties of an integrin. J Exp Med 1992; 175:1457–1465
37. Bowers WE, Berkowitz MR. Differentiation of dendritic cells in cultures of rat bone marrow cells. J Exp Med 1986; 163:872–883.
38. Inaba K, Inaba M, Romani N, Aya H, Deguchi M, Ikehara S, Muramatsu S, Steinman RM. Generation of large numbers of dendritic cells from mouse bone marrow cultures supplemented with granulocyte/macrophage colony-stimulating factor. J Exp Med 1992; 176:1693–1702.
39. Inaba K, Steinman RM, Pack MW, Aya H, Inaba M, Sudo T, Wolpe S, Schuler G. Identification of proliferating dendritic cell precursors in mouse blood. J Exp Med 1992; 175:1157–1167.
40. Bujdoso R, Hopkins J, Young P, McConnell I: Characterization of sheep afferent lymph dendritic cells and their role in antigen carriage. J Exp Med 1989; 170:1285–1302.
41. Kripke ML, Munn CG, Jeevan A, Tang JM, Bucana C. Evidence that cutaneous antigen-presenting cells migrate to regional lymph nodes during contact sensitization. J Immunol 1990; 145:2833–2938.
42. Cumberbatch M, Kimber I. Dermal tumour necrosis factor-alpha induces dendritic cell migration to draining lymph nodes, and possibly provides one stimulus for Langerhans' cell migration. Immunology 1992; 75:257–263.
43. Cumberbatch M, Fielding I, Kimber I. Modulation of epidermal Langerhans' cell frequency by tumour necrosis factor-alpha. Immunology 1994; 81:395–401.

44. Cumberbatch M, Kimber I. Tumour necrosis factor-alpha is required for accumulation of dendritic cells in draining lymph nodes and for optimal contact sensitization. Immunology 1995; 84:31–35.
45. Inaba K, Inaba M, Deguchi M, Hagi K, Yasamizu R, Ikehara S, Muramatsu S, Steinman RM. Granulocytes, macrophages and dendritic cells arise from a common major histocompatibility complex class II-negative progenitor in mouse bone marrow. Proc Natl Acad Sci USA 1993; 90:3038–3042.
46. O'Doherty U, Peng M, Gezelter S, Swiggard WJ, Betjes M, Bhardwaj N, Steinman RM. Human blood contains two subsets of dendritic cells, one immunologically mature and the other immature. Immunology 1994; 82:487–493.
47. Thomas R, Lipsky PE. Human peripheral blood dendritic cell subsets. J Immunol 1994; 153:4016–4028.
48. Havenith CEG, Van Miert PPMC, Breedijk AJ, Beelen RHJ, Hoefsmit ECM. Migration of dendritic cells into the draining lymph nodes of the lung and intratracheal instillation. Am J Respir Cell Biol 1993; 9:484–488.
49. Weissman D, Li Y, Ananworanich J, Zhou LJ, Adelsberger J, Tedder TF, Baseler M, Fauci AS. Three populations of cells with dendritic morphology exist in peripheral blood, only one of which is infectable with human immunodeficiency virus type I. Proc Natl Sci USA 1995; 92:826–830.
50. Kupiec-Weglinski JW, Austyn JM, Morris PJ. Migration patterns of dendritic cells in the mouse. J Exp Med 1988; 167:632–645.
51. Suda T, Callahan RJ, Wilkenson RA, Van Rooijen N, Schneeberger EE. Interferon-gamma reduces Ia+ dendritic cell traffic to the lung. J Leukoc Biol 1996; In press.
52. Roake JA, Rao AS, Morris PJ, Larsen CP, Hankins DF, Austyn JM. Dendritic cell loss from nonlymphoid tissues after systemic administration of lipopolysaccharide, tumor necrosis factor and interleukin 1. J Exp Med 1995; 181:2237–2247.
53. Roake JA, Rao AS, Morris PJ, Larsen CP, Hankins DF, Austyn JM. Systemic lipopolysaccharide recruits dendritic cell progenitors to nonlymphoid tissues. Transplantation 1995; 59:1319–1324.
54. Doerschuk CM, Allard MF, Martin BA, MacKenzie A, Autor AP, Hogg JC. Marginated pool of neutrophils in rabbit lungs. J Appl Physiol 1987; 63:1806–1815.
55. Martin BA, Wiggs BR, Lee S, Hogg JC. Regional differences in neutrophil margination in dog lungs. J Appl Physiol 1987; 63:1253–1262.
56. Gong J, Suda T, Schneeberger EE. Dendritic cells (DC) are enriched in the pulmonary vascular compartment. FASEB J 1995; 9:A1035.
57. Abraham E, Freitas AA, Coutinho AA. Purification and characterization of intraparenchymal lung lymphocytes. J Immunol 1990; 144:2117–2122.
58. Kaltreider HB, Chan MKL. The class-specific immunoglobulin composition of fluids obtained from various levels of the canine respiratory tract. J Immunol 1976; 116:423–429.
59. Sedgwick JD, Holt PG. Induction of IgE-secreting cells and IgE isotype-specific suppressor T cells in the respiratory lymph nodes of rats in response to antigen inhalation. Cell Immunol 1985; 94:182–194.
60. Holt PG, Sedgwick JD. Suppression of IgE responses following inhalation of antigen. Immunol Today 1987; 8:14–15.
61. McMenamin C, McKersey M, Kuhnlein P, Hunig T, Holt PG. Gamma/delta T cells

down-regulate primary IgE responses in rats to inhaled soluble protein antigens. J Immunol 1995; 154:4390–4394.
62. Schild H, Mavaddat C, Litzenberger EW, Davis MM, Bluestone JA, Matis L, Draper RK, Chien Y. The nature of major histocompatibility complex recognition by gamma/delta T-cells. Cell 1994; 76:29–32.
63. Havenith CEG, Breedijk AJ, Betjes MG, Calame W, Beelen RH, Hoefsmit ECM. T cell priming in situ by intratracheally instilled antigen-pulsed dendritic cells. Am J Respir Cell Mol Biol 1993; 8:319–324.
64. Van Haarst JMW, Hoogsteden HC, De Wit HJ, Verhoeven GT, Havenith CEG, Drexhage HA. Dendritic cells and their precursors isolated from human broncho-alveolar lavage: Immunocytologic and functional properties. Am J Respir Cell Mol Biol 1994; 11:244–350.
65. Havenith CEG, Van Haarst JMW, Breedijk AJ, Betjes MGH, Hoogsteden HC, Beelen RHJ, Hoefsmit ECM. Enrichment and characterization of dendritic cells from human bronchoalveolar lavages. Clin Exp Immunol 1994; 96:339–343.
66. Holt PG. Alveolar macrophages. II. Inhibition of lymphocyte proliferation by purified macrophages from rat lung. Immunology 1979; 37:429–436.
67. Holt PG. Down-regulation of immune responses in the lower respiratory tract. The role of alveolar macrophages. Clin Exp Immunol 1986; 63:261–270.
68. Stossel H, Koch F, Kampgen E, Stoger P, Lenz A, Heufler C, Romani N, Schuler G. Disappearance of certain acidic organelles (endosomes and Langerhans cell granules) accompanies loss of antigen processing capacity upon culture of epidermal Langerhans cells. J Exp Med 1990; 172:1471–1482.
69. Elbe A, Schleischitz S, Strunk D, Stingl G. Fetal skin-derived MHC class I+, MHC class II− dendritic cells stimulate MHC class I-restricted responses of unprimed CD8+ T cells. J Immunol 1994; 153:2878–2889.
70. Inaba K, Inaba M, Naito M, Steinman RM. Dendritic cell progenitors phagocytose particulates, including bacillus Calmette-Guerin organisms and sensitize mice to mycobacterial antigens in vivo. J Exp Med 1993; 178:479–488.
71. Kampgen E, Koch F, Heufler C, Eggert A, Gill LL, Gillis S, Dower SK, Romani N, Schuler G. Understanding the dendritic cell lineage through a study of cytokine receptors. J Exp Med 1994; 179:1767–1776.
72. Romani N, Lenz A, Glassel H, Stossel H, Stanzl U, Majdic O, Fritsch P, Schuler G. Cultured human Langerhans cells resemble lymphoid dendritic cells in phenotype and function. J Invest Dermatol 1989; 93:600–609.
73. Kampgen E, Koch N, Koch F, Stoger P, Heufler C, Schuler G, Romani N. Class II major histocompatibility complex molecules of murine dendritic cells: Synthesis, sialylation of invariant chain and antigen processing capacity are down-regulated upon culture. Proc Natl Acad Sci USA 1991; 88:3014–3018.
74. Sertl K, Takemura T, Tschachler E, Ferrans VJ, Kaliner MA, Shevach EM. Dendritic cells with antigen presenting capability reside in airway epithelium, lung parenchyma and visceral pleura. J Exp Med 1986; 163:436–451.
75. Schon-Hegrad MA, Oliver J, McMenamin PG, Holt PG. Studies on the density, distribution and surface phenotype of intraepithelial class II major histocompatibility complex antigen (Ia)–bearing dendritic cells (DC) in the conducting airways, J Exp Med 1991; 173:1345–1356.

76. Plopper CG, Chang AM, Pang A, Buckpitt AR. Use of microdissected airways to define metabolism and cytotoxicity in murine bronchiolar epithelium. Exp Lung Res 1991; 17:197–212.
77. Suda T, Vu Q, Callahan R, Van Rooijen N, Schneeberger EE. Effect of IFN-gamma and dexamethasone on dendritic cell (DC) traffic to the lung. FASEB J 1996; 10: A1007.
78. Pollard AM, Lipscomb MF. Characterization of murine lung dendritic cells: Similarities to Langerhans cells and thymic dendritic cells. J Exp Med 1990; 172:159–167.
79. Holt PG, Haining S, Nelson DJ, Sedgwick JD. Origin and steady-state turnover of class II MHC-bearing dendritic cells in the epithelium of the conducting airways. J Immunol 1994; 153:256–261.
80. Churchill L, Friedman B, Schleimer RP, Proud D. Production of granulocyte macrophage colony-stimulating factor by cultured human tracheal epithelial cells. Immunology 1992; 75:189–195.
81. Christensen PJ, Armstrong LR, Fak JJ, Chen GH, McDonald RA, Toews GB, Paine R. Regulation of rat pulmonary dendritic cell immunostimulatory activity by alveolar epithlelial cell-derived granulocyte macrophage colony-stimulating factor. Am J Respir Cell Mol Biol 1995; 13:426–433.
82. Tazi A, Bouchonnet F, Grandsaigne M, Boumsell L, Hance AJ, Soler P. Evidence that granulocyte-colony-stimulating factor regulates the distribution and differentiated state of dendritic cells/Langerhans cells in human lung and lung cancers. J Clin Invest 1993; 91:566–576.
83. Stanley E, Lieschke GJ, Grail D, Metcalf D, Hodgson G, Gall JA, Maher DW, Cebon J, Sinickas V, Dunn AR. Granulocyte/macrophage colony-stimulating factor–deficient mice show no major perturbation of hematopoiesis by develop a characteristic pulmonary pathology. Proc Natl Acad Sci USA 1994; 91:5592–5596.
84. Dranoff G, Crawford AD, Sadelain M, Ream B, Rashid A, Bronson RT, Dickersin GR, Bachurski CJ, Mark EL, Whitsett JA, Mulligan RC. Involvement of granulocyte-macrophage colony-stimulating factor in pulmonary homeostasis. Science 1994; 264: 713–716.
85. Thepen T, Van Rooijen N, Kraal G. Alveolar macrophage elimination in vivo is associated with an increase in pulmonary immune response in mice. J Exp Med 1989; 170:499–509.
86. Thepen T, McMenamin C, Oliver J, Kraal G, Holt PG. Regulation of immune response to inhaled antigen by alveolar macrophages: differential effects of in vivo alveolar macrophage elimination on the induction of tolerance vs. immunity. Eur J Immunol 1991; 21:2845–2850.
87. Kaltreider HB, Caldwell JL, Byrd PK. The capacity of normal murine alveolar macrophages to function as antigen-presenting cells for the initiation of primary antibody-forming cell responses to sheep erythrocytes in vitro. Am Rev Respir Dis 1986; 133:1097–1104.
88. Lipscomb MF, Pollard AM, Yates JL. A role for TGF-β in the suppression by murine bronchoalvolar cells of lung dendritic cell initiated immune responses. Reg Immunol 1993; 5:151–157.
89. Skoskiewicz MJ, Colvin RB, Schneeberger EE, Russell PS. Widespread and selective

induction of major histocompatibility complex-determined antigens in vivo by gamma interferon. J Exp Med 1985; 162:1645–1664.
90. Schneeberger EE, De Ferrari M, Skoskiewicz M, Russell PS, Colvin RB. Induction of MHC-determined antigens in the lung by interferon- gamma. Lab Invest 1985; 55: 138–144 1986.

11

Regulation of Dendritic Cell Recruitment and Traffic in the Lung

RICHARD L. KRADIN

Harvard Medical School
and Massachusetts General Hospital
Boston, Massachusetts

I. Introduction

The last decade has seen an increasing recognition of the critical role that dendritic cells (DCs) play in mediating the immune response. DCs are derived from a CD45R^{+} common precursor cell in the bone marrow and constitute a morphologically and functionally related family of mononuclear cells (1). They are widely distributed in vivo and have been localized in virtually all mammalian tissues except the brain. In addition, DCs circulate in a pathway that leads from the peripheral blood to tissues and their draining lymph nodes.

DCs have been well characterized as highly effective antigen-presenting cells (APCs) (2). Although there is currently no other recognized functional role for DCs, their presence in large numbers at sites where antigens are unlikely to be deposited, for example, dura mater, perineural sheaths, and the fibrous capsules of the viscera (our unpublished observation), suggests that they may also participate in nonimmunological activities.

In the lung, DCs have been localized in the conducting and respiratory airways, the alveolar interstitium, the adventitia of the pulmonary microvasculature, and the pleura (3–6). In murine species, they are readily distinguished from pulmonary macrophages by their well-developed microvillar cytoplasmic extensions, strong expression of class II major histocompatibility (MHC) antigens, minimal phagocytic capacities, and their ability to present antigen and costimulate the activities of naive and sensitized lymphocytes (3,7). The ability to distinguish DCs from mononuclear phagocytes is complicated in humans by the fact that human macrophages also express high levels of class II MHC (Ia) antigens (8).

Two subsets of pulmonary DCs have been identified in murine species (4,9). A subset of FcR^+C3bi^+ dendritic cells shares phenotypic features with the Langerhans' cell of the skin and represents the majority of the DCs located in the conducting airways (6,10,11). Whereas these FcR^+ airway dendritic cells are excellent APCs, they display relatively inefficient accessory activities for the proliferative responses of T cells. In the pulmonary airways, they are likely to play a critical role in the early capture and processing of inhaled antigens (11). There is substantial evidence to suggest that these FcR^+ cells differentiate under the influence of granulocyte-macrophage colony-stimulating factor (GM-CSF) into FcR^- connective tissue DCs (12,13).

In contrast to the pulmonary airways, the normal lung parenchyma is populated almost exclusively by FcR^- $C3bi^-$ DCs that share phenotypic features with extrapulmonary connective tissue DCs (10). These cells are modestly less efficient than FcR^+ cells in their abilities to process antigen, but they excel as accessory cells in the proliferative responses of T cells (9).

II. Dendritic Cells and Cytokines

A. Cytokine Regulation of DC Maturation

The maturation of DCs is regulated by cytokines. Langerhans' cells mature into connective tissue DCs in the presence of GM-CSF, and tumor necrosis factor-α (TNF-α) and GM-CSF also promote the viability of DCs (12). A subset of blood monocytes appears to differentiate into DCs under the influences of interferon-γ (IFN-γ), interleukin-4 (IL-4), and GM-CSF (13,14).

B. Interferon-γ

IFN-γ, a T helper (h)-1 cytokine, secreted by activated T lymphocytes and natural killer (NK) cells during inflammation, is a potent activator of macrophages with respect to their class II MHC antigen expression and APC activities. However, IFN-γ has little effect on Ia expression by DCs and does not augment their accessory activities (3). Nevertheless, there is evidence to suggest that IFN-γ is a critical mediator of dendritic cell migration in vivo.

Injecting recombinant IFN-γ parenterally into Lewis rats yielded a marked increase in pulmonary DCs (3). In addition, the compartmental distribution of dendritic cells in the lung was altered (Fig. 1), and there was a significant increase in the number of DCs that were intercalated within the airway lining epithelium

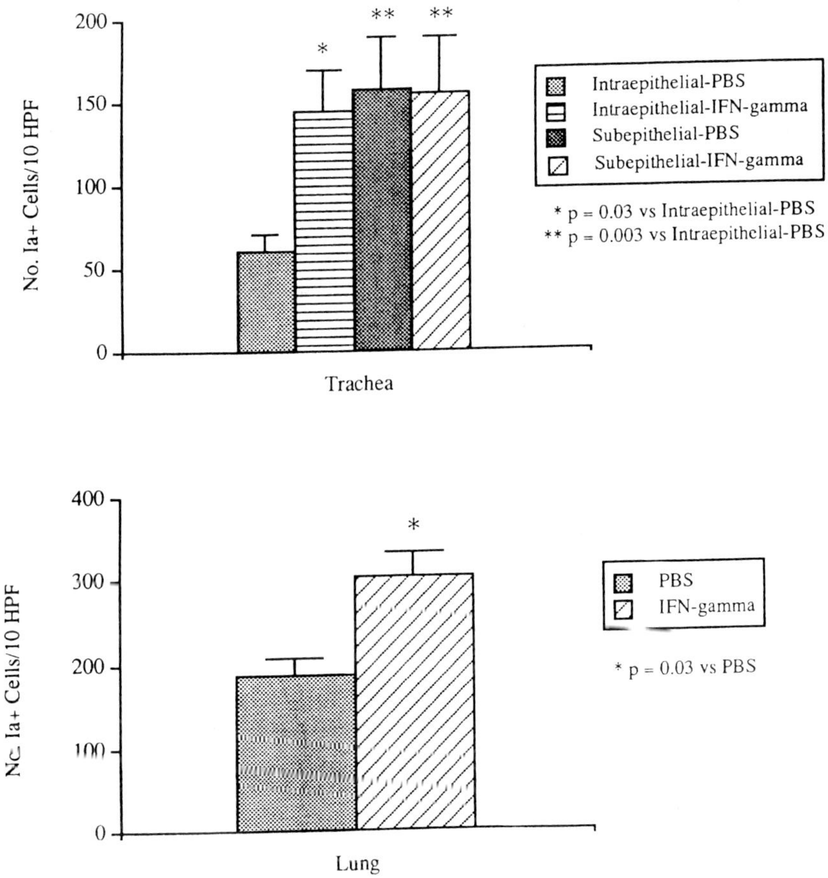

Figure 1 Distribution of Ia (OX-6)$^+$ DCs in lung. Following administration of recombinant murine IFN-γ (2×10^5 U/rat/day) or phosphate-buffered saline (PBS) for 5 days, sections of trachea and lung were incubated with OX-6 and the number of Ia$^+$ cells in tracheal mucosa (upper panel) and alveolar septa (lower panel) was enumerated in situ.

(Fig. 2). This finding was judged either to reflect the translocation of resident DCs within the airway wall or the directed migration of "exudate" DCs from the bronchial circulation to the lung in response to IFN-γ.

C. Interleukin-2

IL-2 is a critical TH1 lymphokine that has recently been shown to promote the accumulation of DCs within the lung (3). It was demonstrated that virtually all

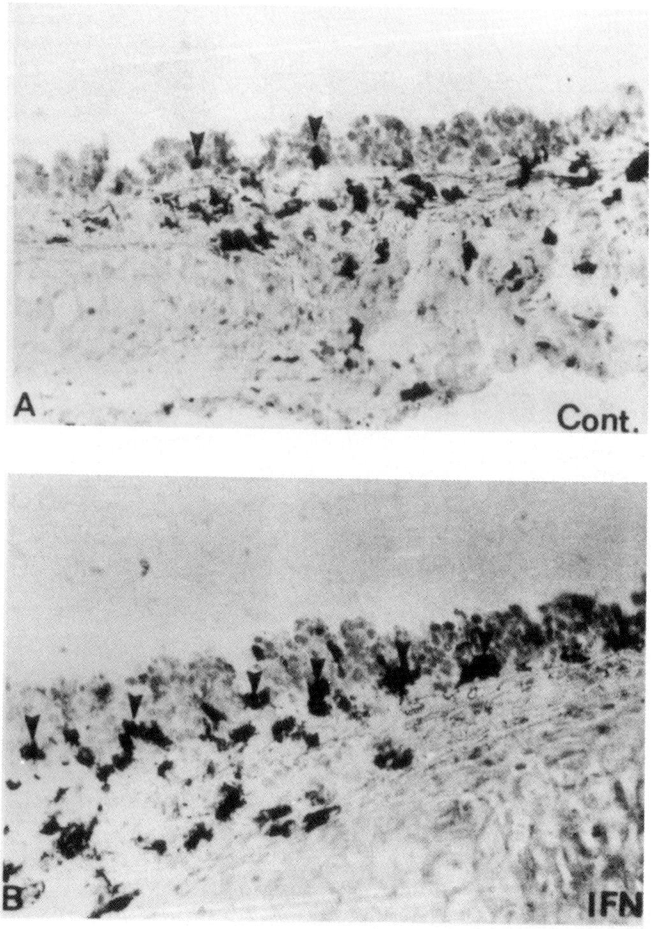

Figure 2 DCs in lung airways following parenteral injection of IFN-γ. (A) Ia^+ DCs are seen in the tracheal mucosa and to a lesser extent among the epithelial lining cells of the saline-treated controls. (B) Administration of IFN-γ for 5 days results in an increase in Ia^+ DCs among the tracheal epithelial cells, whereas the number of Ia^+ cells in the mucosa is not substantially changed. (Peroxidase stain: ×200.)

normal rat DCs express receptors for the low-affinity IL-2 receptor (CD25) and that IL-2 was able to promote directed migration of DCs in Boyden chamber assays. A monoclonal antibody (OX-39) that inhibits the binding of IL-2 to its receptor specifically decreased IL-2–mediated migration of DCs in vitro (Table 1).

Both parenteral and direct intratracheal injection of recombinant IL-2 significantly increased the number of DCs in the lung. The evident site of entry into the lung was the postcapillary venule (Fig. 3). The injection of IL-2 also yielded a significant increase in the percentage of FcR^+ pulmonary DCs, suggesting that these "exudate" DCs were phenotypically immature (15), resembling the Langerhans'-like cells that predominate in the normal pulmonary airways. The accumulation of DCs in tissues was significantly inhibited by administering OX-39 (anti-CD25) intraperitoneally prior to the injection of IL-2. These findings suggest that the migration of DCs, like that of T lymphocytes (16,17), may be mediated by IL-2. To date, the role of other potential chemoattractants for dendritic cells, for example, the CC chemokines (18), has not been evaluated.

The parallel effects of IFN-γ and IL-2 in vivo suggest that Th1 lymphokines play a crucial role in how DCs accumulate in the lung during an inflammatory response. As GM-CSF and TNF-α are also secreted by Th1 lymphocytes, it is posited that a proinflammatory response to inhaled antigen may lead concomitantly to accumulation, increased viability, and differentiation of DCs at sites of inflammation. Variations in cell-response kinetics (18) could potentially allow a single Th1 cytokine (e.g., IL-2) to control the migration of DCs and T lymphocytes differentially during an immune response. It is likely that Th2 lymphokines will be shown to downregulate the motility of DCs directly or by antagonizing Th1 lymphokine production. However, their effects on DC migration have not been directly elucidated.

Table 1 Anti–IL-2R Abrogates IL-2–Induced Motility

Condition	Pretreatment of DCs	Cells/HPF
HBSS	None	80 ± 2
Il-2	None	134 ± 3[a]
IL-2	OX-39 (IL-2R)	73 ± 7
IL-2	OX-22 (CD45RC)	128 ± 4[a]

Dendritic cells were pretreated with an anti–IL-2R mAb, OX-39, (100 μg/ml) or a control mAb, OX-22 (anti-CD45RC, 100 μg/ml) and examined in a Boyden chamber assay for chemotaxis in response to IL-2 (10 U/ml). Data is the mean ± SE of three experiments.
[a]$P = .02$ compared with HBSS control.

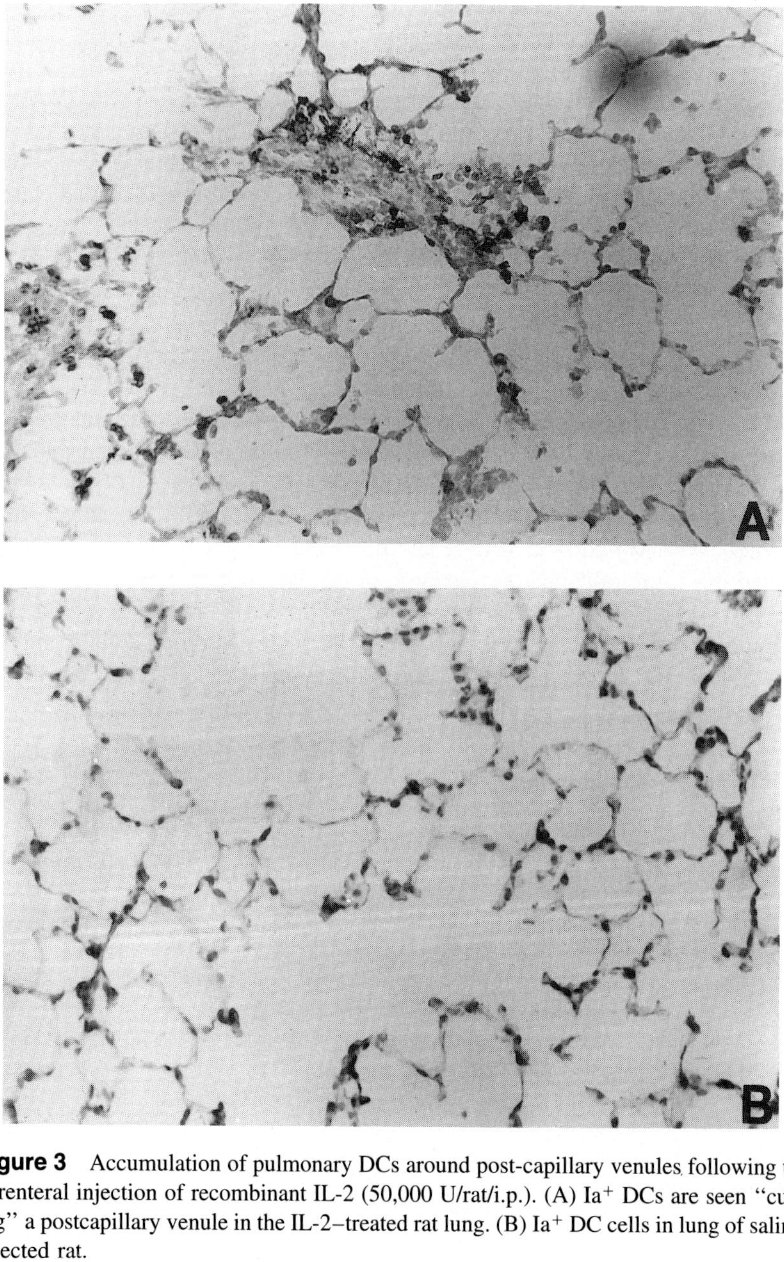

Figure 3 Accumulation of pulmonary DCs around post-capillary venules following the parenteral injection of recombinant IL-2 (50,000 U/rat/i.p.). (A) Ia^+ DCs are seen "cuffing" a postcapillary venule in the IL-2–treated rat lung. (B) Ia^+ DC cells in lung of saline-injected rat.

III. Pulmonary Dendritic Cells During the Response to a Soluble Antigen Challenge

A. Response to Soluble Antigen

McWilliams et al. have demonstrated that acute inflammation leads to the rapid recruitment of DCs to the airway mucosa and subsequently to the draining lymph nodes (19). Recently, the role of DCs in a delayed-type hypersensitivity (DTH) response to inhaled antigen has also been investigated. Lewis rats were injected intratracheally with the soluble antigen, hen egg lysozyme (HEL), at day 0. Thereafter, at regular intervals, DCs were purified from the lung parenchyma and examined for their ability to serve "spontaneously" as APCs for HEL-immune $CD4^+$ lymphocytes (20). Control wells received purified pulmonary DCs supplemented with HEL in order to achieve a maximal response.

DCs purified from HEL-challenged lungs were able to serve as antigen-specific APCs in the absence of added antigen for up to 7 days (Fig. 4). Following that time, no "spontaneous" APC activity was observed for DCs unless HEL was added to the culture wells. DCs purified from the lymph nodes draining the lung showed increased APC activities at day 3 following the intratracheal injection of antigen, suggesting that free antigen had drained to the local lymph nodes or pulmonary DCs had captured antigen and transported it directly to the nodes from the lung. Both pathways may contribute to the transport of antigen in vivo, and it has been proposed that the delivery of antigen bound to DCs may be an effective pathway for sensitizing T cells when the antigen dose is limited (20).

Direct evidence supporting a role for the transport of antigen by DCs has recently been provided by experiments showing that fluorochrome-labeled DCs injected into the trachea can be detected subsequently in the draining lymph nodes (21). In addition, Havenith and coworkers have reported that ex vivo antigen-pulsed DCs delivered via the airways were able to sensitize T cells for a subsequent pulmonary DTH response in vivo (22).

When rats were sensitized to HEL and then injected with HEL intratracheally at day 14, the APC activity mediated by lung DCs rapidly increased to levels seen following the primary exposure to antigen. In addition, a DTH response developed in the lung at 48 hr, as judged by the presence of perivascular lymphocytic inflammation. Immunostaining showed a substantial increase in the number of Ia^+ DCs cuffing pulmonary postcapillary venules at 12–24 hr and prior to the maximal development of the inflammatory response in the lung.

Considered together, these findings suggest that DCs residing in the airways play a critical role in the capture and processing of antigen. These DCs may subsequently participate directly in the transport of inhaled antigen from the lung to the draining lymph nodes and in the sensitization of naive $CD45RA^+$ T cells. In addition, the appearance of "exudate" Ia+ DCs in the lung may be a relatively early event in inflammation that is instrumental in promoting the effector limb

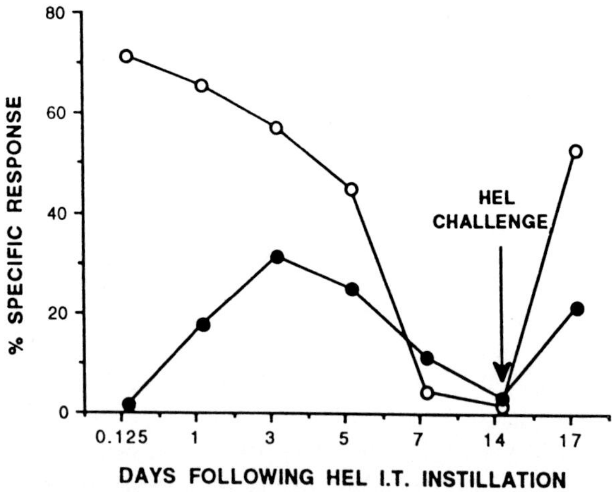

Figure 4 The kinetics of DC-associated APC activities after an airway challenge with HEL. DCs were purified from the lung and draining hilar lymph nodes (LN) at intervals following an airway challenge with HEL. DCs from the lung showed peak APC activity at 3 hr, which subsequently decreased gradually to day 7. ○ APC activity in the LN peaked at day 3 and was detectable until day 14. ● When rats received a second challenge with HEL intrathecally, specific DC-associated APC responses were again stimulated in lung and LN.

of the DTH response by virtue of the ability of DCs to stimulate the expansion of "memory" T cells recruited to the lung during the secondary response to antigen (23,24).

The DTH response is recognized to be mediated primarily by Th1 cytokines (25). In fact, the pulmonary response to antigen closely mimics those changes seen in the lung following the parenteral injection of recombinant IL-2 or IFN-γ in vivo. Th2 cytokines may antagonize these responses (26,27).

IV. Dendritic Cell Responses During the Response to Particulate Antigen

A. Response to Heat-killed *Listeria*

Many of the antigens in the ambient air, including microbes and dusts, are complex particulates. The activities of DCs in the response to a complex particulate antigen, heat-killed *Listeria* (HKL), were examined in an experimental model comparable to the one described above (28). First, it was established that highly purified pulmonary DCs were able to present HKL to immune T cells in vitro (Fig. 5). Next, Lewis rats received graded dosages of HKL (10^7–10^{10}) intratracheally

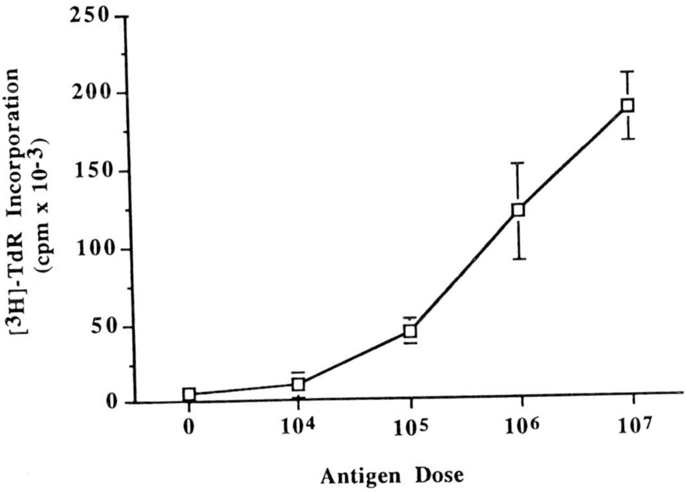

Figure 5 APC activity of pulmonary DCs for HKL in vitro. DCs purified from the lungs of naive rats were able to support the specific proliferation of HKL-immune T cells in vitro in a dose-dependent manner. Data are the mean ± SD of triplicate samples from a representative experiment (three experiments performed).

on day 0, and subsequently DCs were harvested and examined for their ability to serve spontaneously as APCs for immune T cells in vitro. For dosages of $<10^9$ bacteria, no spontaneous APC activity was mediated by lung DCs for up to 7 days (Fig. 6), whereas the same cells served effectively as APCs when HKL was added directly to the culture wells. Despite the absence of APC activity by pulmonary DCs, T-lymphocytes purified from the draining lymph nodes at day 14 were sensitized to HKL, indicating either that free HKL organisms had been transported from the lung to the draining nodes or alternatively that the activities of DCs were depressed specifically within the lung. Bronchoalveolar lavage (BAL) performed on the HKL-challenged rats showed that bacteria were ingested by neutrophils and pulmonary macrophages; few extracellular HKL organisms were detected.

In contrast, when the intratracheal challenge consisted of $>10^9$ HKL, DCs purified from lung at 24 hr showed enhanced specific APC activity for immune T cells (Fig. 7). Examination of the BAL again showed engorgement of phagocytes by HKL; however, now substantial numbers of extracellular organisms were seen in the BAL fluid. When fluorchrome-labeled HKL was injected intratracheally, intra-alveolar phagocytes were seen to contain large numbers of labeled organisms. Extracellular bacteria were detected within the lung interstitium, when the number of injected HKL exceeded 10^9/rat. These findings were interpreted to suggest that particulate bacteria had been effectively sequestered by intra-alveolar

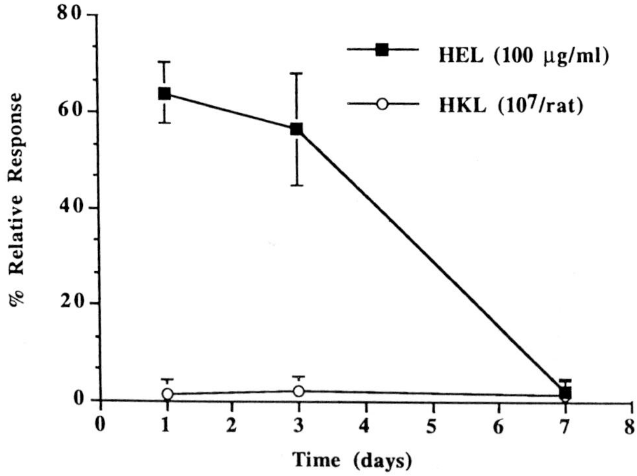

Figure 6 Comparison of APC activity of pulmonary DCs after an intratracheal challenge with either HEL or HKL. APC activity of pulmonary DCs isolated from lung was measured on days 1, 3, and 7 after instillation of either HEL or HKL intratracheally. HEL-challenged rats showed the rapid acquisition of excellent APC activity for HEL-immune T cells that diminished over the interval studied. In contrast, pulmonary DCs from rats receiving HKL (10^7/rat) showed no demonstrable APC activity. Data are expressed as the percentage relative response ± SEM of three experiments.

phagocytes and were not available to interact with interstitial DCs except when the particulate burden exceeded the capacities of phagocytes in the lung.

In order to explore the mechanism of the response to particulates further, soluble HEL was chemically immobilized on latex beads. Pulmonary DCs were demonstrated to present immobilized HEL effectively to immune T cells in vitro. However, when HEL-conjugated beads ("particulate" HEL) were injected intratracheally, DCs purified from the lung failed to stimulate the proliferation of HEL-immune lymphocytes unless soluble HEL was added to the wells.

Particulate HEL was able to sensitize T cells in vivo and a DTH response developed in the lung when the bead-challenged rats subsequently received *soluble*, but not particulate HEL at day 14. These results lend further support to the role that antigen sequestration by phagocytes plays in limiting pulmonary inflammation in vivo.

V. Neural Control of Pulmonary Dendritic Cells

A. DC Interaction with Pulmonary Nerves

Until recently, attention has focused on the role of cytokines as the exclusive regulators of DC functions. However, the orderly distribution of DCs in the

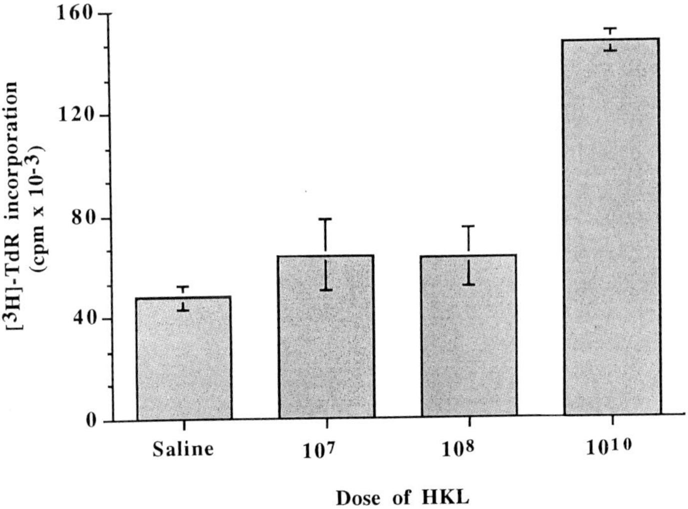

Figure 7 Effect of antigen dose on APC activity of pulmonary DCs after HKL instillation intratracheally. APC activity of pulmonary DCs isolated from the lung 24 hr after introduction of HKL was dose dependent. At relatively low concentrations of HKL (10^7–10^8 organisms/rat), DCs were unable to stimulate [^{3}H]thymidine incorporation into HKL-immune T cells, whereas rats receiving higher doses (10^{10} organisms/rat) showed increased APC activity compared with saline controls. Data are the mean ± SD of triplicate samples from a representative experiment (three experiments performed).

normal lung and the accumulation of these cells following local antigen deposition highlight the fact that DCs are directed in vivo to specific sites in the lung. The peripheral nervous system is an excellent candidate for coordinating the localization of DCs in the lung.

DCs in the lung airways and in the adventitia of the pulmonary vasculature are located in close proximity to unmyelinated tachykinin-producing nerve fibers (29,30). In many instances, mast cells are present in close proximity to nerve fibers (31) and DCs, so that a functionally interactive neuroimmune system of cells may exist in vivo (32). A posited functional linkage between these cells has been supported by the finding that purified DCs specifically bind substance P (SP) (R. Kradin et al., in preparation), a nonadrenergic, noncholinergic peptide (NANC). In addition, SP specifically promotes the motility of pulmonary DCs in vitro (29).

B. Effect of Capsaicin on the Pulmonary Immune Response

In order to determine the effects of unmyelinated nerve fibers on the response to antigen in vivo, Lewis rats were pretreated with capsaicin, a neurotoxin yielded by

Table 2 Putative Regulators of Dendritic Cells in Vivo

Mediator	Effect
IFN-γ (36)	Increased numbers of DCs in lung
	Translocation of DCs into airway epithelium
IL-2 (3)	Chemoattraction in vitro blocked by anti–IL-2R
	Increased numbers of DCs in lung
Substance P (3)	Increased motility of DCs in vitro
	DC accumulation in lung blocked by capsaicin

the red pepper (33). Capsaicin leads to the rapid release of neuropeptides from NANC nerve terminals and anatomical destruction of unmyelinated axonal nerve endings occurs, so that no further release of neuropeptides can be achieved in vivo. Following an HEL challenge intratracheally, rats that had received capsaicin showed no substantial change in the location, number, or functional activities of pulmonary DCs. When rats were secondarily challenged with HEL intratracheally at day 14, the lungs showed a marked decrease in the numbers of Ia^+ DCs centered around pulmonary postcapillary venules, and the magnitude of the pulmonary DTH response was markedly inhibited.

In separate studies, Nilsson et al. observed diminished pulmonary DTH responses following capsaicin treatment (34), as judged by a decrease in BAL lymphocytes capable of reacting specifically to the sensitizing antigen. Recently, Kaltreider and coworkers have shown that inhibiting tachykinin activity in vivo leads to diminished pulmonary immunity (35). This increasing body of evidence suggests that both cytokines and peripheral nonadrenergic noncholinergic nerves (Table 2) play a role in the regulation of DCs in vivo and in the development of pulmonary cell-mediated immunity.

VI. Conclusions

In the last decade, there has been a marked increase in the awareness of pulmonary immunologists that DCs are the dominant APCs of the lung. However, progress in the field of pulmonary DC biology has been hampered by difficulties in purifying these cells from the lung. For this reason, relatively little is known concerning how pulmonary DCs are regulated in vivo. This is an area of critical importance to our understanding of how the lung responds to inhaled antigens, so that in the future, attention will continue to focus on the elucidation of DC biology in pulmonary health and disease.

Acknowledgments

The author wishes to thank Maria Valles for her expert assistance in compiling this manuscript. Some of the research cited here was supported by NIH grant HL-36781.

References

1. Steinman RM. The dendritic cell system and its role in immunogenicity. Annu Rev Immunol 1991; 9:271–296.
2. Steinman R, Witmer-Pack M, Inaba K. Dendritic cells: antigen presentation, accessory function and clinical relevance. In: Kamperdijk, et al, eds. Dendritic Cells in Fundamental and Clinical Immunology. New York: Plenum Press, 1993:1–9.
3. Kradin R, Xia W, Pike M. Pulmonary Dendritic cells show enhanced motility in response to Interleukin-2. (abstr). Am Rev Respir Dis 1993; 147:A473.
4. Pollard A, Lipscomb M. Characterization of murine lung dendritic cells: similarities to Langerhans' cells and thymic dendritic cells. J Exp Med 1990; 172:159–167.
5. Holt PG. Pulmonary dendritic cell populations (review). Adv Exp Med Biol 1993; 329:557–562.
6. Sertl K, Takemura T, Tscachler E, Ferrans VJ, Kaliner MA, Stevach EM. Dendritic cells with antigen-presenting capability reside in airway epithelium, lung parenchyma and visceral pleura. J Exp Med 1986; 163:436–451.
7. Rochester CL, Goodell EM, Stoltenberg JK, Bowers WE. Dendritic cells from rat lung are potent accessory cells. Am Rev Respir Dis 1990; 138:121–128.
8. Brown KA, Bedford P, Macey MG, McCarthy DA, Dumonde DC, Knight SC. Characterization of adhesion molecules on the surface of dendritic cells isolated from human blood. Adv Exp Med Biol 1995; 378:97–99.
9. Kradin R, Xia W, McCarthy K, Schneeberger E. FcR+/- subsets of Ia+ pulmonary interstitial cells display differences in their ability to phagocytize particle and to provide accessory co-stimulation for naive (OX-22+) and sensitized (OX-22−) T cells. Am J Pathol 1993; 142:811–819.
10. Gong J, McCarthy K, Tamatani T, Miyasaka M, Schneeberger E. Intraepithelial airway dendritic cells; a distinct subset of pulmonary dendritic cells obtained by microdissection. J Exp Med 1992; 175:797–807.
11. Holt PG, Schon-Hegrad J, Oliver J, Holt BJ, McMenamin PG. A contiguous network of dendritic antigen-presenting cells within the respiratory epithelium. Int Arch Allergy Appl Immunol 1990; 91:155–159.
12. Witmer-Pack M, Oliver D, Valinsky J, Schuler G, Steinman R. Granulocyte/macrophage colony-stimulating factor is essential for the viability and function of cultured murine epidermal Langerhans' cells. J Exp Med 1987; 166:1481–1498.
13. Tazi A, Bouchonnet F, Grandsaigne M, Boumsell L, Hance A, Soler P. Evidence that granulocyte macrophage-colony-stimulating factor regulates the distribution and differentiated state of dendritic cells/Langerhans' cells in human lung and lung cancers. J Clin Invest 1993; 91:566–576.

14. Xu H, Kramer M, Spengler HP, Peters JH. Dendritic cells differentiated from human monocytes through a combination of IL-4, GM-CSF and IFN-gamma exhibit phenotype and function of blood dendritic cells. Adv Exp Med Biol 1995; 378:75–78.
15. Degwert J, Steckel F, Hoppe U. Antigen presenting and primary in vitro sensitizing capacity of CD1a+ dendritic cells generated from human blood. Adv Exp Med Biol 1995; 378:187–189.
16. Berman J, Cruikshank W, Beer D, et al. Lymphocyte motility and lymphocyte chemotactic factors. Immunol Invest 1988; 17:625–677.
17. Kornfeld H, Berman J, Beer D, Center D. Induction of human T lymphocyte motility by interleukin-2. J Immunol 1985; 134:3887–3890.
18. Hancock W. Chemokines and the pathogenesis of T-cell dependent immune responses. Am J Pathol 1996; 148:681–684.
19. McWilliam A, Nelson D, Thomas J, Holt P. Rapid dendritic cell recruitment is a hallmark of the acute inflammatory response at mucosal surfaces. J Exp Med 1994; 179:1331–1336.
20. Xia W, Pinto CE, Kradin RL. The antigen presenting activities of Ia+ dendritic cells shift dynamically from lung to lymph node following an airway challenge with soluble antigen. J Exp Med 1995; 181:1275–1283.
21. Havenith CE, Van MP, Breedijk AJ, Beelen RH, Hoefsmit EC. Migration of dendritic cells into the draining lymph nodes of the lung after intratracheal instillation. Am J Respir Cell Mol Biol 1993; 9:484–8.
22. Havenith CE, Breedijk AJ, Betjes MG, Calame W, Beelen RH, Hoefsmit EC. T cell priming in situ by intratracheally instilled antigen-pulsed dendritic cells. Am J Respir Cell Mol Biol 1993; 8:319–24.
23. Lipscomb M, Lyons C, O'Hara R, Stein-Streilen J. The antigen-induced selective recruitment of specific T lymphocytes to the lung. J Immunol 1982; 128:111–115.
24. Marathias KP, Preffer FI, Pinto C, Kradin RL. Most human pulmonary infiltrating lymphocytes display the surface immune phenotype and functional responses of sensitized T cells. Am J Respir Cell Mol Biol 1991; 5:470–6.
25. Mosmann TR, Coffman RL. TH1 and TH2 Cells: Different patterns of lymphokine secretion lead to different functional properties. Ann Rev of Immunol 1989; 7: 145–173.
26. Buelens C, Willems F, Pierard G, Delvaux A, Velu T, Goldman M. IL-10 inhibits the primary allogeneic T cell response to human peripheral blood dendritic cells. Adv Exp Med Biol 1995; 378:363–365.
27. Buelens C, Willems F, Delvaux A, et al. Interleukin-10 differentially regulates B7-1 (CD80) and B7-2 (CD86) expression on human peripheral blood dendritic cells. Eur J Immunol 1995; 25:2668–2672.
28. MacLean J, Xia W, Pinto C, Zhao L, Liu H-W, Kradin R. Sequestration of inhaled particulate antigens by lung phagocytes. Am J Pathol 1996; 148:657–666.
29. Kradin R, Xia W, Pinto C, Zhao L-H. Substance P-containing unmyelinated c-fibers are present in immediate proximity to airway dendritic cells and may promote their migration during inflammation. (abstr). Am J Respir Crit Care Med 1994; 149:A267.
30. Barnes P. Non-adrenergic non-cholinergic neural control of human airways. Arch Int Pharmacodyn Ther 1986.
31. Bienenstock J, Perdue M, Blennerhassett M, et al. Inflammatory cells and the epithe-

lium. Mast cell/nerve interactions in the lung in vitro and in vivo. Am Rev Respir Dis. 1988; 138:S31–34.

32. Ioffreda M, Whitaker D, Murphy G. Mast cell degranulation upregulates alpha 6 integrins on epidermal Langerhans cells. J Invest Derm 1993; 101:150–154.
33. Holzer P. Capsaicin: Cellular targets, mechanisms of action, and selectivity for thin sensory neurons. Pharmacol Rev 1991; 43:143–201.
34. Nilsson G, Alving K, Lundberg J, Ahlstedt S. Local immune response and bronchial reactivity in rats after capsaicin treatment. Allergy 1991:304–311.
35. Kaltreider H, Ichikawa S, Byrd P, et al. The role of neuropeptides in modulating immune responses in lung parenchyma of mice. Submitted. 1996.
36. Kradin R, McCarthy K, Xia W, Lazarus D, Schneeberger E. Accessory cells of the lung. I. Interferon-gamma increases Ia+ dendritic cells in the lung without augmenting their accessory activities. Am J Respir Cell Mol Biol 1991; 4:210–218.

12

Role of Pulmonary Epithelial Cells and Dendritic Cells in Regulating Pulmonary Immune Responses

PAUL J. CHRISTENSEN and GALEN B. TOEWS

University of Michigan
Ann Arbor, Michigan

I. Respiratory Epithelium: Pulmonary Microenvironment of Antigen Exposure

The lung is both a site of gas exchange and a vast surface of interaction of the body with the outside world. Elaborate host defenses have evolved to protect the lung from invasion and destruction; however, the delicate structure of the lung which is required for gas exchange is easily damaged by excessive immune and inflammatory responses. Thus, pulmonary immune and inflammatory responses must be tightly regulated locally to balance the competing interests of gas exchange and host defense.

The respiratory epithlelium serves as a barrier separating the external environment from the pulmonary interstitium and capillary bed. For the purposes of this discussion, we have divided the epithelium into airway epithelium (from the trachea to the terminal bronchi) and alveolar epithelium (comprising cells that line the alveoli). The airway epithelium performs many important homeostatic functions in the lung such as mucociliary clearance, ion and fluid transport, gas transport, and cellular proliferation/repair. Similarly, the alveolar epithelium performs homeostatic roles in ion and fluid transport, cellular proliferation and repair, and the maintenance of surface tension through the secretion of pulmonary surfactant. In the course of normal respiration, both types of epithelium are exposed to a wide range of noxious stimuli, including toxic gases, microbes and their by-products, and both organic and inorganic particulate dusts. These agents can initiate local inflammation and immune responses which protect the lung by limiting the exposure to potentially injurious stimuli. However, in some instances, these responses lead to persistent inflammation which can lead to destruction or altered physiology (asthma, sarcoidosis, pulmonary fibrosis).

There is increasing evidence the epithelial cells in the lung play a role in regulating immune and inflammatory response in the lung. A potential control point for pulmonary immune response generation is the regulation of antigen-presenting cell (APC) function. APC activity in the lung is largely attributable to dendritic cells (1–3). Dendritic cells (DCs) respond to local environments by altering their immunostimulatory behavior, often as a result of cytokine stimulation (4–5). Pulmonary DCs behave in a similar manner, upregulating immunostimulatory capacity in response to granulocyte-macrophage colony-stimulating factor (GM-CSF) and interleukin-6 (IL-6) (6). Once pulmonary DCs become loaded with antigen, they most likely move to regional lymph nodes. In the lymph node, the DC can interact with large populations of T lymphocytes, increasing the likelihood of encountering a T lymphocyte which can respond to the specific antigen in question. The signals which initiate and direct this movement also are likely to be delivered locally. Accordingly, the microenvironment in the lung interstitium is likely to be critical in determining the immunostimulatory capacity of pulmonary DCs. Because DCs are located in a subepithelial distribution in the interstitium of the lung, pulmonary epithelial cells are likely to play a major role in defining the local microenvironment for pulmonary DCs. This chapter details the mechanisms by which respiratory epithelial cells can influence the behavior of pulmonary DCs and antigen presentation in the lung.

A. Alveolar Epithelium

The alveolar epithelium of the adult lung is formed by two cell types. Type I cells are large thin cells that cover 93% of the alveolar surface (7). Type II cells cover the remaining 7% of the alveolar surface. In contrast to type I cells, type II cells are compact and cuboidal. Their cytoplasm contains numerous lamellar bodies that store surfactant. In addition to their importance in surfactant biology, type II cells are progenitor cells for the alveolar epithelium (8,9). Attempts to isolate type I cells to study biological functions in vitro have met with limited success. However, established protocols are available for isolating pure populations of type II cells in several species (10,11). The phenotype of type II cells in culture is unstable. Cells undergo changes which cause them to lose certain type II cell characteristics (such as surfactant production); coincident with loss of type II characteristics, the cultured cells begin to express type I cell markers (12,13). This transformation parallels the in vivo situation where type II cells differentiate into type I cells. This phenotype change is a useful model for studying alveolar epithelial cell biology in vitro (14). For the purposes of our discussion, we refer to type II cells which are cultured in vitro as alveolar epithelial cells (AECs).

B. Airway Epithelium

The airway epithelium is a heterogeneous tissue composed of several cell types (15–17). The most abundant cells are ciliated, basal, and goblet cells. Ciliated cells comprise about 50% of the epithelial cell population of the airway (16,17). Cilia on the apical surface of columnar epithelial cells move mucus and trapped particulates to the oropharynx where they are swallowed or expectorated. Basal cells have no luminal contact, and they are classically thought to be precursor cells that may differentiate into either ciliated or goblet cells (18,19). Goblet cells are nonciliated cells that contain dense mucous granules. These cells along with submucosal glands secrete the mucus necessary for normal mucociliary clearance. Airway epithelial cells have been studied in a variety of in vitro systems (20). Evidence presented below demonstrates that these cells, like AECs, are not just passive barrier cells but active participants in immune inflammatory responses (Table 1).

II. Antigen-Presenting Cell/T-Cell Interaction

Immune responses in the lung are likely to be regulated at multiple points. One critical level of regulation occurs at the APC/T-lymphocyte interaction during the initiation of specific immunity. T lymphocytes only recognize processed antigen fragments displayed on the surface of an APC in conjunction with class II major histocompatibility complex (MHC) molecules. To present antigen effectively, a cell must be able to take up proteins and degrade them into relevant antigenic

Table 1 Pulmonary Epithelial Cells Participate in Immune and Inflammatory Processes in the Lung

Function	AECs (reference)	Airway epithelial cells (reference)
Antigen presentation	6	53,55,56
Class II MHC	31–39	31,52–55
IL-1		101,142
IL-6	102	79,81,100,101,140,141
Chemokines	69,108–110,114	65,66,102,133,137–143,145,146,149
GM-CSF	6,90,111,112	77–82,89,140

fragments and express these peptide fragments with class II MHC molecules (21). For optimal activation of resting T cells, additional antigen-nonspecific signals, such as surface-bound molecules or cytokines, are required (22–24) (Fig. 1). Traditionally, the role of an APC has been reserved for bone marrow–derived cells such as DCs, B cells, and monocytes, which constitutively express class II MHC molecules. However, there is mounting evidence that non–bone marrow–derived cells also initiate lymphocyte activation. In the alveolus, AECs are directly exposed to inhaled antigens, have many of the characteristics necessary to function as APCs, and in some instances are capable of initiating immune responses.

A. Initiation of Immune Responses: AECs

Uptake and Processing of Protein Antigens

Type II cells may be capable of processing antigen from the alveolar space. Ferritin-labeled lectins delivered intratracheally to rats have been shown to be taken up and deposited into multivesicular bodies and lysosomes in type II cells (25,26). These organelles contain lysosomal enzymes and have low pH, condi-

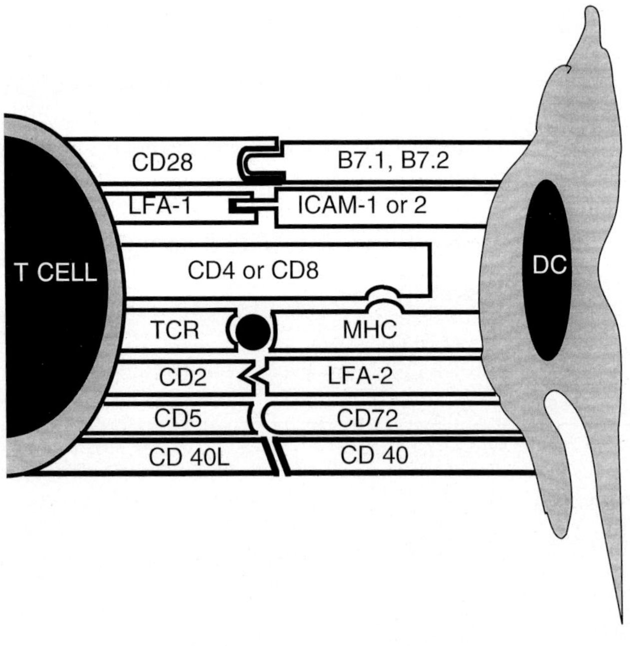

Figure 1 Surface bound molecules required for optimal activation of resting T cells.

tions which have been shown to be important in endosomal processing of antigen by traditional APCs.

Rat AECs and A-549 cells are capable of taking up bacteria (27,28). Surfactant proteins may facilitate the uptake of bacteria and antigens by epithelial cells. Surfactant protein A (SP-A) is a member of a group of chimeric molecules called collectins (29). The structure of these molecules includes a globular head which contains a carbohydrate-recognition domain and a collagen-like tail. SP-A is an integral part of the tubulomyelin framework which stabilizes alveoli. Type II cells are known to have receptors for SP-A. SP-A may facilitate the uptake of bacteria or other pathogens trapped in lipid/SP-A complexes that are cleared by type II cells (29). Lysozyme has a similar intracellular distribution as SP-A in type II cells. Lysozyme is found in lamellar bodies where surfactant is packaged for secretion and in lysosomal fractions where it may undergo degradation (30). During the process of recycling surfactant, type II cells may take up pathogens or particulate antigen bound in surfactant. These antigens could be processed and loaded onto class II MHC molecules. MHC molecules containing antigen then could be expressed on either the Type II cell luminal surface or the basolateral epithelial cell surface. In the latter circumstance, DCs in the interstitium could pick up the processed antigen, move to the lymph node, and present the antigen to T cells. This proposed mechanism would allow dendritic cells to take up and present inhaled antigens without requiring disruption of the epithelial cell barrier. Taken together, these data suggest that alveolar epithelial cells have the capacity to take up and process antigen (Fig. 2).

Expression of Class II MHC Molecules

Class II MHC molecules are expressed on the cell membranes of AECs. In "normal" resected human lung, the expression of class II MHC molecules is limited to a select population of cells, including scattered alveolar cells (31). In an autopsy study of infants, class II MHC molecules were found on type II cells; the expression appeared to be increased during infection (32). In the mouse, class II molecules are not constitutively expressed on type II cells. However, when intraperitoneal injections of interferon-γ (IFN-γ) are given daily for 7 days, high levels of class II MHC antigens are expressed on interstitial macrophages, DCs, and type II cells but not on type I epithelial cells, endothelial cells, or airway epithelial cells (33). Class II molecules were localized to the basolateral surface of the type II cells. The isolation procedure for rat type II cells causes high-level class II expression (>60% of cells) which decays over 48 hr in standard culture to less than 10% of the cells. After 2 days, rat type II cells can be reinduced to express class II molecules with IFN-γ (34,35). In addition, class II MHC expression has been found on alveolar epithelial cells in histological sections taken from humans or animals with inflammatory lung disease (36–39).

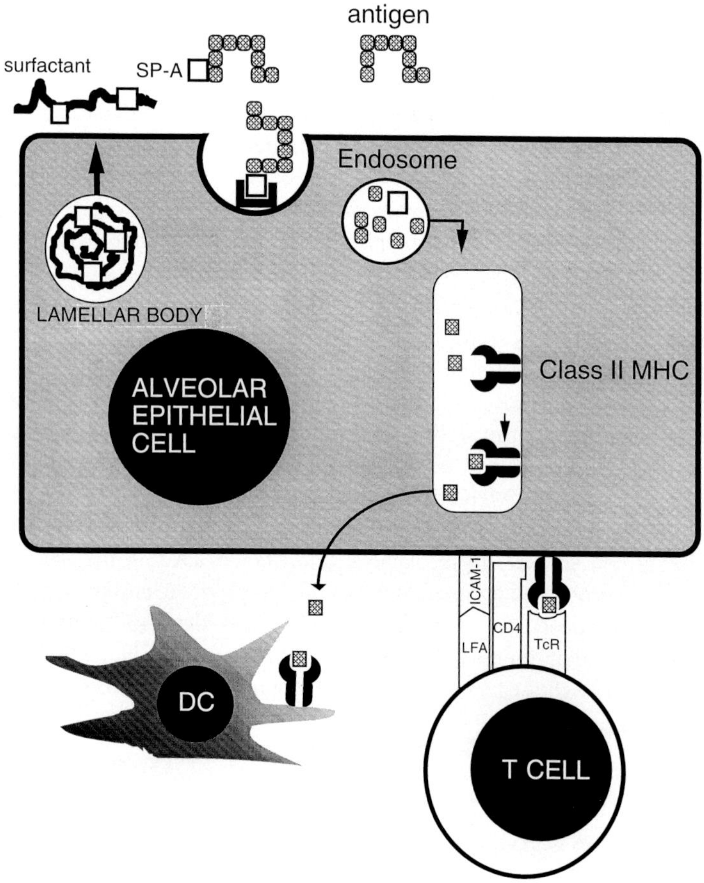

Figure 2 Proposed mechanism of AEC interaction with inhaled antigen (see text for detailed discussion).

Expression of Costimulatory Molecules

AECs interact with inflammatory cells through cell contact–mediated mechanisms. Intercellular adhesion molecule-1 (ICAM-1) is crucially involved in these interactions. This adhesion molecule, which is the natural ligand for the leukocyte integrins leukocyte function–associated factor-1 (LFA-1) and Mac-1, is expressed on the apical surface of type I cells but not type II cells in normal, uninflamed mouse and rat lungs (40,41). In mice that are subjected to hyperoxic insult, ICAM-1 expression is induced on type II cells (42). When rat type II cells are isolated and placed in culture under standard conditions, ICAM-1 expression is rapidly induced as the cells spread over the surface of the plastic dish (41). Because these isolated type II cells also express features characteristic of type I cells, this finding suggests ICAM-1 expression in AECs might be a function of type I cell differentiation. Accordingly, ICAM-1 expression might be regulated in part by signals that control the state of differentiation of these cells. In support of this hypothesis, the induction of ICAM-1 expression by isolated type II cells may be modulated by culture on extracellular matrix preparations that influence cell shape. When cells are cultured under circumstances that preserve cuboidal morphology and the production of components of pulmonary surfactant, ICAM-1 expression is inhibited. Cells that are cultured under conditions that promote spreading maximize ICAM-1 expression (43).

ICAM-1 plays an important role as a costimulatory molecule for T-cell activation. ICAM-1 expression on target cells also results in enhanced killing by cytotoxic lymphocytes. In experiments evaluating the adherence of activated T cells to rat type II cells that have spread in culture (type I like cells), approximately 50% of the lymphocyte binding is inhibited by antibody to ICAM-1 (41). Thus, although ICAM-1 is an important mediator of lymphocyte adherence, it is likely there are other, as yet undescribed, leukocyte adhesion molecules expressed on AECs.

Antigen Presentation

Rat AECs stimulate an allogeneic mixed lymphocyte reaction (MLR), and thus directly influence T-cell activation through mechanisms involving cell-cell contact (6). The potency of AECs as stimulator cells in an MLR was 100-fold less than that of DCs (6). However, their potential significance in vivo cannot be ignored because of the number and location of these cells in the lung. Thus, under some circumstances, AECs can function as APCs and directly activate T cells. Epithelial cells differ in their ability to present antigen (44,45). Keratinocytes can express class II MHC molecules, but they are inefficient APCs that are unable to stimulate resting T cells. However, they can induce a proliferative response in an allospecific T-cell line. In contrast, class II–bearing intestinal epithelial cells can function as accessory cells both for antigen-specific lyphocyte proliferation and for an allogeneic MLR (46,47). These data demonstrate that the accessory cell function of epithelial cells is not uniform and depends on the organ of origin.

Inhibitory Effects on T Cells

The direct effects of AECs on T cells is complex. Crude surfactant, surfactant phospholipids, and purified phosphatidylcholine inhibit T-cell proliferation (48–50). In addition, uncharacterized protein factors secreted by AECs inhibit T-cell proliferation (51). The complexity of the interactions with T cells serves to emphasize the importance of AECs in creating environments that either inhibit or promote immune and inflammatory responses. Tight control of these multiple factors is likely to be critical for normal pulmonary host defense.

Summary

AECs can interact directly with inhaled antigens. AECs express class II MHC molecules and possess the necessary machinery for uptake and processing of antigens. Under some circumstances, AECs can function as APCs in vitro. However, the interactions with T cells are complex and involve both stimulatory and inhibitory factors. The possibility exists that AECs could function as antigen-transporting cells by taking up and processing antigens. These processed antigens might subsequently be deposited in the lung interstitium. Pulmonary dendritic cells in the interstitium might carry these antigen fragments to the lymph node and present them to lymphocytes to initiate specific immune responses. Thus, evidence exists that AEC may be involved in the initiation of specific immunity via direct interactions with inhaled antigens.

B. Initiation of Immune Responses: Airway Epithelial Cells

Airway epithelial cells are also directly exposed to antigen. These cells are capable of presenting antigen directly to T cells in a similar manner to AECs, as discussed above. Class II MHC molecules are expressed on airway cells in vivo (31,52) and in vitro (53–55). Indeed, the high-level bronchocentric expression of class II MHC has been suggested as a target for activated T cells in lung transplants. The interaction between activated T cells and class II–positive airway epithelial cells may result in bronchiolitis obliterans, a pathological lesion in chronic lung transplant rejection (31).

Class II MHC molecules expressed an airway epithelial cells are capable of activating T cells. Human airway epithelial cells isolated from both the lobar bronchus and nasal turbinates were capable of stimulating a MLR (56). The ability to activate T cells persisted when epithelial cells were rigorously depleted of other accessory cells and was dependent on expression of class II MHC, because incubation with anti–class II MHC monoclonal antibodies completely inhibited the response (56). Similarly, human bronchial epithelial cells stimulated with IFN-γ were capable of presenting antigens to autologous T cells albeit significantly less efficient than pulmonary monocytes. Bronchial epithelial cell antigen presentation was a class II MHC–restricted phenomenon (55). Bronchial epithelial cells also enhanced mitogen-induced DNA synthesis in a non–MHC-restricted fashion in T cells subjected to CD3 cross linking. This effect did not require cell-cell contact, as epithelial cell–conditioned media had a similar effect, demonstrating the capacity of airway epithelial cells to affect T-cell proliferation through secreted factors (55). Rat bronchial epithelial cells may also be involved in airway mucosal immunity (53). Primary cultures of rat bronchial epithelial cells did not express class II antigens, but IFN-γ readily induced expression. In addition, when IFN-treated bronchial epithelial cells were pulsed with ovalbumin, they significantly stimulated the proliferation of ovalbumin (OVA)–sensitized T cells. Epithelial cells that were not treated with IFN-γ but were pulsed with OVA did not stimulate T-cell proliferation. In summary, airway epithelial cells express class II MHC antigens and can stimulate T-cell proliferation both through the antigen-specific T-cell receptors and via the secretion of soluble factors. These cells, in a similar manner to AECs, may function as APCs.

III. Pulmonary Dendritic Cells

A. Characteristics of Pulmonary DCs

APC activity in the lung is largely attributable to DCs that are distributed along the airways and throughout the pulmonary interstitium. Pulmonary DCs have been isolated from human, rat, and mouse lungs (1–3, 57,58,60) and possess morphological and functional characteristics similar to those of DCs isolated from the spleen, lymph node, or skin (59). The function of pulmonary DCs can change based on the local microenviroment. They acquire immunostimulatory capacity probably via the expression of costimulatory molecules under the influence of secreted cytokines. Presumably, this activation also primes DCs for movement to the lymph node where efficient antigen presentation can occur. It has been suggested that the activation of pulmonary DCs can occur during the innate immune response or during microenvironmental tissue injury. In this model, the precursor DC (sentinel cell) is located in the periphery of the lung in contact with either alveolar or airway epithelial cells. During the innate immune response, DCs become activated either by common pattern recognition receptors or tissue injury and change their functional capacity (receptor cell). These cells are able to process antigen and migrate to the lymph node. The net result of these processes is a DC (presenter cell) which is fully functional APC capable of initiating an immune response.

Pulmonary DCs can be isolated using flow cytometry to select cells with high Ia expression (60). This method allows the evaluation of the functional characteristics of highly purified pulmonary DCs prior to changes that might occur in cell culture. Freshly isolated DCs respond to exogenous GM-CSF and IL-1 with enhanced immunostimulatory capacity (60). Our laboratory has used this technique to study the effects of AECs on DC functional capacity (6).

B. Role of Epithelial Cells in DC Function

AECs and airway epithelial cells are ideally situated to influence pulmonary DCs. DCs from many tissues respond to local environments by altering their immunostimulatory behavior, often as a result of cytokine stimulation. In the lung, DCs are located in the interstitium adjacent to the basal epithelial surface (61,62). Epithelial cells express certain proteins in a polar fashion; molecules may be transported preferentially to either their apical or basal lateral surfaces. The polar expression of immunologically important molecules may allow AECs and airway epithelial cells to participate in the compartmentalization of immune responses. AECs and airway epithelial cells may influence DCs at multiple levels: (1) by providing signals necessary for the recruitment of DCs to the epithelium; (2) by regulating DC immunostimulatory capacity; (3) by interacting with cells of the innate immune system, mostly local tissue macrophages, to influence DC immunostimulatory capacity; and (4) by processing and/or transporting antigens through the epithelial barrier as described in Section II.A.

Recruitment of Pulmonary DCs

Pulmonary DC precursors originate in the bone marrow and subsequently migrate into the lung. In order to initiate immune responses, DCs must migrate via the afferent lymphatics to draining lymph nodes where they stimulate antigen-specific T lymphocytes (59). Additionally, the rapid recruitment of DCs is a hallmark of inflammation caused by bacteria and their products at the mucosal surface of the lung (63). The molecules that control the migration of DCs in the lung are unknown.

DCs are highly motile in vitro (64). Peripheral blood DCs are stimulated to migrate by C5a, *N*-formyl-Met-Leu-Phe (fMLP), and the CC chemokines monocyte chemoattractant protein (MCP-3), macrophage inflammatory protein-1α (MIP-1α/LD78), and RANTES. MCP-1 had a minor affect on basal migration. IL-8, and IP-10, a CXC chemokine that lacks the ELR motif and is active on monocytes and T cells, but not on neutrophils, were completely inactive in in vitro DC chemotaxis experiments. Thus, the pattern of chemotactic activation of DCs in response to chemokines is distinct from that of neutrophils and to some extent distinct from that of monocytes as well (64).

RANTES, an 8-kDa protein, can be produced by endothelial cells, epithelial cells, fibroblasts, T cells, and macrophages. Primary isolates of human bronchial epithelial cells constitutively produce RANTES, and the human pulmonary type II epithelial cell line A549 produces RANTES following stimulation with tumor necrosis factor-α (TNF-α) or IL-1β (65,66). RANTES mRNA and protein production are stimulated in the human bronchial epithelial cell line BEAS-2B by TNF-α and IFN-γ (67). Infection of bronchial epithelial cells with influenza virus in vitro increases RANTES mRNA expression (68). Following intratracheal installation of lipopolysaccharide (LPS), epithelial cells are a major in vivo source of RANTES protein (69). Although MIP-1α/LD78 is produced predominately by leukocytes, it is also produced by nonleukocytic cells. MIP-1α/LD78 has recently been shown to be produced by epithelial cells in vitro (70). Taken together, these data suggest that epithelial cells could be crucial participants in DC trafficking to the lung by their constitutive or regulated production of DC chemotaxins.

Regulation of DC Immunostimulatory Capacity

GM-CSF

GM-CSF is a immunomodulatory cytokine with pleiotropic effects on immune and inflammatory cells. It belongs to a family of glycoprotein cytokines with potent effects in stimulating proliferation, maturation, and activation of hematopoietic cells (71,72). GM-CSF is required for maximal expression of immunostimulatory capacity by APCs and for primary antibody responses (5,73–76).

Culture of pulmonary dendritic cells in the presence of AECs significantly increased immunostimulatory activity in an allogeneic MLR (6). This enhancement was dependent on the dose of both AECs and DCs. The enhancement of DC-immunostimulatory activity was mediated by soluble factor(s) released from AECs. Furthermore, DC immunostimulatory activity was enhanced when DCs were preincubated with AEC-conditioned medium (CM) prior to the addition of T cells, indicating the AEC medium had a direct effect on DC functional capacity. AECs might secrete factors preferentially from the basal lateral surface that influence DC behavior. To test this postulate, conditioned medium was collected from the apical and basal lateral surfaces of AECs and tested for the ability to enhance DC function. Although both apical and basal lateral AEC-CM enhanced the accessory cell activity of DCs in an allogeneic MLR, basal lateral AEC-CM was more potent in this effect. Previous studies had shown that murine recombinant GM-CSF significantly enhanced the immunostimulatory capacity of pulmonary DCs as stimulator cells in an MLR (60). Accordingly, we used a polyclonal goat antimurine GM-CSF IgG to neutralize the effect of rat AEC-conditioned medium on DC activity. Incubation of AEC-conditioned medium with antimurine GM-CSF antibody prior to the addition of DCs and allogeneic T cells significantly reduced the immunostimulatory activity in the MLR. Treating conditioned medium with control antiserum had no effect. Thus, rat AECs in primary culture secrete factors that influence pulmonary DC behavior. Furthermore, at least part of this activity is due to GM-CSF which is secreted in a polar manner.

Airway epithelial cells also produce GM-CSF. The production of GM-CSF has been noted to be increased in patients with airways disease such as asthma (77–81). GM-CSF production by airway epithelial cells can enhance local DC immunostimulatory capacity (82). In a recent study, intraepithelial DCs isolated from biopsy specimens from patients with atopy were studied for their ability to induce the activation of autologous T lymphocytes in vitro. Incubation of antigen with asthmatic epithelial cells, DCs, and autologous resting $CD4^+$ T cells induced T-cell activation and release of high levels of IL-4 and IL-5. The antigen-presenting activity of DCs was potentiated by asthmatic epithelial cells. This effect was partially abrogated by antibodies against GM-CSF.

GM-CSF is an attractive candidate molecule to control immune responses in the lung. Although a variety of cell types produce GM-CSF, including activated lymphocytes, macrophages, endothelial cells, fibroblasts, and epithelial cells

(83,84), GM-CSF exerts its effects locally and its production is tightly regulated. Under normal conditions, GM-CSF protein is undetectable in human and mouse serum (83). Moreover, the overproduction of GM-CSF in transgenic mice results in accumulations of macrophages, tissue damage, and death (85). Although most cells do not express GM-CSF, production is potently induced by several agonists, including both bacterial products such as LPS and early-response cytokines. Several cell types have been shown to respond to these agonists, including airway epithelial cells (86–89). Preliminary work in our laboratory indicates that rat AECs respond to IL-1 and LPS with the induction of GM-CSF mRNA. Interestingly, SP-A has also been shown to induce GM-CSF activity in rat AECs (90). The regulation of GM-CSF production in pulmonary epithelial cells likely involves cytokines which suppress GM-CSF expression. One potential source of these cytokines are activated T cells recruited to the site of inflammation. Lymphokines IFN-γ and IL-4 suppress GM-CSF production by other cell types. IFN-γ inhibits GM-CSF expression in vascular endothelial cells and synovial cells that have been stimulated with IL-1 (91,92). IL-4 suppresses GM-CSF production in LPS-stimulated monocytes and IL-1–stimulated B cells (93,94). No studies have evaluated the effects of these lymphokines on stimulated pulmonary epithelial cells. However, understanding the regulatory mechanisms involved in the epithelial cell production of GM-CSF is likely to be important in understanding the mechanisms whereby epithelial cells create local environments for the activation of pulmonary DCs.

GM-CSF is a growth factor for which there is considerable variation between species at the levels of both antibody activity and receptor binding (95–98). Although the murine cytokine will stimulate rat cells, antibodies to murine GM-CSF have limited activity against the rat protein. To understand fully the cellular and molecular biology of GM-CSF in rat systems, our laboratory has (1) determined the entire rat GM-CSF coding sequence; (2) expressed and purified functional rat GM-CSF; and (3) generated specific antibodies which recognize and neutralize rat GM-CSF. The rat sequence is 81% homologous to the murine GM-CSF cDNA and 65% homologous to the human GM-CSF cDNA. The predicted protein for rat GM-CSF, excluding the putative leader sequence, is 67% homologous to murine GM-CSF. There is significant dissimilarity between the rat and murine sequences in areas thought to be critical for receptor binding (97,99). A bacterial recombinant GM-CSF has been used to immunize rabbits. The hyperimmunized rabbit serum has successfully been used for immunohistochemistry.

IL-1 and IL-6

IL-1 and IL-6 are two pleiotropic cytokines known to enhance DC immunostimulatory capacity. Airway epithelial cells produce both IL-6 and IL-1 in response to various stimuli (79,81,100,101). AECs have been shown to produce IL-6 (102).

Downregulation of DC Responses

The mechanisms which downregulate DC responses are not understood. DC activation might carry within itself a downregulatory control mechanism. Pulmonary airway DCs have a turnover time of approximately 30 days (103). A similar turnover time has been demonstrated for the DC population of the gut wall (104). Lung DCs have a somewhat longer turnover time, but their turnover time is considerably shorter than that of Langerhans' cells. Accordingly, when a local antigen challenge has been dealt with or local tissue damage has been repaired, a reduced influx of new DC precursors into the peripheral tissue and the resultant reduction of influx of activated DCs into draining lymph nodes may be sufficient to shift the balance of the response back to that of the resting sentinel state.

Specific downregulatory mechanisms may exist. Calcitonin gene–related peptide (CGRP), a major neuropeptide of peripheral sensory nerves, has been shown to inhibit the antigen-presenting function of Langerhans' cells (105). Pretreatment of epidermal Langerhans' cells with CGRP inhibited alloantigen presentation and presentation of the defined protein chicken ovalbumin to an ovalbumin-responsive line. Several other proteins had no effect on Langerhans' cells antigen-presenting capacity, including transforming growth factor-β2 (TGF-β2) and TNF-α. The mechanisms whereby CGRP inhibits Langerhans' cells antigen presentation are not completely understood, but CGRP has been shown to suppress the induction of B7-2 expression by peritoneal macrophages and by a DC line (XS52). CGRP had no effect on Ia expression and no effect on B7-1 expression in these cells (106).

Cells that produce CGRP are generally neural or neuroendocrine cells. However, rat alveolar type II cells and tracheal epithelial serous cells produce CGRP (107). Rat type II alveolar epithelial cells both express and secrete CGRP. Secretion occurs constituitively but can be regulated by a protein kinase C–dependent pathway. Rat type II AECs produce comparable or greater quantities of CGRP than are secreted by other cultured cell types that produce CGRP. The physiological significance of CGRP secretion by alveolar type II cells is unclear. The possibility that CGRP plays a role in the maintenance of DCs as sentinel cells or in the downregulation of DC function following the generation of an immune response are intriguing possibilities yet to be explored.

AEC Interactions with the Innate Immune System

Cytokines

AECs are ideally situated to influence the number and activation state of macrophages via the production of cytokines (6,69,90,108–110). Primary isolates of rat type II cells secrete biologically active GM-CSF; GM-CSF can be localized within the epithelial cells (6,90). GM-CSF has been identified in type II cells in lung tissue (111,112).

GM-CSF is involved in normal alveolar homeostasis. Transgenic knockout mice in which the gene for GM-CSF is disrupted develop abnormal accumulations of pulmonary surfactant resembling pulmonary alveolar proteinosis histologically (113). This accumulation occurs despite normal levels of expression of surfactant apoprotein mRNAs. Accordingly, the pathogenic mechanism likely involves decreased surfactant clearance by alveolar macrophages (AMs) in the absence of GM-CSF stimulation.

MCP-1 is a product of type II cells (108). MCP-1 is a chemoattractant for monocytes. MCP-1 also activates monocytes, inducing the respiratory burst and enhancing tumoricidal activity in vitro. MCP-1 protein, bioactivity, and mRNA have been identified in AECs using both primary cultures of rat type II cells (108) and a human transformed cell line (A549 cells) that displays characteristics of type II cells (110). MCP-1 expression in endothelial cells and A549 cells is induced primarily by the early-response cytokines IL-1 and TNF-α but not directly by bacterial endotoxin. Interestingly, primary isolates of rat type II cells respond directly to LPS with increased expression of MCP-1 mRNA. This suggests that AECs may participate in initiating the response to invading organisms. Although the role of MCP-1 in normal homeostasis or pulmonary pathology has not yet been defined, immunohistochemical staining of lungs from patients with idiopathic pulmonary fibrosis demonstrate increased type II cell–associated MCP-1 (114). Whether this epithelial cell–derived cytokine is a response to cellular derangements in the lung or has a causal role in IPF awaits further investigation.

Evidence has accumulated to indicate that alveolar epithelial cells express a number of other cytokines potentially involved in inflammatory cell recruitment or in the modulation of their activity. The alveolar cell carcinoma line A549 secretes the neutrophil chemotactic protein IL-8 (109), whereas the monocyte chemoattractant RANTES is induced in rat type II cells following stimulation with IL-1 (69). TGF-β comprises an ubiquitous family of cytokines that inhibit lymphocyte and macrophage activation and are produced in large quantities by AECs (115). Finally, IL-3, a cytokine that shares features with GM-CSF and that is importantly involved in eosinophil responses, is a product of rat type II cells in primary culture (90). Thus, through the production of a number of cytokines, AECs play an important role in defining a local environment to recruit and maintain a population of resident macrophages. These macrophages are a critical first line of defense against the invasion of pulmonary pathogens.

Soluble Factors Involved in Alveolar Host Defense—Surfactant Proteins

AECs secrete proteins which play important first-line roles in pulmonary host defense. Surfactant proteins A and D are members of a group of chimeric molecules called the collectins (29). Although SP-A may not function as a true opsonin, SP-A may facilitate the uptake of bacteria or other pathogens trapped in lipid/

SP-A complexes that are cleared by AMs and type II cells (29). In vitro exposure of AMs to SP-A results in enhanced phagocytosis of a number of pathogenic bacteria such as *Staphylococcus aureus* (116) and *Pseudomonas aeruginosa* (117), whereas SP-A directly binds to and opsonizes type a *Haemophilus influenzae* (118). SP-A may also mediate interactions between AMs and virally infected host cells; there is increased binding of SP-A to cells infected with herpes simplex type I via the carbohydrate moiety of the apoprotein (119). SP-A also promotes chemotaxis by AMs (120), increases secretion of GM-CSF and IL-3 (90), and modifies macrophage production of oxidants (121). SP-D, another member of the collectin family of molecules, is produced by freshly isolated type II cells and noncilated bronchiolar Clara cells (122). Unlike SP-A, SP-D does not interact with lipids. However, SP-D mediates agglutination of gram-negative bacteria, suggesting an important role of SP-D in host defense (123).

Complement

The complement system provides a critical defense mechanism against infection (124). Complement-dependent phagocytosis is critical for defense against many microorganisms. This is particularly true for the nonimmune host. Complement also can directly mediate killing of bacteria. Components of the complement system (C5a, C4a, and C3a) are important proinflammatory mediators. These molecules elicit a myraid of biological effects, including smooth muscle spasm, increased vascular permeability, chemotaxis, the induction of cytokine synthesis, oxidative metabolism, and the release of eicosanoids (reviewed in ref. 125). AECs synthesize several complement components, including C5 (126). The synthesis of complement proteins in AECs provides a local source of these proteins for inflammatory responses in the lung. Thus, AECs are a source of several proteins important for first-line defense of the nonimmune lung.

ICAM-1

In the normal lung, ICAM-1 is expressed constitutively on type I cells (40,41). ICAM-1 plays a well-recognized role in host defense, with critical involvement in the recruitment and activation of immune and inflammatory cells (127). Although ICAM-1 may be induced on a variety of pulmonary cells, the function of ICAM-1 on type I AECs has not yet been defined. The high-level constitutive expression of ICAM-1 on the apical surface of type I cells and regulation by differentiation-related signals rather than inflammatory cytokines (43) suggests that ICAM-1 on these cells may play a role in normal homeostasis in the alveolar space. This may involve the definition of the unique immunological environment in the alveolar space. Type I alveolar epithelial cell ICAM-1 may be instrumental in maintaining the distinctive phenotype of AMs when compared with lung interstitial macrophages or resident macrophages from other tissues. We have speculated that

ICAM-1 may be important for the migration of AMs over the surface of type I cells as these macrophages patrol the alveolar spaces. Experiments examining the migration of leukocytes over planar lipids in which adhesion molecules have been incorporated have demonstrated that anchoring of these adhesion molecules is critical for cellular locomotion (128). We have recently performed studies demonstrating ICAM-1 is associated with the cytoskeleton of AECs (129). In this way, ICAM-1 may serve as a "scaffold" to allow efficient macrophage migration.

In summary, there are several mechanisms by which AECs interact with cells important for the innate immune response. First, cytokines that recruit, activate, and stimulate the proliferation of local cells in the innate immune system are produced by AECs. Polar expression of these immunologically important molecules allow AECs to participate in compartmentalization of immune and inflammatory responses by creating microenviroments for specific cells in both the alveolar space (AMs) or interstitium (DCs). Second, AECs produce collectins, SP-A and SP-D, and complement proteins which facilitate the clearance of pathogens by macrophages and in some instances AECs themselves. The production of these proteins promotes the rapid clearance and processing of potential pathogens in the alveolus. Finally, AECs express adhesion molecules such as ICAM-1 which promote close associations and cell-cell communication with local immune and inflammatory cells. Thus, by several different mechanisms, AECs can participate in innate immunity by creating a local environment which is prepared for rapid response to invading pathogens and other inhaled challenges.

Airway Epithelial Cell Interactions with the Innate Immune System

The airway is exposed to a wide range of noxious stimuli which can initiate inflammation. In much the same way as AECs, airway cells may participate in this process by (1) responding directly to noxious stimuli or pathogens by secreting cytokines, (2) acting as paracrine cells amplifying ongoing airway inflammation through the production of cytokines in response to inflammatory mediators, and (3) expressing surface molecules that enhance cell-cell contact between epithelial cells and immune/inflammatory cells. Thus, airway cells are not passive barrier cells but rather are active participants in immune and inflammatory responses.

Cytokines

Airway epithelial cells produces chemokines, lymphocyte chemoattractant factor (LCF), colony-stimulating factors, pleiotropic cytokines, and growth factors (130–139). These molecules are involved in all stages of the inflammatory response, including cellular recruitment (e.g., IL-8, MCP-1 LCF), activation (e.g., GM-CSF, TNF, IL-1), and enhanced survival of recruited cells (e.g., GM-CSF). The production of cytokines by airway epithelial cells can be induced directly by noxious stimuli. Infection of a human bronchial epithelial cell line with respiratory syncytial virus directly stimulates IL-6, IL-8, and GM-CSF (140). The exposure of airway epithelial cells to pollutant gases such as ozone also induces IL-6 and IL-8 (141). Finally, direct exposure of normal human bronchial epithelial cells to asbestos fibers in vitro leads to the secretion of IL-8 and IL-1 (142). Thus, airway epithelial cells can secrete cytokines in direct response to infectious agents, air pollutant gases, and particulate fibers.

Early-response cytokines act on airway cells to amplify inflammatory responses. TNF-α and IL-1 induce epithelial cells to produce a wide array of cytokine products, including GM-CSF, G-CSF, IL-6, IL-8, IL-11, and MCP-1 (24, 133,136,138,139,143). Airway cells also respond to mast cell secretory products such as histamine by producing IL-6 and LCF (134,144). The neutrophil product elastase induces IL-6, IL-8, and GM-CSF (145,146). Hence, airway cells not only produce inflammatory cytokines but are also paracrine targets for inflammatory mediators in the local microenvironment.

IV. Epithelial Cell–Dendritic Cell Interactions: A Link Between Innate and Specific Immune Systems

The lung is the portal of entry for antigenic stimuli ranging from microorganisms to environmental pollutants. These antigens must be recognized and immunological responses mounted without disrupting pulmonary gas exchange. The generation of an immune response following antigen exposure in the lung likely involves the activation or maturation of pulmonary DCs (Fig. 3) (147,148). Precursor DCs (sentinel cells) are located in the periphery of the lung in contact with either alveolar or airway epithelial cells. Following exposure to antigen, DCs become activated. During the activation process, DCs acquire increased immunostimulatory capacity, probably via the expression of costimulatory molecules under the influence of secreted cytokines. Presumably, this activation process also primes DCs from movement to the lymph node where efficient antigen presentation can occur. The net result of the activation process is the maturation of DCs into presenter cells which are fully capable of initiating an adaptive immune response. Activation of pulmonary DCs might occur through two different pathways following antigen exposure. First, microbial products might bind to common microbial pattern recognition receptors present on both epithelial cells and macrophages. Second, microenvironmental tissue injury might initiate the activation of pulmonary DCs. The precise signaling required for this latter pathway could involve cytokines, low molecular weight mediators such as oxygen-free radicals, or changes in the glycosylation of cell surface molecules. Thus, pulmonary epithelial cells likely serve as crucial transducers of information to antigen-presenting cells in the lung. Through their capacity for self and nonself discrimination and via changes induced by injury, pulmonary epithelial cells serve as an essential link

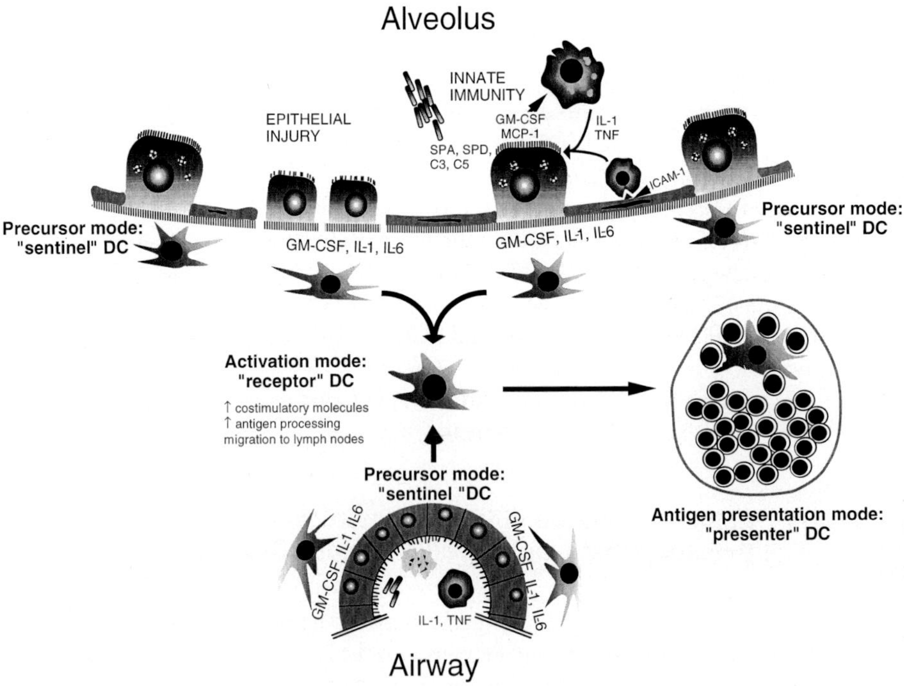

Figure 3 Epithelial cell–dendritic cell interactions: a link between innate and specific immune systems (see text for detailed discussion).

between the innate immune system and specific T-cell immunity. The local milieu created by alveolar and airway epithelial cells plays a crucial role in DC maturation, activation, and migration.

References

1. Rochester C, Goodell E, Stoltenborg J, Bowers W. Dendritic cells from rat lung are potent accessory cells. Am Rev Respir Dis 1988; 138:121–128.
2. Nicod L, Lipscomb M, Toews G, Weissler J. Separation of potent and poorly functional human lung accessory cells based an autofluorescence. J Leukoc Biol 1989; 45:458–465.
3. Pollard A, Lipscomb M. Characterization of murine lung dendritic cells: similarities to langerhans cells and thymic dendritic cells. J Exp Med 1990; 172:159–167.
4. Witmer PM, Olivier W, Valinsky J, Schuler G, Steinman RM. Granulocyte/macrophage colony-stimulating factor is essential for the viability and function of cultured murine epidermal Langerhans cells. J Exp Med 1987; 166:1484–1498.

5. Heufler C, Koch F, Schuler G. Granulocyte/macrophage colony-stimulating factor and interleukin 1 mediate the maturation of murine epidermal Langerhans cells into potent immunostimulatory dendritic cells. J Exp Med 1988; 167:700–705.
6. Christensen P, Armstrong L, Fak J, Chen GH, McDonald RA, Toews GB, Paine R. Regulation of rat pulmonary dendritic cell immunostimulatory activity by alveolar epithelial cell-derived GM-CSF. Am J Respir Cell Mol Biol 1995; 13:426–433.
7. Crapo JD, Barry BE, Gehr P, Bachofen M, Weibel ER. Cell number and cell characteristics of the normal human lung. Am Rev Respir Dis 1982; 126:332–337.
8. Adamson IY, Bowden DH. The type 2 cell as progenitor of alveolar epithelial regeneration. A cytodynamic study in mice after exposure to oxygen. Lab Invest 1974; 30:35–42.
9. Evans MJ, Cabral LJ, Stephens RJ, Freeman G. Renewal of alveolar epithelium in the rat following exposure to NO2. Am J Pathol 1973; 70:175–198.
10. Dobbs L. Isolation and culture of alveolar type II cells. Am J Physiol 1990; 258: L134–L147.
11. Kikkawa Y, Yoneda K. The type II epithelial cell of the lung. I. Method of isolation. Lab Invest 1974; 30:76–84.
12. Dobbs L, Williams M, Gonzalez R. Monoclonal antibodies specific to apical surfaces of rat alveolar type I cells bind to surfaces of cultured, but not freshly isolated type II cells. Biochim Biophys Acta 1988; 970:146–156.
13. Danto S, Zabski S, Crandall E. Reactivity of alveolar epithelial cells in primary culture with Type I cell monoclonal antibodies. Am J Respir Cell Mol Biol 1992; 6:296–306.
14. Paine R, Simon RH. Expanding the frontiers of lung biology through the creative use of pulmonary alveolar epithelial cells in culture. Am J Physiol 1996; 270:L484–L486.
15. Rennard SI, Beckman JD, Robbins RA. Biology of airway epithelial cells. In: Crystal RG, West JB, Barnes PJ, Cherniak NS, Weibel ER, eds. The Lung: Scientific Foundations. New York: Raven Press, 1991:157–167.
16. Mariassy AT. Epithelial cells of the trachea and bronchi. In: Parent RA, ed. Comparative Biology of the Normal Lung. Boca Raton, FL: CRC Press, 1991:63–76.
17. Harkema JR, Mariassy JSG, Hyde DM, Plopper CG. Epithelial cells of the conducting airways: a species comparison. In: Farmer SG, Hay DWP, eds. The Airway Epithelium: Physiology, Pathology, and Pharmacology. New York: M Dekker, 1991: 3–39.
18. Johnson NF, Hubbs AF. Epithelial progenitor cells in the rat trachea. Am J Respir Cell Mol Biol 1990; 3:579–585.
19. Berger H, Wellsh M. Electrolyte transport in the lungs. Hosp Pract 1991; 26:53–59.
20. Gruenert DC, Finkbeiner WE, Widdicombe JH. Culture and transformation of human airway epithelial cells. Am J Physiol 1995; 268:L347–L360.
21. Ziegler K, Unanue ER. Identification of a macrophage antigen-processing event required for I-region–restricted antigen presentation to T-lymphocytes. J Immunol 1981; 127:1869–1875.
22. Mizel S. Interleukin-1 and T cell activation. Immunol Rev 1972; 63:51.
23. Greenbaum LA, Horowitz JB, Woods A, Pasqualini T, Reich E, Bottomly K. Autocrine growth of CD4$^+$ T cells: Differential effects of IL-1 on helper and inflammatory T cells. J Immunol 1988; 140:1555–1560.

24. Elias JA, Trinchieri G, Beck JM, Simon PL, Sehgal PB, May LF, Kern JA. A synergistic interaction of IL-6 and IL-1 mediates the thymocytes stimulating activity produced by recombinant IL-1–stimulated fibroblasts. J Immunol 1989; 142:509.
25. Williams JM, Ransil BJ, Shapiro HM, Strom TB. Accessory cell requirement for activation, antigen expression and cell cycle progression by T lymphocytes. J Immunol 1984; 133:2986–2995.
26. Williams MC. Uptake of lectins by pulmonary alveolar type II cells: subsequent deposition into lamellar bodies. Proc Natl Acad Sci USA 1984; 81:6383–6387.
27. Mody C, Paine R, Shahrabadi MS, Simon RH, Pearlman E, Eisenstein BI, Toews GB. Legionella pneumophila replicates within rat alveolar epithelial cells. J Infect Dis 1993; 167:1138–1145.
28. Chi E, Mehl T, Nunn D, Lory S. Interaction of Pseudomonas aeruginosa with A549 pneumocyte cells. Infect Immun 1991; 59:822–828.
29. Sastry K, Ezekowitz RA. Collectins: pattern recognition molecules involved in first line host defense. Curr Opin Immunol 1993; 5:59–66.
30. Gibson KF, Phadke S. Intracellular distribution of lysozyme in rat alveolar type II epithelial cells. Exp Lung Res 1994; 20:595–611.
31. Glanville AR, Tazelaar HD, Theodore J, Imoto E, Rouse RV, Baldwin JC, Robin ED. The distribution of MHC class I and II antigens on bronchial epithelium. Am Rev Respir Dis 1989; 139:330–334.
32. Peters U, Papadopoulos T, Muller HH. MHC class II antigens on lung epithelial of human fetuses and neonates. Ontogeny and expression in lungs with histologic evidence of infection. Lab Invest 1990; 63:38–43.
33. Schneeberger EE, DeFerrari M, Skoskiewicz MJ, Russell PS, Colvin RB. Induction of MHC-determined antigens in the lung by interferon-γ. Lab Invest 1986; 55: 138–144.
34. Rochat TR, Casale JM, Hunninghake GW. Characterization of type II alveolar epithelial cells by flow cytometry and fluorescent markers. J Lab Clin Med 1988; 112:418–425.
35. Harbeck RJ, Gegen NW, Struhar D, Mason R. Class II molecules on rat alveolar type II epithelial cells. Cell Immunol 1988; 111:139–147.
36. Chang SC, Hsu HK, Perng RP, Shiao GM, Lin CY. Increased expression of MHC class II antigens in rejecting canine lung allografts. Transplantation 1990; 49:1158–1163.
37. Kallenberg CG, Schilizzi BM, Beaumont F, De LL, Poppema S, The TH. Expression of class II major histocompatibility complex antigens on alveolar epithelium in interstitial lung disease: relevance to pathogenesis of idiopathic pulmonary fibrosis. J Clin Pathol 1987; 40:725–33.
38. Romaniuk A, Prop J, Petersen AH, Nieuwenhuis P, Wildevuur CR. Class II antigen expression on bronchial epithelium in rat lung allographs is prevented by cyclosporine treatment. Transplant Proc 1987:218–219.
39. Romaniuk A, Prop J, Petersen AH, Wildevuur CR, Nieuwenhuis P. Expression of class II major histocompatibility complex antigens by bronchial epithelium in rat lung allografts. Transplantation 1987; 44:209–214.
40. Burns A, Takei F, Doerschuk C. Expression of intercellular adhesion molecule

(ICAM-1) on murine pulmonary capillary endothelium and alveolar epithelium. FASEB J 1991; 5:A1056.
41. Christensen P, Kim S, Simon S, Toews G, Paine R. Differentiation-related expression of ICAM-1 by rat alveolar epithelial cells. Am J Respir Cell Mol Biol 1993; 8:9–15.
42. Kang B, Crapo J, Wegner C, Letts L, Chang L. Intercellular adhesion molecule-1 expression on the alveolar epithelium and its modification by hyperoxia. Am J Respir Cell Mol Biol 1993; 9:350–355.
43. Paine RI, Christensen P, Toews G, Simon R. Regulation of alveolar epithelial cell ICAM-1 expression by cell shape and cell-cell interactions. Am J Physiol 1994; 266: L476–L484.
44. Niederwieser D, Auböck J, Troppmair J, Herold M, Schular G, Boeck G, Lotz J, Fritsch P, Huber C. IFN-mediated induction of MHC antigen expression on human keratinocytes and its influence on in vitro alloimmune responses. J Immunol 1988; 140:2556–2564.
45. Umetsu DT, Pober JS, Jabara HH, Fiers W, Yunis EJ, Burakoff SJ, Reiss CS, Geha RS. Human dermal fibroblasts present tetanus toxoid antigen to antigen-specific T cell clones. J Clin Invest 1985; 76:254–260.
46. Bland P, Warren LG. Antigen presentation by epithelial cells of the rat small intestine I. Kinetics, antigen specificity and blocking by anti-Ia antisera. Immunology 1986; 58:1–7.
47. Bland P, Warren L. Antigen presentation by epithelial cells of the rat small intestine II. Selective induction of suppressor T cells. Immunology 1986; 58:9–14.
48. Ansfield MJ, Kaltrieder HB, Benson BJ, Caldwell JL. Immunosuppressive activity of canine pulmonary surface active material. J Immunol 1979; 122:1062–1066.
49. Ansfield MJ, Benson BJ. Identification of the immunosuppressive components of canine pulmonary surface active material. J Immunol 1980; 125:1093–1098.
50. Sitrin RG, Ansfield MJ, Kaltreider HB. The effect of pulmonary surface-active material on the generation and expression of murine B- and T-lymphocyte effector functions in vitro. Exp Lung Res 1985; 9:85–97.
51. Paine III R, Chavis A, Gaposchkind D, Christensen PJ, Mody CH, Toews GB. A factor secreted by a human pulmonary alveolar epithelial-like cell line blocks T cell proliferation between G1 and S phase. Am J Respir Cell Mol Biol 1992; 6:658–666.
52. Hasegawa S, Ockner DM, Ritter JH, Patterson GA, Trulock EP, Cooper JD, Wick MR. Expression of class II major histocompatibility complex antigens (HLA-DR) and lymphocyte subset immunotyping in chronic pulmonary transplant rejection. Arch Pathol Lab Med 1995; 119:432–439.
53. Suda T, Sato A, Sugiura W, Chida K. Induction of MHC class II antigens on rat bronchial epithelial cells by interferon-gamma and its effect on antigen presentation. Lung 1995; 173:127–137.
54. Spurzem JR, Sacco O, Rossi GA, Beckmann JD, Rennard SI. Regulation of major histocompatibility complex class II gene expression on bovine bronchial epithelial cells. J Lab Clin Med 1992; 120:94–102.
55. Mezzetti M, Soloperto M, Fasoli A, Mattoli S. Human bronchial epithelial cells modulate CD3 and mitogen-induced DNA synthesis in T cells but function poorly as

antigen-presenting cells compared to pulmonary macrophages. J Allergy Clin Immunol 1991; 87:930–938.
56. Kalb TH, Chuang MT, Marom Z, Mayer L. Evidence for accessory cell function by class II MHC antigen-expressing airway epithelial cells. Am J Respir Cell Mol Biol 1991; 4:320–329.
57. Nicod L, Lipscomb M, Weissler J, Toews G. Mononuclear cells from human lung parenchyma support antigen-induced T lymphocyte proliferation. J Leuk Biol 1989; 45:336–344.
58. Holt PG, Schon-Hegrad MA, Oliver J. MHC class II antigen-bearing dendritic cells in pulmonary tissues of the rat. Regulation of antigen presentation activity by endogenous macrophage populations. J Exp Med 1988; 167:262–274.
59. Steinman R. The dendritic cell system and its role in immunogenicity. Annu Rev Immunol 1991; 9:271–296.
60. Armstrong LR, Christensen PJ, Paine R, Chen GH, McDonald RA, Lim TK, Toews GB. Regulation of the immunostimulatory activity of rat pulmonary dendritic cells by cell-cell interactions and cytokines. Am J Respir Cell Mol Biol 1993; 11: 682–691.
61. Sertl K, Takemura T, Tschachler E, Ferrans V, Kaliner M, Shevach E. Dendritic cells with antigen-presenting capability reside in airway epithelium, lung parenchyma, and visceral pleura. J Exp Med 1986; 163:436–451.
62. Schon-Hegrad MA, Oliver J, McMenamin PG, Holt PG. Studies on the density, distribution and surface phenotype of intraipithelial class II major histocompatability complex antigen (Ia)–bearing dendritic cells (DC) in the conducting airways. J Exp Med 1991; 173:1345–1356.
63. McWilliam AS, Nelson D, Thomas JA, Holt PG. Rapid dendritic cell recruitment is a hallmark of the acute inflammatory response at mucosal surfaces. J Exp Med 1994; 179:1331–1336.
64. Sozzani S, Sallusto F, Luini W, Zhou D, Piemonti L, Allavena P, Van Damme J, Valitutti S, Lanzavecchia A, Mantovani A. Migration of dendritic cells in response to formyl peptides, C5a, and a distinct set of chemokines. J Immunol 1995; 155:3292–3295.
65. Abdelaziz MM, Devalia JL, Khair OA, Calderon M, Sapsford RJ, Davies RJ. The effect of conditioned medium from cultured human bronchial epithelial cells on eosinophil and neutrophil chemotaxis and adherence in vitro. Am J Respir Cell Mol Biol 1995; 13:728–737.
66. Kwon OJ, Jose PJ, Robbins RA, Schall TJ, Williams TJ, Barnes PJ. Glucocorticoid inhibition of RANTES expression in human lung epithelial cells. Am J Respir Cell Mol Biol 1995; 12:488–496.
67. Stellato C, Beck LA, Gorgone GA, Proud D, Schall TJ, Ono SJ, Lichtenstein LM, Schleimer RP. Expression of the chemokine RANTES by a human bronchial epithelial cell line. Modulation by cytokines and glucocorticoids. J Immunol 1995; 155: 410–418.
68. Matsukura S, Kokubu F, Izumi H, Noda H, Kurokawa M, Tokunaga H, Adachi M. [RANTES expression on human bronchial epithelial cells]. Arerugi 1995; 44: 715–717.
69. VanOtteren GM, Strieter RM, Kunkel SL, Paine R, Greenberger MJ, Danforth JM,

Burdick MD, Standiford TJ. Compartmentalized expression of RANTES in a murine model of endotoxemia. J Immunol 1995; 154:1900–1908.
70. Burdick MD, Kunkel SL, Lincoln PM, Wilke CA, Strieter RM. Specific ELISAs for the detection of human macrophage inflammatory protein-1 alpha and beta. Immunol Invest 1993; 22:441–449.
71. Lieschke GJ, Burgess AW. Granulocyte colony-stimulating factor and granulocyte-macrophage colony-stimulating factor (2). N Engl J Med 1992; 327:99–106.
72. Lieschke GJ, Burgess AW. Granulocyte colony-stimulating factor and granulocyte-macrophage colony-stimulating factor (1). N Engl J Med 1992; 327:28–35.
73. Akagawa KS, Kamoshita K, Tokunaga T. Effects of granulocyte-macrophage colony-stimulating factor and colony stimulating factor-1 on the proliferation and differentiation of murine alveolar macrophages. J Immunol 1988; 141:3383.
74. Chen G, Curtis J, Mody C, Christensen P, Armstrong L, Toews G. Effect of granulocyte-macrophage colony-stimulating factor on rat alveolar macrophage anti-cryptococcal activity in vitro. J Immunol 1994; 152:724–734.
75. Fischer HG, Frosch S, Reske K, Reske-Kunz AB. Granulocyte-macrophage colony stimulating factor activates macrophages derived from bone marrow cultures to synthesis of MHC class II molecules and to augment antigen presentation function. J Immunol 1988; 141:3882–3888.
76. Morrissey PJ, Bressler L, Park LS, Alpert A, Gillis S. Granulocyte-macrophage colony stimulating factor augments the primary antibody response by enhancing the function of antigen presenting cells. J Immunol 1987; 139:1113–1119.
77. Soloperto M, Mattoso VL, Fasoli A, Mattoli S. A bronchial epithelial cell-derived factor in asthma that promotes eosinophil activation and survival as GM-CSF. Am J Physiol 1991:L530–L538.
78. Sousa AR, Poston RN, Lane SJ, Nakhosteen JA, Lee TH. Detection of GM-CSF in asthmatic bronchial epithelium and decrease by inhaled corticosteroids. Am Rev Respir Dis 1993:1557–1561 .
79. Mattoli S, Marini M, Fasoli A. Expression of the potent inflammatory cytokines, GM-CSF, IL6, and IL8, in bronchial epithelial cells of asthmatic patients. Chest 1992:27S–29S.
80. Davies RJ, Wang JH, Trigg CJ, Devalia JL. Expression of granulocyte/macrophage-colony-stimulating factor, interleukin-8 and RANTES in the bronchial epithelium of mild asthmatics is down-regulated by inhaled beclomethasone dipropionate. Int Arch Allergy Immunol 1995; 107:428–429.
81. Marini M, Vittori E, Hollemborg J, Mattoli S. Expression of the potent inflammatory cytokines, granulocyte-macrophage-colony-stimulating factor and interleukin-6 and interleukin-8, in bronchial epithelial cells of patients with asthma. J Allergy Clin Immunol 1992; 89:1001–1009.
82. Bellini A, Vittori E, Marini M, Ackerman V, Mattoli S. Intraepithelial dendritic cells and selective activation of Th2-like lymphocytes in patients with atopic asthma. Chest 1993; 103:997–1005.
83. Gasson JC. Molecular physiology of granulocyte-macrophage colony-stimulating factor. Blood 1991; 77:1131–1145.
84. Kupper TS, Lee F, Birchall N, Clark S, Dower S. Interleukin 1 binds to specific receptors on human keratinocytes and induces granulocyte macrophage colony-

stimulating factor mRNA and protein: a potential autocrine role for interleukin 1 in epidermis. J Clin Invest 1988; 82:1787–1792.
85. Lang RA, Metcalf D, Cuthbertson RA, Lyons I, Sanley E, Kelso A, Kannourakis G, Williamson DJ, Klintworth GK, Gonda TJ, Dunn AR. Trangenic mice expressing a hemopoietic growth factor gene (GM-CSF) develop accumulations of macrophages, blindness and a fatal syndrome of tissue damage. Cell 1987; 51:675–686.
86. Seelentag W, Mermod JJ, Vassalli P. Interleukin 1 and tumor necrosis factor-alpha additively increase the levels of granulocyte-macrophage and granulocyte colony-stimulating factor (CSF) mRNA in human fibroblasts. Eur J Immunol 1989; 19: 209–212.
87. Seelentag WK, Mermod JJ, Montesano R, Vassalli P. Additive effects of interleukin 1 and tumour necrosis factor-alpha on the accumulation of the three granulocyte and macrophage colony-stimulating factor mRNAs in human endothelial cells. Embo J 1987; 6:2261–2265.
88. Horowitz MC, Coleman DL, Flood PM, Kupper TS, Jilka RL. Parathyroid hormone and lipopolysaccharide induce murine osteoblast-like cells to secrete a cytokine indistinguishable from granulocyte-macrophage colony-stimulating factor. J Clin Invest 1989; 83:149–157.
89. Smith SM, Lee DK, Lacy J, Coleman DL. Rat tracheal epithelial cells produce granulocyte/macrophage colony-stimulating factor. Am J Respir Cell Mol Biol 1990; 2:59–68.
90. Blau H, Riklis S, Kravtsov V, Kalina M. Secretion of cytokines by rat alveolar epithelial cells: possible regulatory role for SP-A. Am J Physiol 1994; 266:L148–L155.
91. Hamilton JA, Whitty GA, Royston AK, Cebon J, Layton JE. Cytokine regulation of colony-stimulating factor (CSF) production in cultured human synovial fibroblasts. II. Similarities and differences in the control of interleukin-1 induction of granulocyte-macrophage CSF and granulocyte-CSF production. Blood 1992; 79:1413–1419
92. Akahane K, Pluznik DH. Interferon-gamma destablizes interleukin-1–induced granulocyte-macrophage colony-stimulating factor mRNA in murine vascular endothelial cells. Exp Hematol 1993; 21:878–884.
93. Hamilton JA, Whitty GA, Royston AK, Cebon J, Layton JE. Interleukin-4 suppresses granulocyte colony-stimulating factor and granulocyte-macrophage colony-stimulating factor levels in stimulated human monocytes. Immunology 1992; 76: 566–571.
94. Akahane K, Pluznik DH. Interleukin-4 inhibits interleukin-1 alpha–induced granulocyte-macrophage colony-stimulating factor gene expression in a murine B-lymphocyte cell line via downregulation of RNA precursor. Blood 1992; 79: 3188–3195.
95. Lee F, Yokota T, Otsuka T, Gemmel L, Larson N, Luh J, Arai K, Rennick D. Isolation of cDNA for a human granulocyte-macrophage colony-stimulating factor by functional expression in mammalian cells. Proc Natl Acad Sci USA 1985; 82:4360–4364.
96. Schrimsher JL, Rose K, Simona MG, Wingfield P. Characterization of human and mouse granulocyte-macrophage colony-stimulating factors derived from Escherichia coli. Biochem J 1987; 247:195–199.
97. Shanafelt AB, Johnson KE, Kastelein RA. Identification of critical amino acid

residues in human and mouse granulocyte-macrophage colony stimulating factor and their involvement in species specificity. J Biol Chem 1991; 266:13804–13810.
98. Wingfield P, Graber P, Moonen P, Craig S, Pain RH. The conformation and stability of recombinant-derived granulocyte-macrophage colony stimulating factors. Eur J Biochem 1988; 173:65–72.
99. Brown CB, Hart CE, Curtis DM, Bailey MC, Kaushansky K. Two neutralizing monoclonal antibodies against human granulocyte-macrophage colony-stimulating factor recognize the receptor binding domain of the molecule. J Immunol 1990; 144: 2184–2189.
100. Devlin RB, McKinnon KP, Noah T, Becker S, Koren HS. Ozone-induced release of cytokines and fibronectin by alveolar macrophages and airway epithelial cells. Am J Physiol 1994:L612–L619.
101. Mattoli S, Miante S, Calabro F, Mezzetti M, Fasoli A, Allegra L. Bronchial epithelial cells exposed to isocyanates potentiate activation and proliferation of T-cells. Am J Physiol 1990:L320–L327.
102. Arnold R, Humbert B, Werchau H, Gallati H, Konig W. Interleukin-8, interleukin-6, and soluble tumour necrosis factor receptor type I release from a human pulmonary epithelial cell line (A549) exposed to respiratory syncytial virus. Immunology 1994 82:126–133.
103. Holt PG, Haining S, Nelson DJ, Sedgwick JD. Origin and steady-state turnover of class II MHC-bearing dendritic cells in the epithelium of the conducting airways. J Immunol 1994; 153:256–261.
104. Pugh CW, MacPherson GG, Steer HW. Characterization of nonlymphoid cells derived from rat peripheral lymph. J Exp Med 1983; 157:1758–1779.
105. Hosoi J, Murphy GF, Egan CL, Lerner EA, Grabbe S, Asahina A, Granstein RD. Regulation of Langerhans cell function by nerves containing calcitonin gene-related peptide. Nature 1993; 363:159–163.
106. Asahina A, Moro O, Hosoi J, Lerner EA, Xu S, Takashima A, Granstein RD. Specific induction of cAMP in Langerhans cells by calcitonin gene–related peptide: relevance to functional effects. Proc Natl Acad Sci USA 1995; 92:8323–8327.
107. Hastings RH, Hua XY. Expression of calcitonin gene-related peptide by cultured rat alveolar type II cells. Am J Respir Cell Mol Biol 1995; 13:563–569.
108. Paine R, Rolfe M, Standiford T, Burdick M, Rollins B, Strieter R. MCP-1 expression by rat type II alveolar epithelial cells in primary culture. J Immunol 1993; 150:4561–4570.
109. Standiford TJ, Kunkel S, Basha M, Chensue SW, Lynch JP, Toews GB, Westwick J, Strieter RM. Interleukin-8 gene expression by a pulmonary epithelial cell line- a model for cytokine networks in the lung. J Clin Invest 1990; 86:1945–1953.
110. Standiford T, Kunkel S, Phan S, Rollins B, Strieter R. Alveolar macrophage–derived cytokines induce monocyte chemoattractant protein expression from human pulmonary type II–like epithelial cells. J Biol Chem 1991; 266:9912–9918.
111. Nakata K, Kiyoko S, Fukayama M, Hayashi Y, Kadokura M, Tokunaga T. Granulocyte-macrophage colony-stimulating factor promotes the proliferation of human alveolar macrophages in vitro. J Immunol 1991; 147;1266–1272.
112. Tazi A, Bouchonnet F, Grandsaigne M, Boumsell L, Hance AJ, Soler P. Evidence that granulocyte macrophage-colony-stimulating factor regulates the distribution

and differentiated state of dendritic cells/Langerhans cells in human lung and lung cancers. J Clin Invest 1993; 91:566–576.
113. Dranoff G, Crawford AD, Sadelain M, Ream B, Rashid A, Bronson RT, Dickersin, GR, Bachurski CJ, Mark EL, Whitsett JA, et al. Involvement of granulocyte-macrophage colony-stimulating factor in pulmonary homeostasis. Science 1994; 264:713–716.
114. Antoniades HN, Neville GJ, Galanopoulos T, Kradin RL, Valente AJ, Graves DT. Expression of monocyte chemoattractant protein 1 mRNA in human idiopathic pulmonary fibrosis. Proc Natl Acad Sci USA 1992; 89:5371–5375.
115. Khalil N, O'Connor RN, Flanders KC, Shing W, Whitman CI. Regulation of type II alveolar epithelial cell proliferation by TGF- beta during bleomycin-induced lung injury in rats. Am J Physiol 1994:L498–L507.
116. van Iwaarden F, Welmers B, Verhoef J, Haagsman HP, van Golde LM. Pulmonary surfactant protein A enhances the host-defense mechanism of rat alveolar macrophages. Am J Respir Cell Mol Biol 1990; 2:91–98.
117. Manz-Keinke H, Plattner H, Schlepper SJ. Lung surfactant protein A (SP-A) enhances serum-independent phagocytosis of bacteria by alveolar macrophages. Eur J Cell Biol 1992; 57:95–100.
118. McNeely TB, Coonrod JD. Aggregation and opsonization of type A but not type 8 Hemophilus influenzae by surfactant protein A. Am J Respir Cell Mol Biol 1994; 11: 114–122.
119. van Iwaarden F, van Strijp J, Visser H, Haagsman HP, Verhoef J, van GL. Binding of surfactant protein A (SP-A) to herpes simplex virus type 1–infected cells is mediated by the carbohydrate moiety of SP-A. J Biol Chem 1992; 267:25039–25043.
120. Wright JR, Youmans DC. Pulmonary surfactant protein A stimulates chemotaxis of alveolar macrophage. Am J Physiol 1993:L338–L344.
121. Katsura H, Kawada H, Konno K. Rat surfactant apoprotein A (SP-A) exhibits antioxidant effects on alveolar macrophages. Am J Respir Cell Mol Biol 1993; 9: 520–525.
122. Crouch E, Parghi D, Kuan SF, Persson A. Surfactant protein D: subcellular localization in nonciliated bronchiolar epithelial cells. Am J Physiol 1992:L60–L66.
123. Kuan SF, Persson A, Parghi D, Crouch E. Lectin-mediated interactions of surfactant protein D with alveolar macrophages. Am J Respir Cell Mol Biol 1994; 10:430–436.
124. Tomlinson S. Complement defense mechanisms. Curr Opin Immunol 1993; 5:83–89.
125. Frank MM, Fries LF. The role of complement in inflammation and phagocytosis. Immunol Today 1991; 12:322–326.
126. Strunk RC, Eidlen DM, Mason RJ. Pulmonary alveolar type II epithelial cells synthesize and secrete proteins of the classical and alternative complement pathways. J Clin Invest 1988; 81:1419–1426.
127. Dustin M, Rothlein R, Bhan A, Dinarello C, Springer T. Induction by IL 1 and interferon-g: tissue distribution, biochemistry, and function of a natural adherence molecule (ICAM-1). J Immunol 1986; 137:245–254.
128. Carpen O, Dustin M, Springer T, Swafford J, Beckett L, Caulfield J. Motility and ultrastructure of large granular lymphocytes on lipid bilayers reconstituted with adhesion receptors LFA-1, ICAM-1, and two isoforms of LFA-3. J Cell Biol 1991; 115:861–871.

129. Barton WW, Wilcoxen SE, Christensen PJ, Paine R. Association of ICAM-1 with the cytoskeleton in rat alveolar epithelial cells in primary culture. Am J Physiol. In press.
130. Levine SJ. Bronchial epithelial cell-cytokine interactions in airway inflammation. J Invest Med 1995; 43:241–249.
131. Magnan A, Frachon I, Rain B, Peuchmaur M, Monti G, Lenot B, Fattal M, Simonneau G, Galanaud P, Emilie D. Transforming growth factor beta in normal human lung: preferential location in bronchial epithelial cells. Thorax 1994; 49:789–792.
132. Sacco O, Romberger D, Rizzino A, Beckmann JD, Rennard SI, Spurzem JR. Spontaneous production of transforming growth factor-beta 2 by primary cultures of bronchial epithelial cells. Effects on cell behavior in vitro. J Clin Invest 1992; 90:1379–1385.
133. Nakamura H, Yoshimura K, Jaffe HA, Crystal RG. Interleukin-8 gene expression in human bronchial epithelial cells. J Biol Chem 1991; 266:19611–19617.
134. Bellini A, Yoshimura H, Vittori E, Marini M, Mattoli S. Bronchial epithelial cells of patients with asthma release chemoattractant factors for T lymphocytes. J Allergy Clin Immunol 1993; 92:412–424.
135. Ohta K, Hirohata S, Yamaguchi M, Suzuki N, Ueda T, Ishii A, Shindoh G, Oka T, et al. Interleukin 6/B cell stimulatory factor-II is expressed and released by normal and transformed human bronchial epithelial cells. Biochem Biophys Res Common 1992; 187:596–602.
136. Marini M, Soloperto M, Mezzetti M, Fasoli A, Mattoli S. Interleukin-1 binds to specific receptors on human bronchial epithelial cells and upregulates granulocyte/macrophage colony- stimulating factor synthesis and release. A J Respir Cell Mol Biol 1991 4:519–524.
137. Richman EJ, Jorens PG, Hebert CA, Ueki I, Nadel JA. Interleukin-8: an important chemoattractant in sputum of patients with chronic inflammatory airway diseases. Am J Physiol 1993:L413–L418.
138. Levine SJ, Larivee P, Logun C, Angus CW, Shelhamer JH. Corticosteroids differentially regulate secretion of IL-6, IL-8, and G-CSF by a human bronchial epithelial cell line. Am J Physiol 1993:L360–L368.
139. Becker S, Quay J, Koren HS, Haskill JS. Constitutive and stimulated MCP-1, GRO alpha, beta, and gamma expression in human airway epithelium and bronchoalveolar macrophages. Am J Physiol 1994:L278–L286.
140. Noah TL, Becker S. Respiratory syncytial virus-induced cytokine production by a human bronchial epithelial cell line. Am J Physiol 1993:L472-L478.
141. Devlin RB, McKinnon KP, Noah T, Becker S, Koren HS. Ozone-induced release of cytokines and fibronectin by alveolar macrophages and airway epithelial cells. Am J Physiol 1994:L612–L619.
142. Rosenthal GJ, Germolec DR, Blazka ME, Corsin E, Simeonova P, Pollock P, Kong LY, Kwon J, Luster MI. Asbestos stimulates IL-8 production from human lung epithelial cells. J Immunol 1994; 153:3237–3244.
143. Cromwell O, Hamid Q, Corrigan CJ, Barkans J, Meng Q, Collins PD, Kay AB. Expression and generation of interleukin-8, IL-6 and granulocyte-macrophage colony-stimulating factor by bronchial epithelial cells and enhancement by IL-1 beta and tumour necrosis factor-alpha. Immunology 1992; 77:330–337.
144. Noah TL, Paradiso AM, Madden MC, McKinnon KP, Devlin RB. The response of a

human bronchial epithelial cell line to histamine: intracellular calcium changes and extracellular release of inflammatory mediators. Am J Respir Cell Mol Biol 1991; 5:484–492.

145. Bedard M, McClure CD, Schiller NL, Francoeur C, Cantin A, Denis M. Release of interleukin-8, interleukin-6, and colony-stimulating factors by upper airway epithelial cells: implications for cystic fibrosis. Am J Respir Cell Mol Biol 1993; 9: 455–462.
146. Ruef C, Jefferson DM, Schlegel HS, Suter S. Regulation of cytokine secretion by cystic fibrosis airway epithelial cells. Eur Respir J 1993; 6:1429–1436.
147. Ibrahim MA, Chain BM, Katz DR. The injured cell: the role of the dendritic cell system as a sentinel receptor pathway. Immunol Today 1995; 16:181-186.
148. Toews GB. Pulmonary dendritic cells: sentinels of lung-associated lymphoid tissues. Am J Respir Cell Mol Biol 1991; 4:204–205.
149. Becker S, Koren HS, Henke DC. Interleukin-8 expression in normal nasal epithelium and its modulation by infection with respiratory syncytial virus and cytokines tumor necrosis factor, interleukin-1, and interleukin-6. Am J Respir Cell Mol Biol 1993; 8:20–27.

13

Function of Human Lung Dendritic Cells

LAURENT P. NICOD

University Hospital of Geneva
Geneva, Switzerland

I. Introduction

Human lung dendritic cells (DCs) have many of the features of lung DCs described in previous chapters. As in other species, human DCs form a network in the airway walls and are also abundant around vessels. Under the influence of mediators released by cells such as epithelial cells in their proximity, they form a specialized system for presenting antigens to naive or quiescent T cells and consequently play a central role in the induction of T-cell and B-cell immunity in vivo. In this chapter, various antigen-presenting cells (APCs) present in the lung will be discussed and compared with DCs, the localization and function of which in human lung will also be discussed.

II. Pulmonary Dendritic Cells and Other Accessory Cells

A. Distribution of DCs

When we first stained normal human lung for HLA-DR–positive cells, alveolar macrophages (AMs) were brightly stained. HLA-DR–positive cells were also seen in the interstitium of the lung. Many had a mature macrophage appearance, but others displayed elongated processes similar to those demonstrated by DCs in lymphoid organs. The majority of these DCs were found in the bronchial mucosa either above or below the basement membrane or deep in the lamina propria of the bronchia. In the mucosa below the epithelium, occasional aggregates of lymphocytes of the bronchus-associated lymphoid tissue (BALT) were visualized and were associated with DCs. Many DCs were found in perivascular loose connective tissue and in the pleura (5). These findings were confirmed not only in humans but also in mice (6). Shortly afterward, DCs were shown to constitute a tight network around the airways. This was well shown by Patrick Holt's group. They performed tangential sections of the airways; and by staining for MHC class II, they demonstrated that DC processes formed a network like Langerhans' cells do in the skin (7). However, the majority of these cells do not express CD1a (OKT6) as do Langerhans' cells.

B. Lung APCs: APCs that Constitutively Express MHC Antigens

Before discussing the role and function of DCs in a given organ, such as the lungs, it is important to know what other cells in the local environment might perform a similar function or synergize with DCs.

Alveolar Macrophages

Alveolar macrophages (AMs) are the predominant cells in the alveoli able to take up foreign antigens. Weinberg and Unanue found that only 5% of murine AMs were Ia positive (8). They found that murine AMs were less effective than resident periotoneal macrophages in presenting *Listeria monocytogenes* and stimulating immune T lymphocytes to proliferate (8). This was in contrast to earlier studies by Seeger and Oppenheim, who, working with inbred guinea pigs, found that both AMs and peritoneal exudate macrophages enhanced antigen-induced proliferation (9). It became apparent that there were species differences in the accessory cell function of AMs (10).

Human AMs strongly express both HLA-DR and HLA-DQ. In separate studies, human AMs have been reported both to enhance and to suppress mitogen or antigen-induced immune responses. Toews et al. (11) studied AMs in normal volunteers using accessory cell depleted peripheral blood lymphocytes (PBLs). AMs from 75% of normal volunteers failed to reconstitute antigen-induced lymphoproliferation. AMs from 50% of volunteers suppressed antigen-induced responses when a ratio of AMs to lymphocytes of 1:2 was studied (11). Suppression of T-cell mitogen stimulation by AMs was observed when increasing numbers of AMs were used (12,13), suggesting that AMs might release immunosuppressive factors. AMs efficiently stimulated the proliferation of T-lymphocyte lines and clones in response to soluble antigens (14). This discrepancy could be explained by less stringent requirements to stimulate lymphocytes clones than for the activation of "naive" or "memory" resting T cells. Thus, activation of resting T cells is more complex and requires more specialized APCs. As already discussed in preceding chapters, the evidence is accumulating that DCs are the cells best equipped to deliver the required second signals.

A variety of studies have attempted to determine why AMs are relatively poor accessory cells and many even suppress T-cell proliferation. These various factors are discussed below and are summarized in Table 1.

Studies comparing the release of IL-1 by AMs or monocytes in response to endotoxin indicated that AMs are poor producers of bioactive interleukin-1 (IL-1) as measured by the thymocyte D10 cell line in proliferation assays (15). However, deficiency in IL-1 production does not entirely explain the reduced capacity of AMs to serve as accessory cells. Indeed, even after the addition of exogenous IL-1, the capacity of AMs to stimulate T-lymphocyte proliferation is not reconstituted (14).

IL-1 receptor antagonist (IL-1ra) is also produced by AMs (16). The inhibitor may have a role in the modulation of subtypes of T cells by competing for IL-1 receptors present on T cells and thus blocking the action of IL-1.

AMs release tumor necrosis factor (TNF) inhibitors in the form of soluble receptors for TNF, especially the TNFsR75 (16,17). The potential role of these inhibitors will be discussed later, but their role seems to be rather limited at physiological concentrations on APC–T-cell interactions.

IL-10 is produced by AMs and monocytes and has inhibitory activity, including inhibiting DC T cell interactions, as discussed later (18).

AMs are a source of PGE_2 and hydrogen peroxide, both known to inhibit T-cell proliferation (19).

Table 1 Factors that Can Interfere with the APC Function of Alveolar Macrophages

Factor	Reference
Decreased production of IL-1	15
Release of IL-1 receptor antagonist	16,17
Release of TNF soluble receptors	16
Release of H202	19
Release of PGE_2	19
Decreased LFA-1 expression	20
Production of IL-10	119,120
Lack of B7 family expression	22

AMs have a reduced capacity as compared with monocytes of DCs to bind T lymphocytes. The mechanisms responsible are not clear. A decrease of the density of leukocyte function–associated antigen-1 (LFA-1) (CD11a/CD18) on macrophages, which binds to intracellular adhesion molecule-1 (ICAM-1) (CD54) on T lymphocytes has been proposed (20). It is not yet settled to what extent AMs have a significant decrease in LFA-1 molecules or if they have a decreased capacity to phosphorylate LFA-1 as compared with other APCs. It has been clearly shown that the phosphorylation of LFA-1 is a requirement for its binding to ICAM (21).

A defect of B7-1 (CD10) and B7-2 (DC86) expression has been recently shown. This could be a major defect compared with DCs which can be overcome "in vitro" by using agonist antibodies against their ligand CD28 during antigen presentation (22).

Cleavage of sialic acids can improve the capacity of some cells to serve as APCs, suggesting that electrostatic charges resulting from these residues can prevent antigen presentation. An altered state of glycosylation of major histocompatibility complex (MHC) class II antigens on AMs compared with monocytes has also been suggested (23).

Langerhans' Cells

Langerhans' cells (LCs) have been described in the epidermis since the early 1960s and were shown in 1961 to possess unique ultrastructural organelles known as Birbeck's granules (24). Cells containing these organelles and expressing the antigen CD1a (OKT6) are found constantly in normal mammalian epidermis and in the disease histiocytosis (24). LCs are most likely derived from bone marrow stem cells (25), although the possibility that LCs are derived from tissue macrophages has also been proposed in morphological studies performed after bone marrow transplantation (26). LCs have a fairly long half-life in tissues (25). They are, however, mobile cells, and following antigenic stimulation, they can migrate to regional lymph nodes and central lymphoid organs (27). In the skin, the capacity of LCs to serve as accessory cells is highly dependent on their state of maturation and differentiation. Isolated LCs are poorly phagocytic cells, although they can internalize and process antigens. Freshly isolated epidermal LCs, although capable of activating T-lymphocyte cell lines, lack the capacity to activate naive T cells (27). In vitro culture of LCs, particularly in the presence of IL-1 and granulocyte-macrophage colony-stimulating factor (GM-CSF), can dramatically improve the capacity of these cells to trigger unprimed T cells (28).

Based on these findings, it has been suggested that LCs in the skin serve as "sentinels" which can pick up and process foreign antigens, but initiation of an immune response against a previously unencountered antigen would require their migration to lymph nodes, during which time they "mature" into potent accessory cells, showing features of DCs. Part of this maturation could occur locally in human lung under the influence of factors that still need to be elucidated. Indeed, few CD1a$^+$, Ia$^+$ DCs are found in uninflamed lung as compared with CD1a$^-$, Ia$^+$ DCs present around the airways, the vessels, or in the pleura (5). In inflammatory conditions such as asthma, LCs are severalfold more abundant in the mucosa of patients with this disease, whereas they are unchanged in patients with allergic rhinitis alone (29). The role of these Langerhans-like cells remains to be established in such conditions.

B Cells

In normal human lung, lymphoid aggregates containing B cells are far less well organized than is BALT described in other species, particularly in the chicken, rabbit, and rat.

Small B cells, like DCs, constitutively express class II MHC products, but levels on B cells appear to be one fifth of the level on DCs (30). The level of class II MHC on B cells can be upregulated by lipopolysaccharide (LPS), anti-Ig, IL-1 (31), or antigens (32). Chestnut et al. (33) directly compared fresh B cells, LPS-activated B-cell blasts, and a B-cell lymphoma line as APCs to stimulate KLH-primed T cells and tetanus toxoid T-cell hybridomas. Unstimulated B cells were weak or inactive, but the other B cells were effective APCs. Metlay et al. (34) compared B cells and DCs as stimulators of CD4 T cells in the murine mixed lymphocyte reaction (MLR). B lymphoblasts initiated a primary MLR, whereas small B cells were weak or ineffective stimulators. DCs were 10–30 times more potent than B lymphoblasts in triggering allogeneic T cells and likewise more effective in binding T cells to form clusters.

C. Lung APCs: Nonlymphoid APCs with Inducible MHC Molecules

MHC class II can be expressed by cells that are not part of the immune system following either systemic or locally applied immunological stimulation (35). The functional implication of the expression of HLA-DR molecules by nonlymphoid tissue is not clear. It has been hypothesized that, at least for some organs, DR expression might be involved in antigen presentation or in the recruitment of immune effector cells.

Epithelial Cells

Ciliated bronchial epithelial cells express HLA-DR transcripts (3), and HLA-DR seems to be present as a thin layer of green fluorescence on bronchial cells obtained by suction. On frozen section, we found that this expression was very weak or absent. The addition of interferon-γ (INF-γ) upregulated HLA-DR expression (3). HLA-DR expression is enhanced during lung allograph rejection (36). Human ciliated bronchial epithelial cells are, however, ineffective in stimulating allogeneic T-cell proliferation (3). Thus a "second signal" appears missing from the cells. As for other epithelial cells, in addition to the induction of DR antigen expression, phorbol myristate acetate (PMA) must be added for bronchial cells to stimulate resting allogeneic T cells (3). One cannot rule out that in some clinical conditions, bronchial epithelial cells could act as APCs, particularly for amplifying the activation of already activated T cells. Alternatively, in the in vivo function of MHC class II molecules on epithelial cells may not be for antigen presentation but rather for transport of degraded peptides across the epithelial layer. The expiratory epithelium could act as an immunological "filter." It is also possible that MHC class II antigens modulate the function of plasma cells within the bronchial mucosa, as postulated for other epithelia (37).

Very few data are available on type II pneumocytes. These cells appear not to function as APCs, and they may even suppress T-cell proliferation (38). However, they appear to release factors such as GM-CSF capable of influencing the function of nearby DCs (39).

Fibroblasts

IFN-γ can induce the expression of MHC class II molecules on lung fibroblasts (40) as well as on dermal fibroblasts (4,41). Although fibroblasts can present antigen efficiently to T-cell clones, antigen presentation by fibroblasts to resting T cells requires the addition of exogenous IL-2. Fibroblasts, like bronchial epithelial cells, differ from classic APCs by also lacking the "second signal" necessary to induce the production of IL-2 in resting T cells (41). There is, however, good evidence that fibroblasts process and display antigens (4). Thus, they may be able to maintain the proliferation of already activated T cells.

Endothelial Cells

Endothelial cells appear to be one of the nonlymphoid cells most capable of acting as APCs once they express class II antigens (4). Umbilical endothelial cells are able to activate resting T cells, although weakly (42). No data are available on lung endothelial cells, but, by extrapolation, one would postulate that they not only play a role in T cell migration but may also stimulate resting or memory T cells to respond to antigens.

In summary, among the nonlymphoid cells able to express MHC class II molecules in response to IFN-γ, endothelial cells are the only ones known to be capable of activating resting T cells in response to nominal or alloantigens. However, nonlymphoid APCs may play an important immunological function in the presence of superantigens which do not require processing (43,44) or to maintain the proliferation of previously activated T cells.

III. Dendritic Cells and T-Cell Activation

As previously stated, DCs are much more potent stimulators of T cells than B cells and macrophages. Furthermore, they have the unique capacity to induce primary immune responses. DCs are potent stimulators of a primary MLR in both experimental animals and in humans (45). They are very effective in stimulating T cells to help B cells develop into plasma cells (46). DCs also appear to be the most potent stimulators of CD8 cytotoxic T lymphocytes (CTLs) (47). Both antigen-specific CTLs, as well as antigen nonspecific natural killer (NK) cells can be elicited by DCs (48). DCs can prime an immune response in vivo even in the absence of adjuvants (49).

A. Antigen Presentation by DCs

Presentation of Exogenous Antigens

The immune response against exogenous protein antigens is initiated by the presentation of small antigenic peptides (13–25 amino acids) in the context of MHC class II molecules. A possible restriction in antigen processing by DCs is they may be only weakly phagocytic. This property has been used to purify DCs, in particular human lung DCs, by separating phagocytic macrophages, rich in autofluorescent phagocytic inclusions, from lung DCs which have little or no autofluorescent inclusions (50). Although micropinocytosis (uptake of small vesicles of less than 0.1 μm in diameter) is used by DCs to take up foreign antigens, a cooperative function of macrophages is also possible. Thus, macrophages might degrade particulate antigens and process them for DCs in the vicinity (51).

Recently, human peripheral blood precursors of DCs have been cultured in vitro using GM-CSF and IL-4, which permits DCs to be maintained in a state in which they are competent to capture and process antigen (52). These DCs express mannose receptors for the efficient capture of a variety of antigens. DCs were shown to be capable of a high level of macropinocytosis, taking up large amounts of fluid and concentrating the solutes in a compartment that contained MHC class II molecules and proteases (53). Mannose receptors bind various types of sugars such as mannose, fucose, N-acetyl-glucosamine as well as hydrophobic molecules. Furthermore, the mannose receptors release ligands at low pH and can recycle to the cell surface. In this way, mannose receptors can deliver ligands in consecutive rounds in quantities that far exceed the number of receptors.

Accumulating data indicate that the capacity to internalize and process antigen is a constitutive property of immature DCs. Evidence was obtained in mice that progenitors to DCs can internalize particles, including the mycobacterium, bacillus Calmette-Guérin (BCG) (54). This property is rapidly lost when DCs, after in vivo antigenic challenge, move to T-cell–dependent areas of secondary lymphoid organs. Using human DCs generated in culture with GM-CSF and

IL-4, it was found that TNF-α, CD40L, IL-1, and LPS induced a coordinate series of changes that included a complete downregulation of macropinocytosis and an upregulation of adhesion and costimulatory molecules. These changes occurred within 24–48 hrs and were irreversible (53).

Freshly isolated human blood DCs have also been shown to have a greater capacity to endocytose colloidal gold–labeled bovine serum albumin as compared with cultured, mature DCs. Gold-labeled bovine serum albumin is transported within 30 mins into MHC class II–positive endosomal compartments, presumably the site of antigen degradation and peptide loading onto MHC class II molecules. Thus, DCs appear to use similar subcellular compartments as B cells (55). In rats, pulmonary DCs display APC activity at 3 hrs which lasts up to 7 days after the injection of hen egg lysozyme (HEL) into the trachea. After primary sensitisation, HEL-immune T cells were detected in hilar lymph nodes but not in the lung, implying the need of a lymph node structure to prime T cells. However, a second airway challenge with HEL elicited an antigen-specific pulmonary immune response, probably from circulating preprimed T cells (56). In humans, whether lung DCs present soluble antigens and whether this ability changes in various states with maturation is unsettled. (57).

Presentation of Endogenous Antigens

Endogenous antigens from intracellular pathogens such as viruses typically follow a distinct processing pathway and are expressed on the cell surface in association with class I MHC molecules, permitting their recognition by $CD8^+$ T lymphocytes (58). However, endogenous antigens can sometimes be associated with MHC II molecules (59,60). Splenic DCs can be infected with a number of viruses, including influenza virus or human immunodeficiency virus (HIV), which can prime virus-specific $CD8^+$ cytotoxic lymphocytes (61,62). Infection of human DCs by HIV has been shown to play a role in the pathogenesis of acquired immunodeficiency syndrome (AIDS) either by decreasing the APC function of DCs (63) or by persistently stimulating antibody responses to HIV in seropositive individuals (64).

Presentation of Superantigens

Superantigens share several important features that distinguish them from conventional antigens. Superantigens are potent mitogens for both human and mouse T cells. Thus, they typically stimulate large numbers of T cells bearing particular T-cell receptor (TCR) Vβ gene products. Superantigens appear to bind to sites on MHC class II molecules that are located outside the peptide-binding groove (65).

Microbial superantigens are of major clinical interest, because they cause a spectrum of disease in both humans and animals. For instance, in humans, proteins derived from *Staphylococcus aureus* induce gastroenteritis, toxic shock syndrome, and the scaled skin syndrome. The whole spectrum of diseases induced by superantigens is not yet known. For instance, in rodents, mycoplasmas are known to produce a superantigen-induced, chronic inflammatory polyarthritis (66). It is possible by analogy that mycoplasmas induce airway inflammation in human by similar mechanisms.

Human DCs are 10- to 50-fold more efficient than monocytes or B cells in presenting superantigens and require only trace levels of superantigens to initiate T-cell responses (67). DCs pulsed with 0.005 mM (5 pM) of SEA effectively stimulate T cells at ratios of 1:100-1:1000, whereas monocytes or B cells pulsed with up to 50 mM SEA induced lower responses at APC/T-cells ratios of 1:30. Studies with long-term APCs and T-cells lines suggest that the amount of MHC peptide complexes could be very small (200–600/APC) and the extent of MHC occupancy likewise $\leq 0.1\%$ to stimulate responses (68).

B. DCs and Soluble Costimulatory Signals

Cytokines can either increase the APC function of DCs or can be released by DCs and act directly on T cells.

Cytokines that Enhance APC Function of DCs

The first cytokine described to be able to increase the APC function of DCs was IL-1. Koide et al. observed that IL-1 enhanced the capacity of mice DCs to cluster with T cells before the onset of mitogenesis, but that DCs exhibited little or no IL-1 activity in the D10 lymphoproliferation bioassay (69). IL-1 in combination with GM CSF, macrophage colony-stimulating factor (M-CSF), and IL-3 has been shown to favor proliferation of DCs from CD34$^+$ progenitors in human bone marrow (70). Human CD 34$^+$ hematopoietic progenitor cells do not bear CD1a (a molecule specifically expressed on Langerhans' cells and thymocytes). However, culture for 12 days of these progenitor cells with GM-CSF resulted in the generation of 5–15% CD1a$^+$ cells. The addition of TNF-α to GM-CSF in cultures of human DC progenitors increased the total cell number and the proportion of CD1a$^+$ cells and resulted in a 10- to 20-fold increase in viable CD1a cells with a good capacity to activate allogeneic T cells (71). TNF-α or TNF-β both act at an early proliferative phase of progenitor cell development in order to influence the development of DCs (70). This may explain our failure to grow differentiated DCs from human minced lung using GM-CSF and TNF-α (unpublished data). GM-CSF alone had been shown to promote differentiation and improve survival of human peripheral blood DCs without changing their APC function (72).

DC lines obtained from peripheral blood mononuclear cells and growth with GM-CSF and TNF-α were shown to stimulate allogeneic T cells. However, they were unable to present soluble antigen as compared with the lines obtained from cultures in which both GM-CSF and IL-4 were added. The response of DCs to

TNF-α include a two- to three-fold increase of MHC class II, ICAM-1, LFA-3, CD40, and B-7 antigens as occurs with IL-4, but the expression of invariant chain and FcγRII was decreased. The functional consequence of stimulating DCs with TNF-α was an increased T-cell stimulatory capacity in an MLR but a 10-fold decrease in the presentation of soluble tetanus toxoid (TT) and a 100-fold decreasing presentation of TT-IgG complexes (52). Furthermore, this study indicated that DCs cultured with GM-CSF and IL-4 were 100- to 300-fold more efficient than polyclonal B cells or peripheral blood mononuclear cells (PBMCs) in presenting soluble antigens. DCs activated by GM-CSF and IL-4 were able to present concentrations of TT as low as 10^{-10}M. In the presence of immune complexes, DCs were also more efficient than antigen-specific B cells and were capable of presenting TT at concentrations of 10^{-12} M. Another group succeeded in generating larger numbers of $CD1a^{+}$ cells from human peripheral blood progenitor cells in vitro using GM-CSF and IL-4 as compared with GM-CSF and TNF-α (73).

GM-CSF is expressed by rat alveolar epithelial cells and increases the stimulatory activity of lung DCs in an allogeneic MLR (39). Similar phenomena may occur for human lung DCs, and the local release of IL-4 or TNF-α may have additional crucial importance in regulating both migration and antigen presentation of lung DCs.

The Role of Cytokines Released by DCs in Activating T Cells

Cytokines such as IL-1α or IL-1β (74), IL-6 (75), or TNF-α (76) have all been described as providing T-cell costimulatory signals. For many years, an important hypothesis for the mechanism of T cell activation was that macrophage-derived IL-1 induced IL-2 release and responsiveness. Although active APCs, DCs, produce little or no IL-1 (69,77,78). Using a sensitive enzyme immunoassay for IL-1α and IL-1β, lung DCs demonstrated far less surface-bound and secreted IL-1 than monocytes. Furthermore, IL-1 was produced in relatively small quantities during a DC-induced MLR as compared with monocytes. In addition, blocking antibodies to either IL-1α or IL-1β did not interfere with human lung DC-induced T-cell proliferation. Similar findings were made using human blood DCs (79). However, in contrast to the findings with DCs, the MLR induced by lung AMs was inhibited by blocking antibodies to IL-1 (78).

TNF-α was also shown to enhance anti-CD3–activated T cell proliferation as well as the response of T cells to antigen-bearing paraformaldehyde-fixed APCs (80). The receptor involved in murine as well as human thymocyte stimulation appeared to be the TNFsR-75 (81,82). TNF-α induces IL-2 receptor and gene-regulatory factors, such as c-*fos* or NF-κB in T lymphocytes (83,84). For human lung DCs, TNF-α production is unlikely completely to explain why these cells are such potent APCs, because LPS-stimulated DCs produce less TNF-α than do AMs (78). However, using high doses of a TNF inhibitor, a fusion protein recombinant

TNFsR-p55hγ3, we were able markedly to reduce T-cell proliferation and Th1 and Th2 cytokine production induced by DCs, monocytes, or anti-CD3 antibodies. This study supports the importance of TNF-α as a costimulatory factor in all of these situations (85).

Human peripheral blood DCs appear to be poor producers of IL-6 (86). Blocking antibodies against IL-6 does not decrease a DC-induced MLR , suggesting that this costimulatory signal is unlikely to be of major importance in the APC function of DCs, at least during an alloreaction (78). The precise role of IL-6 as well as of IL-1 in the activation of distinct subtypes of T cells remains to be clarified.

IL-12, a heterodimeric cytokine, has many important biological properties that relate to DC-induced T cell stimulation. It enhances the proliferation of activated NK and T cells and induces the secretion of IFN-γ (87). Mouse DCs produce IL-12, directing the development of Th1 cells from naive $CD4^+$ T cells (88). IL-12 has also been found to induce the production of IFN-γ by neonatal human CD4 T cells (89). The relative capacity of human macrophages or DCs to produce bioactive IL-12 under various stimuli has not been reported. In mice, adding protein antigen to DCs in a particle-absorbed form, as compared with a soluble form, induced de novo synthesis of IL-12 p 40/p35 and MHC class II transcripts and, most importantly, caused augmentation and more prolonged antigen-presentation capacity (90).

C. Adhesion Molecules and DC–T-cell Interactions

When T cells are cultured in the presence of DCs, typical cell aggregates or clusters become apparent (78,91). At this critical stage, the physical association of the two cells seems to be due to an antigen-independent mechanism (92). Subsequently, the specific recognition by the TCR of MHC molecules displaying antigenic peptides is facilitated.

Role of Integrins in DC–T-cell Interactions

Studies of the molecular basis of the interaction between T cells and DCs have shown that adhesion-dependent interactions are partially mediated through binding of integrins. Integrins are heterodimers consisting of noncovalently associated α and β units (93), which mediate both cell-substratum and cell-cell adhesion. For instance, the heterodimer of the α chain (CD11a) and the β 2 chain (CD18) interacts with ICAM-1. This interaction improves the adhesion of interacting DCs and T cells and, in association with TCR activation, delivers costimulatory signals to lymphocytes (94). There are at least seven different β subunits. Interest has focused on the expression and function of β_1- and β_2-integrins on DCs.

The β_1-integrin family includes receptors that bind to extracellular matrix. Analogous to β_2-integrin antibodies, monoclonal antibodies against β_1-integrins affect stimulation of T-cell proliferation initiated by antibodies to CD3 (95). Antibodies to β_1-integrins reduce DC-induced alloreactions to the same extent as monoclonal antibodies against β_2-integrins do, with a synergistic effect when antibodies to both integrins are combined (96). Elevated levels of VLA-5 are found on both DCs and monocytes, and the mean fluorescence intensity is only slightly increased on DCs compared with monocytes. VLA-4 is only weakly present on lung DCs and its ligand, vascular cell adhesion molecule (VCAM), is not detected (96).

Among the β_2-integrins, LFA-1 (CD11a/CD18) is decreased on human lung DCs as compared with blood monocytes. However, its ligand, ICAM-1, is more highly expressed on lung and blood DCs as compared with monocytes (97). In blocking experiments, antibodies against LFA-1 or ICAM-1 only partially decrease T-cell proliferation (96). The low efficacy of these antibodies in blocking an alloreaction could be partly related to the presence of ICAM-2 and/or ICAM-3 on DCs. However, the β_2-integrins are important in DC–T-cell interactions, because monoclonal antibodies against the common β_2 chain or the α chain of CD11c (p150,95) markedly decreases T-cell proliferation. The α chain of CD11c is more highly expressed on DCs than are CD11a or CD11b. In mouse spleen, DCs express the highest levels of CD11c (98) of any leukocyte. In human lung, however, CD11c is expressed to the same extent as it is on monocytes. The same appears true for human blood DCs as compared with monocytes (97).

Role of LFA-3

LFA-3 is highly expressed on peripheral blood and lung DCs compared with monocytes (97). CD2 is absent from DCs. LFA-3 which binds to T-cell CD2 participates in the binding of T cells and DCs, stabilizing cell clusters (99,100). LFA-3 and CD2 also participate in T-cell activation via an interaction with the TCR-CD3 complex (101). The profound inhibition observed when antibodies against either LFA-3 or CD2 are used during DC–T-cell interactions (96) suggests that molecules could be targeted by potential therapeutic agents.

Role of the B7 Family

The importance of the B7 family costimulation in T-cell–APC cross talk was indicated in studies where T-cell stimulation in the absence of B7-CD28 cross linking led to T-cell anergy (102). With the possible exception of keratinocytes (103), B7 expression appears to be limited to cells of hematopoietic origin. Both CD28 and the closely related CTLA-4 molecule bind members of the B7 family of counterreceptors, B7-1(CD80), and B7-2 (CD86).

CD80 was shown to be constitutively expressed by human blood DCs and was increased after binding to allogeneic T cells (104). However, another group found that short-term culture was needed to detect of CD80 and CD86 on human

Table 2 Expression of Important Cell Surface Glycoproteins on Human Lung Dendritic Cells and Blood Monocytes

Glycoproteins	DCs	Monocytes
MHC class I	+++	+
MHC class II	+++	+
β_1-integrins: β_1 chain (CD29)	+++	+
VLA-4 (CD49D)	+	−
VLA-5 (CD49E)	+++	++
VLA-6 (CD49F)	+	−
β_2-integrins: β_2 chain (CD18)	+++	+++
LFA-1 (CD11a)	+	++
CR-3 (CD11b0	+	+
p150/90 (CD11c)	++	++

blood DCs (105). Blocking experiments generally show that CD86 is the predominant functional ligand as compared with CD80 (105). It is possible that the two molecules transmit different signals to T lymphocytes, and perhaps they may even regulate the development of a Th1 versus Th2 response (106). Further, some evidence suggests that interaction with CTLA-4 may mediate a negative signal (107). On human lung DCs, both CD80 and CD86 are expressed to a low degree after 18 hrs of culture. The minimal expression of CD28/CTLA-4 ligands on freshly isolated blood DCs suggests that these cells only acquire full APC function in vivo following exposure to the appropriate cytokine and cellular environment.

Various important surface glycoproteins expressed on human DCs are summarized in Table 2. Their functional importance, as tested with monoclonal antibodies in alloreactions, are indicated in Table 3.

IV. Macrophages and Dendritic Cells

AMs constitute 85% of the cells in the alveoli, whereas DCs account for no more than 1% of the cells in this lung compartment. AMs phagocytose and neutralize foreign organisms and particles that reach distal airways and produce oxygen radicals, cytokines, lysozyme, complement, and many other substances. They can process antigens, but as previously discussed, they are usually poor APCs. In addition, the alveolar milieu, which contains about 10% lymphocytes, appears to be a hostile environment for lymphocytes. BAL lymphocytes have been shown to be hypoactive in proliferative and antibody responses to mitogens and in NK activity as compared with blood-derived lymphocytes. In this regard, several

Table 3 Effect of Antibodies Against (or Soluble Ligands of) Surface Glycoprotein on Dendritic Cell–Induced T-Cell Proliferation

Monoclonal antibody raised against	Inhibition of alloreaction
B7-1,2,3 (using the soluble ligand CTLA-4Ig)	+++ (80–90%)
LFA-3 (CD58)	+++
LFA-2 (CD2)	+++
β_1 chain (CD29)	++ (40%)
β_2 chain (CD18)	++
$\beta_1 + \beta_2$ chains (CD29 and CD18)	+++ (80–90%)
CD11c	++ (70%)
ICAM-1, CD11a, or CD11b	+ (30%)
VLA-4, VLA-5, or VLA-6	0

investigators have found that surfactant lipids have a profound suppressive effect on lymphocyte proliferation and effector function in vivo (108). As previously discussed, the in vivo importance of the suppressive factors released by AMs or of the direct action of surfactant is as yet uncertain, but AM suppression might prevent the lung from experiencing undue inflammation in order to protect gas exchange. Meanwhile, if antigens penetrate the interstitium where most DCs are located and are protected from the alveolar milieu, an immune response should be initiated.

A. Macrophages as Coprocessing Cells

The weak capacity of DCs to phagocytose particles and organisms has already been discussed. This diminished function contrasts with the capacity of macrophages to endocytose particles and degrade them in both human and in animal models (109). AMs and tissue macrophages provide protection against invading organisms or antigens. However, they could also process antigens and release part of them in the vicinity of DCs either in lung tissue or in lymph nodes (110) so that DCs can more readily initiate T-cell proliferation.

B. Effect of AM-Produced IL-1 or TNF Inhibitors on DC Function

AMs not only produce IL-1ra (111), they also release TNF-sR that neutralizes TNF, and can thus regulate immune and inflammatory processes. The final biological result depends on the relative amount of proinflammatory cytokines and the respective inhibitors (112). IL-1ra is a polypeptide of 17-kDa and has 26%

identity with IL-1β and 19% identity with IL-1α (113). IL-1ra apparently has no agonist activity, but it inhibits the functions induced by both IL-1α and IL-1β. The kinetics of the production of IL-1ra, IL-1α, and IL-1β mRNA in AMs reveals that in the presence of PMA, the expression of IL-1α or IL-1β occurs after 15 mins, whereas the expression of IL-1ra mRNA is delayed and detected after 3 hrs (114). To induce IL-1ra mRNA in peripheral blood monocytes, PMA alone is insufficient and a second signal is required such as complexes of IgG or IL-4 (115). Although the anti-inflammatory role of IL-1ra has been clearly demonstrated, evidence was lacking that antigen presentation or an allogeneic reaction induced by DCs was decreased by even high doses of IL-1ra (116). This could be because the occupancy of a few IL-1 receptors at the interface of APCs and T cells is sufficient to provide a second signal. Alternatively, molecules other than IL-1 might transduce similar signals.

Human AMs are capable of the enhanced production of TNF-α, unlike blood monocytes (117). The shedding of TNF-sR may be crucial in maintaining the alveoli free of deleterious inflammatory processes. The presence of TNF inhibitor in the supernatant of AMs has been reported for some time (112). The link between this inhibitory activity and the shedding of TNF-sRs by AMs has been shown by using a specific immunoassay measuring the two soluble fragments of TNF-R, TNF-sR75 and TNF-sR55 (118). After 48 hrs in culture, AMs release similar amounts of each TNF-sR. In the presence of PMA, the shedding of TNF-sR75 from both AMs and monocytes was four- and fivefold higher respectively, than TNF-sR55 after 48 hrs of stimulation. The TNF-sR75 released by monocytes was twice that of AMs. Thus, monocytes produce less TNF-α and more TNF-sR than AMs (16). A 5- to 10-fold excess of TNF-sR can decrease the inflammatory reactions induced by TNF-α or lymphotoxin (TNF-β). However, a thousandfold higher concentration of these inhibitors is required to influence APC–T-cell interactions in vitro. These pharmacological concentrations may be required to block membrane-bound TNF on mononuclear cells. Thus, when DCs are exposed to 10 μg/ml of TNF-sR, their capacity to activate T cells decreases significantly, emphasizing the importance of TNF as secondary signal (85). Nevertheless, it is possible that TNF-sR might influence DC function in vivo, perhaps by preventing maturation in lung tissue.

C. Effect of IL-10 on DCs

IL-10, a cytokine first identified as a product of cloned Th2 cell populations, is produced by many cell types but mainly by monocytes and macrophages. IL-10 belongs to the group of mediators which are able to suppress inflammation. By suppressing synthesis of both proinflammatory cytokines and the expression of surface antigens on APCs (118), IL-10 can decrease alloreactions induced by human DCs and, even more efficiently, decrease the release of inflammatory cytokines during these alloreactions such as TNF-α or lymphotoxins (119). IL-10 downregulates the expression of MHC class II antigens, prevents the expression of B7 (120,121), and may also be responsible for enhancing DC apoptosis (122). IL-10 also prevents T-cell activation by anti-CD3 antibodies in the absence of APCs, implying a direct inhibition of T cells (123,124). Overall, the modulation of IL-10 production by AMs or interstitial lung macrophages must be better understood in order to assess the importance of this cytokine in vivo for the control of lung DC function.

V. Conclusions

DCs are the most powerful APCs in human lung. They form a network around the airways, in the loose connective tissue around the vessels, and in the pleura. Their number and function is tightly regulated by mediators released by several types of cells in their immediate environment, including epithelial cells, monocytes, and macrophages. The growth factor, GM-CSF, in combination with IL-1, TNF-α, and IL-4, is crucial for the maturation of DCs and for their expression of the essential accessory molecules required to facilitate the adherence and activation of resting T cells. Continuing studies on DC regulation in human lung are important in order to come to grips with controlling unwanted hypersensitivity reactions and transplant rejection and yet maintaining effective host responses against infection.

Acknowledgment

This work was supported by grants of the Swiss National Science Foundation and the Lancardis Foundation.

References

1. Barclay AN, Mason DW. Induction of Ia antigen in rat epidermal cells and gut epithelium by immunological stimuli. J Exp Med 1982; 156:1665–1676.
2. Nunez G, Ball EJ, Stastny P. Accessory cell function of human endothelial cells. I/A subpopulation of Ia positive cells is required for antigen presentation. J Immunol 1983; 131:666–673.
3. Rossi GA, Sacco O, Balbi B, Oddera S, Mattioni T, Corte G, Ravazzoni C, Allegra L. Human ciliated bronchial epithelial cells: expression of the HLA-DR antigens and of the HLA-DR alpha gene, modulation of the HLA-DR antigens by gamma-interferon and antigen presenting function in the mixed leukocyte reaction. Am J Respir Cell Mol Biol 1990; 3:431–439.
4. Geppert TD, Lipsky PE. Antigen presentation by interferon-gamma–treated endothelial cells and fibroblasts: differentiation ability to function as antigen-presenting cells despite comparable Ia expression. J Immunol 1985; 135:3750–3762.

5. Nicod LP, Lipscomb MF, Toews GB. Presence of DR-positive dendritic cells in the lung. Clin Res 1986; 34:202A.
6. Sertl K, Takemura T, Tschachler E, Ferrans VJ, Kaliner MA, Shevach EM. Dendritic cells with antigen presenting capability reside in airway epithelium, lung parenchyma and visceral pleura. J Exp Med 1986; 163:436–451.
7. Shon-Hegrad MA, Oliver J, McMenamin PG, Holt PG. Studies on the density, distribution and surface phenotype of intraepithelial class II MHC antigen (Ia)–bearing dendritic cells (DC) in the conducting airways. J Exp Med 1992; 173:1345–1356.
8. Weinberg DS, Unanue ER. Antigen-presentating function of alveolar macrophages: uptake and presentation of Listeria monocytogenes. J Immunol 1981; 126:794–799.
9. Seeger RC, Oppenheim JJ. Synergistic interaction of macrophages and lymphocytes in antigen-induced transformation of lymphocytes. J Exp Med 1970; 132:44–65.
10. Mason R, Austyn J, Brodsky F, Gordon S. Monoclonal antimacrophage antibodies: human pulmonary macrophages express HLA-DR (Ia-like) antigens in culture. Am Rev Respir Dis 1982; 125:586–593.
11. Toews GB, Vial WC, Dunn MM, Guzzetta P, Nunez G, Stastny P, Lipscomb MF. The accessory cell function of human alveolar macrophages in specific T cell proliferation. J Immunol 1984; 132:181–186.
12. Liu MC, Proud D, Schleimer RP, Plaut M. Human lung macrophages enhance and inhibit lymphocyte proliferation. J Immunol 1984; 132:2895–2903.
13. DeShazo RD, Banks DE, Diem JE, Nordberg JA, Baser Y, Bevier D, Salvaggio JE. Bronchoalveolar lavage cell-lymphocyte interactions in normal nonsmokers and smokers. Am Rev Respir Dis 1983; 127:545–548.
14. Lipscomb MF, Lyons CR, Nunez G, Ball EJ, Stastny P, Vial WC, Lem VM, Weissler JC, Miller LM, Toews GB. Human alveolar macrophages: HLA-DR–positive macrophages that are poor stimulators of a primary mixed leukocyte reaction. J Immunol 1986; 136:497–504.
15. Elias JA, Schreiber AD, Gustilo K, Chien P, Rossman MD, Lammie PJ, Daniele RP. Differential interleukin 1 elaboration by unfractionated and density fractionated human alveolar macrophages and blood monocytes: relationship to cell maturity. J Immunol 1985; 135:3198–3204.
16. Nicod LP, Galve-de-Rochemonteix G, Dayer JM. Modulation of IL-1 receptor antagonist and of TNF-soluble receptors produced by alveolar macrophages and blood monocytes. Ann NY Acad Sci 1994; 725:323–330.
17. Moore SA, Strieter RM, Rolfe MW, Standiford TJ, Burdick MD, Kunkel ST. Expression and regulation of human alveolar macrophage–derived interleukin-1 receptor antagonist. Am J Respir Cell Mol Biol 1992; 6:569–575.
18. Springer TA, Dustin ML, Kishimoto TK, Marlin SD. The lymphocyte function-associated LFA-1, CD2, and LFA-3 molecules: cell adhesion receptors of the immune system. Annu Rev Immunol 1987; 5:223-252.
19. Metzger ZVI, Hoffeld JT, Oppenheim JJ. Macrophage-mediated suppression. J Immunol 1980; 124:983–988.
20. Lyons CR, Ball EJ, Toews GB, Weissler JC, Stastny P, Lipscomb MF. Inability of human alveolar macrophages to stimulate resting T cells correlates with decreased antigen specific T cell-macrophage binding. J Immunol 1986; 137:1173–1180.

21. VanKooyk Y, Van de Wiel-Van Kemenade P, Weder P, Kuijpers TW, Figdor CG. Enhancement of LFA-1–mediated cell adhesion by triggering through CD2 or CD3 on T lymphocytes. Nature 1989; 342:811–813.
22. Chelen CJ, Fang Y, Freeman GJ, Secrist H, Marshall JD, Hwang Pt, Frankel LR, DeKruyff RH, Umetsu DT. Human alveolar macrophages present antigen ineffectively due to defective expression of B7 costimulatory cell surface molecules. J Clin Invest 1995; 95:1415–1421.
23. Ferro TJ, Monos DS, Spear BT, Rossman MD, Zmijewski CM, Kamoun M, Daniele RP. Carbohydrate differences in HLA-DR molecules synthesized by alveolar macrophages and blood monocytes. Am Rev Respir Dis 1987; 135:1340–1344.
24. Basset F. Soler P, Wyllie L, Mazin F, Turiaf J. Langerhans' cells and lung interstitium. Ann NY Acad Sci 1976; 278:599–611.
25. Stingl G, Tamaki K, Katz SI. Origin and function of epidermal Langerhans cells. Immunol Rev 1980; 53:149–174.
26. Murphy GF, Messadi D, Fonferko E, Hancock W. Phenotypic transformation of macrophages to Langerhans' cells in the skin. Am J Pathol 1986; 123:401–406.
27. Inaba K, Schuler O, Witmer MD, Valinsky J, Atassi B, Steinman RM. Immunologic properties of purified epidermal Langerhans cells. Distinct requirements for stimulation of unprimed and sensitized T lymphocytes. J Exp Med 1986; 164:606–613.
28. Romani N, Koide S, Crowley M, Witmer-Pack M, Livingstone AM, Fathman CG, Inaba K, Steinman RM. Presentation of exogenous protein antigens by dendritic cells to T cell clones. Intact protein is presented best by immature, epidermal Langerhans cells. J Exp Med 1989; 169:1169–1178.
29. Bellini A. Vittori E, Marini M, Ackerman V, Mattoli S. Intraepithelial dendridic cells and selective activation of TH2-like lymphocytes in patients with atopic asthma. Chest 1993: 103:997–1005.
30. Nussenzweig MC, Steinman RM, Unkeless JC, Witmer MD, Gutchinov B, Cohn ZA. Studies of the cell surface of mouse dendritic cells and other leukocytes. J Exp Med 1981; 154:168–187.
31. Roehm NW, Leibson HJ, Zlotnik A, Kappler J, Marrack P, Cambier JC. Interleukin-induced increase in Ia expression by normal mouse B cells. J Exp Med 1984; 160: 679–694.
32. Monroe JG, Cambier JC. B cell activation. III/B cell plasma membrane depolarization and hyper-Ia antigen expression induced by receptor immunoglobulin crosslinking are coupled. J Exp Med 1983; 158:1589–1599.
33. Chesnut RW, Grey HM. Studies on the capacity of B cells to serve as antigen-presenting cells. J Immunol 1981; 126:1075–1079.
34. Metlay JP, Puré E, Steinman RM. Distinct features of dendritic cells and anti-Ig activated B cells as stimulators of the primary mixed leukocyte reaction. J Exp Med 1989; 169:239–254.
35. Natali PG, DeMartino C, Quaranta V, Nicotra MR, Frezza F, Pellegrino MA, Ferrone S. Expression of Ia-like antigens in normal human nonlymphoid tissues. Transplantation 1981; 31:75–78.
36. Romaniuk A, Prop J, Petersen AH, Wildevuur CR, Niewenhuis P. Expression of class II major histocompatibility complex antigens by bronchial epithelium in rat lung allografts. Transplantation 1987; 44:209–214.

37. Bland P. MHC class II expression by the gut epithelium. Immunol Today 1988; 9: 174–178.
38. Chavis A, Mody CH, Spahr MA, Toews GB, Paine R. Type II cell monolayers reversibly inhibit rat splenocyte proliferation. Am Rev Respir Dis 1990; 141:A91.
39. Armstrong LR, Christensen PJ, Paine R 3rd, Chen GH, McDonald RA, Lim TK, Toews GB. Regulation of the immunostimulatory activity of rat pulmonary interstitial dendritic cells by cell-cell interactions and cytokines. Am J Respir Cell Mol Biol 1994; 11:682–691.
40. Phipps RP, Penney DP, Keng P, Quill H, Paxhia A, Derdak S, Felch ME. Characterization of two major populations of lung fibroblasts: distinguishing morphology and discordant display of Thy 1 and class II MHC. Am J Respir Cell Mol Biol 1989; 1: 65–74.
41. Umetsu DT, Katzen D, Jabara HH, Geha RS. Antigen presentation by human dermal fibroblasts: activation of resting T lymphocytes. J Immunol 1986; 136:440–445.
42. Geppert TD, Lipsky PE. Dissection of the antigen presenting function of tissue cells induced to express HLA-DR by gamma interferon. J Rheumatol 1987, 14:59–62.
43. Irwin MJ, Gascoigne NRJ. Interplay between superantigens and the immune system. J Leukoc Biol 1993; 54:495–503.
44. Nickoloff RM, Mitra RS, Green J, Zheng XG, Shimizu Y, Thompson C, Turka LA. Accessory cell function of keratinocytes for superantigens. J Immunol 1993; 150: 2148–2159.
45. Steinman RM, Gutchinov B, Witmer MD, Nussenzweig MC. Dendritic cells are the principal stimulators of the primary mixed leukocyte reaction in mice. J Exp Med 1983; 157:613–627.
46. Inaba K, Steinman RM. Protein-specific helper T-lymphocyte formation initiated by dendritic cells. Science 1985; 229:475–479.
47. Inaba K, Young JW, Steinman RM. Direct activation of $CD8^+$ cytotoxic T lymphocytes by dendritic cells. J Exp Med 1987; 166:182–194.
48. Young JW, Steinman RM. Dendritic cells stimulate primary human cytolytic lymphocyte responses in the absence of $CD4^+$ helper T cells. J Exp Med 1990; 171: 1315–1332.
49. Inaba K, Metlay JP, Crowley MT, Steinman Rm. Dendritic cells pulsed with protein antigens in vitro can prime antigen-specific, MHC-restricted T cells in situ. J Exp Med 1990; 172:631–640.
50. Nicod LP, Lipscomb MF, Toews GB, Weissler JC. Separation of potent and poorly functional human lung accessory cells based on autofluorescence. J Leukoc Biol 1980.
51. Van Rooijen N. Macrophages as accessory cells in the in vivo humoral immune response: from processing of particulate antigens to regulation by suppression. Semin Immunol 1992; 4:237–245.
52. Sallusto F, Lanzavecchia A. Efficient presentation of soluble antigen by cultured human dendritic cells is maintained by granulocyte/macrophage colony-stimulating factor plus interleukin 4 and downregulated by tumor necrosis factor α. J Exp Med 1994; 179:1109–1118.
53. Sallusto F, Cello M, Danieli C, Lanzavecchia A. Dendritic cells use macropinocytosis and the mannose receptor to concentrate macromolecules in the major

histocompatibility complex class II compartment: downregulation by cytokines and bacterial products. J Exp Med 1995; 182:389–400.
54. Inaba K, Inaba M, Naito M, Steinman RM. Dendritic cell progenitors phagocytose particulates including bacillus Calmette-Guerin organisms, and sensitive mice to mycobacterial antigens in vivo. J Exp Med 1993; 178:479–488.
55. Nijman HW, Kleijmeer MJ, Ossevoort MA, Oorshot VMJ, Vierboom MPM, van de Keur M, Kenemans P, Kast WM, Geuze HJ, Melief CJM. Antigen capture and major histocompatibility class II compartments of freshly isolated and cultured human blood dendritic cells. J Exp Med 1995; 182:163–174.
56. Xia W. Pinto CE, Kradin RL. The antigen-presenting activities of Ia^+ dendritic cells shift dynamically from lung to lymph node after an airway challenge with soluble antigen. J Exp Med 1995; 181:1275–1283.
57. Nicod LP, Lipscomb MF, Weissler JC, Toews GB. Mononuclear cells from human lung parenchyma support antigen-induced T lymphocyte proliferation. J Leukoc Biol 1989; 45:336–344.
58. Sweetser M.T. Morrison LA, Braciale VL, Braciale TJ. Recognition of pre-processed endogenous antigen by class I but not class II MHC-restricted T cells. Nature 1989; 342:180–182.
59. Nuchtern JG, Biddison WE, Klausner RD. Class II MHC molecules can use the endogenous pathways of antigen presentation. Nature 1990; 343:74–76.
60. Guéry J.C, Adorini L. Dendritic cells are the most efficient in presenting endogenous naturally processed self-epitopes to class II restricted T cells. J. Immunol 1995; 154:536–544.
61. Nonacs R, Humborg C, Tam JP, Steinman RM. Mechanisms of mouse spleen dendritic cell function in the generation of influenza-specific, cytolytic T lymphocytes. J Exp Med 1992; 18:219–223.
62. Langhoff E. Terwilliger EF, Bos HJ, Dalland KH, Poznansky MC, Bacon OM, Haseltine WA, Replication of human immunodeficiency virus type 1 in primary dendritic cell cultures. Proc Natl Acad Sci USA 1992; 88:7998–8002.
63. Macatonia SE, Gompels M, Pinching AJ, Patterson S, Knight SC. Antigen presentation by macrophages but not by dendritic cells in human immunodeficiency virus (HIV) infection. Immunology 1992; 75:576–571.
64. Roberts MS, Gompels M, Pinching AJ, Knight SC. Dendritic cells persistently stimulate antibody responses to HIV in seropositive individuals. AIDS 1994; 8: 1097–1101.
65. Dellabona P, Peccoud J, Kappler J, Marrack P, Benoist C, Mathis D. Superantigens interact with MHC class II molecules outside of the antigens groove. Cell 1990; 62: 1115–1121.
66. Cole BC, Ward JC. Mycoplasma as arthitogenic agens. In: The Mycoplasmas. Vol. II. New York: Academic Press, 1979:367–398.
67. Bhardwaj N, Friedman SM, Cole BC, Nisanian AJ. Dendritic cells are potent antigen-presenting cells for microbial superantigens. J Exp Med 1992; 175:267–273.
68. Bhardwaj N, Young JW, Nisanian AJ, Baggers J, Steinman RM. Small amounts of superantigen, when presented on dendritic cells, are sufficient to initiate T cell responses. J Exp Med 1993; 178:633–642.
69. Koide SL, Inaba K, Steinman RM. Interleukin 1 enhances T-dependent immune

responses by amplifying the function of dendritic cells. J Exp Med 1987; 165: 515–530.
70. Reid CDL, Stackpoole A. Meager A, Tikerpae J. Interactions of tumor necrosis factor with granulocyte-macrophage colony-stimulating factor and other cytokines in the regulation of dendritic cell growth in vitro from early bipotent $CD34^+$ progenitors in human bone marrow. J Immunol 1992; 149:2681–2688.
71. Caux C, Dezutter-Dambuyant C, Schmitt D, Banchereau J. GM-CSF and TNF-α cooperate in the generation of dendritic Langerhans cells. Nature 1992; 360: 258–261.
72. Markowicz S, Engleman EG. Granulocyte-macrophage colony-stimulating factor promotes differentiation and survival of human peripheral blood cells in vitro. J Clin Invest 1990; 85:955–961.
73. Mackensen A. Herbst B, Köhler G, Wolff-Vorbeck G, Rosenthal FM, Veelken H, Kulmburg P, Schaefer HE, Mertelsmann R, Lindermann A. Delineation of the dendritic cell lineage by generating large numbers of Birbeck granule–positive Langerhans cells from human peripheral blood progenitor cells in vitro. Blood 1992; 86:2699–2707.
74. Dinarello CA. Biology of interleukin 1. FASEB J 1988; 2:108–115.
75. Wong GG, Clark SC. Multiple actions of interleukin 6 within a cytokine network. Immunol Today 1988; 9:137–139.
76. Yokota S, Geppert TD, Lipsky PE. Enhancement of antigen- and mitogen-induced human T lymphocyte proliferation by tumor necrosis factor-alpha. J Immunol 1988; 140:531–536.
77. McKenzie JL, Prickett TCR, Hart DNJ. Human dendritic cells stimulate allogeneic T cells in the absence of IL-1. Immunol 1989; 67:290–297.
78. Nicod LP, Galve-de Rochemonteix B, Dayer JM. Dissociation between allogeneic T cell stimulation and interleukin-1 or tumor necrosis factor production by human lung dendritic cells. Am J Respir Cell Mol Biol 1990; 2:515–522.
79. Bhardwaj N, Lau LL, Friedman SM, Crow MK, Steinman RM. Interleukin 1 production during accessory cell-dependent mitogenesis of T lymphocytes. J Exp Med 1989; 169:1121–1136.
80. Plaetinck G, Combe MC, Corthesy P, Sperisen P, Kanamori H, Honjo T, Nabholz M Control of IL-2 receptor-α expression by IL-1, tumor necrosis factor, and IL-2. J Immunol 1990; 145:3340–3347.
81. Scheurich P, Thoma B, Ucer U, Pfizenmaier K. Immunoregulatory activity of recombinant human tumor necrosis factor (TNF)-alpha: induction of TNF receptors on human T cells and TNF-alpha–mediated enhancement of T cell responses. J Immunol 1987; 138:1786–1790.
82. Tartaglia LA, Weber RF, Figari IS, Reynolds C, Palladino Jr MA, Geoddel DV. The two different receptors for tumor necrosis factor mediate distinct cellular responses. Proc Natl Acad Sci USA 1991; 88:9292–9296.
83. Tartaglia LA, Rothe M, Hu YF, Geoddel DV. Tumor necrosis factor's cytotoxic activity is signaled by the p55 TNF receptor. Cell 1993; 73:213–216.
84. Grilli M. Chen-Tran A, Lenardo MJ. Tumor necrosis factor alpha mediates a T cell receptor–independent induction of the gene regulatory factor NF-kappa B in T lymphocytes. Mol Immunol 1993; 30:1287–1294.

85. Nicod LP, Isler P, Chicheportiche R, Songeon F, Dayer JM. Production of PGE2 and collagenase is inhibited by the rsTNFR-p55Hhγ3 fusion protein at concentrations hundred fold lower than those decreasing T cell proliferation. European Cytokine Network, in press.
86. Vakkila J, Sihvola M, Hurme M. Human peripheral blood-derived dendritic cells do not produce interleukin 1α, interleukin 1β , or interleukin 6. Scand J Immunol 1990; 31:345–352.
87. Brunda MJ. Interleukin-12. J Leukoc Biol 1994; 55:280–288.
88. Macatonia SE, Hosken NA, Litton M, Vieira P, Hsieh CS, Culpepper JA, Wysocka M, Trinchieri G, Murphy KM, O'garra A. Dendritic cells produce IL-12 and direct the development of TH1 cells from natue CD4+ T cells. J Immunol 1995; 154: 5071–5079.
89. Wu CH, Demeure C, Kiniwa M, Gately M, Delespesse G. IL-12 induces the production of IFN-γ by neonatal human CD4 T cells. J Immunol 1993; 151:1938–1949.
90. Cheider C, Mehlig M, Dienes HP, Reske K. Uptake of microparticle-absorbed protein antigen by bone marrow–derived dendritic cells results in up-regulation of interleukin-1α and interleukin 12 p40/p35 and triggers prolonged, efficient antigen presentation. Eur J Immunol 1995; 25:1566–1572.
91. Flechner ER, Freudenthal PS, Kaplan G, Steinman RM. Antigen-specific T lymphocytes efficiently cluster with dendritic cells in the human primary mixed-leukocyte reaction. Cell Immunol 1988; 111:183–195.
92. Inaba K, Romani N, Steinman RM, An Antigen-independent contact mechanism as an early step in T cell-proliferative response to dendritic cells. J Exp Med 1989; 10:527–542.
93. Hynes RO. Integrins, a family of cell surface receptors. Cell 1987; 48:549–555.
94. Dustin ML, Springer TA. T-cell receptor cross-linking transiently stimulates adhesiveness through LFA- l. Nature 1989; 341:619–624.
95. Van Noesel C, Miedema F, Brouwer M, de Rie MA, Aarden LA, van Lier RAW. Regulatory properties of LFA-1 alpha and beta chains in human T-lymphocyte activation. Nature 1988; 333:850–852.
96. Nicod LP, El Habre F. Adhesion molecules on human lung dendritic cells and their role for T-cell activation. Am J Respir Cell Mol Bio 1992; 7:207–213.
97. Freudenthal PS, Steinman RM. The distinct surface of human blood dendritic cells, as observed after an improved isolation method. Proc Natl Acad Sci USA 1990; 87: 7698–7702.
98. Steinman RM. The dendritic cell system and its role in immunogenicity. Annu Rev Immunol 1991; 9:271–290.
99. King PD, Katz DR. Human tonsillar dendritic cell–induced T cell responses: analysis of molecular mechanisms using monoclonal antibodies. Eur J Immunol 1989; 19:581–587.
100. Hauss P. Selz F, Cavazzana-Calvo M, Fischer A. Characteristics of antigen-independent and antigen-dependent interaction of dendritic cells with DC4$^+$ cells. Eur J Immunol 1995; 25:2285–2294.
101. Breitmeyer JB, Daley JF, Levine HB, Schlossman SF. The T11 (CD2) molecule is functionally linked to the T3/Ti T cell receptor in the majority of T cells. J Immunol 1987; 139:2899–2905.

102. Doherty TM. T-cell regulation of macrophage function C O Immunol 1995; 7: 400–404.
103. Nickoloff BJ, Mitra R, Lee K, Turka L, Green J, Thompson CB, Shimuzu Y. Discordant expression of CD28 ligands, BB-1 and B7 on keratinocytes in vitro and psoriatic cells in vivo. Am J Pathol 1993; 42:1428–1436.
104. Young JW, Koulova L, Soergel SA, Clark EA, Steinman RM, Dupont B. The B7/BB1 antigen provides one of several costimulatory signals for the activation of CD4+ T lymphocytoses by human blood dendritic cells in vitro. J Clin Invest 1992; 90:229–237.
105. McLellan AD, Starling GC, Williams LA, Hock BD, Hart DNJ. Eur J Immunol 1995; 25:2064–2068.
106. Kuchroo VK, Das MP, Brown JA, Ranger AM, Zamvii SS, Sobel RA, Weiner HL, Nabavi N, Glimcher LH. B7-1 and B7-2 costimulatory molecules activate differentially the THI/TH2 development pathways: application to autoimmune disease therapy. Cell 1995; 80:707–718.
107. Gribben JG, Freeman GJ, Boussiotos VA, Rennert P, Jellis CL, Greenfield E, Barber M, Restivo VA Jr, Ke X, Gray GS, Nadler LM. CTLA4 mediates antigen-specific apoptosis of human T cells. Proc Natl Acad Sci USA 1995; 92:811–815.
108. Wilsher ML, Hughes DA, Haslam PL. Immunoregulatory properties of pulmonary surfactant: effect on lung lining fluid on proliferation of human blood monocytes. Thorax 1988; 43:354–359.
109. Witmer-Pack MD, Crowley MT, Inaba K, Steinman RM. Macrophages, but not dendritic cells, accumulate colloidal carbon following administration in situ. J Cell Sci 1993; 105:965–973.
110. Kapsenberg ML, Teunissen MBM, Stiekema FEM, Keizer HG. Antigen-presenting cell function of dendritic cells and macrophages in proliferative T cell responses to soluble and particulate antigens. Eur J Immunol 1986; 16:345–350.
111. Galve-de Rochemonteix B, Nicod LP, Junod AF, Dayer JM. Characterization of a specific 20- to 25-kD interleukin-1 inhibitor from cultured human lung macrophages. Am J Respir Cell Mol Biol 1990; 3:355–361.
112. Galve-de Rochemonteix B, Junod AF, Dayer JM. Fibroblast-alveolar cell interactions in sarcoidosis and idiopathic pulmonary fibrosis: evidence for stimulatory and inhibitory cytokine production by alveolar cells. Eur Respir J 1990; 3:653–664.
113. Carter DB, Deibel MR Jr, Dunn CJ, Tomich CS, Laborde AL, Slightom JL, Berger AE, Bienkowski MJ, Sun FF, McEwan RN, et al. Purification, cloning expression and biological characterization of an interleukin-1 receptor antagonist protein. Nature 1990; 344:633–638.
114. Galve-de Rochemonteix B, Nicod LP, Chicheportiche R, Lacraz S, Baumberger C, Dayer JM. Regulation of interleukin-1ra, interleukin-1 alpha, and interleukin-1 beta production by human alveolar macrophages with phorbol myristate acetate, lipopolysaccharide and interleukin-4. Am J Respir Cell Mol Biol 1993; 8:160–168.
115. Moore SA, Strieter RM, Rolfe MW, Standiford TJ, Burdick MD, Kunkel SL. Expression and regulation of human alveolar macrophage-derived interleukin-1 receptor antagonist. Am J Respir Cell Mol Biol 1992; 6:569–575.
116. Nicod LP, El Habre F, Dayer JM. Natural and recombinant interleukin-1 receptor antagonist does not inhibit human T cell proliferation induced by mitogens, soluble antigens or allogeneic determinants. Cytokine 1993; 4:29–35.

117. Bachwich PR, Lynch JP, Larrick J, Spengler M, Kundel S. Tumor necrosis factor production by human sarcoid alveolar macrophages. Am J Pathol 1986; 125:421–425.
118. Galve-de Rochemonteix B, Nicod LP, Dayer JM. Tumor necrosis factor soluble receptor 75: the principal receptor form released by human alveolar macrophages and monocytes in the presence of interferon γ. Am J Respir Cell Mol Biol 1996; 14:279–287.
119. Nicod LP, El Habre F, Dayer JM, Boehringer N. Interleukin-10 decreases tumor neorosis factor α and β in alloreactions induced by human lung dendritic cells and macrophages. Am J Respir Cell Mol Biol 1995; 13:83–90.
120. De Waal Malefyt R, Haanen J, Spits H, Roncarolo MG, te Velde A, Figdor C, Johnson K, Kastelein R, Yssel H, de Vries JE. Interleukin 10 (IL-10) and viral IL-10 strongly reduce antigen-specific human T cell proliferation by diminishing the antigen-presenting capacity of monocytes via downregulation of class II major histocompatibility complex expression. J Exp Med 1991; 174:915–924.
121. Willems F, Marchant A, Delville JP, Gerard C, Delvaux A, Velu T, de-Boer M, Goldman M. Interleukin-10 inhibits B7 and intercellular adhesion molecule-1 expression on human monocytes. Eur J Immunol 1984; 24(4):1007–1009.
122. Ludewig B, Graf D, Gelderblom HR, Becker Y, Kroczek RA, Pauli G. Spontaneous apoptosis of dendritic cells is efficiently inhibited by TRAP (CD40-ligand) and TNF-α, but strongly enhanced by interleukin-10. Eur J Immunol 1995; 25:1943–1950.
123. Taga K, Mostowski H, Tosato G. Human interleukin-10 can directly inhibit T cell growth. Blood 1993; 81(11):2964–2971.
124. Louden M, Songeon F, Nicod LP. IL-10 reduces TNF-α & TNF-β produced by αCD3-activitated T cells. Am J Respir Crit Care Med 1995; 154:A34.

14

Role of Dendritic Cells and Alveolar Macrophages in Regulating Pulmonary B-Cell Function

DAVID S. WILKES

Indiana University Medical Center
Indianapolis, Indiana

JONATHAN C. WEISSLER

University of Texas Southwestern Medical Center
Dallas, Texas

I. Introduction

Although B lymphocytes are able to function as antigen presenting cells (1), the primary role of B lymphocytes in the lung is to produce immunoglobulins (Ig) important for local defense against microbial pathogens (2). Abnormal production of Ig may also play a pathogenic role in autoimmune diseases and possibly participate in allograft rejection (3,4). Immunoglobulin production represents the final stage of differentiation for B lymphocytes (5). However, several stages precede the development of an Ig-secreting cell. Because macrophages and dendritic cells (DCs) may act on B cells directly or through the stimulation of cytokine release by other cells, it is important to review thoroughly the mechanisms by which B lymphocytes develop into Ig-secreting cells. This chapter discusses B-lymphocyte ontogeny, with descriptions of activation and differentiation (Ig production) as they relate to humoral immune responses in the lung. Additionally, the direct role of macrophages and DCs on B-lymphocyte function in normal and diseased lungs will be discussed.

II. B-Lymphocyte Development

A. Ontogeny

B-lymphocyte ontogeny may be divided into two phases: lymphopoiesis and immunopoiesis, which culminates in the generation of a memory B lymphocyte or plasma cell (6–12).

Lymphopoiesis

Initially, a multipotent stem cell in the fetal liver and bone marrow undergoes several maturational changes known as lymphopoiesis which ultimately results in the production of mature B lymphocytes (6) (Fig. 1). Lymphopoiesis is an antigen-independent and T-lymphocyte–independent process, and it is characterized by several changes in the expression of surface antigens, as well as intracellular Ig gene rearrangements (6,7). Stromal cells, which produce interleukin-7 (IL-7) interacting with stem cells in the microenvironment of the marrow facilitate lymphopoiesis (8,12). IL-7 has been shown to be crucial for B-lymphocyte lymphopoiesis in that anti–IL-7 antibodies completely inhibit intramarrow B-lymphocyte development (8). In addition to IL-7, other cytokines are believed

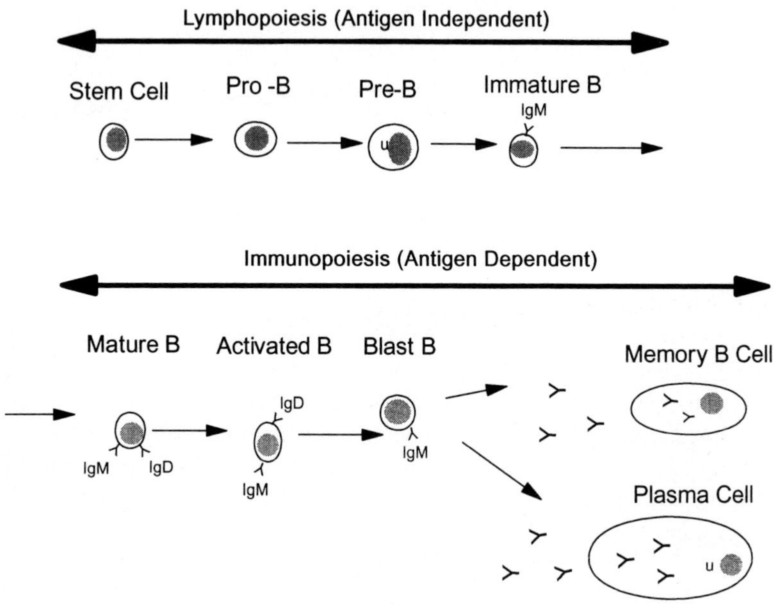

Figure 1 B-Lymphocyte ontogeny.

to be involved in B-lymphocyte lymphopoiesis, including IL-3 (9), IL-11 (12), insulin-like growth factor (IGF-1)(7), granulocyte-macrophage colony-stimulating factor (GM-CSF) (7), and possibly IL-6 (9).

Pro-B lymphocytes, the first developmental stage after a stem cell (6), is characterized by the presence of Ig genes in a germline configuration; that is, the Ig genes have not undergone recombination. The subsequent rearrangement of the germ-line Ig genes results in the expression of intracytoplasmic μ chains, which defines the next phase of development known as the pre-B lymphocyte (6,7). Further rearrangements of Ig genes result in the expression of both μ and light chains leading to the expression of surface IgM which identifies the immature B lymphocyte (see Fig. 1) (6–12). Ultimately, IgD is coexpressed with IgM on the B-lymphocyte membrane which identifies a mature/naive B lymphocyte and the completion of lymphopoiesis (see Fig. 1) (6). Other markers of mature B lymphocytes include the expression of cluster antigens CD19, CD20, CD21 in humans (11) and B220 in mice, among others (13).

Immunopoeisis

At the completion of lymphopoiesis, the mature B lymphocyte enters immunopoiesis. One of two endpoints occurs during immunopoeisis; that is, the development of either plasma cells, which are short-lived and constitutively produce Ig; or memory B lymphocytes (6–8,14), which produce Ig only in response to reexposure to particular antigens (14) (see Fig. 1). Immunopoeisis occurs in secondary lymphoid organs such as spleen, lymph nodes, or tonsils (6). In the lung, these sites could include the draining pulmonary lymph nodes, such as bronchus-associated lymphoid tissue (BALT), hilar lymph nodes, or mediastinal lymph nodes (15). Immunopoeisis is highly regulated and involves the interactions of the mature B lymphocyte with T lymphocytes and/or soluble T-lymphocyte factors such as IL-2 and IL-10 (6,7,16–18), antigen-presenting cells (APCs) such as follicular DCs, interdigitating DCs, macrophages, and antigen (6,7,16). The development into a memory cell or plasma cell appears to involve the cytokines IL-2 and IL-10 plus signaling though CD40 on B lymphocytes (16–18). For example, the engagement of the CD40 ligand (CD40L), which is expressed on T lymphocytes, with CD40 on B lymphocytes, in the presence of IL-2 and IL-10 results in the development of cells with the characteristics of memory B lymphocytes (16). Alternatively, the removal of CD40L under the same conditions results in the development of plasma cells (16). For both memory and plasma cells, activation with antigen results in either the production of IgM (the primary humoral response) (10) or further recombination of the Ig genes (isotype class switch) (10), which results in the expression and production of other Ig isotypes such as IgG, IgA, or IgE (6). Isotype class switch is under the control of several cytokines which are described below. Once the plasma cell or memory B-lymphocyte stage is reached, IgD is no longer expressed (6). Memory B lymphocytes and plasma cells are then widely distributed throughout various organs and are located adjacent to mucosal surfaces within the interstitium or in the draining lymph nodes of the target organs (6). Examination of the human lung B-cell phenotype in cells obtained from mechanically digested lung tissue in our laboratory disclosed an overwhelming predominance of surface IgD-negative B cells in the lung (our unpublished observations), suggesting that the majority of B lymphocytes in human lung are memory B lymphocytes.

B. Activation

B lymphocytes may be activated to undergo proliferation by either T-lymphocyte–independent or T-lymphocyte–dependent mechanisms (5,6). T-lymphocyte–independent B-lymphocyte activation occurs when the B-cell receptor (BCR) is cross linked by antigens in the absence of T lymphocytes (19,20). T-lymphocyte–dependent B-lymphocyte activation refers to the stimulation of B lymphocytes by direct contact with T lymphocytes and soluble factors (21–25).

Antigens that are capable of T-lymphocyte–independent B-lymphocyte activation are known as direct B-lymphocyte mitogens (5a). These include polysaccharides such as lipopolysaccharide (LPS), particles of *Staphylococcus aureus* Cowan strain (SAC), lectins such as pokeweed mitogen, and immobilized anti-immunoglobulin (5a). All of these mitogens have in common the ability to cross link the BCR, which is a surface Ig complexed with Igα and Igβ chains, known as CD79a and b, respectively (19). Unlike the surface Ig chains, the Ig α and β chains have cytoplasmic domains that activate several tyrosine kinases (reviewed in ref. 20) which results in the responding B lymphocyte leaving the G_0 stage and entering the G_1 stage of the cell cycle (20). If the antigenic signal is strong enough, this results in proliferation of the B lymphocyte (20). In the presence of exogenous T-lymphocyte–derived supernatants, or specific cytokines, the B lymphocytes differentiate into Ig-secreting cells (6). It is important to note that the differentiation of B lymphocytes, that is, the transition to an Ig-secreting cell, will not occur in the absence of exogenous cytokine help derived from T lymphocytes (6,10–12). The ability of many of these mitogens directly to activate B lymphocytes may have clinical significance. For example, some of these antigens are major components of the capsules of bacterial pathogens such as pneumococci (polysaccharide capsular antigen) and gram-negative bacteria (LPS) which may be pathogenic in the lung. Therefore, exposure to these pathogens may directly induce a protective immune response in the host by stimulating antigen-specific memory cells or the development of antigen-specific plasma cells in the lung.

In contrast to the direct B-lymphocyte mitogens, B-lymphocyte activation in response to most protein antigens occurs in a T-lymphocyte–dependent manner

(21–25). Examples of T-lymphocyte–dependent B-lymphocyte activators may include sheep red blood cells (26–28), alloantigens (29), and bacterial or viral proteins (30,31). In response to T-lymphocyte–dependent mitogens, B-lymphocytes may become activated by two different methods. First, activation may occur in an antigen and major histocompatibility complex (MHC) nonrestricted manner whereby T lymphocytes are activated by interactions with APCs which in turn activate the B-lymphocyte (21–25). Alternatively, the B lymphocyte may function as an APC and present peptide antigens to T lymphocytes in an MHC class II restricted manner which results in T-lymphocyte activation, costimulatory molecule expression, and cytokine production which in turn facilitates Ig production (32). In either pathway, B-lymphocyte activation requires direct contact with T lymphocytes (6,21). T-lymphocyte membrane-bound signals that are able to activate B lymphocytes include CD40L (10,11,14,17) and intracellular adhesion molecule-1 (ICAM-1) (14,21). In addition, activated T lymphocytes can induce the expression of B7-1 (CD80) (6) and B7-2 (CD86) (6) on antigen-presenting B-lymphocytes which are ligands for CD28/CTLA4 on the responding T lymphocyte (6) and act as costimulators of T-cell proliferation.

III. Role of T-Lymphocyte–Derived Cytokines in B-Lymphocyte Activation, Differentiation, and Control of Immunoglobulin Sublype Production

Activated T lymphocytes produces several cytokines that facilitate the activation and subsequent differentiation of B lymphocytes into Ig-secreting cells (5,6) as well as inducing Ig isotype class switch (33). The following discussion summarizes the role of these cytokines in B lymphocyte function, particularly as they relate to pulmonary humoral immunity.

A. IL-2

IL-2 is produced by activated T helper type 1 lymphocytes (Th1) (34) and is a potent mediator of B-lymphocyte activation, proliferation, and Ig production (35–41). When bound to the intermediate affinity IL-2 receptor, IL-2 induces B-lymphocyte proliferation (39–41). B-lymphocytes that have been activated with SAC or anti-Ig antibodies in the presence of IL-2 secrete IgG, IgA, and IgM (5a,6,42). In response to T-lymphocyte–dependent B-lymphocyte activators, IL-2 also stimulates B-lymphocyte proliferation and Ig production (5a,6,21,22). Additionally, IL-2 may act synergistically with other cytokines such as IL-1, IL-6, and tumor necrosis factor-α (TNF-α), to facilitate Ig secretion from resting B lymphocytes (43). However, IL-2 does not induce Ig isotype class switch in activated B lymphocytes.

B. IL-4

In contrast to IL-2, IL-4 is produced by T helper type 2 T lymphocytes (Th2) (34). Depending on the nature and state of B-lymphocyte activation, IL-4 has divergent effects on B-lymphocyte function. Unlike IL-2, IL-4 does not costimulate with SAC (44) and inhibits IL-2–induced proliferation in B-lymphocytes costimulated through their BCRs (44). However, IL-4 may act as a progression factor in that it induces the proliferation of human B lymphocytes activated with anti-Ig antibodies (45). In murine studies, IL-4 has been suggested to act as a competence factor in that it appears to induce B lymphocytes to be responsive to other T-lymphocyte–derived cytokines that induce proliferation and immunoglobulin secretion (46).

With regard to B-lymphocyte differentiation, IL-4 is able to stimulate Ig production from B-lymphocytes activated by lung alloantigens or SAC plus IL-2 (47,48). In response to SAC plus IL-2, IL-4 induces the production of IgG, IgM, and IgA (47). In addition, IL-4 facilitates IgG production in response to lung alloantigen (48). Stimulation of B lymphocytes through CD40 in the presence of IL-4 results in the selective upregulation of IgE production (49). The effects of IL-4 on B-lymphocyte proliferation and Ig production may be antagonized by interferon γ (IFN-γ) (50).

IL-4 may also act to induce isotype class switch (51,52). For example, Snapper et al. (51,52) reported that IL-4 was able to induce IgG1 and IgE production from surface IgG^- and IgE^- naive murine B lymphocytes that had been activated with LPS. In addition, IL-4 induced IgG4 and IgE production from resting IgG^- and IgE^- human B lymphocytes activated with SAC (53).

In the lung, IL-4 may have a central role in the pathogenesis of asthma (54). Robinson et al. reported the IL-4–producing lymphocytes were found in the bronchoalveolar lavage fluid of asthmatic patients. Data showing that IL-4 stimulates IgE production (49–53) and is produced by mast cells in the lung (55) is further evidence of the role of IL-4 in asthma.

C. IL-5

Although IL-5, a Th2 cytokine (34,54), does not appear to appear to induce proliferation in B lymphocytes by itself, several investigators have shown that IL-5 may act synergistically with other cytokines to induce Ig production (56). For example, Bertolini et al. (57) reported that IL-5 synergizes with IL-4 and IL-2 to induce IgM secretion from SAC-activated human B lymphocytes. Similarly, in mice, IL-5 was shown to be a B-lymphocyte differentiation factor (58) and supports IgA production from IgA-committed B lymphocytes (59,60). Furthermore, IL-5, in combination with transforming growth factor-β (TGF-β), results in enhanced IgA production from surface IgA^+ murine B lymphocytes (60). IL-5 also has synergistic effects with IL-4 in stimulating IgE production, which may have clinical relevance in atopic disorders such as asthma (56). In fact, IL-5 was found in lymphocytes isolated from the bronchi of symptomatic asthmatic patients (54). Although IL-5 induces a critical component of the recombination machinery in normal murine B lymphocytes (61), it does not appear to be able to induce Ig isotype class switch in B lymphocytes.

D. IL-6

IL-6 is produced by monocytes and macrophages as well as T and B lymphocytes (62,64), and it may act in an autocrine fashion on B lymphocytes (64). Although IL-6 is not able to induce proliferation in resting human B lymphocytes (65), it is a potent stimulator of Ig secretion from differentiated B lymphocytes in response to both T-lymphocyte–dependent and T-lymphocyte–independent activators (62–64). For example, IL-6 induces significant IgG production from B lymphocytes stimulated in vivo with pneumococcal polysaccharide antigen (62–64) and facilitates IgG, IgA, and IgM secretion in response to SAC activation (62). In the lung, IL-6 derived from macrophages is believed to play a crucial role in the ability of macrophages to facilitate Ig production in response to certain antigens (66). We have recently reported that deficient IL-6 production from lung allograft macrophages in part accounted for the inability of these macrophages to induce IgG production form allogeneic peripheral blood mononuclear cells (66).

Although, IL-6 does not induce Ig isotype class switch, recent studies indicate that IL-6 stimulates IgA production along mucosal surfaces (67), which suggests that IL-6 may have a critical role in mucosal humoral immunity in the upper and lower respiratory tract. For example, Ramsay et al. (67) reported that IL-6–deficient mice had fewer IgA^+ plasma cells in the lungs than wild-type mice. In addition, IL-6–deficient mice had a markedly impaired ability to mount an IgA response after local immunization with hemagglutinin glycoprotein of the influenza virus (67). Therefore, IL-6 is important in facilitating overall Ig production, and it appears to have a central role in maintaining the integrity of IgA-mediated mucosal immunity in the lung.

E. IL-10

Similar to IL-4 and IL-5, IL-10 is a Th2-type cytokine (68,69). However, unlike IL-4 and IL-5, IL-10 is produced by many cell types other than T helper lymphocytes. Monocytes, macrophages, and B lymphocytes are all sources of IL-10 (68, 69). Although, IL-10 was described initially to be immunosuppressive of many immune cell functions (68,69), IL-10 can either be stimulatory or inhibitory of Ig production (70–73). For example, Punnonen et al. (73) reported that IL-10 was a potent inhibitor of spontaneous IgG, IgA, and IgM production. Other studies have shown that IL-10 is able to inhibit Ig production stimulated by T lymphocyte–independent antigens and IL-5 (71). Furthermore, we recently reported that IL-10 produced by lung allograft alveolar macrophages was partially responsible for the inhibitory effects of these macrophages on IgG production (66). In contrast, after B lymphocytes are stimulated with anti-CD40 antibody (70) or SAC (73), IL-10 vigorously stimulates B-lymphocyte proliferation and IgG, IgA, and IgM production. IL-10 may also act synergistically with IL-6 to stimulate proliferation in CD40-stimulated B lymphocytes (70).

Additionally, IL-10 has recently been demonstrated to induce Ig isotype class switch (74). For example, surface IgD^+IgM^+ B lymphocytes when stimulated with CD40 and IL-10 produced IgG1 and IgG3 (74). Furthermore, IL-10 may synergize with TGF-β to stimulate IgA production, which suggests a possible role for IL-10 in mucosal immunity (75).

F. IL-13

IL-13 is a recently described cytokine produced by T lymphocytes (76) and mast cells (77) which affects B-lymphocyte function similar to that of IL-4 (78,79). For example, IL-13 promotes growth in activated B lymphocytes (78) and IgG4 and IgE production from naive B lymphocytes (78). IL-13 is able to induce IgG4 and IgE production from immature pre-B lymphocytes and IgD^+ human peripheral blood B lymphocytes; thus suggesting an ability to induce an isotype class switch similar to IL-4 (78). The fact that IL-13 is produced by mast cells (80) and is able to stimulate IgE production suggests that it may have a role in the pathogenesis of asthma and other atopic diseases.

G. TGF-β

TGF-β is produced by several cell types, including B and T lymphocytes, monocytes, macrophages, and platelets (81), and it is typically found at sites of active inflammation. The effects of TGF-β on Ig production depend on whether exogenous TGF-β is added to in vitro cell culture systems (82,83) or if TGF-β produced by B lymphocytes is acting in an autocrine fashion (84). Exogenous TGF-β has been shown to be a potent inhibitor of IgG and IgM production from SAC-activated B lymphocytes, whereas having no effect on IgA production (83). We have reported that TGF-β production by human immunodeficiency virus (HIV)–infected alveolar macrophages (AMs) inhibited IgG but not IgA production from surface IgD^+IgM^+ peripheral blood B lymphocytes (85). In contrast, McIntyre et al. (82) reported that TGF-β produced by B lymphocytes may act in an autocrine fashion to induce IgG2b production from LPS-activated B lymphocytes in vitro.

Similar to other cytokines, TGF-β has profound effects on IgG subtype class switch (82). In murine studies, TGF-β induced the production of IgG2b from LPS-activated B lymphocytes (82). In both murine and human studies, TGF-β induces Ig isotype switch to IgA production (83,84,86).

TGF-β likely plays a role in pulmonary humoral immunity. For example, upregulated TGF-β production by HIV-infected AMs may account for the lower than normal concentrations of IgG antibodies in bronchoalveolar lavage (BAL) fluid from HIV-infected individuals (85). In addition, we reported that the production of TGF-β by lung allograft alveolar macrophages was in part responsible for the inability of the allograft macrophages to stimulate IgG production in response to alloantigen (48).

H. INF-γ

IFN-γ is produced by T-lymphocytes (87) and natural killer (NK) cells (87) and may have profound effects on B lymphocyte proliferation and differentiation. Similar to IL-2, IFN-γ is a Th1-type cytokine (34), but it exerts different effects on B lymphocytes (34). In contrast to IL-2, IFN-γ inhibits T-lymphocyte–dependent B-lymphocyte proliferation in vitro (88). Additionally, IFN-γ inhibits IL-4–induced B-lymphocyte proliferation (50) as well as proliferation induced by stimulation through CD40 (89). Conversely, IFN-γ added to IL-4 facilitates CD40-mediated proliferation (89). When added to SAC-activated B lymphocytes at the initiation of culture, IFN-γ enhances proliferation but may be inhibitory if added to B lymphocytes preactivated by SAC (90). In contrast, anti–μ-antibody–activated B lymphocytes proliferated vigorously in the presence of IFN-γ (90). Therefore, depending on the experimental conditions, IFN-γ has variable effects on B-lymphocyte proliferation.

The effects of IFN-γ on Ig production are also variable. In general, IFN-γ alone is not able to induce Ig production (88), but it is able to promote Ig production from murine and human B lymphocytes activated with IL-2 (91). In response to LPS, IFN-γ stimulated the production of IgG2a from murine B lymphocytes (92,93), and it is able to stimulate IgG3 synthesis under other experimental conditions (94). The mechanism by which IFN-γ stimulates IgG2a production is mediated by both isotype class switch from surface IgG2a$^-$ B lymphocytes (92,93) and augmented production form IgG2a-committed cells (93). IFN-γ induced IgG3 production appears to be mediated by isotype class switch mechanisms (94). Conversely, IFN-γ antagonizes IL-4–induced Ig production, and therefore strongly inhibits IgG1 and IgE production (50). In humans, IFN-γ induces IgG2 germline transcripts, suggesting that selective IgG2 production may be stimulated by IFN-γ (95). IFN-γ acts synergistically with IL-6 to stimulate IgG2 production from human peripheral blood mononuclear cells (96), and we have reported that IgG2 production in response to lung alloantigen is solely dependent on IFN-γ (48).

In vivo, IFN-γ appears to stimulate IgG2a and IgG2 production during lung allograft rejection in murine and human studies, respectively (48,97). IFN-γ also appears to be involved in the local antiviral IgG2a response in mice (31).

IV. Review of Immunoglobulin Structure and Function as It Relates to Pulmonary Immunity

In humans, there are five Ig isotypes: IgG, IgA, IgM, IgE, and IgD. Within IgG, there are four subtypes: IgG1, IgG2, IgG3, IgG4. Also, IgA has two subtypes, IgA1 and IgA2 (98). In mice, the IgG subclasses include IgG1, IgG2a, IgG2b, and IgG3 (98). Otherwise, the nomenclature of murine Ig is similar to that in humans. IgD has no known effector functions and is not discussed in this section. IgG, IgA, and IgM are associated with different types of effector function, which are described below.

The structure of an Ig molecule is characterized by two heavy chains (55 kDa) held together by two disulfide bonds and light chains (25 kDa) bound to the heavy chains in the shape of a "y" (99) (Fig. 2). The arms of the Y contain the part of the molecule known as the Fab fragments, which consist of variable (v) and constant (c) regions (99). The variable regions of the Fab determine the antigenic specificity of an Ig molecule (99). The stem of the Y is termed the Fc portion, which is composed of constant regions that can mediate other immune effector systems such as complement binding or attachment to Fc receptors on other immune cells (99). The Fab and Fc regions are linked by a "hinge region" (Fig. 2)

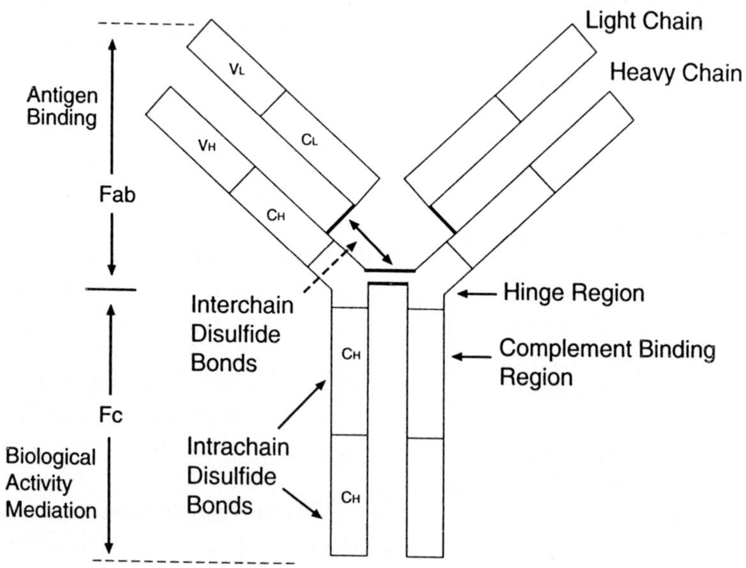

Figure 2 General immunoglobulin structure.

which gives flexibility to the molecule and is believed to be responsible for the ability of the Fab portions of the molecule to bind bivalently to an antigen (98).

Metabolic and effector properties (exclusive of antigen recognition) are determined by the constant regions of the heavy chains. Although the constant regions are highly homologous within an isotype, there exists only 30% homology in the constant regions between isotypes (99). The variable regions exist at the N-terminal portions of both heavy and light chains, and they are crucial for determining antigenic specificity. Because the variable regions must be able to recognize a wide array of antigens, significant differences exist in the peptide sequences of isotypes and subtypes (99).

A. IgG

IgG is a monomeric 150-kDa molecule, and it plays a major role in humoral immunity in the lower respiratory tract in both mice and humans (100–103). IgG subclasses are structurally related, and they show more than 95% amino acid sequence identity between homologous domains (101). In contrast, the hinge regions are only 60–70% homologous and vary greatly in length among the subclasses (101). Therefore, the main structural differences in the IgG subclasses are located in their hinge regions (101), and it has been demonstrated that some of the biological differences in the subclasses, for example, the ability to activate complement, may be related to this area of the molecule (98,101).

The effector functions of these molecules involve several different mechanisms. First, IgG mediates complement activation via the classic pathway known as complement-dependent cellular cytotoxicity, or CDCC (100). In vitro or in experimental animals, IgG3 has a greater ability than IgG1 or IgG2 to bind complement, whereas IgG4 does not bind complement (100). In certain human disease studies, IgG1 and to a lesser extent IgG4 are able to activate complement (104). In mice, IgG2a and IgG2b are far more potent complement activators than IgG1 or IgG3 (105,106).

The other effector functions of IgG antibodies involve the targeting of bound antigen for destruction through phagocytosis by macrophages and granulocytes (98). In addition, IgG antibodies may facilitate the clearance of antigens by antibody-dependent cellular cytotoxicity (ADCC), which is mediated by lymphocytes and NK cells (98). Both of these functions are mediated by the binding of the Fc portion of the IgG molecule to Fcγ receptors of effector cells (98). Although receptors exist for all IgG subtypes (100), the subtypes differ in their ability to bind to Fcγ receptors and, therefore, vary in their ability to induce phagocytosis or ADCC. In humans, IgG1 and IgG3 are most efficient in inducing ADCC, likely because of the affinity of Fcγ receptors on macrophages for these subtypes (100). For example, in vitro studies have demonstrated that IgG3 bound to nearly 30% of AMs in culture and IgG1 bound to another 10% (100). In mice, IgG2a is the most potent mediator of ADCC (105,106).

Reports showing that there is little de novo Ig production in the lower respiratory tract (107), and that under normal conditions local immunoglobulin production is usually suppressed by macrophages in the alveolus (42), suggest that most of the IgG in the lower respiratory tract results from transudation of IgG molecules from serum to alveolar spaces in mature mice (D.S. Wilkes, unpublished data) and adults humans (102). Merrill et al. (102) reported the concentrations of IgG and its subtypes in BAL fluid in adult nonsmoking normal volunteers. The relative concentrations of the IgG subtypes in BAL adjusted for total protein or albumin was such that IgG1 > IgG2 > IgG3 > IgG4 (102). The relative concentrations of the IgG subtypes were essentially similar to serum with the exception of IgG4, which existed in a greater concentration that could be accounted for by total protein or albumin concentrations in BAL fluid (102).

Studies in our laboratory suggest that there appears to be de novo IgG subtype production in the lung during normal murine development. Table 1 shows the concentration of each IgG subtype/milligram of total protein in serum or BAL from mice that were 30, 60, and 90 days old. Thirty-day-old mice are immature and 90-day-old mice are believed to be mature and fully developed (108). In

Table 1 Concentrations of IgG Subtypes in Serum and BAL Fluid of BALB/c Mice

	Age (days		
Ig Subtypes	30	60	90
Serum			
IgG1	3.9 ± 0.4	4.2 ± 0.3	43.1 ± 7.5[a]
IgG2a	0.48 ± 0.04	1.1 ± 0.3	7.4 ± 0.7[a]
IgG2b	0.98a ± 0.2	2.5 ± 1.0	9.6 ± 1.0[a]
IgG3	1.5 ± 0.1	3.9 ± 0.3	15 ± 1.5[a]
Total protein	47.1 ± 5.1	47.9 ± 1.3	41.1 ± 1.3[a]
BAL fluid			
IgG1	26.7 ± 7.6[b]	24.7 ± 7.4[b]	23.0 ± 4.8
IgG2a	4.3 ± 2.1	2.2 ± 0.48[b]	4.4 ± 1.2
IgG2b	15.1 ± 8.0[b]	7.3 ± 2.1[b]	4.7 ± 0.8
IgG3	12.6 ± 5.4[b]	4.3 ± 1.1	7.5 ± 1.3
Total protein	0.016 ± 0.005	0.026 ± 0.006	0.060 ± 0.010[c]

IgG subtype concentrations expressed as micrograms of IgG per milligram protein in serum and BAL fluid. Total protein expressed as milligrams per milliliter in serum and BAL fluid. Data shown as mean ± SEM of 10 mice.
[a]Significantly different from 30- and 60-day values in the serum ($p < .05$).
[b]Significantly different from corresponding values in the serum ($p < .05$).
[c]Significantly different from 30- and 60-day values in BAL ($p < .05$).
Source: D.S. Wilkes, C. Moore, and C.M. Doerschuk, unpublished data.

serum, IgG1, IgG2, IgG2b, and IgG3 concentrations were increased at 90 days relative to 30 or 60 days ($p < .05$). The increases at 90 days were due to an increase in the IgG subtypes, as a slight decrease in total protein in serum at 90 days was not enough to account for the significant increase in the IgG subtypes (Table 1). In contrast, IgG1, IgG2a, IgG2b, and IgG3 levels were comparable in BAL fluid at 30, 60, and 90 days (Table 1; $p > .05$ for each subtype 30 versus 60 versus 90 days. Although the absolute values of the IgG subtypes/milliliter of BAL fluid increased, this was offset by comparable increases in BAL fluid total protein (Table 1). Comparisons of serum and BAL fluid IgG subtype concentrations in Table 1 showed that IgG1 and IgG2b were enriched in BAL fluid relative to serum at 30 and 60 days of age ($p < .05$, IgG1, IgG3 BAL fluid vs serum) and BAL IgG3 was greater than serum at 30 days ($p < .001$ BAL fluid vs serum). In contrast, IgG2a concentrations were equivalent in BAL and serum at 30, 60, and 90 days.

Figures 3 and 4 show the quantitation of IgG subtype concentration in serum and BAL fluid expressed as a percentage of total IgG. In serum, IgG1 comprised the greatest percentage of total IgG at 30 and 90 days ($p < .001$ IgG1 compared with other subtypes), whereas 60 days, IgG1 and IgG3 comprised a similar proportion of total IgG (Fig. 3). The percentage of IgG2a, IgG2b, and IgG3 were variable, and the only consistent comparison was that IgG3 comprised a greater proportion of total IgG than IgG2a at all time points ($p < .044$, IgG3 vs IgG2a at 30, 60, and 90 days).

In BAL fluid, IgG1 comprised the greatest percentage of total IgG at all three time points (Fig. 4.; $p < .001$ IgG1 compared with all other subtypes at 30, 60, or 90 days). The percentage of IgG2a, IgG2b, and IgG3 was similar to all three ages. The disparity of IgG subtype concentrations relative to total protein in murine BAL at different stages of development suggests that some IgG subtypes may be produced locally during different stages of development.

The importance of IgG in human pulmonary immunity is exemplified by the observation that deficiencies of certain subclasses, specifically IgG2, is associated with recurrent sinopulmonary infections and bronchiectasis (109,110). Additionally, we have shown in a murine model that mimics the histology and immunology of lung allograft rejection that only IgG2a production is upregulated locally and deposited in perivascular and peribronchiolar tissues involved in the rejection process (97). In addition, locally upregulated IgG2 production is a marker of rejection in lung allograft recipients (111).

B. IgA

Depending on whether it exists in its monomeric or dimeric form, IgA is a 160- or 400-kDa molecule. It is the most abundant Ig present in secretions (103,112), and in the lung, IgA is the predominant antibody isotype present in the lower respira-

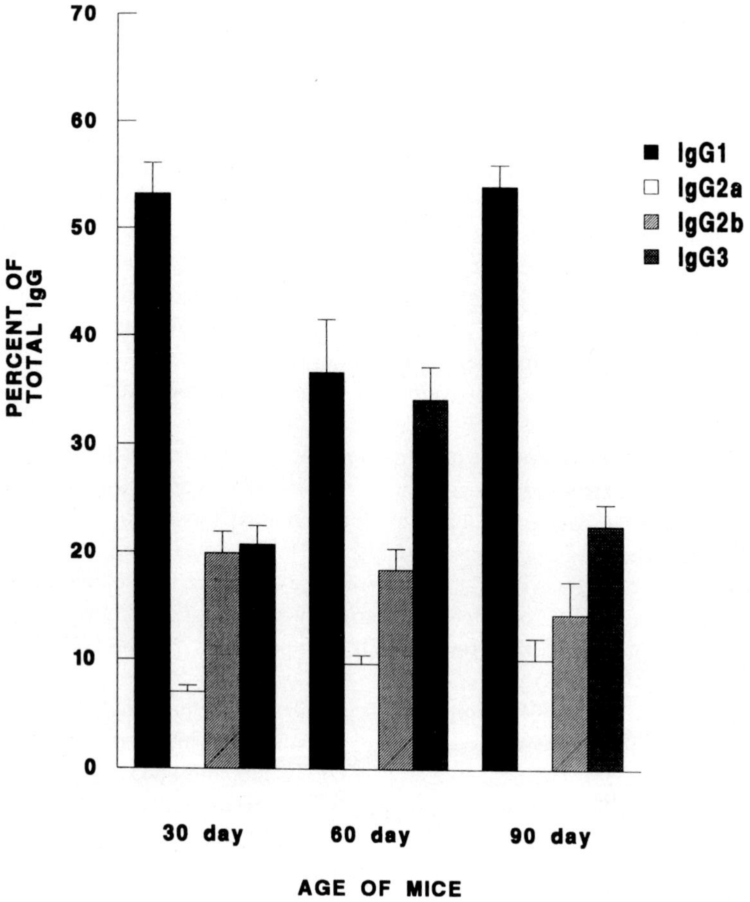

Figure 3 Serum IgG1, IgG2a, IgG2b, or IgG3 levels as a percentage of total IgG BALB/c mice at 30,60, and 90 days of age. Total IgG was calculated as the sum of the individual subtypes. Data are expressed as a mean ± SEM for 10 mice. (From D.S. Wilkes, C. Moore, and C.M. Doerschuk, unpublished data.)

tory tract with the ratio of IgA/IgG equaling approximately 2.5:1.0 (103). However, most of the IgA is localized in the bronchial tubes and in the alveolar spaces, IgG may be the predominant Ig isotype (100,102,103). The fact that the concentration of IgA adjusted for total protein in BAL fluid is greater than that of serum suggests that a large portion of IgA is produced locally in the lung (103). In fact, immunocytochemical studies have shown the presence of IgA-producing cells adjacent to the lamina propria at the basilar surface of airway epithelium (113).

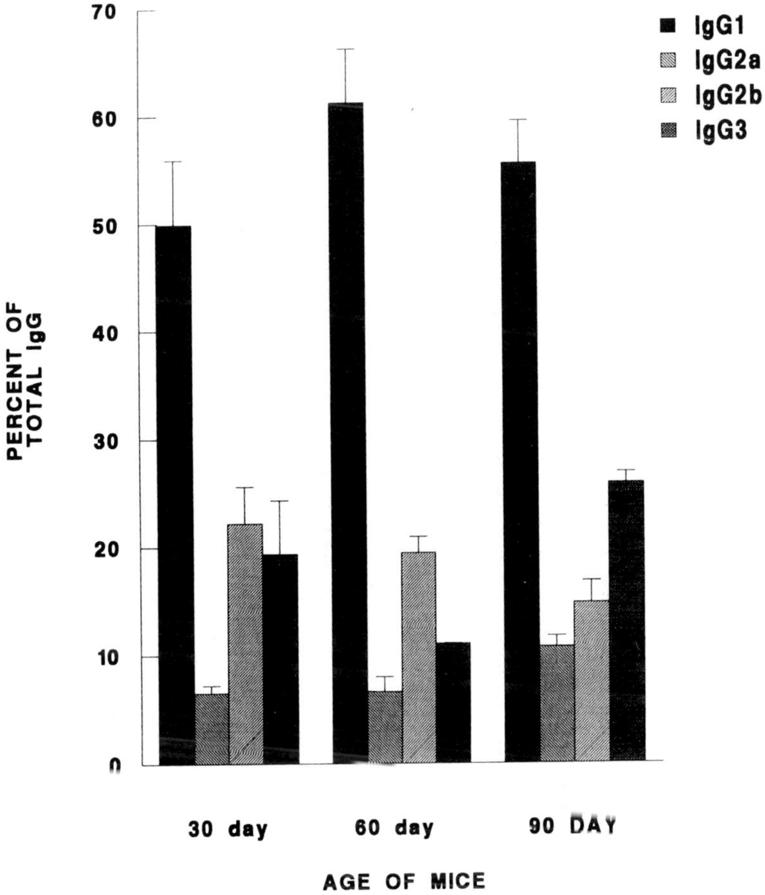

Figure 4 Bronchoalveolar lavage IgG1, IgG2a, IgG2b, or IgG3 levels as a percentage of total IgG in BALB/c mice at 30, 560, and 90 days of age. Total IgG was calculated as the sum of the individual subtypes. Data are expressed as a mean ± SEM for 10 mice. (From D.S. Wilkes, C. Moore, and C.M. Doerschuk, unpublished data.)

Although approximately 90% of the IgA in blood exists in the monomeric form, 50–80% of IgA in lung secretions is dimeric; that is, two molecules of IgA bridged by a J chain (103,112,113). Most of the IgA in lung secretions exists in the secretory form referring to the fact that the dimeric form of IgA is complexed with a heavily glycosylated protein called the secretory component which facilitates transport of the IgA molecule across epithelial cells into the airway lumen (112,113). Human IgA has two subtypes, IgA1 and IgA2 (103,112,113), with

IgA2 further divided into two variants IgA2m[1] and IgA2m[2] (112). IgA1 and IgA2 may exist in the monomeric or dimeric form (112). IgA1 and IgA2 differ in their susceptibility to damage by proteases, and IgA2 is resistant to proteases produced by many pathogenic bacteria. This may have clinical relevance in pulmonary disease in that bacterial proteases may play a role in diminishing the protective effect of opsonizing antibodies in the lower respiratory tract (103). Although IgA1 is the predominant IgA subtype in the lower respiratory tract (66% IgA1 and 33% IgA2 (103), IgA2 resistance to proteolysis and local production suggest it plays a prominent role in local humoral immunity (103).

In contrast to the IgG subclasses, the production of IgA subclasses are not regulated by specific cytokines. However, the total production of IgA may be stimulated by IL-6, IL-5, and TGF-β (60,67,83,84,86), with TGF-β being able to stimulate B lymphocytes selectively to produce IgA antibodies by inducing isotype class switch (60,83,84,86).

Mazanec et al. (114), proposed that there are three mechanisms by which IgA participates inmucosal defense. First, it is generally agreed that the protective function of IgA is to provide an immunological barrier that excludes foreign matter, particularly microorganisms from the body by blocking the adherence of bacteria to epithelial cells in the airway lumen. IgA bound to pathogens may then bind to IgA-specific Fc receptors on AMs which phagocytose the pathogen. Second, the normal transcytosis of IgA from the basilar to apical region of epithelial cells suggests that IgA may be effective at neutralizing intracellular pathogens (114), and this may suggest a role for the intraepithelial neutralization of viruses by this mechanism. Finally, IgA located in proximity to the lamina propria may trap pathogens or antigenic material that may have been absorbed from the airway lumen to the basilar region of epithelium (114). In this way, the IgA may complex with the offending agent and transport the pathogen by transcytosis back to the airway lumen where it can be expelled from the body. IgA may also facilitate local defense by activation of the alternate complement pathway (103).

The importance of IgA in pulmonary humoral immunity has been clearly demonstrated in patients with IgA deficiencies who may experience recurrent sinopulmonary infections (115). This condition is usually associated with other Ig subclass deficiencies, such as IgG2 and IgG4 deficiencies (115). In addition, mucosal IgA deficiency induced as a result of IL-6 deficiency is associated with an increased susceptibility to viral lower respiratory tract infections in mice (67).

C. IgM

IgM is a pentamer or hexamer of individual IgM molecules linked by a J chain (103,116). Consequently IgM is the largest of the immunoglobulin isotypes with a molecular weight of approximately 900 kDa (116). Because of its large size, IgM is not able easily to diffuse into lung secretions from blood. Concentrations of IgM in bronchoalveolar lavage fluid are greater than would be expected for simple diffusion of the molecule into bronchial secretions from serum (103). This is likely explained by the active secretion of IgM by IgM-expressing B lymphocytes present in the bronchial mucosa (103) and the transport of IgM by secretory component across epithelial cells into the airway lumen (103).

The cytokine control of IgM production has been extensively studied. IL-5, IL-6, IL-10, and TGF-β have all been reported to have an effect on IgM production (57,62,70–73,83). In response to several different mitogens, IL-5 promotes IgM secretion (57). For example, Bertolini et al. (57) reported that IL-5 induces SAC-activated B lymphocytes to secrete IgM. Similar data have been reported for IL-6 (62) and IL-10 (70,73). In conjunction with signaling through CD40, IL-10 has been demonstrated to be a potent stimulator of IgM production from both surface IgD^+ and IgD^- B lymphocytes (70). Conversely, TGF-β inhibits IgM production from SAC-activated peripheral blood B lymphocytes (83), and IL-10 inhibits spontaneous IgM synthesis from human peripheral blood mononuclear cells (73).

The effector functions of IgM are linked to its ability to fix complement (103). Although IgM is the first Ig to be detected in response to pathogens in serum (103), its role in local humoral immunity in the lung is well defined. Because IgM exists in much lower concentrations in the lower respiratory tract than IgG or IgA (103), it may play a limited role in pulmonary humoral immunity. However, IgM may replace IgA effector functions in patients with IgA deficiency, which suggests that under certain conditions, IgM may have a role in humoral defenses in the lower respiratory tract (115).

D. IgE

IgE, a 190-kDa molecule, is produced in several organs, including the lungs, skin, and gut (103,117). IgE causes a wide range of responses in the body, including inflammation, coughing, lacrimation, and bronchoconstriction (117), and it appears to be involved in antiparasitic responses (117). Unlike IgG or IgA, IgE does not have any subtypes. Relative to other Igs, IgE comprises a very small fraction of the total Ig pool. For example, serum concentrations of IgE are 50–300 ng/ml compared with that of IgG, which is as high as 10 mg/ml (103). The concentration of IgE in BAL fluid adjusted for total protein suggests active local synthesis of IgE by plasma cells that are present in the bronchial mucosa (103). However, the concentration of IgE in BAL is lower than that of IgG, IgA, or IgM (103).

There are two receptors reported for IgE, FcεRI and FcεRII (117, 118). FcεRI is located on mast cells and basophils, and is termed the high-affinity receptor for IgE (117). FcεRI has a critical role in regulating local responses to allergens as discussed below. The low-affinity receptor for IgE is FcεRII (CD23), and it is expressed on B lymphocytes, macrophages, eosinophils, and NK cells, as well as T lymphocytes, follicular dendritic cells, and Langerhans' cells (118). FcεRII also facilitates the presentation of antigen. For example, antigen bound to IgE on B cells may effectively activate T lymphocytes resulting in further B-lymphocyte activation (118).

Despite the low concentrations of IgE, it is a potent inducer of immune responses to antigens by indirect or direct mechanisms (117). When antigen binds to IgE, which in turn is bound to the FcεRI on mast cells or basophils, IgE may *indirectly* stimulate a vigorous immune response resulting in a hypersensitivity reaction (117). This occurs by antigen-induced cross linking the FcεRI-bound IgE which leads to degranulation of the mast cell or basophil with release of several soluble mediators, particularly histamine, leukotrienes, prostaglandins, IL-4, IL-5, and IL-13 (77,80,117). In addition, signaling through IgE on lung macrophages and mast cells is a powerful stimulus for the release of tumor necrosis factor-α (TNF-α) which can enhance local inflammatory responses to pathogens (119). This response appears to be particularly helpful in the clearance of parasites but may also exacerbate other conditions such as asthma. IgE may also provide *direct* protection against pathogens by antibody-dependent cellular cytotoxicity (ADCC) (103).

There appears to be three signals for selective switching to IgE production in B lymphocytes (49,51–53,118). The first is the cytokine IL-4, which induces recombination of the ε-heavy chains resulting in resting B lymphocytes expressing IgE on their cell membranes (51–53). The second is CD40 on B lymphocytes cell membranes, which in conjunction with IL-4, induces IgE production (49). Third, when FcεRII on APCs, such as follicular DCs, binds to its ligand, CD21, on B lymphocytes in the presence of IL-4, IgE-producing cells are stimulated to become plasmacytes (118). Conversely, cytokines that antagonize IL-4 activity, such as IFN-γ, downregulate IgE synthesis (50). Additionally, recent studies have shown that IL-13 may also stimulate IgE synthesis (79).

The main role of IgE in pulmonary immunity is in the pathogenesis of atopic disorders such as asthma (103,120–122). Elevated serum IgE has been associated with increased bronchial reactivity and increased airways obstruction (120–122). It appears that the elevated serum IgE in these studies serves as a marker of disease activity (120–122). Allergens cross linking IgE on mast cell basophil membranes cause the release of several soluble mediators including histamine, prostaglandins, and leukotrienes which participate in anaphylactic reactions and bronchial reactivity, which indicates that IgE has a clear role in the pathogenesis of atopic disorders (77,80,117). In addition, IgE on mast cell and basophil cell membranes may have a role in further stimulating IgE synthesis (80). For example, allergens that can stimulate mast cells to produce inflammatory mediators, as described above, can also stimulate these cells to produce IL-4 (80), which further facilitates IgE synthesis in B lymphocytes (49).

Although no IgE deficiency states have been reported, overproduction of IgE is associated with disease in humans (123). The hyperimmunoglobulin IgE syndrome (hyperIgE) is a rare condition characterized by recurrent staphylococcal abscesses in different organs and markedly elevated serum IgE levels (123). The serum concentrations of other Igs are normal in this condition (123).

V. Role of Lung Dendritic Cells and Alveolar Macrophages in Pulmonary Humoral Responses

A. Effect of AMs on B-Lymphocyte Responses

AMs are the predominant mononuclear cell in the air spaces of the lung, usually exceeding the population of lymphocytes by a ratio of 10:1. In examining the regulation of lymphocyte function in the alveolus in many systems, a consistent theme has emerged; that is AMs act to downregulate primary lymphocyte responses. Indeed, murine studies involving the in vivo depletion of AMs by the intratracheal administration of liposomes containing a diphosphonate toxic for macrophages resulted in a marked upregulation of antibody production in regional lymph nodes following intrapulmonary administration of antigen (124).

Despite the suppressive nature of AMs, several observations suggest that B-lymphocyte responses can be generated in response to airborne antigen. In a variety of species, antigen-specific IgG, IgA, and IgM are found in regional lymph nodes, blood, and lung following primary intrapulmonary immunization (125–127). However, this appears to result primarily from the development of a primary immune response outside the lung with subsequent recruitment of the antibody-forming cells into the lung (127,128). Conversely, in some systems involving chronic exposure to antigen, the long-term production of antibody appears to be localized in the interstitium (129,130). Although there may be species difference in the function of AMs, the majority of data in humans suggest that de novo Ig synthesis is tightly regulated in the normal air space. Hance and colleagues reported little constitutive Ig production in the alveolar space (107). In contrast, patients with sarcoidosis (131, 132), a disease characterized by the presence of multiple granulomas in the lung, exhibit significant local production of Ig.

Experiments performed with human lung cells have suggested that AMs play a major role in modulating Ig responses. Utilizing whole lung tissue, spleen, or peripheral blood mononuclear cells (PBMs), the response of B lymphocytes to SAC and IL-2 was examined (42). After adjusting for similar numbers of B cells, the amount of IgG produced during in vitro stimulation of lung cells was 10 and 20% that of spleen and blood, respectively. Following depletion of lung macrophages, which were predominantly AMs in the mechanically digested lung cell population, the IgG production of the remaining lung cells was equivalent to that of PBMs.

Using a more direct evaluation of the effect of AMs, the addition of AMs to syngeneic PBMs, a dose-related inhibition of IgG and IgM synthesis to SAC was observed (Fig. 5). However, it was noted that AMs had no effect on the production of IgA. Because AMs are capable of producing a variety of potentially suppressive moieties and AM supernatants were suppressive of IgG production in this system, experiments were conducted which evaluated the mechanism of suppression. The addition of indomethacin and catalase to block the production of arachidonic acid metabolites and oxidants had no effect on AM-mediated suppression. Because TGF-β is capable of inhibiting IgG and IgM, whereas augmenting IgA production (133), it is possible that AM production of TGF-β (134) was involved. However, the addition of blocking antibody to TGF-β also failed to reverse suppression.

It was observed that paraformaldelyde fixed AMs were markedly inhibitory of IgG production in this system. AMs appear to have a unique property compared with other mononuclear phagocytes in that a membrane-bound substance on AMs

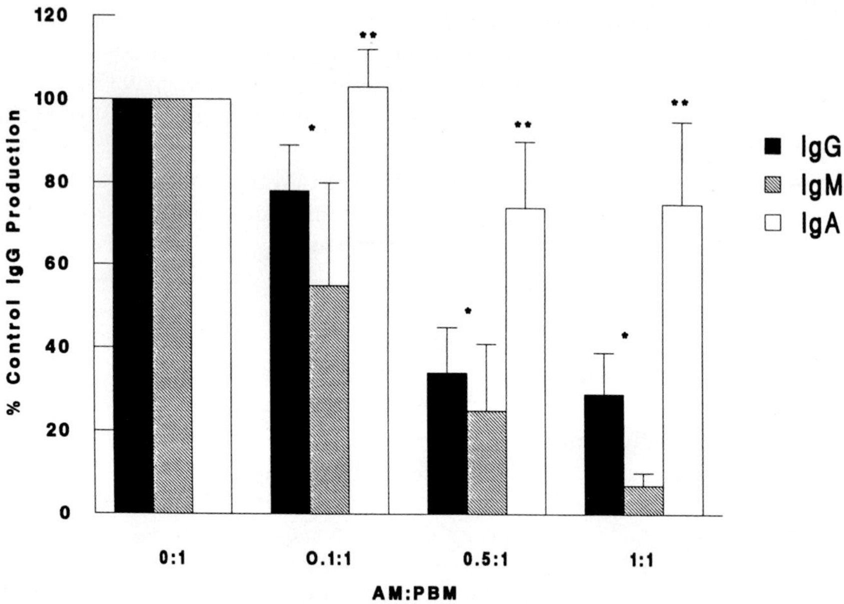

Figure 5 Effects of AMs on Ig production in response to SAC. AMs were incubated in varying ratios with 5×10^4 PBMs in the presence of SAC and 10 U/ml IL-2. Controls were PBM cultured with SAC and IL-2 in the absence of AMs. Culture supernatants were harvested after 6 days and assayed for Ig production by ELISA. Results are the mean ± SEM of seven experiments. $*P < .005$ IgG, IgM versus control. Average Ig production in cultures was 1,100, 900, and 500 ng/ml for IgG (solid bars), IgA (open bars), and IgM (hatched bars), respectively. $**P > .1$ versus control. (From Ref. 42.)

inhibits early events in signal transduction following perturbation of lymphocyte receptors (135). Whether live or fixed, AMs inhibit increases in intracellular calcium concentration in B lymphocytes, T cells, and NK cells in response to cross linkage of surface receptors. Interestingly, in patients with sarcoidosis who demonstrate an active alveolitis and local IgG production in the lung, AMs were permissive of calcium fluxes in response to cross-linked CD3 (136). This observation in concert with others regarding the function of AMs in sarcoid (137,138) suggest that a less suppressive population of AMs is present in patients with sarcoid.

The identity of the membrane-bound substance on AMs responsible for suppressing B-cell responses remains unclear. However, attractive candidates can be found in the surfactant lipids and proteins located in the alveolar space which are also capable of inhibiting lymphocyte calcium fluxes and Ig synthesis (136,139).

B. Effect of Lung DCs on Ig Production

This section briefly reviews the general characteristics of DCs as they relate to B-lymphocyte function. For more extensive reviews of DC phenotype and function, please see other chapters in this text. DCs are located in several different areas of the body, and they are classified into interstitial DCs of the heart, kidney, and lung; Langerhans' cells in the skin and mucous membranes; interdigitating DCs in the thymic medulla and secondary lymphoid tissue (not including germinal center follicular DCs); and blood and lymph DCs (140–144). As recently reviewed by Caux et al. (145), each of these cells is derived from bone marrow with the exception of follicular DCs. Despite varying in their phenotype and function, the primary role of all of these types of DCs is to stimulate cellular immune responses from both naive and memory T lymphocytes (140–144).

Some type of DCs have been demonstrated to be potent inducers of Ig production (146–150). In vitro, murine splenic DCs, or those derived from Peyer's patches have been reported to stimulate Ig production from pre-B lymphocytes and pre-pre-B lymphocytes (148,150). Similarly, murine splenic DCs and human blood DCs induced Ig production from splenic lymphocytes or peripheral blood mononuclear cells, respectively, in a T-lymphocyte–dependent manner (146). In vivo, Sornasse et al. (147) and other investigators reported that antigen-pulsed DCs are able to stimulate Ig production (149). The type of humoral response may depend on the tissue type of DCs utilized to stimulate Ig production (148). For example, splenic DCs induced IgM production, and Peyer's patch DCs stimulated IgG, IgM, and IgA synthesis (148).

Cell contract with T lymphocytes as well as T-lymphocyte–derived cytokines are necessary to induce Ig production (6). Therefore, the production of specific Ig isotypes by DCs from different tissues suggests that these DCs stimulate different cytokine signals form T-lymphocytes specific for the production a given Ig isotype. Depending on the experimental conditions, DCs stimulate IFN-γ production from responding lymphocytes in vitro (48). In vivo, DCs are able to induce a vigorous IFN-γ response, which likely explains how immunization with antigen-pulsed DCs results in IgG2a production in mice (147). The ability of DCs to induce the production of cytokines that stimulate other Ig isotypes, such as IgA or IgM, has not been clearly defined.

In addition to T lymphocyte–derived soluble signals, in vitro studies have shown that DCs must form clusters with T and B lymphocytes to induce Ig production (146–150). The membrane-bound signals necessary for DCs to stimulate Ig production include CD40-CD40 ligand interactions, CD54 (ICAM-1)-CD11a/CD18 (LFA-1) interactions, CD80 (B7-1)/CD86(B7-2)-CD28/CTLA4 interactions, and VLA-4-VCAM-1 interactions (145,151).

In contrast to the DCs described above, follicular dendritic cells (FDCs) are not believed to be derived from leukocytes (145) and reside in specialized areas of lymphoid tissues known as the germinal center (145). Additionally, FDCs appear to be primarily involved in the stimulation of humoral, and not cellular, immune responses (145). The cognate interactions among FDCs, B, and T lymphocytes and the ability of FDCs to capture and express antigens for extended periods of time (152) likely explains their critical role in regulating humoral immunity in the lymph nodes (145,153). In fact, FDCs have been demonstrated to be crucial for the development of memory B lymphocytes and plasmacytes, which are crucial for long-term antibody production (145).

Similar to DCs, several investigators have reported that FDCs stimulate Ig production in vitro in a T-lymphocyte–dependent manner (145,152,153), which requires clustering of FDCs with B and T lymphocytes (152). Also similar to DCs from other tissues T-lymphocyte–derived cytokines have been shown to be crucial for FDC-induced B-lymphocyte activation and Ig production (16). In vitro, IL-2 and IL-10 have been shown to be critical in the generation of memory B lymphocytes and plasma cells (16). Membrane-bound signals on the FDC, B, and T lymphocyte that are important in FDC-induced B-lymphocyte activation and Ig production are similar to those of the DC (145,151). Additionally, FDC–B-lymphocyte interactions mediated through LFA-1/ICAM-1 and VLA-4/VCAM-1 interactions prevents apoptosis in germinal center B lymphocytes (154).

It is important to note that all of the studies examining the ability of DCs and FDCs to induce B-lymphocyte activation and Ig production were performed utilizing DCs and FDCs from nonpulmonary lymphoid tissues. The direct role of pulmonary lymph node DCs and FDCs to induce B-lymphocyte activation and Ig production has not been evaluated. Because pulmonary humoral immune responses to antigens initially occurs in the corresponding draining lymph nodes (15), it may be possible to extrapolate the data regarding interactions between B lymphocytes and nonpulmonary DCs or FDCs to that of interactions between B lymphocytes and pulmonary DCs or FDCs. However, owing to the phenotypic and functional heterogeneity of DCs and FDCs from nonpulmonary tissues, the exact role of DCs and FDCs from pulmonary lymphoid tissues in local humoral immunity may not be easily inferred from studies utilizing nonpulmonary DCs and FDCs.

Effect of Lung Parenchymal DCs on Ig Production

As stated above, DCs can reside within several different locations of solid organs. Exclusive of nodal tissue, recent studies have shown that there are at least two different types of DCs present in the human lung (155); that is, interstitial or parenchymal DCs and airway epithelial DCs. A recent comparison of the two types shows different phenotype and functional capabilities, but both appear to be potent APCs for stimulating T lymphocyte activation in vitro (155). We recently compared the effects of parenchymal lung DCs with AMs on Ig production (48). In response to lung alloantigen, AMs were potent inducers of IgG > IgA production from peripheral blood mononuclear cells (PBMs) in a manner that was dependent of the production of IFN-γ from T lymphocytes in the PBM population (48). In contrast, parenchymal lung DCs were not able to induce IgG or IgA production from PBMs despite stimulating comparable amounts of IFN-γ from T lymphocytes in PBMs (48) and inducing more vigorous proliferation in the responding lymphocyte population (48). An inhibitory effect of DC-stimulated PBMs on Ig production was suggested by experiments showing the release of a soluble suppressive factor in DC-PBM cocultures which decreased IgG production in AM/PBM cultures (48). As reported above, the results of these studies contrast with those of DCs from other anatomical sites which generally stimulate Ig production (146,149). The differences between our studies of lung DCs and those of other organs on Ig production underscores the functional heterogeneity of DCs from different anatomical sites. Indeed, some studies suggest that interstitial DCs may undergo maturational changes after migrating to regional lymph nodes which confers Ig-stimulating properties (124).

VI. Pulmonary Humoral Immunity in Disease States

A. Lung Allograft Rejection

Fung et al. (156) hypothesized that lung allograft rejection involved the interaction of donor lung accessory cells, such as macrophages and DCs, with recipient lymphocytes resulting in upregulating cellular and humoral responses. The fact that rejection episodes diminish in the first 8–12 weeks after transplantation at a time when donor lung macrophages and DCs are replaced by those of the recipient supports this hypothesis (157). We developed a murine model that mimics the histology and immunology of acute lung allograft rejection (97). In this model, when allogeneic (C57BL/6, I-a^b) BAL cells (97% macrophages, 1–2% DCs) were instilled into the lungs of recipient BALBc, (I-a[d]) mice weekly for 4 weeks, a lymphocytic vasculitis and bronchitis analogous to grade 1–2 acute lung allograft rejection, resulted, and locally upregulated Th1 activity in that IL-2, IFN-γ, and IgG2a production were increased in the BAL fluid, but not serum, of recipient mice.

We recently reported that locally upregulated IgG2 production in human allograft BAL fluid, as determined by the ratio of IgG2/IgG1 > 1 in allograft BAL fluid, was associated with acute or chronic lung allograft rejection in patients undergoing lung allograft rejection. These data showed that the ratio of IgG2/IgG1 when >1 in allograft BAL fluid may serve as a marker for lung allograft rejection (111). Similar to our murine studies, only IgG2 was deposited in the perivascular and peribronchiolar tissues of the allograft lung undergoing rejection (D.S. Wilkes, unpublished observations). Significantly, IgG2 production was not increased in the serum of lung allograft recipients undergoing rejection (111).

The fact that upregulated IgG subtype production occurred locally and not systemically in our murine and human studies of lung allograft rejection suggested that B lymphocytes were present in the murine lungs that received allogeneic BAL cells and human lung allografts during rejection. Studies by other investigators and ourselves have confirmed the presence of B lymphocytes at the sites of rejection activity (158,159). In fact, the presence of large numbers B lymphocytes was associated with a refractory response to corticosteriods utilized to treat acute rejection, and this may have been associated the development of chronic rejection known as obliterative bronchiolitis (158). Recent studies from our laboratory utilizing mononuclear cells propagated from human lung allograft tissue revealed that IgG2 is preferentially produced by allograft lung B lymphocytes during acute and chronic rejection (D.S. Wilkes, unpublished observations).

Although Igs are produced locally during lung allograft rejection the antigenic specificity of these IgG2 and IgG2a antibodies in human and murine studies, respectively, are unknown. Antibodies produced in response to allograft rejection are usually referred to as alloantibodies in that they are directed against donor major histocompatibility complex (MHC) antigens (160). However, in our murine (97) and human studies (D.S. Wilkes, unpublished observations), the IgG subtypes were deposited in the perivascular and peribronchiolar connective tissues, which are not rich in MHC class II expression. Therefore, it may be possible that alloantibodies produced during lung allograft rejections may also recognize structures that have some homology to MHC class II proteins. Interestingly, we have shown that type V collagen is recognized as an antigen by IgG2 antibodies produced during lung allograft rejection in humans (D.S. Wilkes, unpublished observations). However, the ability of these anticollagen antibodies to cross react with MHC class II proteins is currently unknown. Unlike studies in cardiac allograft rejection (3), there is no direct evidence that these Igs can induce the pathological changes of the rejection process. Ongoing studies should help to define the specific role of locally produced Ig in the pathogenesis of lung allograft rejection.

B. Role of B Lymphocytes in Human Immunodeficiency Virus Pulmonary Disease

Several studies have shown that patients infected with the human immunodeficiency virus (HIV) are at significantly increased risk for pulmonary infections with encapsulated organisms such as *Streptococcus pneumoniae* (161). Patients at increased risk for infection from these organisms who are not infected with the HIV virus may have defects in the production of opsonizing antibodies, usually of the IgG isotype, specifically IgG2 (162). Interestingly, we have shown that there are decreased levels of IgG in the lower respiratory tract of HIV-infected individuals compared with controls (85). In addition, HIV-infected AMs produce TGF-β which suppresses IgG and IgM production (85). Therefore, it appears that HIV-infected individuals may exhibit local defects in both humoral and as cellular immune defenses.

Acknowledgments

The authors would like to thank Rebecca Synder for her assistance in preparing this manuscript. This work was supported by a Robert Wood Johnson Foundation Award to D.S.W. and the NIH #AR 20852-18 to D.S.W. and NIH HL01797, M01-RR00633 and the James M. Collins Center for Biomedical Research to J.C.W.

References

1. Constant S, Sant'Angelo D, Pasqualini T, Taylor T, Levin D, Flavell R, Bottomly K. Peptide and protein antigens require distinct antigen-presenting cell subsets for the priming of CD4+ T cells. J Immunol 1995; 154:4915–4923.
2. Gyetko MR, Toews GB. Immunology of the aging lung. In: Clinical Chest Medicine, Fein AM, ed. Vol. 14. Pulmonary Disease in the Elderly Patient. Philadelphia: Saunders, 1993:379–391.
3. Russell PS, Chase CM, Winn HJ, Colvin RB. Coronary atherosclerosis in transplanted mouse hearts. II. Importance of humoral immunity. J Immunol 1994; 152: 5135–5141.
4. Reap EA, Sobel ES, Jennette JC, Cohen PL, Eisenberg RA. Conventional B cells, not B1 cells, are the source of autoantibodies in chronic graft-versus-host disease. J Immunol 1993; 151:7316–7323.
5. Banchereau J, Brière F, Liu YJ, Rousset F. Molecular control of B lymphocyte growth and differentiation. Stem Cells 1994; 12:278–288.

5a. Jelinek DF, Lipsky PE. Regulation of human B lymphocyte activation, proliferation, and differentiation. Adv Immunol 1987; 40:1–59.

6. Banchereau J, Rousset F. Human B lymphocytes: phenotype, proliferation, and differentiation. Adv Immunol 1992; 52:125–261.
7. Hentges F. B lymphocyte ontogeny and immunoglobulin production. Clin Exp Immunol 1994; 97 (suppl):3–9.
8. Era T, Nishikawa S, Sudo T, Fu-Ho W, Ogawa M, Kunisada T, Hayashi SI, Nishikawa SI. How B-precursor cells are driven to cycle. Immunol Rev 1994; 137: 35–51.
9. Kincade PW. Molecular interactions between stromal cells and lymphocyte precursors. Semin Immunol 1991; 176:927–935.
10. Tarlinton D. B-cell differentiation in the bone marrow and the periphery. Immunol Rev 1994; 137:203–229.
11. Uckun FM. Regulation of human B-cell ontogeny. Blood 1990; 76:1908–1923.
12. Cumano A, Kee BL, Ramsden DA, Marshall A, Paige CJ, Wu GE. Development of B lymphocytes from lymphoid committed and uncommitted progenitors. Immunol Rev 1994; 137:5–33.
13. Coffman RL, Weissman IL. B220: a B cell-specific member of the T200 glycoprotein family. Nature 1981; 289:681–683.
14. Tsiagbe VK, Linton PJ, Thorbecke GJ. The path of memory B-cell development. Immunol Rev 1992; 126:113–141.
15. Lipscomb MF, Bice DE, Lyons CR, Schuyler MR, Wilkes DS. The regulation of pulmonary immunity. In: Dixon FJ, ed. Advances in Immunology. Vol. 59. Academic Press 1995.
16. Arpin C, Dèchanet J, Van Kooten C, Merville P, Grouard G, Brière F, Banchereau J, Liu YJ. Generation of memory B cells and plasma in vitro. Science 1995; 268: 720–722.
17. Blanchard D, Gaillard C, Hermann P, Banchereau J. Role of CD40 antigen and interleukin-2 in T cell–dependent human B lymphocyte growth. Eur J Immunol 1994; 24:330–335.
18. Burdin N, Van Kooten C, Galibert L, Abrams JS, Wijdenes J, Banchereau J, Rousset F. Endogenous IL-6 and IL-10 contribute to the differentiation of CD40-activated human B lymphocytes. J Immunol 1995; 154:2533–2544.
19. Reth M, Hombach J, Wienands J, Campbell KS, Chien N, Justement LB, Cambier JC. The B-cell antigen receptor complex. Immunol Today 1991; 12:196–201.
20. Scheuermann RH, Uhr JW. Connection between signal transduction components and cellular responses initiated by antigen receptor on B lymphocytes. J Exp Med 1995; 182:903–906.
21. Tohma S, Lipsky PE. Analysis of the mechanisms of T cell-dependent polyclonal activation of human B cells. J Immunol 1991; 146:2544–2552.
22. Croft M, Swain SL. B cell response to fresh and effector T helper cells. Role of cognate T-B interaction and the cytokines IL-2, IL-4 and IL-6. J Immunol 1991; 146: 4055–4064.
23. Van Vlasselaer P, Gascan H, DE Waal Malefyt R, de Vries JE. IL-2 and a contact-mediated signal provided by TCR$\alpha\beta^+$ or TCR$\alpha\sigma^+$ CD4$^+$ T cells induce polyclonal Ig production by committed human B cells. Enhancement by IL-5, specific inhibition of IgA synthesis by IL-4. J Immunol 1992; 148:1674–1684.
24. Hodgkin PD, Yamashita LC, Coffman RL, Kehry MR. Separation of events mediat-

ing B cell proliferation and Ig production by using T cell membranes and lymphokines. J Immunol 1990; 145:2025–2034.
25. Kawakami K, Parker DC. Antigen and helper T lymphocytes activate B lymphocytes by distinct signalling pathways. Eur J Immunol 1993; 23:77–84.
26. McLeod E, Caldwell JL, Kaltreider HB. Pulmonary immune responses of inbred mice. Appearance of antibody-forming cells in C57BL/6 mice after intrapulmonary or systemic immunization with sheep erythrocytes. Am Rev Respir Dis 1978; 118: 561–571.
27. Kaltreider HB, Turner FN. Appearance of antibody-forming cells in lymphocytes from the lower respiratory tract of the dog after intrapulmonary or intravenous immunization with sheep erythrocytes. Am Rev Respir Dis 1976; 113:613–617.
28. Curtis JL, Kaltreider HB. Characterization of bronchoalveolar lymphocytes during a specific antibody-forming cell response in the lungs of mice. Am Rev Respir Dis 1989; 139:393–400.
29. Sherman LA, Chattopadhyay S. The molecular basis of allorecognition. Annu Rev Immunol 1993; 11:385–402.
30. Brandtzaeg P. Session V: Biology and pathogenesis of mucous membrane infections. Humoral immune response patterns of human mucosae: Induction and relation to bacterial respiratory tract infections. J Infect Dis 1991; 165:S167–S176.
31. Nguyen L, Knipe DM, Finberg RW. Mechanism of virus-induced Ig subclass shifts. J Immunol 1994; 152:478–484.
32. Metlay JP, Puré E, Steinman RM. Distinct features of dendritic cells and acti-Ig activated B cells as stimulators of the primary mixed leukocyte reaction. J Exp Med 1989; 169:239–254.
33. Finkelman FD, Holmes J. Lymphokine control of in vivo immunoglobulin isotype selection. Annu Rev Immunol 1990; 8:303–333.
34. Cherwinski HM, Schumacher JH, Brown KD, Mossmann TR. Two types of mouse helper T cell clones. III. Further differences in lymphokine synthesis between Th1 and Th2 clones revealed by RNA hybridization, functionally monospecific bioassays, and monoclonal antibodies. J Exp Med 1987; 166:1229–1244.
35. Karasuyama H, Rolink A, Melchers F. Recombinant interleukin 2 or 5, but not 3 or 4, induces maturation of resting mouse B lymphocytes and propagates proliferation of activated B cell blasts. J Exp Med 1988; 167:1377–1390.
36. Romagnani S, Del Prete G, Giudizi MG, Biagiotti R, Almerigogna F, Tiri A, Alessi A, Mazzetti M, Ricci M. Direct induction of human B-cell differentiation by recombinant interleukin-2. Immunology 1986; 58:31–35.
37. Ralph P, Jeong G, Welte K, Mertelsmann R, Rabin H, Henderson LE, Souza LM, Boone TC, Robb RJ. Stimulation of immunoglobin secretion in human B lymphocytes as a direct effect of high concentrations of IL2. J Immunol 1984; 133:2442–2445.
38. Nakagawa N, Nakagawa T, Volkman DJ, Ambrus JL Jr, Fauci AS. The role of interleukin 2 in inducing Ig production in a pokeweed mitogen-stimulated mononuclear cell system. J Immunol 1987; 138:795–801.
39. Waldmann TA, Goldman CK, Robb RJ, Depper JM, Leonard WJ, Sharrow SO, Bongiovanni KF, Korsmeyer SJ, Greene WC. Expression of interleukin 2 receptors on activated human B cells. J Exp Med 1984; 160:1450–1466.

40. Zubler RH, Lowenthal JW, Erard F, Hashimoto N, Devos R, MacDonald HR. Activated B cells express receptors for, and proliferate in response to, pure interleukin 2. J Exp Med 1984; 160:1170–1183.
41. Muracuchi A, Kehrl JH, Longo DL, Volkman DJ, Smith KA, Fauci AS. Interleukin 2 receptors on human B cells. Implications for the role of interleukin 2 in human B cell function. J Exp Med 1985; 161:181–197.
42. Wilkes DS, Yarbrough WC, Weissler JC. Human alveolar macrophages inhibit immunoglobulin production in response to direct B cell mitogen. Am J Respir Cell Mol Biol 1993; 9:141–147.
43. Tucci A, James H, Chicheportiche R, Bonnefoy JY, Dayer JM, Zubler RH. Effects of eleven cytokines and of IL-1 and tumor necrosis factor-α in a human B cell assay. J Immunol 1992; 148:2778–2784.
44. Jelinek DF, Lipsky PE. Inhibitory influence of IL-4 on human B cell responsiveness. J Immunol 1988; 141:164–173.
45. Clark EA, Shu GL, Lüscher B, Draves KE, Banchereau J, Ledbetter JA, Valentine MA. Activation of human B cells. Comparison of the signal transduced by IL-4 to four different competence signals. J Immunol 1989; 143:3873–3880.
46. Rabin EM, Mond JJ, Ohara J, Paul WE. Interferon-γ inhibits the action of B cell stimulatory factor (BSF)-1 on resting B cells. J Immunol 1986; 137:1573–1576.
47. Splawski JB, Jelinek DF, Lipsky PE. Immunomodulatory role of IL-4 on the secretion of Ig by human B cells. J Immunol 1989; 142:1569–1575.
48. Wilkes DS, Weissler JC. Alloantigen-induced immunoglobulin production in human lung: Differential effects of accessory cell populations on IgG synthesis. Am J Respir Cell Mol Biol 1994; 10:339–346.
49. Spriggs MK, Armitage RJ, Strockbine L, Clifford KN, Macduff BM, Sato TA, Maliszewski CR, Fanslow WC. Recombinant human CD40 ligand stimulates B cell proliferation and immunoglobulin E secretion. J Exp Med 1992; 176:1543–1550.
50. Del Prete G, Maggi E, Parronchi P, Chrétien I, Tiri A, Macchia D, Ricci M, Banchereau J, De Vries J, Romagnani S. IL-4 is an essential factor for the IgE synthesis induced in vitro by human T cell clones and their supernatants. J Immunol 1988; 140:4193–4198.
51. Snapper CM, Pecanha LMT, Levine AD, Mond JJ. IgE class switching is critically dependent upon the nature of the B cell activator, in addition to the presence of IL-4. J Immunol 1991; 147:1163–1166.
52. Snapper CM, Paul WE. B cell stimulatory factor-1 (interleukin 4) prepares resting murine B cells to secrete IgG1 upon subsequent stimulation with bacterial lipopolysaccharide. J Immunol 1987; 139:10–14.
53. Gauchat JF, Lebman DA, Coffman RL, Gascan H, de Vries JE. Structure and expression of germline ε transcripts in human B cells induced by interleukin 4 to switch to IgE production. J Exp Med 1990; 172; 463–473.
54. Robinson DS, Hamid Q, Ying S, Tsicopoulos A, Barkans J, Bentley AM, Corrigan C, Durham SR, Kay AB. Predominant Th2-like bronchoalveolar T-lymphocyte population in atopic asthma. N Eng J Med 1992; 326:298–304.
55. Okayama Y, Petit-Frére C, Kassel O, Semper A, Quint D, Tunon-de-Lara MJ, Bradding P, Holgate ST, Church MK. IgE-dependent expression of mRNA for IL-4 and IL-5 in human lung mast cells. J Immunol 1995; 155:1796–1808.

56. Mahanty S, Nutman TB. The biology of interleukin-5 and its receptor. Cancer Invest 1993; 11:624–634.
57. Bertolini JN, Sanderson CJ, Benson EM. Human interleukin-5 induces staphylococcal A cowan 1 strain-activated human B cells to secrete IgM. Eur J Immunol 1993; 23:398–402.
58. Purkerson JM, Isakson PC. Independent regulation of DNA recombination and immunoglobulin (Ig) secretion during isotype switching to IgG1 and IgE. J Exp Med 1994; 179:1877–1883.
59. Eckmann L, Morzycka-Wroblewska E, Smith JR, Kagnoff MF. Cytokine-induce differentiation of IgA B cells: studies using an IgA expressing B-cell lymphoma. Immunology 1992; 76:235–241.
60. Mclntyre TM, Kehry MR, Snapper CM. Novel in vitro model for high-rate IgA class switching. J Immunol 1995; 154:3156–3161.
61. Mandler R, Chu CC, Paul WE, Max EE, Snapper CM. Interleukin 5 induces Sμ-Sγ1 DNA rearrangement in B cells activated with dextran-anti-IgD antibodies and interleukin 4: a three component model for Ig class switching. J Exp Med 1993; 178: 1577–1586.
62. Bertolini JN, Benson EM. The role of human interleukin-6 in B-cell isotype regulation and differentiation. Cell Immunol 1990; 125:197–209.
63. Hirano T, Akira S, Taga T, Kishimoto T. Biological and clinical aspects of interleukin 6. Immunol Today 1990; 443–449.
64. Kishimoto T, Akira S, Taga T. Interleukin-6 and its receptor: A paradigm for cytokines. Science 1992; 258:593–600.
65. Lue C, Kiyono H, McGhee JR, Fujihashi K, Kishimoto T, Hirano T, Mestecky J. Recombinant human interleukin 6 (rhIL-6) promotes the terminal differentiation of in vivo-activated human B cells into antibody-secreting cells. Cell Immunol 1991; 132:423–432.
66. Wilkes DS, Neimeier M, Mathur PN, Soliman DM, Twigg HL, III, Bowen LK, Heidler KM. Effect of human lung allograft alveolar macrophages on IgG production: Immunoregulatory role of interleukin-10, transforming growth factor-β, and interleukin-6. Am J Respir Cell Mol Biol 1995; 13:621–628.
67. Ramsay AJ, Husband AJ, Ramshaw IA, Bao S, Matthaei KI, Koehler G, Kopf M. The role of interleukin-6 in mucosal IgA antibody responses in vivo. Science 1994; 264; 561–563.
68. De Waal Malefyt R, Yssel H, Roncarolo MG, Spits H, de Vries JE. Interleukin-10. Immunology 1992; 4:314–320.
69. Moore KW, O'Garra A, de Waal Malefyt R, Vieira P, Mosmann TR. Interleukin-10. Annu Rev Immunol 1993; 11:165–190.
70. Burdin N, Van Kooten C, Galibert L, Abrams JS, Wijdenes J, Banchereau J, Rousset F. Endogenous IL-6 and IL-10 contribute to the differentiation of CD40-activated human B lymphocytes. J Immunol 1995; 154:2533–2544.
71. Peçanha LMT, Snapper CM, Lees A, Mong J J. Lymphokine control of type 2 antigen response: IL-10 inhibits IL-5- but not IL-2-induced immunoglobulin secretion by T cell-independent antigens. J Immunol 1992; 148:3427–3432.
72. Peçanha LMT, Snapper CM, Lees A, Yamaguchi H, Mong JJ. IL-10 inhibits T cell–independent but not T cell–dependent responses in vitro. J Immunol 1993; 150: 3215–3223.

73. Punnonen J, de Waal Malefyt R. IL-10 and viral IL-10 prevent IL-4-induced IgE synthesis by inhibiting the accessory cell function of monocytes. J Immunol 1993; 151:1280–1289.
74. Brière F, Servet-Delprat C, Bridon JM, Saint-Remy JM, Banchereau J. Human interleukin 10 induces naive surface immunoglobulin D^+ (slg D^+) B cells to secrete IgG1 and IgG3. J Exp Med 1994; 179:757–762.
75. Defrance T, Vanbervliet B, Brière F, Durand I, Rousset F, Banchereau J. Interleukin 10 and transforming growth factor β cooperate to induce anti CD40–activated naive human B cells to secrete immunoglobulin A. J Exp Med 1992; 175:671–682.
76. de Waal Malefyt R, Figdor CG, Huijbens R, Mohan-Peterson S, Bennett B, Culpepper J, Dang W, Zurawski G, de Vries JE. Effects of IL-13 on phenotype, cytokine production, and cytotoxic function of human monocytes. Comparison with IL-4 and modulation by IFN-γ or IL-10. J Immunol 1993; 151:6370–6381.
77. Burd PR, Thompson WC, Max EE, Mills FC. Activated mast cells produce interleukin 13. J Exp Med 1995; 181:1373–1380.
78. de Vries JE, Zurawski G. Immunoregulatory properties of IL-13: Its potential role in atopic disease. Int Arch Allergy Immunol 1995; 106:175–179.
79. Brinkmann V, Kristofic C. TCR-stimulated naive human $CD4^+45R0^-$ T cells develop into effector cells that secrete IL-13, IL-5, and IFN-γ, but no IL-4, and help efficient IgE production by B cells. J Immunol 1995; 154:3078–3087.
80. Gauchat JF, Henchoz S, Mazzel G, Aubry JP, Brunner T, Blasey H, Life P, Talabot D, Flores-Romo L, Thompson J, Kishl K, Butterfield J, Dahinden C, Bonnefoy JY. Induction of human IgE synthesis in B cells by mast cells and basophils. Nature 1993; 365:340–343.
81. Rodland KD, Muldoon LL, Magun BE. Cellular mechanisms of TFG-β action. J Invest Dermatol 1990; 94:33S–40S.
82. McIntyre T, Klinman D, Rothman P, Lugo M, Dasch J, Mond J, Snapper C. Transforming growth factor β_1 selectivity stimulates immunoglobulin G2b secretion by lipopolysaccharide-activated murine B cells. J Exp Med 1993; 177:1031–1037.
83. Sonoda E, Matsumoto R, Hitoshi Y, Ishii T, Sugimoto M, Araki S, Tominaga A, Yamaguchi N, Takatsu K. Tranforming growth factor β induces IgA production and acts additively with interleukin 5 for IgA production. J Exp Med 1989; 170:1415–1420.
84. Snapper C, Waegell W, Beernink H, Dasch J. Transforming growth factor-β1 is required for secretion of IgG of all subclasses by LPs-activated murine B cells in vitro. J Immunol 1993; 151:4625–4636
85. Twigg HL, Spain BA, Soliman DM, Bowen LK, Heidler KM, Wilkes DS. Impaired IgG production in the lungs of HIV-infected individuals. Cell Immunol 1996; 170: 127–133.
86. Kim P, Kagnoff M. Transforming growth factor β1 increases IgA isotype switching at the clonal level. J Immunol 1990; 145:3773–3778.
87. Chomarat P, Rissoan M, Banchereau J, Miossec P. Interferon γ inhibits interleukin 10 production by monocytes. J Exp Med 1993; 177:523–527.
88. Reynolds D, Boom W, Abbas A. Inhibition of B lymphocyte activation by interferon-γ. J Immunol 1987; 139:767–773.
89. Rousset F, Garcia E, Banchereau J. Cytokine-induced proliferation and immuno-

globulin production of human B lymphocytes triggered through their CD40 Antigen. J Exp Med 1991; 705–710.
90. Defrance T, Aubry J, Vanbervliet B, Banchereau J. Human interferon-γ acts as a B cell growth factor in the anti-IgM antibody co-stimulatory assay but has no direct B cell differentiation activity. J Immunol 1986; 137:3861–3867.
91. Finkelman F, Holmes J. Lymphokine control of in vivo immunoglobulin isotype selection. Annu Rev Immunol; 1990. 8:303–333.
92. Bossie A, Vitetta E. IFN-γ enhances secretion of IgG2a from IgG2a-committed LPS-stimulated murine B cells: implications for the role of IFN-γ in class switching. Cell Immunol 1991; 135:95–104.
93. Snapper C, Peschel C, Paul W. IRN-γ stimulates IgG2a secretion by murine B cells stimulated with bacterial lipopolysaccharide. J Immunol 1988; 140:2121–2127.
94. Snapper C, Mclntyre T, Mandler R, Pecanha L, Finkelman F, Lees A, Mond J. Induction of IgG3 secretion by interferon γ: a model for cell-independent class switching in response to T cell–independent type 2 antigens. J Exp Med 1992; 175: 1367–1371.
95. Kitani A, Strober W, Regulation of Cγ subclass germ-line transcripts in human peripheral blood B cells. J Immunol 1993; 151:3478–3488.
96. Kawano Y, Noma T, Yata J. Regulation of human IgG subclass production by cytokines: IFN-γ and IL-6 act antagonistically in the induction of human IgG1 but additively in the induction of IgG2. J Immunol 1994; 153:4948–4958.
97. Wilkes D, Heidler K, Bowen L, Quinlan W, Doyle N, Cummings O, Doerschuk C. Allogeneic bronchoalveolar lavage cells induce the histology of acute lung allograft rejection, and deposition of IgG2a in recipient murine lungs. J Immunol 1995; 155: 2775–2783
98. Carayannopoulos L, Capra JD. Immunoglobulins: Structure and function. In: Paul WE, ed. 3rd ed. Fundamental Immunology. New York: Raven Press, 1993:283–314.
99. Virella G, Wang AC. Immunoglobulin structure. Immunol Series 1993; 58:75–90.
100. Reynolds H. Immunoglobulin G and its function in the human respiratory tract. Mayo Clin Proc 1988; 63:161–174.
101. Burton D. Immunoglobulin G: functional sites. Mol Immunol 1985; 22:161–206.
102. Merrill W, Naegel G, Olchowski J, Reynolds H. Immunoglobulin G subclass proteins in serum and lavage fluid of normal subjects. Am Rev Respir Dis 1985; 131: 584–587.
103. Burnett D. Immunoglobulins in the lung. Thorax 1986; 41:337–344.
104. Bergamaschini L, Santangelo T, Faricciotti A, Ciavarella N, Mannucci P, Agostoni A. Study of complement-medicated anaphylaxis in humans: the role of IgG subclasses (IgG1 and/or IgG4) in the complement-activating capacity of immune complexes. J Immunol 1996; 156:1256–1261.
105. Miklos K, Tolnay M, Bazin H, Medgyesi G. Antibody mediated lysis of hapten-conjugated target cells by macrophages and by complement: the influence of IgG subclass, antibody and hapten density. Mol Immunol 1992; 29:379–384.
106. Hughes-Jones N, Gorick B, Howard J. The mechanism of synergistic complement-mediated lysis of rat red cells by monoclonal IgG antibodies. Eur J Immunol 1983; 13:641–646.
107. Hance AJ, Saltini C, Crystal RG. Does de novo immunoglobulin synthesis occur on

the epithelial surface of the numan lower respiratory tract? Am Rev Respir Dis 1988; 137:17–24.
108. Redfern R, Rowe FP. Wild rats and mice. UFAW Handbook Care Man Lab Animals. 1987:266–274.
109. Gracia J, Rodrigo M, Morell F, Vendrell M, Miravitlles M, Cruz M, Codina R, Bofil J. IgG subclass deficienceis associated with bronchiectasis. Am J Respir Crit Care Med 1996; 153:650–655.
110. Siber G, Schur P, Aisenberg A, Weitzman S, Schiffman G. Correlation between serum IgG-2 concentrations and the antibody response to bacterial polysaccharide antigens. N Engl J Med 1980; 303:178–182.
111. Wilkes D, Heidler K, Niemeier M, Schwenk G, Mathur P, Breite W, Cummings O, Weissler J. Increased bronchoalveolar IgG1/IgG1 ratio is a marker for human lung allograft rejection. J Invest Med 1994; 42:652–659.
112. Kerr M. The structure and function of human IgA. Biochemistry J 1990; 271:285–296.
113. Soutar CA. Distribution of plasma cells and other cells containing immunoglobulin in the respiratory tract and alveoli. Virchows Archiv (Pathol Anat) 1984; 404: 369–380.
114. Mazanec M, Nedrud J, Kaetzel C, Lamm M. A three-tiered view of the role of IgA in mucosal defense. Immunol Today 1993; 14:430–435.
115. Liblau R, Bach J. Selective IgA deficiency and autoimmunity. Int Arch Allergy Immunol 1992; 99:16–27.
116. Brewer J, Randall T, Parkhouse RM, Corley R. IgM hexamers? Immunol Today 1994; 15:165–168.
117. Sutton B, Gould H, The human IgE network. Nature 1993; 366:421–428.
118. Delespesse G, Sarfati M, Wu CY, Fournier S, Letellier M. The low-affinity receptor for IgE. Immunol Rev 1992; 125:77–97.
119. Ohkawara Y, Yamauchi K, Tanno Y, Tamura G, Ohtani H, Nagura H, Ohkuda K, Takishima T. Human lung mast cells and pulmonary macrophages produce tumor necrosis factor-α in sensitized lung tissue after IgE receptor triggering. Am J Respir Cell Mol Biol 1992; 7:385–392.
120. Sherrill D, Lebowitz M, Halonen M, Barbee R, Burrows B. Longitudinal evaluation of the association between pulmonary function and total serum IgE. Am J Respir Crit Care Med 1995; 152:98–102.
121. Ericsson C, Swartengren M, Mossberg B, Camner P. Bronchial reactivity, lung function, and serum immunoglobulin E in smoking-discordant monozygotic twins. Am Rev Respir Dis 1993; 147:296–300.
122. Martinez F, Wright A, Taussig L, Holberg C, Halonen M, Morgan W, Group Health Medical Associates. Asthma and wheezing in the first six years of life. N Engl J Med 1995; 332:133–138.
123. Geha RS, Leung DYM. Hyper-Immunoglobulin E syndrome. Immunol Rev 1989; 1: 155–172.
124. Holt P, Oliver J, Bilyk N, McMenamin C, McMenamin P, Kraal G, Thepen T. Down regulation of the antigen presenting cell functions of pulmonary dendritic cells in vivo by resident alveolar macrophages. J Exp Med 1993; 177:397–408.
125. Kaltreider HB, Bryd PK, Daughety TW, Shalaby MR. The mechanism of appearance

of specific antibody-forming cells in lungs of inbred mice after intratracheal immunization with sheep erythrocytes. Am Rev Respir Dis 1983; 127:316–321.
126. Bice DE, Harris DL, Hill JO, Muggenburg BA, Wolff RK. Immune responses following localized deposition of antigen in the lung. Am Rev Respir Dis 1980; 122: 755–760.
127. Mason MH, Bice DE, Muggenburg BA. Local pulmonary immune responsiveness after multiple antigenic exposure in the cyomolgus monkey. Am Rev Respir Dis 1985; 132:657–660.
128. Bice DE, King-Herbert AP, Morris MJ, Hanna N, Haley PJ. Inflammation and recruitment of immune cells into the lung. Reg lmmunol 1989; 2:376–384.
129. Bice DE, Weissman DN, Muggenburg BA. Long-term maintenance of localized antibody responses in the lung. Immunology 1989; 74:215–222.
130. Bice DE, Jones SE, Muggenburg BA. Long-term antibody production after lung immunization and challenge: Role of lung and lymphoid tissues. Am J Respir Cell Mol Biol 1993; 8:662–667.
131. Bauer W, Gorny NK, Baumann HR, Morell A. T-lymphocyte subsets and immunoglobulin concentrations in bronchoalveolar lavage of patients with sarcoidosis and high and low intensity alveolitis. Am Rev Respir Dis 1985; 132:1060–1065.
132. Rankin JA, Naegel GP, Schrader CE, Matthay RA, Reynolds HY. Air-space immunoglobulin production and levels in bronchoalveolar lavage fluid of normal subjects and patients with sarcoidosis. Am Rev Respir Dis 1983; 127:442–448.
133. Kehrl JH, Taylor AS, Delsing GA, Roberts AB, Sporn MB, Fauci AS. Further studies of the role of transforming growth factor-β_1 in human B cell function. J Immunol 1989; 143:1868–1874.
134. Roth MD, Golub SH. Human pulmonary macrophages utilize prostaglandins and transforming growth factor β_1 to suppress lymphocyte activation. J. Leukoc Biol 1993; 53:366–371.
135. Yarbrough WC, Wilkes DS Jr, Weissler JC. Human alveolar macrophages inhibit receptor-mediated increases in intracellular calcium concentration in lymphocytes. Am J Respir Cell Mol Biol 1991; 5:411–415.
136. Weissler JC, Mendelson C, Moya F, Yarbrough WC Jr. Effect of interstitial lung disease macrophages on T-cell signal transduction. Am J Respir Crit Care Med 1994; 149:191–196.
137. Lem VM, Lipscomb Mf, Weissler JC, Nunez G, Ball EJ, Stastny P, Toews GB. Bronchoalveolar cells from sarcoid patients demonatrate enhanced antigen presentation. J Immunol 1985; 135:1766–1771.
138. Hance AJ, Douches S, Winchester RJ, Ferrans VJ, Crystal RG. Characterization of mononuclear phagocyte subpopulations in the human lung by using monoclonal antibodies: changes in alveolar macrophage phenotype associated with pulmonary sarcoidosis. J Immunol 1985; 134:284–292.
139. Honda Y, Tsunematsu K, Suzuki A, Akino T. Changes in phospholipids in bronchoalveolar lavage fluid of patients with interstitial lung diseases. Lung 1988; 166: 293–301.
140. Inaba K, Inaba M, Romani N, Aya H, Deguchi M, Ikehara S, Muramatsu S, Steinman R. Generation of large numbers of dendritic cells from mouse bone marrow cultues

supplemented with granulocyte/macrophage colony-stimulating factor. J Exp Med 1992; 176:1693–1702.
141. Inaba K, Steinman RM, Pack MW, Aya H, Inaba M, Sudo T, Wolpe S, Schuler G. Identification of proliferating dendritic cell precursors in mouse blood. J Exp Med 1992; 175:1157–1167.
142. Paglia P, Girolomoni G, Robbiati F, Granucci F, Ricciardi-Castagnoli P. Immortalized dendritic cell line fully competent in antigen presentation initiates primary T cell respones in vivo. J Exp Med 1993; 178:1893–1901.
143. Nicod LP, Lipscomb MF, Weissler JC, Lyons CR, Albertson J, Toews GB. Mononuclear cells in human lung parenchyma: Characterization of a potent accessory cell not obtained by bronchoalveolar lavage. Am Rev Respir Dis 1987; 136:818–823.
144. Pimpinelli N, Santucci M, Romagnoli P, Giannotti B. Dendritic cells in T- and B-cell proliferation in the skin. Dermatol Clin 1994; 12:255–270.
145. Caux C, Liu YJ, Banchereau J. Recent advances in the study of dendritic cells and follicular dendritic cells. Immunol Today 1995; 16:2–4.
146. Inaba K, Steinman RM, Van Voorhis WC, Muramatsu S. Dendritic cells are critical accessory cells for thymus-dependent antibody responses in mouse and in man. Proc Natl Acad Sci USA 1983; 80:6041–6045.
147. Sornasse T, Flamand V, De Becker G, Bazin H, Tielemans F, Thielemans K, Urbain J, Leo O, Moser M. Antigen-pulsed dendritic cells can efficiently induce an antibody response in vivo. J Exp Med 1992; 175:15–21.
148. Saleh F, Spalding DM, Griffin JA. Pre-pre-B cells containing only cytoplasmic B220 can be induced by dendritic cells and T cells to differentiate in vitro. Cell Immunol 1989; 123:44–59.
149. De Becker G, Sornasse T, Nabavi N, Bazin H, Tielemans F, Urbain J, Leo O, Moser M. Immunoglobulin isotype regulation by antigen-presenting cells in vivo. Eur J Immunol 1994; 24:1523–1528.
150. Spalding DM, Griffin JA. Different pathways of differentiation of pre-B cell lines are induced by dendritc cells and T cells from different lymphoid tissues. Cell 1986; 44:507–515.
151. Larsen CP, Ritchie SC, Pearson TC, Linsley PS, Lowry RP. Functional expression of the costimulatory molecule B7/BB1, on murine dendritic cell populations. J Exp Med 1992; 176:1215–1220.
152. MacLennan ICM. Germinal centers. Annu Rev Immunol 1994; 12:117–139.
153. Burton GF, Conrad DH, Szakal AK, Tew JG. Follicular dendritic cells and B cells costimulation. J Immunol 1993; 150:31–38.
154. Koopman G, Keehnen RMJ, Lindhout E, Newman W, Schimizu Y, van Seventer GA, de Groot C, Pals ST. Adhesion through the LFA-1 (CD11a/CD18)-ICAM-1 (CD54) and the VLA-4 (CD49d)-VCAM-1 (CD106) pathways prevents apoptosis of germinal center B cells. J Immunol 1994; 152:3760–3767.
155. Gong JL, McCarthy KM, Telford J, Tamatani T, Miyasaka M, Schneeberger EE. Intraepithelial airway dendritic cells: A distinct subset of pulmonary dendritic cells obtained by microdissection. J Exp Med 1992; 175:797–807.
156. Fung JJ, Zeevi A, Kaufman C, Paradis IL, Dauber JH, Hardesty RL, Griffith B, Duquesnoy RJ. Interactions between bronchoalveolar lymphocytes and macrophages in heart-lung transplant recipients. Hum Immunol 1985; 14:2587–2594.

157. Paradis IL, Marrai M, Zeevi A. HLA phenotype of lung lavage cells following heart lung transplantation. J Heart Transplant 1985; 4:422–425.
158. Yousem SA, Martin T, Paradis IL, Keenan R, Griffith BP. Can immunohistological analysis of transbronchial biopsy specimens predict responder status in early acute rejection of lung allografts? Hum Pathol 1994; 25:525–529.
159. Hasegawa S, Ockner DM, Ritter JH, Patterson GA, Trulock EP, Cooper JD, Wick MR. Expression of class II major histocompatibility complex antigens (HLA-DR) and lymphocyte subset immunotyping in chronic pulmonary transplant rejection. Arch Pathol Lab Med 1995; 119:432–439.
160. Oluwole SF, Tezuka K, Wasfie T, Stegall MD, Reemtsma K, Hardy MA. Humoral immunity in allograft rejection. The role of cytotoxic alloantibody in hyperacute rejection and enhancement of rat cardiac allografts. Transplant 1989; 48:751–755.
161. Janoff EN, Breiman RF, Daley CL, Hopewell PC. Pneumococcal disease during HIV infection: Epidemiologic, clinical, and immunologic perspectives. Ann Intern Med 1992; 117:314–324.
162. Mascart-Lemone F, Gérard M, Libin M, Crusiaux A, Franchioly P, Lambrechts A, Goldman M, Clumeck N. Differential effect of human immunodeficiency virus infection on the IgA and IgG antibody responses to pneumococcal vaccine. J Infect Dis 1995; 172:1253–1260.

Part Four

MACROPHAGES IN LUNG INFECTIONS

15

Mechanisms of Macrophage Recruitment into Infected Lungs

GARY B. HUFFNAGLE and GALEN B. TOEWS

University of Michigan
Ann Arbor, Michigan

I. Introduction

The objective of host defenses in pulmonary infection is, simply stated, to clear the infection from the lungs. The major mechanism by which host defenses are marshalled to clear a pulmonary infection is via the recruitment and activation of effector cells into the lungs. Macrophages are the major effector cells in a number of infections. Although macrophages do possess some constitutive antimicrobial activity, macrophage-activating factors are usually required for an effective response. In some infections, the presence of opsonins is sufficient, but in other pulmonary infections, T-cell–mediated immunity is required for macrophage activation. This chapter focuses on the recruitment and activation of macrophages during a T-cell–mediated immune response to a pulmonary infection using the model of murine pulmonary cryptococcosis (1,2).

II. Murine Pulmonary Cryptococcosis as a Model of Macrophage Recruitment into Infected Lungs

The recruitment of macrophages into the lungs is a hallmark of an effective immune response against *Cryptococcus neoformans*, the etiological agent of cryptococcosis (1–4). *C. neoformans* is an ubiquitous, encapsulated yeast that is the most common cause of fatal fungal infection in patients with acquired immunodeficiency syndrome (AIDS) (5–9). The respiratory tract is the portal of entry for this opportunistic infectious agent, and a protective pulmonary cell-mediated immune response is crucial to eradicate the organism and prevent cryptococcal dissemination from the lungs. Recruited macrophages are the primary effector cells in eliminating *C. neoformans* from the lungs; thus, murine

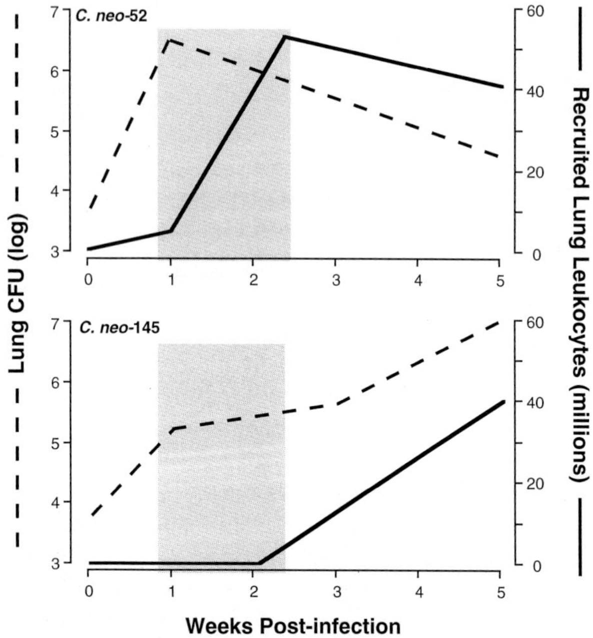

Figure 1 Pulmonary growth and clearance of an opportunistic and a virulent strain of *C. neoformans*. Data shown are the lung CFU in CBA/J mice following i.t. inoculation with 5000 CFU of the opportunistic strain *Cneo-52* or the virulent strain *Cneo-145*. The region highlighted is the period of time when the number of inflammatory cells rapidly increases, resulting in the initiation of pulmonary clearance of *Cneo-52*. The same period of time is highlighted in the response graph for *Cneo-145* to emphasize the difference between the two strains of *C. neoformans*. (Adapted from ref. 14.)

models of pulmonary cryptococcosis provide valuable insights into the mechanisms of macrophage recruitment into the lungs.

The interaction of two factors determines the outcome of a pulmonary infection: (1) the production of virulence factors by the invading microbe and (2) the recognition of the microbe as a pathogen by host defenses and subsequent development of an appropriate host response. The first factor, virulence factor production, is a significant determinant of *C. neoformans* virulence in mice. Strains of *C. neoformans* can be avirulent, virulent, or "opportunistic" (10–13). For example, low-melanin–producing *Cneo-52* is an opportunistic strain of *C. neoformans*, whereas high-melanin–producing *Cneo-145* is a virulent strain (1,12–19). Following intratracheal (i.t.) inoculation of *Cneo-145* into CBA/J mice, the organisms progressively increase in number in the lungs (Fig. 1) and disseminate widely to liver, spleen, kidney, and brain by 14–21 days after inoculation (13,14,19). Mean survival of *Cneo-145*–infected mice is 49 days. A distinctly different pattern is found following i.t. inoculation with *Cneo-52* (1,12–17). The pulmonary burden of *Cneo-52* increases over 1000-fold by day 7, but then it is progressively cleared over the next 5–6 weeks (Fig. 2). Thus, models of both fatal infection and resolving infections are available to dissect the mechanisms of macrophage recruitment and activation in anticryptococcal pulmonary defenses.

A second factor that influences the outcome of a pulmonary cryptococcal

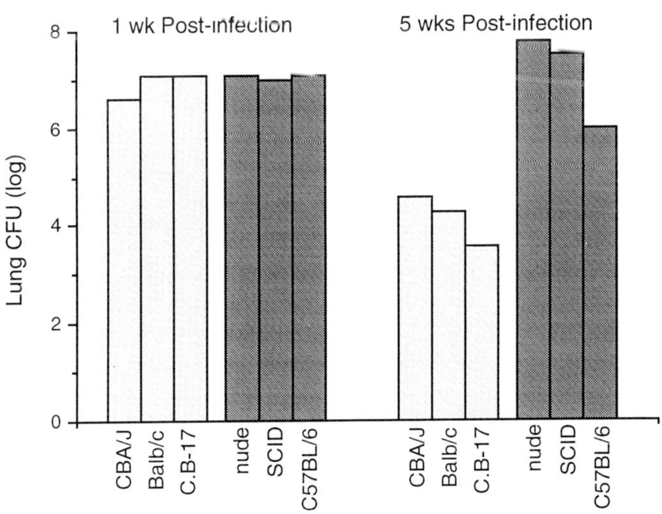

Figure 2 Role of genetics in determining susceptibility of *C. neoformans*. Data shown are the lung CFU in different inbred strains of mice at weeks 1 and 5 after i.t. inoculation with 5000 CFU of *Cneo-52*. (Adapted from refs. 12, 14, 17, and 20.)

infection is the ability of host defenses to recognize the microbe as a pathogen and generate the appropriate host response; that is, protective cell-mediated immunity. Both the genetic background and immune status of the host determine how *C. neoformans* is recognized and immunity develops. Inbred strains of mice can be classified as being either resistant or susceptible to pulmonary infection by opportunistic cryptococcal strains (12,14,17,20,21). For example, CBA/J, BALB/c, and C.B-17 are resistant inbred mouse strains, whereas C57BL/6 and genetically T-cell–deficient mice (nude, SCID) are susceptible (see Fig. 2). Although BALB/c ($H\text{-}2^d$), C.B-17 ($H\text{-}2^d$), CBA/J ($H\text{-}2^k$), and C57BL/6 ($H\text{-}2^b$) mice differ in major histocompatibility (MHC) haplotype, preliminary data indicate that genes outside the MHC are responsible for the differences in cryptococcal susceptibility. However, the salient feature that distinguishes whether a strain of mice clears a pulmonary cryptococcal infection is whether macrophage chemotactic factors and T-cell–derived macrophage-activating factors are generated in the lungs (12–15, 17,20,21).

III. Role of Macrophage Recruitment in Cryptococcal Clearance

Macrophage recruitment in the lungs of resistant CBA/J mice is strikingly different following infection by *Cneo-145* and *Cneo-52* (13,14). Striking differences are noted in the kinetics and magnitude of pulmonary inflammation induced by these two strains (Fig. 3). Macrophage recruitment into the lungs of resistant CBA/J mice dramatically increases coincident with the initiation of pulmonary clearance of the opportunistic strain *Cneo-52* (between weeks 1 and 2) (see Fig. 1). Macrophage numbers peak a week later (week 2), but remain elevated through week 5. Dense, perivascular cuffs of mononuclear cells develop as the inflammatory response progresses (13,17). These are major sites of monocyte extravasation into the lungs from the circulation (Fig. 4). During the clearance of the infection (weeks 2–5), numerous medium-sized macrophages, large foamy macrophages, and multinucleated giant cells fill the alveolar spaces. Although macrophages are the predominant intra-alveolar cell type from weeks 3–5, large numbers of lymphocytes are also present (Fig. 4). After week 2, few extracellular cryptococci are present, and macrophages contain only one or two, if any, ingested encapsulated organisms. Nearly 100% of the mice clear the infection (1). In contrast, few macrophages are recruited into the lungs of CBA/J mice infected with the virulent strain *Cneo-145* (13,14) (Fig. 3). The inflammatory response develops only after 2–3 weeks, and the maximal numbers of macrophages in the lungs are not reached until week 5 (Fig. 3). Numerous readily identifiable encapsulated yeast organisms are found within pulmonary macrophages indicative of an unactivated macrophage that is unable to eliminate ingested cryptococci. In addition, there are large masses of extracellular, heavily encapsulated *C. neoformans* in the alveoli; that is, "cryptococcomas" (13). The pulmonary *Cneo-145* infection is not cleared and

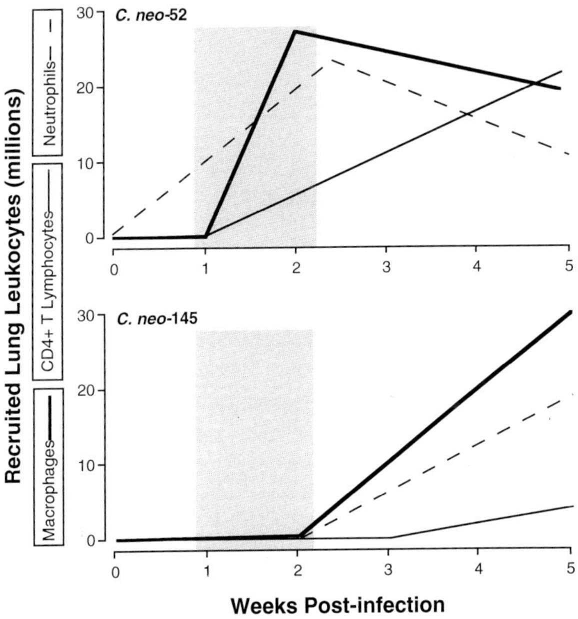

Figure 3 Leukocyte recruitment into the lungs of CBA/J mice infected with either the opportunistic strain *Cneo-52* or the virulent strain *Cneo-145*. The highlighted area is described in the legend for Figure 1. (Adapted from refs. 13 and 14.)

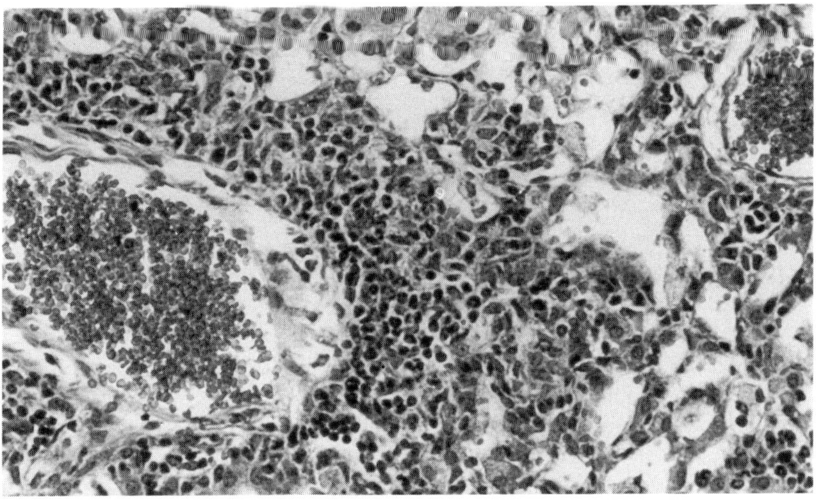

Figure 4 Photomicrograph of the perivascular cuffs (PVCs) of mononuclear cells and the intra-alveolar inflammatory cells at the peak of the inflammatory response to *C. neoformans*. PVCs are a major site of monocyte and lymphocyte extravasation into the lungs from the circulation and are also the first site of contact in the lungs between newly arriving macrophages and T cells.

subsequently disseminates to the brain and other extrapulmonary sites resulting in a uniformly fatal infection (14). Thus, the ability to evade recognition by immune defenses and the resultant attenuation of macrophage recruitment is a mechanism by which an infectious organism, such as *C. neoformans*, can markedly increase its virulence. This virulence mechanism also further underscores the importance of macrophage recruitment in both the containment of *C. neoformans* within the lungs (prevention of dissemination) and the pulmonary clearance of the infection.

IV. Role of T-Cell–Mediated Immunity

Cneo-52 is classified as an opportunistic strain of *C. neoformans* in mice, because pulmonary clearance of *Cneo-52* is dependent on the development of T-cell immunity and the subsequent recruitment and activation of macrophages by immune T cells (2,12,16,17). In genetically T-cell–deficient nude and SCID mice, few macrophages are recruited into the lungs following i.t. inoculation with *Cneo-52* and the infection is not cleared from the lungs (see Fig. 2) (12,21). In immunocompetent resistant strains of mice, clearance (by activated recruited macrophages) does not begin until after the development of antigen-specific delayed-type hypersensitivity (DTH) to the organism (day 7), indicating that T-cell immunity develops before clearance is initiated (12,17). If resistant strains of mice are treated with anti-CD4 and anti-CD8 monoclonal antibody (mAb) at the time of infection with *Cneo-52*, few macrophages are recruited into the lungs and the infection is not cleared directly, demonstrating that T-cell immunity is required for macrophage recruitment and effective clearance of a pulmonary *C. neoformans* infection (16,17).

T cells can orchestrate macrophage Recruitment into the lungs by two mechanisms. First, T cells can directly mediate macrophage recruitment via the production of macrophage chemotactic factors; that is, chemokines, such as macrophage inflammatory protein-1α (MIP-1α) and RANTES (22,23). The biology of MIP-1α and RANTES are discussed in more detail in Section VIII.A below. Second, T cells can produce cytokines that induce the expression of chemokines and adhesion molecules by other cell types in the lungs (e.g., epithelial cells, endothelial cells, fibroblasts, resident leukocytes) (22–25). These T-cell–derived cytokines include interferon-γ (IFN-γ), interleukin-4 (IL-4), tumor necrosis factor-α (TNF-α), and granulocyte-macrophage colony–stimulating factor (GM-CSF) (26). The biology of these cytokines, in relation to macrophage recruitment, is described in more detail in Sections VII.C, VIII.A, and VIII.B below.

Depletion of $CD4^+$ T cells alone significantly reduces macrophage recruitment into the lungs of *Cneo-52*–infected mice (16). Protective T-cell immunity to *C. neoformans* is associated with the production of Th1 type cytokines such as IFN-γ (16,20). This cytokine, along with a number of other T-cell–derived cytokines, is a potent inducer of chemokine production by parenchymal cells in the lungs (Section VII). In addition, activated $CD4^+$ T cell clones directly produce a number of chemokines such as MIP-1α and MIP-1β (27,28). Although both $CD4^+$ and $CD8^+$ T cells are required for protective immunity to *Cneo-52*, depletion of $CD4^+$ T cells has a far greater impact in decreasing macrophage numbers than does depletion of $CD8^+$ T cells (16). Depletion of $CD4^+$ T cells in mice infected with the virulent strain *Cneo-145* also results in an increased extrapulmonary organ colony-forming unit (CFU) and decreased survival (18). Thus, $CD4^+$ T cells are the predominant lymphocyte mediators of macrophage recruitment into the lungs during a *C. neoformans* infection.

The involvement of $CD8^+$ T cells in the protective response to *C. neoformans* is surprising, because this fungus is not considered an intracellular pathogen and, therefore, would not be anticipated to be associated with a class I MHC–mediated response. However, *C. neoformans* can survive in phagosomes of unactivated alveolar macrophages (29) (Fig. 5), possibly providing the intracellular link with the endogenous pathway for antigen processing (30–33). Consistent with this notion is the observation that $CD8^+$ T cells are not required for clearance once T-cell immunity has developed (1). In addition, $CD8^+$ T cells play a role

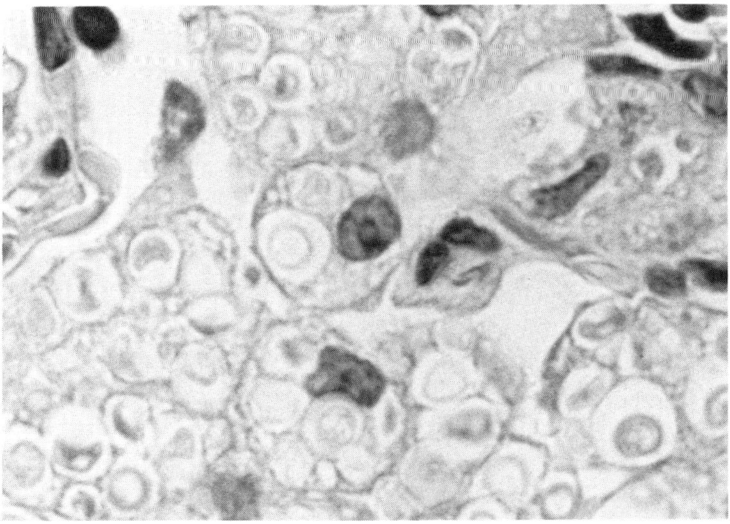

Figure 5 Survival of *C. neoformans* in the phagosomes of unactivated AMs. The clear halo around the ingested cryptococci is the polysaccharide capsule of the organism, which is a clear indication that lysosomal fusion has not occurred.

in the development of anticryptococcal DTH if the mice are immunized with live cryptococci (34), but immunization with soluble cryptococcal antigen results in the generation of $CD8^+$ T cells that suppress DTH responses (reviewed in ref. 3). These results suggest that, similar to the intracellular bacterium *Listeria monocytogenes*, cryptococci must be metabolically active to be processed by the class I MHC pathway (30–33,35). Depletion of $CD8^+$ T cells by anti-CD8 mAb reduces survival of mice infected with *Cneo-145* (19). In the response to infection by the opportunistic strain *Cneo-52*, $CD8^+$ T cells are required for maximal recruitment of $CD4^+$ T cells into the lungs (16). In the absence of $CD8^+$ T cells, $CD4^+$ T cells predominantly produce Th2 type cytokines (IL-4, IL-5, and IL-10) in vitro. The mechanism for the switch of $CD4^+$ T cells to Th2 type cytokines likely involves the $CD8^+$ T cell production of IL-2 and IFN-γ, both of which play a role in macrophage activation and development of Th1 type $CD4^+$ T cells (reviewed in ref. 36). Thus, $CD8^+$ T cells appear to play an indirect role in macrophage recruitment by promoting the development of protective Th1 type $CD4^+$ T-cell immunity to *C. neoformans*.

V. Role of Interleukin-1 and Tumor Necrosis Factor-α as Early Response Cytokines

TNF-α and IL-1 are the two major cytokines produced immediately following both antigen-specific and nonspecific stimulation; hence the designation of TNF-α and IL-1 as "early-response" cytokines (reviewed in refs. 26 and 37). They were originally described as a product solely of monocytes and macrophages (hence the original designation as a "monokine"), but they have now been identified as a product of a number of other cell types as well. TNF-α is also produced by activated T cells, and IL-1 is produced by fibroblasts, smooth muscle cells, and endothelial cells following appropriate stimulation. In the context of an early response to infection, potent stimuli for TNF-α and IL-1 synthesis include phagocytosis and microbial products such as lipopolysaccharide (LPS). Phagocytosis of both encapsulated and unencapsulated cryptococci by monocytes and macrophages stimulates the production of both IL-1 and TNF-α (14,38,39). Thus, IL-1 and TNF-α are two important early-response cytokines that induce a number of similar biological activities, albeit via distinct cell surface receptors.

TNF-α plays a role in a number of events associated with the development of a T-cell–mediated inflammatory response (37,40). In contrast to resting T cells, activated T cells express biologically active TNF-α receptors (41). TNF-α also enhances the expression of high-affinity IL-2 receptors and stimulates T-cell colony formation (41,42). TNF-α can enhance antigen- and mitogen-induced T-cell proliferation and synergize with IL-2 to stimulate T-cell proliferation and production of IFN-γ (41,43). TNF-α is a major cytokine involved in the development of a mixed lymphocyte response and generation of cytotoxic T lymphocytes in response to alloantigen (44,45). Numerous stimuli can induce the production of TNF-α by macrophages and T cells, including LPS, phagocytosis, antigen, reactive oxygen metabolites, and other cytokines. In turn, the production of TNF-α can be inhibited by other factors such as prostaglandin E_2 (PGE_2), IL-4, tumor growth factor-β (TGF-β), and glucocorticosteroids (37,40). Taken together, TNF-α is a cytokine that is increasingly being recognized for its role in mediating numerous types of T-cell–dependent inflammatory and immunological responses.

TNF-α plays a crucial role in the afferent phase of cell-mediated immunity against *C. neoformans*. Following i.t. inoculation of CBA/J mice with *Cneo-52*, levels of TNF-α mRNA and protein in the lungs increase within 2–7 days (G. B. Huffnagle et al., submitted), coinciding with the development of an inflammatory response in the lungs. Because alveolar macrophages (AMs) are the first cells to come in contact with the organism following infection, they are likely the major cellular source of TNF-α early in the infection. In vitro coculture of cryptococci and mouse AMs (or human mononuclear cells) stimulates the production of TNF-α (14,38,39). In contrast, the production of TNF-α in vitro is significantly reduced if the cryptococci contain high levels of melanin or polysaccharide capsule (14,39). The in vivo correlate of these in vitro data is that induction of TNF-α is delayed following infection with the high-melanin–producing strain *Cneo-145* compared with the low-melanin–producing strain *Cneo-52* (Fig. 6). The administration of anti-TNF-α antibodies at the onset of a *Cneo-52* infection blocks the development of protective T-cell–mediated immunity (G. B. Huffnagle et al., J Immunol, in press), thereby preventing macrophage recruitment into the

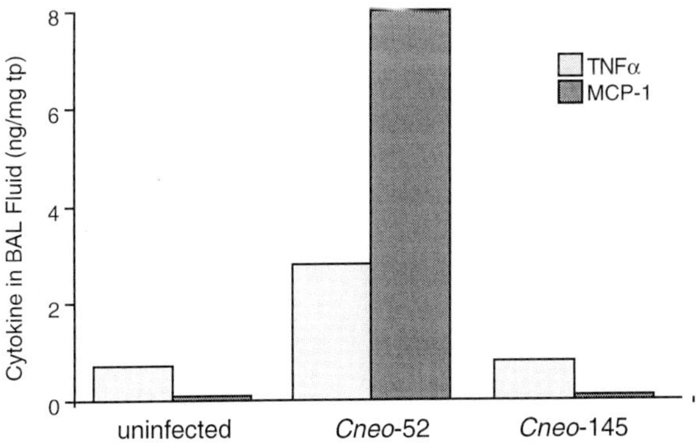

Figure 6 BAL fluid levels of TNF-α and MCP-1 at 2 weeks after i.t. inoculation in CBA/J mice infected with either the opportunistic strain, *Cneo-52*, or the virulent strain, *Cneo-145*. (Adapted from refs. 14 and 15.)

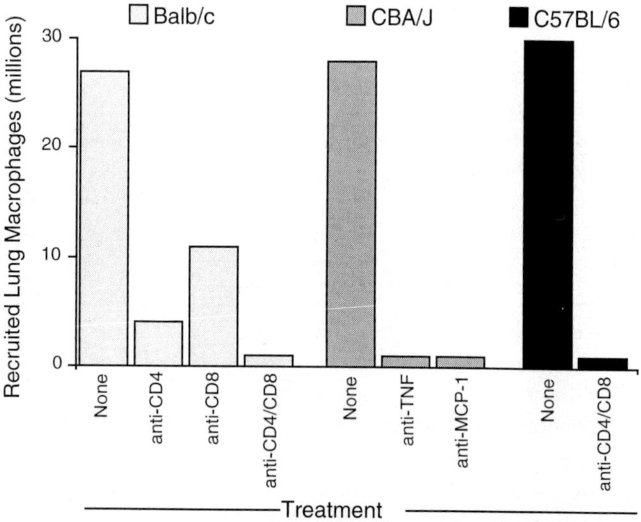

Figure 7 Effect of in vivo administration of various anticytokine and anti–T-cell antibodies on macrophage recruitment into the lungs of resistant and susceptible strains of mice following i.t. inoculation with *Cneo-52*. (Adapted from refs. 15 and 16 and G. B. Huffnagle et al., J Immunol, in press).

lungs (Fig. 7). The decrease in the number of macrophages results in a threefold increase in lung CFU at week 2 and a 10,000-fold increase at week 5. Thus, neutralization of TNF-α bioactivity by injection of anti–TNF-α antibodies changes the pathogenesis of the opportunistic *Cneo-52* into that of the virulent *Cneo-145*. In summary, the production of TNF-α during the afferent phase of the immune response against *C. neoformans* is critical for both the development of protective T-cell immunity and the subsequent recruitment of macrophages into the lungs.

VI. Role of Adhesion Molecules

The migration of monocytes from the circulation through the vascular endothelium into the lung results from a sequential interaction of cytokine-chemokines, cell adhesion molecules, and proteases. Selectins, integrins, and members of the immunoglobulin supergene family all play crucial roles in monocyte-endothelial cell cascades. Adhesion molecule expression on endothelial cells is normally low; but in the context of infection and inflammation, adhesion molecule expression markedly increases as a result of stimulation by either proinflammatory cytokines, such as IL-1 or TNF-α or microbial products, such as LPS. There are a vast number of references on the mechanisms of leukocyte trafficking and a number of excellent reviews (24,46–48). Thus, this section will simply sum up and highlight leukocyte adhesion in the context of monocyte/macrophage recruitment.

The selectin family consists of three closely related cell surface molecules that are differentially expressed by leukocytes (LECAM-1, L-selectin, CD62L), platelets and activated endothelial cells (GMP-140, P-selectin, CD62P), and vascular endothelium (ELAM-1, E-selectin, CD62E) (25,46,49,50). Although structurally similar, the selectins perform adhesive functions that mediate the initial interactions of leukocytes with the endothelium. L-selectin binds to GlyCAM-1 (CD34), MAdCAM-1, or as yet unidentified adhesion molecules on the surface of the endothelial cell and mediates leukocyte rolling on vascular endothelium at sites of tissue injury and inflammation. P-selectin and E-selectin, which are rapidly expressed on the endothelium after activation with inflammatory cytokines, bind myeloid cells via interaction with the ligands sialyl Lewis A (SLe^a) and sialyl Lewis X (SLe^x). The function and role of selectins in mediating leukocyte recruitment into sites of infection and inflammation have been demonstrated using cell culture systems, in vivo blocking with mAb directed against specific selectins, and selectin knockout mice. The relatively weak bonds between monocytes and endothelial cells in selectin-ligand binding slows down rapidly flowing monocytes in the blood vessels and results in monocyte "rolling" along the vascular wall.

Integrins (VLA-4, LFA-1, MAC-1, and P150,95) are heterodimeric proteins (composed of an α and β subunit) that bind to members of the immunoglobulin supergene family (VCAM-1, ICAM-1, ICAM-2, and PECAM) (47,51). Binding of the monocyte integrin VLA-4 (CD49d/CD29) to VCAM-1 slows "rolling" down even further and initiates loose adhesion of the monocyte. Monocyte activation alters integrin conformation and significantly increases the affinity of the integrins for their endothelial ligands. Binding of the monocyte integrins leukocyte function-associated antigen (LFA-1) (CD11a/CD18), MAC-1 (CD11b/CD18), or P150,95 (CD11c/CD18) leads to firm adhesion of monocytes to the endothelium via binding to the endothelial ligands ICAM-1, ICAM-2, or fibrinogen. Blocking of both VLA-4 and LFA-1 function is required to inhibit maximally a DTH reaction demonstrating the complementary role of these two integrins in loose and firm adhesion. Firm adhesion to the endothelium is followed by flattening or spreading of the monocyte. Subsequent transmigration (diapedesis) of the monocyte through endothelial cell junctions is mediated by PECAM-1 (CD31) via the interaction of monocyte PECAM-1 with endothelial cell PECAM-1 (homophilic binding). Diapedesis and migration to an inflammatory site occurs along a chemotactic gradient. There is some evidence that adhesion molecules such as CD44 can bind chemokines and present them on the vascular side of the endothelium, thereby initiating the chemotactic gradient beginning on the vascular side of the endothelium. Chemokines can also initiate monocyte spreading in vitro, suggesting that both adhesion molecules and chemokines are involved in this processs (52). Although no studies have been published looking at the role of adhesion molecules in monocyte recruitment in response to *C. neoformans*, preliminary studies from our laboratory have indicated that VCAM-specific mRNA increases in whole lung total RNA extracts during a *Cneo-52* infection in CBA/J mice, which is consistent with the increased recruitment of macrophages into the lungs.

VII. Role of Urokinase

Following transendothelial migration, monocytes must degrade matrix proteins and traverse tissue planes to enter the alveolar space eventually (reviewed in refs. 53–55). Production of proteases by monocytes is critically important for their migration to inflammatory sites. Substantial evidence implicates the urokinase-type plasminogen activator (uPA)/plasmin system as a central mediator in this process (56,57). Monocytes express uPA, which converts the inactive proenzyme plasminogen to plasmin. This protease has broad substrate specificity (58). A specific receptor for uPA (urokinase receptor, CD87) is present on the cell surface of monocytes. CD87 clusters to the leading front of the cell membrane during cellular migration of monocytes in response to an formyl-methionyl-leucyl-phenylalanine (fMLP) gradient, thus focusing the proteolytic activity of PA in the direction of migration. CD87 is required for monocyte chemotaxis in vitro (59).

The intratracheal inoculation of transgenic urokinase knockout mice (uPA−/−) with *Cneo-52* demonstrates that urokinase production is also important for macrophage recruitment into the lungs during infection by *C. neoformans* (60). uPA−/− mice have significantly fewer macrophages, neutrophils, and $CD4^+$ T cells in the lungs following *Cneo-52* infection that uPA+/+ littermates at weeks 2–3. Consistent with the decrease in the number of macrophages, pulmonary clearance is also attenuated in uPA−/− mice. The increased pulmonary burden of *Cneo-52* in uPA−/− mice results in the destruction of the parenchymal tissue and leads to significantly decreased long-term survival (21% at day 85) compared with *Cneo-52*–infected uPA+/+ mice (84% at day 85). Thus, urokinase appears to play a role in mediating the extravasation of monocyte/macrophages from the circulation and surrounding tissues into the site of cryptococcal infection in the lungs (60).

VIII. Role of Chemokines

A. Overview—Chemokines and Other Chemoactic Factors

The production of chemotactic factors at the site of a pulmonary infection is critical for the effective recruitment of monocyte-macrophages to that site. A number of molecules are chemotactic for macrophages, and these molecules can be categorized into two groups. The first group is chemotactic factors that possess little specificity for recruiting particular leukocyte subpopulations and includes peptides derived from the activation of the complement pathway (primarily C5a), leukotrienes, and bacterial products (e.g., fMLP). The second group is chemotactic factors that possess a high degree of specificity for specific leukocyte subpopulations and includes chemokines, which is a supergene family of small inducible peptides with potent chemotactic activity for specific leukocyte subpopulations (reviewed in refs. 22, 23, and 61–63).

There is only limited evidence that complement-derived peptides play a significant role in macrophage recruitment into infected lungs. C5a is a potent macrophage chemotaxin in vitro, and noninfectious particles, such as asbestos, can activate the alternative complement pathway in vitro, thereby generating C5a (64,65). Rats depleted of complement with cobra venom factor in vivo exhibit a significant reduction in macrophage accumulation following exposure to asbestos fibers. When serum complement levels return to normal, macrophage recruitment also returns to normal (66,67). *C. neoformans* also activates the alternative complement pathway, liberating C3a and C5a (68–70). However, in host defense against *C. neoformans*, complement components predominantly play a role in extrapulmonary defenses rather than in the alveolar space (71,72). C5-deficient mice do not exhibit differences in pulmonary inflammation or fungal clearance following i.t. inoculation with *C. neoformans* (71). Thus, although complement is clearly important for inflammatory cell recruitment into tissues outside the lungs, complement plays only a limited role as a macrophage chemotaxin in the lungs during a *C. neoformans* infection.

Chemokines are an important group of chemotactic factors that are recognized increasingly for their important role in inflammatory processes (reviewed in refs. 22, 23, and 61–63). Chemokines possess a high degree of specificity for the recruitment of macrophages without recruitment of neutrophils. Chemokines have been grouped into two different families based on the position of the first two conserved cysteine residues: the CC and CXC families. The CC chemokines include monocyte chemotactic protein-1 (MCP-1), MCP-2, MCP-3, RANTES, MIP-1α, MIP-1β, and I-309. The CXC chemokines include IL-8, ENA-78, IP-10, MIP-2, PF-4, NAP-2, MGSA/gro-α, and gro-β/γ. With the exception of IP-10 and PF-4, the CXC chemokines are predominantly chemotactic for neutrophils, whereas the CC chemokines are predominantly chemotactic for mononuclear cells. Chemokines are produced by a variety of pulmonary cells, including leukocytes, epithelial cells, endothelial cells, fibroblasts, and smooth muscle cells following stimulation by cytokines or microbial products. In the context of macrophage recruitment, research has predominantly concentrated on the function and regulation of MCP-1, RANTES, and MIP-1α.

MCP-1

MCP-1 was one of the first chemokines characterized at the levels of DNA, protein, and bioactivity. It is a secreted 9- to 13-kDa, glycosylated, heparin-binding protein that was first cloned from mouse fibroblasts as a PDGF-inducible gene termed JE (73). Other investigators later purified and cloned a human monocyte chemotactic protein, the homologue of JE, and variously named it lymphocyte-derived chemotactic factor (LDCF), glioma-derived chemotactic factor (GDCF), monocyte chemotactic-activating factor (MCAF), or smooth muscle cell–derived chemotactic factor (SMC-CF) (74–80). The name monocyte chemotactic protein-1 was subsequently adopted (MCP-1 should not be confused with macrophage cationic protein-1, a member of the defensin family of antimicrobial peptides that also possess chemotactic activity). Although MCP-1 is produced by nearly all the cells in the body, MCP-1 is chemotactic for only monocytes and T lymphocytes (74–77,81,82).

MCP-1 is remarkable, because it is produced by almost all cells or tissues. Stimuli for MCP-1 production include microbial products and a number of cytokines. From the vasculature to the alveoli, cellular sources of MCP-1 include monocytes (74,75,80), endothelial cells (76,83–85), smooth muscle cells (76,79), fibroblasts (73,85–88), epithelial cells (89,90), alveolar macrophages (90,91), and leukocytes recruited to a site of infection (74,75,80,92–94). LPS, IL-1, TNF-α, IFN-γ, and IL-4 can all stimulate MCP-1 production by endothelial cells in vitro (76,83–85). Stimuli for MCP-1 production by fibroblasts include IL-1, TNF-α, and PDGF (73,85,87,88). Supernatants from LPS-stimulated AMs, recombinant IL-1, and recombinant TNF-α all induce MCP-1 production by the type II epithelial cell line A549 in vitro (95). IL-1 and TNF-α can also stimulate MCP-1 expression in nasal, bronchial, and type II epithelial cells (89,90,95). The induction of epithelial cell MCP-1 can be blocked by the glucocorticosteroid dexamethasone; a finding consistent with the anti-inflammatory properties of corticosteroids in treating inflammatory diseases of the lungs. AMs serve as the "gate keeper" of the lungs, and one function of these phagocytic mononuclear cells is to elicit MCP-1 production by parenchymal cells. One mechanism is via the production if IL-1 and TNF-α (95). In addition to inducing MCP-1 production, AMs also produce MCP-1 following in vitro stimulation with LPS, TNF-α, or IL-1 (90,96). Rat AMs recovered from either immune complex-mediated or bleomycin-induced lung injury produce MCP-1 spontaneously in culture (91,97). MCP-1 is produced by macrophage cell lines and peripheral blood monocytes following treatment with cytokines such as IFN-γ (74,75,80). In addition, MCP-1 is elaborated during a mixed leukocyte response in vitro (92). Taken together, MCP-1 is a chemotactic stimulus that can be produced by all the cells in the lungs and lung-associated tissues, making it a "universal signal molecule" for the recruitment of macrophages.

Although increased MCP-1 production is associated with a number of inflammatory diseases, few studies have addressed the role of MCP-1 in antimicrobial host defenses. MCP-1 is critical for the recruitment for macrophages into the lungs of *C. neoformans*–infected mice (15). The administration of neutralizing anti–MCP-1 antibodies in mice blocks the pulmonary inflammatory response to *C. neoformans* and also the granulomatous response to *Schistosoma mansoni* (15,98). MCP-1 has been identified in the pathogenesis of human periodontal disease (99). MCP-1 also stimulates a number of cellular events associated with phagocyte activation (100). In the context of inflammatory diseases, increased MCP-1 levels are detected in patients with idiopathic pulmonary fibrosis (IPF), an end-stage fibrotic disease of the lungs with an unknown etiology and known inflammatory component (101–103). MCP-1 expression is also increased in the lungs of asthmatics (104). Significant levels of pulmonary MCP-1 have been detected in rodent models of bleomycin and immune complex–induced lung injury, IgA-mediated alveolitis, and glucan-induced granulomas (91,96,97,105–108). In summary, the macrophage chemotactic factor MCP-1 plays an important role in host defense against pulmonary infections and in the pathogenesis of pulmonary inflammatory diseases.

RANTES

RANTES was originally identified by subtractive hybridization as a transcript expressed in activated T cells but not in B cells (109). Although the acronym RANTES has been ascribed to regulated on activation normal T expressed and secreted, the protein was named after a character from a science fiction film (61). RANTES is an 8-kDa protein that, similar to MCP-1, can be produced by endothelial cells, epithelial cells, fibroblasts, T cells, and macrophages. RANTES is constitutively produced by cultured primary human bronchial epithelial cells and by the human pulmonary type II epithelial cell line, A549, following stimulation with TNF-α or IL-1β (110,111). TNF-α and IFN-γ synergize in stimulating RANTES mRNA and protein in the human bronchial epithelial cell line, BEAS-2B (112). RANTES induction in these two epithelial cell lines is also glucocorticosteroid, sensitive (110,112). Incubation of bronchial epithelial cells with influenza virus in vitro also increases RANTES mRNA expression (113). Epithelial cells are a major in vivo source of RANTES protein following i.t. instillation of LPS (114). Although IFN-γ, IL-1β , and TNF-α alone cannot induce RANTES expression by endothelial cells (HUVECs), the combination of TNF-α and IFN-γ induces RANTES expression (115). This is partially inhibitable by IL-4 and IL-13 but is unaffected by treatment with IL-10. RANTES is also detectable by immunohistochemistry in endothelial cells from the lymph nodes of patients with tuberculosis or sarcoidosis (116). Cultured dermal fibroblasts and synovial fibroblasts treated with either TNF-α or IL-1 produce RANTES, suggesting that pulmonary fibroblasts may be a potential source of RANTES (117,118). In the context of a developing inflammatory response in the lungs, recruited leukocytes are also an important source of RANTES in the lungs. Peripheral blood T cells, especially $CD8^+CD45RO^+$ T cells, produce significant amounts of RANTES following mitogen or anti-CD3 stimulation (119). In the setting of sarcoidosis and tuberculosis, RANTES is detectable by immunohistochemistry in draining lymph node macrophages (116). Thus, RANTES is produced by both resident and recruited cells in the lungs following stimulation with cytokines such as IL-1 or TNF-α.

To date, there are limited data regarding the role of RANTES in monocyte-macrophage recruitment in vivo and none in the context of infection. However, there is significant correlative data to suggest that RANTES is an important monocyte-macrophage chemotaxin in pulmonary infections. RANTES is chemotactic in vitro for monocytes, eosinophils, and memory $CD4^+$ T cells (109,120–126). The subdermal injection of recombinant human RANTES into dogs elicits the selective migration of both eosinophils and monocytes into the site (122). In SCID mice reconstituted with human T cells, subdermal injection of recombinant human RANTES elicits the migration of memory T cells (121). RANTES has also been implicated in increasing leukocyte adhesion (120,127,128). Passive immunization with anti-RANTES–specific antibodies blocks monocyte and eosinophil recruitment into the lungs of mice inoculated intratracheally with LPS (114). In summary, RANTES is similar to MCP-1, in that it is a chemotactic factor for macrophages and T cells that is produced by both parenchymal and recruited inflammatory cells in the lungs and likely plays a role in the recruitment of macrophages into sites of infection.

MIP-1α

In contrast to MCP-1 and RANTES, MIP-1α is produced predominantly by leukocytes. MIP-1α is an 8-kDa, heparin-binding protein originally identified in culture supernatants from an LPS-stimulated macrophage cell line (129,130). MIP-1α is often noncovalently associated with a related polypeptide MIP-1β (129,130). Although these two cytokines have a high degree of sequence homology and share many bioactivities, MIP-1α induces a more vigorous inflammatory response. MIP-1α is chemokinetic for neutrophils (131) and chemotactic for monocyte-macrophages and lymphocytes (120,121,129–133).

Both microbial products and cytokines induce MIP-1α production by macrophages and neutrophils. Treatment of macrophage cell lines, human peripheral blood monocytes, human AMs, mouse AMs, and rat AMs with LPS induces MIP-1α production (94,130,131,134,135). In addition to LPS from gram-negative bacteria, lipoteichoic acid from *Staphylococcus aureus* and *S. pyogenes* induces MIP-1α production by human peripheral blood monocytes (136). IFN-γ also stimulates MIP-1α production by macrophage cell lines (94,130). IL-1 or phytohemagglutinin (PHA) induces MIP-1α production in human peripheral blood monocytes and AMs (134). Inhibitors of MIP-1α production include PGE and IL-4 (94,134). LPS-stimulated neutrophils are another cellular source of MIP-1α (137). GM-CSF alone does not induce MIP-1α production in neutrophils, but GM-CSF synergizes with suboptimal amounts of LPS to induce MIP-1α (137). Thus, cells of the myeloid lineage (monocytes, macrophages, neutrophils, and eosinophils) are significant cellular sources of MIP-1α.

T lymphocytes are also an important cellular source of MIP-1α. MIP-1α is expressed in activated CD4 Th1 and CD8 T cell lines at high levels and at lower levels in CD4 Th2 cell lines (27,28,138). Human $CD8^+CD45RO^+$ cells produce significant amounts of MIP-1α following stimulation with either anti-CD3 or phorbol myristate acetate (PMA)/ionomycin (119). The expression of MIP-1α by T lymphocytes correlates well with the observation that CD4 Th1 and CD8 T cells but not CD4 Th2 mediate DTH responses; that is, antigen-specific recruitment of macrophages and lymphocytes (36,139,140).

Evidence is emerging that MIP-1α is also produced by nonleukocyte sources and plays a role in cell activation. MIP-1α is produced by pulmonary fibroblasts and epithelial cells in vitro (134,141,142). In contrast to fibroblasts from normal lungs, fibroblasts isolated from the lungs of patients with inflammatory diseases such as IPF produce MIP-1α following stimulation with IFN-γ or IL-10 (142). MIP-1α can stimulate phagocytic cells to produce hydrogen peroxide and release lysosomal enzymes and can also induce fibroblasts, basophils, mast cells, and macrophages to produce other proinflammatory cytokines, such as IL-1, IL-6, and TNF-α (23,94,143,144). Thus, there is in vitro evidence that MIP-1α is produced by and activates cells other than bone marrow culture-derived cells, suggesting a role for MIP-1α in amplifying inflammatory responses in the lungs.

MIP-1α plays a prominent role in a number of pulmonary inflammatory diseases and infections. In transgenic MIP-1α knockout mice, the pulmonary inflammatory response to influenza virus infection is significantly attenuated, demonstrating that MIP-1α is required for a normal inflammatory response to this virus (145). The administration of neutralizing anti–MIP-1α antibodies into mice blocks macrophage recruitment into the lungs following LPS treatment and granuloma formation following injection of schistosomal eggs (146,147). The injection of MIP-1α–specific neutralizing antibodies reduces leukocyte infiltration and associated acute lung injury in rats treated with LPS or IgG immune complexes (148). In asthmatics and IPF patients, MIP-1α levels in the bronchoalveolar lavage (BAL) fluid are significantly elevated (149,150). Alveolar macrophages, interstitial macrophages, and fibroblasts from IPF patients all produce MIP-1α in vitro (142,150). In contrast to normal pulmonary fibroblasts, fibroblasts isolated from patients with IPF produce significant amounts of MIP-1α following stimulation with IL-1β or IFN-γ (142,150). In the animal model of IPF, bleomycin-induced lung injury MIP-1α levels in the BAL fluid increase following treatment with bleomycin (151). The administration of neutralizing anti–MIP-1α antibodies to these mice decreases both macrophage recruitment and the development of the ensuing fibrosis. In summary, MIP-1α is a chemotaxin for monocytes, T cells, and eosinophils and plays an important role in mediating macrophage recruitment into the lungs during infection and inflammatory diseases.

B. MCP-1 and MIP-1α in Pulmonary Cryptococcosis

MCP-1 is one of the major chemotactic factors produced in the lungs during the protective host response to pulmonary *C. neoformans* infection (15). Significant levels of MCP-1 are detectable in the BAL fluid of *Cneo-52*–infected CBA/J mice between 1 and 2 weeks postinfection. The levels of MCP-1 increase concomitantly with the initiation of the inflammatory response and correlate closely with the recruitment of $CD4^+$ T cells and monocytes (Figs. 3 and 8). The administration of neutralizing anti–MCP-1 antibodies in vivo beginning 5 days postinfection results in decreases in the pulmonary recruitment of macrophages (>95%) (see Fig. 7) and $CD4^+$ T cells (76%) and inhibition of cryptococcal clearance at day 15. Neutralization of MCP-1 also results in decreased BAL fluid levels of the proinflammatory cytokines TNF-α and IL-6. The production of MCP-1 and the development of T-cell immunity to *C. neoformans* appear to be interdependent, because $CD4^+$ T-cell recruitment into the lungs is mediated by MCP-1 and the administration of either anti–CD4/CD8 or anti–MCP-1 antibodies block >95% or macrophage recruitment into the lungs (15,16). Thus, MCP-1 plays a critical role in the T-cell–mediated pulmonary inflammatory response to *C. neoformans*.

Preliminary experimental evidence suggests that differences in chemokine induction may be one of the major differences between the host response to pulmonary infection by the opportunistic strain *Cneo-52* and the virulent strain

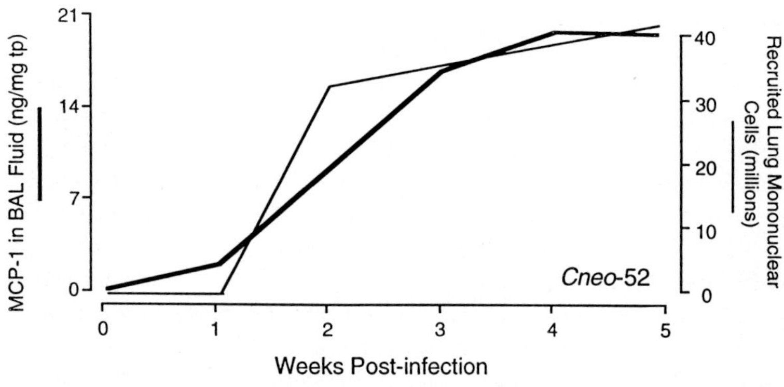

Figure 8 Comparison of mononuclear cell recruitment to the BAL fluid levels of MCP-1 in CBA/J mice following i.t. inoculation with *Cneo-52*. (Adapted from ref. 15.)

Cneo-145. A *Cneo-52*–infected induced significant levels of MIP-1α in the alveoli, whereas a *Cneo-145* infection failed to elicit MIP-1α production. Similarly, *Cneo-52* induced significant amounts of MCP-1, whereas *Cneo-145* did not. Overall, the levels of MCP-1 and MIP-1α in the alveoli following *Cneo-52* and *Cneo-145* infections appeared to correlate with whether these cryptococcal strains could elicit macrophage recruitment into the lungs. Either anti–MCP-1 or MIP-1α alone could decrease macrophage recruitment by >50% in *Cneo-52*–infected mice, suggesting that both of these CC chemokines may be required for macrophage recruitment. In summary, the production of both MCP-1 and MIP-1α in the lungs appears to be critical for macrophage recruitment during a pulmonary *C. neoformans* infection.

C. TNF-α as a Proximal Signal for Chemokine Induction

The phagocytosis of *C. neoformans* by AMs is, alone, insufficient to elicit chemokine production (unpublished observation), suggesting that there must be an event proximal to the induction of chemokines, such as the production of an activating cytokine. TNF-α is a likely candidate, because, in the context of cryptococcal infection, AMs produce TNF-α in vitro on coculture with cryptococci (14) and high levels of TNF-α are expressed before macrophage recruitment is initiated, within the first week of infection (G. B. Huffnagle et al., J Immunol, in press). Although TNF-α is not directly chemotactic, the treatment of *Cneo-52*–infected CBA/J mice with anti–TNF-α serum significantly reduces the recruitment of macrophages into the lungs (G. B. Huffnagle et al., submitted). Anti–TNF-α serum also blocks the increases in MCP-1 and MIP-1α levels in the BAL fluid of *Cneo-52*–infected mice. In CBA/J mice infected with the virulent strain *Cneo-145*, there is a 2-week delay in TNF-α production, MCP-1 and MIP-1α induction, and macrophage recruitment (G. B. Huffnagle et al., submitted). These latter observations provide a correlative link between TNF-α production, chemokine induction, and macrophage recruitment. In other models, anti–TNF-α serum blocks mononuclear cell recruitment in a murine footpad DTH reaction (152). TNF-α can also regulate the production of both CC and CXC chemokines by macrophages in vitro (reviewed in refs. 22, 23, and 61–63). Taken together, these data support the hypothesis that TNF-α is a proximal signal for the induction of the chemokines MCP-1 and MIP-1α during a *C. neoformans* infection.

IX. Final Step: Macrophage Activation

This chapter outlines the events involved in macrophage recruitment into infected lungs. However, as previously stated, the ultimate objective of macrophage recruitment in pulmonary infection is to clear the infection from the lungs. Although recruitment and activation are often concurrent processes, the mechanisms of macrophage recruitment are different from the mechanisms of macrophage activation.

This paradigm is best illustrated in the setting of *Cneo-52* pulmonary infection of genetically susceptible C57BL/6 mice (see Fig. 2). In both CBA/J (resistant) and C57BL/6 (susceptible) mice (12,14), numerous macrophages are recruited into the lungs following *Cneo-52* infection and these macrophages phagocytose the cryptococci. However, the macrophages in C57BL/6 mice are unable to eliminate the ingested organisms. Since AMs can readily phagocytose *C. neoformans* but require an activating signal to cause lysosome-phagosome fusion (29), an activator, such as IFN-γ or GM-CSF (153,154), is missing for activation of the recruited macrophages in C57BL/6 mice (G. B. Huffnagle et al., in preparation).

The same scenario is true for infection with the virulent strain *Cneo-145*. Although macrophage recruitment is significantly delayed following infection with *Cneo-145* compared with *Cneo-52* infection (14), large numbers of macrophages are ultimately recruited into the lungs of *Cneo-145*–infected mice by weeks 4–5 (see Fig. 3). However, these "unactivated" macrophages are unable to eliminate ingested *Cneo-145* organisms, as evidenced by an inability to clear the infection and the large numbers of macrophages that contain numerous fully encapsulated cryptococci (13). If *Cneo-145*–infected mice are coinfected with *Cneo-52*, the host response to the opportunistic strain results in the production of activating factors that stimulate the macrophages to eliminate ingested *Cneo-145* organisms (14). Thus, the clearance of a pulmonary *C. neoformans* infection requires both the recruitment and activation of macrophages.

T cells are important mediators of macrophage activation. The production of IFN-γ (described in more detail in Section IX.B below) by T cells is a major mechanism by which T cells stimulate fungicidal activity in macrophages. Unstimulated murine AMs, monocytes, or other tissue macrophages will phagocytose cryptococci in the presence of opsonins but not kill the organisms (153–157). However, macrophages treated with the T-cell cytokines IFN-γ or GM-CSF exhibit significant fungicidal activity (153,154). GM-CSF, but not IFN-γ, also increases complement-dependent phagocytosis of cryptococci (156). Depletion of $CD4^+$ T cells *in vivo* inhibits multinucleated giant cell formation, which is one mechanism for containing *C. neoformans* in the lungs (158). T-cell–derived cytokines such as IL-3, IL-4, IFN-γ, or GM-CSF can stimulate the fusion of macrophages into multinucleated giant cells *in vitro* (158–162). Taken together, these observations demonstrate that T cells mediate macrophage activation via the production of activating factors such as IFN-γ, TNF-α, and GM-CSF.

A. Role of TNF-α in Macrophage Activation

TNF-α also appears to be an important efferent phase molecule in infections. TNF-α is a potent activator of microbicidal activity in monocytes and macrophages (reviewed in ref. 40). TNF-α also increases the phagocytosis of cryptococci by peritoneal macrophages (156). Treatment of mice with anti–TNF-α antibodies after T-cell–mediated immunity has developed decreases survival following intraperitoneal or intracranial challenge with *C. neoformans* (156,163). Late in a *Cneo-52* infection in uPA−/− mice, macrophages contain numerous intracellular cryptococci, suggesting a defect in macrophage activation in these mice (60). Because TNF-α production has a uPA-dependent component (164), the defect in macrophage activation in uPA−/− mice may be due to reduced TNF-α production in these mice. In CBA/J mice infected with *Cneo-52*, BAL fluid levels of TNF-α and IL-6 at the peak of macrophage recruitment (week 2) are dependent on the production of MCP-1 by cells in the lungs (15), suggesting that there is both afferent phase and efferent phase production of TNF-α.

B. Role of IFN-γ in Macrophage Activation

IFN-γ is both an indirect mediator of macrophage recruitment into the lungs and a direct mediator of macrophage activation. IFN-γ can indirectly mediate macrophage recruitment by increasing adhesion molecule expression on cells and by inducing chemokine production (22–24,165). IFN-γ can mediate macrophage activation by three mechanisms (36,165): (1) directly inducing microbicidal activity in macrophages, (2) increasing MHC class II expression on a number of cell types, including macrophages, and (3) by promoting the development of a Th1 type of T-cell response. The first mechanism directly affects the clearance of a pulmonary infection by increasing the ability of the macrophages to kill ingested organisms or mediate extracellular killing. The second mechanism promotes the recognition of the microbe and subsequent development of a vigorous immune response by increasing the antigen-presenting capacity of the macrophage or other cell. The binding of microbial peptide–bearing MHC II molecules on the macrophage with the appropriate T-cell receptor (TCR)/CD4 complex on the surface of the T cell brings the T cell and macrophage in contact, setting up an extremely efficient mechanism to deliver T-cell–derived activating signals to the macrophage. The third mechanism is a form of an autocrine loop in which IFN-γ promotes the development of T cells that produce more IFN-γ in addition to other macrophage-activating factors, such as IL-2, GM-CSF, TNF-α, and TNF-β; that is, Th1 cells (36,138–140). Altogether, IFN-γ mediates the clearance of pulmonary infections via a multifaceted mechanism that ultimately involves macrophage activation.

In the context of pulmonary cryptococcosis, there are significant data to suggest that the production of IFN-γ is crucial for host defense. IFN-γ stimulates in vitro fungicidal activity in bone culture–derived marrow macrophages, resident peritoneal macrophages, and AMs (154,166,167). IFN-γ also stimulates macrophage-mediated production of small, secreted protein(s) that cause extracellular fungistasis of *C. neoformans* (166). The in vitro production of IFN-γ by cultured murine lung and draining lymph node lymphocytes correlates with genetic susceptibility to *C. neoformans* infection (16,20). Both $CD4^+$ and $CD8^+$ T cells from the lungs of *Cneo-52*–infected BALB/c and C.B-17 mice produce IFN-γ in vitro on coculture with mitogen or cryptococcal antigen (16,20). The production of IFN-γ in these resistant strains of mice is in contrast to the production of IFN-γ in the susceptible mouse strain, C57BL/6. Isolated lung lymphocytes from *Cneo-52*–infected C57BL/6 mice produce little IFN-γ in vitro on coculture with mitogen or cryptococcal antigen (20). Treatment of *Cneo-52*–infected C.B-17 mice with anti–IFN-γ mAb in vivo blocks the induction of nitrous oxide (NO) in pulmonary macrophages and attenuates pulmonary clearance (21). Overall, T cells play two major roles in macrophage-mediated pulmonary clearance of *C. neoformans*: (1) as mediators (directly or indirectly) of macrophage recruitment into the lungs, and (2) as mediators of macrophage activation through the production of activating factors, such as IFN-γ, TNF-α, and GM-CSF.

X. Conclusions

The recruitment and activation of macrophages are major mechanisms by which host defenses are marshalled to clear a pulmonary infection. In a number of infections, the development of T-cell–mediated immunity is required for effective macrophage recruitment and activation. This chapter has discussed the molecular and cellular events of this process in the context of an infection model in which both macrophages and T cells are required to clear the infection from the lungs; that is, murine pulmonary cryptococcosis. This process can be divided into three phases (recognition, recruitment, and response) and is summarized below and in Figure 9 (the numbers in brackets in the text below refer to the number in Fig. 9).

A. Recognition

The first phase is recognition of the infection by host defenses. Initially, many of the invading microbes (*C. neoformans*) are opsonized (likely by surfactant protein A) and phagocytized by AMs. Although unactivated AMs are unable to kill the organism, phagocytosis results in the production of cytokines, such as IL-1 and TNF-α[1]. These cytokines begin the cascade of events that lead to pulmonary inflammation. One role of IL-1 and TNF-α is to induce adhesion molecule expression on endothelial cells [3]. Another, and perhaps the most important, role of IL-1 and TNF-α is in promoting the development of T-cell–mediated immunity [2]. Activated T cells, in turn, provide cytokine signals that also upregulate adhesion molecule expression on endothelial cells [3] and induce chemokine expression by parenchymal cells and leukocytes (including T cells) [8]. The recognition phase is the target of a number of microbial virulence factors, including cryptococcal melanin and polysaccharide.

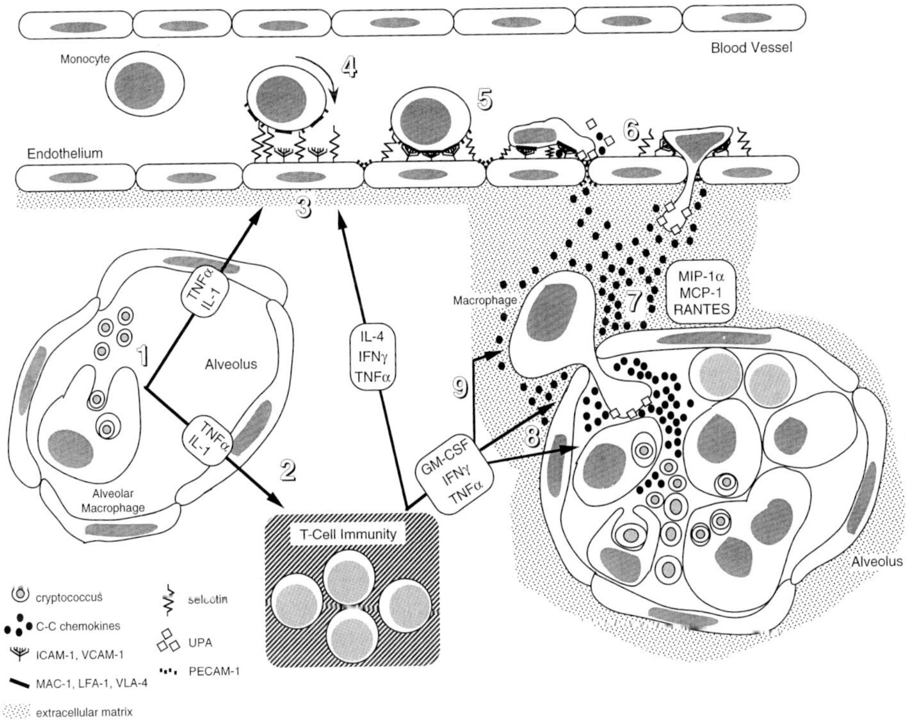

Figure 9 Model summarizing the events involved in the recruitment of macrophages from the circulation to the site of infection in the lungs. The numbers in the figure refer to the more detailed description in Section X of the text.

B. Recruitment

Following recognition of the infection and induction of early-response cytokines, leukocyte recruitment into the lungs is initiated. Macrophage recruitment to the site of infection results from monocyte sequestration (by adhesion to endothelial cells in postcapillary venules) followed by extravasation into the tissue (in response to a chemotactic gradient; i.e., chemokines). Adhesion of monocytes to the endothelium is an ordered process of events involving rolling [4], loose adhesion [5], firm adhesion, and spreading [6]. These events are mediated by the interaction of various adhesion molecules on the surfaces of both monocytes and endothelial cells. Once a monocyte has firmly attached to an endothelial cell, the monocyte flattens and begins to emigrate from the blood vessel by squeezing between adjacent endothelial cells [6]. This latter process is mediated by adhesion molecules, proteases, and chemokines. The emigrating monocytes (and lymphocytes) form a perivascular cuff of cells around the postcapillary venule and the first contact between T cells and monocytes in the lungs occurs. These tissue monocytes, that is, macrophages, continue to migrate along a chemokine gradient through the extracellular matrix and into the alveoli [7].

C. Response

The third phase of pulmonary host defense, the response, is the activation of recruited macrophages and elimination of the infection [9]. Although a number of the same cytokines mediate both recruitment and activation and these processes usually occur concurrently, macrophage recruitment and activation are separate processes. This is best illustrated by *C. neoformans* infection of C57BL/6 mice: Numerous macrophages are recruited and phagocytize cryptococci but do not kill the ingested yeast owing to a defect in production of macrophage-activating cytokines by T cells. It is not clear at what point during the recruitment process macrophage activation by T-cell–derived cytokines occurs. However, activation likely occurs at a number of sites during the trek of the macrophage into the alveolus, including in the perivascular cuffs, extracellular matrix, and alveoli. In conclusion, macrophage recruitment and macrophage–T-cell interactions are crucial for pulmonary host defenses, and this is no more evident than in the clinical setting of immunocompromised patients: Respiratory infections are by far and away the most common type of infection, leading to significant morbidity and mortality in this patient population.

Acknowledgments

These studies have been supported in part by a Parker B. Francis award from The Francis Families Foundation, Kansas City, Missouri, NIH grants 1R29-AI38190, 1R01-HL51082, 1P50-HL46487 SCOR, and a Veterans Administration Merit Review Research Grant.

References

1. Huffnagle GB, Lipscomb MF. Animal model of human disease: Pulmonary cryptococcosis. Am J Pathol 1992; 141:1517–1520.
2. Lipscomb MF, Huffnagle GB, Lovchik JA, Lyons CR, Pollard AM, Yates JL. The role of T lymphocytes in pulmonary microbial defense mechanisms. Arch Pathol Lab Med 1993; 117:1225–1232.
3. Murphy JW. Cryptococcosis. In: Cox RA, ed. Immunology of the Fungal Diseases. Boca Raton, FL: CRC Press, 1989:93–138.
4. Diamond RD. *Cryptococcus neoformans* In: Mandell GL, Douglas RG, Bennett JE,

eds. Principles and Practice of Infectious Disease. New York: Churchill Livingstone, 1990:1980–1989.

5. Kovacs JA, Kovacs AA, Polis M, Wright WC, Gill VJ, Tuazon CU, Gelman EP, Lane HC, Longfield R, Overturf G, Macher AM, Fauci AS, Parillo J, Bennett JE, Masur H. Cryptococcosis in the acquired immunodeficiency syndrome. Ann Intern Med 1985; 103:533–538.
6. Zuger A, Louie E, Holzman R, Simberkoff M, Rahal JJ. Cryptococcal disease in patients with the acquired immunodeficiency syndrome: Diagnostic features and outcome of treatment. Ann Intern Med 1986; 104:234–240.
7. Lahdevirta J, Maury CP, Teppo AM, Repo H. Elevated levels of circulating cachectin/tumor necrosis factor in patients with acquired immunodeficiency syndrome. Am J Med 1988; 85:289–291.
8. Chuck SL, Sande MA. Infections with *Cryptococcus neoformans* in the acquired immunodeficiency syndrome. N Engl J Med 1989; 321:794–799.
9. Gallant JE, Moore RD, Chaisson RE. Prophylaxis for opportunistic infections in patients with HIV infection. Ann Intern Med 1994; 120:932–944.
10. Kagaya K, Yamada T, Miyakawa Y, Fukazawa Y, Saito S. Characterization of pathogenic constituents of *Cryptococcus neoformans* strains. Microbiol Immunol 1985; 29:517–532.
11. Kwon-Chung K, Rhodes J. Encapsulation and melanin formation as indicators of virulence in *Cryptococcus neoformans*. Infect Immun 1986; 51:218–223.
12. Huffnagle GB, Yates JL, Lipscomb MF. T cell–mediated immunity in the lung: a *Cryptococcus neoformans* pulmonary infection model using SCID and athymic nude mice. Infect Immun 1991; 59:1423–1433.
13. Curtis JL, Huffnagle GB, Chen GH, Warnock ML, Gyetko MR, McDonald RA, Scott PJ, Toews GB. Experimental murine pulmonary cryptococcosis: differences in pulmonary inflammation and lymphocyte recruitment induced by two encapsulated strains of *Cryptococcus neoformans*. Lab Invest 1994; 71:113–126.
14. Huffnagle GB, Chen GH, Curtis JL, McDonald RA, Strieter RM, Toews GB. Down-regulation of the afferent phase of T cell–mediated immunity and pulmonary inflammation by a high melanin-producing strain of *C. neoformans*. J Immunol 1995; 155: 3507–3516.
15. Huffnagle GB, Strieter RM, Standiford TJ, McDonald RA, Burdick MD, Kunkel SL, Toews GB. The role of monocyte chemotactic protein-1 (MCP-1) in the recruitment of monocytes and $CD4^+$ T cells during a pulmonary *Cryptococcus neoformans* infection. J Immunol 1995; 155:4790–4797.
16. Huffnagle GB, Lipscomb MF, Lovchik JA, Hoag KA, Street NE. The role of CD4+ and CD8+ T cells in the protective inflammatory response to a pulmonary cryptococcal infection. J Leukoc Biol 1994; 55:35–42.
17. Huffnagle GB, Yates JL, Lipscomb MF. Immunity to a pulmonary *Cryptococcus neoformans* infection requires both CD4+ and CD8 + T cells. J Exp Med 1991; 173:793–800.
18. Mody CH, Lipscomb ML, Street NE, Toews GB. Depletion of CD4+ (L3T4+) lymphocytes *in vivo* impairs murine host defense to *Cryptococcus neoformans*. J Immunol 1990; 144:1472–1477.

19. Mody CH, Chen GH, Jackson C, Curtis JL, Toews GB. In vivo depletion of murine CD8 positive T cells impairs survival during infection with a highly virulent strain of *Cryptococcus neoformans*. Mycopathologia 1994; 125:7–17.
20. Hoag KA, Street NE, Huffnagle GB, Lipscomb MF. Early cytokine production in pulmonary *Cryptococcus neoformans* infections distinguishes susceptible and resistant mice. Am J Respir Cell Mol Biol 1995; 13:487–495.
21. Lovchik JA, Lyons CR, Lipscomb MF. A role for gamma interferon-induced nitric oxide in pulmonary clearance of *Cryptococcus neoformans*. Am J Respir Cell Mol Biol 1995; 13:116–124.
22. Schall TJ. Biology of the RANTES/SIS cytokine family. Cytokine 1991; 3:165–183.
23. Oppenheim JJ, Zachariae COC. Properties of the novel proinflammatory supergene "intercrine" cytokine family. Annu Rev Immunol 1991; 9:617–648.
24. Bevilacqua MP. Endothelial-leukocyte adhesion molecules. Annu Rev Immunol 1993; 11:767–804.
25. Bevilacqua MP, Nelson RM. Selectins. J Clin Invest 1993; 91:379–387.
26. Thomson A, ed. The Cytokine Handbook. London: Harcourt Brace, 1994.
27. Brown K, Zurawski S, Mosmann T, Zuraski G. A family of small inducible proteins secreted by leukocytes are members of a new superfamily that includes leukocyte and fibroblast-derived inflammatory agents, growth factors, and indicators of various activation processes. J Immunol 1989; 142:679–687.
28. Cherwinski HM, Schumacher JH, Brown KD, Mosmann TR. Two types of mouse helper T cell clone. III. Further differences in lymphokine synthesis between Th1 and Th2 clones revealed by RNA hybridization, functionally monospecific bioassays and monoclonal antibodies. J Exp Med 1987; 166:1229–1239.
29. Vecchiarelli A, Dottorini M, Pietrella D, Monari C, Retini C, Todisco T, Bistoni F. Role of human alveolar macrophages as antigen-presenting cells in *Cryptococcus neoformans* infection. Am J Respir Cell Mol Biol 1994; 11:130–137.
30. Engelhard VH. Structure of peptides associated with class I and class II MHC molecules. Annu Rev Immunol 1994; 12:181–207.
31. Germain RN, Margulies DH. The biochemistry and cell biology of antigen processing and presentation. Annu Rev Immunol 1993; 11:403–450.
32. Heemels MT, Ploegh H. Generation, translocation, and presentation of MHC class I–restricted peptides. Annu Rev Biochem 1995; 64:463–491.
33. Shawar SM, Vyas JM, Rodgers JR, Rich RR. Antigen presentation by major histocompatibility complex class I-B molecules. Annu Rev Immunol 1994; 12:839–880.
34. Mody CH, Paine R III, Jackson C, Chen GH, Toews GB. CD8 cells play a critical role in delayed type hypersensitivity to intact *Cryptococcus neoformans*. J Immunol 1994; 152:3970–3979.
35. Brunt LM, Portnoy DA, Unanue ER. Presentation of Listeria monocytogenes to CD8+ T cells requires secretion of hemolysin and intracellular bacterial growth. J Immunol 1990; 145:3540–3546.
36. Mosmann TR, Coffman RL. TH1 and TH2 cells: different patterns of lymphokine secretion lead to different functional properties. Annu Rev Immunol 1989; 7:145–173.
37. Strieter RM, Kunkel SL, Bone RC. Role of tumor necrosis factor alpha in disease states and inflammation. Crit Care Med 1993; 21:S447–S463.
38. Levitz SM, Tabuni A, Kornfeld H, Reardon CC, Golenbock DT. Production of tumor

necrosis factor alpha in human leukocytes stimulated by *Cryptococcus neoformans*. Infect Immun 1994; 62:1975–1981.
39. Vecchiarelli A, Retini C, Pietrella D, Monari C, Tascini C, Beccari T, Kozel TR. Downregulation by cryptococcal polysaccharide of tumor necrosis factor alpha and interleukin-1 beta secretion from human monocytes. Infect Immun 1995; 63:2919–2923.
40. Beutler B, Cerami A. The biology of cachectin/TNF—a primary mediator of the host response. Annu Rev Immunol 1989; 7:625–655.
41. Scheurich P, Thoma B, Ucer U, Pfizenmaier K. Immunoregulatory activity of recombinant human tumor necrosis factor TNF: induction of the TNF receptors on human T cells and TNF-mediated enhancement of T cell responses. J Immunol 1987; 138:1786–1790.
42. Zucali JR, Elfenbein GJ, Barth KC, Dinarello CA. Effects of human interleukin 1 and human tumor necrosis factor on human T lymphocyte colony formation. J Clin Invest 1987; 80:772–777.
43. Yokota S, Gepper TD, Lipsky PE. Enhancement of antigen- and mitogen-induced human T lymphocyte proliferation by tumor necrosis factor alpha. J Immunol 1988; 140:531–536.
44. Shalaby MR, Espevik T, Rice GC, Ammann AJ, Figari IS, Ranges GE, Palladino MA. The involvement of human tumor necrosis factors alpha and beta in the mixed lymphocyte reaction. J Immunol 1988; 141:499–503.
45. Ranges GE, Figari IS, Espevik T, Palladino MA. Inhibition of cytotoxic T cell development by transforming growth factor beta and reversal by recombinant tumor necrosis factor alpha. J Exp Med 1987; 166:991–998.
46. Hogg N, Berlin C. Structure and function of adhesion receptors in leukocyte trafficking. Immunol Today 1995; 16:327–330.
47. Springer TA. Traffic signals for lymphocyte recirculation and leukocyte emigration: the multistep paradigm. Cell 1994; 76:301–314.
48. Albeda SA, Smith CW, Ward PA. Adhesion molecules and inflammatory injury. FASEB J 1994; 8:504–512.
49. Tedder TF, Steeber DA, Chen A, Engel P. The selectins: vascular adhesion molecules. FASEB J 1995; 9:866–873.
50. Lasky LA. Interpreters of cell-specific carbohydrate information during inflammation. Science 1992; 258:964–969.
51. Imhof BA, Dunon D. Leukocyte migration and adhesion. Adv Immunol 1995; 58: 348–416.
52. Tanaka Y, Adams DH, Hubscher S, Hirano H, Siebenlist U, Shaw S. T-cell adhesion induced by proteoglycan-immobilized cytokine MIP-1 beta. Nature 1993; 361: 79–82.
53. Saksela O, Rifkin RDB. Cell-associated plasminogen activation: regulation and physiological functions. Annu Rev Cell Biol 1988; 4:93–126.
54. Vassalli J, Sappino A, Belin D. The plasminogen activator/plamin system. J Clin Invest 1991; 88:1067–1072.
55. Chapman HA, Reilly JJ, Kobzik L. Role of plasminogen activator in degradation of extracellular matrix proteins by live human alveolar macrophages. Am Rev Respir Dis 1988; 137:412–419.

56. Kirchheimer JC, Remold HG. Endogenous receptor-bound urokinase mediates tissue invasion of human monocytes. J Immunol 1989; 143:2634–2639.
57. Estreicher A, Muhlhause J, Carpentier J, Orci L, Vassalli J. The receptor for urokinase type plasminogen activator polarizes expression of the protease to the leading edge of migrating monocytes and promotes degradation of enzyme inhibitor complexes. J Cell Biol 1990; 111:783–792.
58. Dano K, Andreasen PA, Grondahl HJ, Kristensen P, Nielsen LS, Skriver L, Dan K. Plasminogen activators, tissue degradation, and cancer. Adv Cancer Research 1985; 44:139–265.
59. Gyetko MR, Todd RF III, Wilkinson CC, Sitrin RG. The urokinase receptor is required for human monocyte chemotaxis in vitro. J Clin Invest 1994; 93:1380–1387.
60. Gyetko MR, Chen GH, McDonald RA, Goodman R, Huffnagle GB, Wilkinson CC, Fuller JA, Toews GB. Urokinase is required for the pulmonary inflammatory response to *C. neoformans*: A murine transgenic model. J Clin Invest 1996; 97:1818–1826.
61. Schall TJ. Chemokines. In: Thomson A, ed. The Cytokine Handbook. London: Harcourt Brace, 1994:419–460.
62. Strieter RM, Kunkel SL. The immunopathology of chemotactic cytokines. Adv Exp Med Biol 1993; 351:19–28.
63. Kunkel SL, Lukacs N, Strieter RM. Chemokines and their role in human disease. Agents Actions 1995; 46(suppl):11–22.
64. Warheit DB, George G, Hill LH, Snyderman R, Brody AR. Inhaled asbestos activates a complement-dependent chemoattractant for macrophages. Lab Invest 1985; 52:505–514.
65. Marder SR, Chenoweth DE, Goldstein IM, Perez HD. Chemotactic responses of human peripheral blood monocytes to the complement-derived peptides C5a and C5a des Arg. J Immunol 1985; 134:3325–3331.
66. Warheit DB, Hill LH, George G, Brody AR. Time course of chemotactic factor generation and the corresponding macrophage response to asbestos inhalation. Am Rev Respir Dis 1986; 134:128–133.
67. Warheit DB, Overby LH, Gorge G, Brody AR. Pulmonary macrophages are attracted to inhaled particles through complement activation. Exp Lung Res 1988; 14:51–66.
68. Laxalt K, Kozel T. Chemotaxigenesis and activation of the alternative complement pathway by encapsulated and non-encapsulated *Cryptococcus neoformans*. Infect Immun 1979; 26:435–440.
69. Kozel T, Pfrommer G. Activation of the complement system by *Cryptococcus neoformans* leads to binding of iC3b to the yeast. Infect Immun 1986; 52:1–5.
70. Kozel T, Wilson M, Murphy J. Early events in initiation of alternative complement pathway activation by the capsule of *Cryptococcus neoformans*. Infect Immun 1991; 59:3101–3110.
71. Lovchik J, Lipscomb M. Role for C5 and neutrophils in the pulmonary intravascular clearance of circulating *Cryptococcus neoformans*. Am J Respir Cell Mol Biol 1993; 9:617–627.
72. Truelsen K, Young T, Kozel T. In vivo complement activation and binding of C3 to encapsulated *Cryptococcus neoformans*. Infect Immun 1992; 60:3937–3939.

73. Cochran BH, Reffel AC, Stiles CD. Molecular cloning of gene sequences regulated by platelet-derived growth factor. Cell 1983; 33:939–947.
74. Matsushima K, Larsen CG, DuBois GC, Oppenheim JJ. Purification and characterization of a novel monocyte chemotactic and activating factor produced by a human myelomonocytic cell line. J Exp Med 1989; 169:1485–1490.
75. Yoshimura T, Yuhki N, Moore SK, Appella E, Lerman MI, Leonard EJ. Human monocyte chemoattractant protein-1 (MCP-1). Full-length cDNA cloning, expression in mitogen-stimulated blood mononuclear leukocytes, and sequence similarity to mouse competence gene JE. FEBS Lett 1989; 244:487–493.
76. Valente AJ, Graves DT, Vialle VC, Delgado R, Schwartz CJ. Purification of a monocyte chemotactic factor secreted by nonhuman primate vascular cells in culture. Biochemistry 1988; 27:4162–4168.
77. Furutani Y, Nomura H, Notake M, Oyamada Y, Fukui T, Yamada M, Larsen CG, Oppenheim JJ, Matsushima K. Cloning and sequencing of the cDNA for human monocyte chemotactic and activating factor (MCAF). Biochem Biophys Res Commun 1989; 159:249–255.
78. Bottazzi B, Colotta F, Sica A, Nobili N, Mantovani A. A chemoattractant expressed in human sarcoma cells (tumor-derived chemotactic factor, TDCF) is identical to monocyte chemoattractant protein-1/monocyte chemotactic and activating factor (MCP-1/MCAF). Int J Cancer 1990; 45:795–797.
79. Jiang Y, Valente AJ, Williamson MJ, Zhang L, Graves DT. Post-translational modification of a monocyte-specific chemoattractant synthesized by glioma, osteosarcoma, and vascular smooth muscle cells. J Biol Chem 1990; 265:18318–18321.
80. Rollins BJ, Stier P, Ernst T, Wong GG. The human homolog of the JE gene encodes a monocyte secretory protein. Mol Cell Biol 1989; 9:4687–4695.
81 Robinson EA, Yoshimura T, Leonard EJ, Tanaka S, Griffin PR, Shabanowitz J, Hunt DF, Appella E. Complete amino acid sequence of a human monocyte chemoattractant, a putative mediator of cellular immune reactions. Proc Natl Acad Sci USA 1989; 86:1850–1854.
82. Carr MW, Roth SJ, Luther E, Rose SS, Springer TA. Monocyte chemoattractant protein 1 acts as a T-lymphocyte chemoattractant. Proc Natl Acad Sci USA 1994; 91: 3652–3656.
83. Rollins BJ, Yoshimura T, Leonard EJ, Pober JS. Cytokine-activated human endothelial cells synthesize and secrete a monocyte chemoattractant, MCP-1/JE. Am J Pathol 1990; 136:1229–1233.
84. Rollins BJ, Pober JS. Interleukin-4 induces the synthesis and secretion of MCP-1/JE by human endothelial cells. Am J Physiol 1991; 138:1315–1319.
85. Strieter RM, Wiggins R, Phan SH, Wharram BL, Showell HJ, Remick DG, Chensue SW, Kunkel SL. Monocyte chemotactic protein gene expression by cytokine-treated human fibroblasts and endothelial cells. Biochem Biophys Res Commun 1989; 162: 694–700.
86. Yoshimura T, Leonard EJ. Secretion by human fibroblasts of monocyte chemoattractant protein-1, the product of gene JE. J Immunol 1990; 144:2377–2383.
87. Rolfe MW, Kunkel SL, Standiford TJ, Orringer MB, Phan SH, Evanoff HL, Burdick MD, Strieter RM.Expression and regulation of human pulmonary fibroblast–derived monocyte chemotactic peptide-1. Am J Physiol 1992; 263:L536–L545.

88. Larsen CG, Zarhariae CO, Oppenheim JJ, Matsushima K. Production of monocyte chemotactic and activating factor (MCAF) by human dermal fibroblasts in response to interleukin 1 or tumor necrosis factor. Biochem Biophys Res Commun 1989; 160: 1403–1408.
89. Paine R III, Rolfe MW, Standiford TJ, Burdick MD, Rollins BJ, Strieter RM. MCP-1 expression by rat type II alveolar epithelial cells in primary culture. J Immunol 1993; 150:4561–4570.
90. Becker S, Quay J, Koren HS, Haskill JS. Constitutive and stimulated MCP-1, GRO alpha, beta, and gamma expression in human airway epithelium and bronchoalveolar macrophages. Am J Physiol 1994; 271:L278–L286.
91. Brieland JK, Jones ML, Clarke SJ, Baker JB, Warren JS, Fantone JC. Effect of acute inflammatory lung injury on the expression of monocyte chemoattractant protein-1 (MCP-1) in rat pulmonary alveolar macrophages. Am J Respir Cell Mol Biol 1992; 7:134–139.
92. Christensen PJ, Rolfe MW, Standiford TJ, Burdick MD, Toews GB, Strieter RM. Characterization of the production of monocyte chemoattractant protein-1 and IL-8 in an allogeneic immune response. J Immunol 1993; 151:1205–1213.
93. Ham JM, Kunkel SL, Dibb CR, Standiford TJ, Rolfe MW, Strieter RM. Chemotactic cytokine (IL-8 and MCP-1) gene expression by human whole blood. Immunol Invest 1991; 20:387–394.
94. Martin CA, Dorf ME. Differential regulation if interleukin-6, macrophage inflammatory protein-1, and JE/MCP-1 cytokine expression in macrophage cell lines. Cell Immunol 1991; 135:245–258.
95. Standiford TJ, Kunkel SL, Phan SH, Rollins BJ, Strieter RM. Alveolar macrophage-derived cytokines induce monocyte chemoattractant protein-1 expression from human pulmonary type II-like epithelial cells. J Biol Chem 1991; 266:9912–9918.
96. Brieland JK, Flory CM, Jones ML, Miller GR, Remick DG, Warren JS, Fantone JC. Regulation of monocyte chemoattractant protein-1 gene expression and secretion in rat pulmonary alveolar macrophages by lipopolysaccharide, tumor necrosis factor-alpha, and interleukin-1 beta. Am J Respir Cell Mol Biol 1995; 12:104–109.
97. Brieland JK, Jones ML, Flory CM, Miller GR, Warren JS, Phan SH, Fantone JC. Expression of monocyte chemoattractant protein-1 (MCP-1) by rat alveolar macrophages during chronic lung injury. Am J Respir Cell Mol Biol 9:300–305.
98. Chensue SW, Warmington KS, Lukacs NW, Lincoln PM, Burdick MD, Strieter RH, Kunkel SL. Monocyte chemotactic protein expression during schistosome egg granuloma formation. Am J Pathol 1995; 146:130–138.
99. Hanazawa S, Kawata Y, Takeshita A, Kumada H, Okithu M, Tanaka S, Yamamoto Y, Masuda T, Umemoto T, Kitano S. Expression of monocyte chemoattractant protein 1 (MCP-1) in adult periodontal disease: increased monocyte chemotactic activity in crevicular fluids and induction of MCP-1 expression in gingival tissues. Infect Immun 1993; 61:5219–5224.
100. Rollins BJ, Walz A, Baggiolini M. Recombinant human MCP-1/JE induces chemotaxis, calcium flux, and the respiratory burst in human monocytes. Blood 1991; 78: 1112–1116.
101. Standiford TJ, Rolfe MR, Kunkel SL, Lynch J III, Becker FS, Orringer MB, Phan S, Strieter RM. Altered production and regulation of monocyte chemoattractant pro-

tein-1 from pulmonary fibroblasts isolated from patients with idiopathic pulmonary fibrosis. Chest 1993; 103:1215.
102. Iyonaga K, Takeya M, Saita N, Sakamoto O, Yoshimura T, Ando M, Takahashi K. Monocyte chemoattractant protein-1 in idiopathic pulmonary fibrosis and other interstitial lung diseases. Hum Pathol 1994; 25:455–463.
103. Antoniades HN, Nevill GJ, Galanopoulos T, Kradin RL, Valente AJ, Graves DT. Expression of monocyte chemoattractant protein 1 mRNA in human idiopathic pulmonary fibrosis. Proc Natl Acad Sci USA 1992; 89:5371–5375.
104. Sousa AR, Lane SJ, Nakhosteen JA, Yoshimura T, Lee TH, Poston RN. Increased expression of the monocyte chemoattractant protein-1 in bronchial tissue from asthmatic subjects. Am J Respir Cell Mol Biol 1994; 10:142–147.
105. Flory CM, Jones ML, Warren JS. Pulmonary granuloma formation in the rat is partially dependent on monocyte chemoattractant protein 1. Lab Invest 1993; 69: 396–404.
106. Flory CM, Jones ML, Miller BF, Warren JS. Regulatory roles of tumor necrosis alpha and interleukin-1 beta in monocyte chemoattractant protein-1-mediated pulmonary granuloma formation in the rat. Am J Pathol 1995; 146:450–462.
107. Jones ML, Mulligan MS, Flory CM, Ward PA, Warren JS. Potential role of monocyte chemoattractant protein 1/JE in monocyte/macrophage-dependent IgA immune complex alveolitis in the rat. J Immunol 1992; 149:2147–2154.
108. Jones ML, Warren JS. Monocyte chemoattractant protein 1 in a rat model of pulmonary granulomatosis. Lab Invest 1992; 66:498–503.
109. Schall TJ, Bacon K, Toy KJ, Goeddel DV. Selective attraction of monocytes and T lymphocytes of the memory phenotype by cytokine RANTES. Nature 1990; 347: 669–671.
110. Abdelaziz MM, Devalia JL, Khair OA, Calderon M, Sapsford RJ, Davies RJ. The effect of conditioned medium from cultured human bronchial epithelial cells on eosinophil and neutrophil chemotaxis and adherence in vitro. Am J Respir Cell Mol Biol 1995; 13:728–737.
111. Kwon OJ, Jose PJ, Robbins RA, Schall TJ, Williams TJ, Barnes PJ. Glucocorticoid inhibition of RANTES expression in human lung epithelial cells. Am J Respir Cell Mol Biol 1995; 12:488–496.
112. Stellato C, Beck LA, Gorgone GA, Proud D, Schall TJ, Ono SJ, Lichtenstein LM, Schleimer RP. Expression of the chemokine RANTES by a human bronchial epithelial cell line. Modulation by cytokines and glucocorticoids. J Immunol 1995; 155: 410–418.
113. Matsukura S, Kokubu F, Izumi H, Noda H, Kurokawa M, Tokunaga H, Adachi M. (RANTES expression on human bronchial epithelial cells). Arerugi 1995; 44:715–717.
114. VanOtteren GM, Strieter RM, Kunkel SL, Paine R III, Greenberger MJ, Danforth JM, Burdick MD, Standiford TJ, Paine RR. Compartmentalized expression of RANTES in a murine model of endotoxemia. J Immunol 1995; 154:1900–1908.
115. Marfaing-Koka A, Devergne O, Gorgone G, Portier A, Schall TJ, Galanaud P, Emilie D. Regulation of the production of the RANTES chemokine by endothelial cells. Synergistic induction by IFN-gamma plus TNF-alpha and inhibition by IL-4 and IL-13. J Immunology 1995; 154:1870–1878.
116. Devergne O, Marfaing KA, Schall TJ, Leger RM, Sadick M, Peuchmaur M, Crevon

MC, Kim KJ, Schall TJ, Kim KJ. Production of the RANTES chemokine in delayed-type hypersensitivity reactions: involvment of macrophages and endothelial cells. J Exp Med 1994; 179:1689–1694. (Published erratum appears in J Exp Med 1994 Aug 1; 180:775.)

117. Sticherling M, Kupper M, Koltrowitz F, Bornscheuer E, Kulke R, Klinger M, Wilhelm D, Kameyoshi Y, Christophers E, Schroder JM. Detection of the chemokine RANTES in cytokine-stimulated human dermal fibroblasts. J Invest Dermatol 1995; 105:585–591.

118. Rathanaswami P, Hachicha M, Sadick M, Schall TJ, McColl SR. Expression of the cytokine RANTES in human rheumatoid synovial fibroblasts. Differential regulation of RANTES and interleukin-8 genes by inflammatory cytokines. J Biol Chem 1993; 268:5834–5839.

119. Conlon K, Lloyd A, Chattopadhyay U, Lukacs N, Kunkel S, Schall T, Taub D, Morimot C, Osborne J, Oppenheim J, et al. CD8+ and CD45RA+ human peripheral blood lymphocytes are potent sources of macrophage inflammatory protein 1 alpha, interleukin-8 and RANTES. Eur J Immunol 1995; 25:751–756.

120. Taub DD, Lloyd AR, Wang JM, Oppenheim JJ, Kelvin DJ. The effects of human recombinant MIP-1 alpha, MIP-1 beta, and RANTES on the chemotaxis and adhesion of T cell subsets. Adv Exp Med Biol 1993; 351:139–146.

121. Taub D, Conlon K, Lloyd A, Oppenheim J, Kelvin D. Preferential migration of activated CD4+ and CD8+ T cells in response to MIP-1α and MIP-1β. Science 1993; 260:355–358.

122. Meurer R, Van RG, Feeney W, Cunningham P, Hora DJ, Springer MS, MacIntyre DE, Rosen H. Formation of eosinophilic and monocytic intradermal inflammatory sites in the dog by injection of human RANTES but not human monocyte chemoattractant protein 1, human macrophage inflammatory protein 1 alpha, or human interleukin 8. J Exp Med 1993; 178:1913–1921.

123. Kameyoshi Y, Dorschner A, Mallet AI, Christophers E, Schroder JM. Cytokine RANTES released by thrombin-stimulated platelets is a potent attractant for human eosinophils. J Exp Med 1992; 176:587–592.

124. Alam R, Stafford S, Forsythe P, Harrison R, Faubion D, Lett BM, Grant JA. RANTES is a chemotactic and activating factor for human eosinophils. J Immunol 1993; 150:3442–3448.

125. Rot A, Krieger M, Brunner T, Bischoff SC, Schall TJ, Dahinden CA. RANTES and macrophage inflammatory protein 1 alpha induce the migration and activation of normal human eosinophil granulocytes. J Exp Med 1992; 176:1489–1495.

126. Murphy WJ, Taub DD, Anver M, Conlon K, Oppenheim JJ, Kelvin DJ, Longo DL. Human RANTES induces the migration of human T lymphocytes into the peripheral tissues of mice with severe combined immune deficiency. Eur J Immunol 1994; 24:1823–1827.

127. Gilat D, Hershkoviz R , Mekori YA, Vlodavsky I, Lider O. Regulation of adhesion of CD4+ T lymphocytes to intact or heparinase-treated subendothelial extracellular matrix by diffusible or anchored RANTES and MIP-1 beta. J Immunol 1994; 153:4899–4906.

128. Kakazu T, Chihara J, Saito A, Nakajima S. Effect of RANTES on eosinophil adhesion to plates coated with recombinant soluble intercellular adhesion mole-

cule-1 and expression of beta 2-integrin adhesion molecules on eosinophils. Int Arch Allergy Immunol 1995; 1:9–11.
129. Sherry B, Tekamp OP, Gallegos C, Bauer D, Davatelis G, Wlope SD, Masiarz F, Coit D, Cerami A. Resolution of the two components of macrophage inflammatory proteoin 1, and cloning and characterization of one of theose components, macrophage inflammatory protein 1 beta. J Exp Med 1988; 168:2251–2259.
130. Davatelis G, Tekamp OP, Wlope SD, Hermsen K, Luedke C, Gallegos C, Coit D, Merryweather J, Cerami A. Cloning and characterization of a cDNA for murine macrophage inflammatory protein (MIP), a novel monokine with inflammatory and chemokinetic properties. (Published erratum appears in J Exp Med 1989 Dec 1; 170: 2189.) J Exp Med 1988; 167:1939–1944.
131. VanOtteran GM, Standiford TJ, Kunkel SL, Danforth JM, Burdick MD, Abruzzo LV, Strieter RM. Expression and regulation of macrophage inflammatory protein-1 alpha by murine alveolar and peritoneal macrophages. Am J Respir Cell Mol Biol 1994; 10:8–15.
132. Taub DD, Sayers TJ, Carter CR, Ortaldo JR. Alpha and beta chemokines induce NK cell migration and enhance NK-mediated cytolysis. J Immunol 1995; 155:3877–3888.
133. Schall TJ, Bacon K, Camp RD, Kaspari JW, Goeddel DV. Human macrophage inflammatory protein alpha (MIP-1 alpha) and MIP-1 beta chemokines attract distinct populations of lymphocytes. J Exp Med 1993; 177:1821–1826.
134. Standiford TJ, Kunkel SL, Liebler JM, Burdick MD, Gilbert AR, Strieter RM. Gene expression of macrophage inflammatory protein-1 alpha from human blood monocytes and alveolar macrophages is inhibited by interleukin-4. Am J Respir Cell Mol Biol 1993; 9:192–198.
135. Christman JW, Blackwell TR, Cowan HB, Shepherd VL, Rinaldo JE. Endotoxin induces the expression of macrophage inflammatory protein 1 alpha mRNA by rat alveolar and bone marrow-derived macrophages. Am J Respir Cell Mol Biol 1992; 7:455–461.
136. Danforth JM, Strieter RM, Kunkel SL, Arenberg DA, VanOtteren GM, Standiford TJ. Macrophage inflammatory protein-1 alpha expression in vivo and in vitro: the role of lipoteichoic acid. Clin Immunol Immunopathol 1995; 74:77–83.
137. Kasama T, Strieter RM, Standiford TJ, Burdick MD, Kunkel SL. Expression and regulation of human neutrophil-derived macrophage inflammatory protein 1 alpha. J Exp Med 1993; 178:63–72.
138. Fong TAT, Mosmann TR. Alloreactive murine CD8+ T cell clones secrete the Th1 pattern of cytokines. J Immunol 1990; 144:1744–1752.
139. Cher DJ, Mosmann TR. Two types of murine helper T cell clone. II. Delayed-type hypersensitivity is mediated by Th1 clones. J Immunol 1987; 138:3688–3694.
140. Street NE, Mosmann TR. Functional diversity of T lymphocytes due to secretion of different cytokine patterns. FASEB J 1991; 5:171–177.
141. Burdick MD, Kunkel SL, Lincoln PM, Wilke CA, Strieter RM. Specific ELISAs for the detection of human macrophage inflammatory protein-1 alpha and beta. Immunol Invest 1993; 22:441–449.
142. Lukacs NW, Chensue SW, Smith RE, Strieter RM, Warmington K, Wilke C, Kunkel SL. Production of monocyte chemoattractant protein-1 and macrophage inflamma-

tory protein-1 alpha by inflammatory granuloma fibroblasts. Am J Physiol 1994; 144:711–718.

143. Alam R, Forsythe PA, Stafford S, Lett BM, Grant JA. Macrophage inflammatory protein-1 alpha activates basophils and mast cells. J Exp Med 1992; 176:781–786.

144. Fahey T III, Tracey KJ, Tekamp OP, Cousens LS, Jones WG, Shires GT, Cerami A, Sherry B, Fahey TJ. Macrophage inflammatory protein 1 modulates macrophage function. J Immunol 1992; 148:2764–2769.

145. Cook DN, Beck MA, Coffman TM, Kirby SL,Sheridan JF, Pragnell IB, Smithies O. Requirement of MIP-1 alpha for an inflammatory response to viral infection. Science 1995; 269:1583–1585.

146. Standiford TJ, Kunkel SL, Lukacs NW, Greenberger ML, Danforth JM, Kunkel RG, Strieter RM. Macrophage inflammatory protein-1 alpha mediates lung leukocyte recruitment, lung capillary leak, and early mortality in murine endotoxemia. J Immunol 1995; 155:1515–1524.

147. Lukacs NW, Kunkel SL, Strieter RM, Warmington K, Chensue SW. The role of macrophage inflammatory protein1 alpha in *Schistosoma mansoni* egg-induced granulomatous inflammation. J Exp Med 1993; 177:1551–1559.

148. Shanley TP, Schmal H, Friedl HP, Jones ML, Ward PA. Role of macrophage inflammatory protein-1 alpha (MIP-1 alpha) in acute lung injury in rats. J Immunol 1995; 154:4793–4802.

149. Cruikshank WW, Long A, Tarpy RE, Kornfeld H, Carroll MP, Teran L, Holgate ST, Center DM. Early identification of interleukin-16 (lymphocyte chemoattractant factor) and macrophage inflammatory protein 1 alpha (MIP1 alpha) in bronchoalveolar lavage fluid of antigen-challenged asthmatics. Am J Respir Cell Mol Biol 1995; 13:738–747.

150. Standiford TJ, Rolfe MW, Kunkel SL, Lynch J III, Burdick MD, Gilbert AR, Orringer MB, Whyte RI, Strieter RM. Macrophage inflammatory protein-1 alpha expression in interstitial lung disease. J Immunol 1993; 151:2852–2863.

151. Smith RE, Strieter RM, Phan SH, Lukacs NW, Huffnagle GB, Wilke CA, Burdick MD, Lincoln P, Evanoff H, Kunkel SL. Production and function of murine macrophage inflammatory protein-1 alpha in bleomycin-induced lung injury. J Immunol 1994; 153:4704–4712.

152. Bromberg J, Chavin K, Kunkel S. Anti-tumor necrosis factor antibodies suppress cell-mediated immunity in vivo. J Immunol 1992; 148:3412–3417.

153. Chen GH, Curtis JL, Mody CH, Christensen PJ, Armstrong LR, Toews GB. Effect of granulocyte-macrophage stimulating factor on rat alveolar macrophage anticryptococcal activity in vitro. J Immunol 1994; 152:724–734.

154. Mody CH, Tyler CL, Sitrin RG, Jackson C, Toews GB. Interferon-gamma activates rat alveolar macrophages for anticryptococcal activity. Am J Respir Cell Mol Biol 1991; 5:19–25.

155. Bulmer G, Tacker J, Phagocytosis of *Cryptococcus neoformans* by alveolar macrophages. Infect Immun 1975; 11:73–79.

156. Collins HL, Bancroft GJ. Cytokine enhancement of complement-dependent phagocytosis by macrophages: synergy of tumor necrosis factor-alpha and granulocyte-macrophage colony-stimulating factor for phagocytosis of *Cryptococcus neoformans*. Eur J Immunol 1992; 22:1447–1454.

157. Kozel T, Highison B, Stratton C. Localization on encapsulated *Cryptococcus neoformans* of serum components opsonic for phagocytosis by macrophages and neutrophils. Infect Immun 1984; 43:574–579.
158. Hill JO. CD4+ T cells cause multinucleated giant cells to form around *Cryptococcus neoformans* and confine the yeast within the primary site of infection in the respiratory tract. J Exp Med 1992; 175:1685–1695.
159. Abe E, Ishimi Y, Jin CH, Hong MH, Sato T, Suda T. Granulocyte-macrophage colony-stimulating factor is a major macrophage fusion factor present in conditioned medium of concanavalin A-stimulated spleen cell cultures. J Immunol 1991; 147: 1810–1815.
160. McInnes A, Rennick DM. Interleukin 4 induces cultured monocytes/macrophages to form giant multinucleated cells. J Exp Med 1988; 167:598–611.
161. Most J, Neumayer HP, Dierich MP. Cytokine-induced generation of multinucleated giant cells in vitro requires interferon-gamma and expression of LFA-1. Eur J Immunol 1990; 20:1661–1667.
162. Tanaka H, Shinki T, Hayashi T, Jin CH, Miyaura C, Abe E, Suda T. Spermidine-dependent proteins are involved in the fusion of mouse alveolar macrophages induced by 1 alpha,25-dihydroxyvitamin D3 and interleukin 4. Exp Cell Res 1989; 180:72–83.
163. Aguirre K, Havell EA, Gibson GW, Johnson LL. Role of tumor necrosis factor and gamma interferon in acquired resistance to *Cryptococcus neoformans* in the central nervous system of mice. Infect Immun 1995; 63.1725–1731.
164. Sitrin RG, Shollenberger SB, Strieter RM, Gyetko MR. Endogenously produced urokinase activity amplifies tumor necrosis factor-alpha secretion by THP-1 mononuclear phagocytes. J Leukoc Biol 1995. In press.
165. Farrar MA, Schreiber RD. The molecular cell biology of interferon-gamma and its receptor. Annu Rev Immunol 1993; 11:571–611.
166. Flesch IE, Schwamberger G, Kaufmann SH. Fungicidal activity of IFN-gamma-activated macrophages. Extracellular killing of *Cryptococcus neoformans*. J Immunol 1989; 142:3219–3224.
167. Levitz SM, DiBenedetto DJ. Differential stimulation of murine resident peritoneal cells by selectively opsonized encapsulated and acapsular *Cryptococcus neoformans*. Infect Immun 1988; 56:2544–2551.

16

Macrophage Function in Bacterial and Fungal Infections of Newborns

MICHAEL P. SHERMAN

Baylor College of Medicine
Houston, Texas

I. Introduction

The role of alveolar macrophages in preventing pulmonary infection was defined more than three decades ago when the clearance of intrapulmonary bacteria was determined in adult experimental animals following infection with staphylococcal aerosols (1). A decade later, group B streptococci (GBS) emerged as the major cause of neonatal pneumonia (2–4), yet mechanisms utilized by fetal or neonatal lungs to prevent pulmonary infection were ill defined compared with adults. As with congenital pneumonia caused by other pathogens (5,6), GBS disease is the consequence of ascending genital infection, with microbial colonization of amniotic fluid and subsequent bacterial colonization or infection of the fetal lung (7–9). Congenital pneumonia, especially that caused by GBS, presents itself at or shortly after birth with a high incidence of respiratory distress, bacteremia, pulmonary vascular hypertension, and death (8–12). In light of the devastating effects that group B streptococcal infections have on newborns and their families, investigators have spent the past two decades trying to understand the relationship between the function of lung macrophages during the perinatal period and the pathobiology of neonatal pneumonia. This chapter reviews the maturation of lung macrophages and the role they play in the pathobiology of newborn infection. The ease with which newborn alveolar macrophages can be obtained by lung lavage means these cells are the best studied, but few alveolar macrophages are present in human newborn lung. For this reason, most of the information about the developmental aspects of lung macrophages is based on animal studies. In this chapter, differences between human and animal studies will be emphasized whenever they are known. We have divided this subject into two specific areas which include (1) the functional maturation of lung macrophages as it relates to host defense against bacteria and fungi and (2) the function of lung macrophages as it relates to specific pathogens and clinical conditions that are prevalent during the neonatal period. Finally, in each area presented, directions for future research involving neonatal lung macrophages will be discussed so that morbidity and mortality associated with neonatal pneumonia can be reduced in the years to come.

II. Functional Maturation of Lung Macrophages in Newborns

A. Types of Macrophages Identified During Lung Development

Lung macrophages have been divided into bronchial, alveolar, interstitial, intravascular, and pleural types based on their specialized functions in different locations (13). Traveling downward into the lung, bronchial macrophages partly represent alveolar macrophages exiting the lung by way of the mucociliary escalator, but there are also identifiable macrophages that are tightly adherent to respiratory epithelium. These cells remove microorganisms from the large airways (14). Studies involving airway macrophages in developing lungs as the initial phagocytes that would encounter bacteria and fungi colonizing pulmonary tissue have not been undertaken. Alveolar macrophages are the resident phagocytes of the distal air spaces in newborns and adults, and as will be discussed later, the population kinetics and functional activities of these phagocytes differ significantly between birth and later life (15). Although never studied during the perinatal period, subpopulations of alveolar macrophages lavaged from adult lungs express different receptors, have varying cytological and phagocytic characteristics, and manifest distinct mediator release depending on the disease state (16–18). Subpopulations likely exist in neonatal lungs, because numerous cells recovered by lung lavage during the perinatal period have the characteristics of monocytes rather than those of mature alveolar macrophages (19). This finding could provide a partial answer for the reduced antimicrobial activity which newborn alveolar macrophages of rabbits manifest in vitro (20) and in vivo (21) compared with their more mature counterparts. Interstitial macrophages are located in the lung's connective tissue and are recovered by digesting minced pulmonary tissue (22). The numbers of interstitial macrophages can change during the course of different lung diseases (23). These phagocytes share some characteristics with alveolar macrophages, such as particulate ingestion and the release of oxygen intermediates and receptors for the Fc fragment of immunoglobulins (22,24,25). The ability of interstitial macrophages to synthesize DNA and replicate has suggested that they function to review the alveolar macrophage population (26). The role of interstitial macrophages have not been studied in newborn lungs relative to host defense against bacteria and fungi. Besides alveolar and interstitial macrophages, newborns in the order Artiodactyla have pulmonary intravascular macrophages (see ref. 27). The presence of pulmonary intravascular macrophages has also been demonstrated in adult humans (28). These macrophages appear to function as an accessory reticuloendothelial clearance system separate from that of liver and spleen (29,30). Neither the numbers of these cells populating the pulmonary vessels nor their function as intravascular phagocytes have been studied in human newborns. Nevertheless, it is interesting to speculate that if these macrophages were present, they would damage pulmonary endothelium and amplify inflammation when newborns have bacteremia. The final macrophage compartment in the lung, that of the pleural macrophage, is akin to peritoneal macrophages (31). Their presence and role in preventing infections of the pleural space during the neonatal period have not been reported. Thus, much remains to be learned about the different types of macrophages in the newborn lung.

B. Origins of Macrophages During Lung Development

In the rat fetal lung, the macrophage precursors are first detected during the bronchial bud stage (32). As lung development progresses, the pulmonary interstitium, and to a lesser extent the free air spaces, have replicating cells which exhibit surface markers and intracellular characteristics of macrophages. These cells are not monocytes and can be distinguished from a second source of mononuclear phagocytes that enter the alveoli of rabbits just before and following parturition (19). All mammalian species studied (19,33–35), including humans (36,37), have an intra-alveolar influx of mononuclear phagocytes around the time of birth. Thus, the alveolar macrophage population has extremely low numbers until mammals reach term gestation. The monocyte-appearing cells entering the alveoli at birth are distributed to the lung from the bone marrow, whereafter they are capable of at least one more cell division (38), as described for lung macrophage renewal in adults (39). Cellular replication is largely responsible for the increasing alveolar macrophage numbers observed between days 1 and 5 of life (38). For every species studied and reported in the literature, lung lavage has recovered increasing numbers of macrophages as postnatal age advances until a stable adult population is reached. In newborn pigs, pulmonary intravascular macrophage numbers also increase by proliferation within the lung's capillaries postnatally (40). The expansion of interstitial, bronchial, and pleural macrophage subpopulations has not been studied after birth.

C. Newborn Alveolar Macrophage Population and Congenital Pneumonia

The low numbers of alveolar macrophages found in fetal lung are an important explanation for the fulminant course of congenital pneumonia observed in premature infants. We tested this hypothesis by studying group B streptococcal clearance in the lungs of premature and term rabbits (41). These studies examined bacterial proliferation or killing in the lung by quantitatively culturing the homogenized left lung at 0, 4, and 8 hr after aerosol infection. Simultaneously, the numbers of macrophages and neutrophils were determined by lavaging the right lung, recovering the phagocytes by centrifugation, counting their numbers using a hemacytometer, and evaluating cell type via Giemsa-stained cytocentrifuge preparations. When premature and term rabbits were infected with group B streptococcal aerosols, there was significant streptococcal replication at 8 hr in premature lung versus killing of GBS in term lung (Fig. 1). This difference seemed related to an alveolar macrophage population that was 17-fold higher in term rabbit lung. In response to a diminished alveolar macrophage population and intrapulmonary GBS growth, the lungs of premature rabbits had a 13-fold increase in the numbers of lavaged neutrophils compared with that of term animals. In the face of a reduced alveolar macrophage population, the importance of neutrophil recruitment to the premature lung as a second-line phagocytic defense was unmasked by administration of the anti-CD18 monoclonal antibody R15.7. This antibody impaired neutrophil-related, but not macrophage-associated, phagocytosis (Table 1). Premature rabbits receiving the antibody had almost exponential intrapulmonary streptococcal proliferation, whereas term rabbits with the larger alveolar macrophage population had a minimal reduction in bacterial clearance (Fig. 1). Physical removal of intrapulmonary bacteria also did not explain the differences in bacterial numbers cultured from premature and term rabbit lung. Previously, we had shown that physical removal of radiolabeled bacteria accounts for less than 15% of the clearance (loss of viable bacteria) from term rabbit lung following aerosol infection (21). In conclusion, the defective clearance of GBS caused by a diminished alveolar macrophage population in the preterm lung helps clarify why there is an increased incidence and severity of GBS infection in human premature newborns as compared with their term counterparts (6,9,12).

D. Phagocytic Behavior of Newborn Lung Macrophages and Antimicrobial Activity

The airways, air sacs, and vasculature of the immature lung have sessile and mobile macrophages. This is most evident in premature rabbit lung where large numbers of mononuclear phagocytes rapidly enter the alveoli following GBS aerosol infection (41). This phenomenon emphasizes the significance of macrophage movement toward threatening microorganisms, the ingestion of those

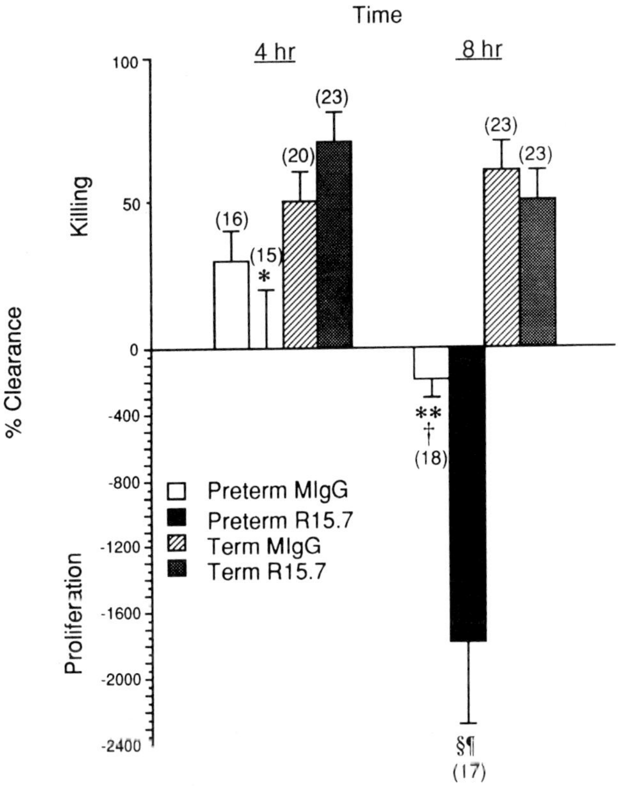

Figure 1 Clearance of pulmonary GBS by preterm or term neonatal rabbits at either 4 or 8 hr after aerosol infection. Time after aerosol infection is displayed above the figure. Percentage of clearance is shown on the vertical axis. Values represent mean ± SEM; numbers of animals studied are in parentheses; key within figure shows gestational age and type of antibody treatment (MIgG, mouse IgG; R15.7, anti-CD18 monoclonal antibody). * = $P < .002$ vs 4 = hr term R15.7, **, $P < .02$ vs 8-hr preterm R15.7; † = $P < .05$ vs 8-hr term control; §, $P < .001$ vs term MIgG and term R15.7; ¶, $P < .002$ vs 4-hr preterm R15.7.

microbes, and their subsequent killing and destruction. The susceptibility of newborns to pneumonia is related not only to a quantitative deficiency of alveolar macrophages in the fetal and newborn lung (41,42) but also is provoked by functional abnormalities in cell movement, microbial uptake, and antimicrobial activity. Chemotactic responses of 1-, 7-, and 28-day-old alveolar macrophages of rabbits have been evaluated in blind-well and agarose assays using endotoxin-activated serum and n-formyl-methionyl-phenylalanine (43). Both assays showed a dramatic increase in the chemotactic responsiveness of alveolar macrophages

Table 1 Effect of Anti-CD18 Monoclonal Antibody (R15.7) on In Vitro Phagocytosis of GBS by Preterm and Term Rabbit Polymorphonuclear Leukocytes (PMNL) and Pulmonary Alveolar Macrophages (PAM)

	PMNL		PAM	
	Control	R15.7	Control	R15.7
% PMNL or PAM with ingested FITC-GBS				
Preterm	56 ± 5 (10)	17 ± 5[a]	34 ± 3 (10)[b]	37 ± 4[b]
Term	57 ± 2 (8)	11 ± 2[a]	70 ± 5 (8)	68 ± 4
No. of FITC-GBS per phagocytic PMNL or PAM				
Preterm	7 ± 1	4 ± 1[c]	10 ± 2	9 ± 2
Term	6 ± 1	3 ± 1[c]	11 ± 1	10 ± 2

Data presented as mean ± SEM (no. of animals). FITC-GBS, fluorescein isothiocyanate–conjugated GBS.
[a] $P < .001$ vs control.
[b] $P < .02$ vs term.
[c] $P < .05$ vs control.

with postnatal maturation regardless of the stimulus used. Kurland and colleagues have shown that directed migration of pulmonary macrophages recovered from <24-hr-old rats and 1-day-old rhesus monkeys was lower than adult cells but reached adult levels by the third and sixth postnatal day, respectively (44,45). Although alveolar macrophages of primates have mature chemotactic behavior by 6 days of age, their ability to kill *Candida albicans* does not reach adult activity until 6 months of age. There is no information describing the responses that lung macrophages of human newborns would have to the plethora of chemoattractant molecules currently reported in the literature, but age-related differences are expected.

Studies of particulate uptake by newborn lung macrophages have resulted in conflicting results. In vitro ingestion of *Candida albicans* by alveolar macrophages of newborn rabbits (43) and humans (46) was equivalent to adult macrophages, whereas newborn lung macrophages of primates had reduced uptake of this yeast (45). Interestingly, adult serum was utilized to opsonize the yeast in each of the aforementioned studies. The use of neonatal serum might have resulted in different findings. In vivo studies (21), using tissue stains to ascertain bacterial ingestion, found reduced phagocytosis of *Staphylococcus aureus* by 1-day-old rabbit alveolar macrophages compared with cells of older animals. In vitro studies (20) using monolayers of newborn and adult alveolar macrophages exposed to *S. aureus* detected no differences in the phagocytic rate or the phagocytic index (bacteria/ingesting macrophage). The preceding in vitro and in vivo studies of rabbit newborn alveolar macrophages that used *S. aureus* as the microbial target employed staining procedures that did not distinguish between surface-adherent and truly internalized bacteria. This problem was overcome in subsequent studies by the use of fluorescein-labeled GBS that were opsonized with either complement (47) or specific antibody (48). To judge the numbers of actual intracellular bacteria, surface-adherent streptococci had their fluorescence quenched by the addition of methylene blue at the end of an experimental period. Intracellular streptococci were protected from the cell-impermeable dye and were brightly visible during fluorescence microscopy. Using this methodology, premature, term, and adult alveolar macrophages ingested GBS equally well when the bacteria were opsonized by the classic complement pathway (47) or type-specific antibody (48). Conversely, GBS opsonized via the alternative complement pathway were poorly ingested by alveolar macrophages of preterm rabbits (47). Other investigations using neonatal rat alveolar macrophages indicate a decreased expression of Fc, C3b, fibronectin, and lectin receptors on the surface of these cells during the first postnatal week (49). Thus, a paucity of phagocytic receptors on fetal or neonatal alveolar macrophages may under certain conditions limit their phagocytic ability. Of special note, the density of the complement and immunoglobulin receptors have never been studied on alveolar macrophages of human infants during lung maturation. Our current knowledge is obviously incomplete, and experimental conditions and species differences make it difficult to draw conclusions regarding defective phagocytic behavior of newborn lung macrophages. Nevertheless, there is the likelihood that mechanisms associated with cell movement and particulate ingestion are reduced in neonatal lung macrophages.

A most interesting area of lung host defense is the role that surfactant-related proteins play in microbial uptake and killing by alveolar macrophages. The two surfactant proteins of interest are, surfactant protein A (SP-A) and surfactant protein D (SP-D), members of the type C lectin or collectin family (50). Alveolar macrophages have receptors for each of these proteins (51,52). These proteins can modulate chemotaxis, the release of reactive oxygen species, and the killing of gram-positive bacteria, gram-negative bacteria, and viruses by alveolar macrophages (50,53–55). In the human fetus and newborn, immunoreactive SP-A has been associated with alveolar macrophages, but these cells has no mRNA for SP-A (56). The role of these two proteins in modulating the phagocytic and metabolic behavior of newborn lung macrophages has not been examined.

E. Maturation of Antimicrobial Mechanisms in Macrophages During Development

The phagocytic killing of bacteria and fungi has often been divided into O_2-dependent and O_2-independent microbicidal mechanisms (57–60). This separation emerged from understanding the mechanisms responsible for chronic granulomatous disease (61). This clinical disorder has several separate inherited defects which result in defective assembly of a membrane-bound NADPH oxidase and is associated with frequent bacterial and fungal infections. This defect results in markedly reduced superoxide anion production by phagocytes. Thus, this genetic disease is characterized by defective oxygen-dependent killing by phagocytes. Because a rapid increase in oxygen consumption occurs when macrophage and neutrophil receptors bind microorganisms and this mechanism is activated, the process is also known as the respiratory burst. Other congenital defects in host defense, such as specific granule deficiency (62,63), may result in O_2-independent microbicidal defects of phagocytes. The killing of ingested microorganisms by phagocytes has become more complicated, however, as more antimicrobial mechanisms are discovered. These mechanisms involve, in part, the production of nitric oxide or peroxynitrite, tryptophan, iron, or other essential nutrient depletion from microbes, and an increasing number of proteins or enzymes that produce membrane pores or cause lysis of the organisms (64). Additional effectors of microbial killing and digestion by lung macrophages will certainly be recognized in the future. This section summarizes our current knowledge of antimicrobial activity by lung macrophages during animal development.

O_2-Dependent Microbicidal Mechanisms of Newborn Lung Macrophages

Production of Reactive Oxygen Intermediates

The reduction of molecular oxygen by alveolar macrophages, including the production of superoxide anion, hydrogen peroxide, and hydroxyl radical (65,66), and the production of singlet oxygen (67) have been implicated in the microbicidal activity of these cells. Compared with neutrophils and eosinophils (68,69), rabbit alveolar macrophages do not produce hypohalous acids (70). In vitro studies of superoxide anion production (48,71,72) or chemiluminescence (73) by neonatal alveolar macrophages show that the respiratory burst is significantly reduced compared with adult cells. These descriptive studies have not been followed up with detailed investigations examining (1) signal transduction defects which result in the assembly of the NADPH oxidase or (2) the hyporesponsiveness of neonatal alveolar macrophages to activating signals such as cytokines or growth factors. Other reasons for the reduced respiratory burst of neonatal alveolar macrophages, however, have been found. Lung expansion at term birth releases large quantities of surfactant into the alveoli, and lamellar vesicles present in the released surface-active material are ingested and remain within neonatal alveolar macrophages for several days after birth (19). Interestingly, Hayakawa et al. (74,75) observed that newborn pulmonary alveolar macrophages (PAMs) incubated with either surfactant or artificial phospholipid vesicles had a suppressed oxidative response to phorbol myristate acetate or opsonized zymosan. The uptake of phospholipid vesicles, whether in vitro (43) or in vivo (76), depressed the ability of newborn alveolar macrophages to kill ingested bacteria or fungi. The findings of Hayakawa and colleagues (74,75) offer yet another explanation for the reduced microbicidal activity of newborn alveolar macrophages. Neonatal lipid-laden alveolar macrophages are akin to those in pulmonary alveolar proteinosis, an adult disease, in which the alveolar macrophages are filled with phospholipid vesicles. Like newborns, patients with pulmonary alveolar proteinosis have an increased susceptibility to pneumonia.

Compared with alveolar macrophages of adults, those lavaged from the lungs of newborn animals do not depend on oxidative phosphorylation as an energy source for the ingestion of bacteria and fungi (48,77). This conclusion is based on a significant inhibition of phagocytosis by cyanide in adult macrophages but not newborn cells. Over the first 7 days of life, neonatal lung macrophages have a substantial increase in the number of mitochondria (19) as these cells adapt to a postnatal alveolar oxygen tension of 150 mm Hg versus a level of 27 mm Hg in the fetal lung. The exact metabolic pathways used as an energy source by neonatal alveolar macrophages during phagocytosis remain undefined. The functional impact of cyanide-insensitive particulate ingestion by neonatal lung macrophages during the immediate postnatal period is also unknown.

Production of Reactive Nitrogen Intermediates

Nitric oxide (NO) is one of the recently recognized microbicidal agents produced by mammalian phagocytes and requires molecular oxygen for its synthesis (78,79). Nitric oxide and citrulline are by-products of L-arginine oxidation. Nitric oxide is metabolized and inactivated by oxidation to its endproducts, nitrite and nitrate. Peroxynitrite, a product of superoxide anion's reaction with nitric oxide, has also been suggested as a bactericidal species produced by macrophages (80,81). Questions have been raised about peroxynitrite production by macrophages, because NADPH oxidase and inflammatory nitric oxide synthase are differentially regulated such that their activities are not simultaneously induced (82). Experimental studies in murine macrophages actually support the concept that either superoxide anion or nitric oxide would be generated based on the stimulus, and superoxide anion or nitric oxide would have separate molecular targets (83). For example, superoxide anion would be important for killing staphylococci, likely through the formation of reactive oxygen species besides superoxide, whereas nitric oxide would inactivate pathogens such as *Leishmania major* (83).

Although nitric oxide–mediated microbiostatic/microbiocidal mechanisms are detectable in murine macrophages, it is not yet clear whether nitric oxide production is operative in human alveolar macrophages (84). We found no evidence that human alveolar macrophages lavaged from healthy adults could be stimulated by lipopolysaccharide (LPS) and interferon-γ to generate nitrite (85), as is the case with rodent alveolar macrophages (86). *Pneumocystis carinii*, however, could stimulate a one- to twofold rise in nitrite production by human alveolar macrophages (85). This level of nitric oxide generation was far below the amounts produced by alveolar macrophages of mice exposed to the same stimuli (85). Despite this controversy, studies have shown that thioglycolate-elicited adult murine macrophages responded to serotypes Ia and III, asialo- and type antigen-deficient mutant strains of the GBS with production of nitric oxide (87). This response is mediated by complement type 3 receptors (CR3). Since hypotension is a major component of GBS disease in newborns, and could be mediated physiologically by nitric oxide production, it is interesting to speculate about its role via macrophage and GBS interactions. In concert with this assumption, we have studied alveolar macrophages lavaged from 3-day-old rats and observed that they produced significantly higher levels of nitrite, nitrate, and citrulline when stimulated by LPS and interferon-γ than did similarly treated 10-day-old or adult macrophages (Fig. 2). We also found that basal or unstimulated nitrite and nitrate production by 3-day-old alveolar macrophages was 10-fold above that of adult cells (data not shown). The heightened production of basal and stimulated nitric oxide generation by 3-day-old alveolar macrophages of rats could not be explained by reduced superoxide anion production in the younger macrophages. Superoxide anion generation by 3-day-old alveolar macrophages of rats was negligible under basal conditions and was equivalent to that produced by older alveolar macrophages following stimulation with phorbol myristate acetate or serum-treated zymosan (Fig. 3). Compared with neonatal alveolar macrophages of rats, newborn alveolar macrophages recovered from rabbits do not release nitric oxide when stimulated with LPS and interferon-γ, whereas their stimulated production of superoxide anion is substantially below that of adult macrophages (72). These variations emphasize the complexity of studying the maturation of lung host defenses in alveolar macrophages of different species. Curiously, the lysozyme content of alveolar macrophages from newborn, infant, and adult rats had an inverse relationship with nitric oxide production.

Finally, it is possible that nitric oxide acts as an immunomodulator of alveolar macrophage maturation or activation in infantile rats rather than as an antimicrobial effector (88). Nitric oxide reduces the expression of message for early growth response gene-1 when rat alveolar macrophages are stimulated with LPS and interferon-γ (89). Early growth response gene-1 is essential for the differentiation of monocytes to macrophages (90), an important event occurring in lung macrophage maturation during the postnatal period. The role that nitric oxide plays in the host defense of newborn lung is problematic at present, and considerable research is needed to clarify our knowledge, particularly with regard to its importance in human neonates.

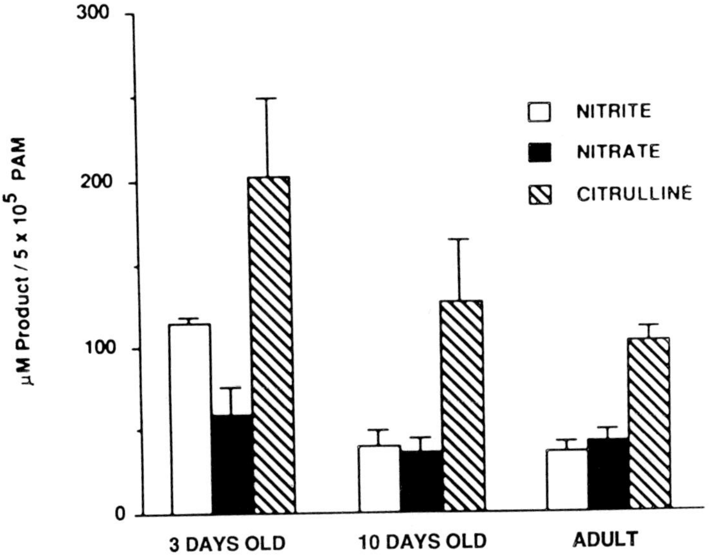

Figure 2 L-arginine oxidation by newborn (3-day), infant (10-day), and adult alveolar macrophages of rats stimulated with LPS (100 ng/10^6 macrophages) and interferon-γ (rat rIFN-γ, 500 U/10^6 macrophages). Age is shown on the horizontal axis, and μM of end-product (nitrite, nitrate or citrulline) is displayed on the vertical axis. The key within the figure describes the type of endproduct. Taken together, nitrite and nitrate production is significantly higher in 3 day-old alveolar macrophages than the adult macrophages (ANOVA, $p < .05$). Data are presented as the mean ± SEM of three separate experiments using lavaged macrophages. The lavaged macrophages from individual litters of rats were pooled and analyzed.

O_2-Independent Microbicidal Mechanisms of Newborn Lung Macrophages

In addition to the depressed respiratory burst that has been observed in newborn alveolar macrophages (48,71–73), there is diminished intracellular concentrations of proteins associated with the O_2-independent killing and destruction of intracellular microorganisms. The content of acid hydrolases (91) and cationic proteins (92), called defensins, in newborn alveolar macrophages of rabbits are remarkably reduced compared with adult cells. Similar to alveolar macrophages of newborn rats, 1-day-old alveolar macrophages of rabbits have extremely low concentrations of lysozyme which progressively rises in macrophages of older animals (91). Since alveolar macrophages are a major source of secreted lysozyme in the alveoli, the rise in lysozyme in postnatal alveolar macrophages parallels its

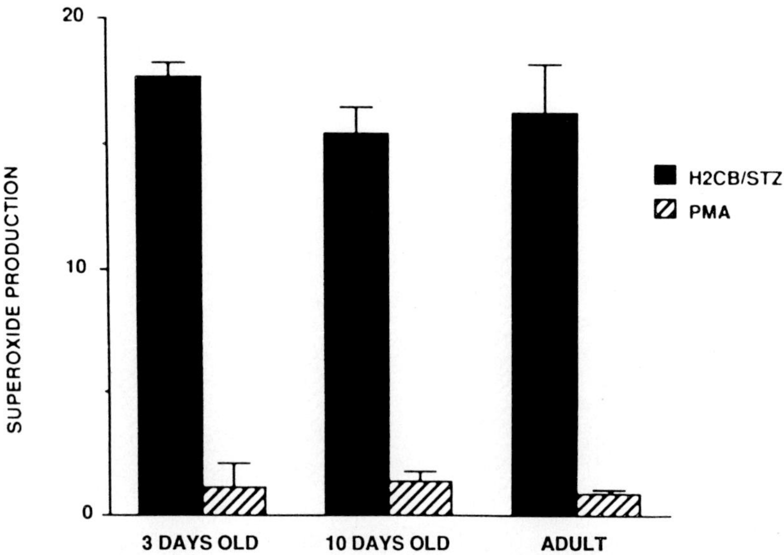

Figure 3 Superoxide anion production by newborn (3-day), infant (10-day), and adult alveolar macrophages of rats stimulated with either serum-treated zymosan after preincubation of cells with dihydrocytochalasin B (H2CB/STZ) or with phorbol myristate acetate (PMA, 100 ng/10^6 macrophages). Age is shown on the horizontal axis, and nanomoles superoxide anion released/10^6 macrophages/10 min is displayed on the vertical axis. The key within the figure describes the type of stimulus. Data are presented as the mean ± SEM of three separate experiments using lavaged macrophages. The lavaged macrophages from individual litters of rats were pooled and analyzed.

appearance in alveolar lining fluid (93). In human newborns, there is a deficiency of lysozyme and lactoferrin in airway secretions (94). To test the importance of lysozyme in the host defense of the neonatal lung, we treated preterm rabbits with surfactant alone or surfactant with lysozyme added followed by infection of these animals with GBS aerosols. The animals that had lysozyme added to surfactant had a significant restriction of GBS growth in the lung compared with preterm rabbits receiving surfactant alone (95). This observation suggests that surfactant could be used as a drug-delivery vehicle to treat neonatal pneumonia and replace host factors, such as lysozyme, that are normally present in the alveolar lining fluid of adults.

The low concentrations of microbicidal proteins in the lysosomes of newborn alveolar macrophages parallels diminished numbers of these vesicles in the phagocytes (19,96). O_2-independent mechanisms include the delivery of lysosomal acid hydrolases, neutral proteases, or lysozyme to the phagosome. The mechanisms of phagolysosomal fusion in newborn alveolar macrophages have only been studied in foals, which often have infection with *Rhodococcus equi* (97). Additional information is needed concerning the effective delivery of lysosomal components to the phagosome of newborn lung macrophages in other species.

The relative importance of O_2-independent mechanisms versus O_2-dependent mechanisms in the killing of microorganisms by newborn lung macrophages has not been studied. Information could be acquired in this area by the use of diphenylene iodonium (98,99). At low concentrations, this flavoprotein inhibitor blocks the production of reactive oxygen and nitrogen species, whereas high concentrations appear to inhibit phagolysosomal fusion. The use of L-arginine analogues to inhibit nitric oxide production alone or the use of diphenylene iodonium to block both reactive oxygen and nitrogen species would define the relative importance of these two agents in microbial killing by macrophages. Using newborn and adult macrophages to ascertain maturational influences, this methodology would likely provide new insight into the mechanisms by which these phagocytes inactivate microorganisms. In conclusion, the mechanisms associated with the killing and destruction of microorganisms by newborn alveolar macrophages are only partly understood, and future investigations must address this lack of knowledge if we are to comprehend the pathobiology of neonatal pneumonia.

F. Cytokines, Adhesion Molecules, Newborn Lung Macrophages, and Antimicrobial Activity

Cytokines and Newborn Lung Macrophages

Alveolar macrophages became activated and enhance the inflammatory response during pulmonary infections (100,101). Cytokines are the mediators of macrophage activation. They are important multifunctional growth factors or biological response modifiers which change cell behavior or initiate cell to cell communication (102). Their action on macrophages may include the control of mitosis, chemotaxis, cytoskeletal arrangement, and immunomodulation. Cytokines can increase or decrease the expression of adhesion molecules on macrophages or alter macrophage interactions with matrix glycoproteins. Macrophages secrete and respond to cytokines. In the lung, interleukin-1β (IL-1β), tumor necrosis factor-α (TNF-α), IL-8, IL-6, and macrophage inflammatory proteins-1 and -2 (MIP-1 and MIP-2) have received the most attention relative to infection (103–105). Little information is available, however, regarding either the secretion of these cytokines by newborn lung macrophages or macrophage activation by them. Increased concentrations of TNF-α and IL-1β have been found in tracheal aspirate specimens from human newborns with pneumonia (106), but the cells producing these cytokines were undefined. Alveolar macrophages lavaged from newborn rabbits produce more IL-1β in culture under basal and LPS-stimulated conditions than do adult alveolar macrophages (107). As might occur in premature human infants hospitalized in intensive care nurseries, infant rats were subjected to severe protein calorie malnutrition, and their alveolar macrophages were assessed for antimicrobial activity, cytokine production, and arachidonic acid metabolism (108). Protein calorie malnutrition did not alter phagocytosis and killing of *Listeria monocytogenes*, the release of hydrogen peroxide and superoxide anion, or the production of IL-1 and TNF by alveolar macrophages. This finding concurs with the low mortality rate seen in premature infants with nosocomial pneumonia as compared with death from congenital pneumonia in intensive care nurseries (109). The secretion of certain cytokines, particularly TNF, has also been associated with lung injury. To test whether inhibiting the secretion of TNF would impair the host defenses of the premature lung, we infected preterm rabbits with GBS and then divided the animals into cohorts that did or did not receive pentoxifylline, an inhibitor of TNF release (110). Streptococcal proliferation was significantly less in the lungs of pentoxifylline-treated animals after 24 hrs than that seen in placebo-treated controls. Pentoxifylline treatment also substantially lowered the content of TNF in the alveolar lining fluid compared with control preterm rabbits at 24 hr after infection. This finding suggested that pentoxifylline might serve as an adjunctive treatment for GBS pneumonia in the newborn. The high concentrations of TNF we observed in the alveolar lining fluid of premature rabbits after GBS infection is consistent with elevated production of TNF by human culture–derived macrophages exposed to fibronection and GBS (111). The preceding information is the extent of our knowledge about the role of cytokines in neonatal pneumonia, although other studies have examined the role of cytokines in the pathogenesis of bronchopulmonary dysplasia (112–114). Cytokine interactions with macrophages in newborn lungs and the secretion of cytokines by neonatal lung macrophages are largely unexplored areas, especially in human subjects.

Adhesion Molecules and Newborn Lung Macrophages

The adhesion molecules consist of three families, selectins, the immunoglobulin supergene family, and integrins (115,116), and they modulate activation, the phagocytosis of opsonized particulates, the antigen-presenting function, the initiation of the respiratory burst, and antibody-mediated cytotoxicity in macrophages (117,118). Almost no information exists on the role of adhesion molecules in neonatal lung infections except for the inhibition of CD18 on neutrophils and macrophages using the monoclonal antibody R15.7 (described in section II.C). Those studies demonstrated the importance of functioning neutrophils in the premature lung which has diminished numbers of alveolar macrophages (41). It is known that its maturation of monocytes to macrophages is characterized by changes in adhesion molecule expression (119). This observation may have implications during the influx of monocyte-appearing macrophages into the lung around the time of birth. The only other literature on the role of adhesion molecules in newborn lungs includes studies in noninfectious pulmonary conditions of the newborn, such as meconium aspiration (120), oxygen toxicity (121), and bronchopulmonary dysplasia (122).

III. Functional Responses of Newborn Lung Macrophages to Microbial Pathogens and Clinical Conditions Common to the Neonatal Period

A. Bacterial and Fungal Infections of the Newborn Lung

Group B Streptococcal Pneumonia

In the 1990s, GBS remains the most common bacterial pathogen responsible for neonatal pneumonia (123). Otherwise, pneumonia caused by GBS occurs only in immunocompromised adults. Neonatal infection is often intrauterine in onset, and symptoms appear during the first 24 hr after birth. The disease is rare as a cause of nosocomial pneumonia. Thus, in human neonates, changes in lung host defenses rapidly become protective against this microorganism. No human studies provide reasons for this observation. Animal studies were the first to examine the pathobiology of GBS pneumonia (124). Less than 1-day-old and 1- to 2-day-old term rabbits were studied at 4 24, 48, and 72 hr after GBS aerosol infection. In younger rabbits, bacteria proliferated at 4, 24, and 48 hr, and thereafter pulmonary inflammation (neutrophil influx) resulted in clearance. In contrast, the intrapulmonary clearance of GBS occurred at the outset in 1- to 2-day old rabbits. Less than 1 day-old rabbits had significantly increased microcolonies of GBS within their alveolar macrophages compared with alveolar macrophages of 1- to 2-day-old rabbits. This finding suggested a bactericidal defect in the alveolar macrophages of younger animals. This resulted in the studies of O_2-dependent and O_2-independent microbicidal mechanisms of newborn animals cited in Section II.E. Those investigations identified immaturity of O_2-dependent and O_2-independent microbicidal systems used by macrophages to kill bacterial pathogens. Other explanations are also likely. Less than 1-day-old rabbits still have low numbers of intra-alveolar macrophages which hinder the lung's ability to kill inhaled GBS (41). Furthermore, alveolar macrophages of newborns contain considerable amounts of ingested phospholipid. Surfactant is released by the lung's expansion at birth (19), and alveolar macrophages ingest and recycle this material (51). We have shown that the in vivo ingestion of phospholipid vesicles by alveolar macrophages of term rabbits impairs their ability to kill inhaled GBS (76). Findings akin to those seen in rabbits are reported in neonatal rats infected with GBS aerosols (125). Martin and colleagues have shown that the type-specific polysaccharide capsule of GBS impairs the initial phagocytosis of these bacteria in the lungs of neonates compared with adults (126). Lastly, the transplacental transfer of type-specific antibody reportedly protects the neonate from fulminant GBS infection (12). GBS are catalase-negative bacteria, and they should be easily killed by hydrogen peroxide generated via the respiratory burst. We have shown that GBS opsonized with antibody, compared with GBS coated with complement alone, initiates a significantly higher production of superoxide anion by alveolar macrophages lavaged from newborn and adult rabbits (48,127). Moreover, premature and term rabbits born to immunized mothers have significantly greater clearance of inhaled GBS than do nonimmune animals (128). Immune newborn rabbits also have infrequent bacteremia after GBS aerosol infection compared with their nonimmune counterparts. Using goat fluorescein-labeled antirabbit IgG antibody, intrapulmonary streptococci can be shown to be opsonized after they are inhaled by immune rabbit pups. Rabbits ordinarily have no natural immunity to GBS unless they are immunized. These findings provide a rationale for immunizing young women with GBS vaccine prior to pregnancy.

Neonatal Pneumonia Caused by Other Bacterial Pathogens

The second most common cause of congenital pneumonia after GBS is *Escherichia coli*, but many other bacteria are also responsible (8,9,109). Today, nosocomial pneumonia in hospitalized neonates is caused predominantly by *Staphylococcus epidermidis* (123), but gram-negative bacteria and *Staphylococcus aureus* still play prominent roles (109,123). Using bacterial aerosol infection, studies have documented that newborn rabbits or rats have a diminished capacity to kill and remove *Staphylococcus aureus* (21,125), *Pseudomonas aeruginosa* (125), and pneumococci (42) from the lung as compared to older animals. In vitro studies of alveolar macrophages obtained by bronchoalveolar lavage of rabbits, lambs, and monkeys reveal age-related impairments in antibacterial activity against *S. aureus* (20,35) and *E. coli* (20). The same factors that are associated with the pathobiology of GBS infections are likely causative in other types of congenital infection. As will be discussed below, hyperoxic injury to neonatal alveolar macrophages (129,130) and the excessive ingestion of phospholipid vesicles by alveolar macrophages initiated by pulmonary oxygen toxicity (131) are mechanisms responsible for the pathogenesis of nosocomial bacterial pneumonia of newborns.

Fungal Infections of the Neonatal Lung

Medical and technical advances in medical care have resulted in the survival of extremely premature infants in the 1990s. The immature immune systems of these infants and the liberal use of corticosteroids for hypotension and lung inflammation have caused significant numbers of these newborns to develop pulmonary colonization with *Candida* (132). Both the immature and damaged intrapulmonary host defenses of these extremely premature infants results in systemic candidiasis.

The susceptibility of newborn lung to colonization with *Candida* is supported by studies in the literature. Alveolar macrophages lavaged from healthy 1-day-old rabbits and 1- to 4-day-old rhesus monkeys demonstrate defects in their chemotactic and candidacidal activities compared with macrophages from older animals or adults (43,45). Both studies utilized cytological methods to assess *Candida* uptake and intracellular killing. The dysfunction in alveolar macrophages from newborn rabbits correlated with large amounts of phagocytosed surfactant-related material, because the candidacidal defect could be produced in 7- and 28-day-old alveolar macrophages by feeding them phospholipid vesicles before the *Candida* challenge. In the only human study that has examined neonatal alveolar macrophage function, we recovered bronchial macrophages from intubated infants and compared them with alveolar macrophages lavaged from nonsmoking, healthy adults (46). These neonatal cells might have been altered, because the infants were receiving supplemental, inspired oxygen. A cytological method was also used to measure ingestion, intracellular germination (fungistatic activity), and killing plus digestion of *C. albicans*. Adult and newborn alveolar macrophages ingested the *C. albicans* at an equivalent rate. Yeast germination was significantly increased in the newborn macrophages, suggesting they lacked a fungistatic mechanism. A similar defect could be produced in newborn alveolar macrophages of rabbits by maintaining them in a hyperoxic environment for 96 hr. However, the human newborn macrophages were able to kill and digest *C. albicans* at a rate equal to adult cells, because intracellular yeast germination rendered the microorganisms more metabolically active and susceptible to inactivation and destruction. These findings indicate that human neonatal alveolar macrophages are immature in their antimicrobial behavior, and their functions can be impaired further by insults such as hyperoxia.

B. Clinical Conditions Affecting Newborn Lung Macrophages

Surfactant, Surfactant Replacement, and Newborn Alveolar Macrophages

At birth, the expansion of the alveoli releases surfactant stored in the type II pneumocyte into the air spaces. This event has profound effects on the function of neonatal lung macrophages. First of all, the alveolar influx and replication of alveolar macrophages following birth appears to be mediated by negatively charged phospholipids. Phosphatidylglycerol strongly stimulates migration and replication of macrophages (133,134). In premature monkeys, increased alveolar macrophage numbers were related to the phospholipid content and the absence of hyaline membrane disease (34). Human premature infants who have received surfactant replacement also have higher macrophage numbers between 3 and 7 days of age compared with untreated newborns (135). An early and consistent finding in the bronchial aspirates of premature infants who will develop bronchopulmonary dysplasia is increased polymorphonuclear leukocytes and diminished numbers of macrophages (36,136). Taken together, these findings suggest that surfactant has a major role in the emergence of the alveolar macrophage population.

In previous sections, it has already been mentioned that the ingestion of phospholipid vesicles impairs the antimicrobial activity of newborn alveolar macrophages (43,45,76,131). We asked another question related to the use of surfactant replacement to treat respiratory distress syndrome. Alveolar macrophages would be markedly reduced in the lungs of prematures, and although surfactant replacement might not affect macrophage-related antibacterial activity under those circumstances, the surfactant itself might act as a nutrient for bacteria. To test this hypothesis, we infected premature rabbits with GBS aerosols and immediately thereafter instilled different, commercially available surfactant preparations into the animal's lungs (137). The streptococci proliferated in the lungs of all premature infants but bacterial growth did not exceed that of sham-treated or untreated preterm rabbits (Fig. 4). Thus, different surfactants used in clinical practice did not act as a primary nutrient for GBS. One preparation, Exosurf, actually restricted the growth of GBS, a finding that was confirmed during in vitro experiments that simulated Exosurf presence in alveolar lining material without macrophages (137). None of the surfactants used in the studies contained SP-A or SP-D. It is possible that the addition of these hydrophilic, macrophage-activating proteins might improve outcomes in neonatal pneumonia if they were added to surfactant and used as adjunctive therapy.

Hyperoxia and Newborn Alveolar Macrophages

Damage to the alveolar macrophage system can reduce innate host defenses in the lung. Newborns with respiratory disorders are often treated with increased concentrations of inspired oxygen. We tested the effects of hyperoxia on the alveolar macrophage function of newborn rabbits and learned that the stimulated respiratory burst of macrophages initially increased after 24 hr of exposure. This was followed by a reduction in superoxide anion generation after 48 hr of treatment (129). Concurrent with the decrease in stimulated respiratory burst activity, there was impaired clearance of inhaled *S. aureus*. The damaged alveolar macrophage thereafter recruited neutrophils to the alveoli 48–96 hr after hyperoxia was begun. An additional adverse effect of hyperoxia which occurred within 24 hr of hyperoxic exposure was that macrophage cell division was significantly blunted. A marked decline in mitotic indices and a substantial reduction in thymidine incorporation by the cells occurred (130). The toxic effects of oxygen on alveolar

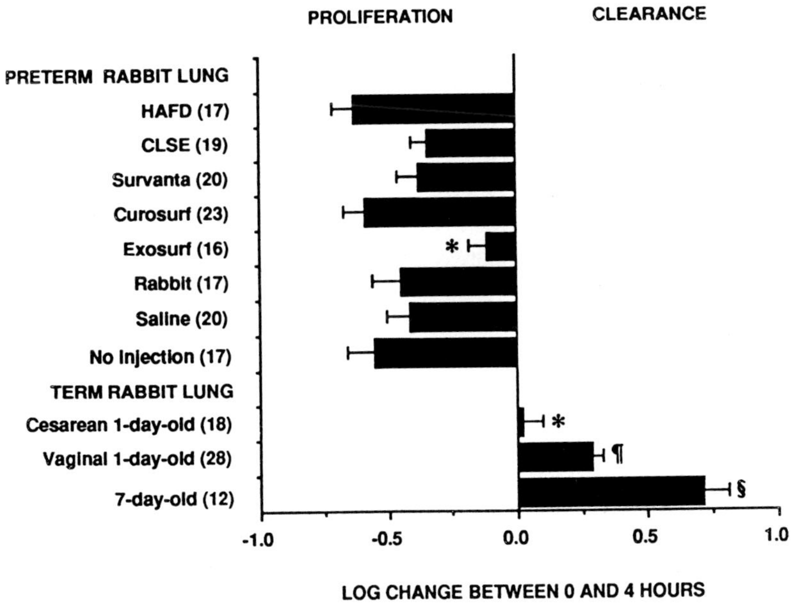

Figure 4 Intrapulmonary clearance of GBS in preterm rabbit lung after surfactant therapy. *Term rabbit lung* refers to rabbits born by cesarean section at 31 days gestation. The histograms on the right side of the figure show a positive $\log_{10}$ change and indicate streptococcal inactivation. Histograms on the left side of the figure show a negative $\log_{10}$ change and represent bacterial growth. Data are presented as mean ± SEM. Numbers in parentheses denote the number of animals studied. $^*P < .05$ for preterm Exosurf Neonatal surfactant and term cesarean rabbit lung versus human amniotic fluid-derived (HAFD) surfactant, Curosurf, and preterm rabbit lung (no injection) (Scheffé F test); $^¶P < .02$ for term vaginal rabbit lung at 1-day versus preterm rabbit lung (no injection) and term cesarean rabbit lung, $^§P < .001$ for 7-day-old rabbits versus preterm (no injection), term (cesarean), and term (vaginal) rabbit lung.

macrophages are likely a major reason why infants on respiratory therapy have an increased incidence of pneumonia.

The use of inhaled nitric oxide to treat pulmonary hypertension of the newborn poses an additional potential insult to newborn alveolar macrophages. Infants receiving inhaled nitric oxide are always breathing increased concentrations of humidified oxygen. The formation of nitrogen dioxide and other oxides of nitrogen occurs readily in this environment, and it has been shown to have detrimental effects on alveolar macrophage function. Inhaled nitric oxide reduces the respiratory burst of piglet alveolar macrophages stimulated with phorbol myristate acetate (138). The biochemical explanation for this effect is that nitric oxide binds to the heme moiety of the NADPH oxidase and inhibits enzymatic activity. In vivo investigations have shown that hyperoxia and inhaled nitric oxide alter surfactant and enhance neutrophil chemotactic activity of bronchoalveolar lavage fluid (139). Clearly, additional research is needed to determine the effects of inhaled nitric oxide on host defenses in the neonatal lung.

Glucocorticosteroids and Perinatal Lung Macrophages

The survival of increasing numbers of extremely premature human neonates is partly due to the recognition that the antenatal administration of steroids matures a number of organ systems. One biochemical effect is augmented surfactant metabolism. The increased survival of these tiny infants can result in chronic pulmonary disease with evidence of significant inflammation. Currently, dexamethasone is a popular drug used to alleviate the inflammation associated with bronchopulmonary dysplasia in these infants. Steroids, however, can have adverse effects on the lung macrophages. In a fetal and neonatal rat model, administration of dexamethasone reduced the number and phagocytic activity of macrophages in the lung (140). Studies indicate that dexamethasone has an additive adverse effect on the ability of the lung to clear inhaled *E. coli* after prolonged hyperoxia (131). Because of the widespread use of glucocorticosteroids to augment lung maturity prior to birth and to reduce inflammation after birth, more attention must be paid to the effects of steroids on the lung's host defenses.

IV. Conclusions

The alveolar surface of the lung is a complex environment which utilizes many mechanisms to defend itself against invading microorganisms. This chapter emphasizes the role of alveolar macrophages in neonatal host defenses. Little attention has been paid to the milieu in which this macrophage exists. This approach has been taken because little information exists regarding how neonatal alveolar macrophages interact with their environment. Information is limited on newborn macrophages in other lung locations or on pulmonary dendritic cells of neonates, although knowledge of dendritic cells in the developing lung is just emerging (141). In the past two decades, we have come to understand some of the reasons why GBS pneumonia occurs in neonates, and we are gaining insight into the prevention and adjunctive therapy of this disease. Nevertheless, our knowledge of newborn lung macrophages has just scratched the surface.

Acknowledgments

The author is the recipient of Public Health Service grant HL 52079.

References

1. Green GM, Kass EH. The role of the alveolar macrophage in the clearance of bacteria from the lung. J Exp Med 1964; 119:167–175.
2. Franciosi RA, Knostmam JD, Zimmerman RA. Group B streptococcal neonatal and infant infections. J Pediatr 1973; 82:707–718.
3. Katzenstein A-L, Davis C, Braude A. Pulmonary changes in neonatal sepsis due to group B β-hemolytic *Streptococcus*: relation to hyaline membrane disease. J Infect Dis 1976; 133:430–435.
4. Hemming VG, McCloskey DW, Hill HR. Pneumonia in the neonate associated with group B streptococcal septicemia. Am J Dis Child 1976; 130:1231–1233.
5. Anderson GS, Green CA, Neligen GA, Newell DJ, Russell JK. Congenital bacterial pneumonia. Lancet 1962; 2:585–587.
6. Naeye RL, Dellinger WS, Blanc WA. Fetal and maternal features of antenatal bacterial infections. J Pediatr 1971; 79:733–739.
7. Baker CJ, Barrett FF. Transmission of group B streptococci among parturient women and their neonates. J Pediatr 1973; 83:919–925.
8. Sherman MP, Goetzman BW, Ahlfors CE, Wennberg RP. Tracheal aspiration and its clinical correlates in the diagnosis of congenital pneumonia. Pediatrics 1980; 65: 258–263.
9. Sherman MP, Chance KH, Goetzman BW. Gram's stains of tracheal secretions predict neonatal bacteremia. Am J Dis Child 1984; 138:848–850.
10. Howard JB, McCracken GH Jr. The spectrum of group B streptococcal infection in infancy. Am J Dis Child 1974; 128:815–818.
11. Shankaran S, Farooki ZQ, Desai R. β-hemolytic streptococcal infection appearing as persistent fetal circulation. Am J Dis Child 1982; 136:725–727.
12. Baker CJ. Summary of the workshop on perinatal infection due to group B streptococcus. J Infect Dis 1977; 136:137–151.
13. Brain JD. Lung macrophages: How many kinds are there? What do they do? Am Rev Respir Dis 1988; 137:507–509.
14. Brain JD, Gehr P, Kavet RI. Airway macrophages. The importance of the fixation method. Am Rev Respir Dis 1984; 129:823–826.
15. Sherman MP, Ganz T. Host defense in pulmonary alveoli. Annu Rev Physiol 1992; 54: 331–350.
16. Shellito J, Kalreider HB. Heterogeneity of immunological function among subfractions of normal rat alveolar macrophages. Am Rev Respir Dis 1984; 129:747–753.
17. Hance AJ, Douches S, Winchester RJ, Ferrans VJ, Crystal RG. Characterization of mononuclear phagocyte subpopulations in the human lung by using monoclonal antibodies: Changes in alveolar macrophage phenotype associated with pulmonary sarcoidosis. J Immunol 1985; 134:284–292.
18. Brannen AL, Chandler DB. Alveolar macrophage subpopulations responsiveness to chemotactic stimuli. Am J Physiol 1988; 132:161–166.
19. Zeligs BJ, Nerurkar LS, Bellanti JA, Zeligs JD. Maturation of the rabbit alveolar macrophage during animal development. I. Perinatal influx into alveoli and ultrastructural differentiation. Pediatr Res 1977; 11:197–208.

20. Zeligs BJ, Nerurkar LS, Bellanti JA. Maturation of the rabbit alveolar macrophage during animal development. III. Phagocytic and bactericidal functions. Pediatr Res 1977; 11:1208–1211.
21. Sherman M, Goldstein E, Lippert W, Wennberg R. Neonatal lung defense mechanisms: A study of the alveolar macrophage system in neonatal rabbits. Am Rev Respir Dis 1977; 116:433–440.
22. Dethloff LA, Lehnert BE. Pulmonary interstitial macrophages: isolation and flow cytometric comparisons with alveolar macrophages and blood monocytes. J Leukoc Biol 1988; 43:80–90.
23. Thet LA, Wrobel DJ, Crapo JD, Shelburne JD. Morphologic aspects of the protection by endotoxin against acute and chronic oxygen-induced lung injury in adult rats. Lab Invest 1983; 48:448–457.
24. Lehnert BE, Valdez YE, Holland LM. Pulmonary macrophages: alveolar and interstitial populations. Exp Lung Res 1985; 9:177–190.
25. Warren JS, Kunkel RG, Johnson KJ, Ward PA. Comparative O_2^- responses of lung macrophages and blood phagocytic cells in the rat. Possible relevance to IgA immune complex induced lung injury. Lab Invest 1987; 57:311–320.
26. Bowden DH, Adamson IYR. The pulmonary interstitial cell as immediate precursor of the alveolar macrophage. Am J Pathol 1972; 68:521–528.
27. Longworth KE, Westgate AM, Grady MK, Westcott JY, Staub NC. Development of pulmonary intravascular macrophage function in newborn lambs. J Appl Physiol 1992; 73:2608–2615.
28. Dehring DJ, Wismar BL. Intravascular macrophages in pulmonary capillaries of humans. Am Rev Respir Dis 1989; 139:1027–1929.
29. Warner AE, Molina RM, Brain JD. Uptake of bloodborne bacteria by pulmonary intravascular macrophages and consequent inflammatory responses in sheep. Am Rev Respir Dis 1987; 136:683–690.
30. Warner AE, Brain JD. The cell biology and pathogenic role of pulmonary intravascular macrophages. Am J Physiol 1990; 258:L1–L12.
31. Zlotnik A, Vatter A, Hayes RL, Blumenthal E, Crowle AJ. Mouse pleural macrophages: characterization and comparison with mouse alveolar and peritoneal macrophages. J Reticuloendothel Soc 1982; 31:207–220.
32. Sorokin SP, McNelly NA, Hoyt RF Jr. CFU-rAM, the origin of lung macrophages, and the macrophage lineage. Am J Physiol 1992; 263:L299–L307.
33. Rothlein R, Gallily R, Kim YB. Development of alveolar macrophages in specific pathogen-free and germ-free Minnesota miniature swine. J Reticuloendothel Soc 1981; 30:483–495.
34. Jacobs RF, Wilson CB, Palmer S, Springmeyer SC, Henderson WR, Glover DM, Kessler L Jr, Murphy JH, Hughes JP, van Belle G, Chi EY, Hodson WA. Factors related to the appearance of alveolar macrophages in the developing lung. Am Rev Respir Dis 1985; 131:548–553.
35. Weiss RA, Chanana AD, Joel DD. Postnatal maturation of pulmonary antimicrobial defense mechanisms in conventional and germ-free lambs. Pediatr Res 1986; 20: 496–504.
36. Ogden BE, Murphy S, Saunders GC, Johnson JD. Lung lavage of newborns with respiratory distress syndrome. Prolonged neutrophil influx is associated with bronchopulmonary dysplasia. Chest 1983; 83:31S–33S.

37. Alenghat E, Esterly JR. Alveolar macrophages in perinatal infants. Pediatrics 1984; 74:221–223.
38. Evans MJ, Sherman MP, Campbell LA, Shami SG. Proliferation of pulmonary alveolar macrophages during postnatal development of rabbit lung. Am Rev Respir Dis 1987; 136:384–387.
39. Van Furth R. Cellular biology of pulmonary macrophages. Int Arch Allergy Appl Immunol 1985; 76(suppl 1):21–27.
40. Winkler GC, Cheville NF. Monocytic origin and postnatal mitosis of intravascular macrophages in porcine lung. J Leukoc Biol 1985; 38:471–480.
41. Sherman MP, Johnson JT, Rothlein R, Hughes BJ, Smith CW, Anderson DC. Role of phagocytes in host defense against group B streptococci in preterm versus term rabbit lung. J Infect Dis 1992; 166:818–826.
42. Coonrod JD, Jarrells MC, Bridges RB. Impaired clearance of pneumococci in neonatal rats. Pediatr Res 1987; 22:736–742.
43. Zeligs BJ, Nerurkar LS, Belianti JA. Chemotactic and candidacidal responses of rabbit alveolar macrophages during postnatal development and the modulating roles of surfactant in these responses. Infect Immun 1984; 44:379–385.
44. Kurland G, Dauber JH. The deficient migration in vitro of lung macrophages from newborn rats normalizes after three days of maturation in situ. J Leukoc Biol 1991; 49:211–213.
45. Kurland G, Cheung AT, Miller ME, Ayin SA, Cho MM, Ford EW. The ontogeny of pulmonary defenses: alveolar macrophages function in neonatal and juvenile rhesus monkeys. Pediatr Res 1988; 23:293–297.
46. D'Ambola JB, Sherman MP, Tashkin DP, Gong H Jr. Human and rabbit newborn lung macrophages have reduced anti-candida activity. Pediatr Res 1988; 24: 285–290.
47. Hall SL, Sherman MP. Intrapulmonary bacterial clearance of type III group streptococcus is reduced in preterm compared with term rabbits and occurs independent of antibody. Am Rev Respir Dis 1992; 145:1172–1177.
48. Sherman MP, Lehrer RI. Oxidative metabolism of neonatal and adult rabbit alveolar macrophages stimulated with opsonized group B streptococci. Infect Immun 1985; 47:26–30.
49. Kradin RL, McCarthy KM, Schneeberger EE. Opsonic receptor function is reduced on the surface of newborn alveolar macrophages. Am Rev Respir Dis 1986; 133: 238–244.
50. van Golde LM. Potential role of surfactant proteins A and D in innate lung defense against pathogens. Biol Neonate 1995; 67(suppl 1):2–17.
51. Oosting RS, Wright JR. Characterization of the surfactant protein A receptor: cell and ligand specificity. Am J Physiol 1994; 267:L165–L172.
52. Miyamura K, Leigh LE, Lu J, Hopkin J, Lopez-Bernal A, Reid KB. Surfactant protein D binding to alveolar macrophages. Biochem J 1994; 300:237–242.
53. van Iwaarden JF, Welmers B, Verhoef J, Haagsman HP, van Golde LM. Pulmonary surfactant protein A enhances host defense mechanism of rat alveolar macrophages. Am J Respir Cell Mol Biol 1990; 2:91–98.
54. Benne CA, Krraijeveld CA, van Strijp JA, Brouwer E, Harmsen M, Verhoef J, van Golde LM, van Iwaarden JF. Interactions of surfactant protein A with influenza A viruses: binding and neutralization. J Infect Dis 1995; 171:335–341.

55. Pikaar JC, Voorhout WF, van Golde LM, Verhoef J, Van Strijp JA, van Iwaarden JF. Opsonic activities of surfactant proteins A and D in phagocytosis of gram-negative bacteria by Alveolar macrophages. J Infect Dis 1995; 172:481–489.
56. Khoor A, Gray ME, Hull WM, Whitsett JA, Stahlman MT. Developmental expression of SP-A and SP-A mRNA in the proximal and distal respiratory epithelium in the human fetus and newborn. J Histochem Cytochem 1993; 41:1311–1319.
57. Thomas EL, Lehrer RI, Rest RF. Human neutrophil antimicrobial activity. Rev Infect Dis 1988; 10(suppl 2):S450–S456.
58. Dyer RM, Majurndar S, Douglas SD, Korchak HM. Bovine parainfluenza-3 virus selectively depletes a calcium-independent, phospholipid-dependent protein kinase C and inhibits superoxide anion generation in bovine alveolar macrophages. J Immunol 1994; 153:1171–1179.
59. Chollet-Martin S, Pasquier C, Marquetty C, Soler P, Hance AJ, Basset F, Hakim J. Cytochrome b-245 in human alveolar macrophages. Am Rev Respir Dis 1985; 132: 836–838.
60. Catterall JR, Sharma SD, Remington JS. Oxygen-dependent killing by alveolar macrophages. J Exp Med 1986; 163:1113–1131.
61. Curnette JT, Whitten DM, Babior BM. Defective superoxide production by granulocytes from patients with chronic granulomatous disease. N Engl J Med 1974; 290: 593–597.
62. Ganz T, Metcalf JA, Gallin JI, Boxer LA, Lehrer RI. Microbicidal/cytotoxic proteins of neutrophils are deficient in two disorders: Chediak-Higashi syndrome and "specific granule deficiency." J Clin Invest 1988; 82:552–556.
63. Johnston JJ, Boxer LA, Berliner N. Correlation of messenger RNA levels with protein defects in specific granule deficiency. Blood 1992; 80:2088–2901.
64. Gebran SJ, Yamamoto Y, Newton C, Tomioka M, Widen R, Klein TW, Friedman H. LPS inhibits the intracellular growth of Legionella pneumophila in thioglycolate-elicited murine peritoneal macrophages by iron-dependent, tryptophan-independent, oxygen-independent and arginine-independent mechanisms. J Leukoc Biol 1995; 57:80–87.
65. Jacobs RF, Locksley RM, Wilson CB, Haas JE, Klebanoff SJ. Interaction of primate alveolar macrophages and Legionella pneumophila. J Clin Invest 1984; 73:1515–1523.
66. Kemmerich B, Rossing TH, Pennington JE. Comparative oxidative microbicidal activity of human blood monocytes and alveolar macrophages and activation by recombinant gamma interferon. Am Rev Respir Dis 1987; 136:266–270.
67. Cadenas E, Daniele RP, Chance B. Low level chemiluminescence of alveolar macrophages: spectral evidence for singlet oxygen generation. FEBS Lett 1981; 123: 225–228.
68. Foote CS, Goyne TE, Lehrer RI. Assessment of chlorination by human neutrophils. Nature 1983; 301:715–716.
69. Mayeno AN, Curran AJ, Roberts RL, Foote CS. Eosinophils preferentially use bromide to generate halogenating agents. J Biol Chem 1989; 264:5660–5668.
70. Goyne TE, Foote CS, Lehrer RI, Sherman MP. Unpublished observations.
71. Sugimoto M, Ando M, Senba H, Tokuomi H. Lung defenses in neonates: effects of bronchial lavage fluids from adult and neonatal rabbits on superoxide production by their alveolar macrophages. J Reticuloendothel Soc 1980; 27:595–606.

72. Sherman MP, Lehrer RI. Superoxide generation by neonatal and adult rabbit alveolar macrophages. J Leukoc Biol 1984; 36:39–50.
73. Zeidler RB, Kim HD. Phagocytosis, chemiluminescence, and cell volume of alveolar macrophages from neonatal and adult pigs. J Leukoc Biol 1985; 37:29–43.
74. Hayakawa H, Myrvik QN, St. Clair RW. Pulmonary surfactant inhibits priming of rabbit alveolar macrophage. Am Rev Respir Dis 1989; 140:1390–1397.
75. Hayakawa H, Giridhar G, Myrvik QN, Kucera L. Pulmonary surfactant phospholipids modulate priming of rabbit alveolar macrophages for oxidative responses. J Leukoc Biol 1992; 51:379–385.
76. Sherman MP, D'Ambola JB, Aeberhard EE, Barrett CT. Surfactant therapy of newborn rabbits impairs lung macrophage bactericidal activity. J Appl Physiol 1988; 65:137–145.
77. Hoffman M, Autor AP. Changes in oxygen consumption and phagocytosis in rat pulmonary macrophages related to animal maturation. Exp Lung Res 1980; 1:313–322.
78. Nathan CF, Hibbs JB Jr. Role of nitric oxide synthesis in macrophage antimicrobial activity. Curr Opin Immunol 1991; 3:65–70.
79. Boockvar KS, Granger DL, Poston RM, Maybodi M, Washington MK, Hibbs JB Jr, Kurlander RL. Nitric oxide produced during murine listeriosis is protective. Infect Immun 1994; 62:1089–1100.
80. Ischiropoulos H, Zhu L, Beckman JS. Peroxynitrite formation from macrophage-derived nitric oxide. Arch Biochem Biophys 1992; 298:446–451.
81. Zhu L, Gunn C, Beckman JS. Bactericidal activity of peroxynitrite. Arch Biochem Biophys 1992; 298:452–457.
82. Bastian NR, Hibbs JB Jr. Assembly and regulation of NADPH oxidase and nitric oxide synthase. Curr Opin Immunol 1994; 6:131–139.
83. Assreuy J, Cunha FQ, Epperlein M, Noronha-Dutra A, O'Donnell CA, Liew FY, Moncada S. Production of nitric oxide and superoxide by activated macrophages and killing of *Leishmania major*. Eur J Immunol 1994; 24:672–676.
84. Cameron ML, Granger DL, Weinberg JB, Kozumbo WJ, Koren HS. Human alveolar and peritoneal macrophages mediate fungistasis independently of L-arginine oxidation to nitrite and nitrate. Am Rev Respir Dis 1990; 142:1313–1319.
85. Sherman MP, Loro ML, Wong V, Tashkin DP. Cytokine and *Pneumocystis carinii* induced L-arginine oxidation by murine and human pulmonary alveolar macrophages. J Protozool 1991; 38:234S–236S.
86. Sherman MP, Aeberhard EE, Wong VZ, Griscavage JM, Ignarro LJ. Pyrrolidine dithiocarbamate inhibits induction of nitric oxide synthase activity in rat alveolar macrophages. Biochem Biophys Res Commun 1993; 191:1301–1308.
87. Goodrum KJ, McCormick LL, Schneider B. Group B streptococcus-induced nitric oxide production in murine macrophages in CR3 (CD11b/CD18) dependent. Infect Immun 1994; 64:3102–3107.
88. Nussler AK, Billiar TR. Inflammation, immunoregulation, and inducible nitric oxide synthase. J Leukoc Biol 1993; 54:171–178.
89. Henderson SA, Lee PH, Aeberhard EE, Adams JW, Ignarro LJ, Murphy WJ, Sherman MP. Nitric oxide reduces early growth response-1 gene expression in rat lung macrophages treated with interferon-γ and lipopolysaccharide. J Biol Chem 1994; 269:25239–25242.

90. Nguyen HQ, Hoffman-Liebermann B, Liebermann DA. The zinc finger transcription factor Egr-1 is essential for and restricts differentiation along the macrophage lineage. Cell 1993; 72:197–209.
91. Nerurkar LS, Zeligs BJ, Bellanti JA. Maturation of the rabbit alveolar macrophage during animal development. II. Biochemical and enzymatic studies. Pediatr Res 1977; 11:1202–1207.
92. Ganz T, Sherman MP, Selsted ME, Lehrer RI. Newborn rabbit alveolar macrophages are deficient in two microbicidal cationic peptides, MCP-1 and MCP-2. Am Rev Respir Dis 1985; 132:901–904.
93. Coonrod JD, Jarrells MC. Impaired antipneumococcal activity of bronchoalveolar lining material of neonatal rats. Infect Immun 1987; 57:359–366.
94. Revenis ME, Kaliner MA. Lactoferrin and lysozyme deficiency in airway secretions: association with the development of bronchopulmonary dysplasia. J Pediatr 1992; 121:262–270.
95. Sherman MP, Hall SL, Johnson JT, Aeberhard EE. Formulating a surfactant therapy for preterm infants who have bacterial pneumonia and respiratory distress syndrome. In: Bevilacqua G, Parmigiani S, Robertson B, eds. Surfactant in Clinical Practice. Singapore: Harwood Academic Publishers, 1992; 103:111.
96. Kradin RL, McCarthy KM, Preffer FI, Schneeberger EE. Flow-cytometric and ultrastructural analysis of alveolar macrophage maturation. J Leukoc Biol 1986; 40:407–417.
97. Zink MC, Yager JA, Prescott JF, Fernando MA. Electron microscopic investigation of intracellular events after ingestion of *Rhodococcus equi* by foal alveolar macrophages. Vet Microbiol 1987; 14:295–305.
98. Robertson AK, Cross AR, Jones OT, Andrew PW. The use of diphenylene iodonium, an inhibitor of NADPH oxidase, to investigate the antimicrobial action of human monocyte derived macrophages. J Immunol Methods 1990; 133:175–179.
99. Langermans JA, Van-der-Hulst ME, Nibbering PH, Hiemstra PS, Fransen L, Van Furth R. IFN-gamma–induced L-arginine–dependent toxoplasmastatic activity in murine peritoneal macrophages is mediated by endogenous tumor necrosis factor-alpha. J Immunol 1992; 148:568–574.
100. Johnson JD, Hand WL, King NL, Hughes CG. Activation of alveolar macrophages after lower respiratory tract infection. J Immunol 1975; 115:80–84.
101. Lamontagne L, Gauldie J, Stadnyk A, Richards C, Jenkins E. In vivo initiation of unstimulated in vitro interleukin-1 release by alveolar macrophages. Am Rev Respir Dis 1985; 131:326–330.
102. Kelley J. Cytokines of the lung. Am Rev Respir Dis 1990; 141:765–788.
103. Huang S, Paulauskis JD, Godleski JJ, Kobzik L. Expression of macrophage inflammatory protein-2 and KC mRNA in pulmonary inflammation. Am J Pathol 1992; 141:981–988.
104. Carre PC, King TE Jr, Mortensen R, Riches DW. Cryptogenic organizing pneumonia: increasing expression of IL-8 and fibronectin genes by alveolar macrophages. Am J Respir Cell Mol Biol 1994; 10:100–105.
105. Xing Z, Jordana M, Kirpalani H, Driscoll KE, Schall TJ, Gauldie J. Cytokine expression by neutrophils and macrophages in vivo: endotoxin induces tumor necrosis factor-alpha, macrophage inflammatory protein-2, interleukin-1 beta, and inter-

leukin-6 but nor Rantes or transforming growth factor-beta 1 mRNA expression in acute lung inflammation. Am J Respir Cell Mol Biol 1994; 10:148–153.
106. Buck C, Gallati H, Pohlandt F, Bartmann P. Increased levels of tumor necrosis factor alpha (TNF-alpha) and interleukin 1 beta (IL-1 beta) in tracheal aspirates of newborns with pneumonia. Infection 1994; 22:238–241.
107. Rozycki HJ. Elevated production of interleukin-1-beta from alveolar macrophages isolated from newborn rabbits. Biol Neonate 1994; 66:93–99.
108. Skerrett SJ, Henderson WR, Martin TR. Alveolar macrophage function in rats with severe calorie malnutrition. Arachidonic acid metabolism, cytokine release, and antimicrobial activity. J Immunol 1990; 144:1052–1061.
109. Webber S, Wilkinson AR, Lindsell D, Hope PL, Dobson SR, Isaacs D. Neonatal pneumonia. Arch Dis Child 1990; 65:207–211.
110. Mah MP, Aeberhard EE, Gilliam MB, Sherman MP. Effects of pentoxifylline on in vivo leukocyte function and clearance of group B streptococci from preterm rabbit lungs. Crit Care Med 1993; 21:712–720.
111. Peat EB, Augustine NH, Drummond WK, Bohnsack JF, Hill HR. Effects of fibronectin and group B streptococci on tumor necrosis factor-alpha production by culture-derived macrophages. Immunology 1995; 84:440–445.
112. Rozycki HJ. Bronchoalveolar interleukin-1 beta in infants on day 1 of life. South Med J 1994; 87:991–996.
113. Pierce MR, Bancalari E. The role of inflammation in the pathogenesis of bronchopulmonary dysplasia. Pediatr Pulmonol 1995; 19:371–378.
114. Sherman MP. The role of macrophages in chronic lung disease of early infancy. In: Bland RD, Coalson JJ, eds. Chronic Lung Disease of Early Infancy. New York: Marcel Dekker, 1996. In press.
115. Pilewski JM, Albelda SM. Adhesion molecules in the lung. An overview. Am Rev Respir Dis 1993; 148:S31–S37
116. Ward PA. Oxygen radicals, cytokines, adhesion molecules, and lung injury. Environ Health Perspect 1994; 102(suppl 10):13–16.
117. Albert RK, Embree LJ, McFeely JE, Hickstein DD. Expression and function of beta 2 integrins of alveolar macrophages from human and nonhuman primates. Am J Respir Cell Mol Biol 1992; 7:182–189.
118. Kang YH, Lee CH, Brummel SE, Newball HH, Forrester J. Effects of endotoxin on expression of VLA integrins by human bronchoalveolar lavage macrophages. J Leukoc Biol 1995; 57:624–634.
119. Prieto J, Eklund A, Patarroyo M. Regulated expression of integrins and other adhesion molecules during differentiation of monocytes into macrophages. Cell Immunol 1994; 156:191–211.
120. Kojima T, Hattoi K, Fujiwara T, Sasai-Takedatsu M, Kobayashi Y. Meconium-induced lung injury mediated by activation of alveolar macrophages. Life Sci 1994; 54:1559–1562.
121. Welty SE, Rivera JL, Elliston JF, Smith CV, Zeb T, Ballantyne CM. Increases in lung tissue expression of intercellular adhesion molecule-1 are associated with hyperoxic lung injury and inflammation in mice. Am J Respir Cell Mol Biol 1993; 9:393–400.
122. Kojima T, Sasai M, Kobayashi Y. Increased soluble ICAM-1 in tracheal aspirates of infants with bronchopulmonary dysplasia. Lancet 1993; 342:1023–1024.

123. Philip AG. The changing face of neonatal infection: experience at a regional medical center. Pediatr Infect Dis J 1994; 13:1098–1102.
124. Sherman M, Goldstein E, Lippert W, Wennberg R. Group B streptococcal lung infection in neonatal rabbits. Pediatr Res 1982; 16:209–212.
125. Martin TR, Rubens CE, Wilson CB. Lung antibacterial defense mechanisms in infant and adult rats: implications for the pathogenesis of group B streptococcal infections in the neonatal lung. J Infect Dis 1988; 157:91–100.
126. Martin TR, Ruzinski JT, Rubens CE, Chi EY, Wilson CB. The effect of type-specific polysaccharide capsule on the clearance of group B streptococci from the lungs of infant and adult rats. J Infect Dis 1992; 165:306–314.
127. Sherman M, Lehrer RI. Superoxide anion production by newborn and adult lung macrophages. Clin Res 1983; 31:123A.
128. Sherman MP, D'Ambola JB, Campbell LA, Anthony BF. Unpublished observations.
129. Sherman MP, Condiotti R. Hyperoxia damages phagocytic defenses of neonatal rabbit lung. J Appl Physiol 1987; 62:684–690.
130. Sherman MP, Evans MJ, Campbell LA. Prevention of pulmonary alveolar macrophage proliferation in newborn rabbits by hyperoxia. J Pediatr 1988; 112:782–786.
131. Sherman MP, Campbell LA, Aeberhard EE, Barrett CT. Hyperoxia (H), dexamethasone (D) and newborn alveolar macrophage (AM) function. Pediatr Res 1987; 21:376A.
132. Rowen JL, Rench MA, Kozinetz CA, Adams JM Jr, Baker CJ. Endotracheal colonization with Candida enhances risk of systemic candidiasis in very low birth weight neonates. J Pediatr 1994; 124:789–794.
133. Lynn WS, Mukherjee C. Motility of rabbit alveolar cells. Role of unsaturated fatty acids. Am J Pathol 1979; 96:663–671.
134. Yui S, Yamazaki M. Relationship of ability of phospholipids to stimulate growth and bind to macrophages. J Leukoc Biol 1987; 41:392–399.
135. Arnon S, Grigg J, Silverman M. Pulmonary inflammatory cells in ventilated preterm infants: effects of surfactant treatment. Arch Dis Child 1993; 69:44–48.
136. Merritt TA, Stuard ID, Puccia JM, Wood B, Edwards DK, Finkelstein J, Shapiro DL. Newborn tracheal aspirate cytology: Classification during respiratory distress syndrome and bronchopulmonary dysplasia. J Pediatr 1981; 98:949–956.
137. Sherman MP, Campbell LA, Merritt TA, Shapiro DL, Long WA, Gunkel JH, Curstedt T, Robertson B. The effect of surfactant preparations used in clinical practice on group B streptococcal pneumonia in preterm rabbits. J Pediatr 1994; 125:939–947.
138. Sherman MP, Moise S, Norberg M, Weatherstone KW, Truog WE. Pulmonary phagocyte responses to inhaled nitric oxide (NO) and hyperoxia (H). Pediatr Res 1995; 37:349A.
139. Robbins CG, Davis JM, Merritt TA, Amirkhanian JD, Sahgal N, Morin FC 3rd, Horowitz S. Combined effects of nitric oxide and hyperoxia on surfactant function and pulmonary inflammation. Am J Physiol 1995; 269:L545–L550.
140. Lortie C, King GM, Adamson IYR. Effects of dexamethasone on macrophages in fetal and newborn rat lung. Pediatr Pulmonol 1990; 8:136–144.
141. Nelson DJ, Holt PG. Defective regional immunity in the respiratory tract of neonates is attributable to hyporesponsiveness of local dendritic cells to activation signals. J Immunol 1995; 155:3517–3524.

17

The Role of Mononuclear Phagocytes in Tuberculosis

LARRY S. SCHLESINGER

University of Iowa
Iowa City, Iowa

I. Introduction

The importance of tuberculosis (TB) as a major infectious cause of worldwide morbidity and mortality is well known (1). Given the ongoing acquired immuno-deficiency syndrome (AIDS) epidemic, the increasing numbers of immuno-compromised individuals, the ease with which drug-resistant strains can continue to disseminate and the problems associated with current antibiotic therapy, the threat of TB likely will be long lived. *Mycobacterium tuberculosis*, the bacterial cause of TB, is one of a diverse group of intracellular pathogens that survives the process of entry into mammalian cells. Multiplication within these cells is critical for disease pathogenesis by these host-adapted microbes. For *M. tuberculosis*, the major host cell reservoir is the mononuclear phagocyte, including both monocytes and macrophages. These cells, along with polymorphonuclear leukocytes, are considered "professional" phagocytes in that they have adapted specifically to engulf and destroy bacteria or other foreign particulate matter. Given the potent microbiocidal mechanisms that these cells possess, *M. tuberculosis* has had to adapt strategies to circumvent adverse host cellular responses during and after entry. On the other hand, tissue macrophages are potentially advantageous host cells, because they are long-lived cells and thus allow these slow-growing bacteria (approximate doubling time of 16–24 hr) to go through several doublings before macrophage death.

Once bacteria begin intracellular multiplication, mononuclear phagocytes serve as antigen-presenting cells in promoting the cellular immune response that leads to protection of the host. Characterization of the molecular determinants of antigen presentation for *M. tuberculosis* molecules by these cells and the regulation of T-cell responsiveness is just beginning to emerge. During infection with *M. tuberculosis*, soluble mediators (host and microbial-derived) play a major role both in the development of the localized protective cellular immune response and in the immunopathogenesis of disease. The regulation of the type and amount of mediator produced dictates in large which of these events dominates.

Although the important role for mononuclear phagocytes in *M. tuberculosis* infection has been known for a long time (reviewed in ref. 2), the resurgence of TB in the United States and worldwide over the past decade has lead to an increase in research activities aimed at further defining the molecular mechanisms underlying many aspects of the *M. tuberculosis*–mononuclear phagocyte interaction. The higher likelihood of TB in AIDS patients has added another layer of complexity in defining this interaction. Although we have become cognizant of the dangers of multidrug-resistant *M. tuberculosis* in the clinical arena, the influence of this phenotype on events in immunopathogenesis remains largely unknown.

This chapter reviews important developments in the *M. tuberculosis*–mononuclear phagocyte interaction. Since *M. tuberculosis* is a respiratory pathogen, the pulmonary alveolar macrophage plays a particularly important role in influencing the clinical outcome during primary infection and reactivation disease in the lung; thus, potential differences between this cell and other mononuclear phagocytes will be discussed. Finally, since human beings serve as the natural reservoir for this pathogen, emphasis will be placed on human phagocytes; however, whenever possible, similarities and differences between these cells and other mammalian phagocyte systems will be discussed.

II. Route of Infection, Including the Unique Role of Lung Macrophages and Lung Immunological Mechanisms

Aerosol droplet nuclei (approximately 0.2–5.0μm) containing small numbers of *M. tuberculosis* reach the alveolus of the lung and are phagocytosed by alveolar macrophages during primary infection. Macrophages that migrate to this tissue site differentiate in the presence of local tissue factors unique to the alveolus, resulting in altered cell physiology (3). Thus, identification of the factors that influence the *M. tuberculosis*–mononuclear phagocyte interaction in this locale is particularly important in furthering our knowledge of disease pathogenesis.

Studies in the mouse model provide evidence for the unique nature of the *M. tuberculosis*–lung macrophage interaction and for the lung's immunological response to this infection in general. In one study, intravenous (IV) inoculation of mice with *M. tuberculosis* resulted in a slowly progressive infection in the lungs despite the development of an effective systemic host immune response capable of resolving infection in the liver, spleen, and kidneys (4). Fewer *M. tuberculosis* colony-forming units given by aerosol were substantially more virulent (faster rate of bacterial growth in the lungs, shorter survival of the mice, and faster development of lung pathology) in a specific mouse strain than much larger numbers inoculated IV even though the number of *M. tuberculosis* organisms given IV resulted in the same number of bacteria initially implanting in the lungs as the number given by aerosol. These results led to the speculation that bacilli reaching the lungs via the respiratory route have a growth advantage over those that enter from the blood during the initial stage of infection. Microorganisms that enter the lung via the respiratory route are ingested by resident alveolar macrophages, whereas those that enter from the blood need to be carried into the organ after being ingested by other phagocytic cell types. Given that resident macrophages appear to be less capable than other macrophages of killing microbial pathogens in general (3), it was further speculated that these cells may enable *M. tuberculosis* to grow more rapidly during the initial stage of infection. Since proinflammatory cytokines produced by lung macrophages in response to ingesting *M. tuberculosis* appear to differ from those produced by infected macrophages elsewhere (see also Section IX), the unique initial inflammatory response in the lung may additionally influence the efficiency by which T-cell–mediated immunity is expressed at this site. Several studies indicate that alveolar macrophages are capable of downregulating the response of T cells to mitogens and specific antigens (5,6) (also see below). Thus, it is possible that this bacterium has evolved specifically to enter via the respiratory route in order to enhance its survival during initial infection.

After intracellular multiplication of *M. tuberculosis* within alveolar macrophages, there is spread of bacilli to regional lymph nodes. The mechanism by which bacilli traverse the alveolus to reach these lymph nodes remains poorly understood. One mechanism may involve the migration of alveolar macrophages containing *M. tuberculosis* across the alveoli and to these nodes. Evidence for this comes from dog studies by Harmsen et al. (7) in which red and green fluorescent-labeled microspheres instilled separately into the trachea were found together within single macrophages in the draining lymph nodes. Also, studies using labeled liposomes phagocytosed by alveolar macrophages revealed that the alveolar macrophages were capable of migrating to the paracortical area of the draining lymph nodes.

Recently, another potential mechanism for the regional spread of bacilli was provided by McDonough et al. (8). In their studies, the virulent *M. tuberculosis* strain, H37Rv, was found to enter A549 human lung epithelial type cells in vitro and was cytotoxic for both polarized and nonpolarized epithelial monolayers as well as for the murine macrophage cell line J774.16. The ability of virulent *M. tuberculosis* to lyse these epithelial cells was enhanced after the bacteria had first been passaged intracellularly in the macrophage cell line. This study suggested that bacteria can directly penetrate damaged alveolar epithelium to gain access to lymph nodes and the circulating system (8).

Following primary pulmonary infection with *M. tuberculosis* in the healthy human, the most common consequence is the development of an effective cellular immune response in which subsequent intracellular bacterial growth is halted and the infection remains subclinical (reviewed in ref. 9). The number of residual viable bacteria following primary infection in the healthy host is unknown. It is generally believed these bacteria become dormant and at some later point (usually with failure of the cellular immune system), bacteria can "reactivate," begin intracellular multiplication, and lead to a disease state. The events that allow for bacterial dormancy within macrophages are largely unexplained. They may involve an adaptation to altered oxygen tension and altered pH (reviewed in ref. 10). *M. tuberculosis* can produce enzymes of anaerobic metabolism, and these are especially prominent when bacilli are tested after direct harvest from host tissues. There is a general shift toward anaerobic pathways in mammalian host grown–derived bacilli as compared with cultured organisms.

III. Potential for Differential Roles for Monocytes, Tissue Macrophages, and Other Cell Types in Handling *M. tuberculosis*

Mononuclear phagocytes originate from a common precursor cell in the bone marrow but differentiate in tissues into long-lived resident macrophages with a variety of phenotypes that are dictated by local tissue factors. In the absence of an inflammatory stimulus, the numbers of resident macrophages greatly exceed the numbers of newly arriving and differentiating monocytes. Thus, predominantly resident macrophages ingest newly arriving mycobacteria as would occur during initial infection of healthy individuals (reviewed in ref. 11). However, as mycobacteria resist killing and form a persistent focus of infection, the numbers of mononuclear phagocytes greatly increase owing to an influx of monocytes from the bloodstream. These monocytes may ingest *M. tuberculosis* shortly after their arrival or may differentiate further in the presence of inflammatory mediators from neighboring cells prior to ingesting *M. tuberculosis*. Thus, interactions between *M. tuberculosis* and the phagocyte will differ depending on the tissue site, the degree of cell differentiation, and the presence of inflammatory mediators. One early study that examined *M. tuberculosis* adherence to guinea pig macrophages illustrates this point (12). Adherence of the virulent H37Rv strain and the attenuated H37Ra strain of *M. tuberculosis* to resident macrophages was comparable using peritoneal macrophages but was greater for H37Rv using alveolar macrophages. In the same study, after immunization with *M. bovis* BCG, adherence of bacteria to alveolar macrophages was markedly increased for H37Rv but was reduced for H37Ra when compared with the resident ("preimmune") macrophages.

The alveolar macrophage is situated at the air-tissue interface in the alveoli and is the first professional phagocyte cell type to encounter inhaled foreign material, including microbes. In vitro studies indicate that alveolar macrophages are inefficient in providing accessory cell activity in vitro compared with macrophages derived from other sources of monocytes (reviewed in ref. 13). In fact, alveolar macrophages have been demonstrated to be efficient suppressors of T-cell activation at high alveolar macrophage to T-cell ratios and can inhibit the presentation of antigens by pulmonary dendritic cells (see below). Thus, pulmonary alveolar macrophages appear to be highly effective in clearing foreign particulate matter, including microbes deposited on the epithelial surface of the lung, and at the same time serve to downregulate activation of the immune system present in lung tissue and in draining lymph nodes. These attributes make them particularly suitable host cells for a host-adapted intracellular pathogen such as *M. tuberculosis*.

Beyond alveolar macrophages, the mononuclear phagocytic system in the lung comprises interstitial macrophages and intravascular macrophages that are believed to come from sequestered monocytes in the pulmonary microcirculation (14). Each of these macrophages likely plays a role in early lung inflammation, and each cell type demonstrates unique physiology, including altered cell-cell and cell-matrix adherence, altered cellular activation (including secretion of soluble mediators), and altered phagocytosis. Differences in the interaction between *M. tuberculosis* and each of these cell types are unknown.

The respiratory tract also contains dendritic cells (DCs) (15,16). Pulmonary DCs possess efficient antigen-presenting capability that can be suppressed by indigenous macrophage populations (13). Dendritic cell progenitor cells can phagocytose mycobacteria and can efficiently present antigens to T cells (17).

Alveolar epithelial cells also contribute to pulmonary inflammation and may play a role in the host response to *M. tuberculosis* (see above) (reviewed in ref. 18). Such alveolar epithelial cells can present antigens to lymphocytes and release a variety of proinflammatory cytokines, chemokines, arachidonic acid metabolites, and complement proteins such as C3 (19), which serves as an opsonin for *M. tuberculosis* (see below). Epithelial cells also produce surfactant proteins and phospholipids that further modulate inflammatory responses and express cell adhesion molecules to which leukocytes can bind. Surfactant protein A produced by epithelial cells upregulates the phagocytosis of *M. tuberculosis* by macrophages (20,21) (see below). Transforming growth factor-β (TGF-β), a cytokine that inhibits lymphocyte and macrophage activation and can enhance intracellular multiplication of *M. tuberculosis* in macrophages (see below), is produced by epithelial cells.

The role of the neutrophil in the host response to mycobacterial infection remains poorly defined (22) but is likely to be involved particularly during the initial inflammatory response (22,23). Phagocytosis of *M. tuberculosis* by neutrophils is accompanied by an oxidative burst (24) and results in bacterial killing by oxygen-dependent and oxygen-independent mechanisms (25,26). Lefkowitz et al. have proposed that the interaction between neutrophils and macrophages can induce a state of chronic inflammation which contributes to the disease state (27). In particular, myeloperoxidase, which is released from neutrophils, can induce macrophages to secrete interleukin-1 (IL-1), interferon-α (IFN-α), and tumor necrosis factor-α (TNF-α) and can enhance the production of reactive oxygen intermediates by these cells.

IV. Determinants of Phagocytosis

A. Mechanism of Entry of *M. tuberculosis* into Phagocytes

Electron microscopy studies have shown that entry of *M. tuberculosis* into human mononuclear phagocytes resembles conventional receptor-mediated phagocytosis (28). We have observed that the levels of phagocytosis in fresh serum of live and heat-killed virulent Erdman strain *M. tuberculosis* by human monocyte-derived macrophages (MDMs) are equivalent (unpublished studies). Thus, the viability of the bacterium does not appear to be a requirement for the phagocytic process itself, although it plays an important role in subsequent events (see below). An *M. tuberculosis* invasion factor has been described in which a cloned fragment of *M. tuberculosis* DNA in *Escherichia coli* conferred enhanced invasiveness of this bacterium into mammalian cells in vitro (29).

B. Major Surface Receptors for Phagocytosis on Monocytes and Macrophages and Their Role in Phagocytosis of Intracellular Pathogens

The earliest interaction between *M. tuberculosis* and the mononuclear phagocyte is binding of the bacterium to the cell surface and subsequent internalization. It has become clear that host-adapted intracellular pathogens utilize several host cell receptors to gain access to their intracellular niche. Recent studies provide evidence for the importance of specific receptor-ligand pathways (single or multiple) and cooperativity between receptors in dictating the fate of phagocytosed intracellular pathogens (30–32). That involvement of specific receptors and groups of receptors can influence the early host cell response has been demonstrated in several studies. As one example, Kobzik et al. compared the oxidative responses of rodent alveolar macrophages during opsonin-dependent and opsonin-independent phagocytosis of latex particles and *Staphylococcus aureus* (33). Alveolar macrophages are known to ingest unopsonized particles avidly when compared with other phagocytic cell types. These investigators found that phagocytosis of opsonized particles (mediated in part by Fc receptors) was accompanied by a detectable oxidative response. In contrast, the uptake of small numbers (1–3) of unopsonized particles per cell led to the inhibition of oxidative activity compared with control cells. Alveolar macrophages ingesting larger numbers of unopsonized particles showed some increase in the oxidative response, but it was much less than that observed with identical numbers of opsonized particles.

These results suggested that alveolar macrophages may ingest the numerous unopsonized inert particles that are inhaled without generation of toxic oxygen metabolites while retaining the capacity to undergo a respiratory burst after ingesting opsonized particles and bacteria. These results also suggested a potential difference in receptor-mediated signal transduction mechanisms. Specifically, the phagocytosis of unopsonized particles may lead to the generation of an inhibitory signal such as elevated cyclic adenosine monophosphate (cAMP) (34,35) which is associated with inhibition of the macrophage respiratory burst.

Apart from receptor-ligand interactions, the fate of the pathogen is dependent on the microbe's unique surface molecules, which either by themselves or by binding to specific host molecules, are also likely to influence the host cell response during and after entry.

Major mononuclear phagocyte receptors that mediate the phagocytosis of intracellular pathogens include Fc receptors, complement receptors (CRs: CR1, CR3, and CR4), and the mannose receptor (MR) (reviewed in ref. 36). In the absence of specific immune antibody, CRs and the MR play major roles (Table 1). CR1 (CD35) binds C3b- and C4b- coated particles and also regulates complement activation (37). CR3 (CD11b/CD18) and CR4 (CD11c/CD18; p150,95) are heterodimers (α- and β-chain) and with leukocyte function–associated antigen (LFA-1) (CD11a/CD18) comprise the β_2 (leukocyte)–integrin family (38,39). In addition to containing binding sites for C3-opsonized particles (primarily C3bi), CR3 and CR4 are involved in cell-cell and cell-surface adhesion (40) and are expressed on other cell types involved in the immune response. CR3 and CR4 contain several extracellular binding domains. In particular, in addition to containing specific binding sites for C3-opsonized particles, CR3 also binds nonopsonized particles and thus can function as a lectin (41).

The phagocytosis of a diverse group of intracellular pathogens is mediated by CRs (42), a family of receptors that has been postulated to play an important role in the early survival of intracellular parasites, because ligation of these receptors by C3-coated particles does not uniformly stimulate the release of oxygen metabolites or inflammatory mediators (43–45) (also see below).

The expression of CRs (particularly CR4) and the MR increases during monocyte differentiation into macrophages, and CR4 and the MR are highly

Table 1 CRs and the MR Mediate Entry of Intracellular Pathogens by Mononuclear Phagocytes[a]

Complement receptors
Mycobacteria: *M. tuberculosis*, *M. leprae*, *M. avium*, *M. bovis* BCG
Other bacteria: *L. pneumophilia*, *L. monocytogenes*, *B. pertussis*
Fungi: *H. capsulatum*, *B. dermatitidis*
Viruses: HIV
Parasites: *Leishmania* sp., T. cruzi
Macrophage MR
Mycobacteria: M. tuberculosis, *M. avium*
Parasites: *Leishmania* sp., *T. cruzi*
Virses: ?HIV

[a]For references and further discussion, see the text.

expressed on alveolar macrophages (46,47). The MR is a C-type lectin that binds glycoconjugates containing mannose and fucose units with highest avidity (reviewed in ref. 48). The extracellular portion of the receptor contains a linear array of eight carbohydrate-recognition domains (CRDs) (49). Macrophage MRs have an internal pool and undergo continual recycling to the cell surface.

Culturing human monocytes in vitro alters their functional capabilities and provides a model for studying mature macrophage function (50,51). This model is particularly relevant to the study of *M. tuberculosis* and other pathogenic mycobacteria. Such MDMs exhibit altered phagocytic capacity (52,53), including the increased expression of existing surface receptors and the expression of new receptors (46,47). The phagocytosis of *M. leprae* increases greater than sixfold by MDMs compared with monocytes when the bacteria to phagocyte ratio is kept the same (54). A similar increase in phagocytosis of *M. tuberculosis* by MDMs occurs.

C. Monocyte and Macrophage Receptors that Mediate Phagocytosis of *M. tuberculosis*

Complement Receptors

The phagocytosis of *M. tuberculosis* by monocytes and MDM is enhanced in the presence of small amounts of fresh (complement sufficient) autologous nonimmune serum which leads to opsonization by activation products of complement protein C3 (C3b and C3bi) (28). Of particular relevance to the pulmonary pathogen *M. tuberculosis*, functional complement proteins are secreted by alveolar macrophages and epithelial cells (3,18) and thus are available to serve as ligands in this setting. Although the *M. tuberculosis* surface glycolipid trehalose dimycolate (cord factor) and tuberculin serve to mediate the activation of complement pathway cascades (55,56), the C3-acceptor molecule(s) on the surface of *M. tuberculosis* has not been identified.

CRs play a major role in the phagocytosis of *M. tuberculosis* by human monocytes and macrophages (28,57,58). CR1 and CR3 on human monocytes mediate the phagocytosis of the virulent Erdman strain of *M. tuberculosis* (28). A combination of two monoclonal antibodies against distinct epitopes on the α-chain of CR3 reduce the phagocytosis of *M. tuberculosis* by 80%. Monoclonal antibodies against the C3bi-binding epitope of CR3 as well as those against other ligand-binding epitopes inhibit phagocytosis. Thus, it is possible that more than one epitope is important in binding *M. tuberculosis* (see below). Fc receptors (in the presence of nonimmune serum) and the β-glucan receptor (59) appear to play little or no role in mediating phagocytosis of *M. tuberculosis* by these cells (28).

The phagocytosis by MDMs of two virulent *M. tuberculosis* strains, Erdman and H37Rv, and the attenuated strain, H37Ra, is enhanced comparably in fresh nonimmune serum. Ingestion of these strains occurs both in the presence and absence of serum and the bacteria are found within phagosomes (57). Thus, reduced phagocytosis is not an explanation for the reduced virulence of the commonly studied H37Ra strain. A combination of monoclonal antibodies against macrophage CR1, CR3, and CR4 inhibits the phagocytosis of all three strains by nearly 80% in the presence of serum (57).

Monoclonal antibodies against CRs also significantly inhibit the phagocytosis of bacteria in the absence of serum (57). This result has at least two explanations. The first is that there is a direct interaction between *M. tuberculosis* surface components and CRs, and the second is that functional C3 protein secreted by the macrophage during the assay binds to the bacterium and serves as the ligand for CRs. Macrophage-derived C3 has been shown to function in this manner (60,61). If the first possibility were true, the binding sites on CRs for C3-opsonized bacteria and unopsonized bacteria may differ. This latter possibility is supported by the antibody studies described above and also by a recent study by Stokes et al. (62) that examined the binding of *M. tuberculosis* to murine macrophage CR3 in a serum-free system. The binding site for *M. tuberculosis* on CR3 was distinct from the C3bi binding site determined by monoclonal antibody–blocking experiments. They also determined that both elicited and activated macrophages bound mycobacteria poorly despite expressing CR3. Thus, binding of *M. tuberculosis* did not correlate with expression of CR3, a finding the investigators hypothesized was related to the functional state of the binding site on CR3. Further evidence for a direct interaction between *M. tuberculosis* components and CRs is provided by a recent study that used the H37Rv strain of *M. tuberculosis* (63).

Macrophage Mannose Receptor and Lipoarabinomannan in the Adherence of Virulent M. tuberculosis *Strains*

CRs mediate phagocytosis of both the virulent Erdman and H37Rv *M. tuberculosis* strains and the attenuated H37Ra strain. In contrast, the macrophage MR has been found to act in concert with CRs to mediate adherence of the virulent

Erdman and H37Rv strains only (57). The outer surface of *M. tuberculosis* and other mycobacteria contains complex acylated carbohydrates and glycolipids (reviewed in ref. 64) that are in a position to mediate initial interactions between *M. tuberculosis* and phagocytes. The major surface lipoglycans are lipoarabinomannan (LAM) and its smaller arabinose-free counterpart lipomannan (LM).

Immunoelectron microscopy studies demonstrate that LAM resembles a capsule surrounding individual bacteria (65). LAM from the Erdman strain and LAM from a rapidly growing avirulent nontuberculous mycobacterial strain (AraLAM) contain acylated core mannan units linked to long arabinose branches. However, Erdman LAM differs from AraLAM in that it contains linear α 1-2–linked oligomannosyl "caps" on the arabinose saccharides at the terminal portions of the molecule (66). These terminal mannosyl units are situated to interact with the MR on macrophages. LAM regulates several macrophage effector functions (64) (also see below).

In studies to evaluate whether LAM serves as a microbial ligand in adherence to MDMs, a model of phagocytosis mediated by LAM was used in which the acyl moieties of LAM are bound to 1-μm hydrophobic microspheres, thus aggregating the lipoglycan on a phagocytic particle (size comparable to a bacterium) and preferentially exposing the glycan units (67). Erdman LAM microspheres adhere more than threefold greater to macrophages than AraLAM or buffer microspheres, and this enhanced adherence is mediated by the MR. Enzymatic removal of the terminal "caps" of Erdman LAM abolishes the enhanced adherence. Preincubation of Erdman *M. tuberculosis* with a monoclonal antibody directed against LAM significantly inhibits the adherence of bacteria to macrophages, indicating a role for Erdman LAM in the adherence of intact bacteria to macrophages. Our recent studies indicate that Erdman LAM–coated microspheres are efficiently phagocytosed by MDMs and reside intracellularly within phagosomes (68). In contrast to macrophages, monocytes demonstrate minimal adherence of Erdman LAM microspheres.

Thus, the binding of the MR on macrophages to the terminal mannosyl "caps" of Erdman LAM on the bacterium is one receptor-ligand interaction involved in the adherence of *M. tuberculosis*. We speculate that the linear array of the terminal mannosyl units of Erdman LAM may be important in enhancing the affinity of this ligand for the MR by clustering several CRDs as described for other glycoconjugates (69). There is not a direct correlation between terminal mannosylation of LAM and virulence. Mycobacteria that vary in virulence in animal models, including H37Rv *M. tuberculosis*, the attenuated H37Ra strain, and *M. bovis* BCG, all contain LAM types with mannosyl caps, although the interior structure of these lipoglycans may differ (70–72). The ability of these LAM types to interact with the MR may also vary (68). Thus, it appears that the precise array, length, and number of caps may be important in determining the interaction with the host cell.

The way in which LAM is presented to the host cell also appears to be critical in these types of studies. During active disease, LAM may be released as a free lipoglycan in various tissue sites. Binding of LAM in solution, a condition in which all domains of the lipoglycan are exposed, to the phagocyte involves the acylated glycosylphosphatidylinositol anchor of the molecule and phagocyte CD14 as well as potentially several portions of the LAM molecule (73–76). AraLAM can signal the host cell through CD14 akin to the means by which lipopolysaccharide (LPS) signals (77).

In addition to the LAM-MR interaction, *M. tuberculosis* strains have been found to possess their own lectins, some of which are specific for mannans and may be involved in bacterial binding to host cells (78). Other recently described *M. tuberculosis* surface molecules that are candidates for microbial ligands in bacterial uptake include glucans and arabinomannans (79).

Thus, *M. tuberculosis*, a highly host-adapted intracellular pathogen, has evolved to utilize several different dominant host cell receptors to mediate its entry into professional phagocytes. This strategy affords the bacterium greater flexibility for entry, since phagocytes differ in expression and/or function of these receptors depending on the tissue site, the degree of cell differentiation, the presence of inflammatory mediators, and the animal model chosen. Whether utilization of specific classes of receptors or binding domains on these receptors influences the earliest host cell responses during entry of *M. tuberculosis* remains to be determined. Although it is clear that CRs and the MR play important roles in the phagocytosis of *M. tuberculosis* by mononuclear phagocytes, competitive inhibitor studies do not completely abolish phagocytosis, leaving open the possibility that other receptors may also participate in uptake either alone or in conjunction with CRs and/or the MR. For example, CD14 has been found to mediate the uptake of unopsonized *M. tuberculosis* by human microglia, resident macrophages of the brain (80).

D. Phagocytosis of *M. tuberculosis* by Alveolar Macrophages

Studies using MDMs provides detailed information that is applicable to human alveolar macrophages (HAMs). However, direct examination of HAMs may provide additional insights. Hirsch et al. (58) found that the phagocytosis of the H37Ra strain of *M. tuberculosis* by HAMs is greater than that by monocytes, and that CR4 plays a particularly important role in phagocytosis by these cells. The phagocytosis of the virulent Erdman strain of *M. tuberculosis* by HAMs also involves CRs (81).

The alveolar type II cell synthesizes and secretes the major components of surfactant, which include phosphatidylcholines, lesser amounts of other phospholipids, and four surfactant-specific proteins: SP-A, SP-B, SP-C, and SP-D (reviewed in ref. 82). SP-A and SP-D belong to a group of proteins called collectins that contain CRDs and homologous collagen-like sequence regions (83). Evidence is accumulating that these molecules play an important role in influencing early host responses to invading microbes in the lung (83). In addition to regulating the level of lung surfactant, SP-A has been found to enhance Fc receptor and CR function (84). We have recently determined that SP-A enhances the phagocytosis of Erdman *M. tuberculosis* by MDM and HAM by a direct interaction between SP-A and the macrophage (20). The degree of enhancement is dependant on the type of SP-A protein used (purified from healthy hosts or patients with alveolar proteinosis or purified recombinant protein) and the way in which SP-A is presented to the host cell (in solution or on a matrix). The carbohydrate moieties of SP-A are critical in the SP-A–macrophage interaction and mannan and antimannose receptor Ab inhibit the enhanced phagocytosis seen with SP-A on the matrix, indicating upregulation of macrophage MR activity. That SP-A may also act as opsonin for *M. tuberculosis* in adherence to murine alveolar macrophages has been described (21). Together these studies implicate SP-A as an important modulator of alveolar macrophage function that results in an enhanced potential for *M. tuberculosis* to gain access to its intracellular niche. SP-A may be particularly important in influencing *M. tuberculosis* phagocytosis (as well as phagocytosis of other pathogens) in disease states such as alveolar proteinosis or human immunodeficiency virus (HIV) infection where SP-A levels are increased (85,86).

E. Receptors and Ligands Mediating Phagocytosis of Other Mycobacteria

Studies of several other pathogenic mycobacteria have demonstrated a major role for CRs and the MR in the phagocytosis by monocytes and macrophages (see Table 1). CRs mediate the phagocytosis of *M. leprae* by mononuclear phagocytes (54,87,88). A role for the MR in the phagocytosis of *M. leprae* has not been found. C3 bound to phenolic glycolipid-1, a major glycolipid on the surface of *M. leprae*, is an important host-derived ligand for the phagocytosis of this bacterium (89). Phagocytosis of *M. avium* by monocytes and macrophages (including alveolar macrophages) is mediated by CR3 and the MR in the presence and absence of serum (90,91). C3 is an important bacterium-bound opsonin for *M. avium*, since *M. avium* adherence to monocytes is enhanced in fresh serum (92), and preincubating serum with antibodies to C3 reduces *M. avium* adherence to macrophages (90). Adherence of *M. bovis* BCG to human monocytes is mediated by CR1 and CR3 (88). Several other receptors and ligands mediate the uptake of *M. avium* by mononuclear phagocytes (90,91,93), including the phagocyte vitronectin receptor (94) and a 68-kDa microbial ligand (95). The involvement of other integrins in the binding of mycobacteria to host cells may be important in determining the cellular tropism of these pathogens (96).

F. Host Molecules that Regulate Phagocytosis of *M. tuberculosis* and Other Mycobacteria

Several host molecules regulate receptor-ligand interactions involving CRs and the MR (such as the surfactant proteins discussed above) and thus may influence the phagocytosis of *M. tuberculosis* and other mycobacteria. Mycobacteria, including *M. tuberculosis*, bind fibronectin (93,97–99). Whether bacterium-bound fibronectin serves as an opsonin for phagocyte fibronectin receptors in the phagocytosis of *M. tuberculosis* is unknown. Fibronectin may also act by upregulating the activity of Fc receptors and CRs on monocytes and macrophages (54,100).

IFN-γ activation of phagocytes downregulates the expression and/or function of CRs and the MR (101,102) and has been found to decrease the adherence of mycobacteria (54,103). In one study, the IFN-γ activation of macrophages decreased adherence but enhanced intracellular multiplication of *M. tuberculosis* (104). Inasmuch as several cytokines and inflammatory mediators, such as prostaglandins, influence the surface expression of phagocyte receptors and receptor-mediated phagocytosis, it is likely that these mediators are important in regulating the level of phagocytosis of *M. tuberculosis* and the early host cell response during different stages of disease. In this regard, IL-4 and prostaglandin E_2 (PGE_2), two mediators that modulate the host response to mycobacterial infection (reviewed in ref. 105), markedly upregulate CR and/or MR expression and function and downregulate phagocyte effector functions (106–108). Treatment of human monocytes with TGF-β1 is reported to decrease phagocytosis of H37Ra *M. tuberculosis* but enhance intracellular growth (109).

Natural antibody to several species of mycobacteria has been found in healthy PPD (purified protein derivative)–negative persons (110). Natural antibody could conceivable play a role in disease pathogenesis for mycobacteria by enhancing C3 deposition onto the bacillus and hence phagocytosis by mononuclear phagocytes (111). Although natural antibody to *M. tuberculosis* exists, its role in regulating phagocytosis of *M. tuberculosis* is unknown.

Later in infection with *M. tuberculosis* or during active disease, lysis of bacterial-laden macrophages leads to the phagocytosis of *M. tuberculosis* by neighboring macrophages in the presence of high-titer immune antibody. High-titer rabbit antimycobacterial immunoglobulin facilitates the multiplication of BCG in the spleens of mice (112). In an in vitro study, rabbit antimycobacterial immunoglobulin had no effect on the phagocytosis of H37Rv *M. tuberculosis* by mouse peritoneal macrophages but enhanced phagosome-lysosome fusion (113) (see below). The role of human immune antibody in influencing receptor-mediated phagocytosis and intracellular survival of *M. tuberculosis* by human cells has not been defined. Costello et al. (114) have reported that the prevention of disseminated forms of TB in childhood correlates with the production of antibody to LAM, which they hypothesize may alter *M. tuberculosis*–macrophage interactions. Heat-stable serum opsonins other than antibody may also mediate the binding of mycobacteria to host cells and influence the host cell response (115). Although resistance to mycobacterial infection in the murine model is strongly dependent on the expression of the Bcg gene product (116) (see below), it may not be at the level of phagocytosis (117,118).

G. Host Cell and Microbial Response to Phagocytosis of *M. tuberculosis*

A diverse group of host-adapted intracellular pathogens has evolved to utilize two major classes of phagocyte receptors, CRs and the MR, in host cell entry. These receptors are expressed to a variable degree on various mononuclear phagocytes and are competent in mediating phagocytosis. Differences in the host cell response to these microbes may relate rather to more subtle differences in the receptor binding epitopes utilized, synergy between receptor classes, unique microbial ligands for these receptors, or the specific host cell studied. For example, CR3 has multiple extracellular binding domains, only some of which appear to trigger a strong oxidative burst (119,120). The phagocytosis of live or heat-killed H37Rv *M. tuberculosis* by human neutrophils, cells that also express CRs, is accompanied by the detectable release of superoxide (24), whereas phagocytosis of *M. tuberculosis* by human monocytes and macrophages is not (121). Although strain differences may account for this last set of observations, subtle differences in binding to CR epitopes, the functional state of CRs, or the use of other receptors (such as the MR) on macrophages may also be a factor.

The entry of *M. tuberculosis* and other intracellular pathogens via the CR pathway has long been postulated to provide the bacterium safe passage into mononuclear phagocytes by allowing it to avoid the toxic consequences of the oxidative burst (43). In this regard, pathogenic mycobacteria as well as some other intracellular pathogens have been found to elicit little or no detectable reactive oxygen intermediates during phagocytosis by mononuclear phagocytes (42,121, 122). We have recently found that the phagocytosis of virulent and attenuated strains of *M. tuberculosis* by MDMs is not accompanied by detectable levels of superoxide anion and hydrogen peroxide in culture supernatants, whereas MDMs primed with IFN-γ do generate detectable (albeit low) levels in response to *M. tuberculosis* (123). Mycobacteria, including *M. tuberculosis*, have surface glycolipids, including LAM, that scavenge toxic oxygen radicals (124–128), providing another mechanism for enhancing the survival of these bacteria during phagocytosis. It remains to be determined whether *M. tuberculosis* downregulates the initiation of the oxidative burst in addition to scavenging the oxygen radicals that are produced. Whether the route of entry influences the oxidative burst in response to *M. tuberculosis* also remains unanswered.

Another potential survival mechanism for *M. tuberculosis* during and shortly following phagocytosis is an alteration in the biosynthesis of specific *M. tuberculosis* proteins that serves to protect the bacterium from host cell microbiocidal mechanisms. In this regard, Alavi and Affronti examined radiolabeled *M. tuberculosis* proteins by SDS-PAGE to identify several polypeptides that appear to be specifically expressed after internalization by the macrophage (129). Plumb and Clark-Curtiss et al. used subtractive cDNA hybridization to identify unique genes expressed by *M. avium* following phagocytosis by the macrophage (130). Other molecular strategies are being utilized (131). The identification of these altered proteins and gene products and their functional consequences await further studies.

V. *M. tuberculosis* and Host Cell Signaling Events

Receptor-mediated signal transduction events in mononuclear phagocytes in response to various stimuli involves the rapid and reversible covalent modification of host cell proteins, usually phosphorylation of either serine, threonine, or tyrosine residues. A receptor may initiate signaling events directly by possessing protein tyrosine kinase (PTK) activity and/or via activation of membrane-associated, nonreceptor PTKs recruited from the cell cytoplasm. These initial rapid events are modulated by phospholipases, calcium mobilization, and protein kinase C (PKC). Surface receptors may also initiate signaling through either various guanosine triphosphate (GTP) binding proteins or via signaling pathways that activate mitogen-activated protein (MAP) kinases. Our knowledge regarding the transmembrane signaling mechanisms that accompany the phagocytosis of *M. tuberculosis* is just beginning to emerge. Stimulation of phospholipase D (PLD) occurs early in the activation of phagocytic leukocytes, and its action generates the lipids phosphatidic acid (PA) and diacylglycerol (132). Kusner et al. (133) have recently demonstrated that phagocytosis of *M. tuberculosis* by MDMs in the presence of fresh nonimmune human serum was associated with a nearly 10-fold increase in PLD enzymatic activity and an increase in its product, PA. The PTK inhibitors genistein and herbimycin A inhibited phagocytosis-associated PLD activity and reduced phagocytosis significantly, as did a competitive inhibitor of PLD, 2,3-diphosphoglycerate. These studies demonstrate that PLD is activated during the phagocytosis of *M. tuberculosis* by human MDMs, and that inhibition of PLD is associated with an inhibition of phagocytosis. The activation of this signal transduction pathway may therefore play an important role in regulating the phagocytosis of *M. tuberculosis* and in modulating subsequent macrophage functions.

Intracellular pathogens such as *M. tuberculosis* may alter host cell signaling pathways during and after phagocytosis in such a way to disrupt normal host cell microbicidal activities and/or phagocyte effector functions (reviewed in ref. 134). In this regard, purified mycobacterial cell wall components have been shown to attenuate mononuclear phagocyte responses to cytokines and other agonists. For example, *M. tuberculosis* sulfatide blocks priming by LPS, IL-1β, TNFα, INF-γ, or muramyl dipeptide for enhanced oxidative burst responses to phorbol esters possibly by altering the phosphorylation of monocyte proteins or reducing PKC enzymatic activity (135,136).

Altered cellular signaling pathways mediated by LAM has been demonstrated in several studies (127,137–140). Specific types of LAM inhibit IFN-γ–induced microbicidal activity and gene transcription, inhibit PKC activity, inhibit the processing of mycobacterial protein antigens by antigen-presenting cells, and suppress T-cell activation (126,127,138,141–143). These studies lend support to the notion that LAM is an important mycobacterial virulence factor. On the other hand, specific LAM types induce the synthesis and release of cytokines and nitric oxide (in murine cells), induce immediate early gene expression, and serve as potent B-cell immunogens (137,140,144–151).

VI. Intracellular Trafficking of *M. tuberculosis* in Mononuclear Phagocytes

As described earlier, following the phagocytosis of *M. tuberculosis* by mononuclear phagocytes, bacteria reside in a membrane-bound phagosome. There is strong evidence that the survival of *M. tuberculosis* within human and murine mononuclear phagocytes over several days in culture is partially related to the absence of phagosome-lysosome fusion and to decreased phagosomal acidification over this time period (152–157). This mechanism of survival appears to be generalizable to other pathogenic mycobacteria (154,158–160). Both virulent strains of *M. tuberculosis* (Erdman and H37Rv) and the attenuated H37Ra strain inhibit fusion (161), although the magnitude of inhibition has been variable with H37Ra (153,155,161).

The incubation of macrophages with killed *M. tuberculosis* results in phagosome-lysosome fusion in these studies (152,157), suggesting that viability is a factor in the inhibition of fusion over time. McDonough et al. recently reported that *M. tuberculosis* initially enters a phagolysosome, then transfers into an isolated phagosome, and finally escapes into the cytoplasm (162). The unusual finding of phagosome-lysosome fusion in their study has been attributed in part to potential differences in the viability, growth conditions, and extent of clumping of the stock *M. tuberculosis* used (157) and in the host cell type studied (155). Although *M. tuberculosis* has been observed to be free in the cytoplasm of phagocytes in some studies (162,163), the majority of evidence indicates that this bacterium resides predominantly within a membrane-bound phagosome.

The *M. tuberculosis* phagosome is only mildly acidified to a pH of ≥6 (154, 156). One potential mechanism for this relates to the recent finding that *M. tuberculosis* phagosomes do not acquire the complete vesicular proton-adenosine triphosphatase (ATPase) pump responsible for phagosomal acidification, possibly because the bacteria selectively inhibit incorporation or influence the rapid removal of the ATPase from the phagosomes (154). The unique composition (see below) of the mycobacterial phagosome is also related to the acquisition of freshly synthesized components (membrane and proteins) budding from the trans-Golgi network. This acquisition is felt to be necessary, because mycobacterial components are trafficked out from the phagosome in budded vesicles with the consequent loss of membrane, and because intraphagosomal bacterial division requires expansion of the phagosomal membrane (155).

The dynamic nature of the formation and maturation of the *M. tuberculosis* phagosome in human monocytes has been highlighted in a recent study by Clemons et al. using quantitative cryosection immunogold electron microscopy (157). They demonstrated that live *M. tuberculosis* retards the maturation of its phagosome along the endosomal-lysosomal pathway in a unique fashion and resides in an endosome-like compartment. This compartment exhibits delayed clearance of major histocompatibility complex (MHC) class I molecules thought to be acquired from the plasma membrane during phagocytosis, relatively intense staining for the transferrin receptor and for MHC class II molecules thought to be acquired from MHC class II–containing vesicles, and relatively weak staining for lysosomal membrane glycoproteins and cathepsin D. The ability of the *M. tuberculosis* phagosome to interact with endosomes allows for antigen processing and presentation and the acquisition of nutrients.

The mechanism(s) that allows for the inhibition of phagosome-lysosome fusion in the case of *M. tuberculosis* is unsettled and may well be multifactorial. Possibilities include (1) involvement of major cell wall constituents (particularly nonprotein molecules), (2) bioactive molecules secreted from viable intraphagosomal bacteria, and/or (3) receptor pathways utilized during phagocytosis. The evidence for the first possibility comes from studies on *M. tuberculosis* sulfatides (anionic trehalose glycolipids) and cord factor, in which these molecules were found to be capable of inhibiting fusion in in vitro models based on charge properties or lipid constituents (164–167).

The evidence for the second possibility comes from studies on ammonia production by the bacterium (168) and more recent studies on the *M. tuberculosis* urease and glutamine synthetase (169,170). The ammonia content of filtrates of *M. tuberculosis* was found to be sufficient to inhibit the lysosomal fusion with yeast-containing phagosomes in murine macrophages (168).

The *M. tuberculosis* urease and glutamine synthetase have recently been purified and characterized (169,170). These enzymes are involved in nitrogen acquisition by the bacterium and have as their by-product ammonia. Since urease production is enhanced in low nitrogen conditions and nitrogen nutrients may be limited in the *M. tuberculosis* phagosome, it is postulated that urease may play a particularly important role in nitrogen acquisition in this locale. *M. tuberculosis* glutamine synthetase plays a role in L-glutamine biosynthesis (a major component of the cell wall of pathogenic mycobacteria) and is produced in culture filtrates as well as within the *M. tuberculosis* phagosome.

The evidence for the third possibility is limited at present, coming from earlier studies on *M. tuberculosis* and extrapolated from studies with other intracellular pathogens and particles (discussed earlier). Armstrong et al. found that opsonizing *M. tuberculosis* with immune antibody enhanced phagosome-lysosome fusion (113), raising as one possibility that entry was directed through Fc receptors which have been found to influence the early intracellular fate (phagosome-lysosome fusion) of *Toxoplasma gondii* (30). As discussed earlier, CRs and the MR play major roles in the phagocytosis of virulent *M. tuberculosis*. Whether entry via one or all of these receptors plays a role in influencing phagosome-lysosome fusion (particularly immediately following phagocytosis) is unknown. Although it appears clear that the intracellular viability of *M. tuberculosis* is important in shaping the unique characteristics of the *M. tuberculosis* phagosome over a period of several hours to days, it remains possible that other factors influence fusion events that normally occur during and immediately following the phagocytosis of pathogens. Understanding these latter events may be particularly important in situations where the number of *M. tuberculosis* encountering the phagocyte is likely to be very low, such as during primary infection in the alveoli of the lung.

VII. Intracellular Growth/Killing of *M. tuberculosis*

A. Intracellular Growth

The growth characteristics of *M. tuberculosis* strains in mononuclear phagocytes have been examined primarily in animal models both in vitro and in vivo. The three most commonly studied strains have been: Erdman, H37Rv, and H37Ra. Erdman and H37Rv cause progressive disease in experimental animals (171,172). H37Rv and H37Ra arose as variants of the original human strain H37 (173). H37Ra is considered "attenuated," because animal studies have demonstrated the limited growth of this bacterium (172).

The difficulties in distinguishing between live and dead *M. tuberculosis* by staining, in maintaining uniform phagocyte monolayers over the course of in vitro studies (174), in disrupting aggregates or clumps of *M. tuberculosis* in stock suspensions or in colony-forming unit (CFU) assays as well as the presence or absence of antibiotics in the culture medium have limited the precision with which intracellular growth can be measured. These limitations have been noted repeatedly in older and more recent studies (175). Beyond this, differences in growth rates relate to the type of mononuclear phagocyte studied (e.g., human vs nonhuman; activated vs nonactivated; lung vs peritoneum).

In an in vitro study using guinea pig peritoneal macrophages, the growth of the H37Rv strain demonstrated a 2-day lag phase followed by logarithmic multiplication (171). In contrast, the H37Ra strain remained stationary over the course of the experiment (5 days). Similarly, in murine peritoneal macrophages, the growth of H37Rv *M. tuberculosis* reveals a lag phase followed by progressive intracellular multiplication and destruction of the monolayer; whereas there is a decline in the growth (apparent killing) of the H37Ra strain between days 0 and 6 followed by variable growth (152,161). In contrast, in murine bone marrow–derived macrophages (BMMϕs), both H37Rv and H3Ra are reported to demonstrate similar growth, which is much lower than that observed in the L-929 fibroblast cell line (153). In this study, evidence was presented that mycobacteria-infected fibroblasts secreted mediators capable of downregulating mycobacteria growth inside macrophages. The ability of *M. tuberculosis* to enter nonprofessional phagocytes in vitro has been noted for some time (176), but the significance of this observation during the course of human infection remains uncertain. The growth of mycobacteria within murine macrophages is regulated by a gene on chromosome 1 (Bcg locus) which encodes for the membrane transporter protein, Nramp (116,177,178). Higher growth of H37Rv *M. tuberculosis* in Bcg^s-derived BMMϕs than Bcg^r-derived BMMϕs has been shown (153). The growth of *M. tuberculosis* strains in human mononuclear phagocytes in vitro has been examined in fewer studies (179,180). The generation time was reported to be approximately 23 hr for Erdman *M. tuberculosis* in one study (over 7 days) (180).

In an in vivo study in guinea pigs infected by the respiratory route, H3Rv *M. tuberculosis* was found to grow rapidly in the lung before achieving a prolonged stationary phase. H3Ra *M. tuberculosis* was also found to grow in the lung over 3 weeks (100-fold lower than H37Rv) before achieving a short stationary phase and then elimination (172). Thus, the "attenuated" H3Ra strain can multiply in the lungs for at least a small amount of time before death depending on the animal model used (172). More recently, the kinetics of growth was studied in the SCID mouse model versus its immunocompetent counterpart (181). In the immunocompetent mouse, virulent bacteria (but not the attenuated H3Ra strain) grew progressively only in the lungs. In SCID mice, virulent bacteria grew in all organs (fastest in the lung). The growth of the attenuated H3Ra strain was not observed in these mice unless the animals were treated with hydrocortisone, which also increased the growth rate of virulent strains (doubling times: Erdman 17.7 hr, H3Rv 17.4 hr, H37Ra 99 hr), suggesting that SCID mice possess a glucocorticoid-sensitive mycobacteriocidal mechanism that normally prevents or reduces *M.*

tuberculosis growth. The growth of *M. tuberculosis* is reduced in Bcg^r mice compared to Bcg^s mice (182). Antibiotic-resistant strains of *M. tuberculosis* demonstrate variable growth rates in the mouse model (183).

B. Intracellular Killing

Antimicrobial activities of professional phagocytes have been characterized as either oxygen-dependent or oxygen-independent. Oxygen-dependent systems include the production of reactive oxygen intermediates (mediated by the phagocyte oxidative burst) and the short-lived reactive nitrogen intermediate, nitric oxide (NO), via oxidation of L-arginine by inducible NO synthase (iNOS) (184). Oxygen-independent microbicidal mechanisms of mononuclear phagocytes include (1) acidification as a result of phagosome-lysosome fusion, (2) lysosomal enzyme activity, (3) nutrient deprivation (including iron), and (4) antimicrobial molecules such as defensins. Although the antimicrobial role of NO in murine mononuclear phagocytes appears clear, its contribution to killing by human mononuclear phagocytes is unresolved.

The mechanisms for killing or growth inhibition of *M. tuberculosis* in human mononuclear phagocytes have been elusive, whereas the mechanisms for growth inhibition in murine macrophages have been shown to include NO, the activation of cells by lymphokines and other cytokines (175,185–187), and the genetic background of the mouse strain using (Bcg^s vs Bcg^r) as noted above. The importance of IFN-γ in the protective cellular immune response in mice to TB has been demonstrated by using IFN-γ knockout mice (188,189). Regardless of the in vitro model used, it has been very difficult to demonstrate complete irradication of viable organisms from any mononuclear phagocyte type. This may relate in part to the impermeable nature of the mycobacterial cell wall which may prevent destruction of intraphagosomal bacilli (190).

The difference between murine and human phagocytes in inhibiting intracellular growth of *M. tuberculosis* after exposure to cytokines has been noted in several studies. As one example, the addition of a crude lymphokine preparation from *M. tuberculosis*-infected human mononuclear cells or recombinant IFN-γ significantly inhibited the growth of *M. tuberculosis* in murine peritoneal cells, whereas the same preparations did not influence overall growth in human alveolar macrophages (187). In one study, the growth of virulent *M. tuberculosis* in recombinant IFN-γ–activated human MDMs was actually enhanced over 5–7 days (104). In contrast to these studies, De Sousa et al. reported that IFN-γ and indomethacin (a potent inhibitor of prostaglandin E_2 synthesis) reduced the growth of *M. tuberculosis* in human MDMs (179); a discrepancy which may relate to assay conditions or the strains of *M. tuberculosis* used. Overall, the data suggest that the IFN-γ activation of human mononuclear phagocytes directly does not significantly reduce the intracellular growth of bacteria, but the possibility remains that IFN-γ may act indirectly through other cell types to influence the intracellular growth by these cells.

Apart from cytokines, calcitriol, the active metabolite of vitamin D, has been reported to inhibit the growth of *M. tuberculosis* in human MDMs in a serum-dependent manner (180,191,192), suggesting an important role for vitamin D (and sunlight) in host defense against *M. tuberculosis*. The mechanism for this observation remains unknown. Apoptosis of monocytes has recently been found to be associated with killing of intracellular *M. bovis* (193).

VIII. *M. tuberculosis*-Infected Mononuclear Phagocytes as Effector and Accessory Cells

Mononuclear phagocytes serve as important antigen-presenting cells for a variety of lymphocyte subsets and other cell types during *M. tuberculosis* infection. In addition to presenting antigens in the context of class I or class II molecules to CD4-positive or CD8-positive T cells, these phagocytes have been recently shown to present nonpeptide antigens (such as mycolic acid or LAM) in the context of CD1b to other αβ T-lymphocyte subsets (194,195). Thus, CD1 presentation of antigens other than peptides to a unique subset of T cells may represent an additional line of antimycobacterial defense, possibly as part of the early innate response to infection.

Tissue macrophages, including those containing intracellular mycobacteria, are long-lived cells that are in constant communication with extracellular matrix proteins as well as other cell types involved in the host cellular immune response. One potential mechanism for the continued survival of *M. tuberculosis* within the macrophage is the avoidance or subversion of the cell's ability to function as an inducer and effector cell in this response. Upregulation or downregulation of expression and/or function of phagocyte surface molecules on *M. tuberculosis*-infected cells may serve to regulate antigen presentation and accessory function, cell migration, and cell to cell and cell to matrix adhesion events (including granuloma formation). Several pathogens have evolved strategies to regulate these events (196).

Pancholi et al. found that macrophages infected with *M. bovis* BCG "chronically" for 7 days stimulated T-cell proliferation poorly, suggesting a defect in antigen presentation potentially related to the sequestration of the antigen within the cell (197). *M. tuberculosis*-containing monocytes have been found to decrease T-cell proliferation to tetanus toxoid, in part due to the reduced expression of class II antigen at very high bacterial loads as well as reduced accessory function (198). A 25-kDa *M. tuberculosis* glycolipoprotein component has been reported to reduce the enhanced expression of class II antigen on monocytes that normally occurs in response to opsonized yeast, LPS, or IFN-γ (199). *M. tuberculosis*-

containing macrophages from a susceptible mouse strain manifest reduced expression of the costimulatory molecule B7 and increased expression of the intracellular adhesion molecule-1 (ICAM-1), resulting in T-cell unresponsiveness which appeared to be linked to prostaglandin synthesis in response to live bacteria (200). Incubation with *M. tuberculosis* or LAM results in an increased expression of ICAM-1 by the myeloid cell line THP-1, in part by secretion of TNF-α (201).

Our ongoing studies in *M. tuberculosis*-infected human macrophages indicate significant decreases in steady-state mRNA levels for the CD11b and CD11c genes of the leukocyte integrins with concomitant changes in the surface expression of these proteins (202). Alterations in expression of the leukocyte integrins in *M. tuberculosis*-infected macrophages could potentially influence interactions with several cell types and matrices as well as influence subsequent endocytic events in these cells. *M. tuberculosis*-infected mononuclear phagocytes were found to have altered Fc receptor function in an earlier study (203). The regulation of phagocyte surface molecules appears to be a general property of mycobacteria-infected cells (204–206).

IX. Inflammatory Mediators Secreted from *M. tuberculosis*-Infected Phagocytes

Mononuclear phagocytes are capable of synthesizing and secreting an array of inflammatory mediators that play a role as chemoattractants, in enhancing inflammation, in granuloma formation, and in bacterial killing. Inflammatory mediators secreted by the *M. tuberculosis*-infected mononuclear phagocyte have been defined as either "protective" (enhance the immunological response to the benefit of the host) or "suppressive" (modulate the immunological response to the benefit of the bacterium). These definitions are not always clearcut. For example, the abundant secretion of a protective mediator (e.g., TNF-α) may lead to many of the signs and symptoms of the disease state in the host. Also, it is clear that these molecules act as part of a cascade of mediators that together determine the immunological response. Thus, the extent and type of mediators produced must create a fine balance in effect (reviewed in ref. 207). The evidence is increasing that host-adapted intracellular pathogens such as *M. tuberculosis* have developed strategies to perturb the fine balance of phagocyte inflammatory mediators released during infection in a way that enhances bacterial survival and allows for persistence in its host niche (reviewed in ref. 208).

Important immunosuppressive molecules secreted by the phagocyte in the case of *M. tuberculosis* include TGF-β, IL-10, and PGE_2 (209). On the other hand, potent immunoprotective mediators include proinflammatory cytokines such as TNF-α and IL-12. The production of immunosuppressive molecules can deactivate macrophages and enhance mycobacterial survival; whereas immunoprotective molecules can inhibit the growth of *M. tuberculosis* and other mycobacteria in macrophages (209). In general, information on the pattern of cytokine synthesis and secretion of TB has been complicated somewhat by attempts to compare data obtained from in vitro macrophage studies, from animal models, and from results obtained from TB-infected patients (210).

TGF-β deactivates host cell oxidant responses, opposes the actions (and production) of proinflammatory cytokines such as IL-1 and TNF-α, and promotes extracellular matrix formation (207,211). TGF-β also inhibits lymphocyte proliferation, reduces the function of cytotoxic T cells and natural killer (NK) cells, blocks the macrophage-activating effects of IFN-γ, and has been found to reduce macrophage killing of intracellular pathogens (207). Infection of mononuclear phagocytes by pathogenic mycobacteria leads to the production of TGF-β (212). PPD enhances the production of TGF-β in monocytes from healthy individuals (213). TGF-β can counteract the antimicrobial effects of TNF-α, which is also secreted by *M. tuberculosis*-infected cells (109). TGF-β production is increased in the monocytes from TB patients and is found in their granulomas (214). In rodent macrophages, TGF-β upregulates the arginase pathway resulting in an increased release of polyamines which reduce macrophage cytotoxicity to tumor cell lines (215). TGF-β also suppresses iNOS activity in macrophages (216).

Mononuclear phagocytes also secrete IL-10 in response to *M. tuberculosis* and LAM, another potent anti-inflammatory cytokine that has inhibitory effects on proinflammatory monokine production (decreasing IL-1α, IL-1β, IL-6, IL-12, and TNF-α [207]). In vitro studies indicate that IL-10 may promote intracellular mycobacterial growth (217,218). Its role in TB is less clearly defined.

PGE_2 is secreted from monocytes exposed to mycobacteria and their products (219,220). Cadranel et al. found that the TNF-α production from the monocytes of TB patients was high and generally correlated with the PGE_2 production, which overall was very low (221). However, in two patients with very high levels of PGE_2, TNF-α was decreased. This latter finding was thought to relate to the fact that at high concentrations, PGE_2 downmodulates TNF-α mRNA expression and hence TNF-α synthesis through a rise in cAMP (222,223). In contrast to these results in TB patients, the in vitro treatment of control peripheral blood mononuclear cells (PBMCs) by PPD was associated with a higher production of PGE_2 than TNF-α. Thus, the investigators concluded that PGE_2-mediated immunoregulation is abnormal in patients with TB. Limiting PGE_2 production could enhance the cellular response, because high levels of PGE_2 are associated with the inhibition of the actions of proinflammatory cytokines and several lymphocyte functions.

Several pathogenic mycobacteria, including *M. tuberculosis*, stimulate increased gene expression and the production of proinflammatory cytokines and chemokines by mononuclear phagocytes which are thought to be important in enhancing effector and accessory functions by this cell (i.e., upregulation of MHC expression and antigen presentation, enhanced T-cell activation, and granuloma formation) and thus are considered "immunoprotective" (224–227). PPD, *M. tuberculosis* culture filtrates, and whole bacilli stimulate TNF-α production from human monocytes and alveolar macrophages in vitro. Several recent studies provide strong evidence for the immunoprotective role of TNF-α in TB infection in mice (208,228), the mechanism of which appears to be related to NO production and granuloma formation (208). TNF-α inhibits mycobacterial growth in macrophage cultures (229). BALB/c mice infected with *M. tuberculosis* and then given exogenous TNF-α have increased survival compared with controls that is abolished by anti–TNF-α antibody. Granuloma formation in mouse liver infected with BCG coincides with local TNF-α synthesis (230). Anti–TNF-α antibody given to these BCG-resistant mice inhibits granuloma formation following BCG challenge and results in a progressive, lethal BCG infection. Although TNF-α is thought to play an important role in granuloma formation (231), it is likely not the only factor, but works in concert with other proinflammatory cytokines and chemokines (see below). The production of TNF-α by mycobacteria may also be cytotoxic to macrophages (232).

Monocytes secrete TNF-α (and other proinflammatory cytokines) in response to both mycobacterial LAM (also see above) and other molecules purified from *M. tuberculosis* (233). AraLAM is a potent stimulus for TNF-α production, whereas LAM from Erdman *M. tuberculosis* is markedly reduced in this respect (145). This downregulation has been hypothesized to be a bacterial adaptation to enhance its virulence. In this regard, in murine macrophages, the secretion of TNF-α appears to be inversely related to the virulence of the mycobacterial strain used (i.e., *M. smegmatis* > H37Ra > H37Rv) (227). In addition, heat-killed virulent H37Rv *M. tuberculosis* induces greater levels of TNF-α than the live bacteria.

TNF-α production has also been examined in patients with TB. Cadranel et al. found that there is an upregulation of TNF-α production and binding to blood monocytes in patients with pulmonary TB (234). Monocytes from TB patients express an increase in TNF-α receptor number compared with controls. The enhanced binding of TNF-α to monocytes enhances the capacity of these cells to generate TNF-α in vitro. However, the amount of TNF-α produced from the monocytes of TB patients relates to the stage of disease (235). Monocytes from patients with chronic refractory TB release lesser amounts of TNF-α than those from patients with newly diagnosed TB. The cerebrospinal fluid of patients with TB has been found to have increased amounts of the soluble receptor forms for TNF and a small biologically active fraction of TNF (236). This finding led to the speculation that Tb and other chronic infectious diseases may be associated with a relatively inadequate production of TNF as a result of the increase in soluble TNF receptors, which may prolong these diseases.

IL-12 production by mononuclear phagocytes is a potent stimulus for INF-γ production (208). Levels of active IL-12 are increased in the pleural effusions of TB patients (237). In the murine model of TB, live *M. tuberculosis* induces IL-12 gene expression in macrophages (238). Treatment of mice with exogenous IL-12 increases resistance to TB and increases the accumulation of macrophages in infected tissues (239,240). Neutralization of IL-12 by antibody in vivo results in loss of granuloma structure. Whether IL-12 acts directly on migrating cells or via potentiation of chemokine responses is unclear.

Phagocytosis and intracellular growth of *M. tuberculosis* in macrophages is followed by the recruitment of monocytes and lymphocyte populations to the site of infection leading to granuloma formation. Chemokines, in particular, MCP-1, a monocyte chemoattractant (241), and IL-8, a T-cell and neutrophil chemoattractant secreted from *M. tuberculosis*-infected cells, likely play an important role in this process. Phagocytosis of *M. tuberculosis* strains and *M. tuberculosis* molecules by murine cells and monocyte cell lines is accompanied by an increase in gene expression and/or secretion of several chemokines (147,242–244). Human blood monocytes stimulated with PPD or *M. tuberculosis* secrete lower levels of IL-8 early compared with those seen with LPS and then demonstrate sustained high levels of MCP-1 secretion over days (245). Early IL-8 secretion by monocytes may contribute to the influx of neutrophils seen in the earliest inflammatory response to *M. tuberculosis*, and MCP-1 may play an important role in the subsequent mononuclear influx (245). Early neutrophil recruitment may also contribute to the recruitment of mononuclear cells, since neutrophils also secrete chemokines (246). These mediators have been found in the pleural effusions of patients with TB (247).

In a recent study, infection of murine macrophages in vitro with several strains of *M. tuberculosis* induced rapid expression of genes incoding four murine chemokines: MIP-1α, MIP-2, IP-10, and MCP-1 (248). In this study, the induction of the expression of several chemokine mRNAs was also found in the lungs of mice following a low-dose aerosol infection. Inasmuch as strains that varied in their capacity to induce chemokine mRNA all grew equally well in the lung, the investigators concluded that factors other than chemokines, such as proinflammatory cytokines, play a more important role in the protective granulomatous response. Thus, together these studies indicate a marked complexity in the interplay between cytokines and chemokines at the site of infection. It appears likely that proinflammatory cytokines and chemokines are secreted during mycobacterial disease in humans, play a role in mobilizing and activating cellular immune responses, and contribute to granuloma formation (150,249).

The local production of macrophage-generated cytokines as well as lymphokines such as IFN-γ in the pleural fluid of patients with TB pleuritis is markedly increased (150,218), and the pattern and amount of cytokines produced differ from those seen in plasma (209). For example, IL-10 concentrations and TH-1 lymphokines are particularly elevated in pleural fluid versus serum (218). The cytokines produced locally in the skin during a PPD-induced delayed-type hypersensitivity (DTH) reaction have also been examined and include TNF-α, IL-1, and IL-6 produced by monocyte-macrophages but also potentially by keratinocytes in this milieu (250). It has been suggested that tissue compartmentalization of cytokine responses may allow for the effective containment of *M. tuberculosis* infection (protective response) at specific sites (such as the pleura), but that these cytokine responses may be inadequate to alter the suppressed systemic responses that can be seen during TB infection (209). The modulation of cytokines and other inflammatory mediators as immunotherapy for TB is under consideration. For example, the systemic "side effects" of the enhanced production of TNF-α in TB patients may be controlled with thalidomide, which inhibits TNF-α synthesis (251).

X. HIV–*M. tuberculosis* Coinfection in Mononuclear Phagocytes

HIV coinfection is a potent risk factor for the development of active TB disease in a person infected with *M. tuberculosis*. Conversely, TB can accelerate the course of HIV disease (209). Macrophage cytokine secretion by *M. tuberculosis* and its constituents can promote HIV expression by latently infected cells (252). Monocytes from TB patients develop a higher productive infection with HIV (253). TB in HIV-infected individuals has been associated with decreased TH-1 responses, potentially induced by immunosuppressive cytokines such as IL-10 produced by mononuclear phagocytes (254). In coinfection models, phagocytosis of *M. tuberculosis* can enhance directly HIV transcription in monocytic cells (255). Preincubation of human monocytes with the HIV envelope glycoprotein gp120 inhibits phagocytosis but enhances intracellular growth of *M. avium* (256). It is of interest that there is enhanced expression of CRs on monocytes from patients with HIV (257), but the precise relationship between this enhanced expression and the development of mycobacterial disease is unknown.

Studies by McDonough suggest that host physiological factors that affect lung epithelial permeability may increase the efficiency with which inhaled bacilli are able to establish infection (8). Such conditions have been reported for HIV-infected individuals and could contribute to the increased susceptibility to *M. tuberculosis* infection and disease in HIV-positive individuals (258).

XI. Conclusions

M. tuberculosis, a highly host-adapted intracellular pathogen, has evolved to utilize a number of strategies to enhance its survival in mononuclear phagocytes (Fig. 1). The receptor-ligand interactions involved in phagocytosis and their

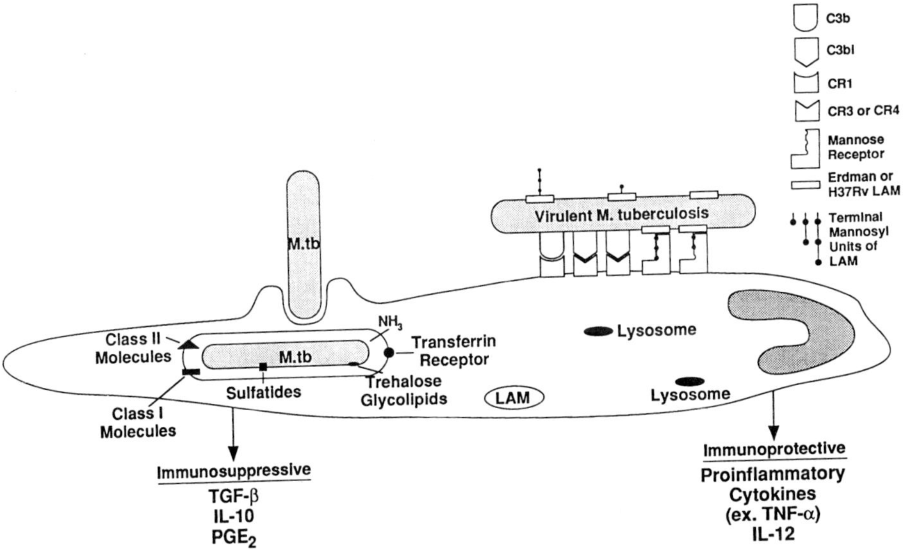

Figure 1 Model of the *M. tuberculosis* (M.tb)–mononuclear phagocyte interaction. M.tb (virulent and attenuated strains) binds C3 fragments that are produced by the activation of complement protein cascades. These fixed C3 fragments are, in turn, bound by CRs (CR1, CR3, and CR4) on the surface of the phagocyte. Alternatively, surface molecules on M.tb may bind to phagocyte CRs directly. Virulent strains of M.tb also bind to the MR. This interaction plays an important role in mediating phagocytosis of bacteria by macrophages which have high MR activity. The terminal manosyl units of M.tb lipoarabinomannan (LAM) serve as ligands for the MR. Following phagocytosis, M.tb multiplies within a mildly acidified (pH $\geqslant$ 6) phagosome that does not fuse with lysosomes and lacks the vesicular proton-ATPase. The phagosome manifests delayed clearance of MHC class I molecules and acquires MHC class II molecules and the transferrin receptor from endosomes. Lysosomal membrane glycoproteins are present in reduced amounts. Ammonia produced in part by the activities of the M.tb urease and glutamine synthetase as well as M.tb surface glycolipids may play a role in the inhibition of fusion of the phagosome with lysosomes. LAM can be seen in vesicles apart from the M.tb phagosome. Several inflammatory mediators are produced from the M.tb-infected phagocyte that are characterized either as "immunosuppressive" or "immunoprotective" (see text).

regulation are complex and are dependent on several variables, including the tissue site, the degree of cell differentiation, the presence of inflammatory mediators, and the model system chosen. The involvement of specific receptor binding domains cooperativity between different receptors and the involvement of specific microbial determinants all will likely be important in defining the early host cell response during and following phagocytosis of *M. tuberculosis*.

Once inside the phagocyte, the survival and growth of *M. tuberculosis* are dependent on several variables, including its intracellular trafficking pathway, its ability to downregulate or overcome the microbiocidal mechanisms it encounters, its ability to access nutrients, the state of "activation" of the cell it encounters, and possibly the genetic background of the host.

Understanding the kinetics and efficacy of the cellular immune response of the host to *M. tuberculosis*-infected phagocytes is particularly complex. There is both modification of the phenotype and function of the infected phagocyte as well as the generation of a specific cascade of inflammatory mediators which result in the recruitment of other cell types to the site of infection. In the human host, the immune response to primary infection is unique in that the majority of infected individuals never develop active disease. Thus, *M. tuberculosis* can continue to survive in its natural ecological niche. Finally, the increasing numbers of immunocompromised patients create an additional challenge in defining the altered immunological response to *M. tuberculosis* infection in these patients and at the same time may provide important clues to immunopathogenesis in the healthy host. Furthering our knowledge of the molecular events underlying the *M. tuberculosis*–phagocyte interaction will continue to be a great challenge but should help to define novel targets for immunotherapy and vaccine development.

Acknowledgments

The author acknowledges research support from the NIH (A133004 and HL51990) and the Department of Veterans Affairs. The author also thanks Deb Nollen for her editorial assistance and Stephen McGowan for his critical review of the manuscript.

References

1. Raviglione MC, Snider DE Jr, Kochi A. Global epidemiology of tuberculosis: Morbidity and mortality of a worldwide epidemic. JAMA 1995; 273:220–226.
2. Chan J, Kaufmann SHE. Immune mechanisms of protection. In: Bloom BR, ed. Tuberculosis: Pathogenesis, Protection, and Control. Washington DC: ASM Press, 1994:389–415.
3. Fels A, Cohn ZA. The alveolar macrophage. J Appl Physiol 1986; 60:353–369.
4. North RJ. *Mycobacterium tuberculosis* is strikingly more virulent for mice when given via the respiratory than via the intravenous route. J Infect Dis 1995; 172:1550–1553.
5. Herscowitz HB. In defense of the lung: paradoxical role of the pulmonary alveolar macrophage. Ann Allergy 1985; 55:634–650.
6. Schauble TL, Boom WH, Finegan CK, Rich EA. Characterization of suppressor function of human alveolar macrophages for T lymphocyte responses to phyto-

hemagglutin: cellular selectivity, reversibility and early events in T cell activation. Am Rev Respir Cell Mol Biol 1993; 8:89–97.
7. Harmsen AG, Muggenburg BA, Snipes MB, Bice DE. The role of macrophages in particle translocation from lungs to lymph nodes. Science 1985; 230:1277–1280.
8. McDonough KA, Kress Y. Cytotoxicity for lung epithelial cells is a virulence-associated phenotype of *Mycobacterium tuberculosis*. Infect Immun 1995; 63:4802–4811.
9. Dannenberg AM Jr, Rook GAW. Pathogenesis of pulmonary tuberculosis: an interplay of tissue-damaging and macrophage-activating immune responses-dual mechanisms that control bacillary multiplication. In: Bloom BR, ed. Tuberculosis: Pathogenesis, Protection, and Control. Washington, DC: ASM Press, 1994:459–483.
10. Wayne LG. Dormancy of *Mycobacterium tuberculosis* and latency of disease. Eur J Clin Microbiol Infect Dis 1994; 13:908–914.
11. Youmans GP. Tuberculosis. Philadelphia: Saunders, 1979.
12. Maxwell KW, Marcus S. Phagocytosis and intracellular fate of *Mycobacterium tuberculosis*: in vitro studies with guinea pig peritoneal and alveolar mononuclear phagocytes. J Immunol 1968; 101:176–182.
13. Thepen T, Kraal G, Holt PG. The role of alveolar macrophages in regulation of lung inflammation. Ann NY Acad Sci 1994; 725:200–206.
14. Staub NC. Pulmonary intravascular macrophages. Annu Rev Physiol 1994; 56:47–67.
15. Steinman RM, Witmer-Pack M, Inaba K. Dendritic cells: antigen presentation, accessory function and clinical relevance. Adv Exp Med Biol 1993; 329:1–9.
16. Steinman RM. The dendritic cell system and its role in immunogenicity. Annu Rev Immunol 1991; 9:271–296.
17. Inaba K, Inaba M, Naito M, Steinman RM. Dendritic cell progenitors phagocytose particulates, including bacillus Calmette-Guerin organisms, and sensitize mice to mycobacterial antigens in vivo. J Exp Med 1993; 178:479–488.
18. Simon RH, Paine R III. Participation of pulmonary alveolar epithelial cells in lung inflammation. J Lab Clin Med 1995; 126:108–118.
19. Strunk RC, Eidlen DM, Mason RJ. Pulmonary alveolar type II epithelial cells synthesize and secrete proteins of the classical and alternative complement pathways. J Clin Invest 1988; 81:1419–1426.
20. Gaynor CD, McCormack FX, Voelker DR, McGowan SE, Schlesinger LS. Pulmonary surfactant protein A mediates enhanced phagocytosis of *Mycobacterium tuberculosis* by a direct interaction with human macrophages. J Immunol 1995; 155: 5343–5351.
21. Downing JF, Pasula R, Wright JR, Twigg HL III, Martin WJ II. Surfactant protein A promotes attachment of *Mycobacterium tuberculosis* to alveolar macrophages during infection with human immunodeficiency virus. Proc Natl Acad Sci USA 1995; 92:4848–4852.
22. Appelberg R, Castro AG, Gomes S, Pedrosa J, Silva MT. Susceptibility of beige mice to *Mycobacterium avium*: Role of neutrophils. Infect Immun 1995; 63:3381–3387.
23. Silva MT, Silva MNT, Appelberg R. Neutrophil-macrophage cooperation in the host defense against mycobacterial infections. Microb Pathog 1989; 6:369–380.
24. May ME, Spagnuolo PJ. Evidence for activation of a respiratory burst in the

interaction of human neutrophils with *Mycobacterium tuberculosis*. Infect Immun 1987; 55:2304–2307.
25. Jones GS, Amirault HJ, Andersen BR. Killing of *Mycobacterium tuberculosis* by neutrophils: a nonoxidative process. J Infect Dis 1990; 162:700–704.
26. Brown AE, Holzer TJ, Andersen BR. Capacity of human neutrophils to kill *Mycobacterium tuberculosis*. J Infect Dis 1987; 156:985–989.
27. Lefkowitz DL, Mills K, Lefkowitz SS, Bollen A, Moguilevsky N. Neutrophil-macrophage interaction: a paradigm for chronic inflammation. Med Hypotheses 1995; 44:58–62.
28. Schlesinger LS, Bellinger-Kawahara CG, Payne NR, Horwitz MA. Phagocytosis of *Mycobacterium tuberculosis* is mediated by human monocyte complement receptors and complement component C3. J Immunol 1990; 144:2771–2780.
29. Arruda S, Bomfim G, Knights R, Huima-Byron T, Riley LW. Cloning of an *M. tuberculosis* DNA fragment associated with entry and survival inside cells. Science 1993; 261:1454–1457.
30. Joiner KA, Fuhrman SA, Miettinen HM, Kasper LH, Mellman I. *Toxoplasma gondii*: fusion competence of parasitophorous vacuoles in Fc receptor–transfected fibroblasts. Science 1990; 249:641–646.
31. Mosser DM, Edelson PJ. The third component of complement (C3) is responsible for the intracellular survival of *Leishmania major*. Nature 1987; 327:329–331.
32. Hoepelman AIM, Tuomanen EI. Consequences of microbial attachment: directing host cell functions with adhesions. Infect Immun 1992; 60:1729–1733.
33. Kobzik L, Godleski JJ, Brain JD. Selective down-regulation of alveolar macrophage oxidative response to opsonin-independent phagocytosis. J Immunol 1990; 144: 4312–4319.
34. Channon J, Leslie C, Johnston RBJ. Zymosan-stimulated production of phosphatidic acid by macrophages: relationship to release of superoxide anion and inhibition by agents that increase intracellular cyclic AMP. J Leukoc Biol 1987; 139:2749.
35. Seyberth HW, Schmidt-Gayk H, Jakobs KH, Hackenthal E. Cyclic adenosine monophosphate in phagocytizing granulocytes and alveolar macrophages. J Cell Biol 1973; 57:567.
36. Wright SD. Receptors for complement and the biology of phagocytosis. In: Gallin JI, Goldstein IM, Snyderman R, eds. Inflammation: Basic Principles and Clinical Correlates. 2nd ed. New York: Raven Press, 1992:477–495.
37. Ross GD. Opsonization and membrane complement receptors. In: Ross GD, ed. Immunobiology of the Complement System. Orlando, FL: Academic Press, 1986: 87–114.
38. Kishimoto TK, Larson RS, Corbi AL, Dustin ML, Staunton DE, Springer TA. The leukocyte integrins. Adv Immunol 1989; 46:149–182.
39. Stewart M, Thiel M, Hogg N. Leukocyte integrins. Curr Opin Cell Biol 1995; 7: 690–696.
40. Van Kooyk Y, Figdor CG. Lymphocyte adhesion mediated by integrins. Res Immunol 1993; 144:709–722.
41. Thornton BP, Vetvicka V, Pitman M, Goldman RC, Ross GD. Analysis of the sugar specificity and molecular location of the β-glucan–binding lectin site of complement receptor type 3 (CD11b/CD18). J Immunol 1996; 156:1235–1246.

42. Horwitz MA. Intracellular parasitism. Curr Opin Immunol 1988; 1:41–46.
43. Wilson CB, Tsai V, Remington JS. Failure to trigger the oxidative metabolic burst by normal macrophages: possible mechanism for survival of intracellular pathogens. J Exp Med 1980; 151:328–346.
44. Wright SD, Silverstein SC. Receptors for C3b and C3bi promote phagocytosis but not the release of toxic oxygen from human phagocytes. J Exp Med 1983; 158:2016–2023.
45. Aderem AA, Wright SD, Silverstein SC, Cohn ZA. Ligated complement receptors do not activate the arachidonic acid cascade in resident peritoneal macrophages. J Exp Med 1985; 161:617–622.
46. Myones BL, Dalzell JG, Hogg N, Ross GD. Neutrophil and monocyte cell surface p150,955 has iC3b-receptor (CR4) activity resembling CR3. J Clin Invest 1988; 82: 640–651.
47. Speert DP, Silverstein SC. Phagocytosis of unopsonized zymosan by human monocyte-derived macrophages: maturation and inhibition by mannan. J Leukoc Biol 1985; 38:655–658.
48. Stahl PD. The mannose receptor and other macrophage lectins. Curr Opin Immunol 1992; 4:49–52.
49. Taylor ME. Recognition of complex carbohydrates by the macrophage mannose receptor. Biochem Soc Trans 1993; 21:468–473.
50. Johnson WD Jr, Mei B, Cohn ZA. The separation, long-term cultivation, and maturation of the human monocyte. J Exp Med 1977; 146:1613–1636.
51. Nakagawara A, Nathan CF, Cohn ZA. Hydrogen peroxide metabolism in human monocytes during differentiation in vitro. J Clin Invest 1981; 68:1243–1252.
52. Newman SL, Musson RA, Henson PM. Development of functional complement receptors during in vitro maturation of human monocytes into macrophages. J Immunol 1980; 125:2236–2244.
53. Wright SD, Silverstein SC. Tumor-promoting phorbol esters stimulate C3b and C3b′ receptor–mediated phagocytosis in cultured human monocytes. J Exp Med 1982; 156:1149–1164.
54. Schlesinger LS, Horwitz MA. Phagocytosis of *Mycobacterium leprae* by human monocyte-derived macrophages is mediated by complement receptors CR2(CD35), CR3(CD11b/CD18), and CR4(CD11c/CD18) and interferon-γ activation inhibits complement receptor function and phagocytosis of this bacterium. J Immunol 1991; 147:1983–1994.
55. Ramanathan VD, Curtis J, Turk JL. Activtion of the alternative pathway of complement by mycobacteria and cord factor. Infect Immun 1980; 29:30–35.
56. Rourke FJ, Fan SS, Wilder MS. Anticomplementary activity of tuberculin: relationship to platelet aggregation and lytic response. Infect Immun 1979; 23:160–167.
57. Schlesinger LS. Macrophage phagocytosis of virulent but not attenuated strains of *Mycobacterium tuberculosis* is mediated by mannose receptors in addition to complement receptors. J Immunol 1993; 150:2920–2930.
58. Hirsch CS, Ellner JJ, Russell DG, Rich EA. Complement receptor-mediated uptake and tumor necrosis factor-alpha–mediated growth inhibition of *Mycobacterium tuberculosis* by human alveolar macrophages. J Immunol 1994; 152:743–753.
59. Czop JK, Kay J. Isolation and characterization of β-glucan receptors on human mononuclear phagocytes. J Exp Med 1991; 173:1511–1520.

60. Wozencraft AD, Sayers G, Blackwell JM. Macrophage type 3 complement receptors mediate serum-independent binding of Leishmania donovani: detection of macrophage-derived complement on the parasite surface by immunoelectron microscopy. J Exp Med 1986; 164:1332–1337.
61. Ezekowitz RAB, Sim RB, Hill M, Gordon S. Local opsonization by secreted macrophage complement components: role of preceptors for complement in uptake of zymosan. J Exp Med 1983; 159:244–260.
62. Stokes RW, Haidl ID, Jefferies WA, Speert DP. Mycobacteria-macrophage interactions: macrophage phenotype determines the nonopsonic binding of *Mycobacterium tuberculosis* to murine macrophages. J Immunol 1993; 151:7067–7076.
63. Hetland G, Wiker HG. Antigen 85C on Mycobacterium bovis, BCG and *M. tuberculosis* promotes monocyte-CR3–mediated uptake of microbeads coated with mycobacterial products. Immunology 1994; 82:445–449.
64. Brennan PJ, Nikaido H. The envelope of mycobacteria. Annu Rev Biochem 1995; 64:29–63.
65. Hunter SW, Brennan PJ. Evidence for the presence of a phosphatidylinositol anchor on the lipoarabinomannan and lipomannan of *Mycobacterium tuberculosis*. J Biol Chem 1990; 265:9272–9279.
66. Chatterjee D, Lowell K, Rivoire B, McNeil MR, Brennan PJ. Lipoarabinomannan of *Mycobacterium tuberculosis*. Capping with mannosyl residues in some strains. J Biol Chem 1992; 267:6234–6239.
67. Schlesinger LS, Hull SR, Kaufman TM. Binding of the terminal mannosyl units of lipoarabinomannan from a virulent strain of *Mycobacterium tuberculosis* to human macrophages. J Immunol 1994; 152:4070–4079.
68. Schlesinger LS, Kaufman TM, Kang BK, Marchiando LK. Host and bacterial determinants that mediate phagocytosis of *M. tuberculosis*. J Cell Biochem 1995; 19B:60.
69. Taylor ME, Bezouska K, Drickamer K. Contribution to ligand binding by multiple carbohydrate-recognition domains in the macrophage mannose receptor. J Biol Chem 1992; 267:1719–1726.
70. Prinzis S, Chatterjee D, Brennan PJ. Structure and antigenicity of lipoarabinomannan from *Mycobacterium bovis* BCG. J Gen Microbiol 1993; 139:2649–2658.
71. Khoo K-H, Dell A, Morris HR, Brennan PJ, Chatterjee D. Inositol phosphate capping of the nonreducing termini of lipoarabinomannan from rapidly growing strains of *Mycobacterium*. J Biol Chem 1995; 270:12380–12389.
72. Venisse A, Berjeaud J-M, Chaurand P, Gilleron M, Puzo G. Structural features of lipoarabinomannan from *Mycobacterium bovis* BCG. Determination of molecular mass by laser desorption mass spectrometry. J Biol Chem 1993; 268:12401–12411.
73. Stokes RW, Speert DP. Lipoarabinomannan inhibits nonopsonic binding of *Mycobacterium tuberculosis* to murine macrophages. J Immunol 1995; 155:1361–1369.
74. Ilangumaran S, Arni S, Poincelet M, Theler J-M, Brennan PJ, Din N, et al. Integration of mycobacterial lipoarabinomannans into glycosylphosphatidylinositol-rich domains of lymphomonocytic cell plasma membranes. J Immunol 1995; 155:1334–1342.
75. Venisse A, Fournié J-J, Puzo G. Mannosylated lipoarabinomannan interacts with phagocytes. Eur J Biochem 1995; 231:440–447.

76. Pugin J, Heumann D, Tomasz A, Kravchenko VV, Akamatsu Y, Nishijima M, et al. CD14 is a pattern recognition receptor. Immunity 1994; 1:509–516.
77. Zhang Y, Doerfler M, Lee TC, Guillemin B, Rom WN. Mechanisms of stimulation of interleukin-1β and tumor necrosis factor-α by *Mycobacterium tuberculosis* components. J Clin Invest 1993; 91:2076–2083.
78. Goswami S, Sarkar S, Basu J, Kundu M, Chakrabarti P. Mycotin: A lectin involved in the adherence of Mycobacteria to macrophages. FEBS Lett 1994; 355:183–186.
79. Ortalo-Magné A, Dupont M-A, Lemassu A, Andersen AB, Gounon P, Daffé M. Molecular composition of the outermost capsular material of the tubercle bacillus. Microbiology 1995; 141:1609–1620.
80. Peterson PK, Gekker G, Hu S, Sheng WS, Anderson WR, Ulevitch RJ, et al. CD14 receptor–mediated uptake of nonopsonized *Mycobacterium tuberculosis* by human microglia. Infect Immun 1995; 63:1598–1602.
81. Schlesinger LS. Roles of complement receptors, the mannosyl-fucosyl receptor, and pulmonary surfactant protein, SP-A in phagocytosis of *Mycobacterium tuberculosis* by human macrophages (abstr). Clin Res 1992; 40:383A.
82. Hawgood S, Clements JA. Pulmonary surfactant and its apoproteins. J Clin Invest 1990; 86:1–6.
83. Haagsman HP. Surfactant proteins A and D. Biochem Soc Trans 1994; 22:100–106.
84. Tenner AJ, Robinson SL, Borchelt J, Wright JR. Human pulmonary surfactant protein (SP-A), a protein structurally homologous to C1q, can enhance FcR- and CR1-mediated phagocytosis. J Biol Chem 1989; 264:13923–13928.
85. Reyes JM, Putong PB. Association of pulmonary alveolar lipoproteinosis with mycobacterial infection. Am J Clin Pathol 1980; 74:478–485.
86. Phelps DS, Rose RM. Inceased recovery of surfactant protein A in AIDS-related pneumonia. Am Rev Respir Dis 1991; 143:1072–1075.
87. Schlesinger LS, Horwitz MA. Phagocytosis of leprosy bacilli is mediated by complement receptors CR1 and CR3 on human monocytes and complement component C3 in serum. J Clin Invest 1990; 85:1304–1314.
88. Launois P, Niang M, Dieye A, Sarthou J-L, Rivier F, Millan J. Human phagocyte respiratory burst by *Mycobacterium bovis* BCG and *M. leprae*: functional activation by BCG is mediated by complement and its receptors on monocytes. Int J Lepr Other Mycobact Dis 1992; 60:225–233.
89. Schlesinger LS, Horwitz MA. Phenolic glycolipid-1 of *Mycobacterium leprae* binds complement component C3 in serum and mediates phagocytosis by human monocytes. J Exp Med 1991; 174:1031–1038.
90. Bermudez LE, Young LS, Enkel H. Interaction of *Mycobacterium avium* complex with human macrophages: roles of membrane receptors and serum proteins. Infect Immun 1991; 59:1697–1702.
91. Roecklein JA, Swartz RP, Yeager H Jr. Nonopsonic uptake of *Mycobacterium avium* complex by human monocytes and alveolar macrophages. J Lab Clin Med 1992; 119:772–781.
92. Swartz RP, Naai D, Vogel CW, Yeager H Jr. Differences in uptake of mycobacteria by human monocytes: a role for complement. Infect Immun 1988; 56:2223–2227.
93. Rao SP, Gehlsen KR, Catanzaro A. Identification of a β_1 integrin on *Mycobacterium avium–Mycobacterium intracellulare*. Infect Immun 1992; 60:3652–3657.

94. Rao SP, Ogata K, Catanzaro A. *Mycobacterium avium–M. intracellulare* binds to the integrin receptor $\alpha_v\beta_3$ on human monocytes and monocyte-derived macrophages. Infect Immun 1993; 61:663–670.
95. Rao SP, Ogata K, Morris SL, Catanzaro A. Identification of a 68 kd surface antigen of *Mycobacterium avium* that binds to human macrophages. J Lab Clin Med 1994; 123:526–535.
96. Byrd SR, Gelber R, Bermudez LE. Roles of soluble fibronectin and β_1 integrin receptors in the binding of *Mycobacterium leprae* to nasal epithelial cells. Clin Immunol Immunopathol 1993; 69:266–271.
97. Ratliff TL, McCarthy R, Telle WB, Brown EJ. Purification of a mycobacterial adhesion for fibronectin. Infect Immun 1993; 61:1889–1894.
98. Abou-Zeid C, Garbe T, Lathigra R, Wiker HG, Harboe M, Rook GAW, et al. Genetic and immunological analysis of *Mycobacterium tuberculosis* fibronectin–binding proteins. Infect Immun 1991; 59:2712–2718.
99. Thole JER, Schöningh R, Janson AAM, Garbe T, Cornelisse YE, Clark-Curtiss JE, et al. Molecular and immunological analysis of a fibronectin-binding protein antigen secreted by *Mycobacterium leprae*. Mol Microbiol 1992; 6:153–163.
100. Pommier CG, Inada S, Fries LF, Takahashi T, Frank MM, Brown EJ. Plasma fibronectin enhances phagocytosis of opsonized particles by human peripheral blood monocytes. J Exp Med 1983; 157:1844–1854.
101. Wright SD, Detmers PA, Jong MTC, Meyer BC. Interferon-γ depresses binding of ligand by C3b and C3bi receptors on cultured human monocytes, an effect reversed by fibronectin. J Exp Med 1986; 163:1245–1259.
102. Mokoena T, Gordon S. Human macrophage activation: modulation of mannosyl, fucosyl receptor activity in vitro by lymphokines, gamma and alpha interferons, and dexamethasone. J Clin Invest 1987; 75:624–631.
103. Toba H, Crowford JT, Ellner JJ. Pathogenicity of *Mycobacterium avium* for human monocytes: absence of macrophage activating factor activity of gamma interferon. Infect Immun 1989; 57:239–244.
104. Douvas GS, Looker DL, Vatter AE, Crowle AJ. Gamma interferon activates human macrophages to become tumoricidal and leishmanicidal but enhances replication of macrophage-associated mycobacteria. Infect Immun 1985; 50:1–8.
105. Barnes PF, Modlin RL, Ellner JJ. T-cell responses and cytokines. In: Bloom BR, ed. Tuberculosis: Pathogenesis, Protection, and Control. Washington, DC: ASM Press, 1994:417–435.
106. Schreiber S, Perkins SL, Teitelbaum SL, Chappel J, Stahl PD, Blum JS. Regulation of mouse bone marrow macrophage mannose receptor expression and activation by prostaglandin E and IFN-γ. J Immunol 1993; 151:4973–4981.
107. Stein M, Keshav S, Harris N, Gordon S. Interleukin 4 potently enhances murine macrophage mannose receptor activity: a marker of alternative immunologic macrophage activation. J Exp Med 1992; 176:287–292.
108. Abramson SL, Gallin JI. IL-4 inhibits superoxide production by human mononuclear phagocytes. J Immunol 1990; 144:625–630.
109. Hirsch CS, Yoneda T, Averill L, Ellner JJ, Toossi Z. Enhancement of intracellular growth of *Mycobacterium tuberculosis* in human monocytes by transforming growth factor-β1. J Infect Dis 1994; 170:1229–1237.

110. Bardana EJJ, McClatchy JK, Farr RS, Minden P. Universal occurrence of antibodies to tubercle bacilli in sera from non-tuberculous and tuberculous individuals. Clin Exp Immunol 1973; 13:65–77.
111. Schlesinger LS, Horwitz MA. A role for natural antibody in the pathogenesis of leprosy: antibody in nonimmune serum mediates C3 fixation to the *Mycobacterium leprae* surface and hence phagocytosis by human mononuclear phagocytes. Infect Immun 1994; 62:280–289.
112. Forget A, Benoit JC, Turcotte R, Gusew-Chartrand N. Enhanced activity of anti-mycobacterial sera in experimental *Mycobacterium bovis* (BCG) infection in mice. Infect Immun 1976; 13:1301–1306.
113. Armstrong JA, Hart PD. Phagosome-lysosome interactions in cultured macrophages infected with virulent tubercle bacilli: reversal of the usual nonfusion pattern and observations on bacterial survival. J Exp Med 1975; 142:1–16.
114. De L, Costello AM, Kumar A, Narayan V, Akbar MS, Ahmed S, Abou-Zeid C, et al. Does antibody to mycobacterial antigens, including lipoarabinomannan, limit dissemination in childhood tuberculosis. Trans R Soc Trop Med Hyg 1992; 86:686–692.
115. Blair AL, Cree IA, Beck JS, Grange JM, Kardjito T. Heat-stable opsonins in tuberculosis and leprosy. FEMS Immunol Med Microbiol 1993; 7:197–204.
116. Stach J-L, Gros P, Forget A, Skamene E. Phenotypic expression of genetically-controlled natural resistance to *Mycobacterium bovis* (BCG). J Immunol 1984; 132: 888–892.
117. Radzioch D, Hudson T, Boule M, Barrera L, Urbance JW, Varesio L, et al. Genetic resistance/susceptibility to mycobacteria: phenotypic expression in bone marrow derived macrophage lines. J Leukoc Biol 1991; 50:263–272.
118. De Chastellier C, Fréhel C, Offredo C, Skamene E. Implication of phagosome-lysosome fusion in restriction of *Mycobacterium avium* growth in bone marrow macrophages from genetically resistant mice. Infect Immun 1993; 61:3775–3784.
119. Ross GD, Cain JA, Lachmann PJ. Membrane complement receptor type three (CR3) has lectin-like properties analogous to bovine conglutinin and functions as a receptor for zymosan and rabbit erythrocytes as well as a receptor for iC3b. J Immunol 1985; 134:3307–3315.
120. Elstad MR, Parker CJ, Cowley FS, Wilcox LA, McIntyre TM, Prescott SM, et al. CD11b/CD18 integrin and a β-glucan receptor act in concert to induce the synthesis of platelet-activating factor by monocytes. J Immunol 1994; 152:220–230.
121. Douvas GS, Berger EM, Repine JE, Crowle AJ. Natural mycobacteriostatic activity in human monocyte-derived adherent cells. Am Rev Respir Dis 1986; 134:44–48.
122. Holzer TJ, Nelson KE, Schauf V, Crispen RG, Anderson BR. *Mycobacterium leprae* fails to stimulate phagocytic cell superoxide anion generation. Infect Immun 1986; 51:514–520.
123. Wayne S, Denning G, Kusner DJ, Kaufman TM, Schlesinger LS. Phagocytosis of virulent and attenuated strains of *M. tuberculosis* by human macrophages does not generate detectable superoxide anion or hydrogen peroxide (abstr). Clin Res 1995; 43:219A.
124. Launois P, Blum L, Dieye A, Millan J, Sarthous JL, Bach MA. Phenolic glycolipid-1 from *M. leprae* inhibits oxygen free radical production by human mononuclear cells. Res Immunol 1989; 140:847–855.

125. Neill MA, Klebanoff SJ. The effect of phenolic glycolipid 1 from *Mycobacterium leprae* on the antimicrobial activity of human macrophages. J Exp Med 1988; 167: 30–42.
126. Chan J, Fujiwara T, Brennan P, McNeil M, Turco SJ, Sibille J-C, et al. Microbial glycolipids: possible virulence factors that scavenge oxygen radicals. Proc Natl Acad Sci USA 1989; 86:2453–2457.
127. Chan J, Fan X, Hunter SW, Brennan PJ, Bloom BR. Lipoarabinomannan, a possible virulence factor involved in persistence of *Mycobacterium tuberculosis* with macrophages. Infect Immun 1991; 59:1755–1761.
128. Vachula M, Holzer TJ, Anderson BR. Suppression of monocyte oxidative response by phenolic glycolipid I of *Mycobacterium leprae*. J Immunol 1989; 142:1696–1701.
129. Alavi MR, Affronti LF. Induction of mycobacterial proteins during phagocytosis and heat shock: A time interval analysis. J Leukoc Biol 1994; 55:633–641.
130. Plum G, Clark-Curtiss JE. Induction of *Mycobacterium avium* gene expression following phagocytosis by human macrophages. Infect Immun 1994; 62:476–483.
131. Mundayoor S, Shinnick TM. Identification of genes involved in the resistance of mycobacteria to killing by macrophages. Ann NY Acad Sci 1994; 730:26–36.
132. Fallman M, Gullberg M, Hellberg C, Andersson T. Complement receptor-mediated phagocytosis is associated with accumulation of PC-deriveid DG in human PMN. J Biol Chem 1992; 267:2656–2663.
133. Kusner DJ, Hall CF, Schlesinger LS. Activation of macrophage phospholipase D is linked to the phagocytosis of *M. tuberculosis* (abstr). Clin Res 1995; 43:356A.
134. Reiner NE. Altered cell signaling and mononuclear phagocyte deactivation during intracellular infection. Immunol Today 1994; 15:374–381.
135. Brozna JP, Horan M, Rademacher JM, Pabst KA, Pabst MJ. Monocyte responses to sulfatide from *Mycobacterium tuberculosis*: inhibition of priming for enhanced release of superoxide, associated with increased secretion of interleukin-1 and tumor necrosis factor α, and altered protein phosphorylation. Infect Immun 1991; 59: 2542–2548.
136. Pabst JJ, Gross JM, Prozna JP, Goren MB. Inhibition of macrophage priming by sulfatide from *Mycobacterium tuberculosis*. J Immunol 1988; 140:634–640.
137. Brown MC, Taffet SM. Lipoarabinomannans derived from different strains of *Mycobacterium tuberculosis* differentially stimulate the activation of NF-kappaB and KBF1 in murine macrophages. Infect Immun 1995; 63:1960–1968.
138. Sibley LD, Hunter SW, Brennan PJ, Krahenbuhl JL. Mycobacterial lipoarabinomannan inhibits gamma interferon-mediated activation of macrophages. Infect Immun 1988; 56:1232–1236.
139. Roach TIA, Chatterjee D, Blackwell JM. Induction of early-response genes KC and JE by mycobacterial lipoarabinomannans: regulation of KC expression in murine macrophages by Lsh/Ity/Bcg (candidate Nramp). Infect Immun 1994; 62:1176–1184.
140. Zhang Y, Broser M, Rom WN. Activation of the interleukin 6 gene by *Mycobacterium tuberculosis* or lipopolysaccharide is mediated by nuclear factors NF-IL6 and NF-kappaB. Proc Natl Acad Sci USA 1994; 91:2225–2229.
141. Moreno C, Mehlert A, Lamb J. The inhibitory effects of mycobacterial lipoarabino-

mannan and polysaccharides upon polyclonal and monoclonal human T cell proliferation. Clin Exp Immunol 1988; 74:206–210.
142. Kaplan G, Gandhi RR, Weinstein DE, Levis WR, Patarroyo ME, Brennan PJ, et al. *Mycobacterium leprae* antigen-induced suppression of T cell proliferation in vitro. J Immunol 1987; 138:3028–3034.
143. Chujor CSN, Kuhn B, Schwerer B, Bernheimer H, Levis WR, Bevec D. Specific inhibition of mRNA accumulation for lymphokines in human T cell line Jurkat by mycobacterial lipoarabinomannan antigen. Clin Exp Immunol 1992; 87:398–403.
144. Moreno C, Taverne J, Mehlert A, Bate CAW, Brealey RJ, Meager A, et al. Lipoarabinomannan from *Mycobacterium tuberculosis* induces the production of tumour necrosis factor from human and murine macrophages. Clin Exp Immunol 1989; 76:240–245.
145. Chatterjee D, Roberts AD, Lowell K, Brennan PJ, Orme IM. Structural basis of capacity of lipoarabinomannan to induce secretion of tumor necrosis factor. Infect Immun 1992; 60:1249–1253.
146. Anthony LSD, Chatterjee D, Brennan PJ, Nano FE. Lipoarabinomannan from *Mycobacterium tuberculosis* modulates the generation of reactive nitrogen intermediates by gamma interferon-activated macrophages. FEMS Immunol Med Microbiol 1994; 8:299–306.
147. Roach TIA, Barton CH, Chatterjee D, Blackwell JM. Macrophage activation: lipoarabinomannan from avirulent and virulent strains of *Mycobacterium tuberculosis* differentially induces the early genes c-fos, KC, JE, and tumor necrosis factor-α. J Immunol 1993; 150:1886–1896.
148. Barnes PF, Chatterjee D, Abrams JS, Lu S, Wang E, Yamamura M, et al. Cytokine production induced by *Mycobacterium tuberculosis* lipoarabinomannan: Relationship to chemical structure. J Immunol 1992; 149:541–547.
149. Dahl KE, Shiratsuchi H, Hamilton BD, Ellner JJ, Toossi Z. Selective induction of transforming growth factor β in human monocytes by lipoarabinomannan of *Mycobacterium tuberculosis*. Infect Immun 1996; 64:399–405.
150. Barnes PF, Fong S-J, Brennan PJ, Twomey PE, Mazumder A, Modlin RL. Local production of tumor necrosis factor and INF-γ in tuberculous pleuritis. J Immunol 1990; 145:149–154.
151. Schuller-Levis GB, Levis WR, Ammazzalorso M, Nosrati A, Park E. Mycobacterial lipoarabinomannan induces nitric oxide and tumor necrosis factor α production in a macrophage cell line: Down regulation by taurine chloramine. Infect Immun 1994; 62:4671–4674.
152. Armstrong JA, Hart PD. Response of cultured macrophages to *Mycobacterium tuberculosis*, with observations on fusion of lysosomes with phagosomes. J Exp Med 1971; 134:713–740.
153. Rastogi N, Labrousse V, Paulo Carvalho de Sousa J. Mycobacterial growth and ultrastructure in mouse L-929 fibroblasts and bone marrow-derived macrophages: Evidence that infected fibroblasts secrete mediators capable of modulating bacterial growth in macrophages. Curr Microbiol 1992; 25:203–213.
154. Sturgill-Koszycki S, Schlesinger PH, Chakraborty P, Haddix PL, Collins HL, Fok AK, et al. Lack of acidification in Mycobacterium phagosomes produced by exclusion of the vesicular proton-ATPase. Science 1994; 263:678–681.

155. Xu S, Cooper A, Sturgill-Koszycki S, Van Heyningen T, Chatterjee D, Orme I, et al. Intracellular trafficking in *Mycobacterium tuberculosis* and *Mycobacterium avium*–infected macrophages. J Immunol 1994; 153:2568–2578.
156. Crowle AJ, Dahl R, Ross E, May MH. Evidence that Vesicles containing living, virulent *Mycobacterium tuberculosis* or *Mycobacterium avium* in cultured human macrophages are not acidic. Infect Immun 1991; 59:1823–1831.
157. Clemens DL, Horwitz MA. Characterization of the *Mycobacterium tuberculosis* phagosome and evidence that phagosomal maturation is inhibited. J Exp Med 1995; 181:257–270.
158. Frehel C, Rastogi N. *Mycobacterium leprae* surface components intervene in the early phagosome-lysosome fusion inhibition event. Infect Immun 1987; 55:2916–2921.
159. Sibley LD, Franzblau SG, Krahenbuhl JL. Intracellular fate of *Mycobacterium leprae* in normal and activated mouse macrophages. Infect Immun 1987; 55:680–685.
160. Frehel C, De Chastellier C, Lang T, Rastogi N. Evidence for inhibition of fusion of lysosomal and prelysosomal compartments with phagosomes in macrophages infected with pathogenic *Mycobacterium avium*. Infect Immun 1986; 52:252–262.
161. Hart PD, Armstrong JA. Strain virulence and the lysosomal response in macrophages infected with *Mycobacterium tuberculosis*. Infect Immun 1974; 10:742–746.
162. McDonough KA, Kress Y, Bloom BR. Pathogenesis of tuberculosis: interaction of *Mycobacterium tuberculosis* with macrophages. Infect Immun 1993; 61:2763–2773.
163. Myrvik QN, Leakes ES, Wright MJ. Disruption of phagosomal membranes of normal alveolar macrophages by the H37Rv strain of *Mycobacterium tuberculosis*. A correlate of virulence. Am Rev Respir Dis 1984; 129:322–328.
164. Goren MB, Hart PD, Young MR, Armstrong JA. Prevention of phagosome-lysosome fusion in cultured macrophages by sulfatides of *Mycobacterium tuberculosis*. Proc Natl Acad Sci USA 1976; 73:2510–2514.
165. Goren MB, Brokl O, Schaefer WB. Lipids of putative relevance to virulence in *Mycobacterium tuberculosis*: correlation of virulence with elaboration of sulfatides and strongly acidic lipids. Infect Immun 1974; 9:142–149.
166. Crowe ML, Spargo BJ, Ioneda T, Beaman BL, Crowe JH. Interaction of cord factor (α,α'-trehalose-6,6′-dimycolate) with phospholipids. Biochim Biophys Acta 1994; 1194:53–60.
167. Spargo BJ, Crowe LM, Ioneda T, Beaman BL, Crowe JH. Cord factor (α,α'-trehalose 6,6′-dimycolate) inhibits fusion between phospholipid vesicles. Proc Natl Acad Sci USA 1991; 88:737–740.
168. Gordon AH, D'Arcy Hart P, Young MR. Ammonia inhibits phagosome-lysosome fusion in macrophages. Nature 1980; 286:79–80.
169. Clemens DL, Lee BY, Horwitz MA. Purification, characterization, and genetic analysis of *Mycobacterium tuberculosis* urease, a potentially critical determinant of host-pathogen interaction. J Bacteriol 1995; 117:5644–5652.
170. Harth G, Clemens DL, Horwitz MA. Glutamine synthetase of *Mycobacterium tuberculosis*: extracellular release and characterization of its enzymatic activity. Proc Natl Acad Sci USA 1994; 91:9342–9346.
171. Hsu HS. The fate of *Mycobacterium tuberculosis* within macrophages of guinea pigs. Am Rev Respir Dis 1971; 103:607–611.

172. Alsaadi A-I, Smith DW. The fate of virulent and attenuated mycobacteria in guinea pigs infected by the respiratory route. Am Rev Respir Dis 1973; 107:1041–1046.
173. Steenken W Jr, Oatway WH Jr, Petroff SA. Biological studies of the tubercle bacillus. III. Dissociation and pathogenicity of the R and S varients of the human tubercle bacillus (H37). J Exp Med 1934; 60:515–540.
174. Warwick-Davies J, Dhillon J, O'Brien L, Andrew PW, Lowrie DB. Apparent killing of *Mycobacterium tuberculosis* by cytokine-activated human monocytes can be an artifact of a cytotoxic effect on the monocytes. Clin Exp Immunol 1994; 96:214–217.
175. Chan J, Xing Y, Magliozzo RS, Bloom BR. Killing of virulent *Mycobacterium tuberculosis* by reactive nitrogen intermediates produced by active murine macrophages. J Exp Med 1992; 175:1111–1122.
176. Sheppard CC. A comparison of the growth of selected mycobacteria in HeLa, monkey kidney, and human amnion cells in tissue culture. J Exp Med 1958; 107: 237–245.
177. Vidal SM, Malo D, Vogan K, Skamene E, Gros P. Natural resistance to infection with intracellular parasites: Isolation of a candidate for Bcg. Cell 1993; 73:469–485.
178. Cellier M, Govoni G, Vidal S, Kwan T, Groulx N, Liu J, et al. Human natural resistance-associated macrophage protein: cDNA cloning, chromosomal mapping, genomic organization, and tissue-specific expression. J Exp Med 1994; 180:1741–1752.
179. De Sousa JPC, Rastogi N. Comparative ability of human monocytes and macrophages to control the intracellular growth of *Mycobacterium avium* and *Mycobacterium tuberculosis*: effect of interferon-γ and indomethacin. FEMS Microbiol Immunol 1992; 89:329–334.
180. Crowle AJ, Ross EJ, May MH. Inhibition by 1,25(OH)-vitamin D of the multiplication of virulent tubercle baccili in cultured human macrophages. Infect Immun 1987; 55:2945–2950.
181. North RJ, Izzo AA. Mycobacterial virulence. Virulent strains of *Mycobacterium tuberculosis* have faster in vivo doubling times and are better equipped to resist growth-inhibiting functions of macrophages in the presence and absence of specific immunity. J Exp Med 1993; 177:1723–1733.
182. Brown DH, Miles BA Zwilling BS. Growth of *Mycobacterium tuberculosis* in BCG-resistant and -susceptible mice: establishment of latency and reactivation. Infect Immun 1995; 63:2243–2247.
183. Ordway DJ, Sonnenberg MG, Donahue SA, Belisle JT, Orme IM. Drug-resistant strains of *Mycobacterium tuberculosis* exhibit a range of virulence for mice. Infect Immun 1995; 63:741–743.
184. Nathan C, Xie Q-W. Regulation of biosynthesis of nitric oxide. J Biol Chem 1994; 269:13725–13728.
185. Flesch I, Kaufmann SHE. Mycobacterial growth inhibition by interferon-gamma–activated bone marrow macrophages and differential susceptibility among strains of *Mycobacterium tuberculosis*. J Immunol 1987; 138:4408–4413.
186. Rook G, Steele J, Ainsworth M, Champion B. Activation of macrophages to inhibit proliferation of *Mycobacterium tuberculosis* comparison of the effects of recombinant γ-interferon on human monocytes and murine peritoneal macrophages. Immunology 1986; 59:333–338.

187. Steele J, Flint KC, Pozniak AL, Hudspith B, Johnson MM, Rook GAW. Inhibition of virulent *Mycobacterium tuberculosis* by murine peritoneal macrophages and human alveolar lavage cells: the effects of lymphokines and recombinant γ-interferon. Tubercle 1986; 67:289–294.
188. Cooper AM, Dalton DK, Stewart TA, Griffin JP, Russell DG, Orme IM. Disseminated tuberculosis in interferon-γ gene-disrupted mice. J Exp Med 1993; 178:2243–2247.
189. Flynn JL, Chan J, Triebold KJ, Dalton DK, Stewart TA, Bloom BR. An essential role for interferon-γ in resistance to *Mycobacterium tuberculosis* infection. J Exp Med 1993; 178:2249–2254.
190. Liu J, Rosenberg EY, Nikaido H. Fluidity of the lipid domain of cell wall from *Mycobacterium chelonae*. Proc Natl Acad Sci USA 1995; 92:11254–11258.
191. Rook GA, Steele J, Fraher L, Barker S, Karmali R, O'Riordan J, et al. Vitamin D, γ-interferon, and control of proliferation of *Mycobacterium tuberculosis* by human monocytes. Immunology 1986; 57:159–163.
192. Denis M. Killing of *Mycobacterium tuberculosis* within human monocytes: activation by cytokines and calcitriol. Clin Exp Immunol 1991; 84:200–206.
193. Molloy A, Laochumroonvorapong P, Kaplan G. Apoptosis, but not necrosis, of infected monocytes is coupled with killing of intracellular bacillus Calmette-Guérin. J Exp Med 1994; 180:1499–1509.
194. Sieling PA, Chatterjee D, Porcelli SA, Prigozy TI, Mazzaccaro RJ, Soriano T, et al. CD1-restricted T cell recognition of microbial lipoglycan antigens. Science 1995; 269:227–230.
195. Beckman EV, Porcelli SA, Morita CT, Behar SM, Furlong ST, Brenner MB. Recognition of a lipid antigen by CD1-restricted αβ T cells. Nature 1994; 372:691–694.
196. Keller R, Keist R, Joller P. Macrophage response to microbial pathogens: modulation of the expression of adhesion, CD14, and MHC class II molecules by viruses, bacteria, protozoa and fungi. Scand J Immunol 1995; 42:337–344.
197. Pancholi P, Mirza A, Bhardwaj N, Steinman RM. Sequestration from immune CD4 T cells of mycobacteria growing in human macrophages. Science 1993; 260: 984–986.
198. Gercken J, Pryjma J, Ernst M, Flad H-D. Defective antigen presentation by *Mycobacterium tuberculosis*-infected monocytes. Infect Immun 1994; 62:3472–3478.
199. Wadee AA, Kuschke RH, Dooms TG. The inhibitory effects of *Mycobacterium tuberculosis* on MHC class II expression by monocytes activated with riminophenazines and phagocyte stimulants. Clin Exp Immunol 1995; 100:434–439.
200. Saha B, Das G, Vohra H, Ganguly NK, Mishra GC. Macrophage–T cell interaction in experimental mycobacterial infection. Selective regulation of co-stimulatory molecules on *Mycobacterium*-infected macrophages and its implication in the suppression of cell-mediated immune response. Eur J Immunol 1994; 24:2618–2624.
201. López Ramírez GM, Rom WN, Ciotoli C, Talbot A, Martiniuk F, Cronstein B, et al. *Mycobacterium tuberculosis* alters expression of adhesion molecules on monocytic cells. Infect Immun 1994; 62:2515–2520.
202. Potts BC, Fogarty SJ, Yi H, Schlesinger LS. *M. tuberculosis* within human macrophages is accompanied by altered levels of β2 integrin steady state mRNA and

surface protein (abstr). Thirtieth US-Japan Joint Conference on Tuberculosis and Leprosy, Fort Collins, Colorado, July 19–21, 1995.
203. Mariano M, Nikitin T, Malucelli BE. Phagocytic potential of macrophages from within delayed hypersensitivity-mediated granulomata. J Pathol 1977; 123:27–33.
204. Kaye PM, Sims M, Feldmann M. Regulation of macrophage accessory cell activity by mycobacteris. II. In vitro inhibition of Ia expression by *Mycobacterium microti*. Clin Exp Immunol 1986; 64:28–34.
205. Tsuyuguchi I, Kawasumi H, Takashima T, Tsuyuguchi T, Kishimoto S. *Mycobacterium avium–Mycobacterium intracellulare* complex-induced suppression of T-cell proliferationin vitro by regulation of monocyte accessory cell activity. Infect Immun 1990; 58:1369–1378.
206. Mshana RN, Hastings RC, Krahenbuhl JL. Infection with live mycobacteria inhibits in vitro detection of Ia antigen on macrophages. Immunobiology 1988; 177:40–54.
207. Schruers J. Cytokines in infectious disease: the battle between host and pathogen. Curr Opin Biotechnol 1993; 4:727–733.
208. Cooper AM, Flynn JL. The protective immune response to *Mycobacterium tuberculosis*. Curr Opin Immunol 1995; 7:512–516.
209. Wallis RS, Ellner JJ. Cytokines and tuberculosis. J Leukoc Biol 1994; 55:676–681.
210. Champsi JH, Bermudez LE, Young LS. The role of cytokines in mycobacterial infection. Biotherapy 1994; 7:187–193.
211. Wahl SM. Transforming growth factor β: the good, the bad, and the ugly. J Exp Med 1994; 180:1587–1590.
212. Bermudez LE. Production of transforming growth factor-β by *Mycobacterium avium*-infected human macrophages is associated with unresponsiveness to IFN-γ. J Immunol 1993; 150:1838–1845.
213. Toossi Z, Young T-G, Averill LE, Hamilton BD, Shiratsuchi H, Ellner JJ. Induction of transforming growth factor β1 by purified protein derivative of *Mycobacterium tuberculosis*. Infect Immun 1995; 63:224–228.
214. Toossi Z, Gogate P, Shiratsuchi H, Young T, Ellner JJ. Enhanced production of TGF-β by blood monocytes from patients with active tuberculosis and presence of TGF-β in tuberculous granulomatous lung lesions. J Immunol 1995; 154:465–473.
215. Boutard V, Havouis R, Fouqueray B, Philippe C, Moulinoux J-P, Baud L. Transforming growth factor-β stimulates arginase activity in macrophages: implications for the regulation of macrophage cytotoxicity. J Immunol 1995; 155:2077–2084.
216. Vodovotz Y, Bogdan C, Paik J, Xie Q, Nathan C. Mechanisms of suppression of macrophage nitric oxide release by transforming growth factor β. J Exp Med 1993; 178:605–613.
217. Bermudez LE, Champsi J. Infection with *Mycobacterium avium* induces production of interleukin-10 (IL-10), and administration of anti-IL-10 antibody is associated with enhanced resistance to infection in mice. Infect Immun 1993; 61:3093–3097.
218. Barnes PF, Lu S, Abrams JS, Wang E, Yamamura M, Modlin RL. Cytokine production at the site of disease in human tuberculosis. Infect Immun 1993; 61:3482–3489.
219. Rostogi N, Bachelet M, De Sousa JPC. Intracellular growth of *Mycobacterium avium* in human macrophages is linked to the increased synthesis of prostaglandin E and inhibition of the phagosome-lysosome fusions. FEMS Microbiol Immunol 1992; 89:273–280.

220. Kleinhenz ME, Ellner JJ, Spagnuolo PJ, Daniel TM. Suppression of lymphocyte responses by tuberculous plasma and mycobacterial arabinogalactan. Monocyte dependence and indomethacin reversibility. J Clin Invest 1981; 68:153–162.
221. Cadranel J, Philippe C, Perez J, Milleron B, Akoun G, Ardaillou R, et al. In vitro production of tumour necrosis factor and prostaglandin E_2 by peripheral blood mononuclear cells from tuberculosis patients. Clin Exp Immunol 1990; 81:319–324.
222. Kunkel SL, Spengler M, May MA, Spengler R, Larrick J, Remick D. Prostaglandin E_2 regulates macrophage-derived tumor necrosis factor gene expression. J Biol Chem 1988; 263:5380.
223. Spengler RN, Spengler ML, Strieter RM, Remick DG, Larrick JW, Kunkel SL. Modulation of tumor necrosis factor-α gene expression. Desensitization of prostaglandin E_2–induced suppression. J Immunol 1989; 142:4346.
224. Friedland JS. Cytokines, phagocytosis, and *Mycobacterium tuberculosis*. Lymphokine Cytokine Res 1993; 12:127–133.
225. Wallis RS, Amir-Tahmasseb M, Ellner JJ. Induction of inteleukin 1 and tumor necrosis factor by mycobacterial proteins: the monocyte western blot. Proc Natl Acad Sci USA 1990; 87:3348–3352.
226. Valone SE, Rich EA, Wallis RS, Ellner JJ. Expression of tumor necrosis factor in vitro by human mononuclear phagocytes stimulated with whole *Mycobacterium bovis* BCG and mycobacterial antigens. Infect Immun 1988; 56:3313.
227. Falcone V, Bassey EB, Toniolo A, Conaldi PG, Collins FM. Differential release of tumor necrosis factor-α from murine peritoneal macrophages stimulated with virulent and avirulent species of mycobacteria. FEMS Immunol Med Microbiol 1994; 8: 225–232.
228. Appelberg R. Protective role of interferon-γ, tumor necrosis factor α and interleukin-6 in *Mycobacterium tuberculosis* and *M. avium* infections. Immunobiology 1994; 191:520–525.
229. Denis M. Involvement of cytokines in determining resistance and acquired immunity in murine tuberculosis. J Leukoc Biol 1991; 50:495–501.
230. Kindler V, Sappino A-P, Grau GE, Piguet P-F, Vassalli P. The inducing role of tumor necrosis factor in the development of bactericidal granulomas during BCG infection. Cell 1989; 56:731–740.
231. Kasahara K, Kobayashi K, Shikama Y, Yoneya I, Kaga S, Hashimoto M, et al. The role of monokines in granuloma formation in mice: the ability of interleukin 1 and tumor necrosis factor-α to induce lung granulomas. Clin Immunol Immunopathol 1989; 51:419–425.
232. Filley EA, Rook GAW. Effect of mycobacteria on sensitivity to the cytotoxic effects of tumor necrosis factor. Infect Immun 1991; 59:2567–2572.
233. Peetermans WE, Raats CJI, Langermans JAM, Van Furth R. Mycobacterial heat-shock protein 65 induces proinflammatory cytokines but does not activate human mononuclear phagocytes. Scand J Immunol 1994; 39:613–617.
234. Cadranel J, Philippe C, Philippe B, Milleron B, Fouqueray B, Mayaud C, et al. Increased expression and occupancy of receptors for tumour necrosis factor on blood monocytes from tuberculosis patients. Clin Exp Immunol 1993; 94:51–56.
235. Takashima T, Ueta C, Tsuyuguchi I, Kishimoto S. Production of tumor necrosis

factor α by monocytes from patients with pulmonary tuberculosis. Infect Immun 1990; 58:3286–3292.

236. Rydberg J, Miorner H, Chandramuki A, Lantz M. Assessment of a possible imbalance between tumor necrosis factor (TNF) and soluble TNF receptor forms in tuberculous infection of the central nervous system. J Infect Dis 1995; 172:301–304.

237. Zhang M, Gately MK, Wang E, Gong J, Wolf SF, Lu S, et al. Interleukin 12 at the site of disease in tuberculosis. J Clin Invest 1994; 93:1733–1739.

238. Flesch IEA, Hess JH, Huang S, Aguet M, Rothe J, Bluethmann H, et al. Early interleukin 12 production by macrophages in response to mycobacterial infection depends on interferon-γ and tumor necrosis factor α. J Exp Med 1995; 181:1615–1621.

239. Cooper AM, Roberts AD, Rhoades ER, Callahan JE, Getzy DM, Orme IM. The role of interleukin-12 in acquired immunity to *Mycobacterium tuberculosis* infection. Immunology 1995; 84:423–432.

240. Flynn JL, Goldstein MM, Triebold KJ, Sypek J, Wolf S, Bloom BR. IL-12 increases resistance of BALB/c mice to *Mycobacterium tuberculosis* infection. J Immunol 1995; 155:2515–2524.

241. Friedland JS. Chemotactic cytokines and tuberculosis. Biochem Soc Trans 1994; 22:310–312.

242. Friedland JS, Remick DG, Shattock R, Griffin GE. Secretion of interleukin-8 following phagocytosis of *Mycobacterium tuberculosis* by human monocyte cell lines. Eur J Immunol 1992; 22:1373–1378.

243. Friedland JS, Shattock RJ, Griffin GE. Phagocytosis of *Mycobacterium tuberculosis* or particulate stimuli by human monocytic cells induces equivalent monocyte chemoattractant protein-1 gene expression. Cytokine 1993; 5:150–156.

244. Zhang Y, Broser M, Cohen H, Bodkin M, Law K, Reibman J, et al. Enhanced interleukin-8 release and gene expression in macrophages after exposure to *Mycobacterium tuberculosis* and its components. J Clin Invest 1995; 95:586–592.

245. Kasahara K, Tobe T, Tomita M, Mukaida N, Shao-Bo S, Matsushima K, et al. Selective expression of monocyte chemotactic and activating factor/monocyte chemoattractant protein 1 in human blood monocytes by *Mycobacterium tuberculosis*. J Infect Dis 1994; 170:1238–1247.

246. Kasama T, Strieter RM, Standiford TJ, Burdick MD, Kunkel SL. Expression and regulation of human neutrophil-derived macrophage inflammatory protein 1α. J Exp Med 1993; 178:63–72.

247. Antony VB, Godbey SW, Kunkel SL, Hott JW, Hartman DL, Burdick MD, et al. Recruitment of inflammatory cells to the pleural space. Chemotactic cytokines, IL-8, and monocyte chemotactic peptide-1 in human pleural fluids. J Immunol 1993; 151:7216–7223.

248. Rhoades ER, Cooper AM, Orme IM. Chemokine response in mice infected with *Mycobacterium tuberculosis*. Infect Immun 1995; 63:3871–3877.

249. Chensue SW, Davey MP, Remick DG, Kunkel SL. Release of interleukin-1 by peripheral blood mononuclear cells in patients with tuberculosis and active inflammation. Infect Immun 1986; 52:341–343.

250. Chu CQ, Field M, Andrew E, Haskard D, Feldmann M, Maini RN. Detection of

cytokines at the site of tuberculin-induced delayed-type hypersensitivity in man. Clin Exp Immunol 1992; 90:522–529.
251. Kaplan G. Cytokine regulation of disease progression in leprosy and tuberculosis. Immunobiology 1994; 191:564–568.
252. Wallis RS, Vjecha M, Amir-Tahmasseb M, Okwera A, Byekwaso F, Nyole S, et al. Influence of tuberculosis on human immunodeficiency virus (HIV-1): enhanced cytokine expression and elevated β-microglobulin in HIV-1-associated tuberculosis. J Infect Dis 1993; 167:43–48.
253. Toossi Z, Sierra-Madero JG, Blinkhorn RA, Mettler MA, Rich EA. Enhanced susceptibility of blood monocytes from patients with pulmonary tuberculosis to productive infection with human immunodeficiency virus type 1. J Exp Med 1993; 177:1511–1516.
254. Zhang M, Gong J, Iyer DV, Jones BE, Modlin RL, Barnes PF. T cell cytokine responses in persons with tuberculosis and human immunodeficiency virus infection. J Clin Invest 1994; 94:2435–2442.
255. Shattock RJ, Friedland JS, Griffin GE. Phagocytosis of *Mycobacterium tuberculosis* modulates human immunodeficiency virus replication in human monocytic cells. J Gen Virol 1994; 75:849–856.
256. Shiratsuchi H, Johnson JL, Toossi Z, Ellner JJ. Modulation of the effector function of human monocytes for *Mycobacterium avium* by human immunodeficiency virus-1 envelope glycoprotein gp120. J Clin Invest 1994; 93:885–891.
257. Palmer S, Hamblin AS. Increased CD11/CD18 expression on the peripheral blood leucocytes of patients with HIV disease: relationship to disease severity. Clin Exp Immunol 1993; 93:344–349.
258. Meignan J, Guillon JM, Denis M, Joly P, Rosso J, Carette MF, et al. Increased lung epithelial permeability of HIV-infected patients with isolated cytotoxic T-lymphocytic alveolitis. Am Rev Respir Dis 1990; 141:1241–1248.

18

Macrophage Function in *Pneumocystis carinii* Infection

HENRY KOZIEL

Harvard Medical School
and Deaconess Hospital
Boston, Massachusetts

JUDD E. SHELLITO

Louisiana State University Medical Center
New Orleans, Louisiana

I. Introduction

Pneumocystis carinii is a unique extracellular opportunistic pulmonary pathogen that produces pneumonia in immunocompromised hosts. In general, infection with this organism remains localized to the alveolar surface of the lungs. At this anatomical locus, alveolar macrophages represent the primary defenders against potential infectious agents. For this reason, characterizing the interplay of alveolar macrophages with *P. carinii* is an important step in understanding the pathogenesis of infection and the pathophysiology of established disease. This in turn may lead to the development of new strategies for the prevention and treatment of *P. carinii* pneumonia based in part on macrophage-directed interventions.

An important role for alveolar macrophages in *P. carinii* pneumonia is suggested by observations from animal models of *P. carinii* pneumonia. The foamy alveolar exudate characteristic of *P. carinii* pneumonia may in part to be composed of coalesced macrophages (1). In the chronic corticosteroid-treated rat model of *P. carinii* pneumonia, microscopic examination of the alveoli of infected animals reveals a paucity of macrophages. However, withdrawal of chronic daily steroid administration results in an influx of alveolar macrophages and the resolution of pneumonia (2). The CD4 T-immunocompetent mice challenged with intratracheal *P. carinii* are able to clear the inoculated organisms, and clearance is accompanied by a twofold rise in alveolar macrophage counts (3). In contrast, CD4 T-lymphocyte–depleted animals challenged with intratracheal *P. carinii* do not demonstrate an influx of macrophages and the animals develop pneumonia. In the SCID mouse model of *P. carinii* pneumonia, the clearance of *P. carinii* in immune reconstituted mice is preceded by marked increases in measurable levels of interleukin-1 and tumor necrosis factor-α, both of which are secretory products of activated alveolar macrophages (4,5). Taken together, these observations support an important role for the alveolar macrophage in the host defense against *P. carinii*. Although much has been recently learned about the specific role of alveolar macrophages in *P. carinii* infection, the mechanisms of interaction are only partially understood.

In studying the pathogenesis of pneumonia, other investigators have examined the interaction of *P. carinii* with alveolar epithelial cells. Current concepts suggest that the interaction of *P. carinii* with alveolar type I epithelial cells results in the death of the epithelial cell and exposure of the alveolar interstium. Although this may permit interaction of *P. carinii* with pulmonary interstitial cells such as dendritic cells, presently there have been no studies which have examined the role of interstitial macrophages or dendritic cells in host defense against *P. carinii* infection.

This chapter focuses on the interaction of *P. carinii* with the predominant alveolar phagocytic cell, the alveolar macrophage. The purpose of this chapter is to review in detail the current understanding for the role of alveolar macrophages in the pulmonary host response to *P. carinii*. This will include examination of the interaction of this organism with specific macrophage receptors, the role of macrophage-specific secretory products (i.e., cytokines), the mechanisms of elimination of this pathogen, and the role of factors present in the alveolar microenvironment (i.e., alveolar lining material). In addition, the impact of human immunodeficiency virus type 1 (HIV-1) infection will be examined because of the high prevalence of *P. carinii* pneumonia in this group of individuals. Efforts will be made to identify areas which have enhanced our understanding of the pathogenesis of *P. carinii* pneumonia and to identify areas which may benefit from further scientific investigation.

II. Biology of *P. carinii*

Taxonomical classification of *P. carinii* remains controversial. For many years, the organism was classified as a protozoon based on its microscopic appearance, lack of growth on fungal media, and response to antiprotozoal drugs. This classification is reflected in the terminology used for the different life forms of the organism (e.g., cyst, trophozoite, sporozoite). However, more recent evidence supports classification of *P. carinii* as a fungus. Analyses of both ribosomal RNA (6,7) and mitochondrial DNA (8) sequences show closer homology of *P. carinii* to fungi than to protozoa. Two microscopic forms of *P. carinii* are commonly identified: the thick-walled cyst (5–8 μm diameter) and the pleomorphic trophozoite (1–4 μm diameter). The wall of the *P. carinii* cyst stains with methenamine silver, toluidine blue, and cresyl echt violet. Cysts may contain up to eight interior forms, often termed sporozoites. Trophozoites stain with Giemsa or Diff-Quik and usually outnumber cysts by a wide margin in clinical specimens. Electron microscopic studies demonstrate that *P. carinii* has a simple structure with a nucleus, poorly preserved mitochondria, and absent or minimal lysosomes and Golgi apparatus (9,10). The relationship between the cyst and trophozoite in terms of the life cycle of *P. carinii* is unknown. Investigators have speculated that trophozoites divide sexually or asexually and mature into cysts. The cysts then rupture to release sporozoites, and each sporozoite then develops into a trophozoite (10). The factors influencing this proposed life cycle have not been identified.

P. carinii is an opportunistic pathogen. Human infection is generally seen only in immunocompromised hosts, although there are case reports of *P. carinii* pneumonia in subjects without apparent recognized risk factors (11,12). Immunosuppressive states permissive for *P. carinii* pneumonia are those where lymphocyte function is profoundly impaired, which include chronic corticosteroid therapy, cancer chemotherapy, chronic lymphocytic leukemia, and malnutrition (reviewed in Ref. 13). However, the prototypical immunosuppressive state associated with *P. carinii* pneumonia is HIV infection, in which *P. carinii* pneumonia accounted for 46% of the acquired immunodeficiency syndrome (AIDS) index diagnoses in 1992 (14,15). In HIV-related *P. carinii* pneumonia, the risk for infection correlates inversely with the number of circulating peripheral blood $CD4^+$ T lymphocytes, justifying the recommendation for prophylactic therapy when the CD4 T-lymphocyte count falls below 200 cells/mm^3 (16) or represent <20% of the total peripheral T lymphocytes.

III. Pathogenesis of *P. carinii* Pneumonia

A. Clinical and Pathological Characteristics

Clinically, HIV-related *P. carinii* pneumonia presents with 25 or more days of dyspnea, fever, and a nonproductive cough in patients with median CD4 T-lymphocyte counts less than 50 cells/mm^3 (17). The duration of symptoms tends to be shorter for no HIV-related *P. carinii* pneumonia (17). Auscultation of the chest may reveal fine crackles but is often normal. The chest radiograph typically shows bilateral interstitial infiltrates, but atypical radiographic patterns have been described, including pneumothorax, diffuse alveolar damage (18), granulomatous disease (19), cavitary pulmonary nodules and cysts (20,21), calcification (22), and prominent interstitial infiltrates with fibrosis (23). Hypoxemia is often present at rest or with exercise (24), serum lactate dehydrogenase (LDH) is often elevated (25), and pulmonary function testing may reveal a reduced diffusing capacity (26).

The histology of *P. carinii* pneumonia is that of an alveolar filling process with little surrounding inflammation. Lung tissue with *P. carinii* pneumonia demonstrates a foamy, eosinophilic exudate that fills and expands alveoli accompanied by proliferation of type II pneumocytes and minimal interstitial infiltration by lymphocytes (27). Involvement of lung tissue may be patchy or diffuse in extent. Small punctate nuclei of trophozoites and sporozoites are visible within the intra-alveolar exudate viewed under oil immersion. Special stains, such as Gomori methenamine silver or toluidine blue, reveal the cyst forms of *P. carinii*, which are crescent or helmet shaped. Electron microscopy has shown the alveolar exudate characteristic of *P. carinii* pneumonia is composed of clumps of cysts and trophozoites in a matrix of cellular debris and fibrin (28). Electron microscopy has further revealed trophozoite forms of *P. carinii* adherent to alveolar epithelial cells but no evidence for intracellular forms of the pathogen (29). Thus, with the exception of organism phagocytosis by macrophages, *P. carinii* is exclusively an extracellular pathogen.

B. Modes of Transmission and Epidemiology

Although an environmental reservoir of *P. carinii* has not been identified, *P. carinii* infection is found worldwide in humans and in a variety of mammalian hosts, including rodents, ferrets, rabbits, horses, dogs, and cats (1). Serological data demonstrate that the majority of humans acquire serum antibodies to *P. carinii* in childhood (30–32). The uniform presence of antibodies to *P. carinii* has been used to support the concept that *P. carinii* pneumonia represents the activation of latent organisms resident in lung tissue, which is analogous to the pathogenesis of tuberculosis reactivation. However, *P. carinii* has not been shown to colonize the respiratory tract in humans, and attempts to find latent organisms in lung tissue from immunocompetent humans have been generally unsuccessful (33,34). Animal to human transmission of infection has never been shown, and surface antigens and DNA of *P. carinii* recovered from lung tissue demonstrate considerable differences between strains infecting different animals and humans (35–37). However, several outbreaks of *P. carinii* pneumonia among immunocompromised humans have been described, suggesting that at least some cases of human *P. carinii* pneumonia are transmitted by inhalation (38,39). It is possible that *P. carinii* may be transmitted from human to human with a very short residence time in normal lung tissue. Alternatively, an ecological niche for *P. carinii* in the environment may yet be identified.

IV. Laboratory Methods for the Investigation of Macrophage–*P. carinii* Interaction

A. Animal Models of *P. carinii* Pneumonia

Much of the current understanding of host defense against *P. carinii* has been developed through experimental models of infection in animals (Table 1). Of the different animal models used, the original and best-studied model is the corticosteroid-treated rat model (40). In this model, conventional rats are treated with corticosteroids injected subcutaneously or added to the drinking water (41). The animals are often given tetracycline to prevent bacterial infection (42). A low-protein diet may also be used to induce a malnourished state (42). Under this regimen, rats will develop *P. carinii* pneumonia and die within 1–2 months, presumably due to progressive respiratory failure. To standardize infection particularly in animals raised under pathogen-free conditions, some investigators instill *P. carinii* organisms directly into the lungs of steroid-treated rats (43). Mice have also been treated with steroids to activate *P. carinii* infection, although the intensity of infection is generally less than that seen in rats (44,45).

Table 1 Major Animal Models of *P. carinii* Infection

Model	Reference	Infection intensity	Advantages	Disadvantages
Steroid rat	40	High	High intensity	Variability; difficult to induce to virus-free animals
Inoculated steroid rat	43	High	High intensity, less variability	May not mimic naturally acquired infection
Steroid mouse	44	Low		Low-intensity infection
Ferret	173		Large animal	Viral infection; few immunological reagents available
Athymic (nude) mouse	49	Low	Immunosuppression not required	Low-intensity infection; variability
SCID mouse	48	High	High intensity; can reconstitute immunity with normal spleen cells	
CD4-depleted mouse	3	High	Selective immune defect; reversible	

In addition to steroid-treated rodents, some investigators have studied *P. carinii* infection in mice with congenital immunodeficiencies or rendered immunodeficient through the administration of specific antibodies. The advantage of these experimental models of *P. carinii* pneumonia is that they utilize more selective forms of immunosuppression than corticosteroid therapy and allow for more sophisticated studies of host resistance to infection. Mice homozygous for the *nu* (nude) mutation are congenitally athymic, have decreased numbers of functional T lymphocytes (46), and can be infected with *P. carinii* (47). The intensity of infection tends to be mild in the nude mouse in comparison with the steroid-treated rat. Mice with the subacute combined immunodeficiency (*scid*) mutation have decreased numbers of both B and T lymphocytes as well as decreased circulating antibodies (46). Colonies of SCID mice develop spontaneous and high-intensity *P. carinii* pneumonia by 5–11 weeks of age (48,49). Established infection in SCID mice can be attenuated by adoptive transfer of spleen cells from normal mice, which provides an experimental opportunity to investigate how selective reconstitution of immunity impacts on host defense against infection (50,51). Finally, studies by Shellito and colleagues show that mice can be selectively depleted of helper T lymphocytes with the administration of monoclonal anti-CD4 antibody. These CD4 T-lymphocyte–depleted mice are susceptible to *P. carinii* pneumonia following intratracheal inoculation with organisms (3,52). This experimental model provides perhaps the most selective immunosuppression permissive for *P. carinii* pneumonia and is also reversible once the administration of the anti-CD4 antibodies is discontinued.

B. In Vitro Methods for Investigating Macrophage–*P. carinii* Interaction

The in vitro study of alveolar macrophage interaction with *P. carinii* is significantly hampered by the lack of a reliable method to propagate the organisms in culture (53). As a consequence, (1) *P. carinii* organisms are not available in large quantities; (2) there is limited availability of "purified" organisms free from host-cell contamination; (3) there is incomplete understanding of metabolic requirements of *P. carinii*; (4) there is a lack of understanding of the *P. carinii* life cycle; and (5) there are technical difficulties in dealing with an intrinsically "sticky" organism, which is present in the infected lungs of animals in large clusters.

Despite these limitations, significant advances in the understanding of macrophage interaction with this opportunistic pathogen have been made using in vitro systems. The majority of these methods utilize rodent alveolar or peritoneal macrophages. As no reliable method currently exists to retrieve *P. carinii* organisms from human lavage specimens in sufficient quantity, *P. carinii* organisms are obtained from steroid-treated rats by tracheal lavage followed by differential centrifugation. Macrophage suspensions or adherent monolayers are allowed to interact with *P. carinii* organisms under experimental conditions. Following removal of nonadherent organisms, various methods have been employed to provide qualitative and quantitative estimates of *P. carinii*–macrophage interaction and to examine the factors which influence the interaction (Table 2).

Although these methods have expanded our understanding of the mechanisms of interaction of *P. carinii* with macrophages, a careful review of the methods should be considered. Each method attempts to measure the ability of macrophages to bind and/or phagocytose *P. carinii*. Frequently, estimates of surface binding versus internalization cannot be established with certainty. Consequently, many of these methods permit only a semiquantitative assessment of the macrophage–*P. carinii* interaction. These limitations should be considered in the interpretation of results reported by investigators.

V. Examination of the Interaction of Macrophages with *P. carinii*

A. Animal Models

Histologically, alveolar macrophages are not numerous in rat *P. carinii* pneumonia and do not appear to be very active in the phagocytosis of the organism (54,55). In human *P. carinii* pneumonia, histopathological examination reveals scanty cellular (lymphocytes, macrophages, plasma cells) infiltrates, sometimes lacking in the alveolar septa (whereas most infiltrates are more noticeable in the perivascular and peribronchial regions [56]). The alveolar lumens are filled with frothy honeycombed material that contains numerous *P. carinii* cysts but only few macrophages and rare neutrophils. Chen and Harmsen (57) investigated the development and resolution of naturally acquired *P. carinii* pneumonia utilizing SCID mice and light and electron microscopy. These investigators observed that the presence of nonlymphoid inflammatory cells in SCID mice is not sufficient to control *P. carinii* infection. Chen and Harmsen observed that the clearance of *P. carinii* from the lungs of reconstituted SCID mice required local recruitment of large numbers of inflammatory cells with an active appearance, and they concluded that intracellular killing of *P. carinii* by phagocytosis does not appear to be a major mechanism of host defense against *P. carinii* infection in this model.

Although these observations may be interpreted to mean that alveolar macrophages may not be important in the host response to *P. carinii*, the apparent paucity of this pulmonary phagocytic cell in the animal model of *P. carinii* pneumonia and in human specimens with active disease may reflect the ability of this opportunistic pathogen to subvert or bypass the host response in susceptable individuals. As noted above, the importance of alveolar macrophages is supported by the observation that following reversal of the immunosuppressive state, investigators observe an influx of alveolar macrophages in the setting of *resolution* of

Table 2 In Vitro Methods for Examination of *P. carinii* Interaction with Macrophages

Macrophage preparations for experimental conditions	*P. carinii* label	Method of detection (ref.)	Stain for *P. carinii*	Determination of binding/phagocytosis	Limitations of assay
Suspensions Fixation of cells following interaction with *P. carinii*	None	Visual (60)	Giemsa	Light microscopy	Observer bias
Adherent No fixation of cells following interaction with *P. carinii*	None	Visual (59)	None	Phase contrast miscroscopy (Sykes-Moore chamber)	Observer bias
	^{51}Cr	Radioactivity (87,174)	None	Gamma counts	Binding only
	*FITC	Fluorescence (112)	None	Plate fluorimetry	Level of detection
Adherent Fixation of cells following interaction with *P. carinii*	None	Visual (58,59)	Giemsa	Light microscopy	Observer bias
	None	Visual (58,59)	Giemsa	Phase contrast microscopy	Observer bias
	None	Photographic (57,61)	None	Transmission electron microscopy	Technically complex
	None	Photographic (58)	None	Scanning electron microscopy	Technically complex
	*FITC	Visual (62)	None	Epifluorescence microscopy	Observer bias

P. carinii pneumonia. Histological examination of animals recovering from infection reveal numerous macrophages accompanied by phagocytosis. Furthermore, perhaps the apparent absence of internalized *P. carinii* in these studies reflects the rapid and efficient clearance of organisms (in the recovery phase) and may reflect the rapid digestion of the organism following engulfment as observed by others (58,59).

B. In Vitro Investigations

Macrophage Receptors

Immunoglobulin (Fc) Receptor

Several investigations have suggested an important role for the macrophage Fc receptor. Masur and Jones (59) observed that immune serum was necessary for the internalization of *P. carinii* cysts by rodent alveolar macrophages. In the absence of serum, freshly explanted macrophages from normal animals only bound *P. carinii* trophozoites to the cell surface. At 2 hr with a multiplicity of 10^6–10^7 organisms per 15 × 10^6 macrophages, approximately 850 *P. carinii* trophozoites were observed to be adherent per 100 macrophages examined by contrast phase microscopy. Some macrophages were observed to have up to 40 adherent organisms. By 72 hr, this number was approximately 700 *P. carinii* per 100 macrophages, without significant changes in the organism morphology or relationship to the macrophage surfaces.

However, phagocytosis occurred promptly with the addition of rabbit anti–*P. carinii* serum, although not with nonimmune rabbit serum. Viewed by inverse phase microscopy in a Sykes-Moore chamber, these investigators observed adherence of *P. carinii* trophozoites only to the surface of macrophage monolayers for up to 120 min in the presence of 20% fetal calf serum or 20% normal rabbit serum. However, within 20 min after adding 20% rabbit anti–*Pneumocystis* serum, each trophozoite was observed to be engulfed by the surface membrane of the alveolar macrophage. A semiquantitative assessment demonstrated that following the addition of 20% immune serum, rat alveolar macrophages with 1–10 adherent organisms ingested virtually all adherent organisms, whereas in macrophages with 30–50 adherent organisms, most cells ingested <75% of adherent organisms. No qualitative differences were observed comparing alveolar macrophages from normal rats with those receiving corticosteroids, and rat peritoneal macrophages behaved in a similar fashion. These investigators concluded that Fc receptors mediated phagocytosis, but they speculated that binding occurred via another receptor.

Taylor and colleagues (60) observed limited phagocytosis of *P. carinii* by elicited rat peritoneal macrophages, although ingestion was greatly enhanced in the presence of immune serum. Following incubation for 10 min at 37°C at a multiplicity of 39:1 (*P. carinii* to macrophage), 36% of macrophages contained at least one internalized *P. carinii* in the presence of 20% immune serum, whereas the number of macrophages with internalized organisms was only 20.5% and 17% in the presence of 1% and 0.2% immune serum, respectively. In these studies, phagocytosis was estimated by lucigenin-amplified chemiluminescence and correlated with light and electron microscopy.

Utilizing 24-hr cultured normal peripheral blood monocytes, Forte and colleagues (61) demonstrated the in vitro ingestion of both trophozoites and cyst forms of *P. carinii*. In the presence of autologous serum, quantitative assessment of the interaction by electron microscopy revealed that by 24 hr of incubation with *P. carinii* at a multiplicity of 10:1 (*P. carinii* to macrophage), 100–450 *P. carinii* were bound to the surface per 100 macrophages, whereas 30–80 *P. carinii* per 100 macrophages were ingested. Over the 24-hr period of exposure to *P. carinii*, 8–38% of macrophages contained ingested *P. carinii*, with microscopic evidence for progressive degradation of the organisms inside macrophage vacuoles. The relative contribution of immunoglobulin was not determined, as experimental conditions omitting autologous serum or utilizing immune serum were not included.

Complement (C3) Receptor

Limited information is available regarding the potential role for complement components in the macrophage recognition of *P. carinii*. In studies by Taylor and colleagues, complement enhanced antibody-mediated opsonization but alone was not an opsonin for *P. carinii* (60). In these studies, only 7% of macrophages ingested at least one *P. carinii* using unopsonized organisms, and 8% of macrophages contained *P. carinii* in the presence of 20% complement. Following incubation for 10 min at 37°C in the presence of 20% immune serum alone, 36% of macrophages contained at least one internalized *P. carinii* organism, whereas 45% of macrophages contained at least one internalized *P. carinii* in the combined presence of 20% immune serum and 20% complement.

Mannose Receptor

Recent investigations demonstrated that binding and ingestion of *P. carinii* can be mediated by the macrophage mannose receptor. The introduction of competitive inhibitors of the mannose receptor (e.g., mannosylated bovine serum albumin or yeast mannan) or the modulation of mannose receptor surface expression significantly reduced the ability of normal alveolar macrophages to bind and internalize *P. carinii* (62). Following a 60-min exposure of adherent alveolar macrophages to fluorescein isothiocyanate (FITC)-labeled *P. carinii* organisms, an attachment index (number of organisms per 100 macrophages) was determined by phase and fluorescence microscopy. In the absence of exogenous opsonins, normal macrophages demonstrated an attachment index of 300–400. In the presence of competitive inhibitors of the mannose receptor (mannosylated bovine serum albumin

[BSA] or mannan), the attachment index was reduced by 85–90%. Redistribution of surface mannose receptor following the adherence of macrophages to mannan-coated glass coverslips resulted in a 90% reduction in the *P. carinii* attachment index. Furthermore, transfection of the nonphagocytic COS-1 cell line with the cDNA for the human mannose receptor rendered the COS-1 cell capable of binding and internalizing *P. carinii* compared with mock transfected cells.

Surfactant Protein A Receptor

A candidate macrophage surfactant protein A (SP-A) receptor has recently been described (63). Although the role of the SP-A receptor in *P. carinii* pneumonia is not yet established, several observations suggest a potential role for this receptor in the pathogenesis of disease. SP-A levels are elevated in the lungs of rats with *P. carinii* pneumonia (64). As detectable levels of SP-A have been measured on the surface of *P. carinii* (65) (SP-A binds to the major surface glycoprotein determinant of *P. carinii* [65]), it would seem reasonable to speculate that a macrophage receptor for SP-A could serve as a means for the binding and phagocytosis of SP-A–coated organisms. However, it is uncertain whether SP-A bound to the surface of *P. carinii* would be recognized by the SP-A receptor in a manner similar to unbound SP-A. One preliminary study utilizing rat alveolar macrophages demonstrated enhanced binding of *P. carinii* in the presence of exogenous SP-A (66). Another preliminary study utilizing human alveolar macrophages observed impaired phagocytosis of *P. carinii* in the presence of excess exogenous SP-A in vitro (67). Currently, no studies are yet available to address whether *P. carinii* specifically interacts with a macrophage SP-A receptor.

Fibronectin Receptor

Pottratz and Martin (68) demonstrated that attachment of *P. carinii* to rat alveolar macrophages occurred through a fibronectin and calcium-dependent mechanism. At a *P. carinii* to macrophage multiplicity of 2:1, approximately 18.9% of the *P. carinii* inoculum remained adherent to alveolar macrophage monolayers by 4 hr at 37°C. Attachment was reduced to 8.3% of the initial inoculum in the presence of antifibronectin antibody or to 2.9% with the addition of the fibronectin cell binding site analogue tetrapeptide Arg-Gly-Asp-Ser (RGDS). Attachment was also reduced to 5.1% in the presence of the calcium ion chelating agent EGTA. Interestingly, fibronectin-mediated attachment did not result in phagocytosis of *P. carinii* as determined by luminol-dependent chemiluminescence measurements.

Other Extracellular Matrix Proteins

Although vitronectin may mediate *P. carinii* attachment to alveolar epithelial cells (69), and laminin may impact on *P. carinii* attachment to epithelial and mesenchymal cells (70), limited information is available regarding the role of vitronectin and laminin in the interaction with alveolar macrophages.

Intercellular Adhesion Molecule-1

Intercellular adhesion molecule-1 (ICAM-1), a member of the immunoglobulin supergene family, is expressed on the surface of alveolar epithelial cells and macrophages. The induction of ICAM-1 by proinflammatory cytokines results in augmented ICAM-1 expression which mediates leukocyte adhesion to epithelial cells. Recently, *P. carinii* was shown to be capable of augmenting ICAM-1 expression on the A549 human alveolar epithelial cell line, and immunohistochemical staining of the lungs of mice with *P. carinii* pneumonia demonstrates increased ICAM-1 expression localized to epithelial cells and alveolar macrophages (71). Although preliminary observations suggest that ICAM-1 does not appear to mediate binding of *P. carinii* (H. Koziel, unpublished data), the precise role for ICAM-1 in *P. carinii* pneumonia has not been established.

P. carinii *Receptors*

Recent investigations describe the presence of specific receptors on the surface of *P. carinii*. Pesanti (72) described a receptor for tumor necrosis factor-α (TNF-α) on the surface of *P. carinii* and postulated that this may potentially serve an important function in host defense. Recently, investigators described an extracellular matrix receptor, similar to laminin, encoded by *P. carinii* organisms and present on the organism's surface (73). Although these receptors may function as a virulence factor, the precise role for these receptors in the pathogenesis of *P. carinii* pneumonia has not been fully elucidated. However, these observations raise potentially important issues regarding the interaction of this extracellular parasite with host cells and host molecules.

Chemotaxis

The response of the alveolar macrophage to an infectious agent typically includes chemotaxis, attachment, and phagocytosis followed by elimination of the agent. Owing to the limited capacity for alveolar macrophage replication in situ, the majority of alveolar macrophages are derived from the general circulation (74). Blood monocytes migrate to the alveolar interstitium in response to chemotactic stimuli. Limited information is available as to the initial event of chemotaxis specifically related to *P. carinii* pneumonia. By electron microscopy, Forte and colleagues (61) observed the movement of normal peripheral blood–derived macrophages toward *P. carinii* accompanied by extension of macrophage pseudopods. By light microscopy, preliminary observations by Blumenfeld and colleagues reported in vitro aggregation of normal monocyte–derived macrophages and alveolar macrophages about human-derived *P. carinii* in the presence of heat-inactivated serum (75). *P. carinii* glycoprotein A (GP-A), the 95- to 120-kDa major surface glycoprotein of *P. carinii*, may serve as a chemotactic factor for normal peripheral blood monocytes and thus function in part in recruiting blood monocytes to the lungs (76).

Binding of P. carinii *to Macrophages*

As described in detail above, several factors may be important in the binding of *P. carinii* to the surface of alveolar macrophages. Binding of this organism to macrophages appears to be calcium dependent and may be potentially mediated through several receptors, including the Fc receptor, mannose receptor, fibronectin receptor, and (possibly) an SP-A receptor. The role of vitronectin, ICAM-1, and potentially other macrophage receptors has not yet been fully investigated. The influence of other exogenous molecules on *P. carinii* binding to macrophages is discussed in detail below.

Phagocytosis of P. carinii *by Macrophages*

Several factors have been described which may be important in the phagocytosis of *P. carinii* by alveolar macrophages (Fig. 1). However, the inherent difficulty in distinguishing with certainty internalized (i.e., phagocytosed) organisms from those bound to the macrophage surface clearly impacts on the ability to study quantitatively this interaction. Thus, the interpretation of the data related to phagocytosis requires consideration of this limitation.

In contrast to Masur and Jones (59), von Behren and Pesanti (58) observed that efficient phagocytosis of *P. carinii* trophozoites occurred in the absence of serum. First, using rats with *P. carinii* pneumonia, these investigators observed various life forms of *P. carinii* attached to freshly explanted alveolar macrophages from these animals. Within 5–6 hr of in vitro incubation of glass-adherent cells, recognizable *P. carinii* trophozoite, sporozoite, and mature cyst forms (MCFs) had essentially disappeared as determined by inverse phase microscopy. Within 24 hr, no identifiable organisms were present either associated with the macrophages or in the culture supernatants.

In subsequent experiments, these investigators demonstrated that alveolar macrophages from normal uninfected rats also engulfed *P. carinii* in the absence of serum, although this required culturing the macrophages for 48 hr prior to exposure to *P. carinii*. Following a 1-hr incubation with exogenous *P. carinii*, only 1% of freshly explanted alveolar macrophages contained MCFs of *P. carinii* and <5% contained any form of the organism. In comparison, following culture for 48 hr, approximately 10% of the cultured macrophages ingested recognizable MCFs, whereas more than 90% were associated with at least one form of *P. carinii*. A quantitative assessment of internalized *P. carinii* MCFs by light microscopy demonstrated approximately 12 *P. carinii* MCFs per 100 alveolar macrophages following a 1-hr exposure at 37°C. The ingestion of *P. carinii* was confirmed by electron microscopy. Experiments performed at 4°C or in the

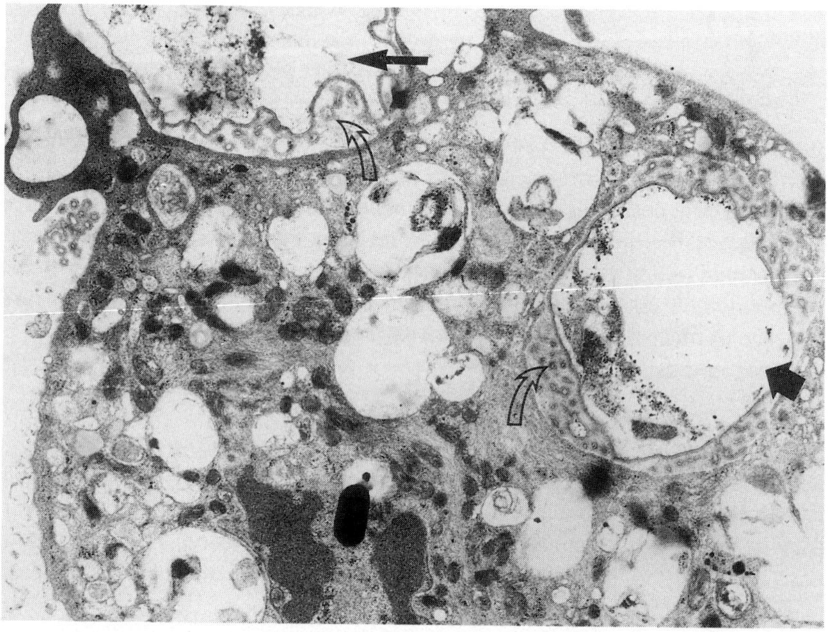

Figure 1 Interaction of a normal human alveolar macrophage with *P. carinii* organisms in vitro. This electron photomicrograph demonstrates one internalized partially degraded trophozoite form of *P. carinii* (short solid arrow) and a second trophozoite form bound to the macrophage surface and undergoing phagocytosis (long solid arrows). External tubular structures (characteristic of *P. carinii*) in cross-section are observed immediately adjacent to the organism cell membrane (curved hollow arrows). The experimental conditions included incubation of adherent alveolar macrophages in the presence of *P. carinii* for 60 min at 37°C in 5% CO_2 in the absence of added exogenous opsonins. Magnification 20,000×. (Photomicrograph provided by Dr. Angeline Warner, Harvard School of Public Health, Boston, MA.)

presence of iodoacetate (inhibitor of phagocytosis) reduced *P. carinii* uptake by 50%. No differences were noted whether alveolar macrophages were isolated by lavage or by mincing of lung tissue. Interestingly, peritoneal macrophages of any age were able to phagocytose *P. carinii* (i.e., peritoneal macrophages did not require preculturing).

Utilizing adherent normal human and rat alveolar macrophages, Ezekowitz and colleagues demonstrated that the macrophage mannose receptor was sufficient for in vitro opsonin-independent phagocytosis of *P. carinii* (62), as described in detail above (see Fig. 1). Utilizing cultured normal peripheral blood monocytes, Forte and colleagues (61) demonstrated in vitro ingestion of both trophozoites and cyst forms of *P. carinii* in the presence of autologous serum. Internalization was confirmed by electron microscopy. Utilizing suspensions of elicited rat peritoneal macrophages, Taylor and colleagues (60) demonstrated 7% of macrophages contained at least one ingested organism, whereas up to 45% of macrophages ingested at least one organism in the presence of serum and complement. Internalization was confirmed by light and electron microscopy.

Thus, within the limitations of the available in vitro assays, current understanding suggests that several macrophage receptors may mediate the phagocytosis of *P. carinii*. The Fc receptor mediates the phagocytosis of immunoglobulin-opsonized organisms (whether opsonized in vivo or through exogenous administration of immunoglobulin in vitro). The complement receptor may enhance Fc receptor–mediated phagocytosis, but it alone does not appear to mediate phagocytosis. The macrophage mannose receptor mediates opsonin-independent phagocytosis of *P. carinii*. Although capable of mediating binding, the fibronectin receptor does not mediate phagocytosis. The relative contribution of each receptor and the potential role of other macrophage receptors (i.e., an SP-A receptor) in mediating phagocytosis of *P. carinii* has not yet been clearly established.

Apparent in Vitro Differences in Macrophage Behavior

As noted by various investigations, the phagocytosis of *P. carinii* by macrophages may be dependent on several factors, including the duration in culture of the macrophages (freshly explanted vs cultured), the source of the macrophages (alveolar vs peritoneal vs monocyte derived), and the presence or absence of exogenous opsonins. The observation of differential behavior of freshly explanted alveolar macrophages compared with cultured alveolar macrophages may reflect the surface changes that occur when alveolar macrophages are removed from an environment rich in alveolar lining fluid to an environment containing serum (77). The macrophage mannose receptor could potentially account for a significant portion of opsonin-independent phagocytosis observed by investigators in prior studies (59), and variable levels of surface expression may potentially account for variable degrees of the phagocytosis of nonopsonized *P. carinii* observed by prior investigators. Blood monocytes and most macrophage cell lines (murine J774e macrophage cell line excluded) do not express functional mannose receptors (78). Thus, the interpretation of results from experiments performed with blood monocytes and certain cell lines in the study of the *P. carinii* interaction with host cells requires consideration of these observations.

Mechanisms of Elimination of P. carinii

As discussed above, resolution of *P. carinii* pneumonia in animal models is accompanied by an influx of macrophages. The mechanism(s) by which macrophages control *P. carinii* infection are not completely understood. Available data suggest that phagocytic cells resident within the alveolar airspaces or responding to an inflammatory stimulus can kill *P. carinii*. However, the mechanism(s) by which macrophages kill *P. carinii* are only partially understood.

Determination of *P. carinii* Viability

Investigations of *P. carinii* killing have been limited by the absence of a reliable assay for distinguishing viable from nonviable organisms. For this reason, interpretation of data regarding "killing" or viability of *P. carinii* require caution, as assays of *P. carinii* viability have not yet been standardized or uniformly accepted.

Early estimates of viability by von Behren and Pesanti assumed that organisms were dead when they lost the characteristic staining properties to Giemsa and Gomori methenamine silver stains (58). Masur and Jones observed destruction of *P. carinii* trophozoites (59) by direct microscopy. Metabolic markers of nucleotide synthesis demonstrated only a low level of incorporation of radiolabeled precursors. Masur and Jones observed uptake of [^{3}H]uridine or [^{3}H]thymidine (but not [^{3}H]leucine) in only 10–20% of trophozoites and cysts (59). This low level of uptake presumably reflected the paucity of energy-generating organelles and/or limited metabolic requirements of the organisms. Pesanti estimated *P. carinii* viability by the conversion of [^{14}C]glucose to [^{14}C]CO_2 by organisms in suspension (79). Recently, β-1,3-glucanase, an enzyme involved in *P. carinii* cell wall synthesis, has been identified (80). The activity of this enzyme correlates directly with the increase in numbers of *P. carinii* organisms in short-term culture and could therefore be useful as a biological marker of viability.

Other investigators have estimated *P. carinii* viability by techniques of incorporating or excluding vital fluorescent dyes (60,81). Exclusion of DNA binding fluorescent dyes can assist in identifying organisms with intact cellular membranes (viable) in contrast to those which have disrupted cell membrane integrity (nonviable) when viewed by epifluorescence microscopy. Others have described the use of phase contrast, Nomarski interference contrast, and bright-field microscopy with oblique illumination in combination with vital dyes to estimate organism viability (82). However, the estimation of *P. carinii* viability by direct visual methods may be difficult because of limited visual sensitivity and the variability of fluorescence staining intensity in mixed life cycle forms of *P. carinii*.

Some of these inherent difficulties in the determination of viability may in part be overcome by utilizing instrumentation with sensitive detection capabilities. Recently, investigators have utilized an adenosine triphosphate (ATP) bioluminescence assay to evaluate *P. carinii* viability (83). This assay employs a luciferin-luciferase ATP-dependent reaction to evaluate the viability of populations of rat derived *P. carinii*. Some investigators have used flow cytometry to characterize a population of *P. carinii*; examining individual organisms that incorporate a fluorescent dye (84,85). Other investigators have adapted a well-described double-labeling viability assay to flow cytometry (86). This assay uses fluorescein diacetate, which is a substrate that is converted to a fluorophore by cell cytoplasmic nonspecific esterase, and propidium iodide, which is a DNA-binding protein excluded by an intact cell membrane Results from this "double-label" assay correlate well with the activity of β-1,3-glucan synthase, a potential biological marker for *P. carinii* viability (80).

Mechanisms of Intracellular Elimination

P. carinii appears to be an obligate extracellular parasite, as a significant intracellular phase in the life cycle of this organism has not been observed or described. The current concept suggests that once internalized by macrophages, *P. carinii* is rapidly degraded. Current methods do not allow evaluation of the viability of *P. carinii* after phagocytosis by macrophages, and current knowledge of the fate of *P. carinii* is based primarily on direct microscopic observation of *P. carinii*–macrophage interaction (58,59).

Masur and Jones observed that within 20 min of adding immune serum, all trophozoites adherent to the surface of alveolar macrophages were engulfed by the surface membrane of the alveolar macrophage. Subsequently, the trophozoite could be recognized within the macrophage phase-lucent vacuole for only 2–4 min, following which the characteristic morphology of the organism was lost. By electron microscopy, only amorphous phase-dense material within the vacuole could be seen, without any recognizable forms of *P. carinii* trophozoites. Von Behren and Pesanti also described rapid digestion of *P. carinii* following phagocytosis (58). Freshly explanted alveolar macrophages from animals with *P. carinii* pneumonia were noted to have various life forms of *P. carinii* attached to the cell surfaces. Within 5–6 hr, all recognizable forms of *P. carinii* had essentially disappeared (presumably ingested by the macrophages), and within 24 hr, no identifiable organisms were present either associated with the macrophages or in the culture supernatants. Following exposure to *P. carinii*, most normal rat alveolar macrophages ingested the organisms by 1 hr (internalized *P. carinii* were confirmed by electron microscopy). Following internalization, degradation was rapid, with eradication complete by 6–8 hr (58). Utilizing cultured normal peripheral blood monocytes, Forte and colleagues (61) demonstrated in vitro ingestion of both trophozoites and cyst forms of *P. carinii*, with microscopic evidence for progressive degradation inside macrophage vacuoles.

Oxidative Burst. A potential role for macrophage-generated oxygen moieties in the killing of *P. carinii* is supported by several investigations. Amplified chemiluminescence assays, measuring the release of reactive oxygen intermediates, have been used by investigators as a surrogate assay for phagocytosis to study host cell *P. carinii* interaction (87). Several investigators have demonstrated the intracellular and extracellular release of hydrogen peroxide and other oxygen moieties by normal macrophages (60,62,88) following in vitro exposure to *P. carinii*. Studies have shown that *P. carinii* is only poorly able to scavenge oxidants

(79) and is susceptible to peroxidative and hydroxyl radical–mediated killing (79,89). Importantly, soluble components of the *P. carinii* cell wall major glycoprotein (GP-A) can stimulate the oxidative burst in normal macrophages (88), suggesting that the interaction of *P. carinii* GP-A with a macrophage receptor (such as the mannose receptor) may represent a significant mechanism for eliciting the oxidative response in normal cells. These observations suggest that phagocytosis followed by a respiratory burst response and the generation of oxygen radicals may be an important route of elimination by normal alveolar macrophages.

Acidic Environment. Pesanti (79) observed that *P. carinii* became metabolically inactive after brief exposure to pH <5 (a pH similar to that which is present within macrophage lysosomes [90]). Recent preliminary studies demonstrated that exposure to an acidic environment (pH = 5) resulted in the loss of *P. carinii* viability as determined by a double-labeling flow cytometry assay (86). Thus, the intracellular environment of phagocyte lysosomes is potentially injurious to *P. carinii*, and normal phagocytes possess biochemical pathways which are capable of killing *P. carinii*.

Nitric Oxide Pathway. A possible role for arginine-dependent nitric oxide pathway–mediated macrophage killing of *P. carinii* has been suggested by several preliminary observations. Sherman and colleagues (91) demonstrated that *P. carinii* induced L-arginine oxidation by murine and human pulmonary alveolar macrophages. Williams and colleagues (92) demonstrated nitrite production by rat alveolar macrophages exposed to *P. carinii* in vitro, whereas the *P. carinii*–mediated release of nitrite was inhibited by corticosteroids. Recent in vivo experiments by Shellito and colleagues (93) observed that tissue expression of the inducible form of nitric oxide synthase (iNOS) mRNA was increased in CD4-depleted mice with *P. carinii* pneumonia in comparison with SCID mice with *P. carinii* pneumonia. However, in these experiments, iNOS enzyme expression did not correlate with either clearance of *P. carinii* organisms or severity of *P. carinii* infection (94). The precise role of the nitric oxide pathway in *P. carinii* pneumonia has not yet been systematically examined.

Mechanisms of Extracellular Elimination

Recent investigations have suggested that killing of *P. carinii* may also occur by extracellular mechanisms. Utilizing a dual-chamber system that permits circulation of culture medium but prohibits direct contact of macrophages with the *P. carinii* organisms, soluble products secreted into the medium by the activated macrophages resulted in the loss of organism viability (95) as measured by conversion of radiolabeled glucose. This extracellular killing may be primarily due to oxidants, as the loss of *P. carinii* viability was inhibited in the presence of catalase and superoxide dismutase.

P. carinii stimulates the in vitro production of TNF-α by human macrophages (96). TNF-α may also be directly toxic to the organisms (95) when viability is measured by the conversion of radiolabeled glucose to CO_2. This unique observation has not been previously reported for other pathogens. However, in a different study measuring *P. carinii* viability utilizing dual-fluorescence labeling and flow cytometry, TNF-α did not have a direct toxic affect on *P. carinii* (89).

The potential role of antibodies and complement to mediate killing of *P. carinii* has been investigated. Although complement may be directly toxic to other pathogens, several studies have not demonstrated a similar adverse effect on *P. carinii* viability. The in vitro exposure of *P. carinii* to fresh complement with immune rat serum (in the absence of macrophages) did not result in death or lysis of the organisms as determined by dye exclusion (97). In the absence of macrophages, immune serum and complement do not appear to have direct adverse effects on *P. carinii* viability as determined by fluorescein diacetate-ethidium bromide staining (60). Antibodies with or without complement do not impair the growth of *P. carinii* in short-term tissue culture (97,98).

VI. Effect of Local Factors in the Alveolar Lining Material

P. carinii infection occurs within the confines of the alveolar lining material (ALM), which includes surfactant components, immunoglobulins, complement, extracellular matrix proteins, serum proteins, and other secretory products of immune and alveolar cells. As the macrophage and *P. carinii* remain in direct contact with the ALM, alterations in the composition of this material may impact on the interaction of the alveolar macrophage with this pathogen. Alterations in the components of the ALM may potentially contribute to the pathobiology of *P. carinii* infection by adversely affecting the functional integrity of alveolar macrophages and other components of the pulmonary host defense system.

A. Antibodies

Children with primary immunodeficiency diseases are at risk for *P. carinii* pneumonia. Prior to the AIDS epidemic, such conditions constituted the underlying illness in 13% of the cases of *P. carinii* pneumonia reported to the Centers for Disease Control and Prevention (CDC) (99,100). However, children with a pure B-lymphocyte defect constitute only a small fraction of immunodeficient patients with *P. carinii* pneumonia. Some investigators have hypothesized that defects in local IgA antibody and complement may have a role in the pathogenesis of the interstitial plasma cell pneumonia form of *Pneumocystis* infection (101,102), although this has not been systematically studied.

Only limited data are available regarding local antibody responses to *P. carinii*. IgG and IgM antibodies have been found in bronchoalveolar lavage

(BAL) fluid of normal adults and a few AIDS patients with pneumocystosis (103). IgG, IgA, and IgM directed against *P. carinii* have been measured in BAL fluid of animals with active infection (104), and IgG, IgM, and IgA have been demonstrated on the surface of human and rat *P. carinii* by immunofluorescence (104,105). Levels of IgG antibodies present in BAL fluid correlate with serum IgG antibody levels (104). The ability of antibodies to function as opsonins is supported by in vitro investigations where immune serum enhanced the in vitro phagocytosis of *P. carinii* by alveolar macrophages (59,60). Despite the frequent presence and the potential function as opsonins, the relative importance of serum and local antibodies in host defense against *P. carinii* is unclear, and the specific role of these opsonins in macrophage processing of *P. carinii* is not known.

B. Surfactant Protein A

Surfactant protein A (SP-A), the major glycoprotein component of surfactant, is a member of a family of C-type lectins. The principal function of SP-A is to reduce alveolar surface tension by enhancing the surface-active properties of the phospholipids (106). However, some investigators have observed that SP-A may serve as an opsonin and participate in local host defenses (107–109). Recent studies characterizing the impact of SP-A on macrophage handling of *P. carinii* have yielded varied results. Preliminary observations demonstrated that exogenous SP-A augmented binding of *P. carinii* to adherent rat alveolar macrophages (66). In comparison, other preliminary observations suggested that exogenous SP-A may impair binding and internalization of *P. carinii* by human alveolar macrophages in vitro (67). SP-A may exert its effect by binding to the mannose-rich glycoprotein A of *P. carinii* (110).

C. Surfactant Protein D

Surfactant protein-D (SP-D) is a collagenous protein synthesized and secreted by alveolar type II epithelial cells. SP-D possesses a calcium-dependent lectin binding domain sharing high sequence homology with other mammalian C-type lectins. Recent observations indicate that SP-D accumulates in the lungs of rats with *P. carinii* pneumonia and binds to *P. carinii* (111). SP-D may bind to the major surface glycoprotein (GP-A) of the organism via a saccharide-mediated mechanism. SP-D is found on the organism in vivo, and it appears to mediate binding to alveolar macrophages in vitro, although it does not significantly affect phagocytosis (111).

D. Extracellular Matrix Proteins

Fibronectin is present in the BAL fluid of normal individuals, and it likely represents a normal component of the alveolar lining material. As discussed above, fibronectin can facilitate the binding of *P. carinii* to alveolar macrophages. Interestingly, fibronectin does not serve the role of a classic opsonin in that it does not facilitate the macrophage ingestion of *P. carinii*. The relative contribution of fibronectin to macrophage binding and its role in the normal host defenses against this opportunistic pathogen are not fully established.

Although other extracellular adhesins such as laminin and vitronectin have been demonstrated to affect binding of the organism to alveolar epithelial cells, their role in the interaction with alveolar macrophages has not yet been fully investigated.

VII. Interaction of Macrophages with Different Life Cycle Forms of *P. carinii*

Most in vitro studies have utilized mixed life cycle forms of *P. carinii*, containing trophozoite, cyst, and intermediate forms. Based on the morphological appearance, investigators commonly identify the trophozoite and cyst forms of *P. carinii*. In general, trophozoites are the predominant form, accounting for 89–99% of the entire population of organisms in specimens obtained from the lungs of infected rodents (59–61,112–114). Some investigators have utilized techniques in an attempt to separate the different life cycle forms of *P. carinii* to examine whether these different forms elicit different responses from macrophages. Masur and Jones described the trophozoite as the predominant form of *P. carinii* in lavage specimens (59), accounting for >99% of the organisms observed by contrast phase microscopy, and thus they focused their investigations primarily on the morphological interactions of the trophozoite form with alveolar macrophages. Von Behren and Pesanti observed no qualitative differences in the uptake of trophozoites, sporozoites, and mature cyst forms (58). Hidalgo and colleagues observed that *P. carinii* cysts elicited a greater respiratory burst response in macrophages (88) than *P. carinii* trophozoites. Thus, the current understanding of the interaction of different life cycle forms of *P. carinii* with macrophages remains rudimentary.

VIII. Role of Cytokines

Cytokines are soluble factors important in controlling tissue inflammation in response to a variety of infections. Cytokines may be products of macrophages that function to modulate the recruitment of inflammatory cells in response to an invading pathogen. Alternatively, cytokines released from nonmacrophage cells, such as lymphocytes, may activate tissue macrophages for increased antimicrobial functions leading to clearance of the infectious challenge. With regard to *P. carinii*, seven cytokines have received investigative attention: TNF-α, interferon-γ (IFN-γ), interleukin-1 (IL-1), interleukin-6 (IL-6), interleukin-8 (IL-8), and macrophage and granulocyte-macrophage colony-stimulating factors (CSFs).

Accumulating evidence implicates TNF in host defense against *P. carinii* infection. In lung tissue, TNF is produced principally by alveolar macrophages. Mononuclear phagocytes do not release TNF constitutively, but when *P. carinii* is added in vitro to normal human monocytes (96) or alveolar macrophages (114, 115), there is stimulated release of TNF. Experimental work by Kolls and colleagues also supports the release of TNF in vivo during murine *P. carinii* infection (5). The release of TNF in response to an acute challenge with *P. carinii* peaks at 3 hr and precedes the influx of inflammatory cells into the lung (5). Both in vitro and in vivo release of TNF are not blocked by polymyxin B, and TNF can be elicited by an acute challenge of *P. carinii* in C3H/HeJ mice, suggesting that the release of TNF in response to *P. carinii* is not secondary to contaminating endotoxin.

The specific stimulus which results in the release of TNF in response to *P. carinii* is not well understood. Work by Hoffman and colleagues has implicated a β-glucan moiety in the organism cyst wall which may mediate the in vitro release of TNF by alveolar macrophages (114). Subsequent data demonstrate that the adhesive proteins, vitronectin and fibronectin, further amplify the TNF response (116).

Thus, it is clear that TNF is elicited by *P. carinii* and that TNF can be detected in the lungs of animals and patients with *P. carinii* pneumonia. The precise role of TNF in pulmonary host defense against this pathogen, however, is less clear. As mentioned above, early data documented binding of TNF to *P. carinii* (72) and direct cytotoxicity when *P. carinii* was incubated with TNF in vitro (95). However, more recent data have not confirmed direct toxicity from TNF (89). Considering the current knowledge of TNF and host defense against other pathogens, rather than killing *P. carinii* directly, it is more likely that TNF serves to activate macrophages or stimulate a cascade of additional cytokines which then lead to the clearance of *P. carinii*.

Experimental approaches to neutralize TNF in vivo indicate that TNF plays a pivotal role in host defense against *P. carinii*. Chen and colleagues demonstrated that the use of an anti-TNF antibody delays resolution of *P. carinii* infection in SCID mice reconstituted with spleen cells, suggesting that endogenous TNF is critical for resolving the infection (57). Moreover, treatment of CD4-depleted, *P. carinii*-infected mice with heat-treated *Escherichia coli*, an inducer of TNF, causes nearly complete resolution of infection (4). The enhanced resolution of infection in this model could be abrogated with anti-TNF antibodies. The mechanism for the beneficial effect of TNF in host defense against *P. carinii* remains unclarified. Some investigators have postulated that pentamidine (115) or steroid-induced (117) decreases in TNF release by alveolar macrophages, and this may explain the improvement in *P. carinii* pneumonia seen when these agents are administered to infected humans.

IFN-γ is predominantly a lymphocyte-derived cytokine which can activate macrophages for enhanced microbiocidal function. Although IFN-γ is not directly lethal to *P. carinii*, it can activate macrophages in vitro to kill the organism (95,118). The evidence supporting an in vivo role for IFN-γ in host defense is conflicting. Ishimine and coworkers found increased release of IFN-γ in lung homogenates for 1–2 weeks after inoculation of normal mice with *P. carinii* organisms, whereas IFN-γ was not detectable in lung homogenates from control mice not exposed to *P. carinii* (118). Because these mice do not develop *P. carinii* pneumonia, this suggests (but does not prove) that the release of this cytokine is involved in host defense against this organism. In further support of this idea, the administration of IFN-γ by osmotic pump to steroid-treated rats (119) or by aerosol to CD4-depleted mice infected with *P. carinii* (120) reduces the intensity of infection. However, in vivo neutralization experiments do not support a major role for IFN-γ in the pathogenesis of *P. carinii* infection in the SCID mouse model. Chen and colleagues demonstrated that in vivo neutralization of IFN-γ with an antibody does not alter clearance of *P. carinii* in T-lymphocyte reconstituted SCID mice (121).

These observations require confirmation in other experimental systems to be certain that sufficient antibody was given completely to neutralize IFN-γ at all stages of infection. Possible mechanisms for IFN-γ to bolster host defense include upregulation of TNF production (122), increased generation of superoxide (123), or increased release of reactive nitrogen intermediates (91,118).

IL-1 is a proinflammatory cytokine which orchestrates inflammation and clearance of a variety of pulmonary pathogens (124). Chen and colleagues have reported that IL-1 activity in the lungs correlates with the clearance *P. carinii* infection in reconstituted SCID mice and that administration of an antibody which prevents binding of IL-1 to its receptor completely abrogates this clearance mechanism (125). These animal data suggest that IL-1 is important in normal host defense against *P. carinii*, but confirmatory evidence in humans is lacking. Huang and colleagues have postulated that corticosteroids may be beneficial in some patients with *P. carinii* pneumonia though suppression of IL-1 release by alveolar macrophages (117).

Data related to the potential importance of IL-6 in host defense against *P. carinii* are relatively sparse. In the reconstituted SCID mouse model, the administration of an antibody to IL-6 has no effect on the clearance of infection (126), but it is associated with more tissue inflammation, suggesting that IL-6 might downregulate the inflammatory response associated with *P. carinii* infection. Preliminary studies with cell lines indicate that IL-6 may augment *P. carinii* attachment to epithelial cells by stimulating the release of fibronectin (127). The precise role and cellular source of IL-6 in *P. carinii* infection in vivo remain to be clarified.

IL-8, a member of the chemokine family of cytokines, is produced by macrophages and functions as a potent chemoattractant for neutrophils into lung tissue (128). Because BAL neutrophilia is associated with more severe *P. carinii* pneumonia and a worse prognosis (129–131), some investigators have hypothesized that *P. carinii* may stimulate the release of IL-8 which then recruits neutrophils into lung tissue. These studies suggest (but do not prove) that IL-8 serves as a signal for the recruitment of neutrophils in response to *P. carinii*. Confirmatory studies in animal systems have yet to be performed. It is likely that IL-8 release is one of the early cytokine responses to *P. carinii*. However, major questions remain regarding the cellular source of BAL fluid IL-8 and the potential influence of IL-8 on other cytokines in response to *P. carinii*. Although alveolar macrophages can produce IL-8 (132), the cytokine is also produced by pulmonary epithelial cells (133), fibroblasts (134), and neutrophils (135).

CSFs are cytokines which stimulate the production of hematopoietic progenitor cells in the bone marrow and serve to regulate increased cell production during states of increased demand, such as infection. In addition to effects on progenitor cells in the bone marrow, CSFs also augment the function of mature hematopoietic cells in the tissues, which may be important in the clearance of an infectious pathogen. The role of CSFs in the pathogenesis of infectious diseases is currently under intense investigation. In spite of this interest, only granulocyte-macrophage colony-stimulating factor (GM-CSF), a cytokine that can activate mature macrophages and neutrophils, has been implicated in *P. carinii* pneumonia. GM-CSF increases the respiratory burst (as measured by chemiluminescence) of human neutrophils incubated with rat *P. carinii* (60). The relevance of these findings to host defense against *P. carinii* is not clear. In animal models, the administration of GM-CSF reduces the intensity of *P. carinii* infection in CD4-depleted mice, perhaps through upregulation of TNF production in the lung (136). This treatment effect for GM-CSF coupled with the previously described treatment effect of IFN-γ (120) suggest that cytokine therapy may offer therapeutic potential for *P. carinii* infection in immunocompromised hosts.

IX. Role of Human Immunodeficiency Virus Infection of Alveolar Macrophages

Pneumocystis carinii pneumonia remains a serious pulmonary complication in persons infected with HIV-1. Despite currently available treatment and prophylaxis, the incidence of *P. carinii* pneumonia has escalated since the onset of the AIDS epidemic (13). *P. carinii* pneumonia occurs in up to 80% of individuals infected with HIV (137), frequently recurs despite prophylaxis, and treatment is frequently complicated by adverse drug reactions or persistent infection despite appropriate therapy.

Despite the recognized impact of HIV infection on the predisposition to *P. carinii*, the mechanism(s) accounting for the high incidence of *P. carinii* pneumonia in persons with AIDS has not been fully identified (138). A progressive reduction in peripheral blood CD4$^+$ T lymphocytes serves to identify individuals at increased risk for developing *P. carinii* pneumonia (16,139). Although the protective role of CD4$^+$ T lymphocytes is further underscored in murine models of *P. carinii* pneumonia (3,51,57), the precise role of CD4$^+$ T-lymphocytes as effector cells in host defense against this pathogen has not been completely defined. The consequences of reduced CD4$^+$ T lymphocytes may include altered B-cell function and the reduced release of cytokines such as INF-γ and IL-2 which in turn may adversely affect natural killer cell and macrophage function.

Although CD4$^+$ T lymphocytes are the main target for HIV, monocyte-macrophages are also susceptible to infection with this virus (140). However, in contrast to lymphocytes, HIV infection of monocyte-macrophages does not result in cell death but rather results in persistence of HIV infection within the cell (140). This persistent HIV infection of alveolar macrophages may contribute to macrophage dysfunction which may then contribute to the frequent pulmonary complications observed in persons with AIDS (141). Alterations of many cellular functions have been reported in monocyte-macrophages obtained from HIV-infected individuals, including decreased chemotaxis (142), decreased class II MHC expression (143), altered IL-1 secretion (144), and altered bactericidal function (145). Macrophage Fc receptor–mediated phagocytosis has been variably described as being impaired (146) and unaffected (147). Although these deficiencies may in part be a direct consequence of viral infection or elaboration of viral products, the precise mechanism for these defects is not known.

A Alveolar Macrophages from HIV-Infected Individuals

Using alveolar macrophages obtained from persons with active AIDS-related lung disorders, preliminary experiments demonstrated an impaired ability to phagocytose *P. carinii* compared with alveolar macrophages from normal individuals (148). Following in vitro incubation at 37°C for 60 min at a multiplicity of 5:1 (*P. carinii*:macrophage), alveolar macrophages from HIV-infected individuals phagocytosed 50% fewer organisms compared with normal cells. This interaction appeared to be pathogen specific, since the HIV-derived cells were able to internalize a phylogenetically related organism, *C. albicans*. At a multiplicity of 5:1, alveolar macrophages form normal individuals were able completely to clear an inoculum of *P. carinii* within 12–24 hr, as determined by plate fluorimetry and electron microscopy. In contrast, alveolar macrophages from HIV-infected individuals fail to eliminate *P. carinii* over the course of 24 hr.

The exact mechanism for the observed reduction in phagocytosis is not yet known. Interestingly, preliminary observations in a small cohort of HIV-seropositive individuals demonstrated that the alveolar macrophage mannose receptor activity is reduced by 50–90% compared with normal individuals (149). Reduced macrophage mannose receptor activity in HIV-infected individuals may in part contribute to the susceptibility to *P. carinii* pneumonia. The mechanism for this observation has not yet been established. The reduced mannose receptor activity may reflect altered surface expression, altered receptor binding affinity, or interference by HIV-specific products. Alternatively, as the mannose receptor cytoplasmic tail is necessary for mediating internalization (150), the observed reduction in the ability of alveolar macrophages from HIV-infected individuals to phagocytose *P. carinii* may reflect altered receptor structure.

As detailed above, several investigators have demonstrated the release of hydrogen peroxide and other oxygen moieties by normal macrophages (60,62,88) following in vitro exposure to *P. carinii*. Recent experiments examined the release of oxygen radicals by alveolar macrophages from HIV-infected individuals following in vitro exposure to *P. carinii* and demonstrated a marked reduction in the respiratory burst response compared with alveolar macrophages from normal individuals (H. Koziel, unpublished data).

B. In Vitro Infection of Normal Alveolar Macrophages

Limited information is available examining the effect of in vitro HIV infection of normal alveolar macrophages with regard to interactions with *P. carinii*. Preliminary experiments with in vitro infection of normal alveolar macrophages with the monocytotropic isolate of HIV-1 (Bal) resulted in a 50% reduction in phagocytosis of *P. carinii* compared with uninfected cells (148). Interestingly, in vitro HIV infection of alveolar macrophages resulted in a 50% reduction in macrophage mannose receptor activity (149). These observations are particularly germane to the observations regarding alveolar macrophages from HIV-infected individuals.

The mechanism(s) for these observations are not yet known. HIV infection of alveolar macrophages may directly or indirectly impact on macrophage effector cell function at several levels. For example, the HIV-associated envelope glycoprotein gp120 inhibits in vitro fungistatic activity of normal alveolar macrophages against *Cryptococcus neoformans* (151). HIV gp120, which is rich in mannose residues (152), is capable of binding to the alveolar macrophage mannose receptor in a competitive manner (149). High levels of HIV gp120 in the alveolar microenvironment could impair the macrophage recognition of *P. carinii* through the mannose receptor pathway. Alternatively, alveolar macrophage function may be affected by altered cytokine release in the alveolar microenvironment.

These preliminary observations suggest that alveolar macrophages from HIV-infected individuals may have impaired phagocytic function against *P. carinii*, which may be pathogen specific, related to HIV infection and may in part reflect impaired macrophage receptor function. Furthermore, alveolar macrophages from HIV-infected individuals may have a reduced respiratory burst response to *P. carinii*, which may impair their ability to eliminate this pulmonary pathogen. As the oxidative-mediated elimination may represent an important mechanism for the killing of *P. carinii*, this abnormality may have profound impact on host defense against this organism. These preliminary observations are the focus of ongoing investigations.

C. Alveolar Lining Material Components in HIV Infection

Surfactant Protein A

SP-A is elevated for unknown reasons in BAL fluid obtained from patients with AIDS-related *P. carinii* pneumonia (153,154), an observation which appears to be independent of corticosteroid use (153) or HIV infection (154). The specific role of SP-A in *P. carinii* pneumonia is not yet established. The elevated SP-A levels may serve as a nonimmune host response in an attempt to combat the organism. Alternatively, this may reflect a mechanism of pathogen subversion of host cell recognition (i.e., by coating the organism to escape immune cell recognition).

Extracellular Matrix Proteins

Although potentially important in the normal macrophage interaction with *P. carinii*, the role of fibronectin and other extracellular matrix proteins has not been investigated in HIV-related disease. Although elevated SP-A levels in BAL fluid have been described in HIV-related *P. carinii* pneumonia, it is unknown whether alterations in fibronectin levels are present in human disease.

Cholesterol

Investigators have described changes in the phospholipid composition of the extracellular milieu that result in altered macrophage membrane lipid composition which reduce fluid-phase pinocytosis and receptor-mediated phagocytosis (155). Thus, altered lipids in the alveolar lining material may impact on the cellular response to *P. carinii*. Examination of the BAL fluid specimens from HIV-infected individuals with active *P. carinii* pneumonia demonstrate altered phosphatidylcholine to cholesterol ratios (156). In vitro exposure of normal alveolar macrophages to increased concentrations of cholesterol resulted in a 33% reduction in phagocytic capacity (although not binding) of *P. carinii* by the macrophages. Thus, alteration of the cholesterol ratio to unfavorable or physiologically disruptive ratios in the alveolar microenvironment may negatively impact on the alveolar macrophage's ability to defend against this opportunistic pathogen and may contribute to the pathogenesis of pneumonia.

D. Cytokines in HIV-Related Disease

Several alterations in the cytokine production and response to specific stimulation have been described in HIV-related disease (141,157). Several investigators observed increased spontaneous release of TNF by alveolar macrophages from HIV-infected individuals (158,159). Krishnan and colleagues measured elevated levels of TNF in the BAL fluid of HIV-infected patients with *P. carinii* pneumonia compared with HIV-infected individuals without pneumonia, and they observed that the TNF levels inversely correlate with cyst burden (159). Data from Kandil and colleagues demonstrated that in vitro infection of human monocyte-derived macrophages with HIV eliminates the release of TNF in response to *P. carinii* (160). These data suggest that infection with HIV may be permissive for infection with *P. carinii* through interference with the macrophage cytokine response to the pathogen.

Circulating $CD4^+$ lymphocytes from patients with AIDS demonstrate a reduced capacity to produce IFN-γ in response to antigenic or mitogenic stimulation (161,162). This observation raises the possibility that in the setting of HIV-related $CD4^+$ T-lymphocyte depletion, *P. carinii* pneumonia may result in part due to the insufficient release of IFN-γ by these cells in lung tissue in response to the infectious challenge.

The release of IL-1 is reduced in macrophages infected with HIV-1 in vitro (144), whereas monocytes (163) and alveolar macrophages (117,164) from AIDS patients may show supranormal release. Kandil and coworkers found that in vitro HIV infection of human monocytes suppresses the release of IL-6 and TNF in response to incubation with *P. carinii* (160). Recent clinical studies found that excess IL-8 is present in BAL fluid from HIV-infected patients with *P. carinii* pneumonia, and the concentration of IL-8 correlates with absolute numbers of lavage neutrophils and with mortality (165–167). However, elevated BAL fluid IL-8 is not specific for *P. carinii* infection, as IL-8 was also present in asymptomatic HIV-infected persons (167).

Although alveolar macrophages release significant amounts of GM-CSF following in vitro stimulation (168), the role of GM-CSF in macrophage function in HIV-related disease is not known. Compared with neutrophils from HIV^+ subjects with no history of *P. carinii* infection, stimulation of blood neutrophils from HIV-infected subjects with a prior history of *P. carinii* pneumonia demonstrated a decreased respiratory burst response following in vitro stimulation with *P. carinii*. This defective respiratory burst response could be corrected with the addition of GM-CSF (169). However, the relevance of these findings to host defense against *P. carinii* is not clear.

Recent observations suggest that in vitro exposure of alveolar macrophages from HIV-infected individuals to certain cytokines may facilitate phagocytosis of *P. carinii*. Preliminary observations suggest that pretreatment of alveolar macrophages from HIV-infected individuals with macrophage CSF (M-CSF) (170) or IL-13 (171) enhances in vitro phagocytosis of *P. carinii*. In both investigations, the enhanced phagocytosis is in part related to increased macrophage mannose receptor expression (170,171). These observations suggest a potential role for cytokine immunomodulation of effector cells as possible adjunct to conventional antimicrobial therapy against *P. carinii* pneumonia.

Thus, several alterations in the cytokine response of immune cells in HIV-related disease (including response to *P. carinii*) have been described. Although these abnormalities likely represent direct consequences of HIV infection or the loss of $CD4^+$ T lymphocytes, the precise mechanism(s) have not yet been elucidated. The importance of better understanding the cytokine response to *P. carinii* resides in the potential application of cytokine-mediated therapy in HIV-related *P. carinii* pneumonia.

X. Future Studies of the *Pneumocystis carinii*–Macrophage Interaction

The available data reviewed above support the concept that alveolar macrophages represent important effector cells against *P. carinii*, and in conjuncture with other immune and nonimmune factors (including intact $CD4^+$ T lymphocytes) likely serve an important role in the pulmonary host defense against this opportunistic pathogen. Although much has been learned over 25 years of investigation, many aspects of the interaction of *P. carinii* with macrophages remain unknown or only partially understood. This includes a limited understanding of the relative importance of specific macrophage receptors, the precise role of host molecules present in the alveolar lining material, the exact mechanism for elimination of *P. carinii*, and the mechanism(s) by which *P. carinii* eludes or subverts macrophage phagocytosis in the setting of active pneumonia in the susceptable host. Clearly, the interaction of the host cellular and humoral immune systems coupled with the complexity of the alveolar microenvironment suggests the host response to *P. carinii* likely represents a complex and coordinated interaction of these immune and nonimmune factors.

Future investigations will be facilitated by the availability of purified organisms. Hopefully, with the development of new systems for culturing *P. carinii* (172), larger quantities of organisms free of host cell contamination will be available. Refining viability assays to assist in the assessment of factors directly injurious to the organisms and to allow better understanding of the life cycle, growth, and metabolic requirements of *P. carinii* will greatly enhance in vitro studies. The observation that disease is generally limited to the lungs suggests that the composition of the alveolar microenvironment may hold certain clues which explain host susceptibility to infection. Additional studies are required to investigate the in vivo clearance of *P. carinii* in animal models to advance the current understanding of factors important in the normal host response to this organism and establish the role of alveolar macrophages. Further investigations are needed of alveolar macrophages from HIV-infected individuals or normal alveolar macrophages infected with HIV in vitro to clarify mechanisms accounting for susceptibility in this particular population of immunocompromised hosts. Finally, the use of animal "knockout" models will likely establish the importance of discrete host factors in macrophage–*P. carinii* interaction and may lead to significant insights regarding disease pathogenesis and suggest potential novel therapies.

References

1. Hughes W. Natural occurrences in animals. Pneumocystis carinii Pneumonitis. Boca Raton, FL: CRC Press, 1987:57–70.
2. Barton E, Campbell W. Pneumocystis carinii in lungs of rats treated with cortisone acetate. Am J Pathol 1969; 54:209–236.
3. Shellito JE, Suzara VV, Blumenfeld W, Beck JM, Steger HJ, Ermak TH. A new model of Pneumocystis carinii infection in mice selectively depleted of helper T lymphocytes. J Clin Invest 1990; 85:1686–1693.
4. Harmsen AG, Chen W. Resolution of Pneumocystis carinii pneumonia in CD4+ lymphocyte-depleted mice given aerosols of heat-treated Escherichia coli. J Exp Med 1992; 176:881–886.
5. Kolls J, Beck J, Nelson S, Summer W, Shellito J. Alveolar macrophage release of tumor necrosis factor during murine Pneumocystis carinii pneumonia. Am J Respir Cell Mol Biol 1993; 8:370–376.
6. Edman JC, Kovacs JA, Mansur H, Santi DV, Elwood HJ, Sogin ML. Ribosomal RNA sequence show Pneumocystis carinii to be a member of the fungi. Nature 1988; 334:519–522.
7. Stringer S, Stringer J, Blase M, Walzer P, Cushion M. Pneumocystis carinii: sequence from ribosomal RNA implies a close relationship with fungi. Exp Parasitol 1989; 68:450–461.
8. Pixley F, Wakefield A, Benerji S, Hopkin J. Mitochonidrial gene sequences show fungal homology for *Pneumocystis carinii*. Mol Microbiol 1991; 5:1347–1351.
9. Bedrossian C. Ultrastructure of Pneumocystis carinii: a review of internal and surface characteristics. Semin Diagn Pathol 1989; 6:212–237.
10. Cushion M, Stringer J, Walzer P. Cellular and molecular biology of Pneumocystis carinii. Int Rev Cytol 1991; 131:59–107.
11. Jacobs J, Libby D, Winters R, Gelmont D, Fried E, Hartman B, Laurence J. A cluster of Pneumocystis carinii pneumonia in adults without predisposing illenesses. N Engl J Med 1991; 324:246–250.
12. Palanage P, Serra P, Di Sabato F, Contini C, Giacovazzo M. Pneumocystis carinii pneumonia in a patient with chronic obstructive pulmonary disease but no evident immunoincompetence. Clin Infect Dis 1994; 19:543–544.
13. Masur H, Lane C, Kovacs JA, Allegra CJ, Edman JC. Pneumocystis pneumonia: from bench to clinic. Ann Intern Med 1989; 111:813–826.
14. Murray J, Mills J. Pulmonary infectious complications of human immunodeficiency virus infection. Part II. Am Rev Respir Dis 1990; 141:1582–1598.

15. Safrin S. Pneumocystis carinii pneumonia in patients with the acquired immunodeficiency syndrome. Semin Respir Infect 1993; 8:96–103.
16. Phair J, Munoz A, Detels R, Kaslow R, Rinaldo C, Saah A. The risk of Pneumocystis carinii pneumonia among men infected with human immunodeficiency virus type 1. N Engl J Med 1990; 322:161–165.
17. Kovacs J, Hiemenz J, Macher A, Stover D, Murray H, Shelhamer J, Lane H, Urmacher C, Honig C, Longo D, Parker M, Natanson C, Parrillo J, Fauci A, Pizzo P, Masur H. Pneumocystis carinii pneumonia: a comparison between patients with the acquired immunodeficiency syndrome and other immunodeficiencies. Ann Intern Med 1984; 100:663–671.
18. Askin F, Katzenstein A. Pnemocystic infection masquerading as diffuse alveolar damage: a potential source of diagnostic error. Chest 1981; 79:420–422.
19. Blumenfeld W, Basgoz N, Owen WJ, Schmidt D. Granulomatous pulmonary lesions in patients with the acquired immunodeficiency syndrome (AIDS) and Pneumocystis carinii infection. Ann Intern Med 1988; 109:505–507.
20. Barrio J, Suarez M, Rodriguez J. Pneumocystis carinii pneumonia presenting as cavitating and noncavitating solitary pulmonary nodules in patients with the acquired immunodeficiency syndrome. Am Rev Respir Dis 1986; 135:1094–1097.
21. Saldana M, Mones J. Cavitation and other atypical manifestations of Pneumocystis carinii pneumonia. Semin Diagn Pathol 1989; 6:273–286.
22. Lee M, Schinella R. Pulmonary calcification caused by Pneumocystis carinii pneumonia. A clinicopathological study of 13 cases in acquired immune deficiency syndrome patients. Am J Surg Pathol 1991; 15:376–380.
23. Gal A, Koss M, Strigle S, Angritt P. Pneumocystis carinii infection in the acquired immunodeficiency syndrome. Semin Diagn Pathol 1989; 6:2387–2395.
24. Stover D, Greeno R, Gagliardi A. The use of a simple exercise test for the diagnosis of Pneumocystis carinii pneumonia in patients with AIDS. Am Rev Respir Dis 1989; 139:1343–1346.
25. Smith R, Ripps C, Lewis M. Elevated lactate dehydrogenase values in patients with Pneumocystis carinii pneumonia. Chest 1988; 93:987–992.
26. Hopewell P. Pneumocystis carinii pneumonia: diagnosis. J Infect Dis 1988; 157: 1115–1119.
27. Watts J, Chandler F. Evolving concepts of infection by Pneumocystis carinii. In: Rosen P, Fechner R, eds. Pathology Annual. Norwalk, CT: Appleton & Lange, 1991: 93–138.
28. Huang S, Marshall K. Pneumocystis carinii infection: a cytologic, histologic and electron microscopic study of the organism. Am Rev Respir Dis 1970; 102:623–629.
29. Vavra J, Kucera K. Pneumocystis carinii: its ultrastructure and ultrastructural affinities. J Protozool 1970; 17:463–472.
30. Pifer L, Hughes W, Stagno S, Woods D. Pneumocystis carinii infection: evidence for high prevalence in normal and immunosuppressed children. Pediatrics 1978; 62: 35–41.
31. Meuwissen J, Tauber I, Leewenberg A, Beckers P, Shiehen J. Parasitologic and serologic observations of infection with Pneumocystis in humans. J Infect Dis 1977; 136:43–49.

32. Peglow S, Smulian G, Linke M, et al. Serologic responses to specific Pneumocystis carinii antigens in health and disease. J Infect Dis 1990; 161:296–306.
33. Peters S, Wakefield A, Sinclair K, Millard P, Hopkins J. A search for Pneumocystis carinii in post mortem lungs by DNA amplification. J Pathol 1992; 166:195–198.
34. Settnes O, Genner J. Pneumocystis carinii in human lungs at autopsy. Scan J Infect Dis 1986; 18:489–496.
35. Kovacs J, Halpern J, Lundgren B, Swan J, Parillo J, Masur H. Monoclonal antibodies to Pneumocystis carinii: identification of specific antigens and characterization of antigenic differences between rat and human isolates. J Infec Dis 1989; 159:60–70.
36. Bauer N, Paulsrud J, Bartlett M, Smith J, Wilde CI. Pneumocystis carinii organisms obtained from rats, ferrets, and mice are antigenically different. Infect Immun 1993; 61:1315–1319.
37. Gigliotti F, Haidaris P, Haidaris C, Wright T, Van Der Meid K. Further evidence of host species-specifica variation in antigens of Pneumocystis carinii using the polymerase chain reaction. J Infect Dis 1993; 168:191–194.
38. Chusid M, Heyrman K. An outbreak of Pneumocystis carinii at a pediatric hospital. Pediatrics 1978; 62:1031–1035.
39. Singer C, Armstrong D, Rosen P, Schottenfeld D. Pneumocystis carinii pneumonia: a cluster of eleven cases. Ann Intern Med 1975; 82:772–777.
40. Frenkel J, Good J, Schultz J. Latent Pneumocystis infections of rats, relapse, and chemotherapy. Lab Invest 1966; 15:1559–1577.
41. Hughes W. Animal models for Pneumocystis carinii pneumonia. J Protozool 1989; 36:41–45.
42. Walzer P. Experimental models of Pneumocystis carinii infections. In: Young L, ed. Pneumocystis carinii Pneumonia: Pathogenesis, Diagnosis, and Treatment. New York: Marcel Dekker, 1984:7–76.
43. Bartlett M, Fishman J, Queener S, Durkin M, Jay M, Smith J. New rat model of Pneumocystis carinii infection. J Clin Microbiol 1988; 26:1100–1102.
44. Walzer PD, Powell RD, Yoneda K. Experimental Pneumocystis carinii pneumonia in different strains of cortisonized mice. Infect Immun 1979; 24:939–947.
45. Bartlett M, Queener S, Durkin M, Shaw M, Smith J. Inoculated mouse model of Pneumocystis carinii pneumonia. J Protozool 1991; 38:130S–131S.
46. Schultz LD, Schweitzer PA, Hall EJ, Sundberg JP, Taylor S, Walzer PD. Pneumocystis carinii pneumonia in scid/scid mice. Curr Top Microbiol Immunol 1989; 152: 243–249.
47. Walzer PD, Schnelle V, Armstrong D, Rosen PP. Nude mouse: a new experimental model for Pneumocystis carinii infection. Science 1977; 197:177–179.
48. Roths JB, Marshall JD, Allen RD, Carlson GA, Sidman CL. Spontaneous Pneumocystis carinii pneumonia in immunodeficient mutant scid mice. Am J Pathol 1990; 136:1173–1186.
49. Walzer P, Kim C, Linke M, Pogue C, Huerkamp M, Chrisp C, Lerro A, Wixson S, Hall E, Shultz L. Outbreaks of Pneumocystis carinii pneumonia in colonies of immunodeficient mice. Infect Immun 1989; 57:62–70.
50. Roths JB, Sidman CL. Both immunity and hyperresponsiveness to Pneumocystis carinii result from transfer of CD4+ but not CD8+ T cells into severe combined immunodeficiency mice. J Clin Invest 1992; 90:673–678.

51. Harmsen A, Stankiewicz M. Requirement for CD4+ cells in resistance to Pneumocystis carinii pneumonia in mice. J Exp Med 1990; 172:937–945.
52. Beck JM, Warnock ML, Curtis JL, Sniezek MJ, Arraj-Peffer SM, Kaltreider HB, Shellito JE. Inflammatory responses to Pneumocystis carinii in mice selectively depleted of helper T lymphocytes. Am J Respir Cell Mol Biol 1991; 5: 186–197.
53. Sloand E, Laughon B, Armstrong M, Bartlett M, Blumenfeld W, Cushion M, Kalica A, Kovacs J, Martin W, Pitt E, Pesanti E, Richards F, Rose R, Walzer P. The challenge of Pneumocystis carinii culture. J Eukaryot Microbiol 1993; 40:188–195.
54. Walzer PD, Powell RD, Yoneda K, Rutledge ME, Milder JE. Growth characteristics and pathogenesis of experimental Pneumocystis carinii pneumonia. Infect Immun 1980; 27:928–937.
55. Yoneda K, Walzer P. The interaction of Pneumocystis carinii with host cells: an ultrastructural study. Infect Immun 1980; 29:692–703.
56. Walzer PD, Kim CK, Cushion MT. Pneumocystis carinii. In: Walzer PD, Genta RM, eds. Parasitic Infections in the Immunocompromised Host. New York: Marcel Dekker, 1989:83–178.
57. Chen W, Mills J, Harmsen A. Development and resolution of Pneumocystis carinii pneumonia in severe combined immunodeficient mice: a morphological study of host inflammatory responses. Int J Exp Pathol 1992; 73:709–720.
58. von Behren LA, Pesanti EL. Uptake and degradation of Pneumocystis carinii by macrophages in vitro. Am Rev Respir Dis 1978; 118:1051–1059.
59. Masur H, Jones TC. The interaction in vitro of Pneumocystis carinii with macrophages and L-cells. J Exp Med 1978; 147:157–170.
60. Taylor MB, Phillips M, Easmon CSF. Opsonophagocytosis of Pneumocystis carinii. J Med Microbiol 1992; 36:223–228.
61. Forte M, Rahelu M, Stubberfield C, Tomkins L, Pithie A, Kumararatne D. In-vitro interaction of human macrophages with Pneumocystis carinii. Int J Experimental Pathol 1991; 72:589–598.
62. Ezekowitz RAB, Williams DJ, Koziel H, Armstrong MYK, Warner A, Richards FF, Rose RM. Uptake of Pneumocystis carinii mediated by the macrophage mannose receptor. Nature 1991; 351:155–158.
63. Chroneos ZC, Shepherd VL. Identification and purification of a high affinity receptor for SP-A. Am J Respir Crit Care Med 1994; 149:A503.
64. Phelps DS, Fishman JA, Rose RM. Surfactant protein A levels in glucocorticoid-immunosupressed rats infected with Pneumocystis carinii. Am Rev Respir Dis 1992; 145:A246.
65. Zimmerman P, Voelker D, McCormack F, Paulsrud J, Martin WI. 120-kD surface glycoprotein of Pneumocystis carinii is a ligand for surfactant protein A. J Clin Invest 1992; 89:143–149.
66. Williams MD, Voelker DR, Martin WJ, II. Surfactant protein A enhances attachment of Pneumocystis carinii to rat alveolar macrophages. Am J Respir Crit Care Med 1994; 149:A289.
67. Koziel H, O'Riordan D, Phelps D, Fishman JA, Armstrong MYK, Richards FF. Surfactant protein-A inhibits binding and internalization of Pneumocystis carinii by alveolar macrophages. Am Rev Respir Dis 1992; 145:A247.

68. Pottratz ST, Martin WJ. Role of fibronectin in Pneumocystis carinii attachment to cultured lung cells. J Clin Invest 1990; 85:351–356.
69. Limper AH, Standing JE, Hoffman OA, Castro M, Neese LW. Vitronectin binds to Pneumocystis carinii and mediates organism attachment to cultured lung epithelial cells. Infect Immun 1993; 61:4302–4309.
70. Fishman JA, Samia JA, Fuglestad J, Rose RM. The effects of extracellular matrix (ECM) proteins on the attachment of Pneumocystis carinii to lung cell lines in vitro. J Protozool 1991; 38:34S–37S.
71. Yu ML, Limper AH. Pneumocystis carinii induces cellular expression of intercellular adhesion molecule 1. Am J Respir Crit Care Med 1995; 151:A794.
72. Pesanti E, Tomicic T, Donta S. Binding of ^{125}I-labelled tumor necrosis factor to Pneumocystis carinii and an insoluble cell wall fraction. J Protozool 1991; 38:28S–29S.
73. Narasimhan S, Armstrong MY, Rhee K, Edman JC, Richards FF, Spicer E. Gene for an extracellular matrix receptor protein from Pneumocystis carinii. Proc Natl Acad Sci USA 1994; 91:7440–7444.
74. Hocking WG, Golde DW. The pulmonary-alveolar macrophage. N Engl J Med 1979; 301:580–645.
75. Blumenfeld W, McCook O, Griffis JM. In vitro aggregation of macrophages around human-derived Pneumocystis carinii. J Protozool 1991; 38:32S–33S.
76. Koziel H, Baik J, Armstrong MYK, Richards FF, Rose RM. Pneumocystis carinii surface glycoprotein-A is a chemotactic factor for normal human blood monocytes. Am Rev Respir Dis 1993; 147:A34.
77. Horio S, Ando M, Ukeshima A, Sugimoto M, Tokuomi H. Surface morphology and function of alveolar macrophages exposed to lung lavage fluids and serum from normal rabbits. J Reticuloendothel Soc 1981; 29:137–152.
78. Stahl PD. The macrophage mannose receptor: current Status. Am J Respir Cell Mol Biol 1990:317–318.
79. Pesanti E. Pneumocystis carinii: oxygen uptake, antioxidant enzymes, and susceptibility to oxygen-mediated damage. Infect Immun 1984; 44:7–11.
80. Williams DJ, Radding JA, Dell A, Khoo K-H, Rogers ME, Richards FF, Armstrong MYK. Glucan synthesis in Pneumocystis carinii. J Protozool 1991; 38:427–437.
81. Kaneshiro ES, Wu YP, Cushion MT. Assays for testing Pneumocystis carinii viability. J Protozool 1991; 38:85S–87S.
82. Ruffolo J, Cushion M, Walzer P. Microscopic techniques for studying Pneumocystis carinii in fresh specimens. J Clin Microbiol 1986; 23:17–21.
83. Chen F, Cushion M. Use of an ATP bioluminescent assay to evaluate viability of Pneumocystis carinii from rats. J Clin Microbiol 1994; 32:2791–2800.
84. DeStefano JA, Sleight RG, Babcock GF, Sramkoski RM, Walzer PD. Isolation of Pneumocystis carinii cysts by flow cytometry. Parasitol Res 1992; 78:179–182.
85. Lapinsky SE, Glencross D, Car NG, Kallenbach JM, Zwi S. Quantification and assessment of viability of Pneumocystis carinii organisms by flow cytometry. J Clin Microbiol 1991; 29:911–915.
86. Armstrong MYK, Koziel H, Rose RM. Arena C, Richards FF. Indicators of Pneumocystis carinii viability in short-term culture. J Protozool 1991; 38:88S–90S.
87. Pottratz ST, Martin WJ. Mechanism of Pneumocystis carinii attachment to cultured rat alveolar macrophages. J Clin Invest 1990; 86:1678–1683.

88. Hidalgo HA, Helmke RJ, German VF, Mangos JA. Pneumocystis carinii induces an oxidative burst in alveolar macrophages. Infect Immun 1992; 60:1–7.
89. Koziel H, Armstrong M, Arena C, Pellitier N, Richards F, Rose R. Tumor necrosis factor-alpha does not affect viability of Pneumocystis carinii in vitro. Am Rev Respir Dis 1993; 147:A33.
90. Gabig TG, Babior BM. The killing of pathogens by phagocytes. Annu Rev Med 1981; 32:313–326.
91. Sherman M, Loro M, Wong V, Tashkin D. Cytokine and *Pneumocystis carinii*–induced L-arginine oxidation by murine and human pulmonary alveolar macrophages. J Protozool 1991; 38:234S–236S.
92. Williams M, Downing J, Martin WI. Dexamethasone inhibits Pneumocystis carinii–mediated nitric oxide production in rat alveolar macrophages (abstr). Am J Respir Crit Care Med 1995; 151:A796.
93. Shellito JE, Kolls JK, Beck JM. Nitric oxide synthase and tissue inflammation during murine P. carinii pneumonia. Am J Respir Crit Care Med 1995; 151:A794.
94. Shellito JE, Kolls JK, Olariu R, Beck JM. Nitric oxide and host defense against Pneumocystis carinii infection in a mouse model. J Infect Dis 1996; 173:432–439.
95. Pesanti EL. Interaction of cytokines and alveolar cells with Pneumocystis carinii in vitro. J Infect Dis 1991; 163:611–616.
96. Tamburrini E, DeLuca A, Ventura G, maiuro G, Siracusano A, Ortono E, Antinori A. Pneumocystis carinii stimulates in vitro production of tumor necrosis factor-alpha by human macrophages. Med Microbiol Immunol 1991; 180:15–20.
97. Pesanti E. In vitro effects of antiprotozoan drugs and immune serum on Pneumocystis carinii. J Infect Dis 1980; 141:755–759.
98. Cushion M, Ruffolo J, Linke M, Walzer P. Pneumocystis carinii: growth variables and estimates in the A549 and WI-38 VA 13 human cells lines. Exp Parisitol 1985; 60:43–54.
99. Walzer P, Schultz M, Western K, Robbins J. Pneumocystis carinii pneumonia and primary immune deficiency diseases of infancy and childhood. J Pediatr 1973; 82:416–422.
100. Walzer P, Perl D, Krogstead D, Rawson P, Schultz M. Pneumocystis carinii pneumonia in the United States: epidemiologic, diagnostic, and clinical features. Ann Intern Med 1974; 80:83–93.
101. Brzosko W, Krawczynski K, Madalinski K, Nowoslawski A. Immunopathologic aspects of Pneumocystis carinii pneumonia in infants as revealed by immunofluorescence and electron microscopy. Natl Cancer Inst Monogr 1976; 43:163–169.
102. Dutz W, Post C, Vessal K, Kohut E. Endemic infantile Pneumocystis carinii infection: the Shiraz study. Natl Cancer Inst Monogr 1976; 43:31–38.
103. Rankin J, Walzer P, Dwyer J, Schraeder C, Enriquez R, Merrill W. Immunologic alterations in bronchoalveolar lavage fluid in the acquired immune deficiency syndrome (AIDS). Am Rev Respir Dis 1983; 128:189–194.
104. Walzer PD, Rutledge ME. Humoral immunity in experimental Pneumocystis carinii infection. I. Serum and bronchial lavage fluid antibody responses in rats. J Lab Clin Med 1981; 97:820–833.
105. Brzosko W, Madalinski K, Crawcznski K, Neweslawski A. Immunohistochemistry in studies of the pathogenesis of pneumocystis pneumonia in infants. Ann NY Acad Sci 1971; 177:156–170.

106. Ryan RM, Morris RE, Rice WR, Ciraolo G, Whitsett JA. Binding and uptake of pulmonary surfactant protein (SP-A) by pulmonary type II epithelial cells. J Histochem Cytochem 1989; 37:429–440.
107. Tenner AJ, Robinson SL, Borchelt J, Wright JR. Human pulmonary surfactant protein (SP-A), a protein structurally homologous to C1q, can enhance FcR- and CR1-mediated phagocytosis. J Biol Chem 1989; 264:13923–13928.
108. Van Iwaarden F, Welmers B, Verhoef J, Haagsman HP, Van Golde LMG. Pulmonary surfactant protein A enhances the host defense mechanism of rat alveolar macrophages. Am J Respir Cell Mol Biol 1990; 2:91–98.
109. Manz-Keinke H, Plattner H, Schlepper-Schafer J. Lung surfactant protein A (SP-A) enhances serum-independent phagocytosis of bacteria by alveolar macrophages. Eur J Cell Biol 1992; 57:95–100.
110. O'Riordan DM, Standing JE, Limper AH. Pneumocystis carinii glycoprotein A binds macrophage mannose receptors. Infect Immun 1995; 63:779–784.
111. O'Riordan D, Standing J, Kwon K, Chang D, Crouch E, Limper A. Surfactant protein D interacts with Pneumocystis carinii and mediates organism adherence to alveolar macrophages. J Clin Invest 1995; 95:2699–2710.
112. Koziel H, Williams DJ, Armstrong MYK, Richards FF, Fishman JA, Ezekowitz RAB, Warner A, Fuglestad J, Rose RM. New rapid method for the study of Pneumocystis carinii interaction with alveolar macrophages. J Protozool 1991; 36 (suppl):173–174.
113. Armstrong M, Richards F. Propogation and purification of rat Pneumocystis carinii in short-term cell cultures. J Protozool 1989; 36:24S–27S.
114. Hoffman OA, Standing JE, Limper AH. Pneumocystis carinii stimulates tumor necrosis factor-alpha release from alveolar macrophages through a beta-glucan-medicated mechanism. J Immunol 1993; 150:3932–3940.
115. Corsini E, Dykstra C, Craig W, Tidwell R, Rosenthal G. Pneumocystis carinii induction of tumor necrosis factor-alpha by alveolar macrophages: modulation by pentamidine isethionate. Immunol Lett 1992; 34:303–308.
116. Neese LW, Standing JE, Olson EJ, Castro M, Limper AH. Vitrinectin, fibronectin, and gp120 antibody enhance macrophage release of TNF-alpha in response to Pneumocystis carinii. J Immunol 1994; 152:4549–4556.
117. Huang Z, Eden E. Effect of corticosteroids on IL1-beta and TNF-alpha release by alveolar macrophages from patients with AIDS and Pneumocystis carinii pneumonia. Chest 1993; 104:751–755.
118. Ishimine T, Kawakami K, Nakamoto A, Saito A. Analysis of cellular response and gamma interferon synthesis in bronchoalveolar lavage fluid and lung homogenate of mice infected with Pneumocystis carinii. Microbiol Immunol 1995; 39:49–58.
119. Shear HL, Valladares G, Narachi MA. Enhanced treatment of Pneumocystis carinii pneumonia in rats with interferon-gamma and reduced doses of trimethoprim/sulfamethoxazole. J Acquir Immune Defic Syndr 1990; 3:943–948.
120. Beck JM, Liggitt HD, Burnette EN, Fuchs HJ, Shellito JE, Debs RJ. Reduction in intensity of Pneumocystis carinii pneumonia in mice by aerosol administration of gamma interferon. Infect Immun 1991; 59:3859–3862.
121. Chen W, Havell E, Harmsen A. Importance of endogenous tumor necrosis factor alpha and gamma interferon in host resistance to Pneumocystis carinii infection. Infect Immun 1992; 60:1279–1284.

122. Burchett S, Weaver W, Westall J, Larsen A, Kronheim S, Wilson C. Regulation of tumor necrosis factor/cachectin and IL-1 secretion in human mononuclear phagocytes. J Immunol 1988; 140:3473–3481.
123. Drath D. Modulation of pulmonary macrophage superoxide release and tumoricidal activity following activation by biological response modifiers. Immunopharmacology 1986; 12:117–126.
124. Dinarello C. Interleukin-1 and its biologically related cytokines. Adv Immunol 1989; 44:153–160.
125. Chen W, Havell EA, Moldawer LL, McIntyre KW, Chizzonite RA, Harmsen AG. Interleukin 1: an important mediator of host resistance against Pneumocystis carinii. J Exp Med 1992; 176:713–718.
126. Chen W, Havell E, Gigliotti F, Harmsen A. Interleukin-6 production in a murine model of Pneumocystis carinii pneumonia: relation to resistance and inflammatory response. Infect Immun 1993; 61:97–102.
127. Pottratz S, Sheldon J. Role of interleukin-6 in attachment of Pneumocystis carinii to alveolar epithelial cells. J Invest Med 1995; 43:479A.
128. Kunkel S, Standiford T, Kasahara K, Strieter R. Interleukin-8: the major neutrophil chemotactic factor in the lung. Exp Lung Res 1991; 17:17–23.
129. Mason G, Hasimota C, Dickman P, Foutty L, Cobb C. Prognostic implications of bronchoalvelar lavage neutrophilia in patients with Pneumocystis carinii pneumonia and AIDS. Am Rev Respir Dis 1989.
130. Jensen B, Lisse I, Gerstoft J, Borgeskov S, Skinhoj P. Cellular profiles in bronchoalveolar lavage fluid of HIV-infected patients with pulmonary symptoms: relation to diagnosis and prognosis. AIDS 1991; 5:527–533.
131. Smith RL, El-Sadir WM, Lewis ML. Correlation of bronchoalveolar lavage cell populations with clinical severity of Pneumocystis carinii pneumonia. Chest 1988; 92:60–64.
132. Strieter R, Chensue S, Basha M, Standiford T, Lynch J, Baggiolini M, Kunkel S. Human alveolar macrophage gene expression of interleukin-8 by tumor necrosis factor-alpha, lipopolysaccharide, and interleukin-1 beta. Am J Respir Cell Mol Biol 1990; 2:321–326.
133. Standiford T, Kunkel S, Basha M, Chensue S, Lynch Jd, Toews G, Westwick J, Strieter R. Interleukin-8 gene expression by a pulmonary epithelial cell line. A model for cytokine networks in the lung. J Clin Invest 1990; 86:1945–1953.
134. Rolfe M, Kunkel S, Standiford T, Chensue S, Allen R, Evanoff H, Phan S, Strieter R. Pulmonary fibroblast expression of interleukin-8: a model for alveolar macrophage-derived cytokine networking. Am J Respir Cell Mol Biol 1991; 5:493–501.
135. Strieter R, Kasahara K, Allen R, Standiford T, Rolfe M, Becker F, Chensue S, Kunkel S. Cytokine-induced neutrophil-derived interleukin-8. Am J Pathol 1992; 141:397–407.
136. Mandujano J, D'Souza N, Nelson S, Summer W, Beckerman R, Shellito J. Granulocyte-macrophage colony stimulating factor and Pneumocystis carinii pneumonia in mice. Am J Respir Crit Care Med 1995; 151:1233–1238.
137. Fauci AS, Macher AM, Longo DL, Lane HC, Rook AH, Masur H, Gelmann EP. NIH Conference. Acquired immunodeficiency syndrome: epidemiologic, clinical, immunologic, and therapeutic considerations. Ann Intern Med 1984; 100:92–106.
138. Martin WJ. Pathogensis of Pneumocystis carinii pneumonia. Am J Respir Cell Mol Biol 1993; 8:356–357.

139. Masur H, Ognibene FP, Yarchoan R, Shelhamer JH, Baird BF, Travis W, Suffredini AF, Deyton L, Kovacs JA, Falloon J, Davey R, Polis M, Metcalf J, Baseler M, Wesley R, Gill VJ, Fauci AS, Lane HC. CD4 counts as predictors of opportunistic pneumonias in human immunodeficiency virus (HIV) infection. Ann Intern Med 1989; 111:223–231.
140. Fauci A. The human immunodeficiency virus: infectivity and mechanisms of pathogenesis. Science 1988; 239:617.
141. Agostini C, Trentin L, Zambello R, Semenzato G. HIV-1 and the Lung. Infectivity, pathogenic mechanisms and cellular immune responses taking place in the lower respiratory tract. Am Rev Respir Dis 1993; 147:1038–1049.
142. Smith PD, Ohura K, Masur H, Lane HC, Fauci AS, Wahl SM. Monocyte function in the acquired immune deficiency syndrome. Defective chemotaxis. J Clin Invest 1984; 74:2121–2128.
143. Belsito D, Sanchez M, Baer R, Valentine F, Thorbecke G. Reduced Langerhans cell Ia antigen and ATPase activity in patients with the acquired immunodeficiency syndrome. N Engl J Med 1984; 310:1279–1282.
144. Roy S, Fitz-Gibbon L, Poulin L, Wainberg M. Infection of human monocyte/macrophages by HIV-1: effect on secretion of IL-1 activity. Immunology 1988; 64: 233–239.
145. Roy S, Wainberg MA. Role of the mononuclear phagocyte system in the development of acquire immunodeficiency syndrome (AIDS). J Leukoc Biol 1988; 43: 91–97.
146. Bender B, Frank M, Lawley T, Smith W, Brinkman C, Quinn T. Defective reticuloendothelial system Fc-receptor function in patients with acquired immunodeficiency syndrome. J Infect Dis 1985; 152:409.
147. Nottet HSLM, de Graaf L, de Vos NM, Bakker LJ, van Strijp JAG, Visser MR, Verhoef J. Phagocytic function of monocyte-derived macrophages is not affected by human immunodeficiency virus type 1 infection. J Infect Dis 1993; 168:84–91.
148. Koziel H, Williams DJ, Fuglestad J, Armstrong MYK, Richards FF, Warner A, Rose RM. Impact of HIV infection on human alveolar macrophage interaction with Pneumocystis carinii. Am Rev Respir Dis 1991; 143:A207.
149. Koziel H, Kruskal BA, Ezekowitz RAB, Rose RM. Human immunodeficiency virus impairs alveolar macrophage mannose receptor function against Pneumocystis carinii. Chest 1993; 103(suppl):111–113.
150. Kruskal BA, Sastry K, Warner AB, Mathieu CE, Ezekowitz RAB. Phagocytic chimeric receptors require both transmembrane and cyoplasmic domains from the mannose receptor. J Exp Med 1992; 176:1673–1680.
151. Wagner RP, Levitz SM, Tabuni A, Kornfeld H. HIV-1 envelope protein (gp120) inhibits the activity of human bronchoalveolar macrophages against Cryptococcus neoformans. Am Rev Respir Dis 1992; 146:1434–1438.
152. Matthews TJ, Weinhold KJ, Lyerly HK, Langlois AJ, Wigzel H, Bolognesi D. Interaction between the human T-cell lymphotropic virus type IIIB envelope glycoprotein gp120 and the surface antigen CD4: role of carbohydrate in binding and cell fusion. Proc Natl Acad Sci USA 1987; 84:5424–5428.
153. Phelps DS, Rose RM. Increased recovery of surfactant protein A in AIDS-related pneumonia. Am Rev Respir Dis 1991; 143:1072–1075.

154. Sternberg RI, Whitsett JA, Hull WM. Baughman RP. Pneumocystis carinii alters surfactant protein A concentrations in bronchoalveolar lavage fluid. J Lab Clin Med 1995; 125:462–469.
155. Mahoney EM, Hamill AL, Scott WA, Cohn ZA. Response of endocytosis to altered fatty acyl composition of macrophage phospholipids. Proc Natl Acad Sci USA 1977; 74:4895–4899.
156. Furlong ST, Gorti R, Catalano PJ, Rose RM. Lipid content of alveolar lining material collected by bronchoalveolar lavage. Improved methods for measuring the major lipid classes. Am Rev Respir Dis 1992; 145:383–387.
157. Agostini C, Sancetta R, Cerutti A, Semenzato G. Alveolar macrophages as a cell source of cytokine hyperproduction in HIV-related interstitial lung disease. J Leukoc Biol 1995; 58:495–500.
158. Israel-Biet D, Cadranel J, Beldjord K, Andrieu JM, Jeffrey A, Even P. Tumor necrosis factor production in HIV-seropositive subjects. Relationship with lung opportunistic infections and HIV expression in alveolar macrophages. J Immunol 1991; 147:490–494.
159. Krishnan VL, Meager A, Mitchell DM, Pinching AJ. Alveolar macrophages in AIDS patients: increased spontaneous tumour necrosis factor-alpha production in Pneumocystis carinii pneumonia. Clin Exp Immunol 1990; 80:156–160.
160. Kandil O, Fishman J, Koziel H, Pinkston P, Rose R, Remold H. Human immunodeficiency virus type 1 infection of human macrophages modulates the cytokine response to Pneumocystis carinii. Infect Immun 1994; 62:644–650.
161. Murray HW. Interferon-gamma, the activated macrophage, and host defense against microbial challenge. Ann Intern Med 1988; 108:595–608.
162. Rudy T, Opelz G, Gerlack R, Daniel V, Schimpf K. Correlation of in vitro immune defects with impaired gamma interferon response in human immunodeficiency virus infected individuals. Vox Sang 1988; 54:92–95.
163. Roux Lombard P, Modoux C, Cruchaud A, Dayer J. Purified blood monocytes from HIV-1 infected patients produce high levels of TNF-alpha and IL1. Clin Immunol Immunopathol 1989; 50:374–384.
164. Eden E, Turino GM. Interleukin-1 secretion by human alveolar macrophages stimulated with endotoxin is augmented by recombinant immune (gamma) interferon. Am Rev Respir Dis 1986; 133:455–460.
165. Benfield T, Vestbo J, Junge J, Nielsen T, Baek Jensen A, Lundgren J. Prognostic value of interleukin-8 in AIDS-associated Pneumocystis carinii pneumonia. Am J Respir Crit Care Med 1995; 151:1058–1062.
166. Benfield T, Van Steenwuk R, Nielsen T, Dichter J, Lipschik G, Jensen B, Shelhamer J, Lundgren J. Interleukin-8 and eicosanoid production in the lung during moderate to sever Pneumocystis carinii pneumonia in AIDS: a role of interleukin-8 in the pathogenesis of P. carinii pneumonia. Respir Med 1995; 89:285–290.
167. Lipschik G, Doerfler M, Kovacs J, Travis W, Andrawis V, Lawrence M, Dichter J, Ognibene F, Shelhamer J. Leukotriene B4 and interleukin-8 in human immunodeficiency virus-related pulmonary disease. Chest 1993; 104:763–769.
168. Agostini C, Trentin L, Zambello R, Builian P, Caenazzo C, Cipriani A, Candrobbi P, Garbisa S, Semenzato G. Release of granulocyte-macrophage colony-stimulating factor by alveolar macrophages in the lung of HIV1-1–infected patients. A mecha-

nism accounting for macrophage and neutrophil accumulation. J Immunol 1992; 149:3379–3385.
169. Laursen A, Rungby J, Andersen P. Decreased activation of the respiratory burst in neutrophils from AIDS patients with prior Pneumocystis carinii pneumonia. J Infect Dis 1995; 172:497–505.
170. Koziel H, Eichbaum Q, Ezekowitz RAB, Sypek J, Armstrong MYK, Richards FF, Rose RM. M-CSF augments human alveolar macrophage host defense against Pneumocystis carinii (abstr). Am Rev Respir Crit Care Med 1994; 149:A289.
171. Koziel H, Eichbaum Q, Armstrong MYK, Richards FF, Minty A, Ferrara P, Caput D, Ezekowitz RAB. IL-13 augments alveolar macrophage phagocytosis of Pneumocystis carinii via up-regulation of mannose receptor (abstr). Clinical Res 1994; 42:157A.
172. Newbury R, Palmer B, Kaltreider H, Beck J. In vitro propagation of viable mouse-derived Pneumocystis carinii. Am J Respir Crit Care Med 1995; 151:A795.
173. Stokes DC, Gigliotti F, Rehg JE, Snellgrove RL, Hughes WT. Experimental Pneumocystis carinii pneumonia in the ferret. Br J Exp Pathol 1987; 68:267–276.
174. Limper AH, Martin WJ. Pathogenesis of Pneumocystis carinii pneumonia: an in vitro model. Chest 1989; 95:180S–181S.

19

Interactions Between Lung Macrophages and Fungi

THOMAS S. HARRISON

Boston University Medical Center Hospital
Boston, Massachusetts

STUART M. LEVITZ

Boston University School of Medicine
Boston, Massachusetts

I. Introduction

In recent years, because of increasing numbers of immunocompromised patients and the widespread use of broad-spectrum antibiotics, the incidence of serious fungal infections has risen dramatically. Those individuals with acquired immunodeficiency syndrome (AIDS) are particularly susceptible to most medically important fungi. In addition, exposure to the endemic mycoses has sharply increased owing to population shifts and increased travel.

With the notable exception of *Candida albicans*, most of the medically important fungi are transmitted by inhalation of airborne conidia. Pulmonary disease often is the primary and predominant clinical manifestation of infection. Bronchoalveolar macrophages are likely to be the first cells of the immune system to interact with these fungi, and they may play several critical roles in host defense: forming a first line of innate cell-mediated immunity; initiating a specific immune response through antigen presentation and cytokine release; and, once activated, serving as final effector cells contributing to the control of infection. In some circumstances, bronchoalveolar macrophages may be able to prevent the initial establishment of infection: Bronchoalveolar macrophages have been shown to kill conidia of *Aspergillus fumigatus*, to prevent the germination of *Rhizopus oryzae* spores, and to inhibit the phase transition of conidia of *Blastomyces dermatitidiis* into yeasts. In contrast to the other fungi discussed, pulmonary infections with *C. albicans* are usually endogenous and hematogenous in origin. In this case, bronchoalveolar macrophages may play a less critical role in protection, although once activated, they have been shown to have anticandidal activity.

Cell-mediated immunity appears to be especially important for host defense against *Cryptococcus neoformans*, and the endemic dimorphic fungi *Histoplasma capsulatum*, *B. dermatitidis*, and *Coccidioides immitis*. Protection against invasive candidiasis, aspergillosis, and mucormycosis is probably more dependent on intact neutrophil function.

In the following sections, interactions of bronchoalveolar macrophages with each of the principal fungal pathogens are discussed. We have tried to emphasize recent developments and molecular mechanisms where these have been explored. *Pneumocystis carinii*, which was recently classified by some as a fungus (1), is discussed in another chapter and will not be discussed here.

II. *Cryptococcus neoformans*

A. Epidemiological and Clinical Features

Cryptococcosis has emerged as a major cause of morbidity and mortality in patients with impaired cell-mediated immunity, especially those with AIDS. Meningitis and, to a lesser extent, pneumonia are the most frequent life-threatening complications of cryptococcosis (2,3). *C. neoformans* is unique among the major fungal pathogens in that it possesses a polysaccharide capsule (reviewed in ref. 4), which is thought to be a major virulence factor.

In clinical specimens, the yeast cell of *C. neoformans*, exclusive of capsule, typically has a diameter of about 5 μm. The size of the surrounding capsule varies greatly, ranging in thickness from less than 1 μm to greater than 30 μm (5). In nature, the yeast cells tend to be smaller and poorly encapsulated (6). Since exposure to *C. neoformans* usually occurs as a consequence of inhalation of airborne organisms, the small size of environmental isolates should favor alveolar deposition. Thus, as with most fungal pathogens, it is likely that bronchoalveolar macrophages form the first line of host defenses against cryptococcosis (7).

B. Binding and Phagocytosis

In the absence of opsonins, most macrophage populations do not bind encapsulated *C. neoformans* (7). Two exceptions are rat and human bronchoalveolar macrophages (8,9). Macrophages generally bind unopsonized acapsular mutants of *C. neoformans* quite avidly. Capsule is thought to inhibit binding by presenting a surface that is not recognized by the phagocyte (10).

Following incubation of encapsulated *C. neoformans* in normal human serum, large amounts of complement, mostly in the form of the cleavage fragment iC3b, gets deposited on the surface of the capsule (11). Macrophages obtained from several different species and anatomical sites have been shown to bind encapsulated *C. neoformans* in the presence of normal serum but not serum lacking a functional complement system (7). Moreover, incubation of *C. neoformans* with purified proteins of the alternative complement pathway results in C3 deposition on the capsular surface and binding by human monocyte-derived macrophages (12). In contrast to C3, immunoglobulins from normal serum contribute relatively little to opsonization (13,14). However, specific anticapsular antibody gets deposited on the capsular surface and is a potent opsonin (15). Mouse-human chimeric antibodies with opsonic and therapeutic activity against *C. neoformans* have been developed (16). Moreover, a vaccine consisting of *C. neoformans* capsular polysaccharide conjugated to tetanus toxoid elicits specific antibodies (17). Both passive administration of antibodies and vaccination have been proposed for clinical trials in persons infected with the human immunodeficiency virus (HIV).

The three defined complement receptors for fragments of C3, CR1 (CD35), CR3 (CD11b/CD18), and CR4 (CD11c/CD18), participate in binding of serum-opsonized *C. neoformans* to monocyte-derived macrophages (14). Tumor necrosis factor-α (TNF-α) and granulocyte-macrophage colony-stimulating factor (GM-CSF) enhance complement receptor–dependent phagocytosis of *C. neoformans* by murine macrophages (18). A T-cell–derived suppressor factor that inhibits binding of *C. neoformans* by murine peritoneal macrophages has been described (19). *C. neoformans* also has been reported to adhere to rat lung epithelial cells (20).

Following binding of a particle to a phagocytic receptor, phagocytosis of the particle may or may not take place, depending on the particular receptor involved and its state of activation. Although complement receptors do not constitutively phagocytize, under certain circumstances, they can be upregulated to do so. The majority of serum-opsonized encapsulated *C. neoformans* are not internalized by human monocyte-derived macrophages in vitro unless specific anticapsular antibody is added (14). In contrast, serum-opsonized acapsular strains are internalized, perhaps because the acapsular strains have ligands (e.g., mannans, β-glucans) exposed on the cell wall that promote phagocytosis. Generally, internalization (but not binding) is thought to be a process requiring energy and cytoskeletal reorganization. However, binding of serum-opsonized encapsulated *C. neoformans* by monocyte-derived macrophages is inhibited by incubation in the cold and by inhibitors of receptor capping and actin polymerization. In contrast, energy is not required for binding of serum-opsonized acapsular *C. neoformans* (14).

C. Inhibition and Killing

Following an intratracheal challenge of mice with *C. neoformans*, multinucleated giant cells surround the fungus. Mice depleted of $CD4^+$ (but not $CD8^+$) T cells did not form multinucleated giant cells and were unable to clear the *C. neoformans* from the lungs, suggesting that $CD4^+$ T cells produce cytokines at the site of infection which promote granuloma formation and activate macrophages (21). Other studies have also supported a role for $CD8^+$ T cells as well as $CD4^+$ cells in in vivo murine pulmonary defenses against *C. neoformans* (22–27). A Th1 immune response with the production of interferon-γ (IFN-γ) which in turn induces nitric oxide appears to be necessary for optimal pulmonary clearance of *C. neoformans* (28). Intraperitoneal injection of IFN-γ into mice enhanced killing of a subsequent intraperitoneal challenge with *C. neoformans* (29). Pulmonary defenses against *C. neoformans* were impaired in mice depleted of the fifth component of complement (C5) (30). A high melanin-producing strain of *C. neoformans* induced a pulmonary inflammatory response in mice that was delayed and relatively ineffective compared with a low melanin–producing strain, suggesting a mechanism by which melanin may act as a cryptococcal virulence factor (31).

In vitro, a variety of effector cells, including macrophages, monocytes, neutrophils, natural killer (NK) cells, and T lymphocytes, have been shown to inhibit or kill *C. neoformans* under defined conditions (32–35). The literature concerning the anticryptococcal activity of macrophages underscores the heterogeneous nature of macrophage populations: Inhibition and killing of *C. neoformans* vary depending on the species of origin, anatomical location, and state of activation of the macrophage (7).

Because bronchoalveolar macrophages are likely to be the first phagocytes to encounter *C. neoformans* following initial exposure to the organism, many cryptococcologists have focused on this cell type. Even in the absence of specific activation, bronchoalveolar macrophages freshly lavaged from mice and rats kill *C. neoformans* (8,36). Paradoxically, encapsulated organisms are more efficiently killed by mouse bronchoalveolar macrophages than are acapsular mutants (36). The antifungal activity of rat bronchoalveolar macrophages diminishes after overnight culture, but it is partially restored by the addition of IFN-γ (37). Growth inhibition, but not killing, of *C. neoformans* by human bronchoalveolar macrophages has been demonstrated (38–40).

Several laboratories have demonstrated killing of opsonized *C. neoformans* by human peripheral blood monocytes (7). Monocytes obtained from patients with cryptococcosis, Hodgkin's disease, or sarcoidosis killed *C. neoformans* as well as control monocytes (41). As was the case with murine bronchoalveolar macrophages (36), human monocytes killed encapsulated strains more efficiently than a capsule-free mutant (35). Cultured human monocytes did not inhibit the growth of encapsulated *C. neoformans* unless the cells were incubated on extracellular matrix proteins (42). However, culture of human blood monocytes and lymphocytes with cryptococcal antigens generated effector cells capable of killing encapsulated *C. neoformans* (43).

The mechanisms by which human mononuclear phagocytes inhibit and kill *C. neoformans* are unknown; however, nitric oxide does not appear to play a role (39). Possibly as a consequence, as opposed to their murine counterparts, human macrophages do not appear to be activated by IFN-γ to kill *C. neoformans*, (43). GM-CSF stimulated human peripheral blood mononuclear cells (PBMCs) to mediate fungistasis (44).

D. Cytokine Secretion and Antigen Presentation

C. neoformans is a potent stimulator of gene expression and the release of the proinflammatory cytokines TNF-α and interleukin (IL)–1β from human monocytes and bronchoalveolar macrophages, apparently via a mechanism involving nuclear translocation of the transcription factor NF-κB (45,46). The production of these cytokines is stimulated by IFN-γ and inhibited by IL-10. Monocytes also make IL-10 when stimulated by *C. neoformans* (46).

Murine macrophages challenged in vitro with *C. neoformans* were poor stimulators of T-cell proliferation unless the fungi were opsonized or an acapsular strain was used (47). Similarly, cryptococcal antigen presentation by human alveolar macrophages to autologous $CD4^+$ T cells was greater using an acapsular strain compared with an encapsulated strain (48).

E. Effects of HIV

Following a 2-hr incubation, PBMCs from persons with AIDS bound and killed *C. neoformans* as well as control cells (49). However, when the incubation was extended to overnight, PBMCs and monocytes obtained from persons with HIV infection had impaired anticryptococcal activity compared with cells from HIV-seronegative donors (50). Normal human monocytes and peritoneal macrophages (but not alveolar macrophages) had temporarily reduced anticryptococcal activity after in vitro infection with a monocytotropic strain of HIV (51). The addition of gp120 (the envelope glycoprotein of HIV-1) to human bronchoalveolar macrophages significantly reduced the capacity of human bronchoalveolar macrophages to phagocytose and inhibit the growth of *C. neoformans* (40). *C. neoformans* induced viral replication in T cells freshly infected with HIV by a mechanism that required monocytic cells but was independent of TNF-α (52). Lymphoproliferation in response to *C. neoformans* progressively wanes during the course of HIV infection (53,54), and this defect is not corrected by the addition of exogenous IL-12 (54). IL-12 release from *C. neoformans*–stimulated monocytes was undetectable regardless of the HIV serological status of the donor unless the cells first were primed with IFN-γ (54,55).

III. *Candida albicans*

A. Epidemiological and Clinical Features

In clinical specimens, *C. albicans* exists in three forms (56). On skin and mucosal surfaces yeasts predominate. They are small (4–6 μm), oval, thin-walled cells that reproduce by budding. Particularly in invasive disease, yeasts also give rise to pseudohyphae and to true septate hyphae. In contrast to the other fungi discussed, *C. albicans* is a normal commensal of humans and other animals. It colonizes the entire gastrointestinal tract and the female genital tract and may be found on diseased skin and in skin folds. Organisms can be isolated from environmental sources, but this probably results from contamination by humans and animals. Most infections are thought to be endogenous in origin, although direct person to person transmission can occur, and growing evidence suggests infection can be acquired from the hospital environment.

Clinical disease usually results when host defenses become impaired. Invasive disease is most commonly associated with neutropenia or defects in neutrophil function, whereas severe or recurrent cutaneous and mucosal disease usually results from defects in cell-mediated immunity, as in patients with AIDS. Pulmonary disease is seen in the setting of disseminated invasive infection and is usually hematogenous in origin, although aspiration of organisms from the oropharynx may also occur. Thus, bronchoalveolar macrophages may play a somewhat different role in pulmonary defenses against *C. albicans* than against fungi that are commonly inhaled directly into the alveoli.

It is also possible that the apparent rarity of aspiration pneumonia due to *C. albicans* reflects the known anticandidal activity of bronchoalveolar macrophages. In mice, large intratracheal inoculations of *C. albicans* were rapidly cleared from the lungs (57). A predominantly neutrophilic infiltrate was seen, but among lavaged cells, both macrophages and neutrophils had ingested organisms. Using carrageenan to impair macrophages and irradiation to deplete neutrophils, Lal et al. concluded neutrophils played the predominant role in defense against intratracheal challenge with *C. albicans* in mice (58). However, in a similar study using cortisone and cyclophosphamide, Sawyer and Harmsen found both cell types were required to clear infection (59).

B. Binding and Phagocytosis

The cell wall of *C. albicans* consists primarily of mannan, glucan, chitin, and protein (60). At least five structural layers have been described. The outermost fibrillar layer is composed of mannan and mannoprotein. The composition and architecture of the cell wall varies with growth conditions and between different forms of the organism. Hyphal forms have been found to have surface proteins, some of which are absent from yeast cells, that can bind the complement products C3d and iC3b, fibrinogen, laminin, and fibronectin.

As with other tissue macrophages, bronchoalveolar macrophages possess a number of receptors, including Fc, complement, mannose, and fibronectin receptors, that could contribute to the binding of *C. albicans* (61). Human monocyte-derived macrophages rapidly phagocytose serum-opsonized *C. albicans*, probably through the combined effects of immunoglobulin and classic and alternative complement activation (62). However, unopsonized organisms were also taken up by monocyte-derived macrophages. This phagocytosis of unopsonized organisms was inhibited in a dose-dependent manner by mannan and mannose–bovine serum albumin (BSA), suggesting a role for the mannose receptor. Cos-1 cells transfected with the mannose receptor cDNA were able to bind and phagocytose *C. albicans* (63). Deletion of the cytoplasmic tail of the receptor resulted in binding but no internalization. Bronchoalveolar macrophages are known to express mannose receptors (64), but in one study, phagocytosis of *C. albicans* by rabbit and human bronchoalveolar macrophages was entirely dependent on opsonization (65). The assay in this case was carried out in suspension and the investigators postulated that adherence may modulate the activity of the mannose receptor.

In a recent study, human vitronectin, a circulating adhesive glycoprotein, was shown to bind to *C. albicans* and increase the binding of the organism to a rat macrophage cell line (66). *C. albicans* also binds to fibronectin (60), an extracellular matrix glycoprotein that has been shown to enhance the attachment of *Pneumocystis carinii* to bronchoalveolar macrophages, although not to mediate phagocytosis (67).

C. Growth Inhibition and Killing

Many studies have demonstrated the ability of macrophages to inhibit and kill *C. albicans* in vitro (recently reviewed in ref. 68), although unstimulated cells have generally shown only modest activity (69). In animal systems, most investigations suggest nonoxidative mechanisms are of primary importance. The anticandidal activity of murine macrophages derived from different anatomical sites and their ability to generate reactive oxygen intermediates were both found to vary. However, these two properties were not well correlated, and catalase and superoxide dismutase failed to inhibit those macrophage populations with the most potent anticandidal activity (70). *C. albicans* has been shown to be less sensitive to the products of oxidative metabolism than a number of pathogenic bacteria (71). However, one early study implicated oxidative mechanisms in the killing of *C. albicans* by activated murine macrophages (72).

In several studies, rabbit bronchoalveolar macrophages demonstrated modest killing of *C. albicans* (73,74). This activity was enhanced by nonspecific stimulation using Freund's complete adjuvant, and it was not reduced by several inhibitors of oxidative metabolism (74). Lysomal extracts from the macrophages inhibited specific amino acid uptake by *C. albicans* and reduced fungal viability (75). These cells were shown to contain two lysosomal cationic proteins, probably defensins, with potent candidacidal activity (76,77). The concentrations of both proteins were increased following treatment with Freud's adjuvant (78).

In one study, human bronchoalveolar macrophages did not kill *C. albicans* after 2 hr of incubation (79). However, killing has been demonstrated after 4–6 hr (80,81). Extracts from cytoplasmic granules had candidacidal activity that was inhibited by trypsin and ethylenediaminetetraacetic acid (EDTA), suggesting a role for cationic proteins in human bronchoalveolar macrophages. Activity was enhanced by the addition of a calcium ionophore, and this enhancement correlated with the increased activity of granule extracts as well as increased superoxide anion production (81). Some evidence suggests that in human macrophages oxidative mechanisms of killing also play a role. Killing of *C. albicans* by human monocyte-derived macrophages was reduced under anaerobic conditions and by catalase and superoxide dismutase (SOD), although not by the inhibitors of the hydroxyl radical and hypohalide formation, thiourea and methionine (82).

Other mechanisms of candidal killing may also operate particularly in activated macrophages, although they have not yet been demonstrated in bronchoalveolar cells. In murine peritoneal macrophages stimulated with IFN-γ, noncationic protein(s) generated in acidic phagolysosomes was implicated in the killing of *C. albicans* (83). In invasive disease, hyphal forms of *C. albicans* too large to be entirely ingested by phagocytic cells may predominate. Murine peritoneal macrophages activated by immunization with *Listeria monocytogenes* killed *C. albicans* hyphae by a serum-independent, contact-mediated mechanism that was not affected by oxygen radical scavengers (84). Supernatants from murine macrophage cell lines cells stimulated with IFN-γ and lipopolysaccharide (LPS) have been shown to inhibit hyphal but not yeast cell growth. Activity was correlated with nitrite production and was almost completely inhibited by the nitric oxide inhibitor N-monomethyl-L-arginine (85). However, the exact role of nitric oxide in macrophage killing of yeast and hyphal forms of *C. albicans* is not yet clear. Some studies have found a correlation between the level of nitric oxide production and the killing capacity of murine macrophages (86,87), whereas a recent report did not (88). The role of nitric oxide in the microbicidal activity of human macrophages is even less certain (89,90).

Several studies have shown that macrophages activated during the course of an immune response to specific and nonspecific stimuli have an enhanced capacity to kill *C. albicans*. Recent studies have focused on the cytokine mediators of this activation. IFN-γ has been found to increase the anticandidal activity of murine peritoneal macrophages (83,91,92) and human monocyte-derived macrophages (93) and bronchoalveolar macrophages (94). A number of mechanisms have been implicated: In one study, increased killing by IFN-γ–treated resident murine peritoneal macrophages required superoxide anion but not the hydrogen peroxide/myeloperoxidase system, although the latter system was implicated in the activation by supernatants from concanavillin A (ConA)–stimulated spleen cells (95). Other studies using murine cells have suggested a role for noncationic protein and nitric oxide (see above) (83,85). In human monocyte-derived macrophages, activation by IFN-γ was associated with increased intracellular free Ca^{2+} and superoxide production on stimulation with *C. albicans*. These increased responses to *C. albicans* were mediated via mannose receptors, although IFN-γ appeared to decrease the numbers of these receptors (93). Bronchoalveolar macrophages from patients with chronic obstructive pulmonary disease were found to have reduced anticandidal activity. IFN-γ treatment increased the anticandidal activity of these cells as well as cells from normal donors (96).

Other cytokines have also been shown to enhance the activity of macrophages against *C. albicans*. Macrophage CSF (M-CSF) increased phagocytosis and killing of *C. albicans* by murine peritoneal exudate macrophages, although not by resident peritoneal macrophages. Increased binding induced by M-CSF was completely inhibited by D-mannose, suggesting activation was mediated by an increase in mannose receptors (97). Other studies have shown increased anticandidal activity of murine macrophages treated with IFN-α (97), IL-1, and TNF-α (92) and increased phagocytosis of *C. albicans* induced by IL-3 (98).

D. Cytokine Production

Most studies to date have focused on the proinflammatory cytokine TNF-α. Whole organisms and mannoprotein, phospholipomannan, oligomannoside, and mannan cell wall components have been shown to stimulate macrophages to produce TNF-α (99–102). In one study, hyphal, but not yeast, forms stimulated TNF-α production in a murine macrophage cell line (103). Mouse bronchoalveolar macrophages secreted TNF-α in response to *C. albicans* mannan (102). The investigators suggested TNF-α could play an important role in pulmonary defense by increasing the antifungal activity of recruited neutrophils (104) as well as that of the macrophages themselves. The release of IL-1β and TNF-α from *C. albicans*–stimulated human PBMCs was inhibited by IL-10. *C. albicans* also stimulated secretion of IL-10 from PBMCs, suggesting a feedback mechanism whereby the host can limit the inflammatory response (46).

IV. *Histoplasma capsulatum*

A. Epidemiological and Clinical Features

H. capsulatum is a dimorphic fungus of worldwide distribution that produces a broad spectrum of disease activity. In the United States, *H. capsulatum* is endemic to the Ohio and Mississippi River valleys and to Puerto Rico, with an estimated 500,000 new infections occurring each year (105). In heavily endemic areas, virtually the entire population is infected (106). The course of infection is mild or even asymptomatic in the vast majority of immunocompetent persons. However, progressive, disseminated infections may ensue, especially in those with impaired cell-mediated immunity. Persons with AIDS are particularly susceptible to disseminated histoplasmosis.

H. capsulatum exists in its native habitat, the soil, in the mycelial phase, which consists of septate branching hyphae that bear spores. Spores are of two types—microconidia, which range in size from 2 to 6 μm, and macroconidia, which range in size from 8 to 14 μm. It is thought that infection occurs following incidental inhalation of airborne spores, especially the microconidia whose size is ideal for deposition in terminal bronchioles and alveoli. In the mammalian host at 37°C, the fungus converts to the yeast phase (106).

In individuals with intact cell-mediated immunity, a granulomatous inflammatory response is generally seen (106,107). Rare complications such as bronchial obstruction, mediastinal fibrosis and pericarditis can result from an overexuberant inflammatory response to the fungus. In contrast, in disseminated histoplasmosis, granulomas tend to be poorly formed or not present and macrophages are heavily parasitized by yeast phase organisms. Immunity in histoplasmosis is thought to be dependent on the development of an effective cellular immune response which activates macrophages to inhibit and kill intracellular yeasts.

B. Binding

H. capsulatum is an intracellular parasite of macrophages. Efforts have focused on defining the mechanisms by which the fungus has evolved to enable it to gain access to and survive in the intracellular environment of the macrophage. The earliest event in intracellular parasitism is binding. Studies by Bullock and Wright identified the CD18 family of adhesion glycoproteins as the major receptor mechanism that mediates attachment of unopsonized *H. capsulatum* yeasts by human monocyte-derived macrophages (108). This family has three members (CD11a/CD18 [LFA-1], CD11b/CD18 [CR3], and CD11c/CD18 [CR4 or p150, 95]), each of which shares a common β-chain (CD18). Inhibition of binding requires downregulation of all three members of the family, suggesting that each of the members of the CD18 complex is independently capable of binding *H. capsulatum* yeast. Optimal binding occurred at 37°C in the presence of divalent cations and did not require Fc, mannose, or β-glucan receptors.

Newman et al. (109) extended these findings by demonstrating that the CD18 complex mediates the binding of *H. capsulatum* microconidia in addition to yeasts. Moreover, human bronchoalveolar macrophages also required the CD18 family to mediate binding of *H. capsulatum* yeasts and microconidia. The identity of the ligand on the surface of the *H. capsulatum* yeasts and microconidia that mediates attachment to macrophages is unknown. Preliminary experiments suggest that it is a heat-stable glycoprotein. Supernatants from a freeze-thaw extract of yeast cells block binding (105). Macrophages from HIV-seropositive persons were profoundly deficient in their capacity to bind *H. capsulatum*. Defective binding was seen at all stages of HIV infection, although it was most profound in persons with less that 200 CD4 cell counts per microliter (110).

Under conditions where opsonins are present, it is likely that complement and/or immunoglobulins get deposited on the surface of *H. capsulatum* and interact with their corresponding receptors (111). Recently, McMahon et al. demonstrated that the extracellular matrix protein laminin binds to *H. capsulatum* in a rapid, specific, and saturable manner (112). A 50-kD laminin binding protein was isolated from a extract of *H. capsulatum* cell wall proteins which was recognized by antibodies directed at the laminin receptor. Macrophages are known to possess laminin receptors, but whether laminin can act as an opsonin for *H. capsulatum* is unknown.

C. Phagocytosis and Intracellular Survival

Phagocytosis (ingestion) of yeasts and microconidia by human monocyte-derived macrophages and bronchoalveolar macrophages occurs rapidly after binding in vitro (109). If analogous events occur in vivo, these findings suggest that phase transition from microconidia to yeast occurs in the intracellular environment of the macrophage. Moreover, as with all the dimorphic fungi, infection could be aborted if the macrophages were able to prevent phase transition.

The role of the macrophage cytoskeleton in binding and phagocytosis has been elucidated. Both binding and phagocytosis are inhibited by cytochalasin D but not by nocodazole, suggesting a requirement for actin but not microtubule polymerization (109).

Studies have been devoted to understanding how *H. capsulatum* survives in the hostile environment of the macrophage. Moreover, as immunity in histoplasmosis is thought to be dependent on the activated macrophage, efforts have also been directed at determining the mechanisms by which the activated macrophage inhibits and kills yeast cells.

On phagocytosis by macrophages, phagosome-lysosome fusion takes place (113,114). Newman demonstrated that human macrophages digest heat-killed yeast cells, whereas live yeasts multiply within the phagolysosome (113). In macrophages simultaneously ingesting both viable and killed yeasts, viable yeasts multiplied, whereas the killed yeasts were digested.

Eissenberg et al. presented evidence that *H. capsulatum* modulates the pH of the phagolysosome (115). In a murine macrophage cell line, vesicles containing live *H. capsulatum* had an approximate pH of 6.5, which was one to two units higher than vesicles containing either killed *H. capsulatum* or zymosan. The higher pH could help *H. capsulatum* avoid host defenses by inactivating lysosomal acid hydrolases. However, as discussed below, raising the pH of the phagolysosome by the administration of chloroquine induces macrophage killing of *H. capsulatum* by preventing the release of iron from transferrin (116). Taken together, these data suggest that survival of *H. capsulatum* within the phagolysosome could depend on a delicately balanced pH.

In murine macrophages, *H. capsulatum* inhibits protein kinase C activity and the respiratory burst (117–119). The effects were seen even with heat-killed yeast cells. However, a respiratory burst is stimulated in human macrophages challenged with *H. capsulatum* (108).

D. Inhibition and Killing

In murine systems, IFN-γ has been reported to activate macrophages for antifungal activity against *H. capsulatum* (120–124). Nitric oxide appears to be the main effector mechanism (121,123,124); although a role for respiratory burst products and defensins has also been postulated (124,125).

However, in human macrophages, which appear to lack significant amounts of nitric oxide and defensins, the critical factors necessary to activate human macrophages to inhibit and kill *H. capsulatum* are incompletely understood. In the absence of cytokine activation, cultured human monocyte-derived macrophages do not inhibit the growth of *H. capsulatum* (126,127). Two groups of investigators failed to note a beneficial effect of IFN-γ on growth of *H. capsulatum* yeasts inside monocyte-derived macrophages (128,129), although a third group did find a dose-dependent beneficial effect (127). A beneficial effect of the colony-stimulating factors IL-3, GM-CSF, and M-CSF has been noted (128,130). Superoxide dismutase inhibited M-CSF–stimulated monocyte-derived macrophage killing of *H. capsulatum*, suggesting a role for reactive oxygen intermediates generated by the respiratory burst (130).

One mechanism by which human activated macrophages could suppress the intracellular growth of *H. capsulatum* is by downregulation of transferrin receptors and the concomitant restriction of intracellular iron. Newman et al. demonstrated that coculture of *H. capsulatum*–infected monocyte-derived macrophages with the iron chelators deferoxamine and VUF 8514 suppressed the fungal growth (116,131). Moreover, chloroquine, which prevents the release of iron from transferrin, induced human monocyte-derived macrophages to kill *H. capsulatum* (116). Chloroquine also was therapeutic in a murine model of histoplasmosis.

H. capsulatum can remain latent in the host for years, only to reactivate when conditions for intracellular replication are favorable, such as might occur during a period of immunosuppression. Eissenberg et al. (132) noted that virulent "rough" strains of *H. capsulatum* destroyed monolayers of the macrophage-like cell line P388D1, whereas "smooth" variants which lacked cell wall α-(1,3)-glucan tended to persist for up to 3 months. Individual strains of *H. capsulatum* were capable of phenotypic switching, leading the investigators to speculate that this process may contribute to intracellular latency.

V. Aspergillosis

A. Epidemiological and Clinical Features

Aspergillosis refers to the broad range of disease states whose etiological agents are members of the genus *Aspergillus* (133). Most cases of human disease are caused by *A. fumigatus* followed in frequency by *A. flavus* and *A. niger*. Aspergilli are ubiquitous, and frequent exposure to the fungus via inhalation of airborne conidia (which are an ideal size for alveolar deposition) undoubtedly occurs. In the susceptible host, conidia germinate into hyphae, the invasion form of the disease. Most human disease occurs in the lung and can be divided into three broad categories: allergic aspergillosis, aspergilloma, and invasive aspergillosis. Allergic disease results from an overexuberant immune response in a hypersensitive or atopic individual, whereas invasive disease is usually seen in those severely immunocompromised as a result of neutropenia, therapy with high doses of corticosteroids, or late-stage AIDS (133–135).

B. Binding and Phagocytosis

A. fumigatus conidia bind to murine bronchoalveolar macrophages in the absence of opsonins via lectin-like attachment sites with characteristics of the mannose receptor (136). Human monocytes bind conidia by a β-glucan–inhibitable receptor (137). *A. fumigatus* conidia bind specifically to human fibrinogen as well as the basement membrane components laminin and fibrinogen (138–141). The significance of these binding interactions is unclear, but it could be an important first step that allows the fungus to establish residence in host tissues. Interestingly, protease(s) produced by *A. fumigatus* degrades both fibrinogen and laminin (138, 141). Thus, fibrinogen and laminin binding could facilitate fungal adherence of conidia, whereas protease production could assist with host cell invasion of the hyphae (135).

C. Inhibition and Killing

The association of invasive aspergillosis with neutropenia and neutrophil dysfunction has prompted research into neutrophil-*Aspergillus* interactions. Whereas *A. fumigatus* conidia resist killing by neutrophils and neutrophil products, hyphae are readily killed, apparently by both oxidative and nonoxidative mechanisms (142–144). Aspergilli produce catalase, an enzyme which breaks down hydrogen peroxide (H_2O_2) to water and oxygen. Evidence for a pathogenic role for catalase is provided by the strong association of aspergillosis with chronic granulomatous disease (145), a disorder associated with a defective respiratory burst and recurrent infections with catalase-positive microbes (146).

The significance of the inability of neutrophils to kill conidia is unclear, as pulmonary macrophages have primary contact with the inhaled conidia, whereas neutrophils are thought to form the second line of defense should germination occur (135,143,144). In mice and rats experimentally infected via the respiratory route, bronchoalveolar macrophages ingest and kill conidia (147). In vitro, human, murine, and rabbit bronchoalveolar macrophages have anticonidial activity (144, 148–150), but murine resident peritoneal macrophages do not (148). Nitric oxide does not appear to be involved in the fungicidal activity of murine or human bronchoalveolar macrophages against *A. fumigatus* conidia (150).

High doses of corticosteroids are a risk factor for invasive aspergillosis. Macrophages treated with corticosteroids in vitro or in vivo have impaired anticonidial activity (143,149,151). The in vitro activity of human mononuclear phagocytes against *A. fumigatus* has been enhanced by GM-CSF, M-CSF, and, in some (but not all) experiments, IFN-γ (149,152,153).

D. *Aspergillus* Virulence Factors

Most environmental strains of *A. fumigatus* and *A. flavus* produce elastase (154, 155). Given that elastin constitutes nearly 30% of lung tissue, elastinolytic activity has been postulated to contribute to the pathogenesis of pulmonary aspergillosis. Elastase-negative strains were relatively avirulent in a mouse model of invasive pulmonary aspergillosis compared with strains selected for high elastase production (154). However, examination of tissue blocks from nine patients with invasive pulmonary aspergillosis revealed no evidence of elastolysis (156). Moreover, the pathogenicity of mutant strains of *A. fumigatus*, created by gene disruption to knockout extracellular elastinolytic activity, was similar to the parental strains in murine models of invasive pulmonary aspergillosis (157–159).

A. fumigatus produces "restrictocin," an RNA nuclease that binds to IgE from allergic patients and IgG from patients with aspergilloma (160,161). Mutants of *A. fumigatus* created by disruption of the gene-encoding restrictocin were indistinguishable from parental strains in their growth in lung tissue and pathogenicity in a murine model of invasive pulmonary aspergillosis (160,162). These results argue against a major role for restrictocin in the pathogenicity of invasive aspergillosis. However, owing to its immunogenicity, restrictocin may still play a role in the immune response to *A. fumigatus*, especially with regard to the immunopathogenesis of allergic aspergillosis (161,162).

VI. *Blastomyces dermatitidis*

A. Epidemiological, Clinical, and Pathological Features

B. dermatitidis is a dimorphic fungus endemic in the southeastern United States and the Mississippi, Ohio, and St. Lawrence River valleys (163). Occasional cases have also been reported from other areas, in particular southern Africa. The mycelial form has been isolated from moist soil of high organic content. Slender septate hyphae give rise to smooth spherical or oval conidia. Infection results from inhalation of conidia, which, within the lung, convert into large, thick-walled yeasts bearing broad-based buds. Pulmonary infection may be subclinical and may resolve spontaneously, or it may give rise to chronic lung involvement and dissemination, most often to the skin, bone, and genitourinary tract.

In a mouse model of lethal infection using intratracheal injection of *B. dermatitidis* conidia, the initial lesion was a focal nodular alveolitis with macrophages and only a few neutrophils (164). By day 5, yeast forms were seen within macrophages and the numbers of neutrophils had increased. After 3–4 weeks, enlarging microabscesses gave rise to the characteristic pyogranulomatous inflammation which is also seen in human disease. Interestingly, in this lethal model coincident with the formation of granulomas was the start of a progressive increase in the number of *B. dermatitidis* colony-forming units recoverable from the lungs.

B. Binding, Growth Inhibition, and Killing

A recent study demonstrated that WI-1, a *B. dermatitidis* cell wall protein, is a major ligand for the binding of *B. dermatitidis* to human monocyte derived macrophages (165). Antibody to WI-1 inhibited binding by 70–80%, and less virulent strains that expressed more WI-1 showed increased binding. A 25–amino acid tandem repeat sequence within WI-1 was shown to mediate binding to the LPS binding sites of CR3 and CD14 on the macrophage.

Mouse and human bronchoalveolar macrophages have been shown to inhibit the phase transition of *B. dermatitidis* conidia into yeast forms, although the conidia were not killed, as shown by the fact that if subsequently incubated at 25°C, they germinated to form hyphae (166,167). The activity of human cells required 5 days of in vitro culture to develop, and it was inhibited by IFN-γ and M-CSF. The mechanism of inhibition was unclear: Inhibition was reproduced by 2.5–25.0 mM hydrogen peroxide but was not significantly reduced by catalase.

A number of studies have examined growth inhibition and killing of *B. dermatitidis* yeast cells by bronchoalveolar macrophages (recently reviewed in refs. 168 and 169). In a 4-hr assay, mouse bronchoalveolar macrophages incubated overnight with 1000 U/ml of IFN-γ or supernatants from ConA-stimulated spleen cells showed modest killing (25% reduction in colony-forming units [cfu]) of *B. dermatitidis* yeasts (170). The administration of IFN-γ in vivo 24 hr prior to harvesting the cells had a similar effect (171). Unstimulated macrophages showed no killing. Experiments with human bronchoalveolar macrophages have used a different assay, but they also suggest that activation is required for antifungal activity (172). Bronchoalveolar macrophages from patients recovering from blastomycosis caused a small reduction in the number of intracellular yeasts over 3 days. In comparison, intracellular yeasts more than doubled in number in bronchoalveolar macrophages from normal donors. Only 12–18% of normal cells and 25–38% of patient cells were reported to contain ingested yeasts. The activity of patient macrophages was thought to result from activation by cytokines secreted by *B. dermatitidis*–specific lymphocytes present in the monolayers. Macrophages from normal donors also inhibited the growth of intracellular yeasts after treament with supernatants from PBMCs stimulated with *B. dermatitidis* antigen.

Mechanisms of killing have only been explored in peritoneal and monocyte-derived macrophages. Using activated murine peritoneal macrophages, killing of *B. dermatitidis* was not enhanced by complement or specific antibody (173), and it was not reduced by inhibitors of oxidative metabolism (174). Contact-mediated killing of extracellular organisms may be important, as suggested by studies of human monocyte-derived macrophages (175).

VII. *Coccidioides immitis*

C. immitis, the causative agent of coccidioidomycosis, is a dimorphic fungus endemic to certain arid and semiarid regions of the Western Hemisphere. The fungus propogates in soil in the mycelial phase that gives rise to infectious arthroconidia. Infection in humans results when aerosolized arthroconidia are inhaled. In the susceptible host, the fungus enters a parasitic life cycle characterized by arthroconidia developing into structures called spherules which in turn reproduce by the formation of internal endospores, each one of which can subsequently develop into a new spherule. The vast majority of individuals develop a protective cell-mediated immune response following exposure to *C. immitis*. However, up to 5.0% of those infected have pulmonary residua and about 0.5% develop disseminated disease. Those with impaired cell-mediated immunity (CMI) are particularly susceptible, although the mechanisms by which the CMI response protects are poorly understood (176).

An intense neutrophil influx is elicited by pulmonary inoculation of arthroconidia and endospores of *C. immitis* (177). However, human neutrophils were relatively ineffective at killing the fungus (178). Peripheral blood monocytes effected modest killing of arthroconidia (178). Endospores phagocytosed by human monocytes developed into spherules, although conversion was retarded by pretreatment of the cells with IFN-γ or TNF-α (179). IFN-γ also had salutory effects on the antifungal activity of murine bronchoalveolar macrophages against *C. immitis* endospores and arthroconidia (180). In experimental murine coccidioidomycosis, genetically resistant mice had increased levels of IFN-γ in lung extracts, whereas susceptible mice had increased levels of IL-4 (181).

VIII. Agents of Mucormycosis

Mucormycosis (zygomycosis) is the common name given to disease caused by fungi of the class Zygomycetes, order Mucorales (182). Disease is typically a result of inhalation of airborne spores and most often manifests itself clinically as sinusitis, rhinocerebral infection, and/or pneumonitis. Like aspergillosis, the pathology is characterized by angioinvasion with resultant hemorrhagic infarction. The major risk factors for mucormycosis are diabetic ketoacidosis, neutropenia, iron overload, and deferoxamine therapy. The striking association between the use of the iron chelator deferoxamine and the occurrence of mucormycosis has led to the hypothesis that the fungi utilize deferoxamine as a siderophore (135).

The experimental induction of diabetes in mice significantly impairs the ability of mice to withstand intrasinus and pulmonary challenge with *Rhizopus oryzae* (183,184). Normal, but not diabetic, bronchoalveolar macrophages inhibited spore germination of *R. oryzae* (151,183). Human neutrophils and neutrophil products can kill *R. oryzae* hyphae (144,151,185,186).

IX. Summary

Cell-mediated immunity, with bronchoalveolar cells probably playing a central role, is of primary importance in host defense against *C. neoformans* and the endemic dimorphic fungi *H. capsulatum*, *B. dermatitidis*, and *C. immitis*. Neutrophils are more important for protection against invasive candidiasis, aspergillosis, and mucormycosis, although bronchoalveolar macrophages may prevent the establishment of infection by *Aspergillus* and *Rhizopus* species by killing or preventing germination of the infective conidia.

The ligands and receptors involved in the binding of macrophages to fungi are being defined. Thus, binding of *C. neoformans* depends on the interaction of complement components, in particular iC3b, deposited on the polysaccharide capsule with macrophage complement receptors. Unopsonized *H. capsulatum* binds to macrophages via the common β-chain, CD18, of CR3, CR4, and LFA-1. WI-1, a cell wall protein of *B. dermatitidis*, binds to CR3 and CD14 on macrophages. Unopsonized *C. albicans* binds via macrophage mannose receptors, whereas complement and immunoglobulin receptors probably contribute to binding of opsonized *C. albicans*. Unopsonized *A. fumigatus* conidia bind via macrophage mannose and/or β-glucan receptors.

The mechanisms whereby bronchoalveolar macrophages inhibit the growth of and kill fungi are less well understood. Cationic proteins such as defensins, reactive oxygen intermediates, iron restriction, and, in mouse cells, nitric oxide have all been implicated as effector mechanisms in different systems. More work is needed to define these mechanisms, especially in human cells. Future studies are also needed to determine what factors influence the pattern of cytokine release from bronchoalveolar macrophages in response to the different fungi. This pattern may be critical in determining whether a protective immune response subsequently develops.

References

1. Edman JC, Kovacs JA, Masur H, Santi DV, Elwood HJ, Sogin ML. Ribosomal RNA sequence shows *Pneumocystis carinii* to be a member of the fungi. Nature 1988; 334:519–522.
2. Levitz SM. The ecology of *Cryptococcus neoformans* and the epidemiology of cryptococcosis. Rev Infect Dis 1991; 13:1163–1169.
3. Diamond RD. *Cryptococcus neoformans*. In: Mandell GL, Bennett JE, Dolin R, eds. Mandell, Douglas and Bennett's Principles and Practice of Infectious Diseases. 4th ed. New York: Churchill Livingstone, 1995:2331–2340.
4. Cherniak R, Sundstrom JB. Polysaccharide antigens of the capsule of *Cryptococcus neoformans*. Infect Immun 1994; 62:1507–1512.
5. Perfect JR. Cryptococcosis. Infect Dis Clin North Am 1989; 3:77–102.
6. Neilson JB, Fromtling RA, Bulmer GS. *Cryptococcus neoformans*: size range of infectious particles from aerosolized soil. Infect Immun 1977; 17:634–638.
7. Levitz SM. Macrophage *Cryptococcus* interactions. In: Zwilling BS, Eisenstein TK, eds. Macrophage Pathogen Interactions. New York: Marcel Dekker, 1994:533–543.
8. Bolanos B, Mitchell TG. Phagocytosis and killing of *Cryptococcus neoformans* by rat alveolar macrophages in the absence of serum. J Leukoc Biol 1989; 46:521–528.
9. Levitz SM, Tabuni A, Wagner R, Kornfeld H, Smail EH. Binding of unopsonized *Cryptococcus neoformans* by human bronchalveolar macrophages: inhibition by a large-molecular-size serum component. J Infect Dis 1992; 166:866–873.
10. Kozel TR, Gotschlich EC. The capsule of *Cryptococcus neoformans* passively inhibits phagocytosis of the yeast by macrophages. J Immunol 1982; 129:1675–1680.
11. Kozel TR, Pfrommer GST. Activation of the complement system by *Cryptococcus neoformans* leads to binding of iC3b to the yeast. Infect Immun 1986; 52:1–5.
12. Kozel TR, Wilson MA, Pfrommer GST, Schlageter AM. Activation and binding of opsonic fragments of C3 on encapsulated *Cryptococcus neoformans* by using an alternative complement pathway reconstituted from six isolated proteins. Infect Immun 1989; 57:1922–1927.
13. Kozel TR, Highison B, Stratton CJ. Localization on encapsulated *Cryptococcus neoformans* of serum components opsonic for phagocytosis by macrophages and neutrophils. Infect Immun 1984; 43:574–579.

14. Levitz SM, Tabuni A. Binding of *Cryptococcus neoformans* by human cultured macrophages. Requirements for multiple complement receptors and actin. J Clin Invest 1991; 87:528–535.
15. Kozel TR, Follette JL. Opsonization of encapsulated *Cryptococcus neoformans* by specific anticapsular antibody. Infect Immun 1981; 31:978–984.
16. Zebedee SL, Koduri RK, Mukherjee J, Mukherjee S, Lee S, Sauer DF, Scharff MD, Casadevall A. Mouse-human immunoglobulin G1 chimeric antibodies with activities against *Cryptococcus neoformans*. Antimicrob Agents Chemother 1994; 38: 1507–1514.
17. Casadevall A, Mukherjee J, Devi SJN, Schneerson R, Robbins JB, Scharff MD. Antibodies elicited by a *Cryptococcus neoformans*–tetanus toxoid conjugate vaccine have the same specificity as those elicited in infection. J Infect Dis 1992; 165: 1086–1093.
18. Collins HL, Bancroft GJ. Cytokine enhancement of complement-dependent phagocytosis by macrophages: synergy of tumor necrosis factor-α and granulocyte-macrophage colony-stimulating factor for phagocytosis of *Cryptococcus neoformans*. Eur J Immunol 1992; 22:1447–1454.
19. Blackstock R, Hall NK, Hernandez NC. Characterization of a suppressor factor that regulates phagocytosis by macrophages in murine cryptococcosis. Infect Immun 1989; 57:1773–1779.
20. Merkel GJ, Scofield BA. Conditions affecting the adherence of *Cryptococcus neoformans* to rat glial and lung cells in vitro. J Med Vet Mycol 1993; 31:55–64.
21. Hill JO. CD4+ T cells cause multinucleated giant cells to form around *Cryptococcus neoformans* and confine the yeast within the primary site of infection in the respiratory tract. J Exp Med 1992; 175:1685–1695.
22. Huffnagle GB, Yates JL, Lipscomb MF. Immunity to a pulmonary *Cryptococcus neoformans* infection requires both CD4+ and CD8+ T cells. J Exp Med 1991; 173: 793–800.
23. Mody CH, Lipscomb MF, Street NE, Toews GB. Depletion of CD4+ (L3T4+) lymphocytes in vivo impairs murine host defense to *Cryptococcus neoformans*. J Immunol 1990; 144:1472–1477.
24. Huffnagle GB, Lipscomb MF, Lovchik JA, Hoag KA, Street NE. The role of $CD4^+$ and $CD8^+$ T cells in the protective inflammatory response to a pulmonary cryptococcal infection. J Leukoc Biol 1994; 55:35–42.
25. Mody CH, Chen GH, Jackson C, Curtis JL, Toews GB. Depletion of murine CD8+ T cells in vivo decreases pulmonary clearance of a moderately virulent strain of *Cryptococcus neoformans*. J Lab Clin Med 1993; 121:765–773.
26. Hill JO, Harmsen AG. Intrapulmonary growth and dissemination of an avirulent strain of *Cryptococcus neoformans* in mice deleted of CD4+ or CD8+ T cells. J Exp Med 1991; 173:755–758.
27. Mody CH, Paine R III, Jackson C, Chen GH, Toews GB. CD8 cells play a critical role in delayed type hypersensitivity to intact *Cryptococcus neoformans*. J Immunol 1994; 152:3970–3979.
28. Lovchik JA, Lyons CR, Lipscomb MF. A role for gamma interferon–induced nitric oxide in pulmonary clearance of *Cryptococcus neoformans*. Am J Respir Cell Mol Biol 1995; 13:116–124.

29. Levitz SM, DiBenedetto DJ. Differential stimulation of murine resident peritoneal cells by selectively opsonized encapsulated and acapsular *Cryptococcus neoformans*. Infect Immun 1988; 56:2544–2551.
30. Lovchik JA, Lipscomb MF. Role for C5 and neutrophils in the pulmonary intravascular clearance of circulating *Cryptococcus neoformans*. Am J Respir Cell Mol Biol 1993; 9:617–627.
31. Huffnagle GB, Chen G-H, Curtis JL, McDonald RA, Strieter RM, Toews GB. Down-regulation of the afferent phase of T cell–mediated pulmonary inflammation and immunity by a high melanin-producing strain of *Cryptococcus neoformans*. J Immunol 1995; 155:3507–3516.
32. Levitz SM, Dupont MP. Phenotypic and functional characterization of human lymphocytes activated by interleukin-2 to directly inhibit growth of *Cryptococcus neoformans* in vitro. J Clin Invest 1993; 91:1490–1498.
33. Levitz SM, Dupont MP, Smail EH. Direct activity of human T lymphocytes and natural killer cells against *Cryptococcus neoformans*. Infect Immun 1994; 62: 194–202.
34. Hidore MR, Nabavi N, Sonleitner F, Murphy JW. Murine natural killer cells are fungicidal to *Cryptococcus neoformans*. Infect Immun 1991; 59:1747–1754.
35. Miller MF, Mitchell TG. Killing of *Cryptococcus neoformans* strains by human neutrophils and monocytes. Infect Immun 1991; 59:24–28.
36. Levitz SM, DiBenedetto DJ. Paradoxical role of capsule in murine bronchoalveolar macrophage-mediated killing of *Cryptococcus neoformans*. J Immunol 1989; 142:659–665.
37. Mody CH, Tyler CL, Sitrin RG, Jackson C, Toews GB. Interferon-γ activates rat alveolar macrophages for anticryptococcal activity. Am J Respir Cell Mol Biol 1991; 5:19–26.
38. Weinberg PB, Becker S, Granger DL, Koren HS. Growth inhibition of *Cryptococcus neoformans* by human alveolar macrophages. Am Rev Respir Dis 1987; 136:1242–1247.
39. Cameron ML, Granger DL, Weinberg JB, Kozumbo WJ, Koren HS. Human alveolar and peritoneal macrophages mediate fungistasis independently of L-arginine oxidation to nitrite or nitrate. Am Rev Respir Dis 1990; 142:1313–1319.
40. Wagner RP, Levitz SM, Tabuni A, Kornfeld H. HIV-1 envelope protein (gp120) inhibits the activity of human bronchoalveolar macrophages against *Cryptococcus neoformans*. Am Rev Respir Dis 1992; 146:1434–1438.
41. Diamond RD, Root R, Bennett JE. Factors influencing killing of *Cryptococcus neoformans* by human leukocytes in vitro. J Infect Dis 1972; 125:367–375.
42. Levitz SM, Farrell TP. Growth inhibition of *Cryptococcus neoformans* by cultured human monocytes: role of the capsule, opsonins, the culture surface, and cytokines. Infect Immun 1990; 58:1201–1209.
43. Levitz SM, Farrell TP, Maziarz RT. Killing of *Cryptococcus neoformans* by human peripheral blood mononuclear cells stimulated in culture. J Infect Dis 1991; 163: 1108–1113.
44. Levitz SM. Activation of human peripheral blood mononuclear cells by interleukin-2 and granulocyte-macrophage colony-stimulating factor to inhibit *Cryptococcus neoformans*. Infect Immun 1991; 59:3393–3397.

45. Levitz SM, Tabuni A, Kornfeld H, Reardon CC, Golenbock DT. Production of tumor necrosis factor-alpha in human leukocytes stimulated by Cryptococcus neoformans. Infect Immun 1994; 62:1975–1981.
46. Levitz SM, Tabuni A, Nong S-H, Golenbock DT. Effects of interleukin-10 on human peripheral blood mononuclear cell responses to *Cryptococcus neoformans*, *Candida albicans*, and lipopolysaccharide. Infect Immun 1996; 64:945–951.
47. Collins HL, Bancroft GJ. Encapsulation of *Cryptococcus neoformans* impairs antigen-specific T-cell responses. Infect Immun 1991; 59:3883–3888.
48. Vecchiarelli A, Pietrella D, Dottorini M, Monari C, Retini C, Todisco T, Bistoni F. Encapsulation of *Cryptococcus neoformans* regulates fungicidal activity and the antigen presentation process in human alveolar macrophages. Clin Exp Immunol 1994; 98:217–223.
49. Washburn RG, Tuazon CU, Bennett JE. Phagocytic and fungicidal activity of monocytes from patients with acquired immunodeficiency syndrome. J Infect Dis 1985; 151:565–566.
50. Harrison TS, Kornfeld H, Levitz SM. The effect of infection with human immunodeficiency virus on the anticryptococcal activity of lymphocytes and monocytes. J Infect Dis 1995; 172:665–671.
51. Cameron ML, Granger DL, Matthews TJ, Weinberg JB. Human immunodeficiency virus (HIV)–infected human blood monocytes and peritoneal macrophages have reduced anticryptococcal activity whereas HIV–infected alveolar macrophages retain normal activity. J Infect Dis 1994; 170:60–67.
52. Orendi JM, Nottet HSLM, Visser MR, Verheul AFM, Snippe H, Verhoef J. Enhancement of HIV-1 replication in peripheral blood mononuclear cells by *Cryptococcus neoformans* is monocyte-dependent but tumour necrosis factor-independent. AIDS 1994; 8:423–429.
53. Hoy JF, Lewis DE, Miller GG. Functional versus phenotypic analysis of T cells in subjects seropositive for the human immunodeficiency virus: a prospective study of in vitro responses to *Cryptococcus neoformans*. J Infect Dis 1988; 158:1071–1078.
54. Harrison TS, Levitz SM. Role of IL-12 in PBMC responses to fungi in persons with and without HIV infection. J Immunol 1996; 156:4492–4497.
55. Harrison TS, Levitz SM. Priming with interferon-gamma restores deficient IL-12 production by PBMC from HIV-seropositive donors. J Immunol 1996. In press.
56. Edwards JE Jr. Candida species. In: Mandell GL, Bennett JE, Dolin R, eds. Mandell, Douglas and Bennett's Principles and Practices of Infectious Diseases. 4th ed. New York: Churchill Livingstone, 1995:2289–2306.
57. Nugent KM, Onofrio JM. Pulmonary tissue resistance to *Candida albicans* in normal and in immunosuppressed mice. Am Rev Respir Dis 1983; 128:909–914.
58. Lal S, Mitsuyama M, Miyata M, Ogata N, Amako K, Nomoto K. Pulmonary defence mechanism in mice. A comparative role of alveolar macrophages and polymorphonuclear cells against infection with *Candida albicans*. J Clin Lab Immunol 1986; 19:127–133.
59. Sawyer RT, Harmsen AG. The relative contribution of resident pulmonary alveolar macrophage and inflammatory polymorphonuclear neutrophils in host resistance to pulmonary infection by *Candida albicans*. Mycopathologia 1989; 108:95–105.

60. Calderone RA, Braun PC. Adherence and receptor relationships of *Candida albicans*. Microbiol Rev 1991; 55:1–20.
61. Fels AO, Cohn ZA. The alveolar macrophage. J Appl Physiol 1986; 60:353–369.
62. Marodi L, Korchak HM, Johnston RB Jr. Mechanisms of host defense against Candida species. I. Phagocytosis by monocytes and monocyte-derived macrophages. J Immunol 1991; 146:2783–2789.
63. Ezekowitz RA, Sastry K, Bailly P, Warner A. Molecular characterization of the human macrophage mannose receptor: demonstration of multiple carbohydrate recognition-like domains and phagocytosis of yeasts in Cos-1 cells. J Exp Med 1990; 172:1785–1794.
64. Stahl PD, Rodman JS, Miller MJ, Schlesinger PH. Evidence for receptor-mediated binding of glycoproteins, glycoconjugates, and lysosomal glycosidases by alveolar macrophages. Proc Natl Acad Sci USA 1978; 75:1399–1403.
65. Rosseau S, Seeger W, Pralle H, Lohmeyer J. Phagocytosis of viable *Candida albicans* by alveolar macrophages: flow cytometric quantification. Am J Physiol 1994; 267:L211–L217.
66. Limper AH, Standing JE. Vitronectin interacts with *Candida albicans* and augments organism attachment to the NR8383 macrophage cell line. Immunol Lett 1994; 42: 139–144.
67. Pottratz ST, Martin WJ II. Mechanism of *Pneumocystis carinii* attachment to cultured rat alveolar macrophages. J Clin Invest 1990; 86:1678–1683.
68. Calderone R, Sturtevant J. Macrophage interactions with *Candida*. Immunol Series 1994; 60:505–515.
69. Stanley VC, Hurley R. The growth of *Candida* species in cultures of mouse peritoneal macrophages. J Pathol 1969; 97:357–366.
70. Decker T, Lohmann-Matthes ML, Baccarini M. Heterogeneous activity of immature and mature cells of the murine monocyte-macrophage lineage derived from different anatomical districts against yeast-phase *Candida albicans*. Infect Immun 1986; 54:477–486.
71. Yamada Y, Saito H, Tomioka H, Jidoi J. Relationship between the susceptibility of various bacteria to active oxygen species and to intracellular killing by macrophages. J Gen Microbiol 1987; 133:2015–2021.
72. Sasada M, Johnston RB Jr. Macrophage microbicidal activity. Correlation between phagocytosis-associated oxidative metabolism and the killing of *Candida* by macrophages. J Exp Med 1980; 152:85–98.
73. Peterson EM, Calderone RA. Growth inhibition of *Candida albicans* by rabbit alveolar macrophages. Infect Immun 1977; 15:910–915.
74. Lehrer RI, Ferrari LG, Patterson-Delafield J, Sorrell T. Fungicidal activity of rabbit alveolar and peritoneal macrophages against *Candida albicans*. Infect Immun 1980; 28:1001–1008.
75. Peterson EM, Calderone RA. Inhibition of specific amino acid uptake in *Candida albicans* by lysosomal extracts from alveolar macrophages. Infect Immun 1978; 21: 506–513.
76. Patterson-Delafield J, Martinez RJ, Lehrer RI. Microbicidal cationic proteins in rabbit alveolar macrophages: a potential host defense mechanism. Infect Immun 1980; 30:180–192.

77. Patterson-Delafield J, Szklarek D, Martinez RJ, Lehrer RI. Microbicidal cationic proteins of rabbit alveolar macrophages: amino acid composition and functional attributes. Infect Immun 1981; 31:723–731.
78. Lehrer RI, Szklarek D, Selsted ME, Fleischmann J. Increased content of microbicidal cationic peptides in rabbit alveolar macrophages elicited by complete Freund adjuvant. Infect Immun 1981; 33:775–778.
79. Kemmerich B, Rossing TH, Pennington JE. Comparative oxidative microbicidal activity of human blood monocytes and alveolar macrophages and activation by recombinant gamma interferon. Am Rev Respir Dis 1987; 136:266–270.
80. Todisco T, Dottorini M, Puliti M, Vecchiarelli A. Comparative oxidative microbicidal activity of human blood monocytes and alveolar macrophages and activation by recombinant gamma interferon [letter]. Am Rev Respir Dis 1988; 137:1246.
81. Vecchiarelli A, Dottorini M, Cociani C, Pietrella D, Todisco T, Bistoni F. Mechanism of intracellular candidacidal activity mediated by calcium ionophore in human alveolar macrophages. Am J Respir Cell Mol Biol 1993; 9:19–25.
82. Thompson HL, Wilton JM. Interaction and intracellular killing of *Candida albicans* blastospores by human polymorphonuclear leucocytes, monocytes and monocyte-derived macrophages in aerobic and anaerobic conditions. Clin Exp Immunol 1992; 87:316–321.
83. Watanabe K, Kagaya K, Yamada T, Fukazawa Y. Mechanism for candidacidal activity in macrophages activated by recombinant gamma interferon. Infect Immun 1991; 59:521–528.
84. Hashimoto T. In vitro study of contact-mediated killing of *Candida albicans* hyphae by activated murine peritoneal macrophages in a serum-free medium. Infect Immun 1991; 59:3555–3561.
85. Blasi E, Pitzurra L, Puliti M, Chimienti AR, Mazzolla R, Barluzzi R, Bistoni F. Differential susceptibility of yeast and hyphal forms of *Candida albicans* to macrophage derived nitrogen-containing compounds. Infect Immun 1995; 63: 1806–1809.
86. Romani L, Puccetti P, Mencacci A, Cenci E, Spaccapelo R, Tonnetti L, Grohmann U, Bistoni F. Neutralization of IL-10 up-regulates nitric oxide production and protects susceptible mice from challenge with *Candida albicans*. J Immunol 1994; 152: 3514–3521.
87. Vazquez-Torres A, Jones-Carson J, Balish E. Candidacidal activity of macrophages from immunocompetent and congenitally immunodeficient mice. J Infect Dis 1994; 170:180–188.
88. Vazquez-Torres A, Jones-Carson J, Balish E. Nitric oxide production does not directly increase macrophage candidacidal activity. Infect Immun 1995; 63:1142–1144.
89. Schneemann M, Schoedon G, Hofer S, Blau N, Guerrero L, Schaffner A. Nitric oxide synthase is not a constituent of the antimicrobial armature of human mononuclear phagocytes. J Infect Dis 1993; 167:1358–1363.
90. Denis M. Human monocytes/macrophages: NO or no NO? J Leukoc Biol 1994; 55: 682–684.
91. Brummer E, Morrison CJ, Stevens DA. Recombinant and natural gamma-interferon activation of macrophages in vitro: different dose requirements for induction of

killing activity against phagocytizable and nonphagocytizable fungi. Infect Immun 1985; 49:724–730.
92. Blasi E, Farinelli S, Varesio L, Bistoni F. Augmentation of GG2EE macrophage cell line-mediated anti-Candida activity by gamma interferon, tumor necrosis factor, and interleukin-1. Infect Immun 1990; 58:1073–1077.
93. Marodi L, Schreiber S, Anderson DC, MacDermott RP, Korchak HM, Johnston RB Jr. Enhancement of macrophage candidacidal activity by interferon-gamma. Increased phagocytosis, killing, and calcium signal mediated by a decreased number of mannose receptors. J Clin Invest 1993; 91:2596–2601.
94. Vecchiarelli A, Todisco T, Puliti M, Dottorini M, Bistoni F. Modulation of anti-Candida activity of human alveolar macrophages by interferon-gamma or interleukin-1-alpha. Am J Respir Cell Mol Biol 1989; 1:49–55.
95. Brummer E, Stevens DA. Candidacidal mechanisms of peritoneal macrophages activated with lymphokines or gamma-interferon. J Med Microbiol 1989; 28:173–181.
96. Vecchiarelli A, Dottorini M, Puliti M, Todisco T, Cenci E, Bistoni F. Defective candidacidal activity of alveolar macrophages and peripheral blood monocytes from patients with chronic obstructive pulmonary disease. Am Rev Respir Dis 1991; 143: 1049–1054.
97. Karbassi A, Becker JM, Foster JS, Moore RN. Enhanced killing of *Candida albicans* by murine macrophages treated with macrophage colony-stimulating factor: evidence for augmented expression of mannose receptors. J Immunol 1987; 139: 417–421.
98. Bleiberg I, Kletter Y, Riklis I, Fabian I. Induction of murine macrophage fungal killing by interleukin 3. Exp Hematol 1989; 17:895–897.
99. Vecchiarelli A, Puliti M, Torosantucci A, Cassone A, Bistoni F. In vitro production of tumor necrosis factor by murine splenic macrophages stimulated with mannoprotein constituents of *Candida albicans* cell wall. Cell Immunol 1991; 134:65–76.
100. Jouault T, Bernigaud A, Lepage G, Trinel PA, Poulain D. The *Candida albicans* phospholipomannan induces in vitro production of tumour necrosis factor-alpha from human and murine macrophages. Immunology 1994; 83:268–273.
101. Jouault T, Lepage G, Bernigaud A, Trinel PA, Fradin C, Wieruszeski JM, Strecker G, Poulain D. Beta-1,2-linked oligomannosides from *Candida albicans* act as signals for tumor necrosis factor alpha production. Infect Immun 1995; 63:2378–2381.
102. Garner RE, Rubanowice K, Sawyer RT, Hudson JA. Secretion of TNF-alpha by alveolar macrophages in response to Candida albicans mannan. J Leukoc Biol 1994; 55:161–168.
103. Blasi E, Pitzurra L, Puliti M, Lanfrancone L, Bistoni F. Early differential molecular response of a macrophage cell line to yeast and hyphal forms of *Candida albicans*. Infect Immun 1992; 60:832–837.
104. Ferrante A. Tumor necrosis factor alpha potentiates neutrophil antimicrobial activity: increased fungicidal activity against *Torulopsis glabrata* and *Candida albicans* and associated increases in oxygen radical production and lysosomal enzyme release. Infect Immun 1989; 57:2115–2122.
105. Newman SL, Bullock WE. Interaction of *Histoplasma capsulatum* yeasts and conidia with human and animal macrophages. Immunol Series 1994; 60:517–532.
106. Bullock WE. *Histoplasma capsulatum*. In: Mandell GL, Bennet JE, Dolin R, eds.

Mandell, Douglas and Bennett's Principles and Practice of Infectious Diseases. 4th ed. New York: Churchill Livingstone, 1995:2340–2353.
107. Deepe GS Jr, Bullock WE. Histoplasmosis. A granulomatous inflammatory response. In: Gallin JI, Goldstein IM, Snyderman R, eds. Inflammation: Basic Principles and Clinical Correlates. 2nd ed. New York: Raven Press, 1992:943–958.
108. Bullock WE, Wright SD. Role of the adherence-promoting receptors, CR3, LFA-1, and p150,95, in binding of *Histoplasma capsulatum* by human macrophages. J Exp Med 1987; 165:195–210.
109. Newman SL, Bucher C, Rhodes J, Bullock WE. Phagocytosis of *Histoplasma capsulatum* yeasts and microconidia by human cultured macrophages and alveolar macrophages. Cellular cytoskeleton requirement for attachment and ingestion. J Clin Invest 1990; 85:223–230.
110. Chaturvedi S, Frame P, Newman SL. Macrophages from human immunodeficiency virus-positive persons are defective in host defense against *Histoplasma capsulatum*. J Infect Dis 1995; 171:320–327.
111. Schnur RA, Newman SL. The respiratory burst response to *Histoplasma capsulatum* by human neutrophils. Evidence for intracellular trapping of superoxide anion. J Immunol 1990; 144:4765–4772.
112. McMahon JP, Wheat J, Sobel ME, Pasula R, Downing JF, Martin WJ II. Murine laminin binds to *Histoplasma capsulatum*. A possible mechanism of dissemination. J Clin Invest 1995; 96:1010–1017.
113. Newman SL, Gootee L, Morris R, Bullock WE. Digestion of *Histoplasma capsulatum* yeasts by human macrophages. J Immunol 1992; 149:574–580.
114. Eissenberg LG, Schlesinger PH, Goldman WE. Phagosome-lysosome fusion in P388D1 macrophages infected with *Histoplasma capsulatum*. J Leukoc Biol 1988; 43: 483–491.
115. Eissenberg LG, Goldman WE, Schlesinger PH. *Histoplasma capsulatum* modulates the acidification of phagolysosomes. J Exp Med 1993; 177:1605–1611.
116. Newman SL, Gootee L, Brunner G, Deepe GS Jr. Chloroquine induces human macrophage killing of *Histoplasma capsulatum* by limiting the availability of intracellular iron and is therapeutic in a murine model of histoplasmosis. J Clin Invest 1994; 93:1422–1429.
117. Wolf JE, Abegg AL, Travis SJ, Kobayashi GS, Little JR. Effects of *Histoplasma capsulatum* on murine macrophage functions: inhibition of macrophage priming, oxidative burst, and antifungal activities. Infect Immun 1989; 57:513–519.
118. Wolf JE, Massof SE, Sherwin JR, Considine RV. Inhibition of murine macrophage protein kinase C activity by nonviable *Histoplasma capsulatum*. Infect Immun 1992; 60:2683–2687.
119. Eissenberg LG, Goldman WE. *Histoplasma capsulatum* fails to trigger release of superoxide from macrophages. Infect Immun 1987; 55:29–34.
120. Wu-Hsieh BA, Howard DH. Inhibition of the intracellular growth of *Histoplasma capsulatum* by recombinant murine gamma interferon. Infect Immun 1987; 55: 1014–1016.
121. Lane TE, Otero GC, Wu-Hsieh BA, Howard DH. Expression of inducible nitric oxide synthase by stimulated macrophages correlates with their antihistoplasma activity. Infect Immun 1994; 62:1478–1479.

122. Lane TE, Wu-Hsieh BA, Howard DH. Iron limitation and the gamma interferon–mediated antihistoplasma state of murine macrophages. Infect Immun 1991; 59: 2274–2278.
123. Zhou P, Sieve MC, Bennett J, Kwon-Chung KJ, Tewari RP, Gazzinelli RT, Sher A, Seder RA. IL-12 prevents mortality in mice infected with *Histoplasma capsulatum* through induction of IFN-gamma. J Immunol 1995; 155:785–795.
124. Brummer E, Stevens DA. Antifungal mechanisms of activated murine broncho-alveolar or peritoneal macrophages for *Histoplasma capsulatum*. Clin Exp Immunol 1995; 102:65–70.
125. Couto MA, Liu L, Lehrer RI, Ganz T. Inhibition of intracellular *Histoplasma capsulatum* replication by murine macrophages that produce human defensin. Infect Immun 1994; 62:2375–2378.
126. Newman SL, Gootee L, Bucher C, Bullock WE. Inhibition of intracellular growth of *Histoplasma capsulatum* yeast cells by cytokine-activated human monocytes and macrophages. Infect Immun 1991; 59:737–741.
127. Brummer E, Kurita N, Yoshida S, Nishimura K, Miyaji M. Killing of *Histoplasma capsulatum* by gamma-interferon–activated human monocyte-derived macrophages: evidence for a superoxide anion-dependent mechanism. J Med Microbiol 1991; 35:29–34.
128. Newman SL, Gootee L. Colony-stimulating factors activate human macrophages to inhibit intracellular growth of *Histoplasma capsulatum* yeasts. Infect Immun 1992; 60:4593–4597.
129. Fleischmann J, Wu-Hsieh B, Howard DH. The intracellular fate of *Histoplasma capsulatum* in human macrophages is unaffected by recombinant human interferon-gamma. J Infect Dis 1990; 161:143–145.
130. Desai G, Nassar F, Brummer E, Stevens DA. Killing of *Histoplasma capsulatum* by macrophage colony stimulating factor-treated human monocyte-derived macrophages: role for reactive oxygen intermediates. J Med Microbiol 1995; 43:224–229.
131. Newman SL, Gootee L, Stroobant V, Van Der Goot H, Boelaert JR. Inhibition of growth of *Histoplasma capsulatum* yeast cells in human macrophages by the iron chelator VUF 8514 and comparison of VUF 8514 with deferoxamine. Antimicrob Agents Chemother 1995; 39:1824–1829.
132. Eissenberg LG, West JL, Woods JP, Goldman WE. Infection of P388D1 macrophages and respiratory epithelial cells by *Histoplasma capsulatum*: selection of avirulent variants and their potential role in persistent histoplasmosis. Infect Immun 1991; 59:1639–1646.
133. Levitz SM. Aspergillosis. Infect Dis Clin North Am 1989; 3:1–18.
134. Khoo SH, Denning DW. Invasive aspergillosis in patients with AIDS. Clin Infect Dis 1994; 19(Suppl 1):S41–S48.
135. Hogan LH, Klein BS, Levitz SM. Virulence factors of medically important fungi. Clin Microbiol Rev 1996; 9:469–488.
136. Kan VL, Bennett JE. Lectin-like attachment sites on murine pulmonary alveolar macrophages bind *Aspergillus fumigatus* conidia. J Infect Dis 1988; 158:407–414.
137. Kan VL, Bennett JE. Beta 1,4-oligoglucosides inhibit the binding of *Aspergillus fumigatus* conidia to human monocytes. J Infect Dis 1991; 163:1154–1156.
138. Tronchin G, Bouchara JP, Larcher G, Lissitzky JC, Chabasse D. Interaction between

Aspergillus fumigatus and basement membrane laminin: binding and substrate degradation. Biol Cell 1993; 77:201–208.
139. Annaix V, Bouchara JP, Larcher G, Chabasse D, Tronchin G. Specific binding of human fibrinogen fragment D to *Aspergillus fumigatus* conidia. Infect Immun 1992; 60:1747–1755.
140. Coulot P, Bouchara JP, Renier G, Annaix V, Planchenault C, Tronchin G, Chabasse D. Specific interaction of *Aspergillus fumigatus* with fibrinogen and its role in cell adhesion. Infect Immun 1994; 62:2169–2177.
141. Larcher G, Bouchara JP, Annaix V, Symoens F, Chabasse D, Tronchin G. Purification and characterization of a fibrinogenolytic serine proteinase from *Aspergillus fumigatus* culture filtrate. FEBS Lett 1992; 308:65–69.
142. Levitz SM, Diamond RD. Mechanisms of resistance of *Aspergillus fumigatus* conidia to killing by neutrophils in vitro. J Infect Dis 1985; 152:33–42.
143. Schaffner A, Douglas H, Braude A. Selective protection against conidia by mononuclear and against mycelia by polymorphonuclear phagocytes in resistance to Aspergillus. Observations on these two lines of defense in vivo and in vitro with human and mouse phagocytes. J Clin Invest 1982; 69:617–631.
144. Levitz SM, Selsted ME, Ganz T, Lehrer RI, Diamond RD. In vitro killing of spores and hyphae of *Aspergillus fumigatus* and *Rhizopus oryzae* by rabbit neutrophil cationic peptides and bronchoalveolar macrophages. J Infect Dis 1986; 154:483–489.
145. Cohen MS, Isturiz RE, Malech HL, Root RK, Wilfert CM, Gutman L, Buckley RH. Fungal infection in chronic granulomatous disease. The importance of the phagocyte in defense against fungi. Am J Med 1981; 71:59–66.
146. Roos D. The genetic basis of chronic granulomatous disease. Immunol Rev 1994; 138:121–157.
147. Schaffner A. Macrophage-Aspergillus interactions. Immunol Series 1994; 60: 545–552.
148. Schaffner A, Douglas H, Braude AI, Davis CE. Killing of Aspergillus spores depends on the anatomical source of the macrophage. Infect Immun 1983; 42:1109–1115.
149. Schaffner A. Therapeutic concentrations of glucocorticoids suppress the antimicrobial activity of human macrophages without impairing their responsiveness to gamma interferon. J Clin Invest 1985; 76:1755–1764.
150. Michaliszyn E, Senechal S, Martel P, de Repentigny L. Lack of involvement of nitric oxide in killing of *Aspergillus fumigatus* conidia by pulmonary alveolar macrophages. Infect Immun 1995; 63:2075–2078.
151. Waldorf AR, Levitz SM, Diamond RD. In vivo bronchoalveolar macrophage defense against *Rhizopus oryzae* and *Aspergillus fumigatus*. J Infect Dis 1984; 150: 752–760.
152. Roilides E, Sein T, Holmes A, Chanock S, Blake C, Pizzo PA, Walsh TJ. Effects of macrophage colony-stimulating factor on antifungal activity of mononuclear phagocytes against *Aspergillus fumigatus*. J Infect Dis 1995; 172:1028–1034.
153. Roilides E, Holmes A, Blake C, Venzon D, Pizzo PA, Walsh TJ. Antifungal activity of elutriated human monocytes against *Aspergillus fumigatus* hyphae: enhancement by granulocyte-macrophage colony-stimulating factor and interferon-gamma. J Infect Dis 1994; 170:894–899.

154. Kothary MH, Chase T Jr, Macmillan JD. Correlation of elastase production by some strains of *Aspergillus fumigatus* with ability to cause pulmonary invasive aspergillosis in mice. Infect Immun 1984; 43:320–325.
155. Rhodes JC, Bode RB, McCuan-Kirsch CM. Elastase production in clinical isolates of Aspergillus. Diagn Microbiol Infect Dis 1988; 10:165–170.
156. Denning DW, Ward PN, Fenelon LE, Benbow EW. Lack of vessel wall elastolysis in human invasive pulmonary aspergillosis. Infect Immun 1992; 60:5153–5156.
157. Smith JM, Tang CM, Van Noorden S, Holden DW. Virulence of *Aspergillus fumigatus* double mutants lacking restriction and an alkaline protease in a low-dose model of invasive pulmonary aspergillosis. Infect Immun 1994; 62:5247–5254.
158. Tang CM, Cohen J, Krausz T, Van Noorden S, Holden DW. The alkaline protease of *Aspergillus fumigatus* is not a virulence determinant in two murine models of invasive pulmonary aspergillosis. Infect Immun 1993; 61:1650–1656.
159. Monod M, Paris S, Sarfati J, Jaton-Ogay K, Ave P, Latge JP. Virulence of alkaline protease-deficient mutants of *Aspergillus fumigatus*. FEMS Microbiol Lett 1993; 80: 39–46.
160. Smith JM, Davies JE, Holden DW. Construction and pathogenicity of *Aspergillus fumigatus* mutants that do not produce the ribotoxin restrictocin. Mol Microbiol 1993; 9:1071–1077.
161. Arruda LK, Mann BJ, Chapman MD. Selective expression of a major allergen and cytotoxin, Asp fI, in *Aspergillus fumigatus*. Implications for the immunopathogenesis of Aspergillus-related diseases. J Immunol 1992; 149:3354–3359.
162. Paris S, Monod M, Diaquin M, Lamy B, Arruda LK, Punt PJ, Latge JP. A transformant of Aspergillus fumigatus deficient in the antigenic cytotoxin ASPFI. FEMS Microbiol Lett 1993; 111:31–36.
163. Kwon-Chung KJ, Bennett JE. Medical Mycology. Philadelphia: Lea & Febiger, 1992.
164. Williams JE, Moser SA, Turner SH, Standard PG. Development of pulmonary infection in mice inoculated with *Blastomyces dermatitidis* conidia. Am J Respir Crit Care Med 1994; 149:500–509.
165. Newman SL, Chaturvedi S, Klein BS. The WI-1 antigen of *Blastomyces dermatitidis* yeasts mediates binding to human macrophage CD11b/CD18 (CR3) and CD14. J Immunol 1995; 154:753–761.
166. Sugar AM, Picard M. Macrophage- and oxidant-mediated inhibition of the ability of live *Blastomyces dermatitidis* conidia to transform to the pathogenic yeast phase: implications for the pathogenesis of dimorphic fungal infections. J Infect Dis 1991; 163:371–375.
167. Sugar AM, Picard M, Wagner R, Kornfeld H. Interactions between human bronchoalveolar macrophages and *Blastomyces dermatitidis* conidia: demonstration of fungicidal and fungistatic effects. J Infect Dis 1995; 171:1559–1562.
168. Klein BS. Immunology of blastomycosis. In: Al-Doory Y, DiSalvo AF, eds. Blastomycosis. New York: Plenum, 1992:133–163.
169. Bradsher RW. Macrophages and Blastomyces dermatitidis. Immunol Series 1994; 60:553–565.
170. Brummer E, Stevens DA. Activation of pulmonary macrophages for fungicidal activity by gamma-interferon or lymphokines. Clin Exp Immunol 1987; 70:520–528.

171. Brummer E, Hanson LH, Restrepo A, Stevens DA. In vivo and in vitro activation of pulmonary macrophages by IFN-gamma for enhanced killing of *Paracoccidioides brasiliensis* or *Blastomyces dermatitidis*. J Immunol 1988; 140:2786–2789.
172. Bradsher RW, Balk RA, Jacobs RF. Growth inhibition of *Blastomyces dermatitidis* in alveolar and peripheral macrophages from patients with blastomycosis. Am Rev Respir Dis 1987; 135:412–417.
173. Brummer E, Sugar AM, Stevens DA. Activation of peritoneal macrophages by concanavalin A or *Mycobacterium bovis* BCG for fungicidal activity against *Blastomyces dermatitidis* and effect of specific antibody and complement. Infect Immun 1983; 39:817–822.
174. Brummer E, Stevens DA. Fungicidal mechanisms of activated macrophages: evidence for nonoxidative mechanisms for killing of *Blastomyces dermatitidis*. Infect Immun 1987; 55:3221–3224.
175. Drutz DJ, Frey CL. Intracellular and extracellular defenses of human phagocytes against *Blastomyces dermatitidis* conidia and yeasts. J Lab Clin Med 1985; 105: 737–750.
176. Stevens DA. *Coccidioides immitis*. In: Mandell GL, Bennett JE, Dolin R, eds. Mandell, Douglas and Bennett's Principles and Practice of Infectious Diseases. 4th ed. New York: Churchill Livingstone, 1995:2365–2375.
177. Savage DC, Madin SH. Cellular responses in lungs of immunized mice to intranasal infection with *Coccidioides immitis*. Sabouraudia 1968; 6:94–102.
178. Ampel NM, Bejarano GC, Galgiani JN. Killing of *Coccidioides immitis* by human peripheral blood mononuclear cells. Infect Immun 1992; 60:4200–4204.
179. Beaman L. Effects of recombinant gamma interferon and tumor necrosis factor on in vitro interactions of human mononuclear phagocytes with *Coccidioides immitis*. Infect Immun 1991; 59:4227–4229.
180. Beaman L. Fungicidal activation of murine macrophages by recombinant gamma interferon. Infect Immun 1987; 55:2951–2955.
181. Magee DM, Cox RA. Roles of gamma interferon and interleukin-4 in genetically determined resistance to *Coccidioides immitis*. Infect Immun 1995; 63:3514–3519.
182. Sugar AM. Mucormycosis. Clin Infect Dis 1992; 14(Suppl 1):S126–S129.
183. Waldorf AR, Ruderman N, Diamond RD. Specific susceptibility to mucromycosis in murine diabetes and bronchoalveolar macrophage defense against Rhizopus. J Clin Invest 1984; 74:150–160.
184. Waldorf AR, Diamond RD. Cerebral mucormycosis in diabetic mice after intrasinus challenge. Infect Immun 1984; 44:194–195.
185. Diamond RD, Clark RA. Damage to *Aspergillus fumigatus* and *Rhizopus oryzae* hyphae by oxidative and nonoxidative microbicidal products of human neutrophils *in vitro*. Infect Immun 1982; 38:487–495.
186. Diamond RD, Krzesicki RK, Epstein B, Jao W. Damage to hyphal forms of fungi by human leukocytes *in vitro*: A possible host defense mechanism in aspergillosis and mucormycosis. Am J Pathol 1978; 91:313–324.

20

Pulmonary Macrophage and Respiratory Virus Interactions

R. CLARK LANTZ, MARK L. WITTEN, and RICHARD J. LEMEN

University of Arizona College of Medicine
Tucson, Arizona

I. Introduction

Viral respiratory infections are major causes of morbidity and mortality in infants and children (1,2) and are frequent precipitants of acute asthmatic exacerbations (3) in patients of all ages. Furthermore, childhood viral respiratory infections may result in the development of chronic lung disease later in life (4). Alveolar macrophages (AMs) are major participants in lung defenses with acute respiratory viral infections and in recovery (5,6); thus, changes in AM function with disease may impact both the short- and long-term consequences. The purpose of this chapter is to review what is known about the interactions of viruses with AMs and to delineate areas for further research.

A. Macrophage Function

AMs located in the alveolar lumen constitute approximately 80–90% of all cells recovered by bronchoalveolar lavage from normal lungs (5,6). They are a key component of lung homeostasis through their role in phagocytosis, antigen presentation, and cell-cell communication. Although the ability of the lung to recognize, detoxify, and remove foreign material involves numerous mechanisms, the initiation of responses to foreign material deposited in the distal lung is centered around AMs (5,6). Furthermore, AMs serve a pivotal role in the pathogenesis of respiratory diseases because of their ability to move and phagocytize particles and to secrete products.

How the AMs responds in health and disease is complex and only partially understood (5,6). Having recognized the presence of foreign material, AMs migrate toward the substances, bind these substances to their plasma membranes, and internalize them. They also internalize soluble molecules from the extracellular fluid by pinocytosis and by other less well-understood pathways. During this process, AMs secrete numerous substances such as cytokines, enzymes, metabolites of arachidonic acid, oxygen metabolites, proteins, free fatty acids, antioxidants, and coagulation factors (5–7) which are used to detoxify the foreign material and to mediate the response of other tissues and cell types to the foreign material.

AMs can also respond in injurious processes that contribute to the development of chronic lung disease (8). Respiratory viruses are known both to enhance and suppress AM functions. These alterations potentially may decrease host resistance to respiratory infections and/or altered immune surveillance as well as contribute to lung injury through the release of inflammatory mediators.

B. Acute Pulmonary Viral Infections

Clinically Significant Pulmonary Viruses

Each year acute viral respiratory infections have enormous impact on the health of children in the United States and worldwide. The most common viral agents are listed in Table 1 (9). Most of the serious respiratory viral infections occur in children less than 5 years of age, especially in infants who are younger than 1 year of age or have preexisting cardiopulmonary diseases. The incidence in the first year of life is 11.4 cases per 100 children and is more common in children of lower socioeconomic status, children exposed to passive smoke, or bottle-fed infants. Approximately 3% of infants with viral infections require hospitalization, and approximately 5000 deaths occur annually (10). Acute lower respiratory infec-

Table 1 Common Causes of Acute Viral Respiratory Infections

Respiratory syncytial virus	Measles
Parainfluenza types I, II, and III	Cytomegalovirus
Adenovirus types I, II, and III	Rhinovirus
Influenza type A and B	Picornaviruses

Source: From ref. 9.

tions with respiratory syncytial virus (RSV) (11), influenza A, parainfluenza (12), rhinoviruses (13–15), and adenoviruses (16) provoke recurrent wheezing illnesses, including classic asthmatic attacks.

Consequences of Acute Pulmonary Infections

Several investigators have suggested a link among genetic factors, severe respiratory infections in infancy, and asthma later in life (17–19). Clearly, little is known about the genetic factors that may control macrophage function between individual cells and between various macrophage functions or how these factors may contribute to the severity of acute respiratory infections. It appears plausible, based on data from animal models, that viral respiratory infections may induce airway injury by upregulation of normal functions of macrophage and other cells important in host inflammatory responses (20).

Several animal models have been developed in order to better understand the mechanisms of virus-induced inflammation (21) and possible interactions with genetic disorders such as asthma (22). Parainfluenza type 2 (23) and adenovirus type 2 (24,25) produced increased airway responsiveness to histamine bronchoprovocation which parallelled the severity of the neutrophilic tissue inflammation and symptoms of the clinical illness in beagle puppies. These viral infections were not associated with long-lasting changes in histamine responsiveness (26), but alterations in lung growth were found, including pulmonary morphometrics and fractal dimensions (27). Unfortunately, even the relationships of airway inflammation and hyperresponsiveness were not clear. Airway hyperresponsiveness with bronchoprovocation were not seen when beagle puppies were infected with *Bordetella bronchiseptica*, an organism that produced a vigorous neutrophilic airway inflammatory response similar to *B. pertussis* in humans (28).

II. Viral Infection of Alveolar Macrophages

A. Binding and Internal Processing of Virus

The interaction of AMs with viruses can be divided into direct and indirect interactions. Direct interactions are those induced by virus-macrophage interactions, whereas indirect interactions refer to macrophage responses to viral infections in other cell types. In this chapter, we mainly discuss the direct interactions of viruses and AMs.

In vitro and in vivo studies have demonstrated that AMs are targeted and infected by respiratory viruses (Fig. 1). Although the mechanism of virus entry into AMs is not known, clues regarding potential mechanisms may be gained from the understanding of how adenovirus enters fibroblasts (29). Receptor-mediated viral endocytosis occurs at specialized regions of the fibroblast membrane called clathrin-coated pits. These membrane areas are indented and may comprise up to

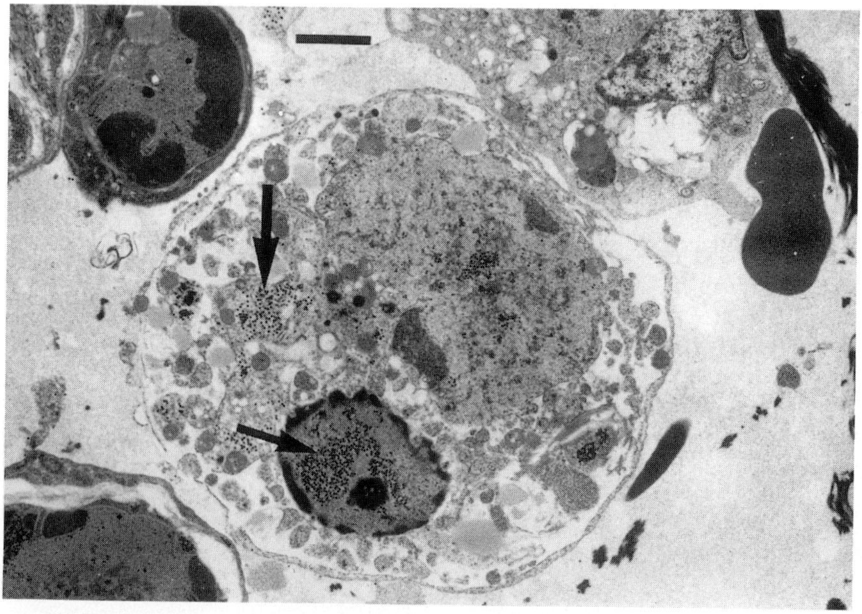

Figure 1 Electron micrograph of a dying large AM, with cytoplasmic and nuclear arrays of virions of adenovirus (arrows). Uranyl acetate-lead citrate stain, bar = 2 μ.

1% of AM surface area. Receptor-virus complexes are internalized as "coated vesicles," or phagosomes. The phagosomes dissociate and fuse with other vesicles to form a larger internal structure, an endosome.

The next step is an endosome-lysosome fusion process (16) (Fig. 2). Foreign matter is transferred to the lysosome and degraded by various proteases functioning in an acidic environment. Recycled products can be released from AMs by exocytosis in the form of small peptides or amino acids or cross from endosomes directly into the cytoplasm.

Electron microscopic studies demonstrate that adenovirus rapidly enters fibroblasts via clathrin-coated pits (29). A few minutes thereafter the virus is observed in endosomes, and after approximately 30 min, adenovirus is found free in the cytosol and eventually becomes associated with the cell nucleus. FitzGerald and coworkers (30) demonstrated by electron microscopic studies that prior to its exit from the endosome, adenovirus causes the endosome membrane to become distorted and wrapped around the virus. After the adenovirus has escaped into the cytosol, fragments of the disrupted endosome membrane are seen close to the viral particles.

Several factors allow adenovirus to escape from the endosome and migrate

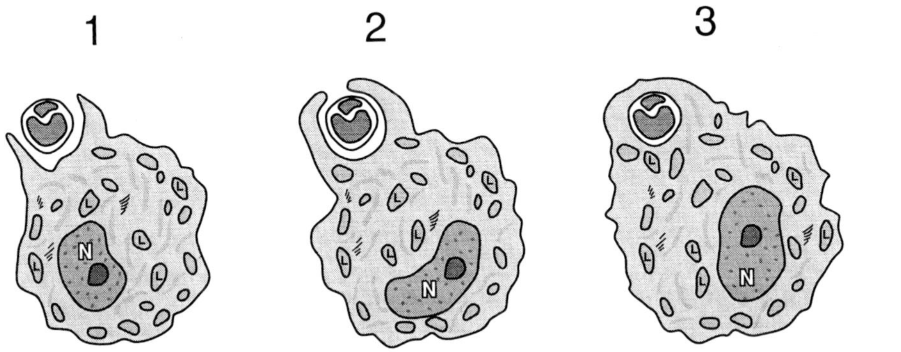

Figure 2 Abbreviations: N, is nucleus; L, lysosome. The following is a theory about how AMs may process respiratory viruses: (1) The AM forms a phagosome to trap a virion. (2) The phagosome encircles the virion and will eventually join with other phagosomes to form an endosome. (3) The endosome begins to fuse with a lysosome. There is evidence that respiratory viruses attenuate the endosome-lysosome fusion process in alveolar macrophages. (From ref. 6.)

toward the nucleus to begin the process of self-replication. The acidic pH within endosomes (approximately pH 5.5) enables the adenovirus to disrupt the membrane of the endosome (31). Another possible mechanism for viral escape from the endosome is the penetration of the endosome membrane lipids by hydrophobic domains of proteins in the adenovirus capsid. Additionally, although the cell membrane is strengthened by actin, myosin, and other cytoskeletal proteins, the endosome membrane lacks these, thus becoming more susceptible to rupture.

Cell maturity influences how AMs interact with respiratory viruses. The AM is a mature form of the monocyte that has migrated into the alveolar septal regions of the lungs. Macrophage precursor cells develop increased phagocytic properties and lysosomal enzyme content and enlarge as they mature into AMs (32–37). There is evidence suggesting that alveolar and blood macrophages take up respiratory syncytial virus (RSV) differently from immature blood monocytes (38). The reasons are not known, but perhaps changes may relate to differences in pH (39). Additionally, restriction of viral infectivity by cultured human blood macrophages is strongly influenced by the number of simultaneous infections, a phenomenon particularly evident with adenovirus type 17 (40). Human sera containing RSV-specific antibodies enhance RSV infection of the U937 macrophage cell line. Human sera enhance viral yield, as measured by the cell-released infectious virus by an average of 50-fold compared with viral infection in the absence of human serum (41). It is thought that RSV-specific antibodies in human serum may have epitopes that can attach readily to virion surface proteins and thus are responsible for the enhancing activity. Therefore, the presence of antiviral antibodies may increase macrophage susceptibility to viral infection.

The AM surface membrane has many receptors, some of which are important in viral uptake and others are important in generalized phagocytic processes and are altered with respiratory viral infections. The Fc receptor which binds the Fc portion of the heavy chain of immunoglobulin G best characterizes this type of alteration that viruses induce.

Approximately 75% of AMs have Fc surface receptors and 60% have C3b receptors. Bovine and ovine alveolar macrophages infected with parainfluenza-3 (PI-3) or RSV and then cultured in vitro have decreased numbers of Fc receptors (42,43). Six days after PI-3 inoculation of calves in vivo, the percentage of AMs with Fc and C3b receptors is significantly reduced (42).

AMs have β_1-receptors which respond to adrenergic stimulation and inhibit AM production of superoxide anion (5). Calves with PI-3 and rhinotracheitis virus infections demonstrate impaired AM β_1-receptor function (44). Treatment with levamisole, a potent antihelminthic, at a dose of 3 mg/kg, partially reverses the virus-induced impairment of the β_1-receptor. It has been suggested that this agent regulates cellular activity by altering intracellular cyclic nucleotide levels (45,46).

B. Incorporation and Detection of Viral Genome in Alveolar Macrophages

Once inside AMs, RSV incorporate their genome, express viral proteins, and replicate. Using cells isolated from bronchoalveolar lavage from RSV infected infants, Midulla et al. (47) have shown that AMs coexpress RSV factor proteins along with HLA-DR, IL-1β, and TNF. Macrophages positive for RSV are seen in adult transplant patients (48) and in an animal model (49). In our laboratory, adenovirus has also been shown to be incorporated into AMs. Using fluorescent immunostaining and electron microscopy, adenovirus virions have been localized to pulmonary epithelial cells, type II alveolar cells, and AMs (see Fig. 1) during the acute phase of the infection (25).

Viral protein and genome can persist in the lung and in AMs for long periods. For instance, RSV can exist in the lung for at least 60 days after infection (50), as evidenced by immunohistochemistry to localize protein and reverse transcriptase-polymerase chain reaction (RT-PCR) for viral RNA. Human AMs exposed in vitro to RSV demonstrate RSV replication, RSV RNA, and protein expression. RSV-exposed macrophages produced and released virus into the supernatants for up to 25 days after infection (51). Immunostaining for RSV RNA is localized to AMs, suggesting this cell type is the source of the RSV RNA. Thus, the persistence of the genome for long periods in macrophages supports the hypothesis that acute respiratory viral infections may produce chronic inflammatory responses, which could in part explain the relationship between viral infections and chronic respiratory diseases such as asthma.

C. Viral Replication in Alveolar Macrophages

Little is known about how AMs support viral replication. Cultured mouse AMs support the growth of Sendai virus as measured by a 10-fold increase in viral titers within the culture medium (52). The synthesis of virus continues for at least 1 month without cytopathic effects. Virus replicates in the cytoplasm of AMs of pigs with experimentally induced African swine fever pneumonitis. These cells subsequently degenerate and release viral antigen (53). AMs phagocytize canine adenovirus type 2 and viral antigen is present in lymphatics of beagle puppies (25). Thus, it is clear that viruses replicate in AMs; however, little is known about the mechanisms of this process and how they may differ between individuals.

Viruses appear selectively to infect AMs. When cultured rabbit AMs are infected with Sendai virus, only 63% of the cells are found to be infected and able to synthesize viral antigens (54). After infection of murine AMs with RSV, only a small amount of infective virus appears in the supernatant, indicating only low levels of replication have occurred (55). However, the level of replication appears to be related to the amount of interferon present. Replication of RSV has also been shown to occur in human AMs (47). Little is known about why respiratory viruses infect some AMs and not others.

III. Direct Effects of Viral Infection on Alveolar Macrophage Function

A. Phagocytosis and Motility

Respiratory infections with RSV, PI-1, PI-3, or influenza are associated with the disruption of several AM functions, which are summarized in Table 2. RSV alters both Fc receptor–mediated and nonopsonized phagocytosis (56,57). The effect of respiratory viral infections seen following both in vivo and in vitro exposures is related to the stage of the infectious process. Early after infection, phagocytosis is decreased, whereas during recovery, phagocytic function is increased (56). Direct infection of AMs leading to a decrease in the sequestration of cobalt oxide

Table 2 Direct Effects of Virus on Macrophage Function

Virus	Phagocytosis	Oxygen radicals	Inflammatory mediators
Respiratory syncytial virus	↓↑	↓	↑
Influenza	↓	↑	↑
Parainfluenza	↓	↓	↑
Adenovirus	—	↑	↑

particles within macrophages is a major factor in the decreased lung clearance of cobalt oxide particles in calves (58). Bovine AMs obtained after viral inoculation are significantly impaired in their ability to phagocytize opsonized *Staphylococcus epidermidis* (42).

Several factors may lead to this defect in phagocytosis. Analysis of phagocytic kinetics in infected bovine AMs demonstrates normal phagocytic receptor numbers but a slower rate of the phagocytic process (59,60). Significant differences between the numbers of control and infected AMs phagocytizing antibody-coated sheep erythrocytes were also observed. Rats infected with PI-1 (Sendai) demonstrated altered cytoplasmic motility (61). This may have been related to alterations in the production of nitric oxide and increased intracellular guanosine 3′,5′-cyclic monophosphate. Additionally, PI-3 virus–infected bovine AMs contained significantly lower concentrations of acid phosphatase, a protease associated with phagocytosis (62). Suppression of phagocytic activity can lead to alterations in macrophage bactericidal activity. Severe influenza infection resulted in maximal suppression of bactericidal activity on day 8 and correlated with impairment of AM phagocytosis. A lesser but significant suppression was seen on the third day following mild infection (63). In spite of diminished phagocytic capacity, macrophages infected with PI-3 appeared able to kill ingested bacteria (42).

B. Radical Production

Several mediators produced by AMs are responsible for effective inflammatory responses against inhaled bacteria. Among these mediators are toxic oxygen radicals with the capability of killing the inhaled pathogens. An alteration in the ability of the AMs to produce oxygen radicals can have severe consequences. Virus-induced decreased production will result in a greater risk of subsequent or simultaneous bacterial infections, whereas overproduction could lead to direct damage of the underlying lung tissue (64).

The particular response of AMs depends on the type of virus that the cells encounter. RSV infection leads to a reduction in reactive oxygen radical production and decreased killing of protozoa (61). Similarly, PI-3 infection reduced phorbol myristate acetate (PMA) and zymosan stimulated production of oxygen radicals in bovine AMs. The suppression required infection with live virus, as heat-inactivated viruses did not alter superoxide production. PI-3 inhibition of superoxide production appears to be mediated through the interaction of the virus with a calcium-independent, phospholipid-dependent kinase. Inhibition of kinase activity and superoxide production occurred at similar viral doses and time frames (65).

Although RSV and parainfluenza virus cause suppression of superoxide production, infection with influenza virus increases the production of superoxide (see Table 2). Mice infected with influenza A have a massive infiltration of leukocytes, mainly polymorphonuclear leukocytes, into alveolar spaces. Cells in the bronchoalveolar fluid (BALF) showed an increased production of superoxide release above control values, and an increased production of radicals was also evident in vivo (66). The increased oxidative stress initiated by influenza infections appears to be due to both the ability of the virus to initiate a strong inflammatory response and a direct effect of the virus on the production of superoxide by macrophages and leukocytes. In vitro exposure to influenza virus can directly activate monocytes and polymorphonuclear neutrophils (PMNs) (67). This activation, combined with recruitment of PMNs and other leukocytes into the alveolar spaces, likely leads to increased injury and insult due to the production of the radicals. Interestingly, although influenza virus can cause increases in superoxide production, infection of avian macrophages leads to the decreased production of nitric oxide (68).

The involvement of free radicals in influenza-induced pathogenesis has been shown by the administration of superoxide scavengers. Administration of superoxide dismutase even 5–8 days after the initiation of an otherwise lethal viral infection prolonged survival of experimental animals (69). Additional evidence of oxidative stress in vivo after influenza infection has been shown by measuring antioxidant enzymes. Infection caused a decrease in total lung glutathione and vitamins C and E. No change in the ratio of oxidized to total concentrations of these antioxidants was seen (70).

Although these data indicate that oxidative stress may play a role in influenza-induced lung injury, this hypothesis is not supported by all investigations. Buffinton et al. (66) found no change in glutathione peroxidase and reductase. Although ascorbic acid concentrations in BALF decreased during the early stages of infection, the levels returned to above normal late in the infections. Also, peroxyl radicals did not cause additional damage in terminally ill infected animals when compared with controls, indicating that the total antioxidant capacity was not different.

The time course of the increased production of oxygen radicals also appears to be dependent on the type of virus. Unlike influenza, where some of the pathogenesis is believed to be due to the production of oxygen radicals, macrophages isolated from animals infected with adenovirus did not show increased superoxide production until 10 days after infections (71) when clinical recovery was in progress. Thus, the increased production of superoxide may indicate macrophage activation associated with the clearing of debris and dead or infected cells. We did not see alterations in macrophage production of superoxide with direct in vitro exposure to adenovirus. Therefore, we believe that increased superoxide production by AMs seen following in vivo infections is most likely due to the indirect effects of the release of activating mediators from other cells recruited to fight the viral infection (i.e., neutrophils or lymphocytes).

C. Inflammatory Mediators

Cytokine Gene Expression and Production

AMs increase the production of many inflammatory mediators when exposed to viruses. Murine macrophages exposed in vitro to RSV produce tumor necrosis factor-α (TNF-α), interleukin-1 (IL-1), and interleukin-6 (IL-6) (55). Human AMs also produce TNF-α in response to RSV, with the level of cytokine production being dependent on the dose of virus administered (72). The overall effect of the virus on cytokine action must be evaluated, however, by examining the entire spectrum of products produced by the infection. For example, whereas RSV induces IL-1 production, AM production of interleukin-1 inhibitor activity in response to RSV is increased (73). Thus, the overall effects on cytokine production will depend on the balance between IL-1 and the IL-1 inhibitor.

Similar effects must also be considered with TNF production and the production of soluble TNF receptors by viral infections (74). Release of these cytokines after in vivo inoculation with RSV has been shown in mice and humans (47,75). Influenza infection has been shown to initiate gene expression for TNF-α, TNF-β, IL-1-α, IL-1-β, IL-2, IL-4, IL-6, IL-8, IL-10, granulocyte-monocyte colony-stimulating factor (GM-CSF), and interferon-γ (76–79). Increases in message were also associated with increases in the production of protein. In general, the time course for the production peaked at the time of greatest infection (between 36 hr and 3 days in mice). The increased cytokine production caused by influenza infection may also be partially attributed to the ability of the virus to cause oxidative stress in lung macrophages. Hemagglutin protein from influenza virus has been shown to induce the activity of the transcription factor NF-κB. Gene expression for several of the cytokines has been shown to be regulated by NF-κB, which can be liberated from its inhibitory subunit by oxidative stress. In the case of influenza infection, activation of gene expression is suppressed by dithiothreitol, suggesting that virus-induced production of oxygen radicals may activate the transcription factor, NF-κB, leading to gene expression.

Eicosinoid Production

Viral infection of AMs infected with respiratory viruses increases the production of prostaglandin E_2 (PGE_2) (80,81), platelet-activating factor (PAF) (82), and thromboxane (83). The increase in PGE_2 may inhibit AM phagocytic ability (81). In normal macrophages, fusion of endosomes with lysosomes is inhibited by increased intracellular levels of cyclic adenosine monophosphate (AMP) (84), and PGE_2 is a potent adenyl cyclase agonist produced by AMs. Thus, exposure to PGE_2 results in increases in intracellular cyclic AMP concentration and diminished endosome-lysosome fusion. As would be expected then, pharmacological blockade of the cyclooxygenase pathway increases AM phagocytic activity. Bacterial killing by PI-3 virus-infected AMs is significantly increased on day 4 of infection by the addition of cyclooxygenase inhibitors 1 hr prior to assay. Additionally, pretreatment of virus-infected AMs with indomethacin, a cyclooxygenase inhibitor, increases the percentage of AMs expressing endosome-lysosome fusion.

The increased production of PAF, a potent inflammatory mediator, and thromboxane, a bronchoconstrictor, in RSV-exposed macrophages may also play a role in the pathogenesis of the viral infections. PAF has been shown to lead to bronchoconstriction, increases in pulmonary resistance, vasoconstriction, and increased vascular permeability. Thus, the overall effects of viral infections on AMs are increased production of proinflammatory mediators and cytokines and suppression of downregulatory mediators, leading to a strong inflammatory response.

Viral Infection of Epithelial Cells

Direct interaction of viruses with macrophages can lead to alterations in macrophage function. In addition, viral infection of other cell types in the lung, in particular epithelial cells, can also lead to the production of mediators that can regulate macrophage function. For example, RSV and influenza and parainfluenza viruses have all been shown to be capable of infecting and replicating in alveolar type II cells (85–87). RSV can induce the production of IL-6, IL-8, GM-CSF, and soluble TNF receptors from bronchial epithelial cells and A549 cells, a human lung epithelial cell line (74). The production of these cytokines by infected epithelium can modulate the inflammation induced by the viral infection. IL-6, for example, plays an important role in B- and T-lymphocyte differentiation and stimulation; IL-8 is a powerful chemoattractant for neutrophils and lymphocytes, and soluble TNF receptors modulate the effects of released TNF. Combined with the proinflammatory cytokines released by macrophages, simultaneous infection of both epithelium and macrophages will result in a strong inflammatory response against the virus.

IV. Time Course and Long-Term Effects of Acute Viral Infections

A. Time Course of Infections

For nonlethal experimentially produced infections, the time course of the inflammation and resolution of the infection are similar between various viruses. Viral plaque-forming units (PFUs) for RSV (88), influenza (89), adenovirus (24), and parainfluenza (23) all begin to rise as early as the first day after infection. The clinical symptoms and pathophysiological effects of the viral infection peak 3–6 days after the infection and resolve by 2–3 weeks postinfection. Inflammatory mediator release induced by the viruses parallels the time course and severity of the infection.

The concentrations of mediators which are produced by AMs in the response to viruses have been measured in BALF during infections. Using a mouse model of influenza infection, investigators have demonstrated increases in IL-1, IL-6, TNF-α, GM-CSF, interferon-γ (IFN-γ), and leukotriene B_4 (LTB_4) (78,89). The exact time course of the presence of these mediators in BALF depended on the dose of the influenza. With a 10 LD_{50} dose, all the mediators began to rise within the first day after infection. Except for IL-6, the concentration of the mediators in BALF peaked between 36 hr and 3 days after infection. IL-6, however, remained elevated throughout the infection. Surprisingly, no increases in the T-cell products, IL-2, IL-3, or IL-4, were seen, nor was PAF changed during the course of the infection. When lower doses were used (89), increases in IL-6, TNF-α, and IFN-γ were still seen. However, the times to peak were later in the infection (between 2 and 4 days). As with the higher infection dose, IL-6 remained elevated throughout the infection. With the lower dose, IL-1 was not detected in the BALF.

Influenza infections can also affect macrophage antibacterial activity (63). Mice inoculated with influenza showed decreased Fc receptor–mediated phagocytosis with fewer cells participating in the phagocytosis as early as day 3 after infection. This led to a suppression in the ability of the macrophages to kill bacteria in vivo, which lasted at least until day 8 postinoculation.

There is also evidence that PI-3–infected AMs can increase arachidonic acid metabolism, specifically the 5- and 12-lipoxygenase enzyme systems (90). Total output of arachidonic metabolites by stimulated AMs is increased by PI-3 viral infection on day 4 postinoculation but not on day 1 postinoculation. The expression of viral antigen and the release of infectious PI-3 virus by AMs are maximal on day 4 postinoculation; the period in which arachidonate metabolism is increased in these cells. This temporal association suggests that viral replication might be required to promote arachidonic acid metabolism. Furthermore, increases in arachidonate metabolite release suggest that greater substrate availability, due to either increased phospholipase A_2 activity or direct virus-membrane interaction, may be responsible for the virus-induced enhancement of these metabolites.

In recent studies from our laboratory (71), we found that AM superoxide production is increased on days 11–14 after infection with canine adenovirus type 2. Thus, increased superoxide production corresponded to the recovery period, since the peak of the clinical illness occurred on days 4–6 after inoculation. We believe that AM superoxide production may be associated with the repair process of the virus-induced lung injury as suggested by others (91). It is possible that earlier in the course of the infections, AMs were overwhelmed by the adenoviral infections and their ability to synthesize mediators was inhibited. On the other hand, we found morphological evidence of AM activation on day 3 after inoculation in other studies (25).

B. Effect on Long-Term Pulmonary Function

As discussed above, it is clear that respiratory viruses alter AM receptor function, phagocytic function, and mediator production. We do not know how these alterations may influence the macrophage-dependent immune responses, but the potential is great for long-term injury to the lungs and, possibly, alterations in normal growth process. Respiratory viruses alter lung function by physiological and pathological processes associated with inflammation (25). We postulate that AMs with an impaired phagocytic ability or bactericidal capacity may not respond adequately to an ongoing viral infection or to bacterial superinfection. In our laboratory, beagle dogs were inoculated with canine adenovirus (CAV2) or sensitized to ragweed antigen followed by inoculation with CAV2 (92). AM bronchoalveolar lavage cell counts were elevated in both ragweed and CAV2 dogs on day 7 following inoculation. AMs from beagles who received both ragweed and CAV2 showed a significant increase over CAV2 alone in their ability to produce superoxide in response to PMA stimulation 3–4 days after inoculation.

An increase in lipoxygenase metabolite production by infected AMs may increase epithelial permeability in the lung, thus allowing aeroallergens and/or other toxins access to the pulmonary interstitium, increase the potential for long-term injury to the lungs, and possibly alter normal lung growth. We demonstrated that prior sensitization to ragweed antigen before inoculation with CAV2 in beagle dogs increased the concentration of LTB_4 in BALF (93). AMs are the chief source of LTB_4 in the lungs (94). However, there was no direct evidence that AMs were the source of LTB_4 in the virus-infected beagle dog model. Both in sheep lung lymph and alveolar fluid, a small increase in protein clearance occured after LTB_4 (2 nmol/ml, 50 ml total) was administered to the caudal lung lobe through a fiberoptic bronchoscope. Staub and coworkers (95) speculated that some of this effect was caused by a direct action of LTB_4. Indeed, many investigators have demonstrated that sulfidopeptide leukotrienes (mainly LTC_4 and LTD_4) can increase vascular permeability (96–99). There is also evidence that prostaglandins, such as PGD_2, may be responsible for the clinical manifestations, such as wheezing, associated with respiratory viral infection (100).

C. Age-Related Differences in Response to Viral Infections

Age-related differences in AMs appear to affect the infectivity of AMs with respiratory viruses. Differences in the populations of lymphocytes and AMs in neonates compared with weanlings may explain the development of alveolar dysplasia and bronchiolar hypoplasia in neonatal rats by 30–110 days after inoculation, whereas weanlings are spared (96,101). Additionally, RSV antigens appear to be detected more frequently in monocytes from children than in those from adults when infected in vitro (102). These findings suggest that children may be highly susceptible to infection of circulating immunocompetent cells, which may suppress host defenses, prolong the infection, and worsen lung injury.

V. Summary

Acute and chronic viral respiratory infections are an important cause of morbidity and mortality worldwide; yet, medical interventions are limited and largely supportive. There is no doubt that AMs play an important role in the pathophysiology of acute viral respiratory infections. AMs are targets of viral infection and release inflammatory mediators which contribute to the tissue injury as well as contribute to host defenses. Disease processes that suppress these functions potentially have severe consequences on health. Very little is known regarding the molecular and genetic factors that may contribute to differences in host responses and chronic sequelae. Furthermore, very little is known regarding pharmacological interventions, such as anti-inflammatory drugs, on AM function in normal and disease states. Important research questions needing attention include understanding how viruses enter cells such as AMs, why some viral infections result in chronic lung damage, and how age influences host defenses.

References

1. Smith JJ, Lemen RJ, Taussig LM. Mechanisms of viral-induced lower airway obstruction. Pediatr Infect Dis 1987; 6:837–842.
2. Denney FW, Collier AM, Henderson FW, Clyde WA Jr. Infectious agents of importance in airways and parenchymal diseases in infants and children with particular emphasis on bronchiolitis. Pediatr Res 1977; 11:234–236.
3. Hudgel DW, Langston L Jr, Selner JC, McIntosh K. Viral and bacterial infections in adults with chronic asthma. Am Rev Respir Dis 1979; 120:393–397.
4. Samet JM, Tager IB, Speizer FE. The relationship between respiratory illness in childhood and chronic air-flow obstruction in adulthood. Am Rev Respir Dis 1983; 127:508–523.
5. Sibille Y, Reynolds HY. Macrophages and polymorphonuclear neutrophils in lung defense and injury. Am Rev Respir Dis 1990; 141:471–501.
6. Witten ML, Grad R, Quan S, Lantz RC, Sobonya RE, Lemen RJ. Effects of respiratory viruses on pulmonary alveolar macrophages. Pediatr Pulmonol 1992; 12: 105–112.
7. Witten ML, Quan SF, Sobonya RE, Bruck D, Devine L, Lemen RJ. Acute cigarette smoke exposure alters lung eicosanoid and inflammatory cell concentrations in rabbits. Exp Lung Res 1988; 14:727–742.
8. Wiltrout RH, Varesio L. Activation of macrophages for cytotoxic and suppressor effector functions. In: Oppenheim JJ, Shevach EM, eds. Immunophysiology: The

Role of Cells and Cytokines in Immunity and Inflammation, New York: Oxford University Press, 1990:365–385.

9. Glezen WP. Viral Pneumonia. In: Chernick and Kendig, eds. Kendig's Disorders of the Respiratory Tract in Children. 5th ed. Philadelphia: Saunders, 1990:394–402.
10. Lemen R. Respiratory syncytial virus and bronchiolitis. Acta Paediatr Sin 1995; 36: 78–85.
11. Frick OL, German DF, Mills J. Development of allergy in children. I. Association with virus infections. J Allergy Clin Immunol 1979; 63:228–241.
12. Gurwitz D, Corey M, Levison H. Pulmonary function and bronchial reactivity in children after croup. Am Rev Respir Dis 1980, 122:95–99.
13. Lemanske RF Jr, Dick EC, Swenson CA, Vrtis RF, Busse WW. Rhinovirus upper respiratory infection increases airway hyperreactivity and late asthmatic reactions. J Clin Invest 1989; 83:1–10.
14. Calhoun WJ, Dick EC, Schwartz LB, Busse WW. A common cold virus, rhinovirus 16, potentiates airway inflammation after segmental antigen bronchoprovocation in allergic subjects. J Clin Invest 1994; 94:2200–2208.
15. Johnston SL. Natural and experimental rhinovirus infections of the lower tract. Am J Respir Crit Care Med 1995; 152:S46–52.
16. Busse WW. Respiratory infections. Their role in airway responsiveness and the pathogenesis of asthma. J Allergy Clin Immunol 1990; 85:671–683.
17. Martinez FD, Morgan WJ, Wright AL, Holberg CJ, Taussig LM. Diminished lung function as a predisposing factor for wheezing respiratory illness in infants. N Engl J Med 1988; 319:1112–1117.
18. Martinez FD, Morgan WJ, Wright AL, Holberg C, Taussig LM. Initial airway function is a risk factor for recurrent wheezing respiratory illnesses during the first three years of life. Am Rev Respir Dis 1991; 143:312–316.
19. Martinez FD. Viral infections and the development of asthma. Am J Respir Crit Care Med 1995; 151:1644–1647.
20. Busse WW. The role of respiratory infections in airway hyperresponsiveness and asthma. Am J Respir Crit Care Med 1994; 150:S77–79.
21. Taussig LM, Lemen RJ, and Ram JS. NHLBI Workshop Summary Models of infectious airway injury in children. Am Rev Respir Dis 1988; 137:979–984.
22. Lemen RJ. Clues to the mechanisms of virus-induced asthma from animal models. Semin Respir Med 1990; 11:321–329.
23. Lemen RJ, Quan SF, Witten ML, Sobonya RE, Ray CG, Grad R. Canine parainfluenza type 2 bronchiolitis increases histamine responsiveness in beagle puppies. Am Rev Respir Dis 1990; 141:199–207.
24. Quan SF, Witten ML, Grad R, Sobonya RE, Ray CG, Dambro NN, Lemen RJ. Acute canine adenovirus 2 infection increases histamine airway reactivity in beagles. Am Rev Respir Dis 1990; 141:414–420.
25. Grad R, Sobonya RE, Witten ML, Quan SF, Ray CG, Devine LC, Lentz LA, Lemen RJ. Localization of inflammation and virions in canine adenovirus type 2 bronchiolitis. Am Rev Respir Dis 1990; 142:691–699.
26. Quan SF, Lemen RJ, Witten ML, Sherrill DL, Grad R, Sobonya RE, Ray CG. Changes in lung mechanics and reactivity with age after bronchiolitis in beagle puppies. J Appl Physiol 1990; 69:2034–2042.

27. Witten ML, McKee JL, Lantz RC, Hays AM, Quan SF, Sobonya RE, Lemen RJ. Fractal and morphometric analysis of lung structures after canine adenovirus-induced bronchiolitis in beagle puppies. Pediatr Pulmonol 1993; 16:62–68.
28. Quan SF, Witten ML, Dambro NN, Lemen RJ. Canine parainfluenza type 2 and Bordetella bronchiseptica produces increased bronchoalveolar lavage thromboxane concentrations in beagle puppies. Prostaglandins Leukot Essent Fatty Acids 1991; 44:171–175.
29. Dales S. Early events in cell-animal virus interactions. Bacteriol Rev 1973; 37: 103–135.
30. FitzGerald DJP, Padmanabhan R, Pastan I, Willingham MC. Adenovirus-induced release of epidermal growth factor and Pseudomonas toxin into the cytosol of KB cells during receptor-mediated endocytosis. Cell 1983; 32:607–617.
31. Tycko B, Maxfield FR. Rapid acidification of endocytic vesicles containing alpha-2 macroglobulin. Cell 1982; 28:643–651.
32. Lehnert BE, Valdez YE, Sebring RJ, Lehnert NM, Saunders GC, Steinkamp JA. Airway intra-luminal macrophages: evidence of origin and comparisons to alveolar macrophages. Am J Respir Cell Mol Biol 1990; 3:377–391.
33. Musson RA, Shafran H, Henson PM. Intracellular levels and stimulated release of lysosomal enzymes from human peripheral blood monocytes and monocyte-derived macrophages. J Reticuloendothel Soc 1980; 28:249–264.
34. Zuckerman SH, Ackerman SK, Douglas SD. Long-term human peripheral blood monocyte cultures: establishment, metabolism, and morphology of primary monocyte-macrophage cell cultures. Immunology 1979; 38:401–411.
35. Newman SL, Musson RA, Henson PM. Development of functional complement receptors during in vitro maturation of human monocytes into macrophages. J Immunol 1980; 125:2236–2244.
36. Willcox MB, Golde DW, Cline MJ. Cytochemical reactions of human hematopoietic cells in liquid culture. J Histochem Cytochem 1976; 24:979–983.
37. Wuest D, Crane R, Rhinehart JJ. Enhancement of Fc receptor function during human monocyte differentiation in vitro. J Reticuloendothel Soc 1981; 30:147–155.
38. Midulla F, Huang YT, Gilbert IA, Cirino NM, McFadden ER Jr, Panuska JR. Respiratory syncytial virus infection of human cord and adult blood monocytes and alveolar macrophages. Am Rev Respir Dis 1989; 140:771–777.
39. White J, Matlin K, Helenius A. Cell fusion by semliki forest, influenza, and vesicular stomatitis virus. J Cell Biol 1981; 89:674–679.
40. Merendino RA, Iannello D, Arena A, Bonina L, Greco V, Mesiti M, Chillemi S, Mastroeni P. Evaluation of macrophage antiviral activity in patients affected by neoplasia. Med Oncol Tumor Pharmacother 1988; 5:191–197.
41. Gimenez HB, Keir HM, Cash P. In vitro enhancement of respiratory syncytial virus infection of U937 cells by human sera. J Gen Virol 1989; 70:89–96.
42. Brown TT Jr, Ananaba G. Effect of respiratory infections caused by bovine herpesvirus-1 or parainfluenza-3 virus on bovine alveolar macrophage functions. Am J Vet Res 1988; 49:1447–1451.
43. Trigo FJ, Breeze RG, Liggitt HD, Evermann JF, Trigo E. Interaction of bovine respiratory syncytial virus and Pasteurella haemolytica in the ovine lung. Am J Vet Res 1984; 45:1671–1678.

44. Ogunbiyi PO, Conlon PD, Black WD, Eyre P. Levamisole-induced attenuation of alveolar macrophage dysfunction in respiratory virus-infected calves. Int J Immunopharmacol 1988; 10:377–385.
45. Anderson R, Glover A, Koornhof HS, Robinson AR. In vitro stimulation of neutrophil motility by levamisole-treated neutrophils. J Immunol 1976; 117:428–432.
46. Cruchaud A, Berney M, Welscher HD. The effects of levamisole on some functions of mouse macrophages after in vitro incubation with rhinovirus. Am Rev Respir Dis 1979; 122:641–666.
47. Midulla F, Villani A, Panuska JR, Dab I, Kolls JK, Merolla R, Ronchetti R. Respiratory syncytial virus lung infection in infants: immunoregulatory role of infected alveolar macrophages. J Infect Dis 1993; 168:1515–1519.
48. Panuska JR, Hertz MI, Taraf H, Villani A, Cirino NM. Respiratory syncytial virus infection of alveolar macrophages in adult transplant patients. Am Rev Respir Dis 1992; 145:934–939.
49. Meehan JT, Cutlip RC, Lehmkuhl HD, Kluge JP, Ackermann MR. Infected cell types in ovine lung following exposure to bovine respiratory syncytial virus. Vet Pathol 1994; 31:229–236.
50. Hegele RG, Hayashi S, Bramley AM, Hogg JC. Persistence of respiratory syncytial virus genome and protein after acute bronchiolitis in guinea pigs. Chest 1994; 105: 1848–1854.
51. Panuska JR, Cirino NM, Midulla F, Despot JE, McFadden ER Jr, Huang YT. Productive infection of isolated human alveolar macrophages by respiratory syncytial virus. J Clinical Invest 1990; 86:113–119.
52. Mills J. Effects of Sendai virus infection on function of cultured mouse alveolar macrophages. Am Rev Respir Dis 1979; 120:1239–1244.
53. Pan IC, Moulton JE, Hess WR. Immunofluorescent studies on chronic pneumonia in swine with experimentally induced African swine fever. Am J Vet Res 1975; 36: 379–386.
54. Watanabe M, Tozawa H, Kumagai K, Ishida N. Specific macrophage immunity to Sendai virus: macrophage aggregation in vitro with Sendai virus by cytophilic antibodies. Infect Immun 1975; 12:324–332.
55. Franke-Ullmann G, Pfortner C, Walter P, Steinmuller C, Lohmann-Matthes ML, Kobzik L, Freihorst J. Alteration of pulmonary macrophage function by respiratory syncytial virus infection in vitro. J Immunol 1995; 154:268–280.
56. Olchowy TW, Ames TR, Molitor TW. Interaction of bovine respiratory syncytial virus with bovine alveolar macrophages in vivo: effects of virus infection upon selected cell functions. Can J Vet Res 1994; 58:42–48.
57. Trigo E, Liggitt HD, Evermann JF, Breeze RG, Huston LY, Silflow R. Effect of in vitro inoculation of bovine respiratory syncytial virus on bovine pulmonary alveolar macrophage function. Am J Vet Res 1985; 46:1098–1103.
58. Slauson DO, Lay JC, Castleman WL, Neilsen NR. Acute inflammatory lung injury retards pulmonary particle clearance. Inflammation 1989; 13:185–199.
59. Slauson DO, Lay JC, Castleman WL, Neilsen NR. Alveolar macrophage phagocytic kinetics following pulmonary parainfluenza-3 virus infection. J Leukoc Biol 1987; 41:412–420.
60. Lehninger AL. Biochemistry, Enzymes: Kinetics and Inhibition. New York: Worth, 1977.

61. Fukushima T, Sekizawa K, Yamaya M, Okinaga S, Satoh M, Sasaki H. Viral respiratory infection increases alveolar macrophage cytoplasmic motility in rats: role of NO. Am J Physiol 1995; 268:L399–406.
62. Liggitt D, Huston L, Silflow R, Evermann J, Trigo E. Impaired function of bovine alveolar macrophages infected with parainfluenza-3 virus. Am J Vet Res 1985; 46: 1740–1744.
63. Nickerson CL, Jakab GJ. Pulmonary antibacterial defenses during mild and severe influenza virus infection. Infect Immun 1990; 58:2809–2814.
64. Maeda H, Akaike T. Oxygen free radicals as pathogenic molecules in viral diseases. Proc Soc Exp Biol Med 1991; 198:721–727.
65. Dyer RM, Majumdar S, Douglas SD, Korchak HM. Bovine parainfluenza-3 virus selectively depletes a calcium-independent, phospholipid-dependent protein kinase C and inhibits superoxide anion generation in bovine alveolar macrophages. J Immunol 1994; 53:1171–1179.
66. Buffinton GD, Christen S, Peterhans E, Stocker R. Oxidative stress in lungs of mice infected with influenza A virus. Free Rad Res Comm. 1992; 16:99–110.
67. Peterhans E, Grob M, Burge T, Zanoni R. Virus-induced formation of reactive oxygen intermediates in cells. Free Radic Res Commun 1987; 3:39–46.
68. Lyon JA, Hinshaw VS. Inhibition of nitric oxide induction from avian macrophage cell lines by influenza virus. Avian Dis 1993; 37:868–873.
69. Oda T, Akaike T, Hamamoto T, Suzuki F, Hirano T, Maeda H. Oxygen radicals in influenza-induced pathogenesis and treatment with pyran polymer-conjugated SOD. Science 1989; 244:974–976.
70. Hennet T, Peterhans E, Stocker R. Alterations in antioxidant defenses in lung and liver of mice infected with influenza A virus. J Gen Virol 1992; 73:39–46.
71. Lantz RC, Grad R, Sobonya R, Brown M, Quan S, Lemen R. Superoxide release corresponds with resolution of adenovirus-induced airway hyperresponsiveness. Am Rev Respir Dis 1990; 141:A689.
72. Panuska JR, Midulla F, Cirino NM, Villani A, Gilbert IA, McFadden ER Jr, Huang YT. Virus-induced alterations in macrophage production of tumor necrosis factor and prostaglandin E2. Am J Physiol 1990; 259:L396–402.
73. Roberts NJ Jr, Prill AH, Mann TN. Interleukin 1 and interleukin 1 inhibitor production by human macrophages exposed to influenza virus or respiratory syncytial virus. Respiratory syncytial virus is a potent inducer of inhibitor activity. J Exp Med 1986; 163:511–519.
74. Arnold R, Humbert B, Werchau H, Gallati H, Konig W. Interleukin-8, interleukin-6, and soluble tumour necrosis factor receptor type I release from a human pulmonary epithelial cell line (A549) exposed to respiratory syncytial virus. Immunology 1994; 82:126–133.
75. Hayes PJ, Scott R, Wheeler J. In vivo production of tumour necrosis factor-alpha and interleukin-6 in BALB/c mice inoculated intranasally with a high dose of respiratory syncytial virus. J Med Vir. 1994: 42:323–329.
76. Pahl HL, Baeuerle PA. Expression of influenza virus hemagglutinin activates transcription factor NF-kappa B. J Virol 1995; 69:1480–1484.
77. Hinder F, Schmidt A, Gong JH, Bender A, Sprenger H, Nain M, Gemsa D. Influenza A virus infects macrophages and stimulates release of tumor necrosis factor-alpha. Pathobiology 1991; 59:227–231.

78. Hennet T, Ziltener HJ, Frei K, Peterhans E. A kinetic study of immune mediators in the lungs of mice infected with influenza A virus. J Immunol 1992; 149:932–939.
79. Houde M, Arora DJ. Human influenza virus neuraminidase, but not hemagglutinin, induces murine macrophage interleukin 1 in vivo and in vitro. Immunol Lett 1989; 22:41–46.
80. Laegreid WW, Liggitt HD, Silflow RM, Evermann JR, Taylor SM, Leid RW. Reversal of virus-induced alveolar macrophage bactericidal dysfunction by cyclooxygenase inhibition in vitro. J Leukoc Biol 1989; 45:293–300.
81. Bielefeldt-Ohmann H, Babiuk LA. Alteration of alveolar macrophage functions after aerosol infection with bovine herpes-virus type 1. Infect Immun 1986; 51: 344–347.
82. Villani A, Cirino NM, Baldi E, Kester M, McFadden ER Jr, Panuska JR. Respiratory syncytial virus infection of human mononuclear phagocytes stimulates synthesis of platelet-activating factor. J Biol Chem 1991; 266:5472–5479.
83. Gershwin LJ, Giri SN, Stewart RS, Chen J. Prostaglandin and thromboxane concentrations in plasma and lung lavage fluids during sequential infection of vaccinated and nonvaccinated calves with bovine respiratory syncytial virus. Am J Vet Res 1989; 50:1254–1262.
84. Lowrie DB, Jackett PS, Aber VR, Carol MEW. Cyclic nucleotides and phagosome-lysosome fusion in mouse peritoneal macrophages. In: Van Furth R, ed. Mononuclear Phagocytes, Functional Aspects. Boston: Martinus Nijhoff, 1980.
85. Castleman WL, Northrop PJ, McAllister PK. Replication of parainfluenza type-3 virus and bovine respiratory syncytial virus in isolated bovine type-II alveolar epithelial cells. Am J Vet Res 1991; 52:880–885.
86. Castleman WL, Northrop PJ, McAllister PK. Replication of parainfluenza (Sendai) virus in isolated rat pulmonary type II alveolar epithelial cells. Am J Pathol 1989; 34:1135–1142.
87. Hahon N, Castranova V. Interferon production in rat type II pneumocytes and alveolar macrophages. Exp Lung Res 1989; 15:429–445.
88. Taylor G, Stott EJ, Hughes M, Collins AP. Respiratory syncytial virus infection in mice. Infect Immun 1984; 43:649–655.
89. Conn CA, McClellan JL, Maassab HF, Smitka CW, Majde JA, Kluger MJ. Cytokines and the acute phase response to influenza virus in mice. Am J Physiol 1995; 268: R78–84.
90. Laegreid WW, Taylor SM, Leid RW, Silflow RM, Evermann JR, Breeze RG, Liggitt HD. Virus-induced enhancement of arachidonate metabolism by bovine alveolar macrophages in vitro. J Leukoc Biol 1989; 45:283–292.
91. Boschert KR, Schoeb TR, Chandler DB, Dillehay DL. Inhibition of phagocytosis and interleukin-1 production in pulmonary macrophages from rats with sialodacryoadenitis virus infection. J Leukoc Biol 1988; 44:87–92.
92. Lantz RC, Witten M, Bice D, Grad R, Bowers M, Quan S, Sobonya R, Lemen R. Superoxide production by pulmonary alveolar macrophages isolated from ragweed sensitized and adenovirus infected beagle dogs. Am Rev Respir Dis 1991; 143: A131.
93. Lemen RJ, Witten ML, Quan SF, Bice DE, Sobonya RE, Leeman S, Lantz RC, Bowers M. Inflammatory mediators are increased in broncho-alveolar lavage fluid (BALF) of normal (N) and ragweed sensitized (RW) beagle puppies with canine adenovirus (CAV2) bronchiolitis. Am Rev Respir Dis 1991; 143:A153.

94. Gerrard JM. Prostaglandins and Leukotrienes: Blood and Vascular Cell Function. New York: Marcel Dekker, 1985.
95. Staub NC, Schultz EL, Koike K, Albertine KH. Effect of neutrophil migration induced by leukotriene B_4 on protein permeability in sheep lung. Fed Proc 1985; 44: 30–35.
96. Leitch AG. The role of leukotrienes in asthma. Ann Acad Med Singapore 1985; 14: 503–507.
97. Austen KF. The role of arachidonic acid metabolites in local and systemic inflammatory processes. Drugs 1987; 33:10–17.
98. Ratnoff WD, Matthay MA, Wong MY, Ito Y, Vu KH, Wiener-Kronish J, Goetzl EJ. Sulfidopeptide-leukotriene peptidases in pulmonary edema fluid from patients with the adult respiratory distress syndrome. J Clin Immunol 1988; 8:250–258.
99. Samuelsson B. Leukotrienes: mediators of immediate hyper-sensitivity reactions and inflammation. Science 1983; 220:568–575.
100. King LS, Fukushima M, Banerjee M, Kang KH, Newman JH, Biaggioni I. Pulmonary vascular effects of prostaglandin D_2, but not its systemic vascular or airway effects, are mediated through thromboxane receptor activation. Circ Res 1991; 68: 352–358.
101. Castleman WL, Owens SB, Brundage-Anguish LJ. Acute and persistent alterations in pulmonary inflammatory cells and airway mast cells induced by Sendai virus infection in neonatal rats. Vet Pathol 1989; 26:18–25.
102. Domurat F, Roberts NJ Jr, Walsh EE, Dagan R. Respiratory syncytial virus infection of human mononuclear leukocytes in vitro and in vivo. J Infect Dis 1985; 152: 895–902.

21

Lung Macrophages in Human Immunodeficiency Viral Infection

HOMER L. TWIGG III

Indiana University Medical Center
Indianapolis, Indiana

I. Introduction

Ever since the initial descriptions of *Pneumocystis carinii* pneumonia in homosexual men (1,2), it has been recognized that pulmonary involvement is frequent in patients infected with the human immunodeficiency virus (HIV). Virtually all HIV-infected individuals will have a pulmonary complication at some point during their illness in the form of opportunistic infections, neoplastic processes, and/or nonspecific lymphocytic infiltrative disorders such as nonspecific interstitial pneumonitis and lymphocytic alveolitis (3–5). Although this could reflect a general systemic immunodeficiency state, in many instances, these processes are unique to or more frequently found in the lung. HIV can affect many cellular functions in the lung resulting in perturbations in normal immune and inflammatory responses (6). This chapter focuses on the effect of HIV on lung macrophages, which are cells critical for the initiation and maintenance of normal immune and inflammatory responses in the lower respiratory tract.

Perhaps nowhere else in the body is the relationship between HIV and host cells more crucial than in the lung. On the air space side, cells in the alveolar space must constantly differentiate potential pathogenic organisms and particles they are exposed to from the vast array of nonspecific antigens and pollutants inhaled with each breath. On the vascular side, lung capillaries serve as a filter for every blood cell or infectious organism found in the circulation. Lung macrophages serve as sentinels against these foreign challenges and are charged with the responsibility of initiating immune and inflammatory responses where appropriate. However, it is critical to point out that not all lung macrophages are alike. It is now clear that substantial functional differences are found between macrophages in the alveolar, interstitial, vascular, airway, and pleural compartments (7).

Because the alveolar macrophage (AM) is the easiest lung macrophage to harvest, it has been the most intensively studied in HIV-infected individuals. After appropriate stimulation, AMs can carry out a wide variety of functions. However, in response to evolutionary pressures to minimize unnecessary inflammation in the lower respiratory tract, these cells have developed relatively poor inflammatory and immunological capabilities. For example, although having exceptional phagocytic abilities, AMs have a poor capacity to mount subsequent inflammatory processes required for killing of intracellular organisms. The AM is also a relatively inactive immune cell compared with other cells of monocyte-macrophage lineage. This does not imply that the AMs cannot participate in immune and inflammatory responses. In certain conditions, usually resulting in cellular activation, the AM can demonstrate potent function in these areas. As will be discussed, HIV infection also results in AM activation, thereby threatening the homeostatic balance between proinflammatory and suppressive responses in the lung. In fact, AM activation may play a central role in the pathogenesis of many of the pulmonary complications seen in these patients.

Finally, virtually no studies exist in which macrophages from other lung compartments have been examined in HIV-infected patients. Studies on lung dendritic cell function in theses individuals are equally sparse. Thus, most of the discussion in this chapter centers around AM function, with reference being made to other macrophage populations where appropriate and literature exists.

II. Infection of Lung Macrophages by HIV

Very early on it was recognized that HIV could be found in the lungs of infected individuals (8,9). Any cell which expresses the CD4 molecule is capable of being infected by HIV. Thus, it is not surprising that lung macrophages are infected with the virus. Whether this represents direct macrophage infection in the lung or influx of HIV-infected monocytes lung is not clear. Clearly AMs can be directly infected with HIV (10). In fact, Rich and colleagues have demonstrated that differentiated mononuclear phagocytes, including AMs, are more susceptible to HIV infection that blood monocytes (11). Thus, if virus is present in the alveolar space, macrophages may be infected in situ. Using coculture techniques where bronchoalveolar lavage fluid (BALF) is cultured with phytohemagglutinin (PHA)–stimulated peripheral blood mononuclear cells (PBMCs), infectious HIV has been isolated from the alveolar space of patients with acquired immunodeficiency syndrome (AIDS) (12,13). In situ hybridization studies also indicate that macrophages within the lung interstitium can be infected with HIV (9). In contrast to infected T cells, macrophages are relatively resistant to the cytopathic effect of HIV (10). Thus, they may serve as a viral reservoir in the lung.

Although it is now accepted that lung macrophages can be infected with HIV, the degree of lung involvement has remained controversial. Differences between studies likely reflect differences in patients studied, cells examined, experimental conditions, and methods used to detect virus. The latter issue is extremely important when evaluating studies on HIV infection in the lung. BAL coculture studies merely document infectious virus in the lung without identifying cell types involved. In situ hybridization studies looking for viral DNA detect latently infected cells, whereas those analyzing viral RNA are examining infectious particles. Immunohistochemical studies merely identify the presence of viral proteins and do not necessarily indicate active infection of positive cells. With these caveats in mind, several generalizations can be made based on available literature (Table 1) (9,12–18). Most patients have evidence of latent viral infection in the lung, as HIV DNA can be found in the majority of clinical BAL samples. In situ hybridization experiments have suggested that 0.1–10.0% of AMs are latently infected with HIV (18). Despite the frequent presence of provirus, HIV RNA is found only rarely in the lungs of infected patients, indicating that in vivo active viral replication is uncommon.

If evidence of active viral replication is rare in the lungs of HIV-infected individuals, why is it so easy to document the presence of virus using coculture or immunohistochemical techniques? Several explanations can be invoked. First, as is discussed in detail later in this chapter, the activation of macrophages directly upregulates HIV replication. Indeed, Lebargy and colleagues demonstrated that after incubation with a combination of tumor necrosis factor-α (TNF-α) and granulocyte-macrophage colony-stimulating factor (GM-CSF), HIV RNA could be documented in 60–80% of AMs (18). These findings are consistent with the wealth of information documenting that cytokines and coinfections can upregulate HIV expression in latently infected cells (19–26). In most coculture studies,

Table 1 Frequency of HIV Infection in the Lung

	Coculture techniques	HIV DNA	HIV RNA	Immunostaining
% BAL samples positive	10–100	46.7–100.0	0	—
% AMs positive	—	0.1–10.0	<0.002	5–90

macrophages are cultured with PHA-stimulated PBMCs which are secreting a variety of macrophage-activating agents. Thus coculture techniques may be detecting upregulation of viral expression in the 0.1–10.0% latently infected cells. Second, immunhistochemical studies may be detecting viral antigen absorbed to the surface of AMs rather that actively infected cells. The third possibility to explain the seemingly easy ability to document HIV infection in the lung (and also the wide variability between studies) involves the patients studied. Several investigators have presented data demonstrating an increased HIV burden in the lung accompanies disease progression (27,28). Most studies looking for evidence of viral infection in the lung have been performed in patients with AIDS, thus representing a population of HIV-infected subjects with a greater likelihood of having pulmonary involvement with HIV.

Although evidence for HIV infection of lung macrophages is strong, data on the involvement of lung dendritic cells are lacking. However, much work has been done on blood and follicular dendritic cells in HIV infection. Blood dendritic cells clearly can be infected by HIV and sustain high levels of viral replication (29,30). Presumably, infection is occurring through the CD4 receptor on these cells. In lymphoid organs, HIV RNA is closely associated with follicular dendritic cells (31–33). Interestingly, viral budding from the cell surface and HIV DNA is not associated with follicular dendritic cells which stain for viral RNA using in situ hybridization techniques (33). This suggests that follicular dendritic cells are not directly infected with HIV but are more likely trapping viral particles on their cell surface, a finding not surprising, since follicular dendritic cells are $CD4^{-}$. Further studies have shown that binding of HIV to follicular dendritic cells is a complement-dependent process (34). These studies suggest that HIV binds to follicular dendritic cells through complement and Fc receptors, both of which are expressed on these cells. On the other hand, lung dendritic cells are thought to be derived from blood dendritic cells, which express CD4 but no Fc or complement receptors (30). Some studies have shown that lung dendritic cells are also Fc receptor negative (35). Since lung dendritic cells more closely resemble blood dendritic cells, it would not be surprising if they too could support HIV replication and thus serve as an important viral reservoir in the lung.

III. Mechanisms of HIV Effects on Macrophages

HIV can effect macrophage function through many potential mechanisms (Table 2). As is described in detail later in this chapter, numerous studies have documented a direct effect of HIV infection on macrophage cytokine secretion and expression of surface molecules. These morphological and functional changes result in significant alterations in the ability of macrophages to participate in inflammatory and immune responses. However, as described above, HIV infec-

Table 2 Mechanisms Through Which HIV May Effect Lung Macrophage Function

Direct effect of HIV infection
on macrophage accessory cell function
on macrophage cytokine secretion
HIV-induced alterations in the lung milieu
decreased $CD4^+$ T cells
potential decreased interferon-γ
increased $CD8^+$ T cells
potential increased interferon-γ
cytotoxic to HIV-infected macrophages
increased secretion of macrophage cytokines
alterations in surfactant
Direct effect of HIV proteins
on cellular (especially T-cell) proliferation
on T-cell and macrophage cytokine secretion

tion in the lung is usually latent, with very little detectable viral RNA or protein found unless AMs are stimulated in vitro (17,18). In addition, the actual number of infected AMs varies with the stage of infection, being low early and increasing as patients progress to AIDS (28). Since AMs demonstrate functional alterations even early in infection, other HIV effects must also be operational.

A second mechanism by which HIV may effect macrophage function is through alterations in the cellular and biochemical makeup of the alveolar and interstitial environment. HIV infection is accompanied by a decrease in the percentage of $CD4^+$ lymphocytes and an increase in the percentage and absolute number of $CD8^+$ lymphocytes in BAL (36–38). Pathological studies have also shown lymphocytic infiltration in the pulmonary interstitium (39,40). Though phenotypic studies are much less complete, at least one study has suggested interstitial lymphocytes are also predominantly $CD8^+$ (40). Since $CD4^+$ memory cells are preferentially infected early in HIV infection (41,42), and since these cells are an important source of interferon-γ (IFN-γ) (43), early HIV infection could be characterized by a relative lack of this important macrophage-activating cytokine. On the other hand, studies show that $CD8^+$ lymphocytes which accumulate in the alveolar space of HIV-infected individuals are HIV-specific cytotoxic T cells (15,44). Since HIV-specific cytotoxic T lymphocytes release IFN-γ when they encounter infected target cells (45), they may make up for the loss of IFN-γ–secreting $CD4^+$ T cells. However, this may be accompanied by killing of HIV-infected AMs (15). Thus, there is a myriad of potential lymphocyte-mediated effects on lung macrophage function in HIV-infected patients. When one adds to this mixture the large number of macrophage-derived cytokines which are produced in increased amounts by HIV-infected macrophages (46), it is easy to see how the lung milieu in HIV infection is vastly different from normal.

A third mechanism by which HIV can alter macrophage function is through effects of viral proteins. The HIV proteins nef, gp120, p24, and tat have all been reported to directly effect cellular function. Nef inhibits NF-κB expression in human T cells (47). Similar inhibition of this important DNA-binding protein in macrophages could have significant effects on cellular activation and cytokine secretion. Soluble gp120 directly impairs accessory cell–dependent T-cell proliferation (48,49). This protein has also been shown to inhibit AM activity against *Cryptococcus neoformans* (50). p24 Impairs IFN-γ induced activation in the monocyte cell line THP1 (51). Finally, the HIV tat protein effects a wide variety of cellular functions from inhibition of T-cell proliferation (52) to induction of cytokine secretion in tissue macrophages and macrophage cell lines (53,54).

IV. Morphological Changes

In addition to an increased number of lymphocytes, one third of HIV-infected individuals have an elevated number of AMs in BAL (55). Whether this is due to in situ proliferation of AMs or recruitment of monocytes into the alveolar space is not known. Agostini and colleagues (56) have shown that AMs from HIV-infected patients spontaneously produce GM-CSF and proliferate in the presence of this factor. On the other hand, monocytes are likely constantly being recruited to the lung in response to the numerous infections seen in these patients. In support of this hypothesis is the finding that AMs from symptomatic HIV-infected patients express increased amounts of the monocyte markers CD11b, CD14, and CD33 compared with normal AMs (57). Evidence suggests that HIV itself can promote differentiation of monocytes into more mature forms (58). It is likely both monocyte recruitment and in situ proliferation account for the increased number of AMs in HIV infection. Since monocytes must traffic through the interstitium to reach the alveolar space, it goes to reason that the number of interstitial macrophages may also be increased, although this remains only speculative.

Once in the lung, HIV-infected AMs appear to have a normal life span. Viability of AMs from normal volunteers and HIV-infected individuals without active pneumonia is similar (59). Ultrastructurally, AMs from these patients appear to have increased numbers of microvilli on the cell surface consistent with a functionally active phagocyte (59). These cells have increased phagocytic vacuoles, some of which contain lymphocytic debris.

As discussed earlier, it is relatively easy to document the presence of viral antigen on the surface of AMs (14,15). In resting cells, this likely does not indicate active infection. However, in the presence of macrophage-activating cytokines or some pulmonary infections, viral replication is upregulated (18–26). In these instances, the presence of viral antigen may truly represent HIV infection in positive cells. Other investigators have shown that viral burden in the lung increases with disease progression (27,28). Thus, it is likely that the ability to detect viral proteins in AMs depends on the stage of disease and the presence or absence of coexisting conditions at the time of study. Whether viral proteins are confined to intracellular vacuoles, as has been demonstrated in other tissue macrophages (60) or are more evenly distributed in the cytoplasm or on the cell surface is not known at this time.

Regardless of the degree of infection by HIV, it is now apparent that monocytes and macrophages isolated from these individuals are activated. This can be demonstrated by the presence of surface molecules associated with cellular activation. Multiple studies have shown that monocytes and macrophages, including AMs, from HIV-infected individuals express increased amounts of major histocompatibility complex (MHC) class II molecules, especially HLA-DR (61–64). Interestingly, the data in monocytes are less consistent than that for macrophages. For example, Petit and colleagues showed that in vitro infection of the monocytic cell line U937 resulted in downregulation of HLA class II molecules (65). However, treatment with IFN-γ increased expression of class II molecules. Thus, monocytes isolated from HIV-infected patients might express variable amounts of HLA-DR depending on the levels of circulating IFN-γ. Expression of β_2-integrins, particularly leukocyte function–associated antigen (LFA-1) (CD11α/CD18), and one of its ligands, intracellular adhesion molecule-1 (ICAM-1) (CD54), are also upregulated in HIV-infected AMs and monocytic cell lines (63,65,66). Other markers of activation such as GM_1 ganglioside (67) and Fc receptors type 1 (68) and III (69) are also upregulated in monocytes from HIV-infected individuals. Whether similar findings occur in tissue macrophages is not known.

Why are macrophages from HIV-infected subjects activated? IFN-γ is one of the most potent macrophage-activating cytokines known. The hypothesis that macrophages in HIV-infected individuals are being activated by IFN-γ at first appears contradictory to standard dogma. Early studies suggested IFN-γ production was impaired in patients with AIDS (70–72), which is likely due to loss of $CD4^+$ memory T lymphocytes responsible for the production of the bulk of this cytokine (43). However, early in HIV infection prior to the loss of memory cells, IFN-γ production by $CD4^+$ T cells is upregulated (73,74). As HIV infection progresses and $CD4^+$ T cells decline, cytotoxic T lymphocytes (CTLs) become an important source of IFN-γ (45). HIV-specific CTLs can be expected to accumulate in areas where there is increased expression of viral antigens. Although the number of infected PBMCs is low and exists mainly in a latent stage (75), tissues containing differentiated macrophages appear to contain substantially more infected cells (15,60). This is especially true in lymphoid organs where active viral replication can be quite high even during the asymptomatic phase of HIV infection (32,33). Increased expression of viral antigens in various organs can be expected to result in the accumulation of HIV-specific CTLs within that tissue. In this regard, $CD8^+$ cells actively producing IFN-γ have been demonstrated in lymph nodes from HIV-infected patients (76). Since HIV-specific CTLs are readily detectable in the lungs of infected subjects (15), they could provide a source of IFN-γ. In support of this hypothesis, Buhl and colleagues have shown that freshly isolated AMs from these individuals express IP-10 gene transcripts, a gene specifically induced by IFN-γ, whereas normal AMs do not (64). Interestingly, these investigators also demonstrated that AMs from these asymptomatic infected individuals released increased amounts of superoxide anion (another sign of macrophage activation) compared with normal AMs. Autologous monocytes did not have similar upregulation of this function. This once again points out the importance of compartmentalized IFN-γ secretion as a potential mechanism for activating tissue macrophages while variably activating blood monocytes. Our laboratory has shown that $CD8^+$ lymphocytes isolated from BAL of HIV-infected patients with lymphocytic alveolitis are a rich source of IFN-γ (77). However, with time even these cells decline in the lungs of HIV-infected patients and are replaced by suppressor $CD8^+$ T cells (78). Thus a scenario can be hypothesized whereby IFN-γ is present in high concentrations in the lungs of HIV-infected individuals till late in the disease process (Fig. 1), resulting in chronic macrophage activation. Early in the disease process, $CD4^+$ cells are the primary source of IFN-γ. Later, $CD8^+$ become an important source of this cytokine. Only very late

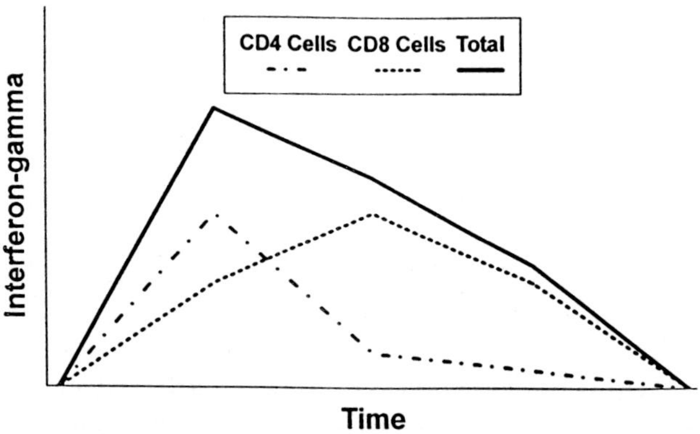

Figure 1 Potential source of IFN-γ in the lungs of HIV-infected subjects. Early in HIV infection, $CD4^+$ memory T cells are an important source of this cytokine. As disease progresses, CD4 counts fall and $CD8^+$ cytotoxic T lymphocytes accumulating in the lung become important sources of IFN-γ. As a result, IFN-γ concentrations have the potential to remain high in the lung of HIV-infected patients until late in the disease process resulting in chronic lung macrophage activation.

in HIV infection does the amount of IFN-γ in the lungs fall to levels inadequate to activate macrophages, leading to the late infectious pulmonary complications of AIDS.

Other surface markers which can be altered in HIV infection include the CD4 molecule. Several studies have shown that CD4 is downmodulated in monocytic cells infected with HIV in vitro (61,79). In contrast, CD4 appears to be upregulated in AMs from HIV-infected individuals (63). Since downmodulation of CD4 is associated with binding of HIV gp120 (49), increased expression of CD4 on AMs may reflect both cellular activation and a relative paucity of free virus in the alveolar milieu. The amount of CD4 on interstitial macrophages, which might be exposed to higher viral concentrations, is unknown. Finally, alterations in the AM mannose receptor have been found which may contribute to an impaired ability to clear *Pneumocystis carinii* in the alveolar space (80).

V. Phagocytosis

Macrophage phagocytosis can be divided into four parts, each of which can be altered by HIV infection (Fig. 2). First, the macrophage must move toward the material to be phagocytized (chemotaxis). Second, it must adhere to the object to be ingested. Third, foreign material must be taken up by the macrophage into phagocytic vacuoles. Finally, the material must be broken down through a variety of cellular digestive mechanisms. Since the predominant phagocytic cell in the lung is the AM and most studies have been performed on this cell population, AMs are the focus of this section. Some phagocytic functions have been examined directly in AMs from HIV-infected subjects. Others have been studied mainly in monocytes or monocyte-derived macrophages and extrapolation to AMs must be done with caution.

A. Chemotaxis

One of the earliest studies performed with monocytes in HIV-infected patients demonstrated impaired chemotaxis in response to N-formyl-methionyl-leucyl-phenylalanine (FMLP), C5a, and the protozoan *Giardia lamblia* when compared with normal monocytes (81). This defect was most profound in patients with AIDS. Other investigators have shown that HIV infection effects the manner in which monocytes interact with extracellular matrix (82). HIV-infected monocytes bound to extracellular matrix more avidly than uninfected cells. Furthermore, infected cells tended to spread out more readily. AMs are less mobile than monocytes but still possess the capability to migrate. Whether the above observations apply to AMs or other tissue macrophages is not known. It should be emphasized though that any analysis of AM chemotaxis in HIV infection will have to separate viral effects on AMs from potential effects on surfactant, which

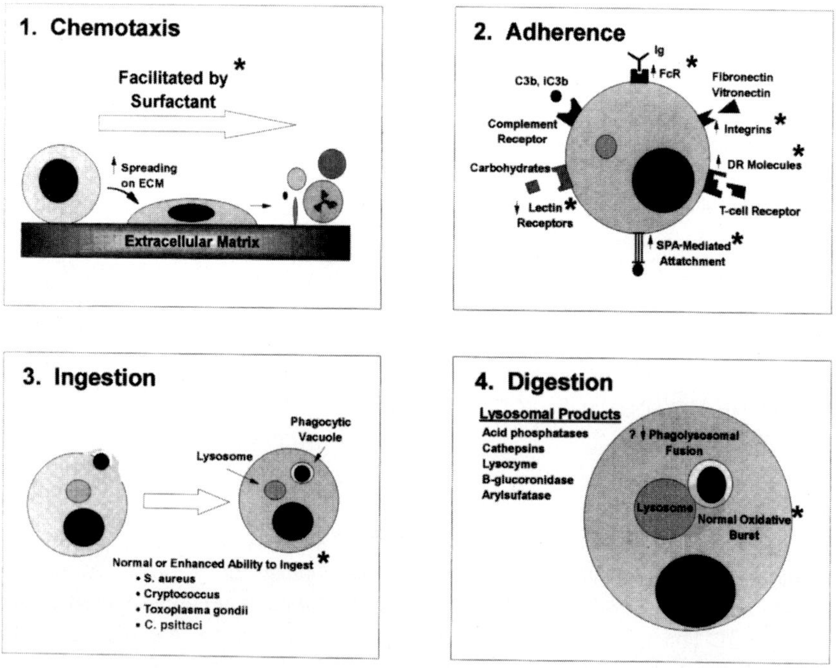

Figure 2 Four steps of AM phagocytosis which may be modified by HIV infection. Information from studies which were done on lung material (i.e., cells, BALF) obtained from HIV-infected subjects is signified by an asterisk (*).

facilitates AM migration (83). This latter point is critical, since recent studies have shown that surfactant protein A, the component of surfactant responsible for enhancing AM chemotaxis (83), is increased in HIV-infected individuals with (84) and without (85) pneumonia.

B. Adherence

Once the AM has migrated into the vicinity of the material to be phagocytized, binding must occur between the particle and cell surface. Many of the molecules described previously which are potentially upregulated on AMs in HIV infection (DR, β_2-integrins, Fc receptors) can promote adherence to other cells or foreign material. In contrast, alterations in other molecules can impair AM binding to antigens (80). Thus, whether AMs from HIV-infected patients bind targets efficiently will depend on the pathway used by AMs to bind the material to be ingested. Many ingestible particles are taken up through Fc and complement receptor–mediated processes. Interestingly, surfactant protein A enhances both Fc and complement receptor–mediated phagocytosis (86). Since surfactant protein A is increased in HIV infection, one might speculate that AMs from these individuals efficiently bind and phagocytize material taken up by these pathways. In support of this hypothesis, Downing and colleagues (85) have shown that *Mycobacterium tuberculosis* binds more avidly to AMs in the presence of BAL fluid from HIV-infected patients compared with normal BAL. Furthermore, removal of surfactant protein A from BAL by immunoprecipitation reduced AM–*M. tuberculosis* binding to normal levels. Surfactant protein A also binds avidly to *Pneumocystis carinii* (87) and thus may be important in AM uptake of this organism. However, uptake of *P. carinii* is also mediated by the macrophage mannose receptor (88), a molecule shown to be altered in HIV infection (80). It must be emphasized that most studies examining AM binding to *P. carinii* have been performed on animal macrophages, normal human AMs, or normal AMs exposed to HIV_{BAL}. Very few studies have examined AMs from HIV-infected patients directly. Thus the question of how well HIV AMs bind this organism remains unresolved.

C. Ingestion

Once the AM has bound the material to be phagocytized, it must ingest it. Although it may seem logical that binding a priori leads to internalization of bound material, this is not always the case. For example, Wagner and colleagues have shown that gp120 inhibits the ingestion of *Cryptococcus neoformans* by normal AMs without altering initial binding of the organism (50). This study emphasizes the need to separate these two important steps when analyzing phagocytosis.

With the above caveat in mind, numerous studies have examined the ability of monocytes and macrophages from HIV-infected patients to phagocytize various organisms. In general, studies have shown normal or decreased monocyte phagocytic function but normal to enhanced AM function; again demonstrating the danger in extrapolating monocyte studies to tissue macrophages. Monocytes from symptomatic HIV-infected patients have an impaired ability to phagocytize *Staphylococcus aureus* (89) and antibody-coated sheep red blood cells (90). Normal monocytes infected in vitro with HIV_{BaL} retain the ability to phagocytize *C. neoformans* (91). In contrast, AMs from HIV-infected patients without pneumonia demonstrate an enhanced ability to phagocytize *S. aureus* (59). Several other studies have shown that the AM phagocytic function is normal in HIV infection (91,92). Only one study has suggested that the AM phagocytic function is impaired in HIV infection (50). This may reflect the organism studied (*C. neoformans*), the cell studied (normal AM), or the conditions examined (exposure to gp120 rather than intact virus). Although the reasons for differential effects of HIV infection on monocyte and the AM phagocytic function remain unclear, one likely explanation involves the cytokine milieu in which each cell must operate.

D. Digestion

The last step in efficient phagocytosis is digestion of the material taken up into phagocytic vacuoles resulting in degradation of inert material and killing of viable organisms. Ingested material is destroyed by lysosomal enzymes, reactive oxygen species, and reactive nitrogen species. Numerous investigators have examined the effect of HIV on monocyte and macrophage oxidative metabolism. Studies on oxidative burst responses in monocytes are conflicting. Several studies demonstrate an impaired ability of monocytes and monocyte-derived macrophages from HIV-infected individuals to produce reactive oxygen species spontaneously or in response to appropriate stimuli (93,94). Other investigators have found this function is intact in HIV-infected subjects (91,92,95,96). Studies in AMs are fewer but consistent. In general, AM oxidative burst potential in HIV-infected patients appears to be normal, especially in the presence of appropriate macrophage-activating cytokines (64,92,97).

Studies on lysosomal function in HIV infection are lacking. One group of investigators found that monocyte phagolysosomal fusion was impaired late in HIV infection (98). Recombinant gp120 induced similar defects in normal monocytes. Whether similar effects are seen in tissue macrophages is not known. Studies looking at lysosomal enzymes in HIV infection are also scarce. Clearly much work needs to be done in this area.

E. Killing

Provided all four steps involved in phagocytosis are intact, the end result should be killing of ingested organisms. Despite the wide discrepancies between some of the studies described above, the end results (killing of organisms) are remarkably uniform. In general, monocytes or monocyte-derived macrophages from HIV-infected patients kill ingested organisms poorly (89,90,95). In contrast, killing of ingested bacteria, parasites, and fungi is largely preserved in AMs (59,92,97). In the one study in which monocytes and macrophages were directly compared, HIV-infected monocytes had reduced anticryptococcal activity, whereas similarly infected AMs retained this function (91). Several mechanisms could explain this dichotomy between monocytes and AMs. Hypothesis center around differences between the vascular compartment and alveolar space with regard to the infectious viral burden, presence of coinfections, and cytokine concentrations. As described earlier, the end result is AM activation (64). Murray has shown that IFN-γ activates HIV-infected AMs to function efficiently as phagocytes (92,97). Thus, it is likely that AMs from HIV-infected patients retain their phagocytic capabilities owing to chronic activation either by IFN-γ, HIV itself, or coinfections present in the lower respiratory tract.

VI. Secretory Products

Macrophages are a rich source of a wide variety of secretory products (99). These products can be loosely grouped into five categories (Table 3): reactive oxygen and nitrogen species, enzymes and proteases, antiproteases and free radical scavengers, metabolites of arachidonic acid, and cytokines. The ability of macrophages from HIV-infected subjects to secrete reactive oxygen species was described above. Very little is known about HIV macrophage function with regard to the next three categories. Only a few studies have been performed in monocytes, and even these present conflicting data. For example, whereas one group found that monocytes from HIV-infected subjects had impaired prostaglandin E_2 (PGE_2) production induced by mitogens (95), another group demonstrated that normal monocytes infected with HIV in vitro produced increased amounts of this prostaglandin (100). In contrast to monocytes, where cyclooxygenase products are the major consequence of arachidonic acid metabolism, AMs preferentially produce lipoxygenase products (101). Interestingly, leukotriene B_4 concentrations have been shown to be elevated in BAL from HIV-infected patients with nonspecific interstitial pneumonitis (102). Phospholipase A_2 concentrations were also elevated in these patients as well as in asymptomatic HIV-positive individuals, suggesting a generalized increase in arachidonic acid metabolism. The cells responsible for these findings were not defined. Thus, it is possible that HIV infection is associated with a generalized increase in arachidonic acid metabolism in monocytes and tissue macrophages. More research is needed in this area.

Table 3 Macrophage Secretory Products

Reactive free radicals	Cytokines
reactive oxygen species	**IL-1**
reactive nitrogen species	**IL-6**
Enzymes and proteases	**IL-8**
Antiproteases and free radical scavengers	**IL-10**
Arachidonic acid metabolites	**GM-CSF**
cyclooxygenase products	**TNF-α**
lipoxygenase products	**TGF-β**

Bolded items have been shown to be produced in normal or increased amounts in the lungs of HIV-infected subjects.

A. Cytokines

In contrast to other secretory products, perhaps no aspect of macrophage biology in HIV infection has been studied as intensely as cytokine production. Virtually every macrophage-derived cytokine studied to date has been found to be secreted in increased amounts in HIV-infected patients. To date, most of the cytokines studied have been proinflammatory in nature, including the interleukins (ILs) IL-1 (28,103), IL-6 (28,103,104), IL-8 (102), granulocyte-macrophage colony-stimulating factor (GM-CSF) (56), and tumor necrosis factor-α (TNF-α) (28,105–108), Importantly, enhanced secretion of these cytokines can occur in the absence of any recognized inflammatory or infectious process as well as in response to low concentrations of nonspecific stimuli (i.e., lipopolysaccharide [LPS], phorbol esters). This implies that macrophages from HIV-infected individuals not only spontaneously synthesize and secrete these factors but also possess the ability to respond in an augmented fashion to stimuli that would otherwise have no or minimal effects on nonactivated cells.

Several explanations can be evoked to explain increased macrophage cytokine secretion in HIV-positive individuals (Fig. 3). It is possible that enhanced cytokine secretion reflects macrophage exposure to pathogens in the lungs of these individuals. Clearly many investigators have found enhanced macrophage cytokine secretion in HIV-infected subjects with pulmonary pathology (56,104,106, 108). However, several studies have shown that enhanced cytokine secretion is seen even in asymptomatic HIV-infected individuals (28,103,105), suggesting this alteration is a direct consequence of HIV infection.

A second explanation for enhanced cytokine secretion demonstrated by macrophages from HIV-infected individuals could be direct stimulation by the virus itself. Merrill et al. demonstrated that HIV could induce the secretion of IL-1 and TNF from monocytes (109). Secretion was induced immediately on exposure of monocytes to heat-inactivated HIV, indicating this effect occurred in the absence of productive viral infection. More recent studies have shown that isolated HIV envelope glycoproteins gp120 and gp160 can induce IL-6 secretion from CD4 T cells and monocytes (110,111). In these studies, cytokine secretion was blocked by soluble CD4, indicating that actual contact between cell surface CD4 and HIV envelope proteins was necessary for stimulation. Studies correlating AM cytokine secretion with pulmonary viral load are conflicting. Israel-Biet and colleagues found that AM TNF secretion was greater in cells shown to express HIV, as documented by coculture techniques with T-cell blasts (105). In contrast, Sierra-Madero et al. found no correlation between viral load determined by polymerase chain reaction (PCR) techniques and the production of TNF-α, IL-1β, or IL-6 (28).

A third explanation for enhanced cytokine secretion observed in HIV macrophages involves the interaction of viral regulatory proteins with the host cell's

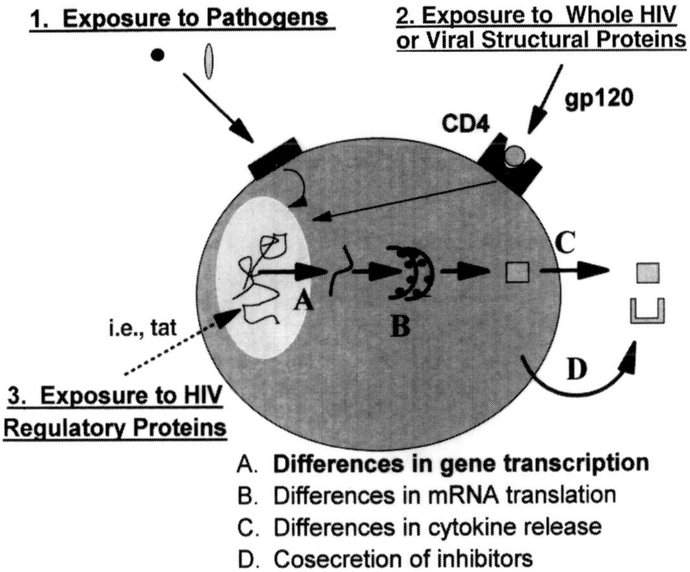

Figure 3 Potential mechanisms for enhanced cytokine secretion by macrophages from HIV-infected individuals. Macrophages may be stimulated to produce cytokines through their normal pathway by usual or opportunistic pathogens, HIV itself, or viral structural proteins. Alternatively, HIV-regulatory proteins can directly initiate cytokine gene transcription. Note that increased cytokine secretion can result from enhanced gene transcription, enhanced mRNA translation and post-translational processing, enhanced secretion of the finished product, or alteration in the secretion of cytokine inhibitors. Increased cytokine gene transcription appears to be the major mechanism whereby HIV upregulates macrophage cytokine production.

genome. Iwamoto et al. have shown that cytomegaloviral immediate-early gene products can induce IL-1 transcription in the monocytic cell line THP-1 (112). Other investigators have suggested that HIV-1 *trans*-activating proteins such as *tat* might likewise induce cytokine gene transcription in HIV-infected cells (113). In support of this hypothesis, recent work has found that THP-1 cells transfected with the HIV-1 *tat* gene demonstrated enhanced unstimulated as well as LPS-induced IL-6 secretion (54) associated with the accumulation of IL-6 mRNA transcripts.

Thus, exposure to infectious organisms associated with AIDS, whole HIV particles, HIV structural proteins, or HIV-regulatory proteins can lead to increases in macrophage cytokine production and secretion. Enhancement can occur at any of a number of steps in cytokine synthesis (see Fig. 3). Most studies documenting enhanced cytokine secretion by HIV-infected macrophages demonstrate concomitant increases in cytokine mRNA (28,56,104,106,113). Thus, increased gene transcription induced by HIV-regulatory proteins may explain increased cytokine secretion by HIV-infected cells. Alternatively, RNA stability could be enhanced. Increased cytokine secretion by HIV-infected macrophages could also reflect enhanced translational and posttranslational machinery in these cells. Secretory processes could be facilitated. However, one study showed no difference in intracellular IL-1β and IL-6 between normal and HIV-infected AMs, suggesting translation and secretion was not affected (103). Finally, the measurement of bioactive cytokines could be altered by cosecretion of cytokine inhibitors.

The above discussion does not consider potential effects HIV may have on cytokine receptors. In fact, one group of investigators has shown that the HIV *tat* gene downmodulates TNF receptors in a human B-cell line (114). In contrast, others have shown that lung T cells from HIV-infected subjects display increased expression of the TNF receptor (115). In another study, Honda and colleagues found that HIV infection induced the release of the soluble IL-6 receptor from a promonocytic cell line (116). Since IL-6 complexed to the soluble IL-6 receptor can bind to gp130 (an IL-6 signal transducing membrane glycoprotein found on many cells, including T cells) and behave as a cytokine agonist (117), secretion of this receptor could serve as another mechanism upregulating the effects of cytokines in HIV infection. Thus, it appears lung T cells in HIV-infected individuals are not only exposed to higher concentrations of cytokines but also have an augmented ability to respond to them through upregulation of cytokine receptors.

Whatever the mechanism, increased macrophage cytokine secretion and their effects on other cells have been implicated as playing a central role in the pathogenesis of HIV-related diseases (118–121). TNF-α in particular has been singled out for its ability to be upregulated by HIV, induce other macrophage cytokines, and cause many of the processes associated with HIV infection such as the wasting syndrome, AIDS encephalopathy, and Kaposi's sarcoma. Not only does HIV induce macrophage cytokines, but these cytokines in turn may upregulate viral expression in other monocyte-macrophages (19–24) or nearby T cells (122–124); thus propagating the virus in an autocrine and/or paracrine manner. This process can be modulated at several points. For example, other infections can either directly upregulate HIV expression (26,105,125) or themselves induce cytokines capable of upregulating viral expression (126).

From the above discussion, it is apparent that BALF from HIV-seropositive patients can be a veritable "cytokine soup" containing increased concentrations of any number of different cytokines. To this hodgepodge one must factor in the effect of cytokines on other lung cells, especially T-cell and AM proliferation. The increased concentration of cytokines and resultant increase in cellular components renders the alveolar space an ideal medium for immune and inflammatory reactions in response to specific and nonspecific antigenic challenges. More importantly, increases in cytokine and cell concentrations can occur spontaneously; thus causing symptoms in the absence of recognized pulmonary disease. It is important to recognize that augmented cytokine secretion and an increased number of immune cells in the alveolar space does not necessarily translate into enhanced immunological function. Clearly HIV infection is characterized by impaired immune responses to various infectious agents. It is likely that cytokines with negative regulatory functions or even cytokine inhibitors are similarly upregulated in the alveolar space. For example, one group of investigators demonstrated that monocytes secreted an IL-1 inhibitor which masked detection of bioactive IL-1 in bioassays (127). Another group found that adherent cells from HIV-infected subjects secreted a 29-kDa protein which inhibits T-cell proliferation by decreasing the expression of the IL-2 receptor and IL-2 secretion (128). Finally, cytokines which have traditionally been felt to have negative inflammatory and immunological effects are also secreted in increased amounts by macrophages from HIV-infected individuals, including transforming growth factor-B (TGF-β) (53) and IL-10 (129).

In summary, the cellular constituents of the lower respiratory tract in HIV-infected individuals are likely to depend on the prevailing cytokine profile in the alveolar milieu, which in turn is affected by many variables, including the presence or absence of specific infectious agents, the degree of AM infection by HIV, and the degree of chronic immune stimulation by potential environmental antigens. The presence of increased cytokine levels and cell numbers in the alveolar space can be expected to have multiple clinical consequences depending on which of these components are upregulated and downregulated.

VII. Immune Function

Besides phagocytosis, the major immune function of macrophages involves their role as accessory cells in the generation of cellular and humoral immune responses. An accessory cell (AC) is a cell which induces lymphoproliferation in the presence of antigens, mitogens, and alloantigens leading to the subsequent production of effector T cells capable of enhancing antibody production or performing cytotoxic functions. Although several cell types possess AC capabilities, monocyte-macrophages and dendritic are classically thought of as the major ACs in the body. In this section, macrophage and dendritic cell function in B-cell and T-cell responses are examined.

A. Accessory Cell–B-Cell Interactions

Despite the presence of a polyclonal hypergammaglobulinemia, HIV-infected patients are predisposed to pulmonary infections with encapsulated bacterial pathogens (130). These infections typically reflect an inability to generate opsonizing antibodies, specifically of the IgG isotype, which are directed against infecting pathogens. Most studies examining immunoglobulin production in HIV-infected patients have been performed in peripheral blood. One group of investigators demonstrated that normal quantities of antigen-specific IgG were produced in HIV-infected individuals immunized with the pneumococcal polysaccharide vaccine (131). IgA responses to the vaccine were below normal. This contrasts with findings from other laboratories demonstrating generalized elevated IgA levels, including secretory IgA, in peripheral blood (132). This latter finding suggests the ability to form antibodies at mucosal sites remains intact. These contrasting data highlight the importance of different antibody responses on exposure to antigen via systemic versus mucosal routes. Thus, information on antibody responses in the lung, which are primarily mucosal in origin, cannot be extrapolated from peripheral blood data but must be examined directly using lung cells and/or lung models.

In vitro, it has been demonstrated that local immunoglobulin production by B lymphocytes depends on stimulation by T-cell cytokines secreted as a result of interactions with antigen-primed accessory cells (133). In the normal lung, this results in the local production of concentrations of IgG greater than IgA. In that the primary opsonizing antibody activity resides in the IgG isotype, it is possible that the increased susceptibility to pulmonary infections caused by encapsulated organisms in HIV-infected patients might be secondary to decreased local IgG concentrations in the lung. AMs induce immunoglobulin production from peripheral blood mononuclear cells (PBMCs) in response to alloantigens (134). Interestingly, dendritic cells, generally felt to be more potent ACs than AMs, are poor inducers of immunoglobulin secretion in PBMCs. IgG production requires IL-4 and IFN-γ (134), whereas IgA production is augmented via IL-5 and TGF-β (135). TGF-β, which is derived from mononuclear phagocytes, has reciprocal activity which results in the inhibition of IgG production. Normally, IgG in the alveolar space is found at a greater concentration than IgA, which in turn is much greater than IgM. In HIV-infected individuals, if the concentration of opsonizing antibodies is decreased in the alveolar space, this might be reflected in a decrease in the concentration of IgG relative to IgA.

Recent studies from our laboratory have shown that despite the ability of AMs from HIV-infected patients to secrete cytokines which enhance B-cell proliferation and differentiation (i.e., IL-6), BAL from these individuals contains significantly decreased IgG concentrations, whereas IgA levels are preserved (136). In that local concentrations of immunoglobulin are the result of AM interactions between T and B lymphocytes, the ability of AMs from HIV-infected patients to stimulate alloantigen-induced immunoglobulin production in normal PBMCs was determined. Although it has been previously shown that normal AMs stimulate much greater IgG than IgA production in response to alloantigens (134), AMs from HIV-infected patients stimulated IgA production without stimulating any IgG or IgM production (Fig. 4). Further studies suggested this may be due to HIV

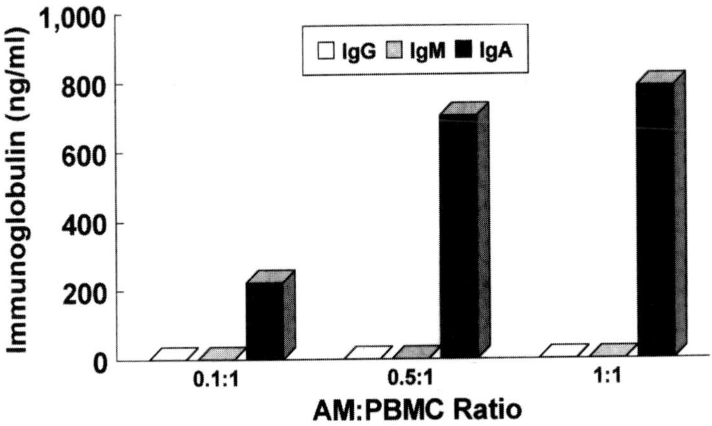

Figure 4 Ability of AMs from HIV-infected patients to stimulate alloantigen-induced immunoglobulin production in normal PBMCs. Fifty thousand PBMCs were cultured with varying ratios of AMs as shown. After 6 days, supernatants were harvested and immunoglobulin levels measured by ELISA. Although it has been previously shown that normal AMs stimulate much greater IgG than IgA production in response to alloantigens, AMs from HIV-infected patients stimulated IgA production without stimulating any IgG or IgM production.

AM overproduction of TGF-β, which has been demonstrated to be a potent inhibitor if IgG production while augmenting IgA production (135,136). These findings are in concordance with those of Kekow et al., who presented data suggesting that TGF-β overexpression was responsible for blunted B-cell responses to mitogens in HIV-infected subjects (137). These data contrast with those of Young and colleagues, who described the presence of elevated IgG and IgA concentrations in the BAL of HIV-infected patients (138). However, all of their patients underwent BAL for diagnostic purposes to evaluate abnormal chest radiographs or clinical symptoms. Lavage was performed specifically in involved areas of the lung documented on chest radiograph. Their patients were found to have an increased number and percentage of neutrophils in BALF fluid, confirming an active ongoing inflammatory process. Thus, part of the elevated immunoglobulin concentrations observed in their study could be explained by the exudation of plasma proteins into the alveolar space. In fact, when immunoglobulin concentrations were expressed as a ratio to albumin concentration, no significant difference was seen between BAL and blood immunoglobulin/albumin ratios, implying that transudation of immunoglobulin rather than local synthesis was operative in their subjects.

B. Accessory Cell–T-Cell Interactions

Cells in the alveolar space are constantly exposed to potential pathogenic organisms and particles inhaled with each breath. Under this evolutionary pressure, AMs, being the most common ACs in the lower respiratory tree, have evolved into relatively poor stimulators of T-cell proliferation compared with blood monocytes (139–141). In contrast, dendritic cells found in the interstitium remain potent ACs in the lung (35,142).

The study of AM AC function in HIV-infected individuals has been an area of intense investigation in our laboratory. AMs from HIV-infected individuals are significantly better ACs compared with normal AMs in stimulating a mitogen response (143). Interestingly, AMs from asymptomatic HIV-infected patients without pulmonary complaints demonstrate the greatest enhancement in AC function, indicating that nonspecific pulmonary inflammation is unlikely to explain these findings. Unlike normal AMs, AMs from HIV-infected subjects are as capable as autologous blood monocytes at stimulating mitogen responses. That AMs from HIV-infected subjects show enhanced AC capabilities should not be surprising, since these cells are activated (64). Individuals with other disease states characterized by cellular activation in the lower respiratory tract (i.e., sarcoidosis) also have AMs with enhanced AC function (144).

The degree of T-cell proliferation induced by ACs depends on the efficiency of AC–T-cell contact and the balance of positive and negative signals secreted by the ACs (145,146). Thus, three potential mechanisms exist to explain enhanced AM AC function in HIV-infected patients (Fig. 5). First, there could be increased secretion of T-cell stimulatory factors, particularly stimulatory cytokines, by HIV AMs. Second, AC–T-cell adherence, through MHC molecules and the T-cell

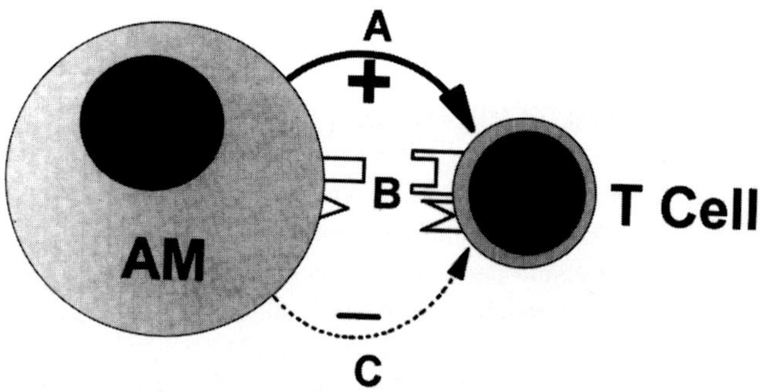

Figure 5 Potential mechanisms of enhanced AM AC function in HIV-infected patients. Enhanced function could reflect (A) increased secretion of T-cell–stimulatory factors, (B) more efficient AM–T-cell contact, and/or (C) decreased secretion of T-cell–inhibitory factors.

receptor or through other receptor-counterreceptor mechanisms, could be more efficient in HIV-infected AMs. Finally, HIV AMs may release less inhibitory factors such as prostaglandins or cytokine inhibitors compared with normal AMs.

With regard to the first possibility, studies have shown that when AM supernatants from normal volunteers and HIV-positive patients are added to mitogen assays using normal AMs as AC lymphoproliferation is enhanced, suggesting normal AMs secrete suboptimal amounts of T-cell–stimulatory factors (103). HIV-positive supernatants were more potent than normal supernatants in this regard, suggesting they contained greater concentrations of these factors. In contrast, both normal and HIV supernatants have negligible effects when added to mitogen assays using AMs from HIV patients as ACs, suggesting HIV AMs are already secreting optimal amounts of T-cell–stimulatory factors. IL-1 and IL-6 are likely candidates for these factors, as they have been shown to enhance AC function (147). As described previously, AMs from HIV-infected patients secrete greater amounts of IL-1 and IL-6 than normal AMs both spontaneously and in response to LPS or mitogenic stimuli (28,103,104). These data demonstrate that secretion of important soluble factors involved in T-cell proliferation is enhanced in AMs from HIV-positive patients. Working in concert, augmented IL-6 and IL-1 production and secretion may in part explain the enhanced AC function demonstrated by AMs from HIV-positive individuals. This conclusion is further supported by data demonstrating that the passage of AM supernatants over protein G immunoaffinity columns coupled with anti–IL-1 and anti–IL-6 antibodies effectively removes the ability of both normal and HIV AM supernatants to enhance mitogen assays (103).

The second possible explanation for enhanced AM AC function in HIV-infected subjects is more efficient AC–T-cell binding. Prior work has shown that AM AC function in normal volunteers correlates with their ability to bind T cells (139). Given the increased expression of HLA-DR and adhesion molecules on the surface of HIV-infected AMs (61–66), one would speculate that these cells possess an enhanced ability to bind T cells. Furthermore, Agostini and colleagues demonstrated increased TNF receptor expression on BAL T cells and increased membrane-bound TNF on AMs, thereby providing another opportunity for augmented AM–T-cell binding (115). However, mitogen-induced T-cell–AM adherence is identical in normal volunteers and HIV-positive patients (148). Furthermore, unlike in normal AMs, an inverse relationship is seen between AM–T-cell binding and AC function in HIV-positive patients, suggesting that close AM–T-cell contact may be detrimental in this setting (Fig. 6). Although these studies suggest factors other than adherence are more important in determining maximal AC function of HIV AMs, AM–T-cell contact is still essential for optimal AC function in HIV-infected patients. This was demonstrated by showing that adhesion molecule antibodies capable of abrogating AM–T-cell adherence also effectively blocked AC function when added at the beginning of culture. When these

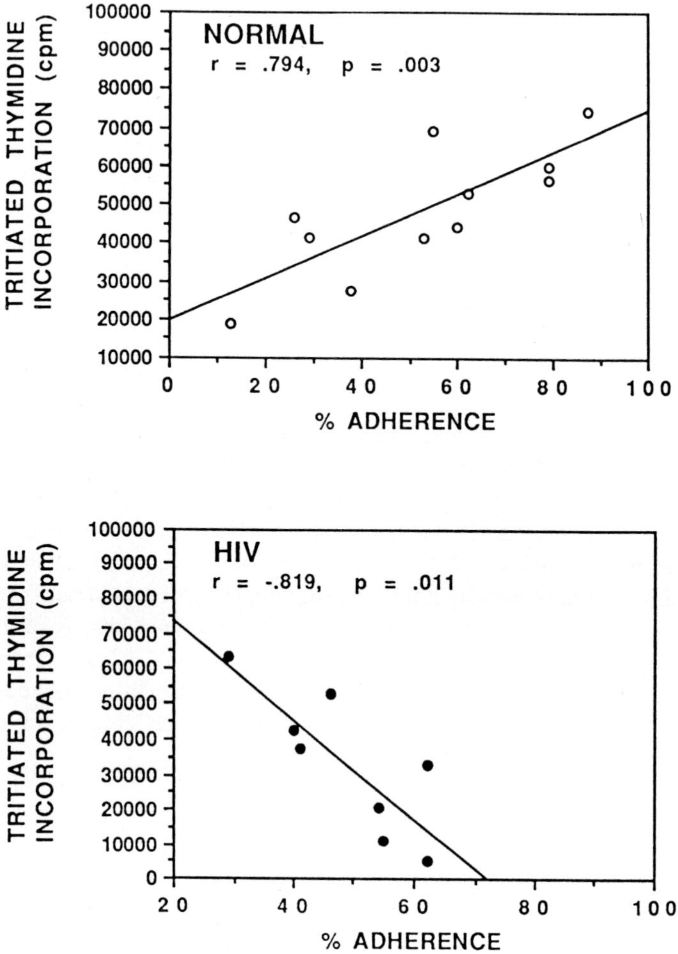

Figure 6 Correlation between AM AC function and AM–T-cell adherence in normal volunteers and HIV-positive patients. T-cell adherence to normal AMs or HIV AMs in the presence of concavallin A (ConA) was determined at 4 hr of culture and correlated with the ability of 25,000 AMs to induce proliferation in 10^5 normal T cells in response to the same mitogens. There was a direct correlation between normal AM AC function in ConA assays and their ability to bind T cells. In contrast, HIV AM AC function in ConA assays was inversely correlated with their ability to bind T cells.

antibodies were added 4 hr after the initiation of culture, accessory cell function in normal volunteers was still completely inhibited. However, significantly less inhibition of accessory cell function was noted in HIV-positive individuals. When antibodies were added after 24 hr of culture, there was a modest inhibitory effect in both groups. These data indicate that augmented AM–T-cell adherence, although still critical for optimal AC function, does not explain the enhanced AC function of HIV-infected AMs.

These findings do not imply that the increased expression of surface adhesion molecules is not facilitating HIV AM AC function. Many receptor-counterreceptor interactions between AC and T lymphocytes serve dual purposes: first, to hold T cells in close proximity to ACs so that secreted AC factors can have an optimal effect and, second, to act as a direct costimulatory signal for T-cell activation (149–154). Thus, increased expression of HLA-DR molecules and β_2-integrins could still contribute significantly to the enhanced AC function of HIV AMs.

Finally, it is possible that HIV AMs are "less inhibitory" than normal AMs. Recent work has shown that normal AMs inhibit increases in intracellular calcium in stimulated T cells (155). Inhibition appears to be mediated by a membrane-bound signal on AMs. Subsequent work has shown that this signal is a membrane-bound surfactant component (156). Since surfactant is altered in HIV infection (84,85), it is possible that surfactant is "less suppressive" in HIV-infected patients. Interestingly, Weissler and colleagues found that AMs isolated from patients with idiopathic pulmonary fibrosis, another immune-mediated disease characterized by activation of AMs, were less suppressive than normal AMs (156). Other investigators have found that GM-CSF abrogates the natural inhibitory properties of normal AMs (157). This might have important effects on AMs from HIV-infected patients where secretion of this cytokine is upregulated (56). Other factors such as prostaglandins and nitric oxide can inhibit AC function (158,159). It is possible that AM production of these mediators is altered in HIV infection. Finally, it is possible that the secretion of inhibitory cytokines or cytokine inhibitors is downregulated in HIV infection.

Based on studies of AM–T-cell adhesion and AM cytokine secretion, a model for AM AC function in normal volunteers and HIV-positive patients can be proposed (Fig. 7). In normal volunteers, it is important to maintain AM T-cell adherence so that T cells may be held in close proximity to the cytokine-secreting macrophage. Since normal alveolar macrophages secrete relatively small amounts of important T-cell cytokines such as IL-1 and IL-6, T cells that are not in close proximity to AMs will not proliferate even if they have been previously activated. A corollary to this would be that the more T cells that remain in contact with AMs, the more T cells will proliferate. In contrast, HIV AMs secrete greater amounts of IL-1 and IL-6. Therefore, cells that have been activated but are no longer in contact with AMs are still able to proliferate in the presence of these T-cells–stimulatory cytokines. The greater the cytokine secretion, the greater this sphere

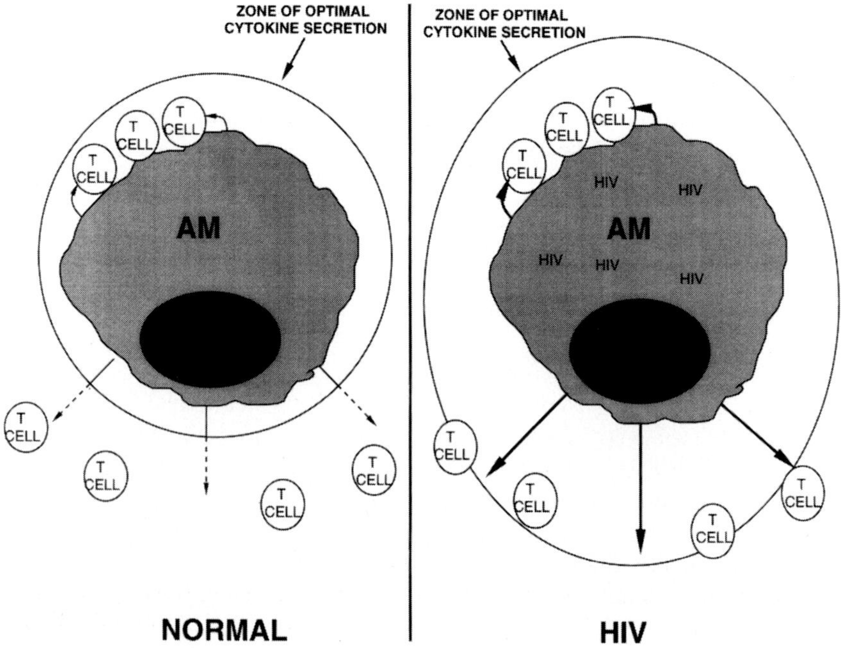

Figure 7 Proposed model of AM AC function in normal volunteers and HIV-positive patients. Initial AM–T-cell contact is required in both groups to initiate cellular immune responses. However, owing to enhanced secretion of T-cell–stimulatory cytokines (i.e., IL-1, IL-6) by HIV AMs, a zone of optimal cytokine secretion extends further from the HIV AM microenvironment than from normal AMs. Thus, more T cells that have been activated and subsequently disengaged from ACs can be optimally stimulated by AM cytokines in HIV-positive patients. Note that AM–T-cell contact is potentially detrimental in HIV-positive individuals if passage of HIV is facilitated from the infected AMs to uninfected T cells.

of influence extends. In contrast to normal volunteers, AM T-cell adherence may actually be detrimental in HIV-positive patients owing to viral passage from infected macrophages to noninfected T cells.

Studies on the effect of HIV on dendritic cell AC function are scarce and conflicting. As described earlier, both peripheral blood dendritic cells as well as follicular dendritic cells can be infected by HIV (29–34). Several investigators have examined the effect of HIV infection on blood dendritic cell AC function. Cameron and colleagues found that blood dendritic cells from HIV-infected patients were able to stimulate a mixed leukocyte reaction (MLR) in normal allogeneic $CD4^+$ T-cell clones (160). In contrast, the ability of blood dendritic cells from these patients to stimulate an MLR in normal PBMCs or antigen-specific responses in autologous cells was impaired in a study by Macatonia et al. (161). Differences in dendritic cell isolation techniques and in the lymphocyte responder population used may explain these contrasting results. Interestingly, though dendritic cell AC function was impaired in the latter study, macrophage function was preserved. The ability of tissue dendritic cells from HIV-infected individuals to stimulate T-cell proliferation is unknown at this time. Regardless of whether T-cell proliferation is induced, studies have shown transmission of HIV from dendritic cells to uninfected $CD4^+$ T cells is highly efficient (162).

VIII. Implications

What are the theoretical implications of enhanced AC function of lung macrophages in HIV-infected patients? There are five potential sequelae whereby enhanced macrophage or dendritic cell AC function in HIV-infected patients could ultimately lead to $CD4^+$ T-cell depletion in the lung (Fig. 8). The first potential sequela is formation of syncytia between infected AC and T cells (143). This process would be facilitated by the increased expression of β2-integrins on HIV-infected macrophages (163,164). Second, macrophages or dendritic cells may infect $CD4^+$ T cells directly during the generation of an immune response (33, 162,165–167). Third, increased T-cell proliferation as a result of enhanced secretion of macrophage cytokines makes T cells more susceptible to infection by free virus (168). Fourth, as described previously, these same cytokines upregulate HIV replication in latently infected $CD4^+$ T cells and other macrophages; thus increasing the amount of free virus in the lung. Finally, enhanced AC-$CD4^+$ T-cell interactions may ultimately lead to the increased production of $CD8^+$ cytotoxic T lymphocytes (CTLs) active against viral antigens which can then kill virus-infected macrophages and $CD4^+$ T cells. The alveolar environment may be especially suitable for generating a CTL response for reasons already described; namely, enhanced macrophage activation and the potential for greater exposure to HIV than occurs in the vascular compartment.

All the responses described above would be expected to result in depletion of $CD4^+$ T cells in the lung (169,170). Recent emphasis has focused on lymphoid organs as the predominant site where HIV propagation and destruction of $CD4^+$ T cells occur. If enhanced AM AC function reflects processes going on in other tissue macrophages, including those in lymphoid organs, then systemic $CD4^+$ T-cell depletion would be facilitated as well through these mechanisms. Indirect support for this hypothesis comes from the observation that an inverse relationship exists between AM AC function and peripheral blood $CD4^+$ T-cell counts or the CD4/CD8 ratio in HIV-infected patients (143).

Besides contributing to the depletion of $CD4^+$ T cells, HIV-infected lung

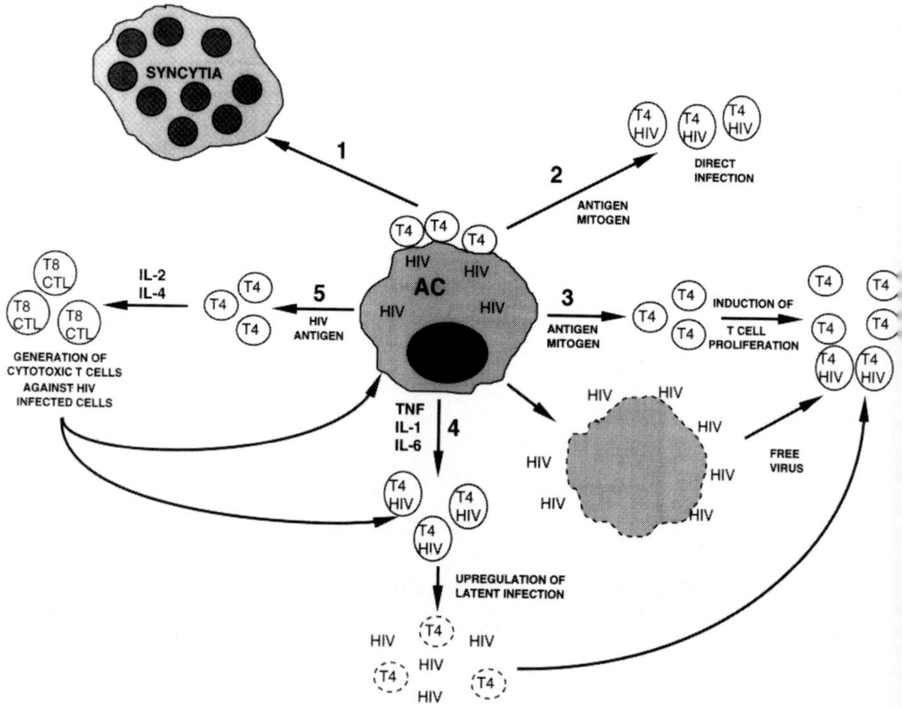

Figure 8 Five potential mechanisms by which enhanced AM AC function may lead to $CD4^+$ T-cell depletion in the lung: (1) formation of syncytia; (2) infection of uninfected T cells through AM–T-cell contact; (3) induction of proliferation in uninfected T cells, thus rendering them more susceptible to infection by free virus; (4) upregulation of viral replication in latently infected T cells; and (5) enhanced ability to facilitate formation of HIV-specific cytotoxic T cells.

macrophages may play a central role in the development of the $CD8^+$ lymphocytic alveolitis so prevalent in these individuals (37,38). Two major mechanisms exist through which T cells may accumulate in the alveolar space: recruitment from the vascular compartment and in situ lymphoproliferation. Recent studies have shown that alveolar $CD8^+$ lymphocytes from HIV-infected subjects proliferate spontaneously and in response to IL-2 but not IL-4 (171). Furthermore, proliferation waned with time unless autologous AMs were added to cultures. Finally, these $CD8^+$ lung lymphocytes secreted IL-2 (171) and IFN-γ (77), a Th1-like cytokine profile consistent with a CTL profile (172). Thus, there is good evidence that in situ proliferation is at least one mechanism accounting for lymphocytic alveolitis in HIV-infected patients. One could hypothesize that lymphocytic alveolitis is

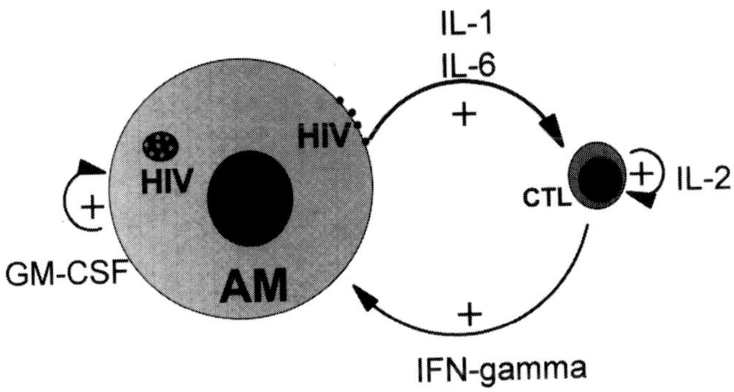

Figure 9 Potential mechanism for the development of lymphocytic alveolitis in HIV infection. Lymphocytic alveolitis could be mediated by chronically infected and activated AMs which have augmented abilities to secrete T-cell–stimulatory cytokines such as IL-1 and IL-6 spontaneously and in response to nonspecific stimuli resulting in enhanced AC function. Responding activated CTLs then secrete IL-2, which may perpetuate T-cell proliferation in the absence of further AC involvement. Furthermore, IFN-γ secretion by CTLs may feed back and further activate AMs. Increased GM-CSF secretion further enhances AM AC function. What perpetuates this response is the apparent inability of HIV-specific CTLs to clear infected cells, especially AMs, from the alveolar space.

mediated by chronically infected and activated AMs which have augmented abilities to secrete T-cell–stimulatory cytokines such as IL-1 and IL-6 spontaneously and in response to nonspecific stimuli (Fig. 9) resulting in enhanced AC function. Responding activated CTLs then secrete IL-2, which may perpetuate T-cell proliferation in the absence of further AC involvement. Furthermore, IFN-γ secretion by CTLs may feed back and further activate AMs. Enhanced GM-CSF further augments AM AC function (157). What perpetuates this response is the apparent inability of HIV-specific CTLs to clear infected cells, especially AMs, from the alveolar space.

In support of this immune hypothesis concerning the development of lymphocytic alveolitis is the recent finding that the incidence of lymphocytic alveolitis is reduced in HIV-infected subjects who smoke (Fig. 10) (173). Smoking has been shown to have immunosuppressive effects on AM function both in decreasing AC function (174) and in impairing cytokine secretion (175–177). In concordance with this data, AMs from HIV-infected patients were found to have a reduced ability to stimulating mitogen-induced lymphoproliferation and secrete IL-1 and IL-6 when compared with AMs from nonsmoking patients (173). It is tempting to speculate that cigarette smoking impairs the generation of cellular immune responses in the lungs of HIV-infected individuals and the subsequent development of lymphocytic alveolitis.

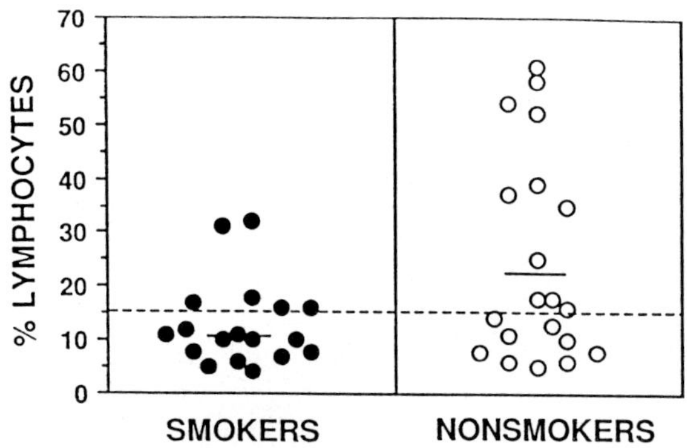

Figure 10 Percentage of lymphocytes in BAL of smoking (dark circles) and nonsmoking (white circles) HIV-positive individuals. BAL from smoking HIV-infected individuals contained a lower percentage of lymphocytes than BAL from nonsmokers (smokers 12.9 ± 1.9%, nonsmokers 24.7 ± 4.3%, $p<0.05$). The incidence of lymphocytic alveolitis, defined as percentage of lymphocytes in BAL ⩾ 15% (dotted line) was significantly lower in HIV-positive smokers (6 of 18 individuals, or 33%) compared with nonsmokers (11 of 20 individuals, or 55%).

IX. Summary

Macrophages and dendritic cells are crucial components of the immune and inflammatory responses which occur in the lung. Alteration in the function of these important cells can have significant detrimental effects on lung function. On the one hand, this may occur through impairment of appropriate responses to pathogens. Alternatively, immune and inflammatory responses may be unnecessarily exuberant, leading to impairment of normal lung functions. Most HIV-infected individuals will have pulmonary complications during the course of their illness. The pathogenesis of many of these complications involves an exuberant immune and/or inflammatory response either to infecting organisms, nonspecific stimuli, or HIV itself. Activated macrophages play a central role in mediating these responses. Macrophages and dendritic cells are natural reservoirs for the virus. Thus, the lower respiratory tract in HIV-infected individuals contains a mixture of activated macrophages, dendritic cells, T cells, stimulatory cytokines, infectious organisms, nonspecific inhaled antigens, and HIV itself (cell-associated or latent). Given this rich medium for immune and inflammatory responses, it is not surprising that pulmonary complications are so common in HIV infection even in the absence of recognized infections. A better understanding of how HIV affects lung macrophages and dendritic cells will allow the development of novel therapeutic strategies directed at the pulmonary complications of HIV infection.

References

1. Gottlieb MS, Schroff R, Schanker HM, Weisman JD, Fan PT, Wolf RA, Saxon A. *Pneumocystis carinii* pneumonia and mucosal candidiasis in previously healthy homosexual men: evidence of a new acquired cellular immunodeficiency. N Engl J Med 1981; 305:1425–1431.
2. Masur H, Michelis MA, Greene JB, Onorato I, Vande Stouwe RA, Holzman RS, Wormser G, Brettman L, Lange M, Murray HW, Cunningham-Rundles S. An outbreak of community-acquired *Pneumocystis carinii* pneumonia: initial manifestation of cellular immune dysfunction. N Engl J Med 1981; 305:1431–1438.
3. Murray JF, Garay SM, Hopewell PC, Mills J, Snider GL, Stover DE. Pulmonary complications of the acquired immunodeficiency syndrome: an update. Am Rev Respir Dis 1987; 135:504–509.
4. Murray JF, Mills J. Pulmonary infectious complications of human immunodeficiency virus infection: Part I. Am Rev Respir Dis 1990; 141:1356–1372.
5. Murray JF, Mills J. Pulmonary infectious complications of human immunodeficiency virus infection: Part II. Am Rev Respir Dis 1990; 141:1582–1598.
6. Agostini C, Trentin L, Zambello R, Semenzato G. HIV-1 and the lung: infectivity, pathogenic mechanisms, and celluar immune responses taking place in the lower respiratory tract. Am Rev Respir Dis 1993; 147:1038–1049.
7. Brain JD. Lung Macrophages: how mancy kinds are there? What do they do? Am Rev Respir Dis 1988; 137:507–509.
8. Gartner S, Markovits P, Markovitz DM, Kaplan MH, Gallo RC, Popovic M. The role of mononuclear phagocytes in HTLV-III/LAV infection. Science 1986; 233:215–219.
9. Chayt KJ, Harper ME, Marselle LM, Lewin EB, Rose RM, Oleske JM, Epstein LG, Wong-Staal F, Gallo RC. Detection of HTLV-III RNA in lungs of patients with AIDS and pulmonary involvement. JAMA 1986; 256:2356–2359.
10. Salahuddin SZ, Rose RM, Groopman JE, Markham PD, Gallo RC. Human T lymphotropic virus type III infection of human alveolar macrophages. Blood 1986, 68:281–284.
11. Rich EA, Chen ISY, Zack JA, Leonard ML, O'Brien WA. Increased susceptibility of differentiated mononuclear phagocytes to productive infection with human immunodeficiency virus–1 (HIV-1). J Clin Invest 1992; 89:176–183.
12. Dean NC, Golden JA, Evans LA, Warnock ML, Addison TE, Hopewell PC, Levy JA. Humna immunodeficiency virus recovery from bronchalveolar fluid in patients with AIDS. Chest 1988; 93:1176–1179.
13. Linnemann CC, Baughman RP, Frame PT, Floyd R. Recovery of human immunodeficiency virus and detection of p24 antigen in bronchoalveolar lavage fluid from adult patients with AIDS. Chest 1989; 96:64–67.
14. Clarke LM, Sierra MF, Shahabbudin M, Cummings RV, Pellegrino MG, Steiner P, Volsky DJ, Nowakowski M. Markers of human immunodeficiency virus infection in pediatric bronchoalveolar lavage samples. Ann Clin Lab Sci 1992; 22:290–299.
15. Plata F, Autran B, Martins LP, Wain-Hobson S, Raphael M, Mayaud C, Denis M, Guillon JM, Debre P. AIDS virus-specific cytotoxic T lymphocytes in lung disorders. Nature 1987; 328:348–351.

16. Clarke JR, Krishnan V, Bennett J, Mitchell D, Jeffries DJ. Detection of HIV-1 in human lung macrophages using the polymerase chain reaction. AIDS 1990; 4:1133–1136.
17. Rose RM, Krivine A, Pinkston P, Gillis JM, Huang A, Hammer SM. Frequent identification of HIV-1 DNA in bronchoalveolar lavage cells obtained from individuals with the acquired immunodeficiency syndrome. Am Rev Respir Dis 1991; 143: 850–854.
18. Lebargy F, Branellec A, Deforges L, Bignon J, Bernaudin JF. HIV-1 in human alveolar macrophages from infected patients is latent in vivo but replicates after in vitro stimulation. Am J Respir Cell Mol Biol 1994; 10:72–78.
19. Biswas P, Poli G, Kinter AL, Justement JS, Stanley SK, Maury WJ, Bressler P, Orenstein JM, Fauci AS. Interferon γ induces the expression of human immunodeficiency virus in persistently infected promonocytic cells (U1) and redirects the production of virions to intracytoplasmic vacuoles in phorbol myristate acetate-differentiated U1 cells. J Exp Med 1992; 176:739–750.
20. Kazazi F, Mathijs JM, Chang J, Malafiej P, Lopez A, Dowton D, Sorrell TC, Vadas MA, Cunningham AL. Recombinant interleukin 4 stimulates human immunodeficiency virus production by infected monocytes and macrophages. J Gen Virol 1992; 73:941–949.
21. Poli G, Bressler P, Kinter A, Duh E, Timmer WC, Rabson A, Justement JS, Stanley S, Fauci AS. Interleukin 6 induces human immunodeficiency virus expression in infected monocytic cells alone and in synergy wiht tumor necrosis factor α by transcriptional and post-transcriptional mechanisms. J Exp Med 1990; 172:151–158.
22. Poli G, Kinter A, Justement JS, Kehrl JH, Bressler P, Stanley S, Fauci AS. Tumor necrosis factor α functions in an autocrine manner in the induction of human immunodeficiency virus expression. Proc Natl Acad Sci USA 1990; 87:782–785.
23. Mellors JW, Griffith BP, Ortiz MA, Landry ML, Ryan JL. Tumor necrosis factor-α/cachectin enhances human immunodeficiency virus type 1 replication in primary macrophages. J Infect Dis 1991; 163:78–82.
24. Kalter DC, Nakamura M, Turpin JA, Baca LM, Hoover DL, Dieffenbach C, Ralph P, Gendelman HE, Meltzer MS. Enhanced HIV replication in macrophage colony-stimulating factor–treated monocytes. J Immunol 1991; 146:298–306.
25. Israel-Biet D, Cadranel J, Even P. Human immunodeficiency virus production by alveolar lymphocytes is increased during pneumocystis carinii pneumonia. Am Rev Respir Dis 1993; 148:1308–1312.
26. Golden MP, Kim S, Hammer SM, Ladd EA, Schaffer PA, DeLuca N, Albrecht MA. Activation of human immunodeficiency virus by herpes simples virus. J Infect Dis 1992; 166:494–499.
27. Jeffrey AA, Israel-Biet D, Andrieu JM, Even P, Venet A. HIV isolation from pulmonary cells derived from bronchoalveolar lavage. Clin Exp Immunol 1991; 85: 488–492.
28. Sierra-Madero JG, Toossi Z, Hom DL, Finegan CK, Hoenig E, Rich EA. Relationship between load of virus in alveolar macrophages from human immunodeficiency virus type 1–infected persons, production of cytokines, and clincal status. J Infect Dis 1994; 169:18–27.
29. Langhoff E, Terwilliger EF, Box HJ, Kalland KH, Poznansky MC, Bacon OML,

Haseltine WA. Replication of human immunodeficiency virus type 1 in primary dendritic cell cultures. Proc Natl Acad Sci USA 1991; 88:7998–8002.

30. Langhoff E, Haseltine WA. Infection of accessory dendritic cells by human immunodeficiency virus type 1. J Invest Dermatol 1992; 99:89S–94S.
31. Spiegel H, Herbst H, Niedobitek G, Foss HD, Stein H. Follicular dendritic cells are a major reservoir for human immunodeficiency virus type 1 in lymphoid tissues facilitating infection of CD4+ T-helper cells. Am J Pathol 1992; 140:15–22.
32. Pantaleo G, Graziosi C, Demarest JF, Butini L, Montroni M, Fox CH, Orenstein JM, Kotler DP, Fauce AS. HIV infection is active and progressive in lymphoid tissue during the clinically latent stage of disease. Nature 1993; 362:355–358.
33. Embretson J, Zupancic M, Ribas JL, Burke A, Racz P, Tenner-Racz K, Hasse AT. Massive covert infection of helper T lymphocytes and macrophages by HIV during the incubation period of AIDS. Nature 1993; 362:359–362.
34. Joling P, Bakker LJ, Van Strijp JAG, Meerloo T, de Graaf L, Dekker MEM, Goudsmit J, Verhoef J, Schuurman HJ. Binding of human immunodeficiency virus type-1 to follicular dendritic cells in vitro is complement dependent. J Immunol 1993; 150:1065–1073.
35. Nicod LP, Lipscomb MF, Weissler JC, Lyons CR, Albertson J, Toews GB. Mononuclear cells in human lung parenchyma: characterization of a potent accessory cell not obtained by bronchoalveolar lavage. Am Rev Respir Dis 1987; 136:818–823.
36. Wallace JM, Barbers RG, Oishi JS, Prince H. Celluar and T-lymphocyte subpopulation profiles in bronchoalveolar lavage fluid from patients with acquired immunodeficiency syndrome and pneumonitis. Am Rev Respir Dis 1984; 130.786–790.
37. Guillon JM, Autran B, Denis M, Fouret P, Plata F, Mayaud CM, Akoun GM. Human immunodeficiency virus–related lymphocytic alveolitis. Chest 1988; 94:1264–1270.
38. Agostini C, Poletti V, Zambello R, Trentin L, Siviero F, Spiga L, Gritti F, Semenzato G. Phenotypical and functional analysis of bronchoalveolar lavage lymphocytes in patients with HIV infection. Am Rev Respir Dis 1988; 138:1609–1615.
39. Travis WD, Fox CH, Devaney KO, Weiss LM, O'Leary TJ, Ognibene FP, Suffredini AF, Rosen MJ, Cohen MB, Shelhamer J. Lymphoid pneumonitis in 50 adult patients infected with human immunodefiiency virus: lymphocytic interstitial pneumonitis versus nonspecific interstitial pneumonitis. Hum Pathol 1992; 23:529–541.
40. Morris JC, Rosen MJ, Marchevsky A, Teirstein AS. Lymphocytic interstitial pneumonia in patients at risk for the acquired immune deficiency syndrome. Chest 1987; 91:63–67.
41. Schnittman SM, Lane HC, Greenhouse J, Justement JS, Baseler M, Fauci AS. Preferential infection of $CD4^+$ memory T cells by human immunodeficiency virus type 1: Evidence for a role in the selective T-cell functional defects observed in infected individuals. Proc Natl Acad Sci USA 1990; 87:6058–6062.
42. van Noesel CJM, Gruters RA, Terpstra FG, Schellekens PTA, van Lier RAW, Miedema F. Functional and phenotypic evidence for a selective loss of memory T cells in asymptomatic human immunodeficiency virus-infected men. J Clin Invest 1990; 86:293–299.
43. Sanders ME, Makgoba MW, Sharrow SO, Stephany D, Springer TA, Young HA, Shaw S. Human memory T lymphocytes express increased levels of three cell

adhesion molecules (LFA-3, CD2, and LFA-1) and three other molecules (UCHL1, CDw29, and Pgp-1) and have enhanced IFN-γ production. J Immunol 1988; 140: 1401–1407.

44. Agostini C, Zambello R, Trentin L, Feruglio C, Masciarelli M, Siviero F, Poletti V, Spiga L, Gritti F, Semenzato G. Cytotoxic events taking place in the lung of patients with HIV-1 infection. Am Rev Respir Dis 1990; 142:516–522.
45. Jassoy C, Harrer T, Rosenthal T, Navia BA, Worth J, Johnson RP, Walker BD. Human immunodeficiency virus type-1-specific cytotoxic T lymphocytes release gamma interferon, tumor necrosis factor alpha (TNF-α) and TNF-B when they encounter their target antigens. J Virol 1993; 67:2844–2852.
46. Twigg HL. Bronchoalveolar lavage fluid in HIV-infected patients: "cytokine soup." Chest 1993; 104:659–661.
47. Niederman TMJ, Garcia JV, Hastings WR, Luria S, Ratner L. Human immunodeficiency virus type 1 nef protein inhibits NF-κB induction in human T cells. J Virol 1992; 66:6213–6219.
48. Chirmule N, Kalyanaraman VS, Oyaizu N, Slade HB, Pahwa. Inhibition of functional properties of tetanus antigen-specific T-cell clones by envelope glycoprotein gp120 of human immunodeficiency virus. Blood 1990; 75:152–159.
49. Cefai D, Ferrer M, Serpente N, Idziorek T, Dautry-Varsat A, Debre P, Bismuth G. Internalization of HIV glycoprotein gp120 is associated with down-modulation of membrane CD4 and $p56^{ick}$ together with impairment of T cell activation. J Immunol 1992; 149:285–294.
50. Wagner RP, Levitz SM, Tabuni A, Kornfeld H. HIV-1 envelope protein (gp120) inhibits the activity of human bronchoalveolar macrophages against *Cryptococcus neoformans*. Am Rev Respir Dis 1992; 146:1434–1438.
51. Nong Y, Kandil O, Tobin EH, Rose RM, Remold HG. The HIV core protein p24 inhibits interferon-γ–induced increase of HLA-DR and cytochrome b heavy chain mRNA levels in the human monocyte-like cell line THP1. Cell Immunol 1991; 132: 10–16.
52. Viscidi RP, Mayur K, Lederman HM, Frankel AD. Inhibition of antigen-induced lymphocyte proliferation by tat protein from HIV-1. Science 1989; 246:1606–1608.
53. Zauli G, Davis BR, Re MC, Visani G, Furlini G, La Placa M. tat protein stimulates production of transforming growth factor-B1 by marrow macrophages: a potential mechanism for human immunodeficiency virus-1-induced hematopoietic suppression. Blood 1992; 80:3036–3043.
54. Iwamoto GK, Konicek SA, Twigg HL. Modulation of accessory cell function and interleukin-6 production by the HIV-1 tat gene. Am J Respir Cell Mol Biol 1994; 10:580–585.
55. Agostini C, Zambello R, Trentin L, Poletti V, Spiga L, Gritti F, Cipriani A, Salmaso L, Cadrobbi P, Senenzato G. Prognostic significance of the evaluation of broncho-alveolar lavage cell populations in patients with HIV-1 infection and pulmonary involvement. Chest 1991; 100:1601–1606.
56. Agostini C, Trentin L, Zambello R, Bulian P, Caenazzo C, Cipriani A, Cadrobbi P, Garbisa S, Semenzato G. Release of granulocyte-macrophage colony-stimulating factor by alveolar macrophages in the lung of HIV-1-infected patients. J Immunol 1992; 149:3379–3385.

57. Wasserman K, Subklewe M, Pothoff G, Banik N, Schell-Frederick E. Expression of surface markers on alveolar macrophages from symptomatic patients with HIV infection as detected by flow cytometry. Chest 1994; 105:1324–1334.
58. Roulston A, D'Addario M, Boulerice F, Caplan S, Wainberg MA, Hiscott J. Induction of monocytic differentiation and Nf-κB activities by human immunoodeficiency virus 1 infection of myelomonoblastic cells. J Exp Med 1992; 175:751–763.
59. Musher DM, Watson DA, Nickeson D, Gyorkey F, Lahart C, Rossen RD. The effect of HIV infection on phagocytosis and killing of staphylococcus aureus by human pulmonary alveolar macrophages. Am J Med Sci 1990; 299:158–163.
60. Meltzer MS, Skillman DR, Gomatos PJ, Kalter DC, Gendelman HE. Role of mononuclear phagocytes in the pathogenisis of human immunodeficiency virus infection. Annu Rev Immunol 1990; 8:169–194.
61. Mann DL, Gartner S, LeSane F, Blattner WA, Popovic M. Cell sufrace antigens and function of monocytes and a monocyte-like cell line before and after infection with HIV. Clin Immunol Immunopathol 1990; 54:174–183.
62. Koethe SM, Carrigan DR, Turner PA. Increased density of HLA-DR antigen on monocytes of patients infected with the human immunodeficiency virus. J Med Virol 1989; 29:82–87.
63. Agostini C, Trentin L, Zambello R, Bulian P, Garbisa S, Cipriani A, Cadrobbi P, Semenzato G. Alveolar macrophages in HIV-1 infection express accessory molecules, activation markers, and release increased biological response modifiers. Chest 1993; 103(suppl 2):108S–111S.
64. Buhl R, Jaffe HA, Holroyd KJ, Borok Z, Roum JH, Mastrangeli A, Wells FB, Kirby M, Saltini C, Crystal RG. Activation of alveolar macrophages in asymptomatic HIV-infected individuals. J Immunol 1993; 150:1019–1028.
65. Petit AJC, Terpstra FG, Miedema F. Human immunodeficiency virus infection down-regulates HLA class II expression and induces differentiation in promonocytic U937 cells. J Clin Invest 1987; 79:1883–1889.
66. Rossen RD, Smith CW, Laughter AH, Noonan CA, Anderson DC, McShan WM Hurvitz MY, Orson FM. HIV-1 stimulated expression of CD11/CD18 integrins and ICAM-1: a possible mechanism for extravascular dissemination of HIV-1-infected cells. Trans Assoc Am Physicians 1989; 102:117–130.
67. Auci DL, Chice SM, Durkin HG, Murali MR. Constitutive productioon of PAI-II and increased surface expression of GM_1 ganglioside by peripheral blood monocytes from patients with AIDS: evidence of monocyte activation in vivo. J Leukoc Biol 1992; 52:282–286.
68. Capsoni F, Minonzio F, Ongari AM, Colombo G, Rizzardi GP, Bonara P, D'Arminio-Monforte A, Zanussi C. Increased expression of IgG Fc receptor type I on neutrophils and monocytes from HIV-infected subjects. Clin Exp Immunol 1992; 90: 175–180.
69. Allen JB, Wong HL, Guyre PM, Simon GL, Wahl SM. Association of circulating receptor FcγRIII-positive monocytes in AIDS patients with elevated levels of transforming growth factor-B. J Clin Invest 1991; 87:1773–1779.
70. Murray HW, Rubin BY, Masur H, Roberts RB. Impaired prdouction of lymphokines and immune (gamma) interferon in the acquired immunodeficiency syndrome. N Engl J Med 1984; 310:883–889.

71. Murray HW, Hillman JK, Rubin BY, Kelly CK, Jacobs JL, Tyler LW, Donelly DM, Carriero SM, Godbold JH, Roberts RB. Patients at risk for AIDS-related opportunistic infections: clinical manifestations and immpaired gamma interferon production. N Engl J Med 1985; 313:1504–1510.
72. Maggi E, Macchia D, Parronchi P, Mazzetti M, Ravina A, Milo D, Romagnani S. Reduced production of interleukin 2 and interferon-gamma and enhanced helper activity for IgG synthesis by cloned $CD4^+$ T cells from patients with AIDS. Eur J Immunol 1987; 17:1685–1690.
73. Rinaldo C, Piazza P, Wang Y, Armstrong J, Gupta P, Ho M, Petteway S, Reed D, Lyter D, Kingsley L. HIV-1-specific production of IFN-γ and modulation by recombinant IL-2 during early HIV-1 infection. J Immunol 1988; 140:3389–3393.
74. Fan J, Bass HZ, Fahey JL. Elevated IFN-γ and decreased IL-2 gene expression are associated with HIV infection. J Immunol 1993; 151:5031–5040.
75. McElrath MJ, Steinman RM, Cohn ZA. Latent HIV-1 infection in enriched populations of blood monocytes and T cells from seropositive patients. J Clin Invest 1991; 87:27–30.
76. Emilie D, Peuchmaur M, Maillot MC, Crevon MC, Brousse N, Delfraissy JF, Dormont J, Galanaud P. Production of interleukins in human immunodeficiency virus-1–replicating lymph nodes. J Clin Invest 1990; 86:148–159.
77. Spain BA, Soliman DM, Konicek SA, Iwamoto GK, Twigg HL. Interferon-gamma production is upregulated in lung lymphocytes from asymptomatic HIV-infected individuals with lymphocytic alveolitis (abstr). Clin Res 1994; 42:425A.
78. Joly P, Guillon JM, Mayaud C, Plata F, Theodorou I, Denis M, Debre P, Autran B. Cell-mediated suppression of HIV-specific cytotoxic T lymphocytes. J Immunol 1989; 143:2193–2201.
79. Geleziunas R, Bour S, Boulerice F, Hiscott J, Wainberg MA. Diminution of CD4 surface protein but not CD4 messenger RNA levels in monocytic cells infected by HIV-1. AIDS 1991; 5:29–33.
80. Koziel H, Kruskal BA, Ezekowitz RAB, Rose RM. HIV impairs alveolar macrophage mannose receptor fucntion against pneumocystis carinii. Chest 1993; 103 (suppl 2):111S–112S.
81. Smith PD, Ohura K, Masur H, Lane HC, Fauci AS, Wahl SM. Monocyte function in the acquired immune deficiency syndrome: defective chemotaxis. J Clin Invest 1984; 74:2121–2128.
82. Dhawan S, Vargo M, Meltzer MS. Interactions between HIV-infected monocytes and the extracellular matrix: increased capacity of HIV-infected monocytes to adhere to and spread on extracellular matrix associated with changes in extent of virus replication and cytopathic effects in infected cells. J Leukoc Biol 1992; 52:62–69.
83. Wright JR, Youmas DC. Pulmonary surfactant protein A stimulates chemotaxis of alveolar macrophages. Am J Physiol 1993; 264:L338–L344.
84. Phelps DS, Rose RM. Increased recovery of surfactant protein A in AIDS-related pneumonia. Am Rev Respir Dis 1991; 143:1072–1075.
85. Downing JF, Pasula R, Wright JR, Twigg HL, Martin II WJ. Surfactant protein A promotes attachment of *Mycobacterium tuberculosis* to alveolar macrophages during infection with human immunodeficiency virus. Proc Natl Acad Sci USA 1995; 92:4848–4852.

86. Tenner AJ, Robinson SL, Borchelt J, Wright JR. Human pulmonary surfactant protein (SP-A), a protein structurally homologous to C1q, can enhance FcR- and CR1-mediated phagocytosis. J Biol Chem 1989; 264:13923–13928.
87. Zimmerman PE, Voelker DR, McCormack FX, Paulsrud JR, Martin II WJ. 120-kd surface glycoprotein of *Pneumocystis carinii* is a ligand for surfactant protein A. J Clin Invest 1992; 89:143–149.
88. Ezekowitz RAB, Williams DJ, Koziel H, Armstrong MYK, Warner A, Richards FF, Rose RM. Uptake of pneumocystis carinii mediated by the macrophage mannose receptor. Nature 1991; 351:155–158.
89. Szkaradkiewicz A. Phagocytosis and microbicidal capacity of human monocytes in the course of HIV infection. Immunol Lett 1992; 33:145–150.
90. Capsoni F, Minonzio F, Ongari AM, Rizzardi GP, Lazzarin A, Zanussi C. Monocyte-derived macrophage function in HIV-infected subjects: in vitro modulation by rIFN-γ and rGM-CSF. Clin Immunol Immunopathol 1992; 62:176–182.
91. Cameron ML, Granger DL, Matthews TJ, Weinberg JB. Human immunodeficiency virus (HIV)–infected human blood monocytes and peritoneal macrophages have reduced anticryptococcal activity wheras HIV-infected alveolar macrophages retain normal activity. J Infect Dis 1994; 170:60–67.
92. Murray HW. Macrophage activation in the acquired immunodeficiency syndrome. In: Steinman RM, North RJ, eds. Mechanisms of Host Resistance to Infectious Agents, Tumors and Allografts. New York: Rockefeller University Press, 1986: 321–332.
93. Muller F, Rollag H, Froland SS. Reduced oxidative burst responses in monocytes and monocyte-derived macrophages from HIV-infected subjects. Clin Exp Immunol 1990; 82:10–15.
94. Spear GT, Kessler HA, Rothberg L, Phair J, Landay AL. Decreased oxidative burst activity of monocytes from asymptomatic HIV-infected individuals. Clin Immunol Immunopathol 1990; 54:184–191.
95. Roux-Lombard P, Aladjem D, Balavoine JF, Chofflon M, Despont JP, Hirschel B, Jeannet M, Kapanci Y, Lang R, Toccanier MF, Voinier B, Wilhelm A, Dayer JM, Cruchaud A. Altered functions of peripheral blood monocytes in homosexual males and intravenous drug users with persistent generalized lymphadenopathy. Eur J Clin Invest 1986; 16:262–270.
96. Dukes CS, Matthews TJ, Weinberg JB. Human immunodeficiency virus type 1 infection of human monocytes and macrohages does not alter their ability to generate an oxidative burst. J Infect Dis 1993; 168:459–462.
97. Murray HW, Gellene RA, Libby DM, Rothermel CD, Rubin BY. Acitvation of tissue macrophages from AIDS patients: in vitro response of AIDS alveolar macrophages to lymphokines and interferon-γ. J Immunol 1985; 135:2374–2377.
98. Pittis MG, Sternik G, Sen L, Diez RA, Planes N, Pirola D, Estevez ME. Impaired phagolysomal fusion of peripheral blood monocytes from HIV-infected subjects. Scand J Immunol 1993; 38:423–427.
99. Nathan CF. Secretory products of macrophages. J Clin Invest 1987; 79:319–326.
100. Foley P, Kazazi F, Biti R, Sorrell TC, Cunningham AL. HIV infection of monocytes inhibits the T-lymphocyte proliferative reponse to recall atnigens, via production of eicosanoids. Immunology 1992; 75:391–397.

101. Balter MS, Toews GB, Peters-Golden M. Different patterns of arachidonate metabolism in autologous human blood monocytes and alveolar macrophages. J Immunol 1989; 142:602–608.
102. Lipschik GY, Doerfler ME, Kovacs JA, Travis WD, Andrawis VA, Lawrence MG, Dichter JR, Ognibene FP, Shelhamer JH. Leukotriene B_4 and interleukin-8 in human immunodeficiency virus-related pulmonary disease. Chest 1993; 104:763–769.
103. Twigg HL, Iwamoto GK, Soliman DM. Role of cytokines in alveolar macrophage accessory cell function in HIV-infected individuals. J Immunol 1992; 149:1462–1469.
104. Trentin L, Garbisa S, Zambello R, Agostini C, Caenazzo C, Di Francesco C, Cipriani A, Francavilla E, Semenzato G. Spontaneous production of IL-6 by alveolar macrophages of HIV-1 seropositive patients. J Infect Dis 1992; 166:731–737.
105. Israel-Biet D, Cadranel J, Beldjord K, Andrieu JM, Jeffrey A, Even P. Tumor necrosis factor production in HIV-seropositive subjects: relationship with lung opportunistic infections and HIV expression in alveolar macrophages. J Immunol 1991; 147:490–494.
106. Agostini C, Zambello R, Trentin L, Garbisa S, Francia Di Celle P, Bulian P, Onisto M, Poletti V, Spiga L, Raise E, Foa R, Semenzato G. Alveolar macrophages from patients with AIDS and AIDS-related complex constitutively synthesize and release tumor necrosis factor alpha. Am Rev Respir Dis 1991; 144:195–201.
107. Cox RA, Anders GT, Cappelli PJ, Johnson JE, Blanton HM, Seaworth BJ, Treasure RL. Production of tumor necrosis factor-alpha and interleukin-1 by alveolar macrophages from HIV-1–infected persons. AIDS Res Hum Retroviruses 1990; 6:431–441.
108. Millar AB, Miller RF, Foley NM, Meager A, Semple SJG, Rook GAW. Production of tumor necrosis factor-α by blood and lung mononuclear phagocytes from patients with human immunodeficiency virus-related lung disease. Am J Respir Cell Mol Biol 1991; 5:144–148.
109. Merrill JE, Koyanagi Y, Chen ISY. Interleukin-1 and tumor necrosis factor α can be induced from mononuclear phagocytes by human immunodeficiency virus type 1 binding to the CD4 receptor. J Virol 1989; 63:4404–4408.
110. Oyaizu N, Chirmule N, Ohnishi Y, Kalyanaraman VS, Pahwa S. Human immunodeficiency virus type 1 envelope glycoproteins gp120 and gp160 induce interleukin-6 production in $CD4^+$ T-cell clones. J Virol 1991; 65:6277–6282.
111. Clouse KA, Cosentino LM, Wehl KA, Pyle SW, Robbins PB, Hochstein HD, Natarajan V, Farrar WL. The HIV-1 gp120 envelope protein has the intrinsic capacity to stimulate monokine secretion. J Immunol 1991; 147:2892–2901.
112. Iwamoto GK, Monick MM, Clark BD, Auron PE, Stinski MF, Hunninghake GW. Modulation of interleukin 1 beta gene expression by the immediate early genes of human cytomegalovirus. J Clin Invest 1990; 85:1853–1857.
113. Yamato K, El-Hajjaoui Z, Simon K, Koeffler HP. Modulation of interleukin-1β RNA in monocytic cells infected with human immunodeficiency virus-1. J Clin Invest 1990; 86:1109–1114.
114. Pocsik E, Higuchi M, Aggarwal BB. Down-modulation of cell surface expression of p80 form of the tumor necrosis factor receptor by human immunodeficiency virus-1 *tat* gene. Lymphokine Cytokine Res 1992; 11:317–325.
115. Agostini C, Zambello R, Trentin L, Cerutti A, Enthammer C, Facco M, Milani A,

Sancetta R, Garbisa S, Semenzato G. Expression of TNF receptors by T cells and membrane TNF-α by alveolar macrophages suggests a role for TNF-α in the regulation of the local immune responses in the lung of HIV-1-infected patients. J Immunol 1995; 154:2928–2938.
116. Honda M, Yamamoto S, Cheng M, Yasukawa K, Suzuki H, Saito T, Osugi Y, Tokunaga T, Kishimoto T. Human soluble IL-6 receptor: its detection and enhanced release by HIV infection. J Immunol 1992; 148:2175–2180.
117. Taga T, Hibi M, Hirata Y, Yamasaki K, Yasukawa K, Matsuda T, Hirano T, Kishimoto T. Interleukin-6 triggers the association of its receptor with a possible signal transducer, gp130. Cell 1989; 58:573–581.
118. Matsuyama T, Kobayashi N, Yamamoto N. Cytokines and HIV infection: is AIDS a tumor necrosis factor disease? AIDS 1991; 5:1405–1417.
119. Merrill JE. Cytokines and retroviruses. Clin Immunol Immunopathol 1992; 64: 23–27.
120. Rosenberg ZF, Fauci AS. Immunopathogenic mechainsms of HIV infection: cytokine induction of HIV expression. Immunol Today 1990; 11:176–180.
121. Butera ST. Cytokine involvement in viral permissiveness and the progression of HIV disease. J Cell Biochem 1993; 53:336–342.
122. Israel N, Hazan U, Alcami J, Munier A, Arenzana-Seisdedos F, Bachelerie F, Israel A, Virelizier JL. Tumor necrosis factor stimulates transcription of HIV-1 in human T lymphocytes, independently and synergistically with mitogens. J Immunol 1989; 143:3956–3960.
123. Clouse KA, Powell D, Washington I, Poli G, Strebel K, Farrar W, Barstad P, Kovacs J, Fauci AS, Folks TM. Monokine regulation of human immunodeficiency virus-1 expression in a chronically infected human T cell clone. J Immunol 1989; 142: 431–438.
124. Kinter AL, Poli G, Fox L, Hardy E, Fauci AS. HIV replication in IL-2-stimulated peripheral blood mononuclear cells is driven in an autocrine/paracrine manner by endogenous cytokines. J Immunol 1995; 154:2448–2459.
125. Brinchmann JE, Gaudernack G, Thorsby E, Vartdal F. Staphylococcal exotoxin superantigens induce human immunodeficiency virus type 1 expression in anturally infected $CD4^+$ T cells. J Virol 1992; 66:5924–5928.
126. Clouse KA, Robbins PB, Fernie B, Ostrove JM, Fauci AS. Viral antigen stimulation of the production of human monokines capable of regulating HIV1 expression. J Immunol 1989; 143:470–475.
127. Berman MA, Sandborg CI, Calabia BS, Andrews BS, Friou. Interleukin 1 inhibitor masks high interleukin 1 production in acquired immunodeficiency syndrome. Clin Immunol Immunopathol 1987; 42:133–140.
128. Ammar A, Sahraoui Y, Tsapis A, Bertoli AM, Jasmin C, Georgoulias V. Human immunodeficiency virus–infected adherent cell–derived inhibitory factor (p29) inhibits normal T cell proliferation through decreased expression of high affinity interleukin-2 receptors and production of interleukin-2. J Clin Invest 1992; 90:8–14.
129. Denis M, Ghadirian E. Dysregulation of interleukin 8, interleukin 10, and interleukin 12 release by alveolar macrophages from HIV type 1-infected subjects. AIDS Res Hum Retroviruses 1994; 10:1619–1627.
130. Janoff EN, Breiman RF, Daley CL, Hopewell PC. Pneumococcal disease during HIV

infection: epidemiologic, clinical, and immunologic perspectives. Ann Intern Med 1992; 117:314–324.

131. Mascart-Lemone F, Gerard M, Libin M, Crusiaux A, Franchioly P, Lambrechts A, Goldman M, Clumeck N. Differential effect of human immunodeficiency virus infection on the IgA and IgG antibody responses to pneumococcal vaccine. J Infect Dis 1995; 172:1253–1260.
132. Vincent C, Cozon G, Zittoun M, Mellquist M, Kazatchkine MD, Czerkinsky C, Revillare JP. Secretory immunoglobulins in serum from human immunodeficiency virus (HIV)-infected patients. J Clin Immunol 1992; 12:381–388.
133. Jelinek DF, Lipsky PE. Ther roles of T cell factors in activation, cell cycle progression, and differentiation of human B cells. J Immunol 1985; 134:1690–1696.
134. Wilkes DS, Weissler JC. Alloantigen-induced immunoglobulin production in human lung: differential effects of accessory cell populations on IgG synthesis. Am J Respir Cell Mol Biol 1994; 10:339–346.
135. Sonoda E, Matsumoto R, Hitoshi Y, Ishii T, Sugimoto M, Araki S, Tominaga A, Yamaguchi N, Takatsu K. Transforming growth factor B induces IgA production and acts additively with interleukin 5 for IgA production. J Exp Med 1989; 170:1415–1420.
136. Twigg HL, Spain BA, Soliman DM, Bowen LK, Heidler KM, Wilkes DS. Impaired IgG production in the lungs of HIV-infected individuals. Cell Immunol 1996; 170: 127–133.
137. Kekow J, Wachsman W, McCutchan JA, Gross WL, Zachariah M, Carson DA, Lotz M. Transforming growth factor-Band suppression of humoral immune responses in HIV infection. J Clin Invest 1991; 87:1010–1016.
138. Young KR, Rankin JA, Naegel GP, Paul ES, Reynolds HY. Bronchoalveolar lavage cells and proteins in patients with the acquired immunodeficiency syndrome. Ann Intern Med 1985; 103:522–533.
139. Lyons CR, Ball EJ, Toews GB, Weissler JC, Stastny P, Lipscomb MF. Inability of human alveolar macrophages to stimulate resting T cells correlates with decreased antigen-specific T cell-macrophage binding. J Immunol 1986; 137:1173–1180.
140. Lipscomb MF, Lyons CR, Nunez G, Ball EJ, Stastny P, Vial W, Lem V, Weissler J, Miller LM, Toews GB. Human alveolar macrophages: HLA-DR-positive macrophages that are poor stimulators of a primary mixed leukocyte reaction. J Immunol 1986; 136:497–504.
141. Toews GB, Vial WC, Dunn MM, Guzzetta P, Nunez G, Stastny P, Lipscomb MF. The accessory cell function of human alveolar macrophages in specific T cell proliferation. J Immunol 1984; 132:181–186.
142. Kradin RL, McCarthy KM, Gifford J, Schneeberger EE. Antigen-independent binding of T-cells by dendritic cells and alveolar macrophages in the rat. Am Rev Respir Dis 1989; 139:207–211.
143. Twigg HL, Lipscomb MF, Yoffe B, Barbaro DJ, Weissler JC. Enhanced accessroy cell function by alveolar macrophages from patients infected with the human immunodeficiency virus: potential role for depletion of $CD4^+$ cells in the lung. Am J Respir Cell Mol Biol 1989; 1:391–400.
144. Lem VM, Lipscomb MF, Weissler JC, Nunez G, Ball EJ, Stastny P, Toews GB. Bronchoalveolar cells from sarcoid patients demonstrate enhanced antigen presentation. J Immunol 1985; 135:1766–1771.

145. Chatila TA, Schwartz DH, Miller R, Geha RS. Requirement for mitogen, T cell–accessory cell contact, and interleukin 1 in the induction of resting T-cell proliferation. Clin Immunol Immunopathol 1987; 44:235–247.
146. Kawakami K, Yamamoto Y, Kakimoto K, Onoue K. Requirement for delivery of signals by physical interaction and soluble factors from accessory cells in the induction of receptor-mediated T cell proliferation. J Immunol 1989; 142:1818-1825.
147. Ruppert J, Peters JH. IL-6 and IL-1 enhance the accessory activity of human blood monocytes during differentiation to macrophages. J Immunol 1991; 146:144–149.
148. Twigg HL, Soliman DM. Role of alveolar macrophage-T cell contact in accessory cell function in HIV-infected individuals. Am J Respir Cell Mol Biol 1994; 11: 138–146.
149. Hynes RO. Integrins: a family of cell surface receptors. Cell 1987; 48:549–554.
150. Arnaout MA. tructure and function of the leukocyte adhesion molecules CD11/CD18. Blood 1990; 75:1037–1050.
151. Dougherty GJ, Hogg N. The role of monocyte lymphocyte function-associated antigen 1 (LFA-1) in accessory cell function. Eur J Immunol 1987; 17:943–947.
152. Haverstick DM, Gray LS. Lymphocyte adhesion mediated by lymphocyte function–associated antigen-1. I. Long term augmentation by transient increases in intracellular cAMP. J Immunol 1992; 149:389–396.
153. Damle NK, Klussman K, Linsley P, Aruffo A. Differential costimulatory effects of adhesion molecules B7, ICAM-1, LFA-3, and VCAM-1 on resting and antigen-primed CD4+ T lymphocytes. J Immunol 1992; 148:1985–1992.
154. Van Seventer GA, Shimizu Y, Horgan KJ, Shaw S. The LFA-1 ligand ICAM-1 provides an important costimulatory signal for T cell receptor–mediated activation of resting T cells. J Immunol 1990; 144:4579–4586.
155. Yarbrough WC, Wilkes DS, Weissler JC. Human alveolar macrophages inhibit receptor-mediated increases in intracellular calcium concentration in lymphocytes Am J Respir Cell Mol Biol 1991; 5:411–415.
156. Weissler JC, Mendelson C, Moya F, Yarbrough WC. Effect of interstitial lung disease macrophages on T-cell signal transduction. Am J Respir Crit Care Med 1994; 149:191–196.
157. Bilyk N, Holt PG. Inhibition of the immunosuppressive activity of resident pulmonary alveolar macrophages by granulocyte/macrophage colony-stimulating factor. J Exp Med 1993; 177:1773–1777.
158. Metzger Z, Hoffeld JT, Oppenheim JJ. Macrophage-mediated suppression. I. evidence for participation of both hydrogen peroxide and prostaglandins in suppression of murine lymphocyte proliferation. J Immunol 1980; 124:983–988.
159. Denham S, Rowland IJ. Inhibition of the reactive prolifeartion of lymphocytes by activated macrophages: the role of nitric oxide. Clin Exp Immunol 1992; 87:157–162.
160. Cameron PU, Forsum U, Teppler H, Granelli-Piperno A, Steinman RM. During HIV-1 infection most blood dendritic cells are not productively infected and can induce allogeneic CD4+ T cells clonal expansion. Clin Exp Immunol 1992; 88: 226–236.
161. Macatonia SE, Gompels M, Pinching AJ, Patterson S, Knight SC. Antigen presentation by macrophages but not by dendritic cells in human immunodeficiency virus (HIV) infection. Immunology 1992; 75:576–581.

162. Cameron PU, Freudenthal PS, Barker JM, Gezelter S, Inaba K, Steinman RM. Dendritic cells exposed to human immunodeficiency virus type-1 transmit a vigorous cytopathic infection to CD4+ T cells. Science (1992; 257:383–387.
163. Valentin A, Lundin K, Patarroyo M, Asjo B. The leukocyte adhesion glycoprotein CD18 participates in HIV-1–induced syncytia formation in monocytoid and T cells. J Immunol 1990; 144:934–937.
164. Pantaleo G, Butini L, Graziosi C, Poli G, Schnittman SM, Greenhouse JJ, Gallin JI, Fauci AS. Human immunodeficiency virus (HIV) infection in CD4+ T lymphocytes genetically deficient in LFA-1: LFA-1 is required for HIV-mediated cell fusion but not for viral transmission. J Exp Med 1991; 173:511–514.
165. Mann DL, Gartner S, LeSane F, Buchow H, Popovic M. HIV-1 transmission and function of virus-infected monocytes/macrophages. J Immunol 1990; 144:2152–2158.
166. Lewis DE, Yoffe B, Bosworth CG, Hollinger FB, Rich RR. Human immunodeficiency virus-induced pathology favored by cellular transmission and activation. FASEB J 1988; 2:251–255.
167. Lyerly HK, Cohen OJ, Weinhold KJ. Transmission of HIV by antigen presenting cells during T-cell activation: prevention by 3′-azido-3′-deoxythymidine.l AIDS Res Hum Retroviruses 1987; 3:87–94.
168. Folks T, Kelly J, Benn S, Kinter A, Justement J, Gold J, Redfield R, Sell KW, Fauci AS. Susceptibility of normal human lymphocytes to infection with HTLV-III/LAV. J Immunol 1986; 136:4049–4053.
169. Pantaleo G, Graziosi C, Fauci AS. The immunopathogenesis of human immunodeficiency virus infection. N Engl J Med 1993; 328:327–335.
170. Fox CH, Cottler-Fox M.The pathobiology of HIV infection.Immunol Today 1992; 13:353–356.
171. Spain BA, Soliman DM, Sidner RA, Twigg HL. Enhanced proliferation and IL-2 secretion by lung lymphocytes from HIV-infected subjects. Am J Physiol 1995; 269: L498–L506.
172. Fong TAT, Mosmann TR. Alloreactive murine CD8+ T cell clones secrete the Th1 pattern of cytokines. J Immunol 1990; 144:1744–1752.
173. Twigg HL, Soliman DM, Spain BA. Impaired alveolar macrophage accessory cell function and reduced incidence of lymphocytic alveolitis in HIV-infected patients who smoke. AIDS 1994; 8:611–618.
174. Laughter AH, Martin RR, Twomey JJ. Lymphoproliferative responses to antigens mediated by human pulmonary alveolar macrophages. J Lab Clin Med 1977; 89: 1326–1332.
175. Brown GP, Iwamoto GK, Monick MM, Hunninghake GW. Cigarette smoking decreases interleukin 1 release by human alveolar macrophages. Am J Physiol 1989; 256:C260–C264.
176. Soliman DM, Twigg HL. Cigarette smoking decreases bioactive interleukin-6 secretion by alveolar macrophages. Am J Phyisol 1992; 263:L471–L478.
177. McCrea KA, Ensor JE, Nall K, Bleecker ER, Hasday JD. Altered cytokine regulation in the lungs of cigarette smokers. Am J Respir Crit Care Med 1994; 150:696–703.

Part Five

MACROPHAGES AND DENDRITIC CELLS IN NONINFECTIOUS LUNG DISEASES

22

Macrophages and Dendritic Cells in Allergic Asthma

MARY F. LIPSCOMB and C. RICHARD LYONS

University of New Mexico School of Medicine
Albuquerque, New Mexico

I. Introduction

Preceding chapters have documented the important roles of lung macrophages and dendritic cells (DCs) in regulating immunity and inflammation in the lung, and, therefore, one would suspect that these cells play important roles in the initiation of immune responses that lead to asthma. This chapter defines asthma and outlines current concepts of the immunological mechanisms in allergic asthma; discusses immunoregulatory determinants of protective versus allergic responses; describes results from animal models that have ascribed roles to lung macrophages and DCs in initiating allergic lung disease; and finally lists and discusses the most likely lung macrophage products that mediate the acute and late phase asthmatic responses.

II. Asthma: An Immunologically Mediated Inflammatory Disease

Asthma is a heterogeneous disease: and as such is both difficult to define and unequivocally diagnose. The following characteristics are widely accepted as being essential to asthma: (1) reversible, obstructive airway disease; (2) bronchial and bronchiolar inflammation; and (3) airway hyperresponsiveness (AHR), with the latter being defined as an increased sensitivity to bronchoconstrictive agents (1–3). Bronchial and bronchiolar inflammation demonstrates increased numbers of mast cells, basophils, eosinophils, lymphocytes, and plasma cells in the airway wall and is accompanied by fibrosis (including subepithelial fibrosis previously described as increased basement membrane thickening) and smooth muscle cell hyperplasia (1,4). During acute asthmatic episodes, bronchoconstriction, mucous hypersecretion, congestion and edema, and a further increase in leukocyte infiltrates accompanied by evidence of eosinophil and mast cell degranulation and epithelial cell necrosis are superimposed on the chronic inflammatory changes (4). It is probable that lung DCs and macrophages play important roles in initiating and regulating the immune response that causes asthma in the predisposed individual and that pulmonary macrophages participate in the secretion of proinflammatory molecules that enhance both the acute asthmatic response and the attendant chronic inflammation.

A. Pathogenesis

Recent studies implicate dysregulated immune responses to inhaled antigens as the underlying cause of allergic (or atopic) asthma (reviewed in refs. 3 and 5–7). The immune response in normal, healthy individuals is a finely tuned balance between responses mediated by two subsets of CD4 T cells. One subset, the Th1 subset, regulates the development of cell-mediated immunity and opsonizing antibody production and is considered essential for the resolution of many infectious diseases. In contrast, the Th2 subset amplifies antibody formation with the production of specific antibody isotypes, including IgG4 and IgE in humans (IgG1 and IgE in mice) and promotes the recruitment of inflammatory cells in which eosinophils are a prominent feature. Although protection against infection may be mediated through the production of Th2-regulated antibody production, IgE and eosinophilia are associated with allergic disease. The effects of Th1 and Th2 cells on immune responses are regulated by the unique set of cytokines which they secrete. Furthermore, some of these secreted cytokines together with others secreted by cells in the environment regulate the differentiation of precursor T cells into one of the two subsets.

Allergic asthma in children and young adults is linked to the development of Th2-mediated responses to common aeroallergins. Evidence for a shift in a Th1/Th2 balance in asthmatics includes (1) elevated total IgE serum levels, (2) immediate skin hyperreactivity (an IgE/mast cell–dependent response), (3) peripheral blood and pulmonary eosinophilia, and (4) the presence in lavage fluids and bronchial mucosa of activated T cells that express cytokines characteristic of the Th2 response (3,5). In regard to blood and tissue eosinophilia, the Th2 cytokine, interleukin-5 (IL-5), supports recruitment and activation of eosinophils in the bronchi (8,9). Since nonasthmatic individuals typically show little evidence of an active Th2 response to inhaled allergens, it is highly likely that a Th2-enhancing environment develops early in life in the lungs of individuals who will develop asthma as children. Once asthma develops, it is also quite possible that the Th2 lung milieu perpetuates Th2 responses to new aeroantigens.

B. T-Cell Subset Regulation: Basics

What is known about the development of antigen-specific Th1 and Th2 cells? The two T-cell subsets are derived from a common precursor, the naive or precursor T cell (Thp; reviewed in refs. 10 and 11) (Fig. 1). T-cell development is antigen dependent and is influenced by cytokines that predominate during the transition from Thp to mature T cells and likely occurs via a transitional cell, the uncommitted, activated Th0 cell. Th0 cells are capable of synthesizing a wide variety of cytokines characteristic of both Th1 and Th2 cells. However, as discussed, differentiated T cells secrete distinct subsets of cytokines. Th1 cells secrete IL-2, interferon-γ (IFN-γ), and lymphotoxin (tumor necrosis factor β [TNF-β]). Th2 cells secrete IL-4, IL-5, IL-10, and IL-13. In mice, both T-cell subsets secrete granulocyte-macrophage colony-stimulating factor (GM-CSF), IL-3, and TNF-α. A debate continues as to how sharply defined Th1 and Th2 subsets truly are, because studies of early immune responses show that there are immune responses without a sharp distinction in cytokine profiles; and T-cell clones from humans and experimental animals often secrete cytokines characteristic of transitional or mixed cell types (12). However, ample evidence from experimental animals and humans supports this enormously useful conceptual Th1/Th2 framework.

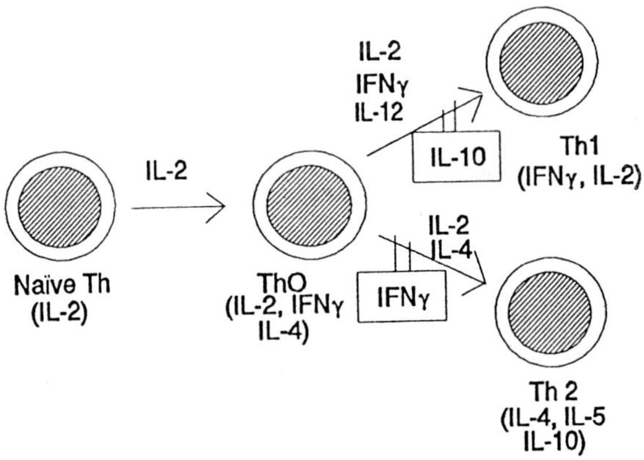

Figure 1 A model for T-cell subset development. (See text.)

A model for cross regulation of T-cell subsets development is shown in Fig. 1, a version modified from the review of Street and Mosman (10). Non-lymphoid sources of key regulatory cytokines are not shown. However, in addition to T cells, DCs, macrophages, and many other resident and recruited leukocytes may secrete the regulatory cytokines INF-γ, IL-12, and IL-4. For example, IL-12 is released by DCs and macrophages and supports development of a Th1 response (13,14). IL-10 may also be secreted by macrophages and suppresses their own production of IL-12, thus inhibiting Th1 development (15). IL-4, a critical factor in Th2 development, is synthesized by mast cells and basophils (16,17), as well as by fully mature Th2 cells. Indeed, the ability of mast cells and basophils to synthesize a broad array of cytokines typical of Th2 cells suggests that they may be important contributors in establishing and maintaining an inappropriate Th2 response in asthma. Whether a Th1 or Th2 response develops may also be determined by the antigen-presenting cell (APC). Thus, DCs tend to drive Th1 responses, whereas B cells tend to support Th2 responses (reviewed in ref. 18). The mechanisms by which DCs and B cells mediate this differential regulation are uncertain, but they may reflect the capacity of APCs to either make relevant cytokines or induce other cells to secrete them. We shall return to the roles of DCs and macrophages in regulating these responses.

C. Immune Regulation in the Lung: Basics

How does the lung respond to an inhaled neoantigen? This subject has been extensively covered in several chapters, particularly in the section on DCs (see Chapters 3, 8, 11–14). Briefly, the immune response may be divided into an afferent phase, a central processing phase, and an efferent phase. The afferent phase refers to the uptake and processing of antigen. Central processing includes the presentation of antigen to T cells and the subsequent clonal expansion and differentiation of T and B lymphocytes. Finally, the efferent phase encompasses both emigration and effector function.

Typically, particulate antigens that penetrate to the deep lung are removed via alveolar macrophage (AM) uptake and/or via the mucociliary elevator. Some antigen may penetrate the epithelium and be processed by resident DCs within the bronchial epithelium or in peribronchial connective tissue, whereas some antigen may enter lymphatics directly. It is likely that the antigen most relevant to the development of an immune response is carried to the regional lymph nodes by DCs, although resident lymph node B cells may also present antigen. As central processing proceeds in the lung-associated lymph nodes (LALNs), Thp cells are stimulated to become Th0 cells and subsequently driven by predominating regulatory cytokines to become either Th1 or Th2 cells. In the presence of appropriate T-cell cytokines, B cells can respond to native antigen by proliferating and differentiating into Ig-secreting plasma cells.

In the normal lung, inhaled antigens are largely cleared by AMs; and when an immune response is elicited, it is usually not a Th2 response. In atopic asthmatics, central processing likely occurs in a LALN bathed in cytokines from a lung milieu rich in Th2 and mast cell–derived cytokines, including IL-4. In infants and young children, a Th2 milieu may be the result of an aberrant response to a viral infection or environmental exposure to pollutants, particularly to cigarette smoke. Thus, the likelihood exists that once a Th2 lung milieu is established, exposure to new aeroantigens will also elicit an allergic response. The combination of IL-4 and the interaction of B-cell surface CD40 with CD40 ligand on Th2 cells favors class switching to IgE, resulting in the secretion of antigen-specific IgE into the blood (19,20). Both T- and B-cell lymphoblasts leave LALNs via efferent lymphatics, enter the bloodstream, and emigrate into inflamed lungs, utilizing appropriate adhesion molecules. If the recruited, clonally expanded T cells encounter antigen on APCs in the lung, they will expand further and/or secrete Th2 cytokines. Once a primary immune response is established within the regional nodes, repeated antigen delivery to the lung should stimulate antigen specific lymphocytes that were recruited to the lung to proliferate and amplify the immune response.

D. Pulmonary Immune Regulation in Asthmatics

The discussion above suggests that a cytokine milieu in the lung could determine the type of primary immune response to an inhaled antigen by flooding the lymphatics with regulatory cytokines that reach the LALNs where the immune response occurs. But what is it that causes a Th2-enhancing cytokine environment to be established in the individual destined to develop asthma? Young children may wheeze in response to infection and yet not develop Th2 responses and asthma. A recent study helped to clarify the distinction between transient and persisting wheezing illnesses in infants and young children (21). In this study, a persisting wheezing illness was associated with elevated IgE levels and a history of maternal asthma. Although a history of maternal asthma was linked to the emergence of a wheezing illness associated with atopy, underlying mechanisms of inheritance are still poorly defined. Other studies have shown that immature lung development, infection, and pollution are all statistically linked to persisting wheezing illnesses and to atopic asthma (summarized in ref. 22). One can only postulate the role of each of these risk factors for asthma, and important questions include: What determines the lung milieu in early infancy and childhood that causes immune responses to allergens to elicit a Th2 response? What is the role of genetics? What role does the developmental status of the lung and/or the maturity of the pulmonary immune system play? What effects do inhaled pollutants, including tobacco smoke, and lung infection have on the lung environment to drive a local immune response toward a Th2 response? And, finally, what causes

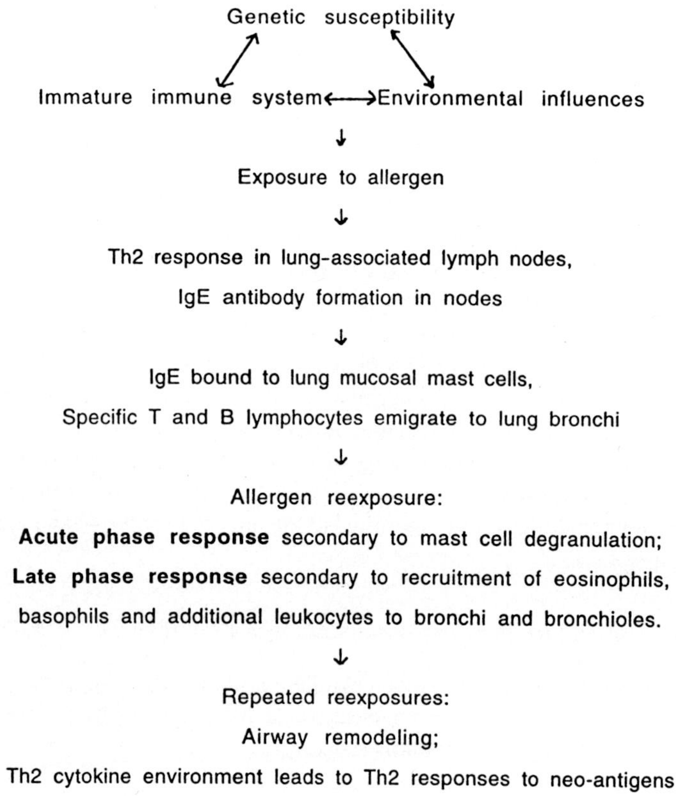

Figure 2 The sequence of events in the development of asthma.

this early Th2 response to persist? These intriguing questions are more easily addressed in the context of a model (Fig. 2) that illustrates the relationship of predisposing factors in the development of asthma.

E. Genes, Immunological Immaturity, and the Environment

The tendency of atopic asthma to "run in families" has led to many epidemiological studies attempting to define the patterns of inheritance for familial asthma (23,24). No single gene determines the inheritance of atopy, although recent data have shown a linkage to the chromosome encoding a number of cytokines that regulate Th1/Th2 responses. There is strong evidence linking elevated total IgE levels and bronchial responsiveness to the short arm of human chromosomal segment 5q close to loci that encode the Th2 cytokines IL-4 and IL-5 (25,26). One postulate is that abnormal transcription of the gene for IL-4 promotes the development of the asthmatic response. Linkage of IgE antibodies to specific allergens has been mapped to the major histocompatibility complex (MHC) locus (27,28). Thus, certain MHC class II allotypes may be capable of binding and presenting allergenic peptides, whereas others are not. Another genetic locus linked to asthma is on 11q12 where the gene for the β subunit of the high-affinity FcεRI receptor resides. A single amino acid substitution in this gene product has been linked to atopy, but this has not been confirmed (29).

The role for the developmental stage of the lung and immune system in the initiation of a Th2 response is not clear. In many experimental models, continuous exposure of normal, adult animals to antigen via the airways resulted in either tolerance or production of antibodies of nonpathogenic significance (30). However, severe allergic asthma is primarily a disease acquired in childhood and usually develops in children who have a history of recurrent wheezing illness as infants. One suggestion is that the immaturity of the lung immune system in infancy may predispose to the development of dysregulated immune responses. Consistent with this notion, Holt et al. demonstrated that lung DCs were immature in newborn rats (31) and suggested that in early childhood, immature lung APCs might enhance the development of a Th2 response (32). We will return to this concept.

A poor correlation has repeatedly been found between wheezing illness of very young children and the presence of elevated IgE. An important corollary is that wheezing illnesses in early childhood in many children does not necessarily indicate that allergic asthma will develop (21). A potential explanation for the wheezing in many young children is that bronchial hyperresponsiveness can be caused by nonspecific inflammation or a Th1-mediated inflammatory response and may not require an inhaled allergen specifically to stimulate IgE sensitized mast cells. A murine model of asthma has shown that indeed AHR can occur in the presence of a Th1 response (33). However, chronicity of inflammation in a subset of predisposed children with early childhood wheezing may eventually lead to a dominate Th2 response in the lung to allergens in the environment and the establishment of asthma. Alternatively, if a Th2 response fails to develop, either because tolerance to allergens occurs or because the appropriate lung environment to drive a Th2 response never develops, the wheezing illness stops.

What is the role of the environment? A link exists between early childhood exposure to infections and pollutants, especially tobacco smoke, and the development of asthma (34,35). However, the link is not absolute. For example, recent studies found a lower incidence of asthma in eastern European cities with heavy air pollution than in western European countries (36). Environmental factors are likely to be only one component in the development of atopic asthma, albeit an important one, particularly when influenced by a genetic predisposition. Furthermore, little is known about the influence of tobacco smoke on initiation of a pulmonary Th2 immune response.

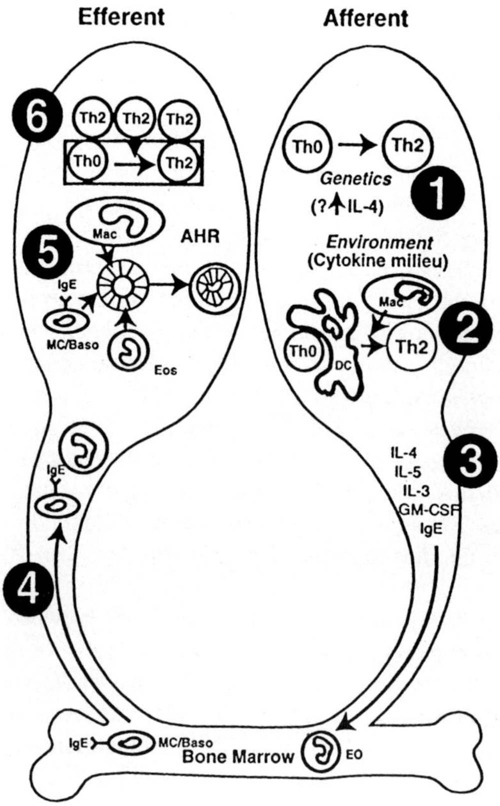

Figure 3 A conceptual framework for the initiation of a Th2 immune response in a naive lung (afferent stage of response, on the right) and the elicitation of an acute response in the inflamed asthmatic lung (efferent stage of response, on the left). (1) On the right, genetics and the environment within the lung (the cytokine milieu predominantly) tend to drive the response to an inhaled antigen toward a Th2 response. (2) The primary immune response actually takes place in the lung-associated lymph nodes (LALNs) following emigration of the antigen-bearing DCs from the lung, although shown here for simplicity as within the whole lung. Note that the cytokines secreted by alveolar macrophages (Mac) might either affect the DCs before they emigrate from the lung or produce sufficient cytokines to flood afferent lymphatics that drain to the LALNs (3) Allergen-specific IgE-secreting B cells enter the medullary sinusoid and secrete IgE, which then reaches the bloodstream. At least some of the antigen-specific T and B lymphoblasts also reach the systemic circulation via efferent lymphatics. Furthermore, cytokines (IL-4, IL-5, IL-3, GM-CSF) generated in the LALNs likely also enter the bloodstream and perfuse the bone marrow to stimulate the expansion and differentiation on basophils and eosinophils. (4) Thus, increased numbers of basophils and eosinophils in the blood are available for recruitment to the lung. Both IgE and lymphoblasts enter the lung where mast cells are sensitized, and B lymphocytes and

F. Conceptual Framework for Asthma

Figure 3 summarizes some of the concepts discussed above and constructs a model for understanding the initial development of a Th2 response in a naive individual (naive lung) which will eventually lead to the clinical expression of allergic asthma. In the naive lung and its draining lymph nodes, genetics may favor an IL-4–dependent Th2 response to neoantigens by controlling (1) alterations in cytokine gene expression that promote Th2 development and (2) expression of MHC class II antigen on APCs which determine whether an important allergenic epitope will be appropriately presented to responding T cells. The external environment, including pollutant exposures or infection, may create an internal lung environment (with altered cytokine milieu) which may influence how the immune response develops. Furthermore, the developmental stage of the immune system, including the maturation of APCs, may also determine whether Th0 cells will develop into Th2 cells in response to new aeroantigens. The production of antigen-specific IgE is an important consequence of the Th2 response. IgE secreted into the bloodstream can subsequently bind to FcIgεR on lung mast cells so that on reexposure to the aeroantigen, these receptors are cross linked causing release of mast cell products and recruitment of eosinophils and basophils to the lung. Cytokines that are produced by eosinophils, mast cells, basophils, and macrophages in response to reexposure to allergens contribute to the Th2-like cytokine profile in the lung environment. These cytokines act as a positive feedback to perpetuate the development and proliferation of Th2 lymphocytes that become lung residents. Thus, in this model, repeated exposures to inhaled allergens establish a persistent Th2 environment in the lung (asthmatic lung), which eventually may lead to fixed airway disease.

A number of important questions remain unanswered and will be discussed in following sections. First, what is the interrelationship of antigen presentation by lung DCs and immunoregulation by resident lung macrophages in responses to neoantigens in the lung that leads to the development of a Th2 response in the lung. Second, once a Th2 response is present in the lung, do DCs and macrophages play roles in the acute asthmatic attack? Specifically, is there evidence that pulmonary macrophages secrete factors that recruit lymphocytes, eosinophils, or basophils into the lung and/or increase AHR?

Th2 cells take up residence in airway walls. (5) These events create a setting so that on reexposure to allergen, mast cells, recruited eosinophils and basophils, and resident and recruited macrophages are activated to initiate bronchospasm (and long-standing airway hyperreactivity [AHR], plasma exudation, mucous secretion, and recruitment of additional inflammatory cells). (6) Antigen-specific T cells in the lung (after antigen presentation on residual lung DCs) add to this inflammatory response by secreting Th2 cytokines that amplify the local inflammatory response as well as drive immune responses to new aeroallergens toward a Th2 response.

III. Factors in Th2 Lymphocyte Differentiation

The ability of a host to respond with an appropriate immune response likely evolved in order to protect against infectious agents (concepts discussed in refs. 37–40). Th1 responses are particularly important in viral and other intracellular infections, whereas Th2 responses are protective in helminthic infections. Host mechanisms that determine whether a Th1 response or a Th2 response will occur are tightly regulated. If either type response continued unabated, the host would experience inappropriate, persisting immune responses with tissue destruction, chronic fibrosis, and organ dysfunction. Whether a Th1 or a Th2 response develops is determined by at least four factors: the character, dose, and route of entry of the antigen; characteristics of the APCs; the genetic makeup of the responding CD4 T cell; and cytokines in the milieu. Cytokines at the site of the developing immune response seem to have overriding importance. Indeed, the final common pathway for the first three factors may well be their effect on generating regulatory cytokines. In addition to considering what regulates the development of subsets from antigen-stimulated naive T cells, one must also consider the possibility that plasticity in CD4 T cells might exist such that at least some committed effector cells may return to being Th0 cells. T cells that are initially primed become effector cells producing the cytokines that define their subset and function in the effector limb of the immune response. However, when antigen stimulation is discontinued, effector T cells become quiescent and together with antigen-specific primed T cells that have failed to commit to either a Th1 or Th2 cell maintain residence in the immunized host as memory cells. Evidence exists that at least some of these memory T cells on restimulation by antigen are again subject to cytokine influences to drive them from the Th0 subset along either a Th1 or Th2 pathway (41). This concept has important implications for vaccination as well as for therapeutics in asthma. The possibility exists that after a period of allergen absence, reexposure might be coupled with appropriate suppressors of Th2-stimulating cytokines to shift what was originally a Th2 response to a less damaging one. In addition, and perhaps most important, anergy induction is possible. The natural progression of a change in the type of immune response may explain why some true childhood asthmatics may not continue to experience asthma into adulthood.

A. Cytokines and Other Secreted Factors that Influence Subset Differentiation

Many lung cells are capable of producing cytokines that can determine the type of immune response to inhaled allergens. IL-12, IFN-γ, and IL-4 are the dominant factors determining the direction of an immune response, but a number of other factors may also play a modulating role. In addition to other cytokines such as

IL-1, TGF-β, and GM-CSF, hormones and prostaglandins may exert effects. However, for these additional factors, the demonstration of a role in immunoregulation is often dependent on the experimental system. Furthermore, these factors generally exert their effects by enhancing or suppressing the secretion of the three final effector cytokines, IL-12, IFN-γ, or IL-4.

Interferon-γ–inducing factor (IGIF) is a newly described, but potentially very important, cytokine in immune regulation. IGIF can enhance a Th1 response in the absence of IL-12 (42), because it is an extremely potent stimulus for IFN-γ production in T cells. It induced higher levels of IFN-γ from anti-CD3 stimulated Th1 clones than IL-12. IGIF production was stimulated in Kupffer and peritoneal exudate cells by in vivo exposure to *Proprionibacterium acnes*, which did not induce IL-12. These studies suggest that IGIF, induced by certain infectious agents that fail to induce IL-12, is an alternative cytokine for stimulating a Th1 response. Furthermore, since peritoneal exudate cells were primarily macrophages in these studies, it is possible that lung macrophages might also make IGIF when appropriately stimulated and thus drive lung Th1 responses.

Table 1 lists important factors that have effects on T-cell subset development. Important cell sources are also indicated. A quick perusal clearly implicates macrophages as an important source of factors that both enhance and suppress a Th1 response. Macrophages are the principal source of IL-12 and can produce this cytokine on exposure to a number of agents, including lipopolysaccharide (LPS), selected intracellular infectious agents or their products, including *Listeria monocytogenes*, extracts from *Staphylococcus aureus*, or a hydrophobic molecule from *Toxoplasma gondii*, and is amplified in the presence of IFN-γ (43; reviewed in ref. 40). They are also a principal source of IL-10. Sources of IL-4, other than Th2 T cells, are mast cells, basophils, eosinophils, and low-frequency α/β T cell subsets which are either NK1.1$^+$, CD4$^+$ or CD4$^-$/8$^-$ and γδ T cells (44–49). It is easy to speculate that the lung, which contains both macrophages and mast cells, is in a position to respond to environmental antigens with secretion of either IL-12 or IL-4. Furthermore, γδ T cells and natural killer (NK) cells are also present in the lung and can amplify the IL-12–secreting response of resident macrophages to intracellular microbes by secreting IFN-γ (18,50,51).

What are the possible effects on immunoregulation in the lung of cytokines other than IL-12, IFN-γ, IGIF, and IL-4? IFN-α likely stimulates the development of Th1 responses (38) by enhancing the production of IL-12 by macrophages. IL-13 binds to many of the same receptors as IL-4 and has many properties in common (52). Although IL-13 does not directly stimulate T cells like IL-4 does, it promotes the proliferation and differentiation of B cells and the secretion of immunoglobulin E. Furthermore, it is upregulated in mononuclear cells in the lavage fluids of allergen-challenged asthmatics (53).

IL-10, secreted by macrophages, B cells, and Th2 T cells, is one of the most important cytokines in downregulating a Th1 response, and thus allows a Th2

Table 1 Factors Influencing Th1 Versus Th2 Subset Development and Predominant Cell Sources

Function and Factors	Major cell sources	Ref.[a]
Th1-enhancing factors		
IL-12	Monocyte, macrophages, dendritic cells, B cells	13,37,65
IFN-γ	NK cells, γδ T cells, CD8 T cells, Th1 cells	37,49
IGIF[b]	Macrophages, Kupffer's cells	42
IL-2	T cells, NK cells	60
IFN-α	Macrophages, fibroblasts	38
TGF-β	Macrophages, epithelial cells	38,60
IL-1β	Macrophages, endothelial cells	38
DHEA[c]	Adrenal cortical cells	65
Th1-suppressing factors		
IL-10	Macrophages, Th2 cells, B cells	55–56
IL-4	See below	57
PGE_2	Macrophages, epithelial cells	62,63
TGF-β	See above	59,60
Th2-enhancing factors		
IL-4	Mast cells, basophils, eosinophils, some T cells	44–47,49
IL-13	Th2 cells	52,53
IL-1	See above	38
Progesterone	Ovarian cells	65
Th2-suppressing factors		
IFN-γ	See above	See above, 107
IL-1a/IL-1b	See above	See above, 61
Il-12	See above	See above, 62
Substance P	Nerve, neuroendocrine cells	64

[a]References are representative and not inclusive.
[b]Interferon-γ–inducing factor.
[c]Dehydroepiandrostenone.

response to develop (15,54). Splenic DCs stimulate IFN-γ secretion by both CD4 and CD8 T cells in a primary allogeneic mixed leukocyte reaction (MLR). IL-10 failed to inhibit DC-induced lymphoproliferation in this system, but it did inhibit IFN-γ production (55), although it was not determined whether this was through an effect on decreasing IL-12 production by DCs. Thus, although IL-10 does not inhibit a DC-stimulated primary T-cell responses, it can prevent the development of a Th1 response, allowing these important APCs to stimulate a Th2 response. Furthermore, IL-12 secretion by lung macrophages is downregulated by IL-10 and to a lesser extent by IL-4 (56,57).

There is ample evidence that TGF-β has immunomodulatory effects on the

T-cell response. TGF-β has been reported to block T-cell proliferation (58) and contribute to a shift toward Th2-type responses both directly and by stimulating IL-10 (59). However, TGF-β can also enhance Th1 responses (38,60) by enhancing endogenous IFN-γ secretion and blocking IL-4 production. TGF-β is made by AMs and could readily play a role in the development of asthma, but whether it enhances or suppresses development of this disease is not clear. Recent studies have shed light on the confusing role of TGF-β in experimental studies. The role of this cytokine in determining Th1 versus Th2 responses is dependent on the mouse strain (60); another piece of evidence that genetics can play an important role in the type of immune response that develops. In these studies, the mechanism was a differential susceptibility of stimulated T cells from different mouse strains to the TGF-β–mediated suppression of secretion of IL-2, a cytokine essential for expansion of Th1 clones.

IL-1 is another cytokine, made primarily by macrophages, for which apparently conflicting data are also available. Romagnani reviewed the evidence for IL-1 as enhancing Th2 development; likely functioning as a growth enhancer of Th2 cells (38). On the other hand, using human peripheral blood mononuclear cells Sandborg et al. showed that IL-1α and IL-1β both selectively inhibited early IL-4 expression, although in these studies, proliferation of the responder T cells was not inhibited (61). The differences are undoubtedly due to the systems studied (both used human T cells), but the reasons for differences remain to be identified.

Prostaglandin E_2 (PGE_2) and hormones have also been studied in relationship to T-cell subset regulation. PGE_2 inhibits mRNA expression and the secretion of IL-2 and IFN-γ, and allows a Th2 response to be expressed when added to cultures of anti-CD3–stimulated naive T cells (62). It also prevents Th1 cytokine secretion in committed T cells (63). PGE_2 suppression of a Th1 response is likely mediated through differential effects on cyclic adenosine monophosphate (cAMP) elevations in Th1 versus Th2 cells, although precise mechanisms are not clear. Both macrophages and follicular DCs (unique cells that reside in lymphoid follicles with the capacity to trap immune complexes and stimulate antigen specific B cells) synthesize PGE_2 and thus could play roles in generating a Th2 response in the lung. An important corollary is that indomethacin, a prostaglandin inhibitor, might circumvent the development or amplification of Th2 responses if it were administered at an appropriate time in the development of a primary immune response or at rechallenge with an antigen that has initiated a Th2 response previously.

Evidence exists that the nervous system may play a role in modulating the immune response, and numerous studies have examined this concept. Two neuropeptides detected in the lung and available in LALNs at the time of a developing immune response are substance P (SP) and vasointestinal peptide (VIP). Both can dampen ongoing responses. In studies in preimmunized mice, subcutaneous injections of SP at the time of peak IgE responses suppressed those responses within 48 hr in an isotype-specific, dose-dependent manner (64). When spleen cells from immunized mice were cultured for 5 days with the immunizing antigen, memory IgE antibody responses were induced in vitro. Inclusion of either SP or VIP, but not other neuropeptides, suppressed IgE responses. Furthermore, SP-mediated, but not VIP-mediated, suppression of IgE responses was abrogated by inclusion of anti–IFN-γ in cultures, suggesting that SP stimulated IFN-γ secretion.

Finally, a number of hormones have also been implicated in the modulating immune response. (See Table 1; briefly discussed in ref. 65.)

B. Character, Dose, and Route of Antigen Exposure

The structural and functional properties of the antigen that initiates a response can exert an effect on the development of a Th1 or Th2 response. Peptides constructed to differ slightly from the relevant antigenic epitope (altered peptide ligand [APL]) can result directly in either partial activation or anergy of T cells (66,67), suggesting that carefully crafted peptides from allergens could be useful in interfering with Th2 responses. Specific molecules from infectious agents may also preferentially stimulate Th1 versus Th2 responses through unclear mechanisms (39,65), but possibly reflecting stimulation of regulatory cytokines from APCs or other accessory cells. For example, LPS containing organisms would be expected to elicit Th1 responses, because LPS delivers a strong stimulatory signal for IL-12 secretion by macrophages. Live versus dead intracellular organisms have also been shown to have differential effects on some, but not all, cytokines that are secreted that could influence an immune response (68). Proteins with proteolytic activity are known to be quite allergenic, perhaps because they can directly trigger IL-4 release from mast cells or because they cleave important regulatory molecules that could lead to initiation or persistence of a Th2 response. The major allergenic component of the house dust mite *Dermatophagoides pteronyssinus*, *Der p* I, has the ability to selectively cleave the low-affinity IgE receptor, CD23, from B cells (69). Binding of IgE immune complexes to CD23 leads to feedback inhibition of both B-cell proliferation and differentiation for IgE synthesis (70). Thus, loss of CD23 from enzymatic cleavage would allow unregulation B-cell expansion and IgE secretion. Furthermore, CD23-knockout mice demonstrate sustained elevated levels of IgE antibodies to T-dependent antigens (71); again illustrating the important role of this surface molecule in downregulating B cells. Therefore, cleavage of CD23 by *Der p* I should enhance a Th2 response and offers an explanation of why this house dust mite antigen, more than many other inhaled protein, is an important cause of asthma. Furthermore, this observation suggests that macrophages and recruited monocytes in the respiratory tract, which secrete proteases with similar abilities to cleave CD23, might play enhancing roles in driving Th2 responses, particularly when respiratory antiproteases are inactivated. Cigarette smoke is known to inactivate the major respiratory tract antiprotease (72), and tobacco smoke exposure is a known risk factor for the development of asthma in young children (35).

The dose of antigen is important (39). Hosken et al. used an in vitro system of naive T cells from αβ T-cell receptor (TCR) transgenic mice (in which the TCR recognized ovalbumin [OVA] peptide when presented on the MHCII antigen, IA^d) to show that at very low or very high doses of antigen, Th2 responses predominated. Th1 responses occurred at intermediate doses (73). Antibody to IL-4 blocked the very low and very high dose effects, indicating that endogenous IL-4 production drove the Th2 response. Interestingly, this antigen dose effect occurred whether highly purified splenic DCs or B cells were used as APCs, results that conflict with the results of others (see below). Similar results were obtained for low-dose (Th2 response) and high-dose allergens (Th1 response) in studies using human peripheral blood cells stimulated with allergens (74). These studies suggest that the dose of allergen exposure could affect lung responses. Furthermore, infections in the lung that alter the permeability of antigens in the airways to the underlying lung parenchyma to be taken up by lung DCs or translated directly to lymph nodes could alter primary immune responses.

The route of antigen can also play a role in determining the type immune response. The development of tolerance characterizes the response of small antigens given intravenously. Protective vaccination of C57BL/6 mice against *Schistosoma mansoni* failed when the skin-stage schistosomula were given by intravenous or intramuscular routes, but it was successful when administered by the intradermal route (75). Failure was associated with the development of macrophage-suppressive activity and was ascribed at least partially to the secretion of TGF-β. It is worth recalling that Th1 immune responses in C57BL/6 mice are susceptible to suppression by TGF-β. Clearly, the route of antigen administration can determine the accessory cells that interact with the antigen and the role of macrophage suppression and thereby affect the type of immune response generated.

C. Antigen-Presenting Cell Type

Several investigators have examined the role of the type of APC on Th2 responses. Immune responses to aeroallergins are likely to result from antigen translocated on lung DCs to LALNs, and B cells may act as APCs to amplify these responses. Furthermore, once immune responses are ongoing in the lung, any Ia-positive cell may function as an APC for recently activated T cells (76). Despite studies to the contrary (73), the majority of studies suggest that DCs and macrophages tend to stimulate Th1 responses, whereas activated B-cell antigen presentation favors a Th2 response (77–79). Using allergen-stimulated peripheral blood human T cells, Secrist et al. also found a trend to more IL-4 secretion by T cells stimulated with activated B cells as APCs and more IFN-γ when peripheral blood monocytes were added to cultures (74). The mechanisms might reflect the capacity of DCs and macrophages to make IL-12 and the failure of B cells to make this cytokine, although this hypothesis has not been systematically examined. However, emerging evidence indicates that costimulatory molecules, that is, CD40 and CD80, expressed on APCs that interact with specific ligands on responder T cells can help determine the type of immune response that ensues. Differential regulation and display of these molecules could determine differences in immune regulation among APCs.

The expression of the costimulatory molecules CD80 (B7-1), CD86 (B7-2) (80), and CD30 ligand (CD30L) (81) plays a critical role the development of Th2 cells. Mice deficient in CD28 appear to have intact cell-mediated immunity but impaired humoral responses, suggesting an overall greater dependence of Th2 responses on B7-CD28 interactions than Th1 responses (39,82). Whether CD80 and CD86 have different effects on subset development continues to be studied. For example, in an in vivo study in an allergic encephalomyelitis (EAE) murine model, anti-CD80 antibodies blocked a Th1 response and decreased disease severity, but anti-CD86 antibodies increased disease severity (80). These findings suggest that CD80 interactions favor a Th1 response that favors disease, whereas CD86 favors a Th2 response.

CD30, a member of the TNF/nerve-grown receptor superfamily, was initially shown to be preferentially expressed by human Th2 clones. Recent studies suggest that stimulation of CD30 might influence T-cell subset differentiation. Culture of either Th0 or Th2 clones with an agonistic anti-CD30 monoclonal antibody (Ab) enhanced proliferation and cytokine secretion (81). Furthermore, stimulation of peripheral blood lymphocyte with anti-CD3 and the agonistic anti-CD30 resulted in the preferential development of T-cell lines and clones showing a Th2-like profile of cytokine secretion. Early blockade of the CD30 ligand–CD30 interaction shifted the development of T cells toward the opposite (Th1-like) phenotype. Whether cytokines or other stimuli can differentially regulate CD30 ligand expression on B cells and DCs cells and thus affect their capacity to stimulate subset differential is not known.

D. Genetic Background of Responder CD4 Lymphocytes

One determinant of the type of immune response that occurs in the lung may be the genetic propensity of CD4 T cells to respond to antigens with an excessive production of IL-4. The linkage of elevated IgE levels (23) and bronchial hyperresponsiveness (26) in humans to genes on chromosome 5q where the IL-4 gene is found is compatible with this view. Evidence using inbred mouse strains also suggests this possibility. In a *Leishmania major* model, BALB/c mice are sensitive to subcutaneous infection, develop an expanding local infection with systemic spread, and die. In contrast, C57BL/6 mice are relatively resistant, resolve the local infection, and do not develop systemic disease. BALB/c mice develop a Th2 response to *L. major*, whereas resistant C57BL/6 mice develop a protective Th1 immune response (83). Treatment of BALB/c mice with antibodies to IL-4 resulted in the development of a Th1 response with cure (84). In vitro studies have demonstrated that BALB/c splenocytes have an innate propensity to develop Th2 responses as expressed by diminished IFN-γ secretion and increased IL-4 in response to mitogen (74). The question was whether differing responses in BALB/c and C56BL/6 mice was a property of the antigen, the APCs, the responder T cell, or cytokines made by other cells in the milieu. An important role for the genetic makeup of the responder T cell was found in determining which subset developed in response to the defined antigen OVA (85). The studies exploited the OVA T-cell receptor (TCR) transgenic mouse system discussed above in which the transgene was bred onto both resistant (B10.D2) and sensitive (BALB/c) H2-compatible mouse strain backgrounds so that APCs (including cells from the splenic milieu) from either strain could be used to present antigen to the opposite strain. Using this system, the investigators found that in the absence of exogenous cytokines, splenic APCs did not determine the cytokine secretion profiles of the OVA-stimulated, transgene-positive T cells. Rather, BALB/c T cells developed primarily a Th2 cytokine profile whether stimulated with antigen on BALB/c or B10.D2 APCs. Nevertheless, the addition of IL-12 and antibodies to IL-4 drove the BALB/c T cells to a Th1 phenotype, suggesting that T-cell–derived endogenous IL-4 might be responsible for this effect and that blocking IL-4 and adding a powerful Th1-enhancing cytokine could overcome the genetic propensity. This conclusion is important, because it predicts that T cells responding to either allergens or infectious agents that do not stimulate cells in the lymphoid environment to produce Th1- or Th2-enhancing cytokines might follow a genetically predetermined default pathway.

E. Possible Effects of Confounding Infections

An important link exists between the development of asthma in children and pulmonary infections, particularly viral infections. Inflammation associated with infection could increase the permeability of the allergen across the epithelium, allowing it to be more immunogenic. Furthermore, infection might create a cytokine milieu so that concurrent allergen exposure might cause the host to respond with a Th2 response rather than a Th1 or null response. Thus, in an infection that generates a Th2 response with concomitant secretion of IL-4 by T cells and mast cells and IL-10 by T cells and macrophages, exposure to a coincident allergen might be tipped toward a Th2 response. What types of infections might initiate a Th2 response?

An enormous literature has developed on the role of Th1 responses and associated IFN-γ secretion in resolving infections (reviewed in ref. 38), particularly in experimental infections in mice (18,37,40). The most exhaustively studied is *L. major*, as previously discussed. Other infections, including many which can involve the lungs such as *Mycobacterium*, *Candida albicans*, *Crypto-*

coccus neoformans, and *Histoplasma capsulatum* also stimulate Th1 responses. Murine helminthic infections such as *Trichuris muris*, *Nippostrongylus braziliensis*, and *Heligmosomoides polygyrus* (39) stimulate Th2 responses in mice, and evidence indicates humans respond with Th2 responses in a large number of helminthic infections as well, including toxocariasis, filariasis, strongyloidiasis, and onchocerciasis. None of these corresponds to common pulmonary infections that children develop in early childhood. In contrast, viral infections typically depend on Th1 responses for resolution. One could reasonably ask whether children with wheezing illnesses during viral infections and who later develop asthma respond to these infections with Th2 rather than Th1 responses as a result of genetic predisposition or accompanying environmental insults as discussed above. Measurements of cytokines secreted from virus-specific T cells from these children would be important to assess.

IV. Role for Lung Dendritic Cells and Macrophages in the Initiation of Lung Immunity Leading to Asthma: Lessons from Animal Models of Asthma

Evidence for a role for lung DCs and macrophages in the initiation of immune responses in the lung has come from many sources, including analysis of tissue from postmortem examinations, biopsies, blood and lavage fluids, and human cells studied ex vivo. One of the most powerful ways to study the pathogenesis of asthma is to use experimental models. A recent publication summarized some of the highlights of a workshop held in Prague, June, 1993, on animal models of asthma. Models have been developed in primates, dogs, sheep, cats, rabbit, guinea pigs, rats, and mice. Pros and cons for each species that has been used to explore facets of this complex disease were discussed. The general consensus was that each model was imperfect (86). However, most models expressed AHR and some degree of eosinophilia, although the degree of AHR and the dependence on eosinophilia for AHR varied among species (87,88). Larger animals are particularly useful for the greater ease of performing lung physiology and the capacity to do repeated sampling and physiological studies. Rats and mice offer the advantage of better defined immune systems with a greater availability of reagents to manipulate the model. In these smaller animals, adoptive transfer techniques are also more feasible to better define the roles of T and B cells and their subsets.

Mice are particularly useful, because large numbers of reagents are available for in vitro and in vivo manipulations and the quantities of monoclonal antibodies or recombinant cytokines necessary for each animal is relatively small as compared with amounts necessary for larger animals. Another important advantage is the availability of many inbred strains, including some with either absent T cells or mast cells and others with variable sensitivities to induction of

AHR and the development of Th2 responses. Mice that have been genetically manipulated offer unique possibilities. Transgenic mice may (1) overexpress cytokines in all tissues, (2) overexpress products in specific cells or organs by manipulating the promoter used to control transgene expression, or (3) completely fail to produce specific cytokines or their receptors ("knockouts").

A number of investigators have developed murine models that mimic various components of human asthma. Particularly useful for studying the regulation of eosinophilia and AHR that correlates with eosinophilia are those that initially immunize intraperitoneally or subcutaneously with antigen in alum (89,90) or with sheep red blood cells via the peritoneum (91) and then elicit AHR with an aerosol challenge of the immunizing antigen. However, models that utilize aerosol or intranasal deposition of antigens rather than the foregoing routes are more physiological mechanisms for immunization. Repeated OVA given to mice led to a transient initial development of IgE antibodies, but tolerance developed by 7 weeks (30). Tolerance was shown in several strains of mice and was adoptively transferred by γδ T cells, possibly through their ability to respond specifically to the antigen by secreting IFN-γ (92). Nevertheless, the aerosol model offers the advantage of studying mechanisms for the development of IgE and to explore whether inflammation and altered lung physiology are present in the early time period prior to tolerance developing.

Gelfand and his colleagues have exploited an aerosol immunization model using BALB/c mice exposed to daily nebulized OVA over a 10-day period. The mice developed an OVA-specific IgE response, immediate cutaneous reactivity to OVA, and increased AHR as measured by contraction of tracheal muscle in response to electrical stimulation (93). When IFN-γ was nebulized at the same time as OVA, serum anti-OVA IgE and immediate skin reactivity were decreased and AHR failed to develop. There was evidence of a Th1 response with increased serum anti-OVA Ig2a. Intraperitoneal delivery of IFN-γ coincident with OVA aerosol exposures was ineffective in blocking the Th2 response, suggesting that the route used to deliver an immunomodulating agent is critical. Other important aspects of this model are that IgE production and AHR could be adoptively transferred with lymphocytes from LALNs but not splenocytes. Instead, adoptive transfer of splenic $CD8^+$ T cells suppressed ongoing responses, which was related to their capacity to secrete IFN-γ (94). These latter results are compatible with the observations of McMennamin et al. discussed above, because the tolerizing IFN-γ–secreting γδ T cells were also $CD8^+$ (92).

In summary, animal models have been useful in defining aspects of the early initiation of asthma and complement the evidence from humans that regulation of Th2, IgE-generating responses are likely pivotal in the development of asthma. What then is the evidence from these models that lung macrophages and DCs play roles in the development of asthma? Chapter 8, by Georg Kraal, Broug, van Iwaarden, Persoons, and Thepen describes their work and that of others on the importance of AM in down-regulating pulmonary immune response, including IgE production (95) by several suppressive mechanisms, including production of nitric oxide (NO) (96). Chapters 3 and 9–14 discuss the role of DCs in immunoregulation in general. In brief, drawing on our previous discussion, lung DCs undoubtedly carry antigen to LALNs to initiate an immune response. They have the potential of secreting IL-12 to initiate a Th1 response, and this suggests that if an antigen is presented solely by these cells in draining nodes, a Th1 response is more likely. Holt and coworkers, using rats, have shown lung DCs in infants are immature (97) and may not function to initiate effective immune responses (32). One can speculate that in infants, particularly in the presence of high levels of allergens, repeated viral infections and/or exposure to environmental pollutants, free antigen (unprocessed by DCs) might reach LALNs and be presented by resident B cells to initiate Th2 rather than Th1 responses. Alternatively, immature DCs could participate but might not have the appropriate costimulatory molecules (perhaps B7-1) to stimulate a Th1 response. Finally, infections and environmental pollutants might alter macrophage-suppressive functions or recruit less suppressive monocytes from the peripheral blood so that the ability of lung macrophages to block the development of immune responses is lost. Additional studies are needed to test these hypotheses.

V. Role for Lung Macrophages in the Effector Limb of the Asthmatic Response: Secretion of Proinflammatory Factors

The previous sections have focused on the role of lung macrophages and DCs in the initiation of the Th2 immune responses that leads to asthma. This section will discuss the role of macrophages in the effector limb of asthma, emphasizing the possible role of AMs and interstitial macrophages in the acute and chronic inflammatory responses that characterize the asthmatic lung. The role of DCs in the effector limb of the Th2 response would be to act as APCs for recruited CD4 T cells to stimulate an allergen-specific response and will not be further discussed.

As in any inflammatory response, the changes in acute asthma include vasodilatation, increased vascular permeability, and leukocyte recruitment. In addition, bronchoconstriction and increased mucous secretion more specifically characterize asthma. In time, chronic changes develop and include airway wall thickening with increased connective tissue, smooth muscle hypertrophy, and neovascularization. All of these changes are regulated by secreted factors. Although mast cells, basophils, and lymphocytes are of major importance in this regard, lung macrophages undoubtedly play a role in both acute and chronic inflammatory changes (98).

Bochner et al. reviewed many of the important factors in mediating acute

Table 2 Factors Made by Lung Macrophages that Mediate the Acute Asthmatic Response

Cause vasodilatation and increase permeability: PAF,[a] LTC4, LTD4, LTEA[b]
Upregulate vascular adhesion molecules; i.e., ICAM,[c] VCAM[d]: TNF-α, IL-1
Chemoattract lymphocytes, eosinophils, basophils and/or monocytes: LTB4, C5a, TNF-α, CC chemokines: (MCP-1, MIP-1α, and MIP-1β)
Prime and activate mast cells, basophils, and/or eosinophils: LTB4, C5a, TNF-α, CC chemokines (MCP-1, MIP-1α)
Bronchoconstrict, either directly or through stimulating neural network: PAF, LTC4, LTD4, LTE4, thromboxane
Stimulate mucus secretion: LTC4, LTD4, LTE4
Cause epithelial damage: Reactive oxygen radicals
Stimulate growth of smooth muscle, fibroblasts, and/or vessels: TGF-β, PDGF[e]

[a]Platelet-activating factor.
[b]LT = leukotrienes.
[c]Intracellular adhesion molecule.
[d]Vascular cell adhesion molecule.
[e]Platelet-derived growth factor.
Source: Summarized from refs. 99 and 100.

asthma (99), and Baggiolini and Dahinden reported on the role of chemokines in allergic reactions (100). From the factors listed in these reviews, Table 2 was constructed to include those factors that are made by macrophages and to indicate their likely role in asthma. Most factors directly cause the indicated effect, but others can act indirectly by activating other cells in the lung environment, including stimulating nerves to cause bronchoconstriction. Factors that mediate the asthmatic response are numerous, and redundancy is so great that an antagonist of any single factor is unlikely to be completely effective at preventing asthma. Similarly, because so many lung cells are capable of making many of these factors, it is difficult to attribute primary responsibility to any one cell type.

Evidence that macrophages play a role in asthma is that they are both increased in tissues of patients with asthma and show evidence of activation (98). In one study, bronchial biopsies were obtained from 14 patients with recently diagnosed asthma and four controls (101). Significantly increased numbers of mast cells, eosinophils, lymphocytes, and macrophages were all present in the epithelium of patients as compared with controls. Significantly increased numbers of all of these cells, except for mast cells, were also present in the lamina propria together with increased numbers of plasma cells. In addition to demonstrating an increase in macrophages, these studies were important to document that inflammation was present in asthmatic lungs even early in the course of the disease.

A special comment about one of the two subfamilies of chemokines, the so-called CC, or β, chemokines, is worthwhile because of their potent recruiting and activating properties for critical effector cells in asthma; that is, eosinophils, basophils, lymphocytes, and monocytes (see also Chapter 7.). Chemokines are inducible, novel small proteins that have four conserved cysteines which form disulfide bonds (100). The two subfamilies are distinguished by the arrangement of the first two cysteines; they are adjacent in the CC chemokines and separated by one amino acid in the CXC, or α, chemokines. IL-8 is the classic CXC chemokine and has particularly potent activity for neutrophils. On the other hand, the CC chemokines, such as monocyte chemotactic proteins (MCPs) MCP-1, MCP-2, and MCP-3, RANTES, and macrophage inflammatory proteins (MIPs) MIP-1α and MIP-1β, have potent activity for monocytes as well as basophils, eosinophils, and lymphocytes. CC chemokines can both prime, recruit, and activate these important effector cells in asthma. MIP-1α and MIP-1β have been shown to have differential abilities to attract lymphocyte subsets; MIP-1α attracts CD8 T cells and B cells, and MIP-1β is chemotactic for CD4 T cells (102,103). Therefore, MIP-1α and MIP-1β could play important roles in the recruitment of IgE-secreting B cells and Th2 T cells, respectively. Not only do macrophages and monocytes respond to the CC chemokines, they also make many of them, including MIP-1α, and MIP-1β, and MCP-1. A role for these chemokines has been found in asthma. Following antigenic challenge of asthmatics, potent lymphocyte-recruiting activity was found in BAL (104). Although the majority of the activity was ascribed to IL-16, high levels of MIP-1α were also found and could have come from macrophages. MCP-1 has also been detected in bronchial biopsies of asthmatics and controls, especially in bronchial epithelium, but also in subepithelial macrophages (105).

Chronic airway thickening in asthma may cause worsening of asthma and eventually lead to fixed airway disease. At a meeting on future directions in asthma research sponsored by the American Lung Association and American Thoracic Society in 1993, scientists were convened to discuss important issues of the relationship of inflammation and asthma. One group addressed issues related to nonbronchospastic airway obstruction in asthma (106). James Hogg, University of British Columbia, using a morphometric analysis of lung tissue from patients with asthma, chronic bronchitis, and controls showed that asthmatic lungs were characterized by a uniform thickness of airway walls as compared with patients with chronic bronchitis or controls. The thickening reflected an expansion of smooth muscle, interstitium, and vessels. In an attempt to assess whether these changes were the result of an excess production of growth factors, immunochemical staining detected no differences in TGF-β and PDGF in asthmatics when compared with chronic bronchitic patients or normal controls. Nevertheless, it is possible that over time these growth factors, both of which can be made by macrophages, are intermittently expressed in asthmatics and contribute to these chronic airway changes.

VI. Conclusions

A burgeoning literature is providing a renewed enthusiasm about the possibilities of understanding the underlying causes of asthma and constructing novel strategies to prevent and treat this widespread disease. Immunologists have provided the basis for this understanding by dissecting mechanisms for regulating immune responses to environmental antigens. Pulmonary scientists, many who have been studying asthma for years, have embraced this knowledge and are applying it to their own systems. Undoubtedly, lung DCs and macrophages are pivotal cells in both initiating immunity to allergens and driving acute and chronic inflammatory responses in the asthmatic lung. With a greater understanding of their role in these processes, ways will be developed to manipulate their involvement to help control asthma.

Acknowledgments

The authors gratefully acknowledge discussions with colleagues who helped shape our ideas about mechanisms in asthma and the role of dendritic cells and macrophages in this disease, particularly Drs. David Bice, Mark Schuyler, Barbara Masten, and Julie Wilder. Studies of ours which are cited here were funded by grants from the National Institutes of Health: AI21951, AI21278, HL29543, and MO1 RR00633.

References

1. Djukanovic R, Roche WR, Wilson JW, Beasley CRW, Twentyman OP, Howarth PH, Holgate ST. State of the art: mucosal inflammation in asthma. Am Rev Respir Dis 1990; 142:435–457.
2. McFadden ER Jr, Gilbert IA. Asthma. N Engl J Med 1992; 327:1928–1937.
3. Paul W. Determination of the lymphokine-producing phenotype of CD4+ T cells: a potential control point in allergic and infectious diseases. Ann Intern Med 1994; 121:698–708.
4. Lipscomb M. The Respiratory System. In: Kumar V, Cotran R, Robbins S, eds. Basic Pathology. Philadelphia: Saunders, 1992; 389–391.
5. Romagnani S. Regulation of the development of type 2 T-helper cells in allergy. Curr Opin Immunol 1994; 6:838–846.
6. Barnes P. Cytokines as mediators of chronic asthma. Am J Respir Crit Care Med 1994; 150:S42–S49.
7. Ricci M, Matucci A, Rossi O. Recent advances in the pathogenetic mechanisms and genetic aspects of atopic diseases. ACI News 1994; 6:103–108.
8. Kroegel C, Virchow J-C, Luttmann W, Walker C, Warner JA. Pulmonary immune

cells in health and disease: the eosinophil leucocyte (Part I). Eur Respir J 1994; 7:519–543.
9. Devos R, Plaetinck G, Cornelis S, Guisez Y, der Heyden JV, Tavernier J. Interleukin-5 and its receptor: a drug target for eosinophilia associated with chronic allergic disease. J Leukoc Biol 1995; 57:813–819.
10. Street NE, Mosmann TR. Functional diversity of T lymphocytes due to secretion of different cytokine patterns. FASEB J 1991; 5:171–177.
11. Swain SL, Bradley LM, Croft M, Tonkonogy S, Atkins G, Weinberg AD, Duncan DD, Hedrick SM, Dutton RW, Huston G. Helper T-cell subsets: phenotype, function and the role of lymphokines in regulating their development. Immun Rev 1991.
12. Kelso A. Th2 and Th2 subsets: paradigms lost? Immunol Today 1995; 16:374–379.
13. Hall SS. IL-12 at the crossroads. Science 1995; 268:1432–1434.
14. Trinchieri G. Interleukin-12 and interferon-γ; Do they always go together? Am J Pathol 1995; 147:1534–1538.
15. Mosmann TF. Properties and functions of interleukin-10. Adv Immunol 1994; 56: 1–26.
16. Schwartz LB. Mast cells: function and contents. Curr Biol 1994; 6:91–97.
17. Marshall JS, Bienenstock J. The role of mast cells in inflammatory reactions of the airways, skin and intestine. Curr Biol 1994; 6:853–859.
18. Lipscomb MF, Bice DE, Lyons R, Schuyler MR, Wilkes D. The regulation of pulmonary immunity. Adv Immunol 1995; 59:369–455.
19. Gascan H, Gauchat J-F, Roncarolo M-G, Yssel H, Spits H, de Vries JE. Human B cell clones can be induced to proliferate and to switch to IgE and IgG4 synthesis by interleukin 4 and a signal provided by activated CD4+ T cell clones. J Exp Med 1991; 173:747.
20. Spriggs MK, Armitage RJ, Strockbine L, Clifford KN, Macduff BM, Sato TA, Maliszewski CR, Fanslow WC. Recombinant human CD40 ligand stimulated B cell proliferation and immunoglobulin E secretion. J Exp Med 1992; 176:1543–1550.
21. Martinez FD, Wright AL, Taussig LM, Holberg CJ, Halonen M, Morgan WJ, Associates GHM. Asthma and wheezing in the first six years of life. N Engl J Med 1995; 332:133–138.
22. Society AT. Early childhood asthma: What are the questions? Workshop on Childhood Asthma. Salem, MA. Am J Respir Clin Care Med 1995; 151:S1–S42.
23. Marsh DG, Meyers DA, Bias WB. The epidemiology and genetics of atopic allergy. N Engl J Med 1981; 305:1551–1559.
24. Blumenthal M. Searching for the asthma and allergy gene—the current status of our knowledge. ACI News 1995; 7:15–19.
25. Marsh D, Neely J, Breazeale D, Ghosh B, Freidhoff L, Ehrlich-Kautzky E, Schou C, Krishnaswamy G, Beaty T. Linkage analysis of IL4 and other chromosome 5q31.1 markers and total serum immunoglobulin E concentrations. Science 1994; 264: 1152–1156.
26. Postma DS, Bleecker ER, Amelung PJ, Holroyd KJ, Xu J, Panhuysen CIM, Meyers DA, Levitt RC. Genetic susceptibility to asthma-bronchial hyperresponsiveness coinherited with a major gene for atopy. N Engl J Med 1995; 333:894–900.
27. Zwollo P, Ehrlich-Kautzky E, Ansari A, Scharf S, Erlich H, Marsh D. Molecular

studies of human immune response genes for the short ragweed allergen, Amb a V: sequencing of HLA-D second exon in responders and non-responders. Immunogenetics 1991; 33:141–151.
28. Matsushita S, Muto M, Suemura M, Saito Y, Sasazuki T. HLA-linked nonresponsiveness to Cryptomeria japonica pollen antigen. J Immunol 1987; 138:109–115.
29. Cookson W, Sharp P, Faux J, Hopkin J. Linkage analysis between immunoglobulin E responses underlying asthma and rhinitis and chromosome 11q. Lancet 1989; 1: 1292–1295.
30. Holt PG, Batty JE, Turner KJ. Inhibition of specific IgE responses in mice by pre-exposure to inhaled antigen. Immunology 1981; 42:409–417.
31. Nelson DJ, McMenamin C, McWilliam AS, Brenan M, Holt PG. Development of the airway intraepithelial dendritic cell network in the rat from class II major histocompatibility (Ia)–negative precursors: differential regulation of Ia expression at different levels of the respiratory tract. J Exp Med 1994; 179:203–212.
32. Holt P. Developmental immunology and host defense: kinetics of postnatal maturation of immune competence as a potential etiologic factor in early childhood asthma. Am J Respir Crit Care Med 1995; 151:S11–S13.
33. Garssen J, Nijkamp F, Ban Der Bliet H, Van Loveren H. T-cell-mediated induction of airway hyperreactivity in mice. Am Rev Respir Dis 1991; 144:931–938.
34 Peden DB. The effect of air pollution in asthma and respiratory allergy—the American experience. ACI News 1995; 7:19–23.
35. Halken S, Høst A, Nilsson L, Taudorf E. Passive smoking as a risk factor for development of obstructive respiratory disease and allergic sensitization. Allergy 1995; 50:97–105.
36. Bråback L, Breborowicz A, Dreborg S, Knutsson A, Pieklik H, Björkstén B. Atopic sensitization and respiratory symptoms among Polish and Swedish school children. Clin Exp Allergy 1994; 24:826–835.
37. Garside P, Mowat AM. Polarization of Th-cell responses: a phylogenetic consequence of nonspecific immune defence? Immunol Today 1995; 16:220–223.
38. Romagnani S. Lymphokine production by human T cells in disease states. Annu Rev Immunol 1994; 12:227–257.
39. Pearce EJ, Reiner SL. Induction of Th2 responses in infectious diseases. Curr Opin Immunol 1995; 7:497–504.
40. Biron DA, Gazzinelli RT. Effects of IL-12 on immune responses to microbial infections: a key mediator in regulating disease outcome. Curr Opin Immunol 1995; 7:485–496.
41. Bradley LM, Yoshimoto K, Swain SL. The cytokines IL-4, IFN-γ and IL-12 regulate the development of subsets of memory effector helper T cells in vitro. J Immunol 1995; 155:1713–1724.
42. Okamura H, Tsutsui H, Komatsu T, Yutsudo M, Hakura A, Tanimoto T, Torigoe K, Okura T, Nukada Y, K. H, et al. Cloning of a new cytokine that induces IFG-γ production by T cells. Nature 1995; 378:88–91.
43. Hsieh C-S, Macatonia SE, Tripp CS, Wolf SF, O'Garra A, Murphy KM. Development of Th1 CD4+ T helper cells through IL-12 produced by *Listeria*-induced macrophages. Science 1993; 260:547–549.

44. Bradding P, Feather I, Howarth P, Mueller R, Roberts J, Britten K, Bews J, Hunt T, Okayama Y, Heusser C, et al. Interleukin 4 is localized to and released by human mast cells. J Exp Med 1992; 176:1381–1386.
45. Brunner T, Heusser CH, Dahinden CA. Human peripheral blood basophils primed by interleukin 3 (IL-3) produce IL-4 in response to immunoglobulin E receptor stimulation. J Exp Med 1993; 177:605–611.
46. Nonaka M, Nonaka R, Woolley K, Adelroth E, Miura K, Okhawara Y, Glibetic M, Nakano K, O'Byrne P, Dolovich J, et al. Distinct immunohistochemical localization of IL-4 in human inflamed airway tissues. J Immunol 1995; 155:3234–3244.
47. do Carmo Leite-de-Moraes M, Herbelin A, Machavoine F, Vicari A, Gombert J-M M, Papiernik M. MHC class I–selected CD4-CD8-TCR-$\alpha\beta$+ T cells are a potential source of IL-4 during primary immune response. J Immunol 1995; 155: 4544–4550.
48. Yoshimoto T, Paul WE. CD4pos, NK1.1pos T cells promptly produce interleukin 4 in response to in vivo challenge with ani-CD3, Nk1.1 T cells promptly produce interleukin 4 in response to in vivo challenge with ani-CD3. J Exp Med 1994; 179: 1285–1295.
49. Ferrick DA, Schrenzel MD, Mulvania T, Hsieh B, Ferlin WG, Lepper H. Differential production of interferon-γ and interleukin-4 in response to Th1- and Th2-stimulating pathogens by $\gamma\delta$ T cells in vivo. Nature 1995; 373:255–257.
50. Stein-Streilein J, Bennett M, Mann D, Kumar V. Natural killer cells in mouse lung: Surface phenotype, target preference, and response to local influenze virus infection. J Immunol 1983; 131:2699–2704.
51. Abraham E, Freitas A, Coutinho A. Purification and charcterization of intraparenchymal lung lymphocytes. J Immunol 1990; 144:2117–2122.
52. Aversa G, Punnonen J, Cocks BG, Waal Mallefyt RD, Vega F, Zurawski SM, Zurawski G, De Vries JE. An interleukin 4 (IL-4) mutant protein inhibits both IL-4 or IL-13-dependent human IgG4 and IgE synthesis and B cell proliferation: support for a common component shared by IL-4 and IL-13 receptors. J Exp Med 1992; 178: 2213–2218.
53. Huang S-K, Xiao H-Q, Kleine-Tebbe J, Paciotti G, Marsh DG, Lichtenstein LM, Liu MC. IL-13 expression at the sites of allergen challenge in patients with asthma. J Immunol 1995; 155:2688–2694.
54. de Waal Malefyt R, Abrams J, Bennett B, Figdor C, de Vries J. Interleukin 10(IL-10) inhibits cytokine synthesis by human monocytes: an autoregulatory role of IL-10 produced by monocytes. J Exp Med 1991; 174:1209–1220.
55. Macatonia SE, Doherty TM, Knight SC, O'Garra A. Differential effect of IL-10 on dendritic cell-induced T cell proliferation and IFN-gamma production. J Immunol 1993; 150:3755–3765.
56. D'Andrea A, Aste-Amezaga M, Valiante NM, Ma X, Kubin M, Trinchieri G. Interleukin-10 (IL-10) inhibits human lymphocyte IFN-γ production by suppressing natural killer cell stimulatory factor/IL-12 synthesis in accessory cells. J Exp Med 1993; 178:1041–1048.
57. Chensue SW, Ruth JH, Warmington K, Lincoln P, Kunkel SL. In vivo regulation of macrophage IL-12 production during type 1 and type 2 cytokine-mediated granuloma formation. J Immunol 1995; 155:3546–3551.

58. Lipscomb MF, Pollard AM, Yates JL. A role for TGF-beta in the suppression by murine bronchoalveolar cells of lung dendritic cell initiated immune responses. Reg Immunol 1993; 5:151–157.
59. Maeda H, Shiraishi A. TGF-β contributes to the shift toward Th2-type responses through direct and IL-10-mediated pathways in tumor-bearing mice. J Immunol 1996; 156:73–78.
60. Hoehn P, Goedert S, Germann T, Koelsch S, Jin S, Palm N, Ruede E, Schmitt E. Opposing effects of TGF-β_2 on the Th1 cell development of naive CD4+ T cells isolated from different mouse strains. J Immunol 1995; 155:3788–3793.
61. Sandborg CI, Imfeld KL, Zaldivar F Jr., Wang Z, Buckingham BA, Berman MA. IL-4 expression in human T cells is selectively inhibited by IL-1α and IL-1β. J Immunol 1995; 155:5206–5212.
62. Katamura K, Shintaku N, Yamauchi Y, Fukui T, Ohshima Y, Mayumi M, Furusho K. Prostaglandin E2 at priming of naive CD4+ T cells inhibits acquisition of ability to produce IFN-γ and IL-2, but not IL-4 and IL-5. J Immunol 1995; 155:4604–4612.
63. Masuda E, Naito Y, Arai K, Arai N. Expression of lymphokine genes in T cells. The Immununologist 1993; 1:198–203.
64. Carrucci J, Auci D, Herrick C, Durkin H. Neuropeptide-mediated regulation of hapten-specific IgE responses in mice I. Substance P-mediated isotype-specific suppression of BPO-specific IgE antibody-forming cell resonses induced in vivo and in vitro. J Leukoc Biol 1995; 57:110–115.
65. Romagnani S. Regulation of the development of type 2 T helper cells in allergy. Curr Opin Immunol 1994; 6:838–846.
66. Fasler S, Aversa G, Terr A, Thestrup-Pedersen K, de Vries JE, Yssel H. Peptide-induced anergy in allergen-specific human Th2 cells results in lack of cytokine production and B cell help for IgE synthesis. J Immunol 1995; 155:4199–4206.
67. Sloan-Lancaster J, Allen PM. Significance of T-cell stimulation by altered peptide ligands in T cell biology. Curr Opin Immunol 1995, 7:103–109,
68. Zhan Y, Cheers C. Differential induction of macrophage-derived cytokines by live and dead intracellular bacteria in vitro. Infect Immun 1995; 64:720–723.
69. Hewitt CRA, Brown AP, Hart BJ, Pritchard DI. A major house dust mite allergen disrupts the immunoglobulin E network by selectively cleaving CD23: innate protection by antiproteases. J Exp Med 1995; 182:1537–1544.
70. Luo H, Hofstetter H, Bancherauu J, Delespesse G. Cross linking of CD23 antigen by its natural ligand (IGE) or by anti-CD23 antibody prevents B lymphocytes proliferation and differentiation. J Immunol 1991; 146:2122–2129.
71. Yu P, Kosco-Vilbois M, Richards M, Kohler G, Lamers M. Negative feedback regulation of IgE synthesis by murine CD23. Nature 1994; 369:753–756.
72. Evans M, Pryor W. Cigarette smoking, emphysema and damage to alpha-1-proteinase inhibitor. Am J Physiol 1994; 266:L593–L611.
73. Hosken NA, Shibuya K, Heath AW, Murphy KM, O'Garra A. The effect of antigen dose on CD4+ T helper cell phenotype development in a T cell receptor–αβ-transgenic model. J Exp Med 1995; 182:1579–1584.
74. Secrist H, DeKruyff RH, Umetsu DT. Interleukin 4 production by CD4+ T cells from allergic individuals is modulated by antigen concentration and antigen-presenting cell type. J Exp Med 1995; 181:1081–1089.

75. Williams ME, Caspar P, Oswald I, Sharma HK, Pankewycz O, Sher A, James SL. Vaccination routes that fail to elicit protective immunity against *Schistosoma mansoni* induce the production of TGF-β, which down-regulates macrophage antiparasitic activity. J Immunol 1995; 154:4693–4700.
76. Lyons CR, Ball EJ, Toews GB, Weissler JC, Stastny P, Lipscomb MF. Inability of human alveolar macrophages to stimulate resting T cells correlates with decreased antigen-specific T cell–macrophage binding. J Immunol 1986; 137:1173–1180.
77. Gajewski TF, Pinnas M, Wong T, Fitch FW. Murine Th1 and Th2 clones proliferate optimally in response to distinct antigen-presenting cell populations. J Immunol 1991; 146:1750–1758.
78. DeKruyff RH, Fang Y, Umetsu DT. IL-4 synthesis by in vivo primed keyhole limpet hemocyanin-specific CD4+ T cells. J Immunol 1992; 149:3468–3476.
79. Rossi-Bergmann B, Müller I, Godinho EB. Th1 and Th2 T-cell subsets are differentially activated by macrophages and B cells in murine leishmaniasis. Infect Immun 1993; 2266–2269.
80. Kuchroo VK, Das MP, Brown JA, Ranger AM, Zamvil SS, Sobel RA, Weiner HL, Nabavi N, Glimcher LH. B7-1 and B7-2 costimulatory molecules activate differentially the Th1/Th2 developmental pathways: Application to autoimune disease therapy. Cell 1995; 80:707–718.
81. Del Prete G, De Carli M, E'Elios MM, Daniel KE, Almerigogna F, Alderson M, Smith CA, Thomas E, Romagnani S. CD30-mediated signaling promotes the development of human T helper type 2-like T cells. J Exp Med 1995; 182:1655–1661.
82. Shahinian A, Pfeffer K, Lee KP, Kundig TM, Kishihara K, Wakeham A, Kawai K, Ohashi PS, Thompson CB, Mak TW. Differential T cell costimulatory requirements in CD28-deficient mice. Science 1993; 261:609–612.
83. Heinzel FP, Sadick MD, Holaday BJ, Coffman RL, Locksley RM. Reciprocal expression of interferon gamma or interleukin 4 during the resolution or progression of murine leishmaniasis. J Exp Med 1989; 169:59–72.
84. Sadick MD, Heinzel FP, Holaday BJ, Pu RT, Dawkins RS, Locksley RM. Cure of murine leishmaniasis with anti-interleukin 4 monoclonal antibody. Evidence for a T cell–dependent, interferon γ–independent mechanism. J Exp Med 1990; 171: 115–127.
85. Hsieh C-S, Macatonia SE, O'Garra A, Murphy KM. T cell genetic background determines default T helper phenotype development in vitro. J Exp Med 1995; 181: 713–721.
86. Chung K. Usefulness of animal models in asthma research. Eur Respir Rev 1995; 5: 184–187.
87. Abraham W. Animal models of late bronchial responses. Eur Respir Rev 1995; 5: 211–217.
88. Tu Y-P, Larsen G, Irvin C. Utility of murine systems to study asthma pathogenesis. Eur Respir Rev 1995; 5:224–230.
89. Kung T, Jones H, Adams G III, Umland S, Kreutner W, Egan R, Chapmen R, Watnick A. Characterization of a murine model of allergic pulmonary inflammation. Int Arch Allergy Immunol 1994; 105:83–90
90. Kurup V, Choi H, Murali P, Coffman R. IgE and eosinophil regulation in a murine model of allergic aspergillosis. J Leukoc Biol 1994; 56:593–598.

91. Gavett S, Chen X, Finkelman F, Wills-Karp M. Depletion of murine CD4+ T lymphocytes prevents antigen-induced airway hyperreactivity and pulmonary eosinophilia. Am J Respir Cell Mol Biol 1994; 10:587–593.
92. McMenamin C, Pimm C, McKersey M, Holt P. Regulation of IgE responses to inhaled antigen in mice by antigen-specific γδ T cells. Science 1994; 265:1869–1871.
93. Lack G, Renz H, Saloga J, Bradley K, Loader J, Leung D, Larsen G, Gelfan E. Nebulized but not parenteral IFN-γ decreases IgE production and normalizes airways function in a murine model of allergen sensitization. J Immunol 1993; 152: 2536–2554.
94. Renz H, Lack G, Saloga J, Schwinzer R, Bradley K, Loader J, Kupfer A, Larsen G, Gelfand E. Inhibition of IgE production and normalization of airways responsiveness by sensitized CD8 T cells in a mouse model of allergen-induced sensitization. J Immunol 1994; 152:351–360.
95. Thepen T, Van Rooijen N, Kraal G. Alveolar macrophage elimination in vivo is associated with an increase in pulmonary immune resonse in mice. J Exp Med 1989; 170:499–509.
96. Holt P, Oliver J, Bilyk N, McMenamin C, McMenamin P, Kraal G, Thepen T. Downregulation of the antigen presenting fell function(s) of pulmonary dendritic cells in vivo by resident alveolar macrophages. J Exp Med 1993; 177:397–407.
97. Nelson D, Holt P. Defective regional immunity in the respiratory tract of neonates is attributable to hyporesponsiveness of local dendritic cells to activation signals. J Immunol 1995; 155:3517–3524.
98. Lee T. Interactions between alveolar macrophages, monocytes, and granulocytes: implications for airway inflammation. Am Rev Respir Dis 1987; 135:S14–S17.
99. Bochner B, Undem B, Lichtenstein L. Immunological aspects of allergic asthma. Annu Rev Immunol 1994; 12:295–335.
100. Baggiolini M, Dahinden C. CC chemokines in allergic inflammation. Immunol Today 1994; 15:127–133.
101. Laitinen L, Laitinen A, Haahtela T. Airway mucosal inflammation even in patients with newly diagnosed asthma. Am Rev Respir Dis 1993; 147:697–704.
102. Taub D, Conlon K, Lloyd A, Oppenheim J, Kelvin D. Preferential migration of activated CD4+ and CD8+ T cells in response to MIP-1a and MIP-1b. Science 1993; 260:355–358.
103. Schall T, Bacon K, Camp R, Kaspari J, Goeddel D. Human macrophage inflammatory protin a (MIP-1α) and MIP-1β chemokines attract distinct populations of lymphocytes. J Exp Med 1993; 177:1821–1825.
104. Cruikshank W, Long A, Tarpy R, Kornfeld H, Carroll M, Teran L, Holgate S, Center D. Early identification of Interleukin-16 (lymphocyte chemoattractant factor) and macrophage inflammatory protein 1α (MIP1α) in bronchoalveolar lavage fluid of antigen-challenged asthmatics. Am J Respir Cell Mol Biol 1995; 13:738–747.
105. Sousa A, Lane S, Nakhosteen J, Yoshimura T, Lee T, Poston R. Increased expression of the monocyte chemoattractant protein-1 in bronchial tissue from asthmatic subjects. Am J Respir Cell Mol Biol 1994; 10:142–147.
106. Boucher R, Gilbert I, Hogg J, King M, Knowles M, Nadel J. Nonmuscular airway obstruction and asthma. Am J Respir Crit Care Med 1995; 152:408–410.

107. McMenamin C, Holt P. The natural immune response to inhaled soluble protein antigens involves major histocompatibility complex (MHC) class I–restricted CD8+ T cell–mediated but MHC class II-restricted CD4+ T cell–dependent immune deviation resulting in selective suppression of immunoglobulin E production. J Exp Med 1993; 178:889–899.

23

Macrophages in the Pneumoconioses

Asbestosis, Silicosis, and Coal Workers' Pneumoconiosis

WILLIAM N. ROM

New York University Medical Center
New York, New York

I. Introduction

Alveolar macrophages (AMs) constitute an important host defense mechanism against inhaled inorganic particulates and fibers. Following respiratory ciliary clearance, particles may impact on airway bifurcations or settle by sedimentary action onto the surface of large airways. Fibers deposit by interception dependent on the width of the fiber rather than the length for those that are propelled like a spear in the inhaled breath. AMs consequently process smaller particles <10 μm in size and those that are <1 μm that reach alveolar space by Brownian motion. Fibers that are narrow but long (high-aspect ratio fibers >10–20 μm) may be phagocytosed by multiple AMs. Thus, AMs function to clear the alveolar spaces.

II. Alveolar Macrophage Particle Burdens in Exposed Workers

The particle burden of AMs reflects the past type of exposure and approximately correlates with particle burden. Bronchoalveolar lavage (BAL) provides a powerful research tool to sample the alveolar spaces to obtain a particle burden index

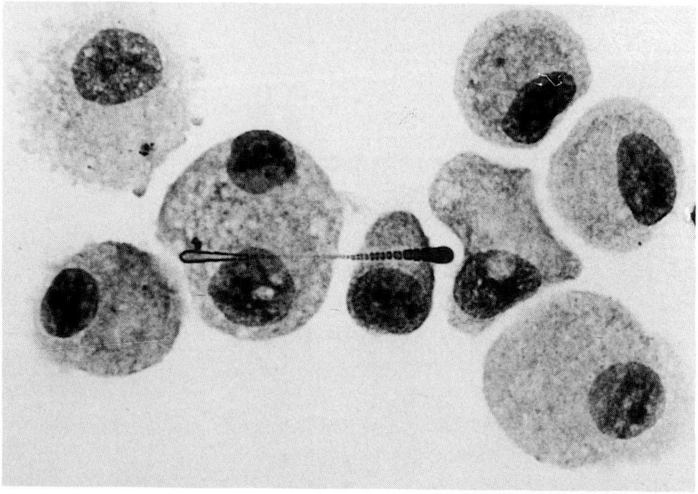

(A)

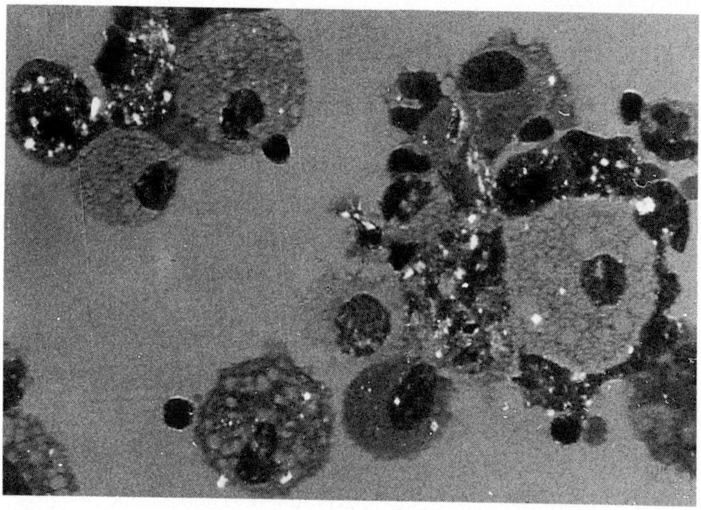

(B)

Figure 1 Light microscopic evaluation of particles in cytocentrifuge preparations of alveolar macrophages. Wright-Giemsa stain; magnification ×400. (A) AMs from nonsmoking asbestos insulator reveals a binucleate macrophage phagocytosing a beaded asbestos body. (B) Birefringent particles are seen in polarized light micrograph of AMs from a silica-exposed subject. (C) Coal and other mineral particles seen in AMs from a coal miner. (Courtesy of T. Takemura and V. Ferrans, MD.)

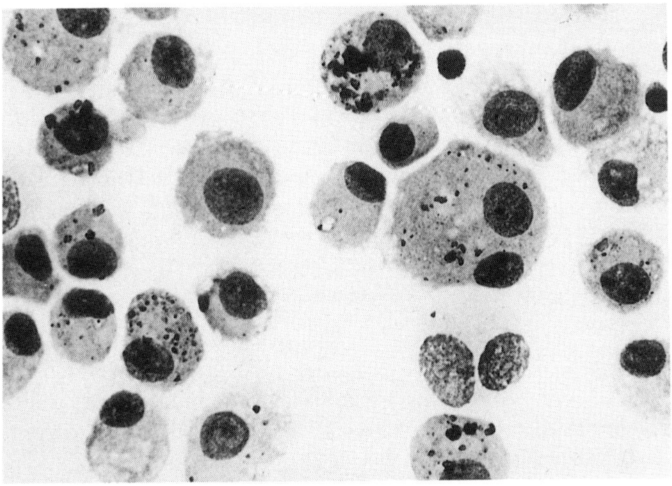

(C)

(per 10^6 AMs or per milliliter of BAL fluid recovered) of past in vivo exposure. Coupled with scanning/transmission electron microscopy (S/TEM), energy-dispsersive x-ray analysis (EDXA), and electron diffraction, particles can be identified, sized, and enumerated.

In a study of 43 inorganic dust-exposed individuals compared with 16 normal volunteers, all life-long nonsmokers or exsmokers for >5 years, Takemura and colleagues reported significantly increased numbers of particles in the dust-exposed individuals compared with normal controls (1). The percentages of AMs containing particles were asbestos-exposed 37 ± 4%, silica-exposed 57 ± 4%, coal-exposed 74 ± 4%, normal 16 ± 1%, $p < .01$; all comparisons to normals (Fig. 1). The asbestos-exposed workers (n = 19) had worked for 32 ± 2 years in the building trades as insulators, boilermakers, or steamfitters. The silica-exposed workers (n = 10) had worked for 11 ± 4 years as foundry workers, pottery workers, or quartz mill or were exposed to calcined diatomaceous earth. The coal miners (n = 11) worked at the coal face in Appalachian coal mines. The percentage of AMs containing birefringent particles was significantly higher among each of the dust-exposed individuals: asbestos 14 ± 2%, silica 52 ± 6%, coal 65 ± 4%, and normal 6 ± 1%. The percentage of particle-laden macrophages were not significantly different in currently exposed workers versus those who had not been exposed for 1 or more years. Coal mine dust particles could be found in AMs of coal miners who had retired 20 years earlier, and the asbestos workers had not been exposed to asbestos since the early 1970s. Dust-exposed workers with restrictive lung function (vital capacity, total lung capacity, or diffusing capacity

<80% predicted) did not have a greater proportion of AMs with particles than did those with normal pulmonary function for each of the three groups.

To develop a particle burden index per 10^6 AMs, AMs were digested with bleach and particles collected on a 0.45-μ pore size Millipore filter to be analyzed by S/TEM with electron diffraction and energy-dispersive x-ray spectroscopy (2) (Fig. 2). Comparing 28 asbestos, coal, or silica-exposed individuals to 8 normals, there was a fourfold increase in particle burden ($3.9 \pm 0.0 \times 10^6$ particles in the dust-exposed workers versus $1.0 \pm 0.1 \times 10^6$ particles in the normals, $p < .02$). The total particle counts per 10^6 AMs were asbestos-exposed AMs $2330 \pm 850 \times 10^3$, coal-exposed AMs $3582 \pm 1324 \times 10^3$, silica-exposed AMs $6938 \pm 1245 \times 10^3$, and normals $928 \pm 319 \times 10^3$. Interestingly, coal miners had a variety of minerals in their AMs: talc, kaolin, mullite, mica, feldspar, aluminum, titanium, biotite, and calcium. Silica was predominantly found in the silica-exposed workers and one third of the particles were <0.1 μm in diameter. At least one third

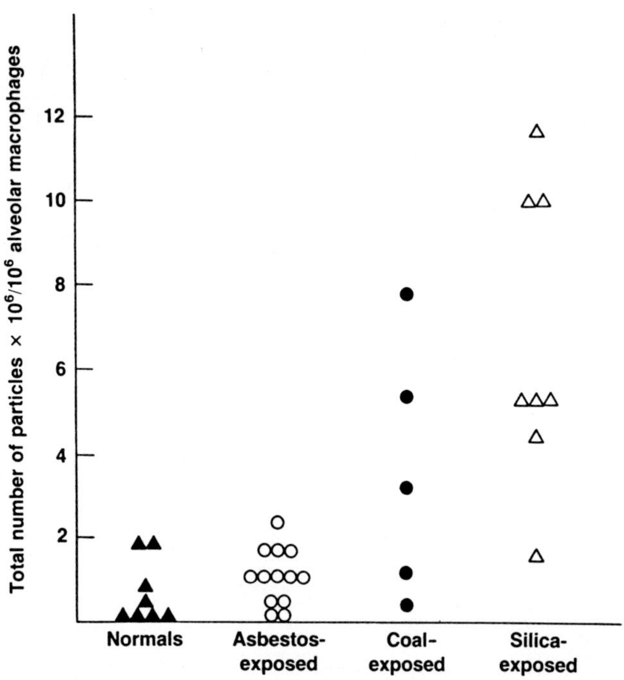

Figure 2 Total number of particles per 10^6 AMs in normal or inorganic dust-exposed individuals. All were nonsmokers or exsmokers for >5 years. There was a significant increase in particles from the silica-exposed ($p < .01$) and coal-exposed ($p < .05$) subjects than normal controls using analytical optical electron microscopy. (From ref. 2.)

of the coal-exposed macrophages contained silica, especially <0.5 μm in size. The presence of silica particles in coal miners' AMs suggests that silica might be responsible for the macrophage activation inducing lung injury and fibrosis in this disorder. Coal miners averaged >700 iron and potassium-containing aluminum silicates per cell that were <1 μm in size. Coal particles were usually >0.5 μm in diameter. Importantly, half of the inorganic dust particles were <0.2 μm in diameter, which is below the resolution of the light microscope.

In collaboration with Andrew Churg, we analyzed asbestos fibers/10^6 AMs in 24 US asbestos workers exposed as insulators or boilermakers and 8 normals. Both chrysotile and amosite were identified (chrysotile 28,163 ± 290 fibers versus 1,114 ± 47 in normals, $p <. 001$, amosite 4658 ± 225 fibers versus 300 ± 294 in normals, $p < .01$). There was approximately one chrysotile fiber per 35 AMs and one amosite fiber per 215 AMs. Chrysotile was found in 20 of 24 asbestos workers' AMs, amosite in 13 of 24, tremolite in 2 of 24, and crocidolite in none (Fig. 3). Two controls had small numbers of chrysotile or tremolite, and one had amosite. The asbestos workers had on average 3308 ± 1173 asbestos bodies/10^6 AMs, which was similar to the mean number of amosite fibers per 10^6 AMs. There was a significant correlation between the number of asbestos bodies observed by

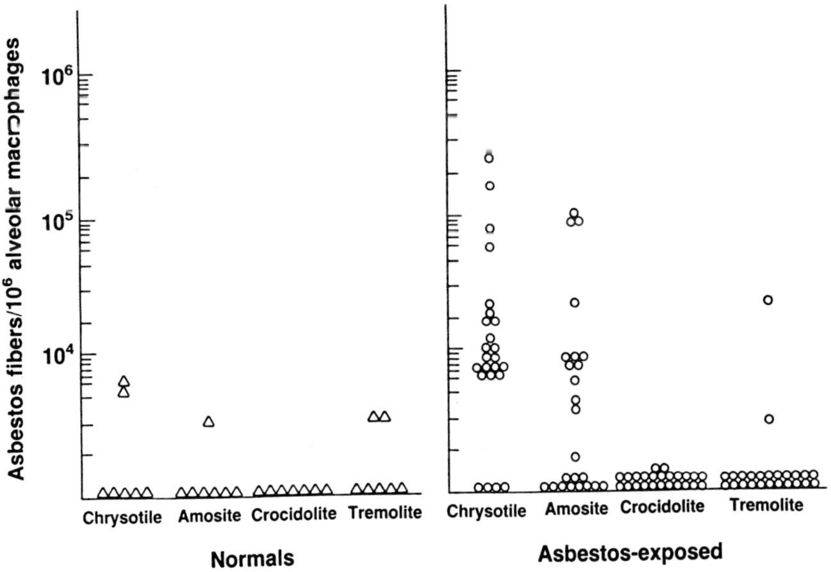

Figure 3 Number of asbestos fibers per 10^6 AMs identified as chrysotile, amosite, crocidolite, or tremolite using analytical optical electron microscopy in normal versus asbestos-exposed subjects. (From ref. 2.)

S/TEM (energy optical analysis) and the asbestos exposure index $r = 0.4, p < .05$). The asbestos exposure index was the number of years exposed times the intensity of exposure with job intensity graded as 4 = asbestos insulator, 3 = boilermaker, 2 = sheet metal worker, and 1 = bystander asbestos exposure.

In France, 18 asbestos workers had asbestos fibers and bodies (ABs) evaluated: 6 had $>10^3$ AB/ml, whereas 8 had fewer than 20 AB/ml, and in 6 of 9 workers, these were predominantly amphiboles (3). Ten of 14 workers had $>10^4$/ml fibers seen with the analytical TEM. In a large series of 563 BAL procedures, 10–30 ml BAL fluid was mixed with equal volumes of bleach and filtered on 0.45 mm Millipore filters for identification of AB/ml BAL fluid with the phase contrast microscope (4). The presence of >1 AB/ml reflected occupational exposure with log mean 120.5 AB/ml seen in radiographic asbestosis (range 0–42,600 AB/ml). Only 7% of white-collar workers and 18% of blue-collar workers had ⩾ 1 AB/ml. Those with pleural disease, lung cancer, mesothelioma, or merely asbestos exposure had considerable overlap with geometric means 4.8–8.5 AB/ml. Of 49 patients with more than 100 AB/ml, 30 had asbestosis, 8 had pleural disease, 13 had lung cancer or mesothelioma, 2 were exposed to chrysotile in a brake-lining factory, and 1 had no apparent disease. The investigators reported a statistically significant correlation ($r = 0.23, p < .01$) between AB counts and the duration of exposure. In bystander asbestos-exposed workers, predominantly sheet metal workers and plumbers, AB/ml BAL ranged from 0 to 34 (5). There were only 3 of 71 workers with ⩾ 10 AB/ml, and there were no positive correlations with indices of exposure, chest radiograph, or pulmonary function. Teschler and colleagues, however, found a positive correlation ($r = 0.75, p < .001$) between BAL and lung tissue AB on 23 paired samples from 20 patients occupationally exposed to asbestos (6). There was a significant correlation for amphiboles, but not chrysotile, which is consistent with the well-known fact that chrysotile fibers occur in bundles that splinter over time and are more readily cleared or broken up in lung and that fewer ferruginous bodies form on chrysotile. The aspect ratio of fibers was 50% higher in the lung tissue than in BAL; asbestos fiber dimensions in tissue were twice as long but had similar widths as BAL fibers.

Johnson and colleagues reported on 12 inorganic dust-exposed patients and 10 patients with no known history of dust exposure using energy dispersive x-ray microanalysis (7). They detected 18 types of particles including fibrous forms of asbestos and mullite; platy forms (e.g., chlorite, talc, mica, and kaolinite); and nonfibrous forms (e.g., chromium/calcium, iron, copper, nickel, zinc, tungsten, titanium, and silicon). Talc was found in large amounts in two patients with talc pneumoconiosis, and tungsten was the predominant particle in two patients exposed to hard metal dust. Copper, zinc, and chromium were detected in those grinding metal alloys or in welding. Kaolinite and mica were frequently detected in the cigarette smokers. Mullite is rarely found naturally and usually emanates from refractory materials and in fly ash from coal-fired power stations. Cobalt may

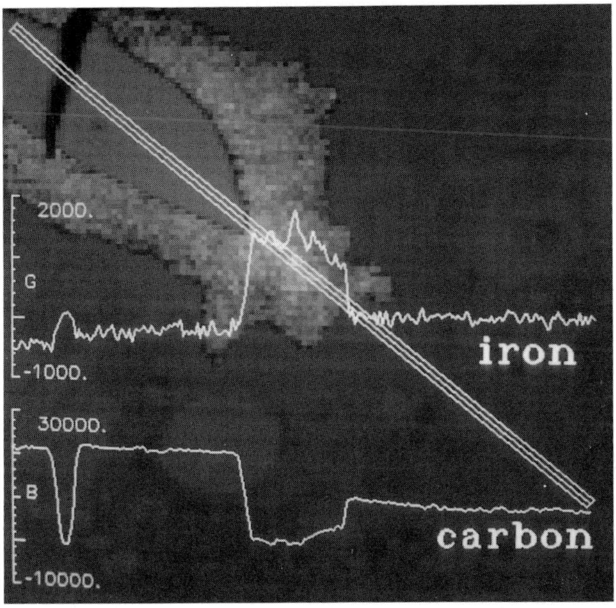

Figure 4 Energy-dispersive x-ray analysis using transmission electron microscopy. Note that the electron probe scans a particle surrounded by iron followed by a carbon peak in a macrophage from a coal miner. (Courtesy of R. Leapman and C. Fiori.)

be detected in lung tissue or BAL from hard metal workers recently after exposure, because cobalt is dissolved in biological fluids. Particles were localized within phagolysosomes either as single particles or aggregated masses. Some particles were enveloped only by a limiting membrane without other lysosomal components (Fig. 4).

III. Alveolar Macrophage Morphology Following Exposure to Dusts

Morphological changes observed by light and S/TEM of AMs exposed to inorganic dusts have been observed over three decades leading to the concept that particles activate AMs. To evaluate morphological effects on inorganic particles, Takemura and colleagues reported on BAL of 43 nonsmokers exposed to asbestos, silica, or coal compared with 16 normal nonsmokers (1) (Fig. 5). Multinucleated AMs were more than twice as common (7 vs 3%) among all dust-exposed groups and contained 2–4 nuclei with some cells containing more than 20 nuclei. Mitoses

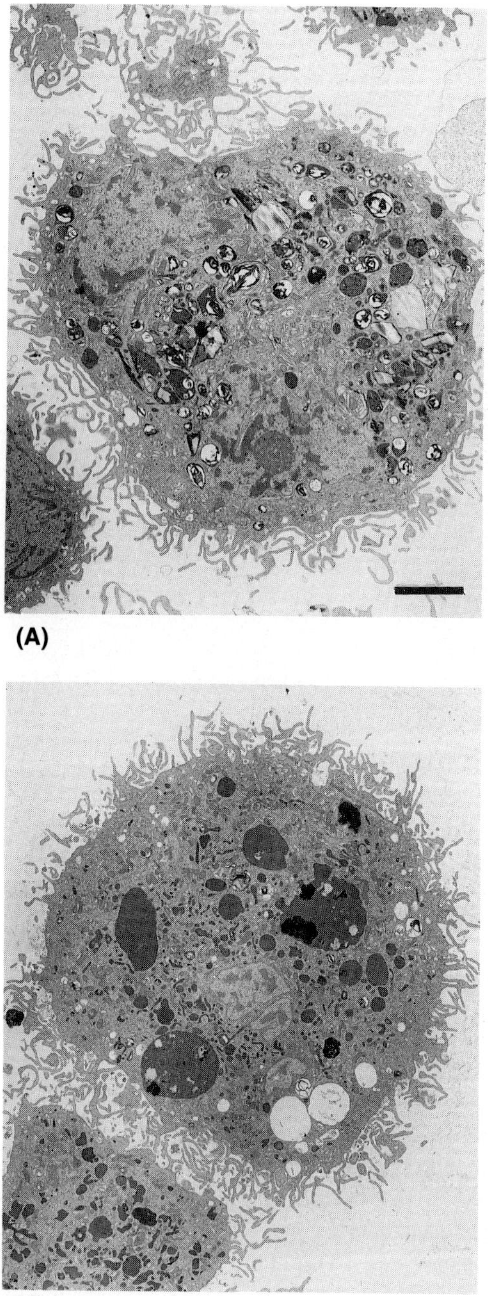

(A)

(B)

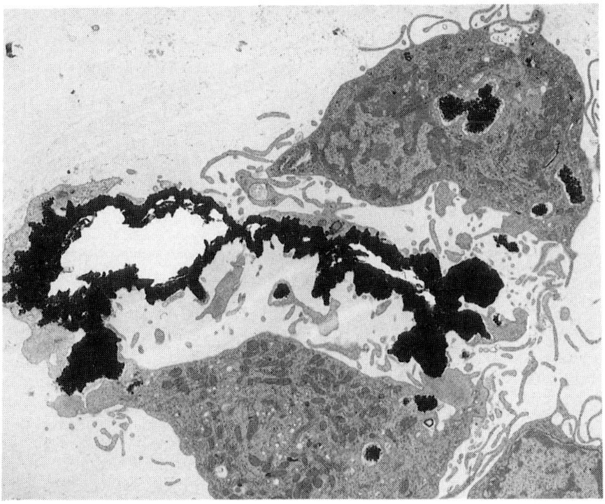

(C)

Figure 5 Transmission electron micrographs of AMs from subjects exposed to inorganic dusts. (A) Several rectangular particles within phagolysosomes are enveloped by a limiting membrane in an AM from a coal miner (×4200 bar is 1 μm). (B) Several highly electron-dense irregularly shaped particles are present in phagolysosomes of an asbestos insulator (×4200). (C) Asbestos body recovered from an asbestos-exposed subject measures 11 μm in length and is engulfed by two macrophages that have thin cytoplasmic processes covering the particle. In addition, there are fragmented particles in their cytoplasm (×7000). (From ref. 1.)

of the AMs were seen in 3.7 vs 1.4% in unexposed controls ($p < .001$), which is consistent with enhanced in situ replication. The size of the AMs assessed by diameters was not different between dust-exposed and normal individuals.

S/TEM showed striking differences in AMs recovered by BAL from dust-exposed subjects (Fig. 6). Rough surfaces, with filopodia (up to 5–6 μm in length), blebs, and rufflings were found more frequently (asbestos-exposed subjects 75 ± 3%, silica-exposed subjects 80 ± 6%, coal-exposed subjects 72 ± 7%, normals 26 ± 6%, $p < .05$, all comparisons to normal values). The mean number of filopodia per entire cell profile (i.e., the cell circumference by TEM) was greater in each dust-exposed group compared with normals. The number of pinocytotic vesicles in the cell surfaces was increased in asbestos and silica-exposed subjects. The number of lysosomes and phagolysosomes per unit area of cytoplasm were increased approximately twofold in each of the three groups of inorganic dust-exposed subjects. The increased lysosomes are consistent with animal models of dust inhalation where lysosomal enzymes such as acid phosophatase and β-glucu-

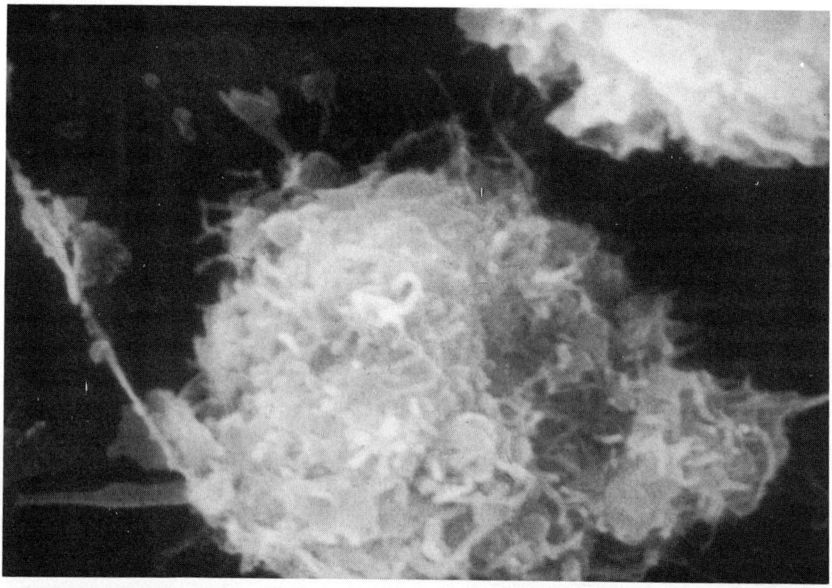

Figure 6 Scanning electron micrograph of AM from an asbestos worker demonstrating filopodia engulfing an asbestos fiber. (Courtesy of T. Takemura and V. Ferrans, MD.)

roridase are increased in BAL (8,9). Mitochondria occasionally showed a dark matrix and an irregular arrangement of cristae, but they were not significantly increased in number in any of the three dust-exposed groups. In some cells, the rough endoplasmic reticulum showed an increase in the numbers of cisterns. The cisterns were occasionally dilated, and lameller complexes were sometimes present. Golgi bodies were well developed. Intracytoplasmic vesicles measuring 0.1–3.0 μm in diameter with a lucent content that were enveloped by a single limiting membrane were increased (asbestos 69 ± 10%, silica 79 ± 3%, coal 60 ± 13%, normal 41 ± 21%, $p < .05$, all comparisons to normal values). Many AMs from dust-exposed subjects had large phagolysosomes containing electron-lucent materials, cell debris, tubular myelin, and cytoplasmic fragments. Concentric lamellae (myelin figures) were common in AMs (asbestos 53 ± 9%, silica 70 ± 17%, coal 42 ± 11%, normal 26 ± 3%; $p < .01$, all comparisons to normal). Subplasmalemmal linear densities forming intercellular junction-like structures connecting two AMs or cell to cell attachments between AMs or AMs to lymphocytes were especially common in asbestos-exposed subjects. These subplasmalemmal densities may represent a network of actin filaments which function in the phagocytosis of particles by activated macrophages (10). Lipid droplets were seen frequently in silica-exposed AMs (42 ± 15% vs 25 ± 14% in normals, $p < .01$). These lipid droplets may result from an interaction between silicic acid and phospholipid and/or activation of phospholipase A by silica particles (11). Intermediate filaments (80–100 Å in diameter) formed broad bundles in the cytoplasm of 82 ± 8% of AMs from the normals compared with 15 ± 3% of the inorganic dust-exposed subjects ($p < .01$). The high frequency of bundles of intermediate filaments in AMs from the normal group may correspond to the resting phase of macrophages, whereas the dispersed pattern of filaments in the exposed groups may correspond to the activated state. Intermediate filaments of macrophages are thought to be related to phagocytosis and cell movement, and both actin and myosin appear to concentrate rapidly at sites of particle phagocytosis (11,12). These studies with TEM suggest that fragmentation, digestion, and dissolution of large particles occurs during phagocytosis or during residence in phagolysosomes. Particles in phagolysosomes are frequently found in association with iron-containing granules regardless of the type of particle, and occasionally particles could be found free in the cytoplasm.

Miller has exposed rats to quartz or crocidolite asbestos to study AM morphology (12). Increased numbers of mitochondria and lysosomes, indicative of macrophage activation, were noted with crocidolite exposure, and variation in AM size was noted with silica exposure, which differed from changes seen with human exposure. Miller noted an increase in Fc receptors for immunoglobulin on macrophages following inorganic dust exposure. IgG1 subclass added to AMs in vitro enhanced their ability to release superoxide anion when stimulated with chrysotile or crocidolite asbestos. Apparently the asbestos fibers adsorb IgG by the Fab domains and cross link the FcR on the macrophage (13). Cross linking of FcR is known to stimulate superoxide release from macrophages (14). Greenberg and colleagues have demonstrated that phagocytosis of IgG-coated particles involves the actin-based cytoskeleton, since F-actin accumulates beneath forming phagosomes (15). They also demonstrated that tyrosine phosphorylation coincides temporally with the appearance of F-actin beneath phagocytic cups. Tyrosine kinase inhibitors (e.g., genistein) was able to inhibit the focal accumulation of F-actin beneath IgG-opsonized particles, and they inhibited the ingestion of particles as well (16). The 68-kDa tyrosine phosphoprotein paxillin colocalized with F-actin beneath nascent phagosomes (17). The γ subunit of FcRI and FcRIII were shown to undergo tyrosine phosphorylation during FcR-mediated phagocytosis. Coating asbestos fibers with vitronectin enhanced phagocytosis in pleural mesothelial cells via the integrin $\alpha v\beta 5$ receptor (18).

Miller demonstrated that there was a 50% increase in lymphocytes attached to dust-exposed AMs, and these lymphocytes proliferated after 3–4 days in in vitro culture (12). Miller exposed rats to crocidolite by inhalation for 6 months and found that AMs demonstrated irregular outlines with many cytoplasmic processes that touched or interdigitated closely with neighboring cells. The AMs showed large nuclei with chromatin marginated against the nuclear membrane and prominent nucleoli. Many lysosomes and phagolysosomes were noted as well as clumps of AMs with indistinct cytoplasmic borders consistent with fusion and aggregation in the process of becoming giant cells. Lemaire reported a rat model of intratracheal injection of short or long chrysotile exposure that resulted in an increase in multinucleated giant cells that preceded and accompanied the development of fibrosis (19).

Morgenroth evaluated human AMs in situ from a 35-year-old asbestos-exposed bricklayer who had had a lung biopsy. AMs had increased lysosomes and fragmented AB, and fibers were seen within the finely granular dense matrix of phagosomes (20). Phagocytosis was observed with trough-shaped depressions in the area where the particles contacted the cell and narrow cytoplasmic appendices extended over the particles. Morgenroth observed large needle-shaped asbestos fibers measuring up to 50.0 μm in length and 0.7 μm in width freely present in the alveolar space that would be difficult to retrieve with BAL. Groups of macrophages were observed coming into close contact with the fibers, and where the cytoplasm of AMs and asbestos came into immediate contact, the fine and dense basic structure of the asbestos had been transformed to a cloudy, finely granular material. No asbestos fibers were observed in alveolar epithelial cells.

In the formation of the asbestos or ferruginous body, small accretions of iron are observed adjacent to the fiber within the phagolysosome (21). The iron has the morphological appearance of hemosiderin containing ferric hydroxide with structural characteristics of ferritin molecules (22). The ferritin core is approximately 60 Å and contains ferric oxyhydroxide, hydrous ferric oxides, acid mucopolysaccharides, calcium, and phosphorus. The fiber is coated concentrically with iron-containing protein leaving a small clear area between the coating and the limiting membrane. Asbestos bodies form on the longer fibers and preferentially on amphiboles rather than chrysotile, suggesting perhaps a protective mechanism against this more toxic fiber type.

IV. Alveolar Macrophage Activation Following Particle Exposure

In addition to the morphological features of AM activation in the pneumoconioses, AMs are increased in numbers and release mediators of lung injury and fibrosis consistent with an AM alveolitis. The source of these AMs includes increased recruitment of blood monocytes, since AMs of asbestos-exposed subjects react at a significantly increased percentage with monoclonal antibodies against monocyte surface antigens than AMs from nonsmoking controls (23). In addition, there is enhanced in situ replication as demonstrated by autoradiography of AMs labeled with [^{3}H]thymidine. Chemotactic factors for monocytes include platelet-derived growth factor (PDGF), and inhaled asbestos in rats activates a complement-dependent chemoattractant for macrophages (24,25). In addition to a macrophage alveolitis, AMs release chemotactic factors for neutrophils, and the percentage of neutrophils recovered by BAL in dust-exposed subjects is usually increased, especially if there is more advanced parenchymal fibrosis. Two important AM chemotactic factors are leukotriene B_4 (LTB_4) generated by arachidonic acid metabolism and interleukin-8 (IL-8). Garcia and colleagues demonstrated that AMs from asbestosis patients released more LTB_4 than asbestos-exposed individuals without interstitial lung disease, and that spontaneous LTB_4 release correlated with BAL neutrophils (26). We have shown that spontaneous IL-8 release by AMs from those with asbestos-induced lung disease (n = 18) is significantly increased over controls, and that gene expression was increased 2.7-fold as measured by RT-PCR (27). In vitro experiments with the myelomonocyte cell line THP-1 cells demonstrated that two different types of asbestos, chrysotile and crocidolite, could stimulate the release of IL-8, and that a nonfibrogenic control dust, titanium dioxide, could not.

BAL neutrophils play a far more significant role in the alveolitis of coal workers' pneumoconiosis (CWP) than in asbestosis and silicosis, because CWP is classically associated with focal emphysema and lung destruction as well as fibrosis. In a BAL study of 17 nonsmoking West Virginia coal miners with CWP, there was a significant increase in neutrophils recovered compared with normal controls or healthy smokers (28). Short-term culture of AMs from these individuals revealed a significant increase in the spontaneous release of superoxide anion and hydrogen peroxide, reactive oxidant species capable of injuring the delicate alveolar walls surrounding the areas of coal dust deposition. Quantification of neutrophil elastase complexed to α1-antitrypsin in lung BAL fluid by ELISA revealed a significant increase in those with CWP compared with normals or normal smokers. Enhanced protease activity and reactive oxidant species in areas of coal dust deposition and AM accumulation likely contribute to the focal emphysema noted in the respiratory bronchioles of lung sections from long-term coal miners.

BAL was used to characterized the alveolitis and the mechanisms of fibrosis in the pneumoconioses due to asbestos (n = 18), coal (n = 15), and silica (n = 6) (29). Gallium-67 lung scans semiquantitated into units were higher than normals in all instances consistent with activated AMs. Total cells recovered by BAL were significantly increased, and the predominant cells recovered were AMs. The spontaneous release of the reactive oxidant species superoxide anion and hydrogen peroxide was increased two- to threefold over controls (Fig. 7). Importantly, AMs spontaneously released significantly increased amounts of fibronectin and fibroblast growth factor. Fibronectin is a key component of extracellular matrix and is a chemotractant for fibroblasts providing a framework for intra-alveolar fibrosis and budding observed in the interstitial lung disorders. The AM fibroblast growth factor originally described as a progression factor by Bitterman and

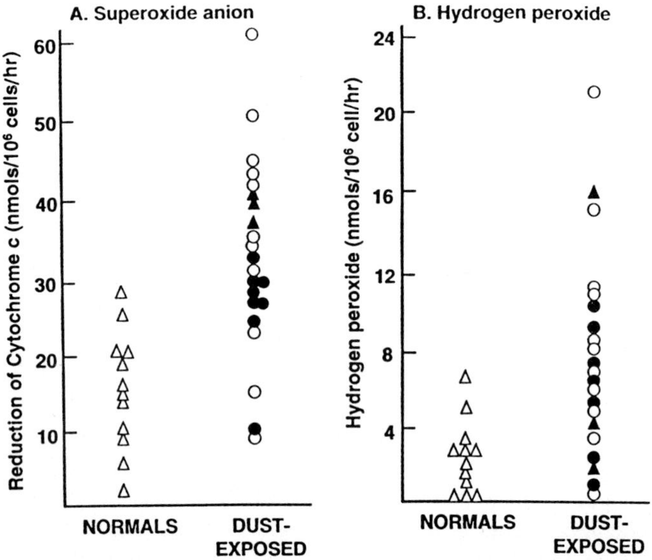

Figure 7 Spontaneous release of oxidants by inflammatory cells recovered from the lower respiratory tract of persons with interstitial lung disease caused by the chronic inhalation of inorganic dust. (A) Spontaneous release of superoxide anion as measured by the reduction of cytochrome-*c*/10^6 cells per hour. Shown are data for normals (open triangles) and persons with asbestosis (open circles), coal workers' pneumoconiosis (closed circles), and silicosis (closed triangles). (B) Spontaneous release of hydrogen peroxide as measured by the reduction of phenol red horseradish peroxidase/10^6 cells per hour. Symbols are the same as in (A). (From ref. 29.)

colleagues (30) was subsequently identified to be a large 25-kDa tissue-type of insulin-like growth factor-I (IGF-I) by copurification of AM supernatant and recombinant IGF-I by column chromatography, blocking of fibroblast growth stimulation by anti–IGF-I, competition with labeled IGF-I for the IGF-I receptor, and gene expression of IGF-I in AMs recovered from asbestosis patients by BAL (31). Dust-exposed individuals with respiratory impairment and parenchymal fibrosis had further increases in BAL neutrophils, fibronectin, and AM fibroblast growth factor release compared to those without respiratory impairment (32). PDGF, a 31-kDa molecular weight peptide growth factor that provides competence growth activity early in G_1, was found to be increased in AM supernatants from five asbestosis patients compared to seven controls using a smooth muscle chemotaxis assay (33).

Macrophages are a major source of cytokines. Animal models of silicosis demonstrated the accumulation of tumor necrosis factor-α (TNF-α) in silicotic nodules by the second week following exposure, and anti–TNF-α could inhibit the development of the lesions (34). ELISA measurement of AM supernatants collected from nine asbestosis patients versus five normal controls revealed a significant threefold increase over controls for TNF-α and a threefold increase for IL-1β. TNF-α is a mitogen for fibroblast proliferation synergizing with PDGF, and IL-1β upregulates the gene for the PDGF-A chain in fibroblasts (35). Gene expression was upregulated as shown by Northern analysis (Fig. 8). Peripheral blood monocytes stimulated in vitro with amosite, crocidolite, or chrysotile asbestos upregulated TNF-α and IL-1β mRNA in AMs and stimulated the release of their respective proteins into cell supernatants (Figs. 9, 10). Either cytokine could stimulate human lung fibroblasts to upregulate genes for collagen types I and III and fibronectin in short-term in vitro culture.

In asbestosis, several investigators have reported increased lymphocytes

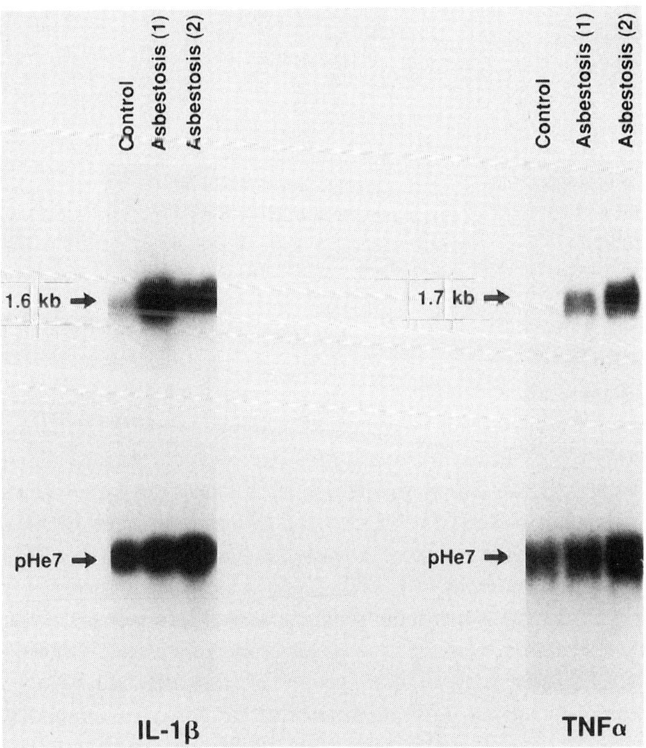

Figure 8 Northern analysis of IL-1β and TNF-α mRNA in AMs from two individuals with asbestos exposure and an unexposed control. Both IL-1β and TNF-β genes were upregulated compared with control, and the pHe7 housekeeping gene demonstrates equal loading of RNA in each lane. (From ref. 35.)

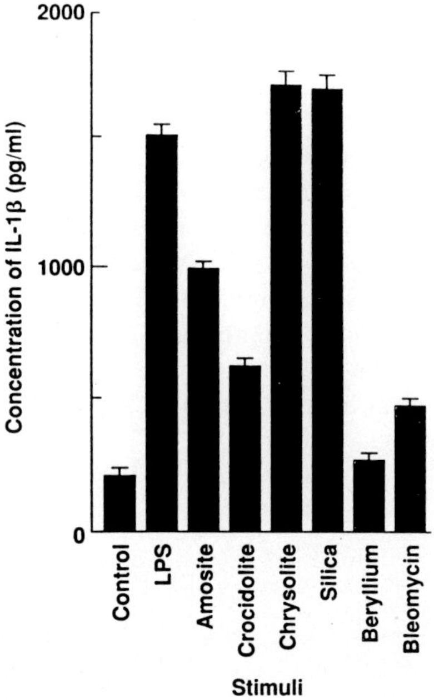

Figure 9 Stimulated release of IL-1β from peripheral blood monocytes (PBMs). PBMs were isolated and cultured in the presence of stimuli for 24 hr and IL-1β measured with ELISA. All three types of asbestos and silica were strong stimuli of IL-1β release. (From ref. 35.)

recovered in the BAL, especially subjects who had pleural plaques (36–38). These lymphocytes have been reported to be $CD3^+$ and $CD4^+$ with an increased $CD4^+/CD8^+$ ratio. In one series of 17 asbestos-exposed individuals all with pleural disease and at least an International Labor Organization (ILO) category $\geqslant 1/0$ parenchymal opacities, the $CD4^+/CD8^+$ ratio was significantly increased in BAL (36). The $CD4^+$ lymphocytes were activated in that they had a threefold increased surface expression of HLA-DR, and 4 of 13 BAL specimens spontaneously released IL-2 into the supernatants. The spontaneous release of interferon-γ (IFN-γ) by BAL cells over 24 hr was increased in 19 of 25 asbestos-exposed individuals, and open lung biopsy specimens from the individuals revealed a patchy chronic interstitial infiltrate consisting primarily of lymphocytes with a few plasma cells (36). The role of activated lymphocytes in the pathogenesis of the pneumoconioses is not settled, but IFN-γ can activate AMs to release oxidants, cytokines such as IL-1β and TNF-α, growth factors (PDGF), and matrix molecules (fibronectin) (36).

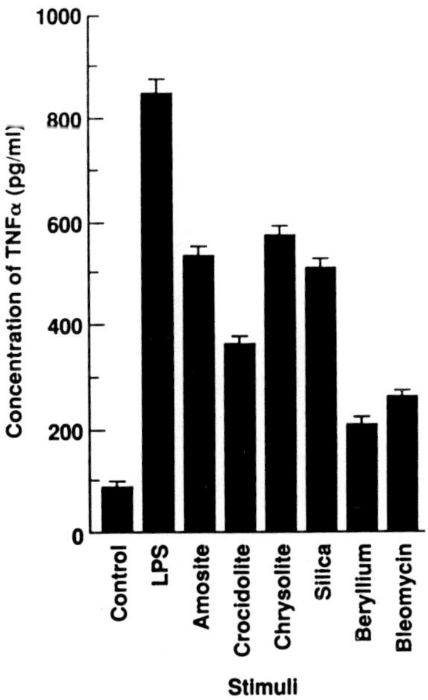

Figure 10 Stimulated release of TNF-α from peripheral blood monocytes (PBMs). PBMs were isolated and cultured in the presence of stimuli for 24 hr and TNF-α measured with ELISA. All three types of asbestos and silica were strong stimuli of TNF-α release. (From ref. 35.)

We have previously demonstrated that the nuclear transcription factor, NF-IL6, mediates transcription of the IL-1β and IL-6 genes by TNF-α (39,40). In addition, NF-IL6 is a transcription factor for IL-8, and TNF-α stimulates IL-8 release. The NF-IL6 transcription factor is a member of a family of nuclear factors termed C/EBP for CCAAT/enhancer binding proteins primarily represented in acute phase proteins from the liver (41). NF-IL6 has a nine base pair consensus sequence T(T or G)NNGNAA(G or T). Both the IL-6 and TNF-α promoter elements contain one or more NF-κB sequences as well. We have shown that these transcription factors act independently of each other or orientation in the IL-1β and IL-6 genes. The C/EBP nuclear factors are characterized by a leucine zipper motif that has five leucines with one in every seventh position. These form an α-helix three-dimensional structure that can form homo- or heterodimers. The leucine zipper structure is followed by the DNA binding domain. The homodimers with NF-IL6β or heterodimers with NF-κB enhance transcriptional activation of cytokine genes (42). The NF-κB p50 as well as the p65 subunit and NF-IL6 proteins interact in the leucine zipper domain forming heterodimers that may synergize; for example, in the transcriptional activation of the IL-8 and IL-6 genes

(43,44). NF-IL6 has been cloned and is expressed in lung tissue after exposure to lipopolysaccharide (LPS), IL-1β, TNF-α, or IL-6 (45). NF-IL6 increases on maturation of U937 cells and in peripheral blood monocytes following stimulation with LPS (46). Cotransfection of NF-IL6 with the NF-κB subunit resulted in strong synergistic activation of an IL-6 promoter-reporter construct (43). TNF-α stimulates dissociation of NF-κB from its inhibitor in the cytoplasm and thus stimulating migration of the active form into the nucleus (45). Interestingly, cell adhesion to fibronectin, a glycoprotein prominent in pneumoconiotic lesions, activates the NF-κB p50/p65 heterodimer in fibroblasts and smooth muscle cells (47). This is dependent on the fibronectin RGD (arginine-glycine-aspartic acid) domain, and cell-matrix interaction is required for cytokine (IL-1β) stimulation of fibroblast gene products, including collagen type I. Thus, the lower respiratory tract may be a localized site of an acute phase response in the alveolitis of the pneumoconioses and other interstitial lung disorders. In this regard, NF-IL6 and NF-κB nuclear factors in AMs may play a central role in directing the inflammatory response orchestrated by cytokines IL-8, IL-1β, TNF-α, and IL-6 which is then followed by an epithelial cell and fibroblast response involving peptide growth factors and the release of matrix macromolecules leading to fibrosis.

References

1. Takemura T, Rom WN, Ferrans VJ, Crystal RG. Morphological characterization of alveolar macrophages from individuals with occupational exposure to inorganic particles. Am Rev Respir Dis 1989; 140:1674–1685.
2. Rom WN, Churg A, Leapman R, Fiori C, Swyt C. Evaluation of alveolar macrophage particle burden in individuals occupationally—exposed to inorganic dusts. J Aerosol Med 1990; 5:543–556.
3. Jaurand MC, Gaudichet A, Atassi K, Sebastien P, Bignon J. Relationship between the number of asbestos fibers and the cellular and enzymatic content of bronchoalveolar fluid in asbestos exposed subjects. Bull Eur Physiopathol Respir 1980; 16:595–606.
4. DeVuyst P, Dumortier P, Moulin E, Yourassowsky N, Yernault JC. Diagnostic value of asbestos bodies in bronchoalveolar lavage fluid. Am Rev Respir Dis 1987; 136: 1219–1224.
5. Schwartz DA, Galvin JR, Burmeister LF, Merchant RK, Dayton CS, Merchant JA, Hunninghake GW. The clinical utility and reliability of asbestos bodies in bronchoalveolar fluid. Am Rev Respir Dis 1991; 144:684–688.
6. Teschler H, Friedrichs KH, Hoheisel GB, Wiak G, Soltner U, Thompson AB, Konietzko N, Costabel U. Asbestos fibers in bronchoalveolar lavage and lung tissue of former asbestos workers. Am J Respir Crit Care Med 1994; 149:614–645.
7. Johnson NF, Haslam PL, Dewar A, Newman-Taylor AJ, Turner-Warwick M. Identification of inorganic dust particles in bronchoalveolar lavage macrophages by energy dispersive x-ray microanalysis. Arch Environ Health 1986; 41:133–144.

8. Miller K. The effects of asbestos on macrophages. CRC Crit Rev Toxicol 1978; 5: 319–354.
9. Kagan E, Oghiso Y, Hartmann D-P. The effects of chrysotile and crocidolite asbestos on the lower respiratory tract: analysis of bronchoalveolar lavage constituents. Environ Res 1983; 32:382–397.
10. Kawanami O, Ferrans VJ, Crystal RG. Subplasmmalemmal linear densities in cells of the mononuclear phagocyte system in lung. Am J Pathol 1980; 100:131–150.
11. Painter RG, McIntosh AT. The regional association of actin and myosin in macrophages following inhalation of crocidolite asbestos and quartz dusts: an overview. Environ Res 1979; 20:162–182.
12. Miller K. Alterations in the surface-related phenomena of alveolar macrophages following inhalation of crocidolite asbestos and quartz dusts: An overview. Environ Res 1979; 20:162–182.
13. Perkins RC, Schekle RK, Holian A. *In vitro* bioactivity of asbestos for the human alveolar macrophage and its modification by IgG. Am J Respir Cell Mol Biol 1991; 4:532–537.
14. Aida Y and Onone K. Triggering of the superoxide generation of macrophages by cross-linking of Fc receptors. J Biochem 1984; 95:1067–1072.
15. Greenberg S, el Khoury J, di Virgilio F, Kaplan EM, Silverstein SC. Ca(2+)-independent F-actin assembly and disassembly during Fc receptor-mediated phagocytosis in mouse macrophages. J Cell Biol 1991; 113:757–767.
16. Greenberg S, Chang P, Silverstein SC. Tyrosine phosphorylation is required for Fc receptor-mediated phagocytosis in mouse macrophages. J Exp Med 1993; 177. 529–534.
17. Greenberg S, Chang P, Silverstein SC. Tyrosine phosphorylation of the gamma subunit of Fc gamma receptors, p72syk, and paxillin during Fc receptor-mediated phagocytosis in macrophages. J Biol Chem 1994; 269:3897–3902.
18. Boylan AM, Sanan DA, Sheppard D, Broaddus VC. Vitronectin enhances internalization of crocidolite asbestos by rabbit pleural mesothelial cells via integrin $\alpha v\beta 5$. J Clin Invest 1995; 96:1987–2001.
19. Lemaire I. Characterization of the bronchoalveolar cellular response in experimental asbestosis. Am Rev Respir Dis 1985; 131:144–149.
20. Morgenroth K. Cellular reaction in the human lung caused by inhalation of asbestos dust over long periods. Beitr Path Bd 1973; 148:199–210.
21. Suzuki Y, Churg J. Structure and development of the asbestos body. Am J Pathol 1969; 55:79–107.
22. Morgan A, Holmes A. The enigmatic asbestos body: its formation and significance in asbestos-related disease. Environ Res 1985; 38:283–292.
23. Spurzem JR, Saltini C, Rom WN, Winchester RJ, Crystal RG. Mechanisms of macrophage accumulation in the lungs of asbestos-exposed individuals. Am Rev Respir Dis 1987; 136:276–280.
24. Mornex J-F, Martinet Y, Yamauchi K, Bitterman P, Grotendorst GR, Crystal RG. Spontaneous expression of the c-*sis* gene and release of a platelet-derived growth factor-like molecule by human alveolar macrophages. J Clin Invest 1986; 78:61–66.
25. Warheit DB, George G, Hill LH, Snyderman R, Brody AR. Inhaled asbestos activates

a complement dependent chemotactic factor for macrophages. Lab Invest 1985; 52: 505–514.
26. Garcia JGN, Griffith DE, Cohen AB, Callahan KS. Alveolar macrophages from patients with asbestos exposure release increased levels of leukotriene B_4. Am Rev Respir Dis 1989; 139:1494–501.
27. Broser M, Zhang Y, Aston C, Harkin T, Rom WN. Elevated interleukin-8 in the alveolitis of individuals with asbestos exposure. Int Arch Occup Environ Health 1996; 68:109–114.
28. Rom, WN. Basic mechanisms leading to focal emphysema in coal workers' pneumoconiosis. Environ Res 1990; 53:16–28.
29. Rom WN, Bitterman PB, Rennard SI, Cantin AC, Crystal RG. Characterization of the lower respiratory tract inflammation of non-smoking individuals with interstitial lung disease associated with chronic inhalation of inorganic dusts. Am Rev Respir Dis 1987; 136:1429–1434.
30. Bitterman PB, Rennard SI, Hunninghake GW, Crystal RG. Human alveolar macrophage growth factor for fibroblasts: regulation and partial characterization. J Clin Invest 1982; 70:806–822.
31. Rom WN, Basset P, Fells G, Nukiwa T, Trapnell BC, Crystal RG. Alveolar macrophages release an insulin-like growth factor I–type molecule. J Clin Invest 1988; 87: 1685–1693.
32. Rom WN. Relationship of inflammatory cell cytokines to disease severity in individuals with occupational inorganic dust exposure. Am J Ind Med 1991; 19:15–27.
33. Rom WN. Activated Alveolar Macrophages from Individuals with Asbestosis Release Peptide Growth Factors. In: Harris CC, Lechner JF, Brinkley BR, eds. Cellular and Molecular Aspects of Fiber Carcinogenesis. Cold Spring Harbor, NY: Cold Spring Harbor Laboratory Press, 1991: 103–114.
34. Piguet PF, Collart MA, Gran GE, Sappino AP, Vassalli P. Requirement of tumour necrosis factor for development of silica-induced pulmonary fibrosis. Nature 1990; 344:245–247.
35. Zhang Y, Lee TC, Guillemin B, Yu MC, Rom WN. Enhanced interleukin-1β and tumor necrosis factor-α release and mRNA expression in macrophages from idiopathic pulmonary fibrosis or following asbestos exposure. J Immunol 1993; 150: 4188– 4196.
36. Rom WN and Travis WD. Lymphocyte-macrophage alveolitis in nonsmoking individuals occupationally exposed to asbestos. Chest 1992; 101:779–786.
37. Wallace JM, Oishi JS, Barbers RG, Batra P, Aberle DR. Bronchoalveolar lavage cells and lymphocyte phenotype profiles in healthy asbestos-exposed shipyard workers. Am Rev Respir Dis 1989; 139:33–38.
38. Sprince NL, Oliver LC, McLoud TC, Eisen EA, Christiani DC, Ginns LC. Asbestos exposed and asbestos-related pleural and parenchymal disease. Am Rev Respir Dis 1991; 143:822–828.
39. Zhang Y and Rom WN. Regulation of the interleukin-1β gene by mycobacterial components and lipopolysaccharide is mediated by two NF-IL6-like motifs. Mol Cell Biol 1993; 13:3831–3837.
40. Zhang Y, Broser M, and Rom WN. Activation of the interleukin-6 gene by *Mycobac-*

terium tuberculosis or lipopolysaccharide is mediated by NF-IL6 and NF-κB. Proc Natl Acad Sci USA 1994; 91:2225–2229.

41. Akira S, Isshiki H, Sugita T, Tanabe O, Kinoshita S, Nishio Y, Nakajima T, Hirano T, Kishimoto T. A nuclear factor for IL-6 expression (NF-IL6) is a member of a C/EBP family. EMBO J 1990; 9:1897–1906.
42. Kinoshita S, Akira S, Kishimoto T. A member of the C/EBP family, NF-IL6β, forms a heterodimer and transcriptionally synergizes with NF-IL6. Proc Natl Acad Sci USA 1992; 89:1473–1476.
43. Matsusaka T, Fujikawa K, Nishio Y, Mukaida N, Matsushima K, Kishiimoto T, Akira S. Transcription factors NF-IL6 and NF-κB synergistically activate transcription of the inflammatory cytokines, interleukin 6 and interleukin 8. Proc Natl Acad Sci USA 1993; 90:10193–10197.
44. Stein B and Baldwin AS. Distinct mechanisms for regulation of the interleukin-8 gene involve synergism and cooperativity between C/EBP and NF-κB. Mol Cell Biol 1993; 13:7191–7198.
45. Akira S and Kishimoto T. IL-6 and NF-IL6 in acute-phase response and viral infection. Immunol Rev 1992; 127:25–50.
46. Natsuka S, Akira S, Nishio Y, Hashimoto S, Sugata T, Isshiki H, Kishimoto T. Macrophage differentiation specific expression of NF-IL6, a transcription factor for interleukin-6. Blood 1992; 79:460–466.
47. Qwarnstrom EE, Ostberg CO, Turk GL, Richardson CA, Bomsztyk K. Fibronectin attachment activates the NF-κB p50/p65 heterodimer in fibroblasts and smooth muscle cells. J Biol Chem 1994; 269:30765–30768.

24

Dendritic Cells and Macrophages in Lung Transplantation

JOCHUM PROP and AALZEN DE HAAN

University Hospital Groningen
Groningen, The Netherlands

TADASHI UYAMA

School of Medicine
University of Tokushima
Tokushima City, Japan

I. Introduction

In this chapter, we discuss how dendritic cells and macrophages can affect the outcome of lung transplantation, particularly in regard to acute rejection, chronic rejection, transplantation tolerance, and infection.

A. Dendritic Cells and Macrophages in the Transplanted Lung

Transplantation of a lung implies transplantation of all the dendritic cells and macrophages that are in the lung at the time of explantation from the donor. After transplantation these cells are gradually replaced by cells of host origin. As a consequence, on transplantation donor-type dendritic cells and macrophages are embedded in the "self"-microenvironment of the lung transplant, but in a "nonself" host with an incompatible major histocompatibility complex (MHC), because at present, donor and recipient are not prospectively MHC matched in clinical lung transplantation. After a certain period, the reverse situation exists: The replacing dendritic cells and macrophages from the host are localized in an organ with a microenvironment of nonself MHC.

How the function of pulmonary dendritic cells and macrophages is affected by transplantation is largely unknown. In normal, nontransplanted lungs, MHC molecules play an important role in the interactions of cells of the immune system. Because after lung transplantation pulmonary dendritic cells and macrophages are MHC incompatible with either the host or the lung, this is likely to complicate their functional position in the pulmonary defense system.

Of particular interest for the management of transplant recipients is the influence of pulmonary dendritic cells and macrophages on the rejection process. Several clinical and experimental studies have addressed adverse and beneficial effects of dendritic cells and macrophages on rejection. Because rejection in combination with infection is the major cause of death of patients after lung transplantation, a better understanding of dendritic cell and macrophage function in transplanted lungs may help to improve the life of lung transplant recipients.

B. Clinical Lung Transplantation

The first clinical lung transplantation was performed in 1963 followed by some 40 lung transplants in the subsequent 15 years. These early lung transplants had little clinical success, partly due to infection and over-immunosuppression (1). With the introduction of cyclosporine, a more selective immunosuppressive drug, a new era of lung transplantation started in the 1980s (2). Today, lung transplantation is widely used for the treatment of end-stage lung disease. According to the Registry of the International Society for Heart and Lung Transplantation (3), almost 7000 lung transplants (including combined heart-lung transplants) have been performed worldwide through January 1996, with over 1150 in 1995. In the past 15 years, the results of lung transplantation have been improving steadily, but they are still behind those of heart transplantation, as shown by Figure 1 with data from the Registry Report (4).

Better immunosuppressive treatment of acute rejection has improved the early survival of lung transplant recipients, but still many patients die within a few years after transplantation. The most common cause of late morbidity and mortality after lung transplantation is bronchiolitis obliterans, a chronic process that affects the small airways of the lungs. The prognosis of bronchiolitis obliterans after lung transplantation remains poor; the patients are severely limited in their daily life by dyspnea and usually die within a couple of years if they are not suitable for retransplantation. Bronchiolitis obliterans is the end stage of a process that starts with injury of the epithelium and leads to a spectrum of airway changes ranging from bronchiolar inflammation and mild submucosal scarring to complete obliteration of the bronchioles (5). The etiology of bronchiolitis obliterans is likely multifactorial. One factor is pulmonary infection, which occurs more frequently after lung transplantation than after heart transplantation (6,7). Other factors in the etiology of bronchiolitis obliterans are repeated episodes of acute rejection and chronic rejection. Dendritic cells and macrophages are involved in all of these factors.

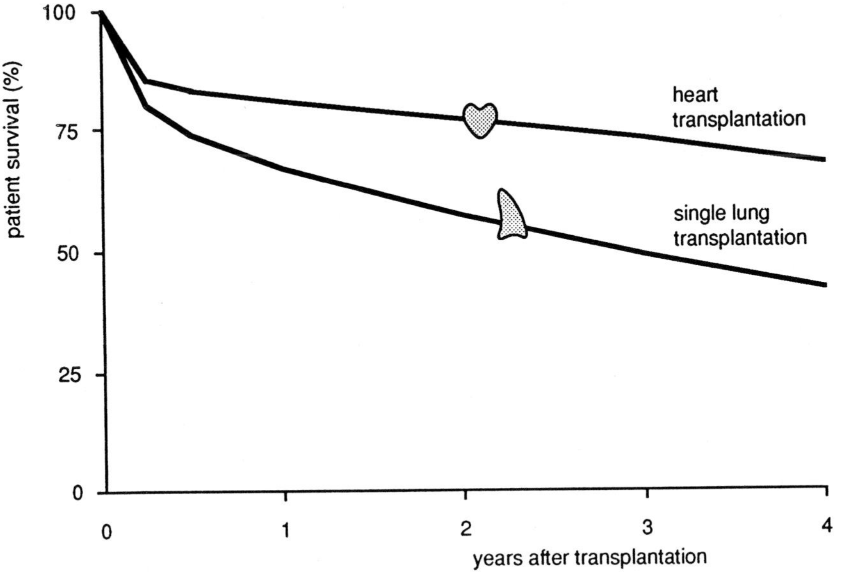

Figure 1 Comparison of patient survival after single lung transplantation and heart transplantation. Lung transplant recipients have a lower survival than heart transplant recipients in the immediate postoperative period and the survival curves continue to diverge in the following 4 years. (Drawn from data extracted from reports of the Registry of the International Society for Heart and Lung Transplantation [3,4].)

II. Acute Rejection

A. Pathology of Lung Rejection

Properties of dendritic cells and macrophages make it conceivable that they are involved in acute rejection in different ways. Like other immune responses, acute rejection is induced by antigen-presenting cells (APCs) such as dendritic cells, presenting the allogeneic MHC antigens of the donor. APCs sensitize donor-specific lymphocytes that proliferate and release cytokines recruiting additional lymphocytes. Donor-specific effector cells are generated, including cytotoxic T lymphocytes, which clear antigen-bearing allogeneic cells, and non-specific cells, such as macrophages, may also play a role in acute rejection. These effector mechanisms lead to destruction of the graft.

Acute rejection can be induced in two ways (Fig. 2). One way is by activation of lymphocytes inside the transplant; this is called peripheral sensitization. The lymphocytes activated by APCs in the transplant discontinue migration, accumulate, and form infiltrates, an important phenomenon recognized by the pathologist in biopsies from the transplant. The second way to induce rejection, central sensitization, is when lymphocytes are activated by APCs in the lymphoid

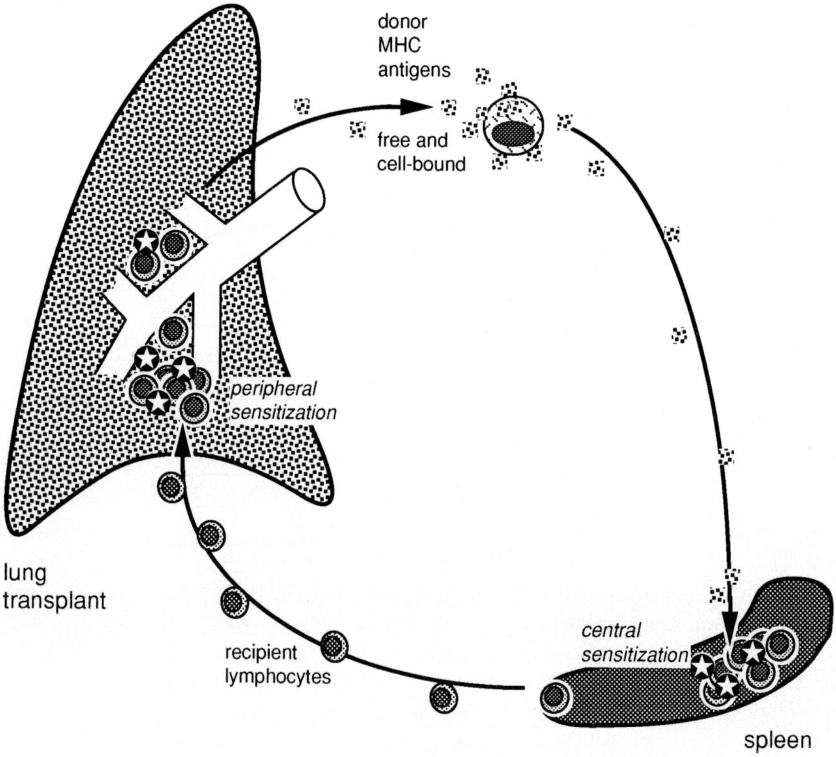

Figure 2 Peripheral and central sensitization of alloreactive lymphocytes. In peripheral sensitization, recipient lymphocytes passing through the lung transplant are activated by donor MHC alloantigens. In central sensitization, recipient alloreactive lymphocytes are activated inside the central lymphoid tissues such as spleen and lymph nodes. This activation is induced by donor MHC alloantigens that are released from the transplant as free molecules or bound to donor passenger cells (lymphocytes, macrophages, dendritic cells) and that are translocated through the blood or lymph. The asterisks (✪) represent activation of lymphocytes.

tissues of the recipient. Central sensitization, although important, is rarely recognized, because biopsies from lymphoid tissue of patients are not obtained.

To analyze the possible role of dendritic cells and macrophages in acute rejection of lung transplants, it is necessary to describe the pathology of acute lung rejection in order to illustrate the initiation of acute rejection and progression to lung destruction. Pathologically, sequential phases of acute rejection are recognized. The phenomena of acute rejection in the lymphoid tissue within the transplanted lung and in the spleen of the recipient is also discussed here.

Rejection Phases

For use in clinical studies, a histological grading system of acute rejection divided into five rejection grades has been described by the Lung Rejection Study Group of the International Society for Heart and Lung Transplantation (5). This standardized classification of acute lung rejection allows comparison between studies performed in different centers and helps to define criteria for treatment of rejection in patients. The clinical grading system is not meant to be an analysis of the rejection process, and it provides no evidence for a role of dendritic cells and macrophages in acute rejection.

Animal studies are more suitable for analysis of the rejection process. Our research group has investigated the histopathology of acute lung rejection in great detail in rats (8,9). After transplantation of the left lung, acute rejection was not treated with immunosuppressive drugs to permit investigation of the rejection process, including the complete destruction of the graft (despite rejection of the left lung, the rats survived because they retained their normal right lung).

In the lung allografts, four sequential phases of acute lung rejection were characterized (8,9): latent, vascular, alveolar, and destructive phase. In the latent phase immediately after transplantation, there is no evidence of rejection. In the vascular phase, lymphocytes infiltrate first in the perivenous region and then in periarterial and peribronchial region. Initially, these infiltrates are small and form a ring of one or two cell layers around venules, most of which are still unaffected. Later the infiltrates grow and form thick cuffs around almost all vessels and airways. The infiltrating cells are mainly lymphocytes, many showing signs of blast transformation or proliferation. It is not until the alveolar phase that alveolar septa and the alveolar spaces are infiltrated, predominantly by lymphocytes. Macrophages are among the infiltrating cells, but their numbers—assessed by bronchoalveolar lavage—do not increase significantly (10), and they are outnumbered by lymphocytes in this phase of acute rejection (11). The final destructive phase is characterized by necrosis of the pulmonary parenchyma, destroying the alveolar architecture of the lung. The alveolar spaces become filled with an admixture of lymphocytes, macrophages, erythrocytes, neutrophils, and fibrinous edema. Neutrophils are the most predominant infiltrating cells in the destructive phase.

Acute Rejection in Bronchus-Associated Lymphoid Tissue and Spleen

The first signs of acute rejection occur in the bronchus-associated lymphoid tissue (BALT) of the lung transplant. BALT is submucosal lymphoid tissue along the bronchial tract (12). It is easily recognized in rats; in humans, BALT consists of "diffuse collections of subepithelial lymphoid tissue" (13). Its existence in the normal human lung has even been disputed (14), but it becomes more prominent in diseases with chronic infection of the airways (15). Dendritic cells and macrophages are a consistent part of BALT structure. High endothelial venules and efferent lymph vessels provide evidence for the participation of BALT in the recirculation of lymphocytes.

After lung transplantation, an immunoproliferative response—with pyroninophilic immunoblasts and cells in mitosis—is seen in BALT, which develops more quickly and is more widespread than is the response in the perivascular infiltrates. In rats, we found marked immunological activity in the BALT as early as 1 day after allotransplantation of lungs. This can be explained by the characteristics of BALT (Fig. 3). After lung allotransplantation, BALT maintains its place in the recirculation pathways of lymphocytes (8). When we injected labeled lymphocytes intravenously in the recipient rat at the time of the lung transplantation, more than 0.5% of these cells were found in the BALT 24 hr later, which is very much like in BALT of normal lungs. Compared with early perivascular infiltrates, BALT is the main site of lymphocyte influx in the allograft (8). Using an immunoperoxidase staining specific for recipient cells, we found impressive numbers of recipient lymphocytes in the BALT of the graft in 1 or 2 days after transplantation (16). The swift immune response locally in the transplant's BALT is likely to be a manifestation of peripheral sensitization of recipient lymphocytes.

In the spleen of the recipient, a similarly swift immune response develops. Already 1 day after lung allotransplantation in rats, a typical proliferative immune response was found in the spleen (8). It reached a maximal intensity a few days later, when massive lymphocyte proliferation involved all periarterial lymphocyte sheaths, which are the T-cell areas of the spleen. Then a B-cell response also became obvious with migrating plasma cells in the red pulp. The spleen is an important organ in the recirculation pathways of lymphocytes and other lymphoid cells. A large number of donor lymphoid cells migrated from the transplanted lung into the spleen (and other lymphoid tissues of the recipient) (8,16). This is the opposite route compared with infiltration of recipient lymphocytes into the BALT. The immune response in the spleen is likely a manifestation of central sensitization of lymphocytes.

B. Dendritic Cells and Acute Rejection

In the pathological description of acute rejection, most emphasis has been on lymphocytes, whereas a role for dendritic cells and macrophages is rarely discussed. To understand the role of dendritic cells in acute rejection, we have to take into account the distribution and function of dendritic cells in lung transplants and include evidence from studies in other organ transplants. Furthermore, we have to distinguish the influence of dendritic cells from the donor and that of dendritic cells from the recipient.

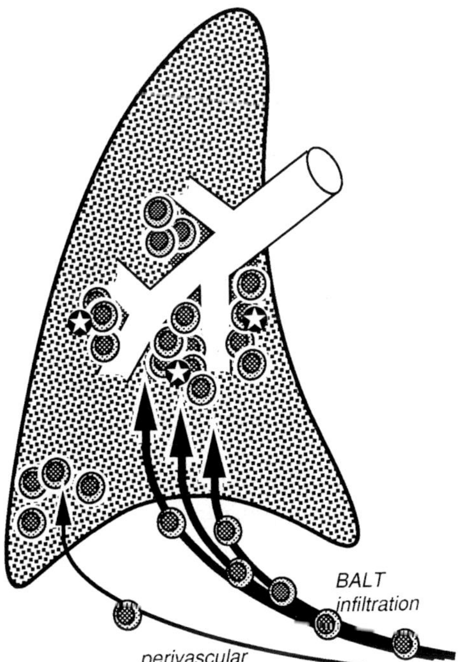

Figure 3 Early signs of rejection in BALT caused by infiltrating recipient lymphocytes. BALT maintains its place in the recirculation pathways of lymphocytes after allotransplantation of the lung. During the first couple of days after lung transplantation, the number of recipient lymphocytes that migrate to BALT is much larger than the number of cells that form perivascular infiltrates. In addition, as soon as they are inside BALT, alloreactive lymphocytes are strongly activated by abundant professional APCs, shown as asterisk (✪).

Distribution and Function of Donor Dendritic Cells

Dendritic cells are dispersed throughout the lung parenchyma of the mouse, rat, and humans (17,18). One population of interstitial dendritic cells is found in perivascular and peribronchial tissue and within alveolar septal walls (18). A second population of dendritic cells is present in BALT as lymphoid dendritic cells. After transplantation of the lung, donor dendritic cells presumably migrate from the transplant into the spleen and other lymphoid tissues of the recipient. Dendritic cells are able to leave the lung as do lymphocytes (8) despite the interruption of lymphatic vessels (19). A similar migration of donor dendritic cells to the spleen was shown after transplantation of allogeneic hearts in mice (20).

Dendritic cells are potent stimulators of immune responses. As accessory cells, dendritic cells are more potent than macrophages and B cells in stimulating resting T cells (21). After transplantation, dendritic cells efficiently present MHC alloantigens to alloreactive T cells and provide second signals for activation and proliferation. If the dendritic cells are from the donor, they present donor MHC alloantigens without a need to first process these antigens (Fig. 4). This is called the direct presentation of alloantigens and is the most powerful stimulation of alloreactive cells.

Stimulation of Acute Rejection by Donor Dendritic Cells

The pathology of early lung rejection only partly agrees with the concept of peripheral sensitization. Thus, immune activation in BALT and perivascular and peribronchial tissue is consistent with this concept. But strikingly absent, despite the large number of dendritic cells in the septal walls, is any response in the alveolar compartment of the lung. This observation was confirmed in experiments where acute rejection of lung transplants was suppressed by treatment of the recipient rats with low-dose cyclosporine. These lungs showed perivascular and peribronchial infiltration (22), whereas the alveolar compartment remained unaffected (23). These findings can be explained in three ways. The first explanation might be that dendritic cells play no role in the peripheral sensitization of lympho-

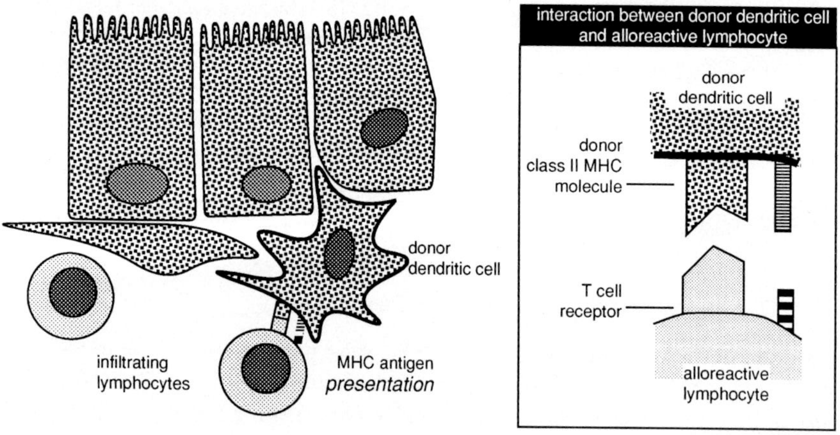

Figure 4 Direct presentation of alloantigens by donor dendritic cells in a lung transplant. Donor dendritic cells present their class II MHC molecules to alloreactive lymphocytes. In the antigen-specific interaction between these cells, the T-cell receptor binds to donor class II MHC molecules on the dendritic cell, shown in the inset. A nonspecific second signal that is necessary for lymphocyte activation is provided by interaction of accessory molecules, represented by the hatched bars.

cytes inside the transplant. In this line of reasoning, nonprofessional APCs such as endothelial cells could account for activation of lymphocytes. However, this does not satisfactorily answer the question of why septal walls remain free of lymphocytic infiltration, because septa contain endothelial cells. The second explanation might be that T-cell activation by dendritic cells is downregulated in septa by macrophages. Such a suppression of dendritic cell function by macrophages was found by Holt et al. in rats (24), and this regulatory mechanism may continue after allotransplantation. The third explanation might be that the intraseptal dendritic cells belong to a population of cells with less prominent antigen-presenting properties. The staining pattern of these dendritic cells has been shown to be different from peribronchial dendritic cells (25), but how this relates to their function is unknown.

Central sensitization of lymphocytes as was observed in the spleen may reflect specific migration patterns of dendritic cells out of the transplanted lung. It should be noted that immune responses against exogenous antigens deposited in nontransplanted lungs tend to remain limited to the draining lymph nodes after translocation of antigens in macrophages (26). The large response in the spleen early during acute rejection could be augmented by migrating donor lymphocytes (8), although lymphocytes have poor antigen-presenting capacities compared with dendritic cells.

That donor dendritic cells stimulate acute rejection, whether peripheral or central, has been proven for different organs in animal studies. Reduction of donor dendritic cells in the organs before transplantation have reduced or prevented rejection. Islets of Langerhans were depleted of dendritic cells by culturing in high oxygen: when the cultured islet were transplanted, they survived longer than freshly isolated islets (27). Vascularized donor organs could be depleted of (donor) dendritic cells before transplantation by pretreatment of the donor with cyclophosphamide and total body irradiation or by transplantation via an immunosuppressed intermediate recipient. Very elegant studies in rats were performed by Lechler and Batchelor, who retransplanted kidney allografts from intermediate recipients into secondary untreated recipients of the same rat strain as the first recipient but allogeneic compared with the donor rat (28). They found that the retransplanted kidneys survived much longer than normal kidney transplants, because they had lost their donor dendritic cells. Most convincing for the role of dendritic cells in this study was the finding that rejection of the retransplanted kidneys was restored if donor strain dendritic cells were administered at the time of retransplantation. Using a similar experimental design, we retransplanted lungs from intermediate recipients into naive recipient strain rats (Fig. 5). These rat lung transplants, depleted of donor dendritic cells, were rejected more slowly than normal lung transplants (29). All these studies support the concept that donor dendritic cells strongly stimulate acute rejection by direct interaction with alloreactive lymphocytes.

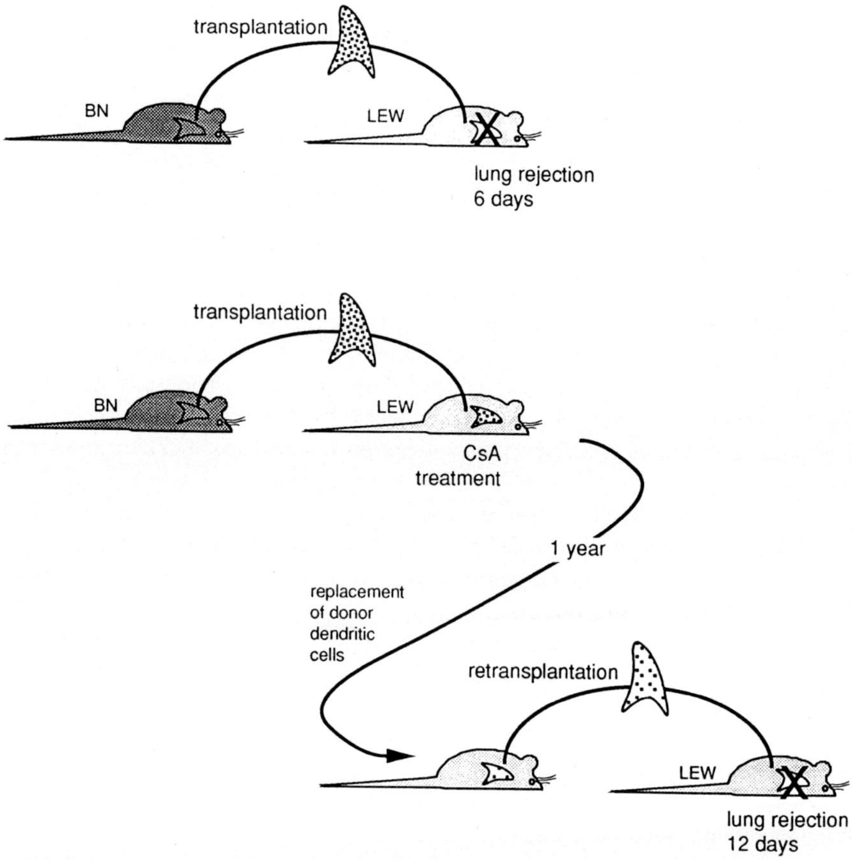

Figure 5 Slow rejection of retransplanted lungs. Normal BN lung transplants are rejected in 6 days by LEW recipient rats. If rejection is prevented by treatment with cyclosporine (CsA), the BN donor dendritic cells in the transplants are gradually replaced by dendritic cells from the LEW recipient. When these lungs, depleted of donor dendritic cells, are retransplanted into new, untreated LEW recipients, they are rejected more slowly than normal lung transplants.

Replacement of Donor Dendritic Cells by Recipient Dendritic Cells

When acute rejection of lung transplants is prevented for some time by immunosuppression, the dendritic cells of the donor are gradually replaced by dendritic cells from the recipient. We examined the immigration of recipient dendritic cells into lung allografts in a rat lung transplantation model (25). Left lung allografts transplanted from BN donor rats into fully mismatched LEW recipient rats and treated with a short course of cyclosporine survived indefinitely without acute rejection episodes for up to 1 year (30). In these lungs, we investigated the replacement of dendritic cells at time points between 3 days and 6 months after transplantation. Recipient dendritic cells were identified in the lung transplant based on their specific staining (with monoclonal antibodies reacting with class II MHC antigens of the LEW rat) and morphology. Few recipient dendritic cells were found in the transplant on day 3, but at 1 week after transplantation, most dendritic cells around vessels and airways were of recipient origin (Fig. 6). Striking is that the population of dendritic cells in the alveolar septa is replaced

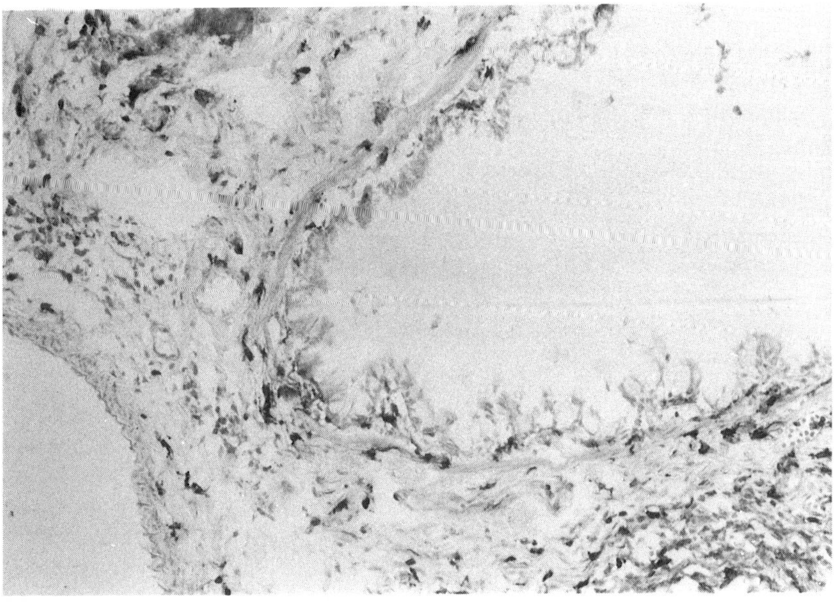

Figure 6 Recipient dendritic cells in the bronchus 1 week after transplantation. In immunosuppressed rats, dendritic cells from the recipient replace the original donor dendritic cells. The figure shows a bronchus in a BN rat lung 1 week after transplantation in an immunosuppressed LEW recipient. Recipient dendritic cells are found in the submucosa staining positively with a monoclonal antibody specific for the LEW class II MHC antigens.

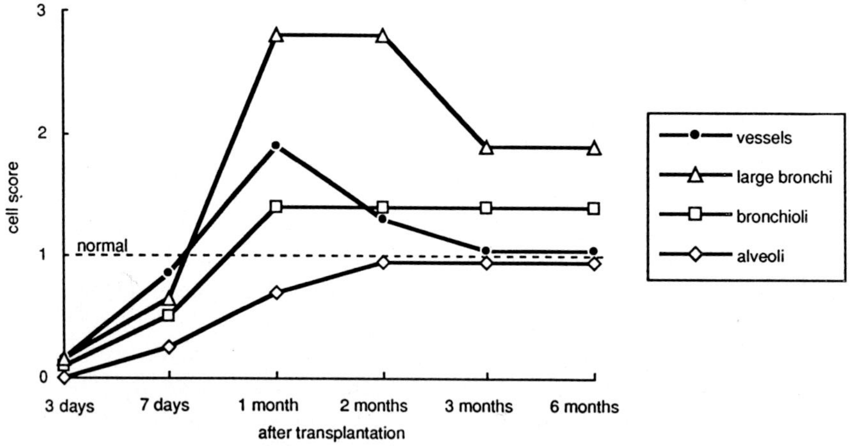

Figure 7 Replacement of donor dendritic cells in different structures of the transplanted lung. Recipient dendritic cells begin to migrate into the vessels, bronchi, and bronchioles by 7 days after transplantation. At 1 month after transplantation, the highest density of dendritic cells is in large bronchi, which persists for over 6 months. Migration of dendritic cells into alveolar septa is slow as compared with migration into bronchi and vessels. A cell scoring system was used, where 0, no recipient dendritic cells; 1, nearly the same number of dendritic cells as in normal lung; 2, number of dendritic cells increased more than normal with some aggregation; 3, dendritic cells in very high numbers and dense aggregation.

relatively slowly; here donor dendritic cells could be found up to 1 month after transplantation (Fig. 7).

A similar rapid turnover of pulmonary dendritic cells was found by Holt et al. (31). They investigated dendritic cell populations in the tracheal mucosa of x-irradiated rats reconstituted with congenic bone marrow. In this radiation chimera model, they showed replacement of dendritic cells in the epithelium of the conducting airways by cells from the donor bone marrow. Replacement of dendritic cells was also described for other organs, such as the kidney where dendritic cells appeared between days 7 and 10 after bone marrow transplantation (32). After xenotransplantation, dendritic cell replacement has also been reported in liver and heart. Valdivia et al. found dendritic cells in the center of the triads near bile ducts in liver xenografts from 30 days after transplantation (33).

Dendritic cells which migrate to organs originate in the bone marrow (31,32). However, only immature MHC class II–negative dendritic cells migrate into nonlymphoid tissues. In mice, systemic administration of bacterial lipopolysaccharide caused immature dendritic cells to migrate into the heart. On transplantation of these hearts, the dendritic cells progenitors matured into MHC class II–positive cells (34). When mature dendritic cells were injected intravenously in mice during acute rejection of cardiac or skin allografts, they failed to migrate into the transplants (35). Although T cells infiltrated the cardiac transplant in large numbers, the mature dendritic cells migrated to the spleen.

In our lung transplantation experiments, we have not performed injections of mature recipient-type dendritic cells. Therefore, we do not know whether MHC class II–positive dendritic cells in the airways and alveolar structures migrated as progenitors of dendritic cells or as mature dendritic cells. Different migration patterns of immature and mature dendritic cells may relate to specific functions (Fig. 8). Thus, immature dendritic cells likely have the capacity to process antigens acquired after migration into nonlymphoid tissues. Mature dendritic cells, on the other hand, are better antigen-presenting cells and migrate to lymphoid tissue where they stimulate lymphocytes to immune responses.

Stimulation of Acute Rejection by Recipient Dendritic Cells

When lungs are depleted of donor dendritic cells before transplantation, acute rejection can be induced by dendritic cells of the recipient via indirect antigen presentation. This antigen presentation is called "indirect," because the dendritic cells process the MHC alloantigens and express the processed peptides or fragments on their own MHC molecules (Fig. 9). Indirect presentation of alloantigens stimulates alloreactive recipient T cells and can induce acute rejection. Sherwood et al. transferred recipient cells presenting donor alloantigens into syngeneic recipients and showed that these cells were able to sensitize the recipients (36), as shown by an accelerated rejection of donor-type skin transplants. Benham et al. also showed in vascularized organ grafts that the indirect presentation of donor antigens contributed to rejection (37). They retransplanted long-term surviving kidney allografts, which were devoid of donor dendritic cells, into the secondary recipient rats. The survival of the retransplanted kidneys was prolonged in control naive recipient rats, because of the lack of direct MHC antigen presentation. Rejection progressed normally, however, if the recipient rats had been preimmunized with donor class I peptides. In this model, recipients were sensitized by their own dendritic cells presenting donor antigens indirectly.

The strength of the acute rejection process stimulated by recipient dendritic cells varies. In some studies, transplants are not rejected at all when rejection is induced only via the indirect antigen-presentation pathway. In most studies, however, the transplanted organs are rejected more slowly than normal acute rejection (28). We found that rejection of retransplanted lungs, which was induced via the indirect pathway, progressed more slowly (29). These observations support the idea that indirect presentation of alloantigens results in a weaker form of acute rejection that might be suppressed more easily by conventional immunosuppression.

The concept that indirect presentation of alloantigens by recipient dendritic

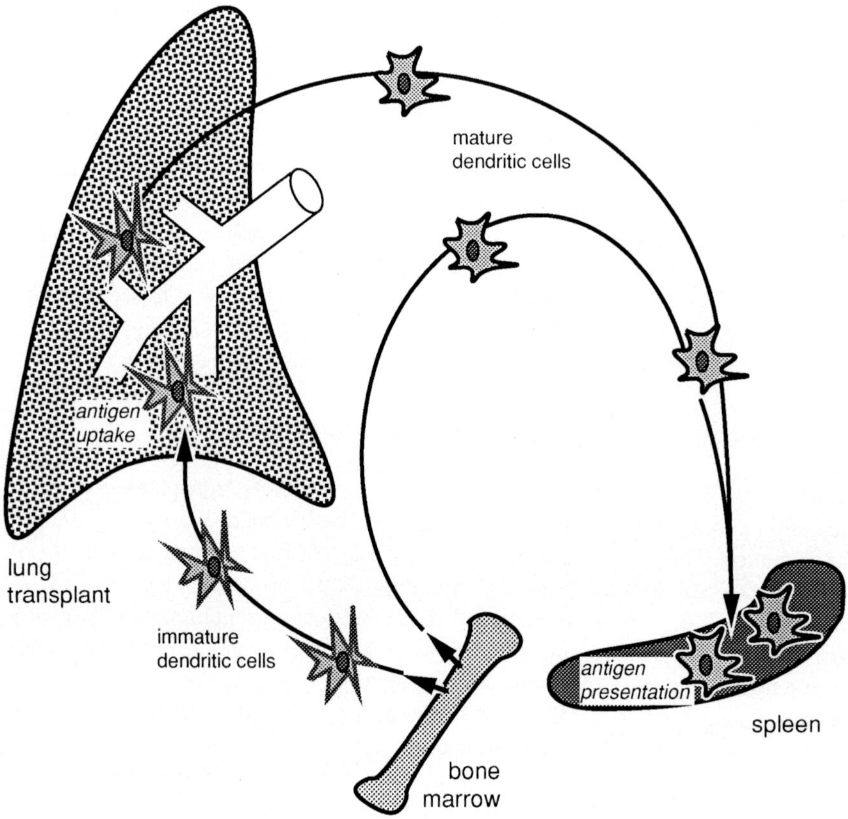

Figure 8 Migration and function of immature and mature dendritic cells. Dendritic cells are released from the bone marrow into circulation. Immature dendritic cells migrate into nonlymphoid tissues (e.g., the lung). The main function is to process antigens. Mature dendritic cells, which have taken up antigens in nonlymphoid tissue while immature, migrate to lymphoid tissue where they function as antigen-presenting cells.

cells results in a weak form of acute rejection might be in agreement with the clinical experience that the dosage of immunosuppressive drugs can be decreased with time after transplantation. Immediately after transplantation, high dosages are needed to prevent activation of alloreactive lymphocytes by donor dendritic cells via the direct pathway. When the donor cells are replaced gradually by cells from the recipient, the shift toward the indirect pathway of lymphocyte activation and a lower immunosuppressive drug dose are sufficient to prevent acute rejection. This idea has been taken one step further with the hypothesis that monitoring the suppression of the indirect pathway can be used to evaluate effective immuno-

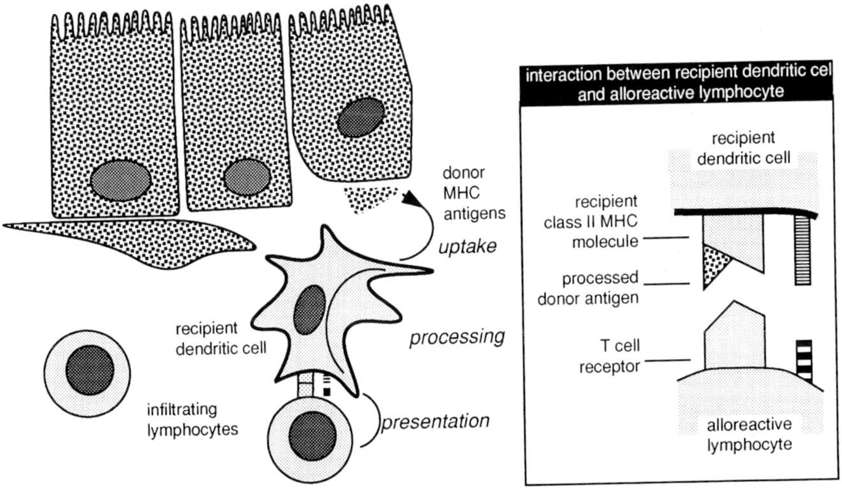

Figure 9 Indirect presentation of alloantigens by recipient dendritic cells in a lung transplant. Recipient dendritic cells stimulate alloreactive lymphocytes by presenting donor alloantigens in the context of their own class II MHC molecules. Before the dendritic cell can present the donor alloantigen, it has to take up MHC molecules that are shed by the cells from the transplanted lung, and process the molecules into peptides. The T-cell receptor of the alloreactive lymphocyte interacts with this complex of processed donor MHC peptide and recipient class II MHC, shown in the inset. A second signal, represented by the hatched bars, is essential for activation of lymphocytes.

suppressive treatment of patients (38). Following this hypothesis the in vitro responses of lymphocytes from kidney recipients were analyzed. It was found that when recipient lymphocytes became responsive to alloantigens presented via the indirect pathway, the patients were at risk for acute rejection.

C. Macrophages and Acute Rejection

Since the early days of lung transplantation, the role of macrophages in the rejection of lung transplants has been an area of research (39). Alveolar macrophages can be obtained repeatedly from the transplants by bronchoalveolar lavage with little discomfort for the patients. It was suggested that the function of lavaged macrophages in interacting with lymphocytes might be used to monitor acute rejection in the lung transplant.

Extensive research on alveolar macrophages after clinical lung transplantation was done in Pittsburgh. Patients were lavaged frequently after transplantation to analyze the in vitro responses of macrophages and lymphocytes. In many cases, a spontaneous proliferation of lymphocytes was found in cultures of cells lavaged in the early postoperative period. This was explained by the composition of the lavaged cells with lymphocytes and macrophages being of different origin. Thus, the lymphocytes were from the recipient, whereas the macrophages were from the donor (40). Donor macrophages were detectable for months and even years after transplantation (41). These donor macrophages were thought to act as antigen-presenting cells to stimulate a macrophage-lymphocyte reaction in vitro but possibly also in vivo. Whether this is an important contribution to the induction of acute rejection has not been proven. The alveolar macrophage is generally a poor stimulator of alloresponses in vitro, and it is hard to exclude that the observed responses were caused by contaminating dendritic cells in the lavaged cells.

If macrophages play a role in acute rejection, they are most likely involved in the effector phase of graft destruction. Their numbers, as assessed by broncho-alveolar lavage, may not increase significantly during acute rejection (10), but they can be activated by the immunological rejection process and hence cause more damage in the transplant. We investigated the activation of macrophages in a study analyzing the production of oxygen radicals by alveolar macrophages lavaged from rat lung transplants during the course of untreated acute rejection (11). However, instead of an increase, we found a decrease of oxygen radical production while rejection progressed. This does not exclude the possibility that activated macrophages contribute to acute lung rejection, for macrophages can express a variety of other effector functions on stimulation. An important effector molecule released by activated macrophages is nitric oxide, particularly in rodents, that causes the destruction of some microorganisms and neoplastic cells. The synthesis of nitric oxide by the enzyme nitric oxide synthase can be inhibited by aminoguanidine. It has been shown that this treatment suppresses acute rejection (42) and prolongs survival of lung allografts (43), indicating that macrophages may play a role in acute rejection.

III. Chronic Rejection

Chronic rejection is a poorly defined process in organ transplants. In fact, chronic rejection in clinical lung transplantation is usually defined by a decrease of graft function. Therefore, chronic graft dysfunction might be a better description of the process (44). The pathophysiology is indicative of an inflammatory process that can be caused both by immune (rejection) and nonimmune (ischemia, infection) mechanisms. Macrophages are thought to be particularly important in the development of chronic rejection (44).

In lung transplants, the process of chronic rejection seems to be aimed mainly at the airways, whereas in most other organ transplants, it is aimed at the vessels. Although vasculopathy is seen in lung transplants, the chronic dysfunction results from bronchiolitis obliterans (5). The ambivalence between "rejection" and "dysfunction" is also reflected by the fact that two classification systems are used to grade the severity of chronic airway disease after transplantation. One system is based on the pathological description of bronchiolitis obliterans in biopsies from the lung transplant (5). The second system is based on the decrease of lung function and was introduced to avoid the problem that adequate pathology is often lacking (45). The cause of bronchiolitis obliterans is multifactorial, but rejection certainly is one of the most important factors. Despite our poor understanding of the underlying process, there is evidence that both dendritic cells and macrophages play a role in its development.

A. Dendritic Cells and Chronic Graft Dysfunction

Accumulation of Dendritic Cells in Lung Allografts

Pulmonary dendritic cells increase in number in different lung diseases. They increase in response to smoking (46) and fibrosing process (47), and they are also found to be increased in some tumors like eosinophilic granuloma (48) and lung cancer (49). Dendritic cells might migrate and increase to elicit T-cell–mediated reactions in lesions of those diseases. In lung transplants, increased numbers of recipient dendritic cells have been observed along the airways in humans (50) and rats (22,25) with long-standing transplants.

Chronic Rejection and Bronchiolitis Obliterans

It has been suggested that recipient dendritic cells are involved in chronic rejection in long-surviving transplants and contribute to the development of bronchiolitis obliterans. In rats, we found that recipient dendritic cells formed aggregates in the submucosa of the large airways (25) (Fig. 10). Interestingly, this was the site where airway damage resembling bronchiolitis obliterans was observed (22). Besides aggregates of dendritic cells, focal infiltrates of lymphocytes were found in the submucosa. The bronchial epithelium covering these infiltrates of dendritic cells and lymphocytes persistently expressed class II MHC antigens. Although these long-surviving lung transplants were not rejected acutely, the recipient rats were able to respond strongly in an in vitro mixed lymphocyte reaction against donor lymphocytes (51). All these observations are consistent with a chronic rejection process in the transplanted lung with the airways as the target. In patients, a similar increase in the number of dendritic cells in bronchial epithelium and submucosa (50) and the expression of class II MHC antigens on bronchial epithelium (52) have been reported in association with bronchiolitis obliterans in lung transplants.

The following role of recipient dendritic cells in the development of bronchiolitis obliterans was postulated by Uyama et al. (22). Recipient dendritic cells that replace the donor cells in the submucosa take up MHC class II antigens expressed by the bronchial epithelium. They process the alloantigens and present them to recipient alloreactive lymphocytes. On activation, the lymphocytes pro-

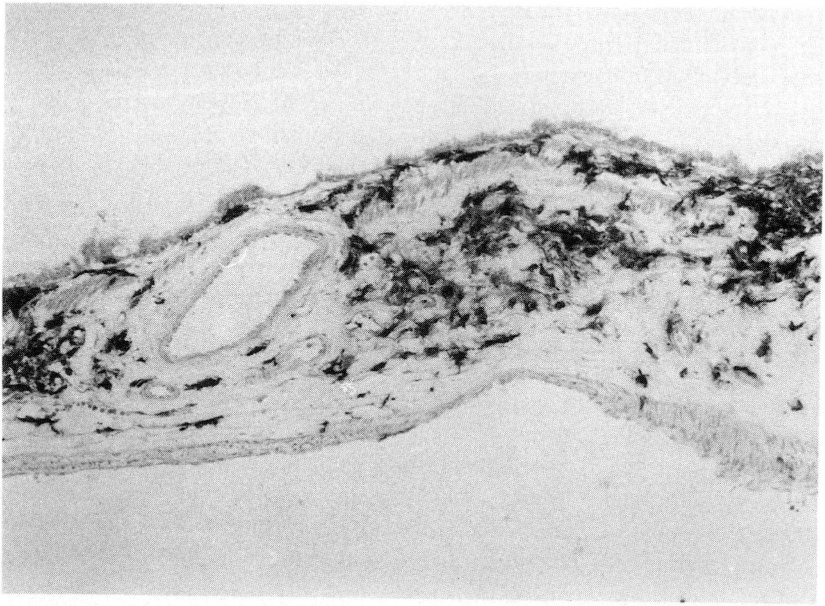

Figure 10 Accumulation of dendritic cells in bronchus of a rat lung 6 months after transplantation. Dendritic cells, shown by staining with a recipient-specific monoclonal antibody, are increased in number and form focal aggregates in the submucosa of the bronchus. The aggregates often correlate with sites where late airway damage resembling bronchiolitis obliterans is observed.

duce cytokines, hence inducing a chronic rejection response and starting a vicious cycle: more MHC antigens are expressed by the adjacent epithelium, more dendritic cells are attracted that process and present the abundant alloantigens, and more lymphocytes are recruited and subsequently activated by the antigen-presenting dendritic cells. In this way, a inflammatory response is initiated that damages the bronchial tissue and, in particular, the epithelium. Finally, an excessive repair mechanism with fibrosis and granulation tissue causes the lumen of the airway to be occluded by scar tissue. This process can be very focal, which indicates that it can be triggered by local stimuli. Viral infections are among the possible triggers of bronchiolitis obliterans after lung transplantation in animals (53) and in humans (54).

B. Macrophages and Chronic Graft Dysfunction

Macrophages have received little attention in studies of chronic lung dysfunction. In lesions of bronchiolitis obliterans, they are not present in conspicuous numbers

(55). Macrophages have been investigated in experimental models of obliterative airway disease resembling bronchiolitis obliterans after lung transplantation. Macrophages were present in tracheal implants that showed immune-mediated injury and obliteration with increased expression of fibroblast growth factor (56), and thus could contribute to immune responses in airway disease that cause obliterative tissue growth after lung transplantation.

In chronic rejection of other organs, macrophages are associated with the development of graft vasculopathy. In rat kidney transplants, macrophages are found in large numbers in vessels and glomeruli with chronic rejection (44). In heart transplants, macrophages together with smooth muscle cells and lymphocytes are found in lesions of graft arteriosclerosis (57). The supposed mechanism is that in response to ongoing allogeneic stimuli, macrophages produce factors that enhance smooth muscle cell proliferation, leading to obstruction of the vascular lumen. Two differences are striking when comparing these organs with lung transplants. First, vascular changes as seen in heart and kidney are less prominent than the airway damage in lung transplants. Second, fibroblasts rather than smooth muscle cells are the proliferating cells that obliterate the lumen. These differences could indicate different immunological responses in the causation of obstruction of vessels versus airways, or that the effects of the same chronic rejection process largely depends on the local cellular environment of the target tissue.

IV. Tolerance and Chimerism

A. Tolerance

Besides activating rejection, donor leukocytes can also deliver tolerogenic signals to the recipient's immune system after lung transplantation. As mentioned above, donor lung macrophages, and particularly donor dendritic cells, are potent stimulators of alloreactive (T-cell) responses mediating rapid graft rejection. With immunosuppressive therapy, however, the responses of recipient T cells to sensitizing signals are prevented, whereas responses to tolerizing signals may be facilitated (58). The outcome is graft acceptance and a state of tolerance. This is achieved by a specific change in the recipient's immune system; in time, reactivity against donor antigen is specifically decreased, whereas reactivity against unrelated (third-party) antigens remains unchanged. After human lung transplantation, the development of donor antigen-specific tolerance or hyporeactivity was demonstrated by a decreased in vitro response of recipient peripheral blood lymphocytes to donor antigen in a mixed lymphocyte reaction (59,60).

The mechanism of tolerance induction after (lung) transplantation and the (donor) cell types involved in this process have not yet been identified. The alveolar macrophage is one possible tolerizing cell that persists for several weeks or months in the lung after transplantation (41), allowing time for interaction with the recipient's immune system. However, to date, no evidence exists to support this hypothesis. In contrast, the alveolar macrophage has been implicated as a cell that sensitizes the recipient's immune system (40).

B. Microchimerism

Another mechanism of tolerance induction and long-term graft acceptance can be donor microchimerism; that is, the persistence for months or even many years of donor-derived cells in the recipient outside the graft (61). Recent studies in kidney and liver transplant recipients have suggested that the establishment of donor cell microchimerism is facilitated by immunosuppressive therapy. In liver transplant recipients, donor-derived chimeric cells were widely distributed and were found in heart, kidney, skin, intestine, and lymph nodes. Based on their morphology and strong expression of MHC class II products, these cells were proposed to be dendritic cells (61). Although dendritic cells are potent stimulators of allo-responses, with immunosuppressive therapy their antigen-presentation function may become ineffective, resulting in T-cell anergy instead of stimulation (58). Alternatively, it has been speculated that unperturbed immature donor dendritic cells from nonlymphoid tissue—in contrast to mature dendritic cells from lymphoid tissue (see Fig. 8)—may deliver tolerogenic signals (62). In support of this view, in vitro studies have shown that murine bone marrow–derived dendritic cells progenitors can render alloreactive T lymphocytes hyporesponsive (63).

The transplanted lung contains large numbers of leukocytes, including dendritic cells, such that microchimerism might be established after lung transplantation. If the degree of microchimerism is substantial, tolerance induction will ensue. Several groups have now shown microchimerism after lung transplantation in rats and humans (60,64–66). The identity of the chimeric cell remains unknown, because "in situ hybridization" or "polymerase chain reaction" (PCR) techniques were employed to detect donor cells, and these techniques do not allow a precise identification of the cells bearing the alloantigens (60,64–66). However, based on the migratory behavior of dendritic cells after solid organ transplantation (20), it is tempting to speculate that dendritic cells contribute to the donor-type chimeric cells.

Interestingly, the degree of microchimerism after human lung transplantation appears to correlate with donor antigen–specific hyporesponsiveness (60). Although these results point to a role of microchimerism in the induction of tolerance, the causal relationship between microchimerism and tolerance induction remains to be established. It is unclear whether chimeric cells induce tolerance or do they simply persist owing to other tolerizing mechanisms? Studies of Bushell et al. (67) have shown that the establishment of microchimerism is not essential for tolerance induction in mice. In this species, short-term exposure of the recipient to donor leukocytes (without development of microchimerism) is a condition which already facilitates tolerance induction (67). Despite the fact that the role of microchimerism in tolerance induction is unclear, clinical trials are presently being undertaken in which the establishment of microchimerism is augmented. This is achieved by infusion of donor bone marrow cells at time of organ transplantation (68). Preliminary data indicate that such an approach may result in an enhancement of allograft acceptance with fewer episodes of acute rejection, although prevention of chronic rejection remains to be established (69).

V. Lung Defense and Infection

Whether dendritic cells and macrophages in donor lungs are able to function normally in the defense of the lung against invading organisms is largely unknown. MHC incompatibility of cells from donor and recipient can hamper the interaction of cells during immune responses, such as dendritic cells, lymphocytes, and cells of the lung parenchyma. The observation that infections occur more frequently after lung transplantation than after heart transplantation (6,7) suggests that the defense in the lung transplant is indeed insufficient. The following discusses whether the structure of the pulmonary defense system is abnormal or that responses against intrapulmonary antigens are inadequate after lung transplantation.

A. Damage of the Pulmonary Defense System

Immediately after transplantation of the lung, the lymphatic vessels from the lung to the draining lymph nodes are interrupted. It takes these lymph vessels about 10 days to regenerate in dogs and rats (19,70). During this period, carbon particles injected into the lung are not translocated into the lymph nodes (19). It is difficult to assess whether the defense system inside allografted lungs is intact. BALT in rat lungs is structurally abnormal. In long-term surviving lung allografts, BALT was small and depleted of lymphocytes (71). Furthermore, BALT and surrounding tissue showed fibrosis (72). No uptake of particles from the bronchiolar lumen into the BALT was observed, perhaps as a result of the extensive fibrosis. Because lungs transplanted between identical rats did not show these changes, it is concluded that the defense system in the lung transplant is damaged by low-grade rejection (72).

B. Function of the Pulmonary Defense System

In some respects, the pulmonary defense system of the transplanted lung was found to respond inadequately against antigens in the airways. The antibody response to intratracheally injected sheep red blood cells (SRBCs) was decreased significantly in allografted lungs compared with hilar stripped or isografted lungs

(73), and responses against respiratory viral infections were also abnormal (74). During the first month after transplantation, antibody responses in allografted lungs recovered more slowly to normal values than in isografted lungs. It is remarkable that in other studies it was shown that dendritic cells in the allografted lungs are replaced by recipient cells about 1 month after transplantation (25). During this period, the immunological mechanisms responsible for presentation of antigens by dendritic cells might not function effectively because of MHC incompatibility between donor dendritic cells and recipient lymphocytes. In long-term surviving lung allografts, however, the antibody responses to SRBCs were similar to that in isografted lungs or normal lungs. Apparently after complete redistribution of recipient-type dendritic cells, a normal antibody response can be induced by the dendritic cells in the transplanted lung.

An important question is whether responses against viruses return to normal. Viral antibody responses remained low (73), but a cytotoxic T-cell response might be more important. If cytotoxic T cells are generated by recipient dendritic cells presenting the viral antigens to recipient T cells, a population of cytotoxic T cells restricted by the recipient MHC would result. These cells could not recognize infected graft cells, because viral antigens would be expressed in association with donor MHC molecules. The consequence would be that the immune system would be unable to clear virus from the transplant. In recipients of lung transplants, it is uncertain to what extent responses against pulmonary viral infections are affected by inadequate interactions by the local defense system or by the immunosuppressive treatment.

VI. Conclusions

In patients, a significant proportion of morbidity after lung transplantation is related to rejection and infection. As was shown above, these two conditions are influenced by dendritic cells and macrophages. Strategies to improve the outcome of lung transplantation cannot ignore the role of these cells. For example, dendritic cells affect the balance between rejection and tolerance; therefore, protocols might be aimed at shifting this balance toward tolerance. A curious effect has been observed after combined transplantation of the heart and lungs (75). In that situation, rejection of the heart was reduced significantly, and the improvement could be attributed to the effect of the lymphoid cells transplanted with the lung (76). Because rejection is induced primarily by dendritic cells, modification of the immunogenicity of dendritic cells may result in concurrent improved survival of the lung transplant.

An approach to reduce at least the systemic side effects of immunosuppressive treatment might be to apply drugs locally in the transplant by inhalation. In experimental studies, inhalation of cyclosporine was effective in preventing rejection (77). Such a local treatment, however, will have immediate effects on the pulmonary dendritic cells and macrophages. It remains to be shown whether these effects are beneficial, resulting in less rejection, or detrimental, with associated increased infection.

Local application of surfactant in the lung transplant might also be an effective inhibitor of rejection. Surfactant has a variety of immunomodulatory effects, particularly affecting the macrophages, but also affects pulmonary fibroblasts (78). Intrapulmonary immune responses can be suppressed by surfactant treatment (79). The effects of surfactant on posttransplant lung function has been investigated after lung transplantation in animals (80,81), but its effect on immune responses in lung transplants needs further investigation.

Acknowledgments

This work was supported by grants from the Netherlands Asthma Foundation and the "Ziekenfonds Raad."

References

1. Wildevuur CR, Benfield JR. A review of 23 human lung transplantations by 20 surgeons. Ann Thorac Surg 1970; 9:489–515.
2. Reitz BA, Wallwork JL, Hunt SA, Pennock JL, Billingham ME, Oyer PE, Stinson EB, Shumway NE. Heart-lung transplantation. Successful therapy for patients with pulmonary vascular disease. N Engl J Med 1982; 306:557–564.
3. International Society for Heart and Lung Transplantation. Slide Registry Series, 1996
4. Hosenpud JD, Novick RJ, Breen TJ, Keck B, Daily P. The registry of the international society for heart and lung transplantation: Twelfth official report—1995. J Heart Lung Transplant 1995; 14:805–815.
5. Yousem SA, Berry GJ, Brunt EM, Chamberlain D, Hruban RH, Sibley RK, Stewart S, Tazelaar HD. A working formulation for the standardization of nomenclature in the diagnosis of heart and lung rejection: lung rejection study group. J Heart Transplant 1990; 9:593–601.
6. Dummer JS, White LT, Ho M, Griffith BP, Hardesty RL, Bahson HT. Morbidity of cytomegalovirus infection in recipients of heart or heart-lung transplants who have received cyclosporine. J Infect Dis 1985; 152:1182–1191.
7. Zenati M, Dowling RD, Dummer JS, Armitage JM, Kormos RL, Hardesty RL, Griffith BP. Influence of the donor lung on the development of early infections in lung transplant recipients. J Heart Transplant 1990; 9:5002–5509.
8. Prop J, Wildevuur CRH, Nieuwenhuis P. Lung allograft rejection in the rat II. Specific immunological properties of lung grafts. Transplantation 1985; 40:126–131.
9. Prop J, Tazelaar HD, Billingham ME. Rejection of combined heart-lung transplants in rats: function and pathology. Am J Pathol 1987; 127:97–105.

10. Prop J, Wagenaar-Hilbers JPA, Petersen AH, Wildevuur CRH. Diagnosis of rejection in rat lung allografts by bronchoalveolar lavage. Transplant Proc 1987; 19:3779–3780.
11. Oosterhoff Y, Noordhoek JA, Petersen AH, Kauffman HF, Postma DS, Prop J. There is no activation of O_2^- production by alveolar macrophages and neutrophil polymorphonuclear leukocytes in rat lung transplants during reimplantation response and acute rejection. Am Rev Respir Dis 1992; 145:1155–1159.
12. Bienenstock J, McDermott MR, Befus AD. The significance of bronchus-associated lymphoid tissue. Bull Eur Physiopathol Respir 1982; 18:153–177.
13. Bienenstock J, Johnston N, Perey DYE. Bronchial lymphoid tissue I. Morphologic characteristics. Lab Invest 1973; 28:686–692.
14. Pabst R, Gehrke I. Is the bronchus-associated lymphoid tissue (BALT) an integral structure of the lung in normal mammals, including humans? Am J Respir Cell Mol Biol 1990; 3:131–135.
15. Sato A, Chida K, Iwata M, Hayakawa H. Study of bronchus-associated lymphoid tissue in patients with diffuse panbronchiolitis. Am Rev Respir Dis 1992; 146:473–478.
16. Prop J, Kuijpers K, Petersen AH, Bartels HL, Nieuwenhuis P, Wildevuur CRH. Why are lung allografts more vigorously rejected than hearts? Heart Transplant 1985; 4:433–436.
17. Sertl K, Takemura T, Tschachler E, Ferrans VJ, Kaliner MA, Shevach EM. Dendritic cells with antigen-presenting capability reside in airway epithelium, lung parenchyma and visceral pleura. J Exp Med 1986; 163:436–451.
18. Holt PG, Schon-Hegrad MA. Localization of T cells, macrophages and dendritic cells in rat respiratory tract tissue: implications for immune function studies. Immunology 1987; 62:349–356.
19. Wang F, Winter JB, Dam M, Groen M, Petersen AH, Wildevuur CRH, Prop J. Influence of interrupted pulmonary lymph drainage on antibody responses in hilar-stripped lungs. J Heart Lung Transplant 1992; 11:S215–S220.
20. Larsen CP, Morris PJ, Austyn JM. Migration of dendritic leukocytes from cardiac allografts into host spleens: a novel pathway for initiation of rejection. J Exp Med 1990; 171:307–314.
21. Inaba K, Steinman RM. Resting and sensitized T lymphocytes exhibit distinct stimulatory requirements for growth and lymphokine release. J Exp Med 1984; 160:1711–1735.
22. Uyama T, Winter JB, Groen G, Wildevuur CRH, Monden Y, Prop J. Late airway changes caused by chronic rejection in rat lung allografts. Transplantation 1992; 54:809–812.
23. Bullard AG, Petersen A, Wildevuur CRH, Prop J, Crapo J. Lung transplantation in the rat. a morphometric analysis of the gas exchange region. Transplantation 1991; 52: 443–449.
24. Holt PG, Schon-Hegrad MA, Oliver J. MHC class II antigen-bearing dendritic cells in pulmonary tissue of the rat. Regulation of antigen presentation activity by endogenous macrophage population. J Exp Med 1988; 167:262–274.
25. Uyama T, Winter JB, Sakiyama S, Monden Y, Groen G, Prop J. Replacement of dendritic cells in the airways of rat lung allografts. Am Rev Respir Dis 1993; 148: 760–767.

26. Bice DE, Shopp GM. Antibody responses after lung immunization. Exp Lung Res 1988; 14:133–155.
27. Bartlett ST, Naji A, Silvers WK, Barker CF. Influence of culturing on the functioning of major-histocompatibility-complex–compatible and incompatible islet grafts in diabetic mice. Transplantation 1983; 36:687–690.
28. Lechler R, Batchelor JR. Restoration of immunogenicity to passenger cell–depleted kidney allografts by the addition of donor strain dendritic cells. J Exp Med 1982; 155:31–41.
29. Prop J, Nieuwenhuis P, Wildevuur CRH. Lung allograft rejection in the rat I. Accelerated rejection caused by graft lymphocytes. Transplantation 1985; 40:26–30.
30. Westra AL, Prop J, Kuijpers KC, Wildevuur CRH. A paradox in heart and lung rejection. Transplantation 1990; 49:826–828.
31. Holt PG, Haining S, Nelson DJ, Sedgwick JD. Origin and steady-state turnover of class II MHC-bearing dendritic cells in the epithelium of the conducting airways. J Immunol 1994; 153:256–261.
32. Leszczynski D, Renkonen R, Hayry P. Localization and turnover rate of rat renal dendritic cells. Scand J Immunol 1985; 21:355–360.
33. Valdivia LA, Demetris AJ, Langer AM, Celli S, Fung JJ, Starzl TE. Dendritic cell replacement in long-surviving liver and cardiac xenografts. Transplantation 1993; 56:482–484.
34. Roake JA, Rao AS, Morris PJ, Larsen CP, Hankins DF, Austyn JM. Systemic lipopolysaccharide recruits dendritic cell progenitors to nonlymphoid tissues. Transplantation 1995; 59:1319–1324.
35. Larsen CP, Barker H, Morris PJ, Austyn JM. Failure of mature dendritic cells of the host to migrate from the blood into cardiac or skin allografts. Transplantation 1990; 50:294–301.
36. Sherwood RA, Brent L, Rayfield LS. Presentation of alloantigens by host cells. Eur J Immunol 1986; 16:569–574.
37. Benham AM, Sawyer GJ, Fabre JW. Indirect T cell allorecognition of donor antigens contributes to the rejection of vascularized kidney allografts. Transplantation 1995; 59:1028–1032.
38. Mulluk SC, Clerici M, Via CS, Weir MR, Kimmel PL, Shearer GM. Correlation of in vitro CD4+ T helper cell function with clinical graft status in immunosuppressed kidney transplant recipients. Transplantation 1991; 52:284–291.
39. Drews J, Shimada K, White PH, Benfield JR. The pulmonary alveolar macrophage in lung transplantation. Transplantation 1974; 17:319–322.
40. Fung JJ, Zeevi A, Kaufman C, Paradis IL, Dauber JH, Hardesty RL, Griffith B, Duquesnoy RJ. Interactions between bronchoalveolar lymphocytes and macrophages in heart-lung transplant recipients. Hum Immunol 1985; 14:287–294.
41. Paradis IL, Marrari M, Zeevi A, Duquesnoy RJ, Griffith BP, Hardesty RL, Dauber JH. HLA phenotype of lung lavage cells following heart-lung transplantation. J Heart Transplant 1985; 4:422–425.
42. Worrall NK, Lazenby WD, Misko TP, Lin T-S, Rodi CP, Manning PT, Tilton RG, Williamson JR, Ferguson TB. Modulation of in vivo alloreactivity by inhibition of inducible nitric oxide synthase. J Exp Med 1995; 181:63–70.
43. Shiraishi T, DeMeester SR, Worrall NK, Ritter JH, Misko TP, Ferguson TB, Cooper

JD, Patterson GA. Inhibition of inducible nitric oxide synthase ammeliorates rat lung allograft rejection. J Thorac Cardiovasc Surg 1995; 110:1449–1460.
44. Azuma H, Tilney NL. Immune and nonimmune mechanisms of chronic rejection of kidney allografts. J Heart Lung Transplant 1995; 14:S136–S142.
45. Cooper JD, Billingham M, Egan E, Hertz M, Higenbottam T, Lynch J, Mauer J, Paradis I, Pallerson A, Smith C, Trulock E, Vreim C, Yousem SA. A working formulation for the standardization of nomenclature and for clinical staging of chronic dysfunction in lung allografts. J Heart Lung Transplant 1993; 12:713–716.
46. Soler P, Moreau A, Basset F, Hance AJ. Cigarette smoking induced changes in the number and differentiated state of pulmonary dendritic cell/Langerhans cells. Am Rev Respir Dis 1989; 139:1112–1117.
47. Hammar S, Bockus P, Remington F, Bartha M. The widespread distribution of Langerhans cells in pathologic tissues. An untrastructural and immunohistochemical study. Hum Pathol 1986; 17:894–905.
48. Basset F, Soler P, Hance AJ. The Langerhans cell in human pathology. Ann NY Acad Sci 1986; 465:324–339.
49. Furukawa T, Watanabe S, Kodama T. T zone histiocytes in adenocarcinoma of the lung in relation to postoperative prognosis. Cancer 1985; 56:2651–2656.
50. Yousem SA, Ray L, Paradis IL, Dauber JA, Griffith BP. Potential role of dendritic cells in bronchiolitis obliterans in heart-lung transplantation. Ann Thorac Surg 1990; 49:424–428.
51. Sakiyama S, Uyama T, Tanida N, Fukumoto T, Nagasawa H, Himeno K, Monden Y. Pathogenesis of late airway changes in long-term surviving lung allografts. Am J Respir Crit Care Med 1994; 150:1137–1141.
52. Taylor PM, Rose ML, Yacoub MH. Expression of MHC antigens in normal human lungs and transplanted lungs with obliterative bronchiolitis. Transplantation 1989; 48:506–510.
53. Tazelaar HD, Prop J, Nieuwenhuis P, Billingham ME, Wildeouur CRH. Airway pathology in the transplanted rat lung. Transplantation 1988; 45:864–869.
54. Burke CM, Glanville AR, Macoviak JA. The spectrum of cytomegalovirus infection following human heart-lung transplantation. J Heart Transplant 1986; 5:267–272.
55. Yousem SA. Pathology of heart-lung transplantation. In: Cooper DKC, Novitzky D, eds. The Transplantation and Replacement of Thoracic Organs. Dordrecht, The Netherlands: Kluwer, 1990:315–324.
56. Huang X, Reichenspurner H, Shorthouse R, Cao W, Berry G, Morris R. Heterotopic tracheal allograft transplantation: a new model to study the cellular and molecular events causing obliterative airway disease in rats (abstr). J Heart Lung Transplant 1995; 14:S53.
57. Russell ME, Wallace AF, Hancock WW, Sayegh MH, Adams DH, Sibinga NE, Wyner LR, Karnovsky MJ. Upregulation of cytokines associated with macrophage activation in the Lewis to F344 rat chronic cardiac rejection model. Transplantation 1995; 59:572–578.
58. Thompson S. Immunosuppressive drugs and the induction of transplantation tolerance. Transpl Immunol 1994; 2:263–270.
59. Reinsmoen NL, Bolman RM, Savik K, Butters K, Matas AJ, Hertz MI. Improved long-term graft outcome in lung transplant patients who have donor-antigen specific hyporeactivity. J Heart Lung Transplant 1994; 13:30–37.

60. Reinsmoen NL, McSherry C, Jackson AJ, Hertz MI, Bolman RM, Matas AJ. Circulating donor antigen and response to donor antigen as predictors of long-term lung and kidney transplant success. Transplant Proc 1995; 27:205–206.
61. Starzl TE, Demetris AJ, Murase N, Ildstad S, Ricordi C, Trucco M. Cell migration, chimerism, and graft acceptance. Lancet 1992; 339:1579–1582.
62. Steinman RM, Inaba K, Austyn JM. Donor-derived chimerism in recipients of organ transplants. Hepatology 1993; 17:1153–1156.
63. Lu L, McCaslin D, Starzl T, Thomson AW. Bone marrow–derived dendritic cells progenitors (NDLC 145^+, MHC class II^+, B7-1^{dim}, B7-2^-) induce alloantigen-specific hyporesponsiveness in murine lymphocytes. Transplantation 1995; 60:1539–1545.
64. Kubit V, Sonmez-Alpan E, Zeevi A, Paradis I, Dauber JH, Iacono A, Keenan R, Griffith BP, Yousem SA. Mixed allogeneic chimerism in lung allograft recipients. Hum Pathol 1994; 25:408–412.
65. Knoop C, Andrien M, Defleur V, Antonine M, de Francquen P, Goldman M, Estenne M. Detection of blood chimerism after lung and heart-lung transplantation. The superiority of nested as compared with standard polymerase chain reaction amplification. Transplantation 1994; 58:1335–1338.
66. Hara M, Shibata K, Hirasawa A, Iwaya M, Okuyama S, Tsujimoto G, Kimura H, Koga Y. Microchimerism and lung allograft acceptance. Transplant Proc 1995; 27: 1580–1582.
67. Bushell A, Pearson TC, Morris PJ, Wood KJ. Donor-recipient microchimerism is not required for tolerance induction following recipient pretreatment with donor-specific transfusion and anti-CD4 antibody. Transplantation 1995; 59:1367–1371.
68. Rao AS, Fontes P, Zeevi A, Rugeles M, Shapiro R, Jordan M, Corry R, Dodson F, Rilo HLR, Rastellini C, Aitouche A, Rosner G, Trucco M, Demetris AJ, Fung JJ, Starzl TE. Enhancement of donor cell chimerism in whole organ allograft recipients by adjuvant bone marrow transplantation. Transplant Proc 1995; 6:3387–3388.
69. Ricordi C, Karatzas T, Selvaggi G, Neri J, Fernandez H, Ruiz MP, Lanetzky E, Kong SS, Webb M, Bottino R, Quian T, Ciancio G, Carreno M, Cirocco R, Garcia R, Olsen L, Esquinazi V, Miller J, Schiff E, Tzakis AG. Enhanced allograft acceptance by multiple infusions of donor bone marrow in humans. Transplant Proc 1995; 27:3381.
70. Eraslan S, Turner MD, Hardy JD. Lymphatic regeneration following lung reimplantation in dogs. Surgery 1964; 56:970–973.
71. Uyama T, Prop J, Petersen AH, Winter JB, Romaniuk A, Westra AL, Wildevuur CRH. Reduction of the lymphoid tissue in long-term surviving rat lung allografts. Transplant Proc 1990; 22:2013–2104.
72. Winter JB, Prop J, Groen M, Petersen AH, Uyama T, Meedendorp B, Wildevuur CRH. Defective bronchus-associated lymphoid tissue in long-term surviving rat lung allografts. Am J Respir Crit Care Med 1995; 152:1367–1373.
73. Winter JB, Groen G, Petersen AH, Wildevuur CRH, Prop J. Reduced antibody responses after immunization in rat lung transplants. Am Rev Respir Dis 1993; 147: 664–668.
74. Winter JB, Groen M, Welling S, van der Logt K, Wildevuur CRH, Prop J. Inadequate antibody response against respiratory viral infection in long-surviving rat lung allografts. Transplantation 1995; 59:1583–1589.
75. Baldwin JC, Oyer PE, Stinson EB, Starnes VA, Billingham ME, Shumway NE.

Comparison of cardiac rejection in heart and heart-lung transplantation. J Heart Transplant 1987; 6:352–356.

76. Westra AL, Petersen AH, Prop J, Wildevuur CRH. The combi-effect: Reduced rejection of the heart by combined transplantation with the lung or spleen. Transplantation 1991; 52:952–955.
77. Zenati M, Duncan AJ, Burckart GJ, Shaper M, Yousem SA, Griffith BP, Casarotto D. Immunosuppression with aerosolized cyclosporine for prevention of lung rejection in a rat model. Eur J Cardiothorac Surg 1991; 5:266–272.
78. Thomassen MJ, Antal JM, Barna BP, Divis LT, Meeker DP, Wiedemann HP. Surfactant downregulates synthesis of DNA and inflammatory mediators in normal lung fibroblasts. Am J Physiol 1996; 270 (Lung Cell Mol Physiol 14):L159–L163.
79. Richman PS, Batcher S, Catanzaro A. Pulmonary surfactant suppresses the immune lung injury response to inhaled antigen in guinea pigs. J Lab Clin Med 1990; 116: 18–26.
80. Erasmus ME, Petersen AH, Hofstede G, Haagsman HP, Bambang Oetomo S, Prop J. Surfactant treatment before reperfusion improves the immediate function of lung transplants in rats. Am J Respir Crit Care Med 1996; 153:665–670.
81. Novick RJ, Veldhuizen RAW, F. P. Lee J, Sandler D, Lewis JF. Exogenous surfactant therapy in thirty-eight hour lung graft preservation for transplantation. J Thorac Cardiovasc Surg 1994; 108:259–268.

25

Induction of Tumoricidal Activity in Lung Macrophages by Liposome-Encapsulated Synthetic Immunomodulators

JERALD J. KILLION, EUGENIE S. KLEINERMAN, and ISAIAH J. FIDLER

The University of Texas
M.D. Anderson Cancer Center
Houston, Texas

I. The Biology of Tumor Metastasis

The major barrier to the eradication of primary and metastatic tumors centers on the heterogeneous nature of neoplastic disease; that is, tumors contain subpopulations of cells that are able to subvert host defenses and recruit infiltrating cells that supply needed growth factors and blood supply. Moreover, metastatic lesions can become autonomous with respect to the homeostatic mechanisms of normal tissues (1–3). Neoplastic transformation involves genetic alterations, such as the activation or dysregulation of oncogenes (4). There are probably immediate selection pressures for cells that are able to circumvent normal growth control mechanisms and these cells give rise to diverse progeny that can subvert or even usurp homeostatic mechanisms. The continual evolution of neoplasms eventually favors the emergence of cell subpopulations with metastatic potential owing to their expression of receptors for growth factors and the secretion of angiogenic molecules and enzymes that allow the cells to invade surrounding stromal tissue. Host cells can modulate these events. For example, organ-specific fibroblasts influence the invasive properties of tumor cells (5), which are facilitated by the secretion of degradative enzymes (6–8). Monocyte and lymphocyte infiltration of the tumor may suppress the growth of some cells yet stimulate the growth of others (9).

By the time of diagnosis of the primary tumor, malignant neoplasms contain multiple cell populations that are heterogeneous with respect to a variety of biological properties, including cell surface characteristics, antigenicity, immunogenicity, growth rate, karyotype, sensitivity to chemotherapeutic drugs, and the ability to invade and metastasize (2,10–14). The existence of these subpopulations of cells presents a dilemma for the clinician. The emergence of drug-resistant variants during or subsequent to chemotherapy has been documented, as have differences in the response of primary and metastatic tumors to therapeutic agents (2). The obstacle to chemotherapy that heterogeneity represents may also affect the success of immunotherapy or the use of biological response modifiers (2,15–17). Biological heterogeneity is equally prominent among tumor cells in metastatic lesions (12,18).

The process of cancer metastasis consists of a series of sequential, interlinked steps (10). Eventually, within the primary tumor, metastatic tumor cells will dominate the proliferative fraction of growing tumor, and these cells will be able to invade the host stroma, blood vessels, and lymphatics surrounding the primary neoplasm owing to their secretion of degradative enzymes and motility (6,7,19). Tumor cells may aggregate with themselves or with circulating blood cells prior to arrest in distant capillary beds via expression of specific adhesion molecules. Subsequent to arrest, tumor cells multiply and destroy the surrounding vessel, penetrate the endothelial basement membrane, or follow migrating white blood cells. The survival, invasion, and growth of malignant cells at particular secondary sites also involves their response to tissue- and organ-specific factors. Tumor cells recognize tissue-specific motility factors, respond to organ-specific growth factors, and must establish adequate vasculature for continued growth (20–22). The successful treatment of cancer is, in reality, the treatment of metastatic disease, which will require an understanding of tumor-host relationships that may be unique even between patients that have the same histological origin of their disease. Therapeutic strategies that center on molecular targeting of the multiple steps within the metastatic cascade, such as biosynthetic pathways associated with expression of growth factors and adhesion molecules, secretion of proteolytic and collagenolytic enzymes, release of angiogenic substances, and subversion of host defenses within the tumor microenvironment (23–26) offer new approaches to inhibition of tumor growth and metastasis. New therapies based on stimulation of natural immune mechanisms, however, must be able to circumvent tumor heterogeneity, overcome the continual emergence of cells that are resistant to conventional cytotoxic therapies, and provoke host-mediated defenses that favor eradication of metastatic tumor cells.

II. Macrophage Infiltration into Tumors

To eradicate tumors in vivo, macrophages must first infiltrate the lesions. The presence of inflammatory macrophages in growing tumors is maintained through recruitment of circulating monocytes (27–34) and, in certain tumors, the proliferation of mononuclear phagocytes (33,35). One mediator of monocyte recruitment to inflammatory sites and neoplasms is monocyte chemotactic activating factor (MCAF; also known as monocyte-chemotactic protein, tumor cell–derived chemotactic factor, or JE gene product) (36–39). In monocytes, MCAF also stimulates an increase in cytosolic calcium levels, superoxide anion production, lysosomal enzyme release, and production of interleukins (ILs) IL-1 and IL-6 (39,40), and regulates surface expression of adhesion molecules, as well as tumoricidal activation (41).

The distribution pattern of host inflammatory cells into tumors has been described (42). In regressing murine sarcomas, tumor-associated macrophages (TAMs) are found throughout the tumors, whereas in progressing sarcomas, TAMs are confined to the periphery of the tumor (43). The induction time of murine tumors varies with their macrophage content: The faster-growing tumors are those with fewer infiltrating macrophages (34). Small tumors are infiltrated with a relatively large number of macrophages, whereas large tumor are not (30,31). Several investigators have suggested possible mechanisms for the decreased recruitment of macrophages into larger tumors. These large tumors and metastases may outstrip the host's ability to mount a macrophage response (34). Reduced macrophage responsiveness or effects of the geometry of large solid tumors may also decrease macrophage invasion in to tumors (30,31). Other investigators have demonstrated that soluble products of tumor cells from several different cell lines impair macrophage recruitment in normal animals (27,44,45). Other indepedent studies have also concluded that the accumulation of macrophages in tumors does not necessarily correlate with the metastatic properties or immunogenicity of the tumors (28,46,47). Actually, the presence of noncytotoxic (nonactivated) macrophage in neoplasms could enhance tumor growth (35,48). Macrophages (and lymphocytes) produce many diffusible growth, angiogenic, and cytotoxic factors (49). According to the type and level of such mediators, TAMs may, therefore, enhance or inhibit the growth of neoplasms.

III. Macrophage Recognition of Tumor Cells

Activated macrophages can destroy syngeneic, allogeneic, and xenogeneic tumor cells but leave normal nontumorigenic cells unharmed, suggesting that histocompatibility and tumor-specific antigens are not involved in this recognition (50).

Cell cycle duration and replication of endogenous C-type viruses or associated fluctuations in antigen expression were also ruled out as factors in macrophage recognition of tumor cells, since tumor cells with different cell cycling times and with various levels of viral expression were all recognized and lysed by activated macrophages (51). In addition, differences in metastatic potential, chromosome number, or resistance to chemotherapeutic agents did not appear to be important factors for macrophage recognition of tumor cells (52–54).

The ubiquitous role of macrophages in homeostasis suggests that macrophages recognize numerous molecules via a wide variety of plasma membrane receptors. For example, macrophages are able to recognize certain sugar residues on tumor cells by specific cell surface lectins (55). An N-acetylgalactosamine/galactose (GalNAc/Gal)–specific receptor on liver macrophages mediates the attachment of desialyated cells to the liver macrophages (56,57). Various simple monosaccharides can inhibit macrophage–tumor cell binding and subsequent cytotoxicity in some systems (58–60). Mouse macrophage binding to concanavalin A–resistant cells is inhibited by D-mannose but not by galactose or N-acetylglucosamine (58). In contrast, the binding of murine macrophages to allogeneic lymphoma cells is inhibited by N-acetylgalactosamine but not by D-mannose (60). These results suggest that macrophages can recognize sugar sequences that may vary among different tumor cells.

Macrophages can also recognize complex asparagine-linked oligosaccharides on tumor cells. Treatment of murine mastocytoma cells with the oligosaccharide-processing inhibitors 1-deoxynojirimycin or 1-deoxymannojirimycin resulted in glycoproteins that lacked the complex structure of N-linked oligosaccharides and in a decreased level of macrophage-mediated lysis (61). The inhibition of tumor cell–macrophage binding by saccharides suggests a role for membrane lectins on the tumor cell and macrophage cell surfaces; for example, a 175-kDa mannose/fucose receptor on the macrophage membrane surface (62). Moreover, endogenous lectins of 14.5 and 34.0 kDa that recognize β-galactosides have been identified on the surfaces of different tumor cells and host effector cells (62). The cell surface expression of carbohydrate on tumor cells has been used to isolate immunogenic subpopulations of tumor cells from mice with leukemia (63,64), and this display of membrane carbohydrates, especially the galactosyl-like saccharides, correlates with the immunogenic properties of tumor cells and their value as therapeutic vaccines (65). Macrophages have plasma membrane receptors that interact with extracellular proteins as well as with adhesion molecules, such as the integrins (55). Macrophages recognize laminin and fibronectin, two basement membrane proteins. The adhesion of macrophages to laminin or fibronectin aids diapedesis into damaged issues (66–70).

Although carbohydrates and proteins have long been known to play a part in cellular adhesion, an increasing body of evidence suggests that phospholipids are also involved in macrophage recognition of target cells (50). All cell membranes are composed of a phospholipid bilayer containing four major phospholipids, the choline phospholipids, phosphatidylchoine (PC) and sphingomyelin, and the aminophospholipids, phosphatidylethanolamine (PE) and phosphatidlyserine (PS). In addition, the distribution of phospholipids in the bilayer membrane is extremely asymmetrical. PC and sphingomyelin are preferentially positioned in the outer leaflet of the lipid bilayer, whereas PE is preferentially distributed in the inner leaflet (71,72). PS displays absolute asymmetry in its distribution; normal cells contain PS only in the inner leaflet of the cell membrane. Experiments based on several lines of evidence now suggest that when PS is expressed on the outer leaflet, it serves as a recognition molecules for macrophages (73–75).

Macrophage recognition of PS in the outer leaflet of the red blood cell (RBC) membrane was demonstrated by inserting fluorescent PS into the plasma membrane. Binding and uptake of these RBCs by macrophages was fivefold higher than binding of untreated RBCs or of those with inserted PC (73). Similar findings were obtained in vivo. RBCs containing a fluorescent analogue injected into the circulation of syngeneic mice were rapidly removed from the circulation (74). This recognition of external PS may form the molecular basis of removal of pathological and aged RBCs by cells of the reticuloendothelial system (RES).

Macrophage recognition of PS is also correlated with their increased binding to tumor cells. The levels of PS on the outer leaflet of murine erythroleukemia cells (MELCs) and differentiated murine erythroleukemia cells (dMELCs) were measured: MELCs contained a greater density of PS on the outer leaflet than dMELCs (76). As MELCs differentiated, they progressively lost PS from the outer leaflet of the plasma membrane; these cells were bound less by macrophages (54,76). Similar data were derived by examining the interaction of human monocytes with three human tumorigenic cell lines and one normal human epidermal keratinocyte line. The three tumor cell lines exhibited elevated levels of external leaflet PS compared with normal cells; the PS levels directly correlated with monocyte binding to the target cells as well as with monocyte-mediated cytotoxicity (77).

Collectively, these results indicate that activated macrophages recognize and destroy neoplastic cells without injuring nontumorigenic cells. The mechanism for this recognition is nonimmunological and requires cell to cell contact. Tumoricidal macrophages can discriminate between neoplastic and normal cells by a process that is independent of major histocompatibility antigens, tumor-specific antigens, cell cycle, or transformation phenotype. Moreover, resistance to tumoricidal macrophage-mediated lysis is rare. Because activated macrophages can destroy phenotypically diverse tumor cells, including cells resistant to killing by other host defense mechanism and by anticancer drugs, macrophage-directed therapy represents a potential strategy for elimination of biologically heterogeneous metastatic cells.

IV. Activation of Macrophages to the Tumoricidal State

A. Classic Activation Pathways

The activation of macrophages results in noncytotoxic macrophages acquiring tumoricidal properties (78–81). There are two major pathways to produce macrophage activation for tumor cell cytotoxicity in vivo. Macrophages can become activated subsequent to interaction with microorganisms or their products such as endotoxins and bacterial cell wall components. Although this interaction is very common, intentional attempts to activate macrophages systemically by administering microorganisms or their products were limited by side effects such as allergic reactions and granuloma formations (81). For this reason, the field did not progress until the discovery of muramyl dipeptide (MDP), a defined component of the bacterial cell wall (82,83). Muramyl dipeptide is a water-soluble, low molecular weight synthetic moiety of N-acetyl-muramyl-L-alanyl-D-isoglutamine; it has potent effects on a variety of host defense cells, including macrophages (82,84). A synthetic, lipophilic MDP derivative, N-acetyl-muramyl-L-alanyl-D-isoglutamyl-L-alanyl-2-(1′,2′-diplamitoyl)-sn-glycero-3′-phosphoryl-ethylamide (MTP-PE), has also been synthesized (85). Although muramyl peptides influence several macrophage functions in vitro, comparable effects have not been observed in vivo, because they are rapidly cleared after parenteral administration (86,87). Even when injected at very high doses, muramyl peptides fail to induce significant macrophage-mediated antitumor activity (88).

Macrophages can also be activated by their interaction with lymphokines. Mitogen- and antigen-stimulated T cells release diffusible mediators that interact with specific receptors on target cells. A family of lymphokines, referred to as macrophage-activation factors (MAFs) (89–91) are able to render macrophages tumoricidal. These include the interferons (INFs)—IFN-γ, which is able to prime macrophages for activation following the induction of a second signal (such as suboptimal concentrations of MDP), IFN-α, and IFN-β—which also prime macrophages in a synergistic fashion with IFN-γ (92,93). There is evidence that IFN-γ–mediated priming of macrophages renders them sensitive to tumoricidal activation by tumor necrosis factor (TNF) and IL-1, demonstrating that interactions of lymphocyte and macrophage products (interferons and interleukins) with macrophage receptors can lead to both transitory and sustained tumoricidal activity (94). Studies indicated that the priming stimulus delivered to macrophages by IFN-γ (or complete activation by MAF) involves the binding and subsequent internalization of cell surface receptors (95), a process that requires a sustained presence of ligand. The tumoricidal properties of lymphokine-activated macrophages are short lived, and the cells become refractory to reactivation by free lymphokines (96). For these reasons, the systemic administration of soluble lymphokines does not result in the sustained level of tumoricidal activity needed as a potential therapeutic modality.

B. Activation of Macrophages by Liposomes Containing Immunomodulators

Advances in liposome technology may provide a mechanism for activating macrophages in situ. Considerable attention has been focused on the use of synthetic phospholipid vesicles, liposomes, to target various drugs to different organ sites in vivo (88,97,98). Most attempts to target liposomes to solid neoplasms have failed, because these phospholipid vesicles are unable to extravasate from the circulation and are rapidly taken up by circulating and fixed phagocytic cells. We have utilized this natural fate of liposomes to target immunomodulators encapsulated within liposomes to cells of the RES, thus activating these phagocytic cells to the tumoricidal state (99,100). Preferential delivery of the liposome-encapsulated substances can be achieved in vivo, because mainly phagocytic cells are exposed to the liposome-entrapped agents. This natural localization pattern also allows efficient targeting of liposomes and their contents to various macrophage compartments in the body.

Through the systematic evaluation of multilamellar vesicles (MLVs) with different phospholipid compositions, we and others have identified certain classes of phospholipids that are preferentially recognized by macrophages. The inclusion of negatively charged phospholipids such as phosphatidylserine or phosphatidylglycerol in MLVs consisting of phospatidylcholine enhances their uptake by macrophages (100). Indeed, liposomes composed of PC-PS result in a significant accumulation by all cells of the RES (91,101–103).

C. In Situ Activation of Murine Macrophages

Since the lung is a major site of disseminated metastatic disease, we identified those liposomes with increased efficacy for localization to the lung parenchyma following intravenous injection of radiolabled liposomes. Large MLVs (>0.1 μm) were retained in the lungs more efficiently than were small unilammelar liposomes of identical lipid composition. In addition, liposomes of the same structural class were more efficiently retained in the lungs when they contained negatively charged lipids (PS) than when they contained only neutral lipids (PC) (97,104).

Subsequent studies indicted that murine blood monocytes can extravasate out of the circulation and differentiate into alveolar macrophages (AMs). This was shown by the intravenous injection of PC-PS (7:3 molar ratio) MLVs that contained trace amounts of a fluorescent lipid analogue into mice and recovery of AMs by lung lavage 24 hr later (105). About 12–15% of the total AM population (as identified by the formation of rosettes between opsonized erythrocytes and macrophages) contained fluorescent MLVs, suggesting that the PC-PS liposomes were effective vehicles for the delivery of encapsulated compounds to blood monocytes (106).

The intravenous injection of PC-PS MLVs containing MDP or MTP-PE immunomodulators results in the in situ activation of AMs. However, any potential therapeutic modality for metastatic disease macrophages must provide sustained activity. In order to assess whether AMs activated by liposomes containing MDP maintained their tumoricidal properties over an extended period of time, the AMs were harvested and assayed for tumoricidal activity 1, 2, 3, or 4 days after infusion of a single dose of MLVs containing MDP, MTP-PE, or a combination of both of these agents. On day 1, AMs harvested from all of the mice that received MLVs with hydrophilic MDP exhibited less than 20% cytotoxicity against the murine tumor target cells; at every other time point, AMs harvested from mice that received MLV containing MTP-PE consistently exhibited significantly higher levels of tumoricidal activity (up to 55% cytolysis of target cells) (107–109).

V. Immunotherapy of Cancer Metastasis with Liposome-Encapsulated Macrophage Activators

A. Therapy of Murine Melanoma Lung Metastases

The systemic activation of macrophages by liposomes containing allogeneic MAF (22,), MDP (107–109), MTP-PE (108–111), recombinant IFN-γ (rIFN-γ) with MTP-PE (112), or MDP with MAF (113) results in the destruction of well-established lung and lymph node metastases produced by primary subcutaneous melanoma. The major tumor model for these studies has been the B16-BL6 melanoma cell line, which is syngeneic to C57BL/6 mice. After implantation in the footpad, this tumor metastasizes to lymph nodes and the lungs in more than 90% of the mice (114). Each of the C57BL/6 mice was given an intrafootpad injection of melanoma cells, and 4–5 weeks later, when the tumors had reached 10–12 mm, the leg-bearing the tumor, including the popliteal lymph node, was amputated. Three days later, the mice were injected intravenously with MLVs containing immunomodulators or a placebo preparation. Both test and control groups were treated twice weekly for 4 weeks (eight IV injections). Once again, PC-PS liposomes were used, because they are not toxic (97) and because they are arrested efficiently in the lungs, as well as in organs of the RES following IV injection (97,100,104).

Spontaneous metastases in the lungs and lymph nodes were well established at the time liposome treatment began (many individual metastases could be seen macroscopically at this time). All mice treated with saline, free-form lymphokines, free-form MDP, or liposomes containing placebo died by day 90 of the experiment; that is, 60 days after the amputation of the tumor-bearing leg. Significantly, 70% of the mice injected intravenously with liposome-encapsulated lymphokines and 60% of the mice injected with liposome-encapsulated MDP were alive when the experiments were terminated at 200 days. In this tumor system, we estimate that the metastases contained 10^7 cells at the beginning of the liposome treatment. Since the median survival time of treated mice was 40–50 days, we speculated that the residual tumor burden of the surviving mice must have been reduced to less than 10 viable cells. Further studies revealed that the degree of efficacy depended on the timing of the first treatment with liposome-encapsulated MTP-PE. In animals initially inoculated in the footpad with B16-BL6 melanoma cells, the first injection of liposomes was given either 3, 7, or 10 days after hind limb amputation. The results indicated that optimal therapy consisted of early treatment beginning 3 days after surgery when tumor burden was smallest. Moreover, in a separate study, a group of mice with metastatic melanoma was treated with varying schedules of liposome administration. The most effective schedule for treatment of lung metastases was twice weekly injections for 4 weeks; 68% of the mice were alive on day 200 of the study (114).

Collectively, these results demonstrate that repeated the administration of the liposomes results in maximal therapeutic benefit and that micrometastases are most susceptible to this form of therapy. Successful treatment of metastases by the intravenous injection of liposomes containing immunomodulators has been reported for several tumor host models, including mouse fibrosarcoma (113,115–117), melanomas (118,119), lung carcinoma (120), and colon carcinoma (121).

B. Therapy of Spontaneous Lung Metastasis of Canine Osteogenic Sarcoma

Like the disease in humans, the natural behavior of canine osteosarcoma involves progressive growth of the primary tumor, with the dog dying of distant visceral (lung) metastasis (122,123). Surgical resection of the primary neoplasm produces a median survival time of 3–6 months, with only 10% of the dogs surviving 1 year or longer. In a randomized study of the therapeutic efficacy of liposome-encapsulated MTP-PE, 27 dogs were studied (124). All animals had histological confirmation of osteosarcoma and complete amputation of the affected limb. Immediately following surgery, dogs were randomized to receive either liposomes containing MTP-PE or liposomes containing saline (placebo) in twice weekly IV infusions for 8 weeks. The dose per treatment was 2 mg/m^2 of MTP-PE (in 500 mg of phospholipids). Fourteen dogs were treated with active compound and 13 dogs received the saline placebo. The median survival time of this control groups was 77 days; a survival that does not differ from treatment with surgery alone (122,123). In marked contrast, the median survival of dogs treated with liposome–MTP-PE was 222 days. Moreover, four dogs in this group were alive and free of metastasis 1 year following surgery. The treatment had no toxic effects and was well tolerated. Of course, as with any therapy based on immunomodulation, the major limitation of eradication of metastases by macrophages is tumor burden (28, 114,125). It is possible that in some dogs, the tumor burden exceeded the amount that could be destroyed by macrophages. In this as well as in other preclinical studies (114,122,126), it was recommended that systemic macrophage activation follow cytoreductive therapies.

VI. Multimodality Treatment of Lung Metastases in Murine Models

A. Combination of Local x-Irradiation and Liposome MTP-PE

At least two factors limit the treatment of metastases by systemic activation of macrophages. First, the activated macrophages may not sufficiently infiltrate metastases, and second, the tumor burden may exceed the capability of the available activated macrophages. For these reasons, we performed a study to determine whether the combination of local thoracic x-irradiation (LTI) combined with systemic macrophage activation would improve the efficacy for the treatment of large lung metastases (127). Ionizing radiation produces a logarithmic decrease in tumor cell survival (128), thus reducing metastatic burden. Radiation also produces blood vessel injury and inflammation in the irradiated tissues, and this inflammation may enhance the migration of macrophages to the irradiated tissue. We injected mice with a murine fibrosarcoma that resulted in a high tumor burden of lung metastases and all of the untreated mice died. As anticipated with this large tumor burden, treatment with liposomes containing MTP-PE did not decrease the median number of lung metastases nor significantly increase overall survival. Treatment of the mice with LTI alone resulted in 32% long-term survivors, but the best therapeutic result was obtained in mice given LTI and liposome MTP-PE twice weekly (for 4 weeks); 60% of these mice were alive on day 120 of the study.

B. Sequential Therapy with Chemotherapeutic Drugs and Liposome MTP-PE

The outcome of macrophage-tumor interaction in vivo depends on factors such as the ratio of macrophages to cancer cells, the activation state of the macrophages, and the rate of tumor cell proliferation. Because the major limitation to treatment of disseminated metastases by systemic macrophage activation appears to be tumor burden, macrophage-directed therapeutic regimens will have to be used in combination with other cytoreductive treatments such as chemotherapy. Obviously, for such a strategy to succeed, it must not be toxic to the macrophage system (129). Because drug-mediated toxicity is likely to be a function of the administered dose, we investigated the pharmacokinetic variables that might interfere with subsequent macrophage activation.

We studied whether macrophages of mice pretreated with doxorubicin (DXR) could be activated by intravenous infusion of liposome MTP-PE. A dose-dependent decrease in white blood cells was found in mice after intravenous injection of DXR (4–15 mg/kg). In mice given <10 mg/kg, the decrease was temporary (about 2 weeks). Macrophages harvested from the peritoneal cavity of mice surviving DXR administration of up to 15 mg/kg were tumoricidal (130,131).

After single or multiple injections of DXR, mice were given intravenous or intraperitoneal injections of liposome MTP-PE. On the following day, the cytotoxic activity of the in situ activated macrophages (alveolar or peritoneal exudate cells) was determined in culture against syngeneic melanoma cells. Under these conditions, the systemic administration of DXR did not compromise the activation of murine macrophages by liposome MTP-PE. Such combination therapy for metastases is doubly attractive, because it provides a mechanism for the reduction of large tumor burdens followed by elimination of residual, resistant tumor cells. This approach is currently being tested in clinical trials with liposome MTP-PE.

C. Combination of Liposome MTP-PE with INF-γ or IL-2

Primary renal cancers can be removed surgically, but the majority of patients still die from metastases that are resistant to conventional systemic therapies. Renal cancer may be a good candidate for therapy based on the in situ activation of macrophages, because it is resistant to most chemotherapeutic strategies and there are reports of successful immunological intervention in this disease.

A new macrophage activator, CGP-31362, is a synthetic lipoprotein analogue of a gram-negative bacterial cell wall component (N-hexadecanoyl-S-[(2R)-3-didodecanoyloxpropyl]-L-cysteinyl-L-alanyl-D-isoglutaminyl-glycyl-tarine sodium salt). This compound is a potent activator of macrophages and produces superior tumoricidal activity compared with that observed for MTP-PE (132–135). Liposome CGP-31362 is able to activate macrophages at a lower liposome concentration than that required by liposome MTP-PE, optimal tumoricidal activity produced by liposome CGP-31362 does not require the presence of IFN-γ. Moreover, macrophages activated with liposome CGP31362 release a wider range of diffusible mediators than those treated with liposome MTP-PE (132,136). We evaluated the therapeutic efficacy of this immunomodulator for inhibition of the growth of spontaneous lung metastases using the murine renal cell carcinoma (RENCA) model (137,138).

BALB/c mice with RENCA implanted in their kidney were injected intravenously with liposome CGP-31362; 1 day later, AMs were harvested that were found to be highly cytotoxic in vitro against RENCA target cells. Maximal cytotoxic activity of the macrophages was induced by injection of about 8 μg/dose of active CGP-31362, and a single injection produced activation of macrophages that lasted for up to 4 days. Repeated intravenous injections produced continuous antitumor activity in alveolar macrophages. To study the lipopeptide's effects on metastasis, the tumorous kidneys of BALB/c mice were removed by nephrectomy about 10 days after the initial implantation of RENCA tumor and groups of mice were injected with empty (control) or liposome CGP-31362 (twice a week for up to 4 weeks). Therapy with repeated injections of liposome CGP-31362 significantly reduced the number of lung metastases in nephrectomized mice. Optimal therapy was achieved by combining the use of intravenous liposome CGP-31362 with subcutaneous administration of recombinant mouse IFN-γ. In addition, administration of the macrophage activator prior to removal of the primary tumor by nephrectomy was superior to postoperative therapy alone.

In the next set of studies, we evaluated the possibility that the systemic administration of liposomes containing the synthetic macrophage activators CGP-31362 or MTP-PE to mice that were simultaneously receiving injections of IFN-γ or IL-2 would lead to enhanced regression of spontaneous lung metastases. The kidneys of BALB/c mice were injected with RENCA cells, and 10 days later, the kidney with local tumor was surgically resected. These mice were then randomized to treatment groups and received injections of liposomes containing either CGP-31362 or MTP-PE alone, subcutaneous IFN-γ or IL-2 alone, or the mixed combinations of therapy using both macrophage activation plus administration of cytokine. The results showed that whereas administration of either cytokine alone did not significantly reduce the median number of lung metastases in treated mice compared with the controls (about 50 lung nodules), the use of either macrophage activator alone did significantly reduce the number of lung tumor nodules. But the combination therapy was superior, and with the use of liposome 31362 plus IL-2, one half of the mice were cured of this aggressive, metastasizing renal carcinoma. In addition, regional and local recurrence at the site of excision of the primary tumor were not observed in mice treated with the combination therapy.

VII. Oral Administration of Liposome MTP-PE for Macrophage-Directed Therapy of Spontaneous Lung Metastases of Murine Renal Cell Carcinoma

Early studies from our laboratory demonstrated the ability of unencapsulated (free-form) MTP-PE to activate alveolar macrophages in mice after repeated oral feedings (139). It was not known whether encapsulation of MTP-PE into multilamellar liposomes would enhance the distribution and delivery of this activator to tissue macrophages after oral administration. However, liposomes composed of only phosphatidylcholine (POPC) are phagocytosed by macrophages and are stable in the presence of serum (100). We therefore evaluated the in vivo modulation of macrophage function by oral administration of POPC-MTP-PE (140). These liposomes were rapidly absorbed in the intestine and reached the systemic circulation within 4 hr (about 60% of the lipid material was no longer in the form of POPC, as determined by radiolabeled lipid using thin-layer chromatography of extracted tissues). AMs harvested from the lungs of mice 24 hr after a single oral feeding were tumoricidal toward the RENCA cells. However, as observed for intravenous administration, this activation was gradually lost after about 3 days. Repeated daily feedings with POPC-MTP-PE generated sustained activation of the AMs (interestingly, activation of peritoneal macrophages to the tumoricidal state required at least 3 daily feedings in contrast to the single feeding that resulted in activation of the AMs). In vitro studies demonstrated the release of tumor necrosis factor-α (TNF-α) and IL-6 by macrophages activated by POPC-MTP-PE in the presence of IFN-γ. IL-1 and nitric oxide were not induced by this liposomal preparation. Daily administration of POPC-MTP-PE after establishment of experimental metastases in the lungs of BALB/c mice inhibited the growth of this tumor and confirmed the potential role of long-term therapy with this new immunomodulator.

The preclinical evaluation of these synthetic macrophage activators established the fundamental parameters necessary for their use in an adjuvant setting (their likely clinical use). The composition of liposomes and route of administration influenced their tissue distribution and uptake by cells of the RES. The optimal formulation of active compound-to-lipid ratio was determined, as well as the schedule dependence and their use in combination with other treatment modalities, including both immunomodulators and cytotoxic drugs, was studied using orthotopic animal models that accurately modeled the in vivo complexity associated with therapy of metastatic disease.

VIII. Clinical Trials

A. Phase I (Toxicity, Dosage, and Biological Effects)

The principles of immunomodulation mediated by activation of macrophages with liposome-encapsulated muramyl peptides motivated careful evaluation of this form of therapy in cancer patients. The ultimate strategy was for potential use against metastatic disease, especially in light of the previous therapeutic efficacy of liposome MTP-PE against lung metastasis, as observed in the preclinical animal tumor models. It was first demonstrated that blood monocytes of patients with local or disseminated colorectal carcinoma could be activated in vitro to become tumoricidal after incubation with either free-form or liposome-encapsulated immunomodulators and these monocytes secreted TNF and IL-1 (141). However, potential therapeutic regimens to stimulate the antitumor properties of macrophages would have to be used in combination with conventional modalities such as chemotherapy. It was shown therefore that in vitro incubation of doxorubicin hydrochloride (DXR) (0.5–500.0 ng/ml) for 1 hr prior to activation of human blood monocytes with MLV-MTP-PE did not inhibit the ability of these cells to become tumoricidal (142). Monocytes obtained from osteosarcoma patients treated with 75 mg/m^2 DXR gave similar results and strengthened the rationale for this form of immunotherapy as adjuvant treatment for residual, metastatic disease. Studies were then designed to evaluate the biological properties (toxicity and maximum-tolerated dose) of MLV-MTP-PE in a multicenter phase I trial in cancer patients. The study included 28 adults with histological confirmation of metastatic cancer who were unresponsive to all previous therapies. The criteria for patient eligibility, evaluation of disease, and monitoring of clinical parameters have been reported in detail (143,144), and the study will be summarized here regarding clinical principles for phase I studies using liposome-encapsulated immunomodulators. Stable, pharmaceutical preparations of freeze-dried liposomes containing MTP-PE were prepared by Ciba-Geigy, Ltd. (Basel, Switzerland). The clinical preparation (PC-PS ratio 7:3; 4 mg MTP-PE per 1200 μmol lipid per vial) was the same formulation shown to have potent biological activity in the murine and canine studies described above. Liposome MTP-PE was administered IV over 1 hr twice weekly for 9 weeks. The first three patients in the study received liposome MTP-PE at a dose of 0.05 mg/m^2 MTP-PE. After 3 weeks of biweekly infusions, the dose was escalated to the next level (0.1 mg/m^2) and subsequently to a third and final level of 0.25 mg/m^2 MTP-PE. When no major toxicity was observed, an additional three patients were entered at a dose one level higher than the initial starting level for the previous group and were escalated every 3 weeks. The doses of MTP-PE thus administered were up to 12 mg/m^2.

The major findings from this phase I trial were: At the dose and schedule described, liposome MTP-PE was safe; dose-limiting side effects were chills and fever (80% of the patients), fatigue (60%), nausea and vomiting (55%), and headaches (20%). Hypo- or hypertension was also observed, but these were self-limiting. In four patients, liposome MTP-PE treatment was deescalated owing to severe malaise and recurrent fever exceeding 38.8°C. Toxicity was not cumulative, and the maximal tolerated dose of liposome-MTP-PE was 6 mg/m^2. Similar to preclinical studies, we observed a consistent increase in absolute white blood cells and granulocytes. This finding was significant in patients infused with $\geqslant 2$ mg/m^2 of MTP-PE liposomes regardless of whether, at 24 hr after infusion, data were analyzed alone or combined with data obtained at 72 and 96 hr after infusion. This could have been due to production of colony-stimulating factors (CSFs) by activated monocytes, in particular granulocyte-CSF (G-GSF) given either alone or in combination with IL-1β (145,146).

The intravenous administration of liposome MTP-PE activated tumoricidal properties in the monocytes of 24 of the 28 patients. This increased antitumor activity could not be explained by simple fluctuations in cytotoxic function, because monocyte-mediated antitumor cytotoxicity in both normal donors and cancer patients did not change when sequentially studied over time (144,147). Activated human monocytes are known to produce numerous cytokines such as IL-6, TNF, IL-1, and granulocyte-monocyte CSF (GM-CSF) (148). We observed increases in serum IL-1 in 6 of 10 patients within 1 hr postinfusion. Elevations in

C-reactive protein, β_2-microglobulin, and ceruloplasmin were frequently observed. Absolute tumor regressions were not observed.

B. Phase II Trials of Liposome MTP-PE for Recurrent Osteosarcoma Lung Metastases

We completed a phase II trial using liposome MTP-PE in patients with recurrent (drug-resistant) osteosarcoma lung metastases. The rationale for the use of liposome MTP-PE in this group of patients was as follows: These patients have a disease-free survival of less than 1 year following surgical excision of all visible and palpable lung metastases. Eighty percent of the patients relapse by 1 year after surgery (149). Salvage chemotherapy regimens have been largely unsuccessful in extending this disease-free interval, emphasizing the need for new therapeutic strategies. Presumably the patients have a subpopulation of tumor cells that is either resistant to chemotherapy or in a location where adequate drug levels cannot be achieved. Since activated macrophages can kill drug-resistant cells (129), we reasoned that perhaps activated macrophages would eradicate the residual cells that were undetectable at the time of surgery.

All patients entered in this trial had histologically proven osteosarcoma and pulmonary metastases that had either developed during adjuvant chemotherapy or that were present at diagnosis and had persisted despite chemotherapy. Prior to entry into the study, each patient's primary tumor had to be resected and the patient rendered disease-free by surgical excision of pulmonary tumors, as established by x-ray. Chest x-ray, lung computed tomography scan, and bone scan were obtained 2–4 weeks postoperatively to document the clinical status of the patients. Liposome MTP-PE was administered intravenously twice weekly for 12 weeks at 2 mg/m^2 over 1 hr. Initially, 12 patients were treated on this regimen. Because of histological changes observed in pulmonary lesions removed from 6 of these patients (see below), the next 16 patients received liposome MTP-PE twice weekly for 12 weeks and then once weekly for an additional 12 weeks for a total treatment course of 24 weeks.

Patients were treated in the outpatient clinic; none required hospitalization. Toxic reactions were limited to those as described above for the phase I trials. Serial blood samples were collected from 16 patients after liposome MTP-PE administration and were assayed for cytokine levels (TNF, IL-6, IFN-γ, neopterin, and C-reactive protein) (150). Circulating TNF and IL-6 were induced shortly after liposome MTP-PE infusion. TNF levels peaked 1–2 hr after infusion, whereas peak IL-6 levels were detected at 2–3 hr. Neopterin levels (a specific marker of macrophage activation) and C-reactive protein increased at 24 hr after infusion. Of the first 12 patients, 7 had disease recurrence within 6 weeks of completing liposome therapy in the area of surgery that had been performed immediately prior to entry into the study. The nodules were excised from six of these patients and submitted for pathological examination. The specimens were compared with the lesions resected immediately before the liposome therapy. All six patients showed histological changes in the pulmonary tumors (151). In three patients, peripheral fibrosis developed around the tumors, as did inflammatory cell infiltration and neovascularization. In contrast, lesions resected following chemotherapy and immediately prior to liposome MTP-PE therapy exhibited either central necrosis, with viable tumors cells at the periphery of the tumor and no inflammatory response or absence of both necrosis and inflammation.

After liposome MTP-PE therapy, case 4 showed early evidence of the fibrotic changes described for the first three cases. Two cases showed a change in the malignant characteristics of the entire tumor from high-grade before therapy to low grade. In case 5, the histological subtype of the pulmonary lesion also changed from an osteoblastic to a fibroblastic subtype. Case 6 showed hemorrhagic necrosis but no fibrosis. However, such findings are often observed in rapidly growing tumors and are not necessarily attributable to liposome MTP-PE therapy. A seventh patient suffered recurrence in the lung 9 months after completing liposome MTP-PE therapy and had no change in tumor characteristics. This was not unexpected in light of probable resolution of the fibrotic reactions within the lesion during extended time off of therapy.

None of the osteosarcoma lung metastases resected prior to liposome MTP-PE therapy exhibited peripheral fibrosis, neovascularization, or host inflammatory cells. Some of these lesions exhibited a zone of central necrosis characteristic of the effects of chemotherapy. The periphery of these lesions, however, was occupied by viable tumor cells growing into compressed normal lung tissue.

In addition to peripheral fibrosis and neovascularization, infiltration of the tumor with chronic inflammatory cells was observed after therapy with liposome MTP-PE in all five cases. Immunohistochemistry studies confirmed that the cells mainly consisted of histiocytes-macrophages. Once again, these were not found in chemotherapy-treated or in untreated lesions. The implication of these results is that liposome MTP-PE altered the host response so that inflammatory cells infiltrated the tumor regions. We interpret the peripheral fibrosis to be the consequence of cellular infiltration of macrophages.

What we observed in these patients is reminiscent of the appearance of pulmonary tuberculosis (152). Initially, the lesion is walled off and slow necrosis proceeds from the outside, so that the lesion is replaced by fibrous tissue. Eradication of tuberculosis by chronic inflammation is a slow process, and viable bacilli can persist for months. Thus, our initial choice of a 12-week course of therapy may have been insufficient. Giving liposome MTP-PE for 6–12 months might have increased peripheral fibrosis and tumor necrosis. For this reason, the next 16 patients entering the study received liposome MTP-PE twice weekly for 12 weeks and then weekly for an additional 12 weeks.

To assess the efficacy of both the 12- and 24-week treatment, the disease-free intervals of the two treatment groups were compared with those of a historical control group of osteosarcoma patients who had relapsed with lung metastases following multiagent chemotherapy and had been rendered disease-free by surgery. Subsequently, all patients had received chemotherapy. The disease-free intervals were calculated from the day of surgery. The study endpoint was the time of disease recurrence after surgery (disease free). The mean time to relapse for the historical control group was 4.5 months; by 1 year after surgery, 85% of the patients had relapsed. Of the 12 patients receiving liposome MTP-PE for only 12 weeks, 2 patients had no evidence of disease for 12 and 19 months before having a recurrence in the lung; 1 patient had no evidence of disease for 8 months before recurrence, in this case, in bone (the lung was free of disease); and 1 patient remained free of disease at 50 months. However, the average disease-free interval for this group of patients, 6.8 months, was not significantly improved over that observed for the control group (153).

For the 16 patients receiving 24 weeks of liposome MTP-PE therapy, the median disease-free interval significantly exceeded that of the control group (9.0 months vs 4.5 months). At 6 months, 75% of the 24-week treatment group were free of disease compared with 43% of the control group. However, this number rapidly dropped once therapy was stopped, and by 210 months (4 months after the completion of therapy), only 49% of these patients had no evidence of disease. These numbers are too small to be conclusive, but the results suggest that more prolonged therapy could benefit some patients. The case of administering liposome MTP-PE in an outpatient setting and the minimal toxicity of this agent make extension of therapy feasible (especially in light of the preclinical data that demonstrated the biological and antitumor activity of liposome-encapsulated MTP-PE given orally). We were unable to demonstrate a correlation between plasma cytokine levels or the level of monocyte tumoricidal activity generated during this therapy with clinical outcome. Therefore, at the present time, we do not have a biological assay that could serve to discriminate between potential responders and nonresponders. However, this clinical preparation has biological activity in the cancer patients who have undergone extensive therapy (and failed). We, therefore, are attempting to add this form of therapy with earlier therapeutic planning for treatment of osteosarcoma patients.

C. Phase IIb Trial of Liposome MTP-PE: Combination with Chemotherapy in Patients with Recurrent Osteosarcoma Lung Metastases

This trial was based on the rationale that the biological activity of liposome MTP-PE could be exploited during the time of tumor reduction by chemotherapy in patients with recurrent osteosarcoma. This study was primarily undertaken to determine the tolerability of this combination therapy and whether ifosfamide

(IFX) chemotherapy would alter or suppress the in vivo immune response to liposome MTP-PE (154). Patients had histologically proven osteosarcoma and pulmonary metastases that either developed during adjuvant chemotherapy or were present at diagnosis, persisted despite chemotherapy, and recurred following surgical excision. Stratum A patients were rendered clinically free of disease within 4 weeks of study entry prior to receiving combination therapy. IFX was administered at 1.8 g/m^2 for 5 days every 21 days for up to eight cycles. Liposome MTP-PE was administered twice weekly for 12 weeks and then once weekly for 12 weeks. Once the cycle of combination therapy was defined as 5 days of IFX and 3 weeks of liposome MTP-PE therapy, stratum B patients had measurable disease at study entry that was judged to be amenable to surgical resection. These patients received three cycles of combination therapy prior to surgery to judge clinical and histological response. Postoperatively, patients received an additional five cycles. A total of nine patients were entered into the protocol: six on stratum A and three on stratum B. Levels of the inflammatory cytokines and tumoricidal activity of blood monocytes were measured as well as complete blood counts. No increase in the toxic side effects of IFX was demonstrated when administered with liposome MTP-PE, nor were delays in IFX administration due to neutropenia. The side effects of liposome MTP-PE were not increased in this combination regimen. There was no interference of the chemotherapy on the immune parameters measured following liposome MTP-PE therapy, and conversely no effect of immunomodulation on the cytotoxic effects of IFX. Elevations of serum C-reactive protein, plasma neopterin, IL-6, IL-8, and TNF-α following combination therapy were similar to those observed in patients treated with liposome MTP-PE alone. Monocyte-mediated tumoricidal activity was elevated 24 and 72 hr following therapy, which is similar to what has been reported following administration of liposome MTP-PE. Tumor specimens obtained from stratum B patients showed the histological characteristics consistent with a "chemotherapy effect,"; that is, dead, amorphous, acellular osteoid with cell dropout. In addition, fibrosis accompanied by an infiltration of chronic inflammatory cells (i.e., histiocytes and lymphocytes) at the periphery of the tumor nodules was observed in many of the lesions, evidence of a biological response to the liposome MTP-PE therapy. These preliminary results suggest that this form of combination therapy may be appropriate for recurrent disease.

IX. Cellular Mechanisms of Macrophage-Directed Antitumor Therapy

A. Macrophage-Mediated Tumoricidal Activity

The tumor microenvironment is a dynamic milieu due to the local secretion of growth factors, regulatory molecules, and cytokines from tumor cells, host stromal and parenchymal cells, and tumor-infiltrating cells. Macrophages are envisioned easily to traffic to local regions of inflammation and sites of altered tissue architecture due to progressive tumor growth. Immunomodulation within this environment provokes undefined cellular responses that alter the activation state and immune reactivity of tumor infiltrating cells.

Classic activation of macrophages to a tumoricidal state requires strong transmembrane signals associated with the internalization of endotoxin-like products and the activation state of the macrophage is modulated (sensitized) by lymphokines, such as IFN-γ and IL-2. In the absence of these strong activation costimuli, the macrophages that do infiltrate progressively growing tumors probably serve as a rich source of growth factors, angiogneic molecules, and a variety of growth-promoting and tissue-modeling substances (49).

Liposome-encapsulated macrophage activators (MTP-PE and lipoprotein CGP-31362) are rapidly engulfed by blood monocytes and fixed macrophages, and in the presence of costimulatory signals, these cells become activated, tumoricidal, and able to induce further immune reactivity against tumor cells. The antitumor effects of macrophage-directed immunomodulation can be considered as twofold: nonspecific tumoricidal properties subsequent to activation and a specific immune component that leads to T-cell recruitment and immunological memory. The direct cytotoxicity of activated macrophages does not require an intact T-cell system (133). This conclusion is based on studies showing that subsequent to intravenous injection of liposome MTP-PE, activation of alveolar macrophages occurs in immunosuppressed mice and in nude mice (133). The direct lytic effects of activated macrophages are well established. Tumoricidal macrophages release a myriad of inflammatory and reactive molecules (49), some of which are directly toxic to tumor cells (nitric oxide, TNF-α, and IL-1), and other monokines that initiate and participate in cytokine cascades within the microenvironment. For example, liposome MTP-PE rapidly upregulates mRNA for IL-1α, IL-1β, TNF-α, IL-6, and IL-8 in human monocytes, a process that is sustained for up to 72 hr (in contrast to rapid upregulation and downregulation observed for LPS stimulation) (155). Cytokines of infiltrating leukocytes can directly affect macrophage function. For example, AMs are sensitive to IL-4 and IL-10, which suppresses their tumoricidal activation (156,157), although in other systems, IL-4 can enhance activation while upregulating Ia antigens (158). IL-4 and IL-10 tend to steer antigen-driven lymphocyte reactivity toward strong humoral immune responses and suppress T-cell function, activities that may or may not favor tumor growth. The direction of T-cell responses (cell-mediated and humoral) are sensitive to relative concentrations of cytokines that favor expansion of the helper cell populations, Th1 (IL-2, IFN-γ) and Th2 (IL-4, IL-5, IL10) (159). In this context, it is interesting that the use of Th1-associated cytokines, IL-2, and IFN-γ enhanced the therapeutic benefit of liposome-encapsulated macrophage activators. Hence, it is likely that substantial modulation of immune events take place in the microenvironment of macrophage-mediated tumor cell killing, with potential activation of T-cell immune reactivity.

B. Systemic Immunity Due to Tumoricidal Activation of Macrophages

Two lines of evidence indicate that alterations in the parameters of systemic immunity can occur following tumoricidal activation of macrophages by liposome MTP-PE (with or without IL-2). First, interstitial natural killer (NK) cell activity in the lung and liver was augmented in normal mice given an intravenous infusion of liposome MTP-PE (160). Interestingly, this augmentation was not observed by direct incubation of lymphocytes with liposome MTP-PE in vitro, suggesting that indirect pathways lead to augmented NK activity in vivo. Moreover, this heightened NK activity of pulmonary-derived lymphocytes was accompanied by the resistance of the treated mice to experimental lung metastases of murine fibrosarcoma (this resistance was abolished by pretreatment with the anti-NK antibody asialo GM1). NK function is also enhanced by intraperitoneal administration of IL-2 or intravenous administration of liposome CGP-31362 in normal BALB/c mice (133), and this activity was significantly greater in mice treated with both agents. This additive effect on NK activity was not diminished in normal mice by the administration of asialo GM1 but was abolished in nude mice, suggesting a role for both NK and T cells in the efficacy of this combination therapy (133).

The location of the macrophages may also determine their interaction with lymphocytes. AMs macrophages can suppress IL-2–mediated lymphokine-activated killer (LAK) cell function, whereas blood monocytes potentiate the induction of LAK cells (161). Perhaps this reflects the release of different cytokines by macrophages that are influenced by their site of residence or relative state of activation. Macrophages are the primary source of IL-12, the potent LAK cell–inducing cytokine (162) and are also the primary source for chemotactic molecules and proliferative/differentiation molecules, such as the colony-stimulating factors (49).

A second line of evidence to support the concept of establishment of systemic immunity subsequent to macrophage-directed therapy is that mice cured of lung metastasis by systemic administration of liposome-encapsulated macrophage activators plus IL-2 were immune to a rechallenge with murine RENCA cells (138). Macrophages are also antigen-presenting cells (APCs) to T helper populations, and present antigens in the context of class II (Ia) self-antigens. This APC activity is enhanced in the presence of IFN-γ (a priming stimulus for activation with liposome MTP-PE), owing to upregulation of Ia expression (163). This function of macrophages together with the presence of other leukocyte populations in the microenvironment may result in enhanced host immune responsiveness to metastatic disease.

X. Conclusions: Macrophage Activation for Therapy of Metastatic Cancer

The continuous emergence of variant subpopulations of tumor cells from both the primary and metastatic lesions implies that the successful therapy of disseminated cancer will have to include a modality that will overcome the cellular heterogeneity of tumors that ultimately lead to resistance to therapy. Macrophages activated to become tumoricidal by interaction with liposomes containing immunomodulators may be one such treatment modality. Tumoricidal macrophages recognize and destroy neoplastic cells but not normal cells in vitro and in vivo. Moreover, evidence indicates that macrophage destruction of tumor cells is not associated with the development of tumor cell resistance.

Intravenously administered liposomes are cleared from the circulation by phagocytic cells and the endocytosis of these liposomes containing appropriate immunomodulators generates cytotoxic macrophages in situ. Multiple administration of such liposomes has been shown to bring about eradication of lung metastases in several murine tumor systems and in spontaneous canine osteosarcoma. These host antitumor responses do not require the participation of functional T lymphocytes or NK cells, although there is evidence for the activation of systemic immune responses following macrophage-directed therapy. However, the destruction of established metastases after the multiple systemic administration of liposome-encapsulated immunomodulators seems to be directly mediated by activated macrophages. The antitumor response of activated macrophages is enhanced when lymphokines, such as IL-2 or IFN-γ, are added to this form of therapy. Liposomes containing MTP-PE can activate tumoricidal properties in blood monocytes of cancer patients. The results of clinical trials indicate that at the minimum dose sufficient to produce biological effects, these liposomes are well tolerated. The cytolytic effects of macrophages in vivo are regulated by factors such as the ratio of macrophages to target cells, the state of activity of the macrophages, and the rate of tumor cell proliferation. The treatment of metastases by macrophages, therefore, would be most successful when the metastases are small or slow to proliferate and are infiltrated by a large number of tumoricidal macrophages. However, the major limitation to treatment of disseminated metastases by the systemic activation of macrophages appears to be the tumor burden. Thus, the therapeutic stimulation of macrophage-mediated antitumor activity will probably be unsuccessful for the treatment of large or numerous metastatic nodules. For this reason, potential therapeutic regimens designed to stimulate host immunity must be used in combination with other treatment modalities, such as chemotherapy, in order to reduce the tumor burden; the activated macrophages may then lyse tumor cells that are resistant to the chemotherapy. A heterogeneous disease such as cancer cannot be treated by a single modality: Systemically activated macrophages should add one more modality to our regimens for the eradication of cancer metastases.

References

1. Fidler IJ, Hart RI. Biological diversity in metastatic neoplasms: origins and implications. Science 1982; 217:998–1003.
2. Fidler IJ, Balch CM. The biology of cancer metastasis and implications for therapy. Curr Prob Surg 1987; 24:131–233.
3. Killion JJ, Fidler IJ. The biology of tumor metastasis. Semin Oncol 1989; 16:106–115.
4. Bishop MJ. The molecular genetics of cancer. Science 1987; 235:305–311.
5. Fabra A, Nakajima M, Bucana CD, Fidler IJ. Modulation of the invasive phenotype of human colon carcinoma cells by organ specific fibroblasts of nude mice. Differentiation 1992; 52:101–110.
6. Sloane BF, Honn KV. Cysteine proteinase and metastasis. Cancer Metastasis Rev 1984; 3:249–265.
7. Jones PA, DeClerck YA. Extracellular matrix destruction by invasive tumor cells. Cancer Metastasis Rev 1982; 1:289–319.
8. Liotta LA, Thorgeirsson UP, Garbisa S. Role of collagenases in tumor cell invasion. Cancer Metastasis Rev 1982; 1:277–288.
9. Prehn RT. The immune reaction as a stimulator of tumor growth. Science 1972; 176: 170–171.
10. Poste G, Fidler IJ. The pathogenesis of cancer metastasis. Nature 1980; 283:139–146.
11. Heppner G. Tumor heterogeneity. Cancer Res 1984; 44:2259–2265.
12. Nicolson GL. Generation of phenotypic diversity and progression in metastatic tumors. Cancer Metastasis Rev 1984; 3:25–42.
13. Killion JJ, Kollmorgen GM. Isolation of immunogenic tumor cells by cell-affinity chromatography. Nature 1976; 259:674–676.
14. Giavazzi R, Miller L, Hart IR. Metastatic behavior of an Adriamycin-resistance murine tumor. Cancer Res 1983; 43:5055–5083.
15. Kerbel RS. Implications of immunological heterogeneity in tumors. Nature 1979; 280:358–360.
16. Killion JJ. Immunotherapy with tumor subpopulations. II. Therapy of drug-resistant L1210 leukemia and EL4 lymphoma. Cancer Immunol Immunother 1979; 5:21–26.
17. Malkovska V, Sondel PM, Malkovsky M. Tumor immunotherapy. Curr Opin Immunol 1989; 1:883–890.
18. Fialkow PJ. Clonal origin of human tumors. Annu Rev Med 1979; 30:135–176.
19. Liotta LA. Tumor invasion and metastases. role of the extracellular matrix: Rhoads Memorial Award lecture. Cancer Res 1986; 46:1–7.
20. Fidler IJ. Critical factors in the biology of human cancer metastasis: Twenty-eighth G.H.A. Clowes Memorial Award lecture. Cancer Res 1990; 50:6130–6138.
21. Folkman J, Klagsburn M. Angiogenic factors. Science 1987; 235:442–447.
22. Fidler IJ, Ellis LM. The implications of angiogenesis for the biology and therapy of cancer metastasis. Cell 1994; 79:185–188.

23. Altomonte M, Gloghini A, Bertola G, Gasparollo A, Carbone A, Ferrone S, Maio M. Differential expression of cell adhesion molecules CD54/CD11a and CD58/CD2 by human melanoma cells and functional role in their interaction with cytotoxic cells. Cancer Res 1993; 53:3343–3348.
24. Hujanen ES, Vaisanen A, Zheng A, Tryggvason K, Turpenniemi-Mi-Hujanen T. Modulation of M_r 72,000 and M_r 92,000 type-IV collagenase (gelatinase A and B) gene expression by interferons alpha and gamma in human melanoma. Int J Cancer 1994; 58:582–586.
25. Berasconi S, Matteucci C, Sironi M, Conni M, Colotta F, Mosca M, Colombo N, Bonazzi C, Landoni F, Corbetta G, Mantovani A, Allavena P. Effects of granulocyte-monocyte colony-stimulating factor (GM-CSF) on expression of adhesion molecules and production of cytokines in blood monocytes and ovarian cancer-associated macrophages. Int J Cancer 1995; 60:300-307.
26. O'Brien T, Cranston D, Fuggie S, Bicknell R, Harris AL. Different angiogenic pathways characterize superficial and invasive bladder cancer. Cancer Res 1995; 55:510–513.
27. Evans R, Lawlor E. Macrophage content and immunogenicity of C57BL/6 and BALB/c ByJ methycholanthrene-induced sarcomas. Int J Cancer 1980; 26:831–837.
28. Evans R. Macrophages and neoplasms: new insights and their implications in tumor immunobiology. Cancer Metastasis Rev 1982; 1:227–243.
29. Bugelski PJ, Kirsh R, Poste G. New histochemical method for measuring intratumoral macrophages and macrophage recruitment into experimental metastases. Cancer Res 1983; 43:5493–5501.
30. Bugelski PJ, Cotwon SP, North SM, Kirsh R, Nicolson GM, Poste G. Macrophage content of spontaneous metastases at different stages of growth. Cancer Res 1987; 47:4141–4147.
31. Bugelski PJ, Kirsh R, Ruscarmo C, Cotwin SP, Poste G. Recruitment of exogenous macrophages into metastases at different stages of tumor growth. Cancer Immunol Immunother 1987; 24:93–98.
32. Acero R, Polentaretti N, Boffazzi H, Alberti S, Rizza A, Mantovani A. Effect of hydrocortisone on the macrophage content, growth, and metastases of transplanted murine tumors. Int J Cancer 1984; 33:95–100.
33. Evans R, Cullen RT. In situ proliferation of intratumor macrophages. J Leukoc Biol 1984; 35:561–568.
34. Normann SJ. Macrophage infiltration and tumor progression. Cancer Metastasis Rev 1985; 4:277–291.
35. Mantovani A. Tumor-associated macrophages. Curr Opin Immunol 1990; 2: 689–694.
36. Matsushima K, Larsen CG, DuBios GC, Oppenheim JJ. Purification and characterization of a novel monocyte chemotactic and activating factor produced by a human myelomonocytic cell line. J Exp Med 1989; 169:1485–1492.
37. Leonard EJ, Yoshimura T. Human monocyte chemoattractant protein-1 (MCP-1). Immunol Today 1990; 11:97–100.
38. Oppenheim JJ, Zachariae COC, Mukaida N, Matsushima K. Properties of novel proinflammatory supergene intercrine cytokine family. Annu Rev Immunol 1991; 9: 617–624.

39. Rollins BJ. JE/MCP-1: an early response gene encodes a monocyte specific cytokine. Cancer Cells 1991; 3:517–525.
40. Jiang Y, Beller DI, Frendl G, Graves DT. Monocyte/chemoattractant protein-1 regulates adhesion molecule expression and cytokine production in human monocytes. J Immunol 1992; 148:2423–2429.
41. Singh RK, Fidler IJ. Synergism between human recombinant monocyte chemotactic and activating factor and lipopolysaccharide for activation of antitumor properties in human blood monocytes. Lymphokine Cytokine Res 1993; 12:285–291.
42. Bucana CD, Fabra A, Sanchez R, Fidler IJ. Different patterns of macrophage infiltration into allogeneic murine and xenogeneic human neoplasms growing in nude mice. Am J Pathol 1992; 141:1125–1236.
43. Russell SW, Gillespie GY. Nature, function and destruction of inflammatory cells in regressing and progressing Moloney sarcomas. J Reticuloendoth Soc 1977; 22: 159–168.
44. Fauve RM, Hevin B, Jacob H, Gaillard JA, Jacob F. Anti-inflammatory effects of murine malignant cells. Proc Natl Acad Sci USA 1974; 71:4052–4056.
45. Snyderman R, Pke MC, Blaylock BI, Weinstein P. Effect of neoplasms on inflammation: depression of macrophage accumulation after tumor implantation. J Immunol 1976; 116:585–589.
46. Talmadge JE, Key M, Fidler IJ. Macrophage content of metastatic and nonmetastatic rodent neoplasms. J Immunol 1981; 126:2245–2248.
47. Russell PI, Phillips J, Allan W, Hume DA. Antigenic variation and macrophage infiltration of human bladder tumors xenografted into nude mice. J Leukoc Biol 1988; 43:335–342.
48. Mantovani A, Bottazzi B, Colatta F, Sozzani S, Ruco L. The origin and function of tumor associated macrophages. Immunol Today 1992; 13:265–270.
49. Nathan CF. Secretory products of macrophages. J Clin Invest 1987; 79:319–326.
50. Fidler IJ, Schroit AJ. Recognition and destruction of neoplastic cells by activated macrophages: discrimination of altered self. Biochim Biophys Acta 1988; 948: 151–173.
51. Fidler IJ. Recognition and destruction of target cells by tumoricidal macrophages. Is J Med Sci 1978; 14:177–191.
52. Fidler IJ, Cifone MA. Properties of metastatic and nonmetastatic cloned subpopulations of an ultraviolet light-induced murine fibrosarcoma of recent origin. Am J Pathol 1979; 97:633–648.
53. Killion JJ, Radinsky RR, Dong Z, Fishbeck R, Whitworth P, Fidler IJ. The immunogenic properties of drug-resistant murine tumor cells do not correlate with expression of the MDR phenotype. Cancer Immunol Immunother 1993; 36:381–386.
54. Pak CC, Fidler IJ. Activated macrophages distinguish undifferentiated-tumorigenic from differentiated-nontumorgenic murine erythroleukemia cells. Differentiation 1989; 41:49–55.
55. Gordon S, Perry VH, Rabinowitz S, Chung L, Rosen H. Plasma membrane receptors of the mononuclear phagocyte system. J Cell Sci 1988; 9:1–26.
56. Kolb-Bachofen V, Schlepper-Scharfer J, Roos P, Hulsmann D, Kolb H. GalNAc/ Gal-specific rat liver lectins: their role in cellular recognition. Biol Cell 1984; 51:219–226.

57. Oda S, Sato M, Toyoshima S, Osawa T. Binding of activated macrophages to tumor cells through a macrophage lectin and its role in macrophage tumoricidal activity. J Biochem 1989; 105:1040–1043.
58. Pohajdak B, Lee K, Sugawara I, Miller V, Wright JA, Greenberg AH. Comparative analysis of natural killer cell and macrophage recognition of concanavalin A–resistant Chinese hamster ovary cells: role of membrane oligosaccharides. J Natl Cancer Inst 1986; 2:257–267.
59. Gjoen T, Seljelid R, Kolset SO. Binding of metastatic colon carcinoma cells to liver macrophages. J Leukoc Biol 1989; 45:362–369.
60. Brunda MJ, Wiltrout RH, Holden HT, Baresio L. Selective inhibition by monosaccharides of tumor cell cytotoxicity mediated by mouse macrophages, macrophage-like cell lines, and natural killer cells. Int J Cancer 1983; 31:373–379.
61. Mercurio AM. Disruption of oligosaccharide processing in murine tumor cells inhibits their susceptibility to lysis by activated mouse macrophages. Proc Natl Acad Sci USA 1986; 83:2609–2613.
62. Lotan R, Raz A. Endogenous lectins as mediators of tumor cell adhesion. J Cell Biochem 1988; 37:107–117.
63. Killion JJ, Wallenbrock MA, Rogers III JA, Kollmorgen GM, Sansing WA, Cantrell JL. Tumorigenicity and the expression of cell-surface carbohydrates. Nature 1976; 261:54–56.
64. Killion JJ. Reduced tumorigenicity of A-10 adenocarcinoma in the presence of immunogenic A-10 tumor cells. Cancer Immunol Immunother 1977; 3:93–95:
65. Killion JJ. The immunotherapeutic value of an L1210 tumor vaccine depends upon the expression of cell-surface carbohydrates. Cancer Immunol Immunother 1977; 3: 87–91:
66. Martin DW, Reece M, Reese AC. Effect of plasma fibronectin, macrophages, and glycosaminoglycans on tumor cell growth. Cancer Invest 1984; 2:339–345.
67. Raynor RH, Reese AC. Effect of fibronectin on macrophage-induced tumor cell cytostasis. Oncology 1984; 41:420–426.
68. Huard TK, Raney JI, Wood JFB, Wieha MA. A potential role for the extracellular matrix glycoprotein laminin on macrophage-tumor cell interactions. Int J Cancer 1985; 36:511–517.
69. Ruoslahti E, Pierschbacher MD. New perspectives in cell adhesion: LRGD and integrins. Science 1987; 238:491–497.
70. Garcia-Prado A, Ferreira OC, Valinsky J, Biano C. Fibronectin receptors of mononuclear phagocytes: binding characteristics and biochemical isolation. Exp Cell Res 1989; 181:420–431.
71. Gordesky SE, Marinetti GV. The asymmetric arrangement of phospholipids in the human erythrocyte membrane. Biochem Biophys Res Commum 1973; 50:1027–1031.
72. Zwaal RFA, Hemker HC. Blood cell membranes and haemostasis. Haemostasis 1982; 11:12–39.
73. Tanaka Y, Schroit AJ. Insertion of fluorescent phosphatidlyserine into the plasma membrane of red blood cells. Recognition by autologous macrophages. J Biol Chem 1983; 258:11335–11341.
74. Schroit AJ, Madsen JW, Tanaka Y. In vivo recognition and clearance of red blood

cells containing phosphatidylserine in their plasma membranes. J Biol Chem 1985; 260:5131–5138.
75. Ratner S, Schroit AJ, Vinson SB, Fidler IJ. Analogous recognition of phospholipids by insect phagocytes and mammalian macrophages. Proceed Soc Exp Biol Med 1986; 182:272–276.
76. Connor J, Bucana CD, Fidler IJ, Schroit AJ. Differentiation-dependent expression of phosphatidylserine in mammalian plasma membranes: quantitative assessment of outer leaflet lipid by prothrombinase complex formation. Proc Natl Acad Sci USA 1989; 86:3184–3188.
77. Utsugi T, Schroit AJ, Conner J, Bucana CD, Fidler IJ. Elevated expression of phosphatidylserine in the outer membrane leaflet of human tumor cells and recognition by activated human blood monocytes. Cancer Res 1991; 51:3062–3066.
78. Hibbs Jr JB, Lambert, LH, Remington, JS. Control of carcinogenesis: a possible role for the activated macrophage. Science 1972; 177:998–1000
79. Hibbs Jr JB. Discrimination between neoplastic and nonneoplastic cells in vitro by activated macrophages. J Natl Cancer Inst 1974; 53:1487–1492.
80. Hibbs Jr JB. Heterocytolysis by macrophages activated by bacillus Calmette-Guerin: lysosome exocytosis into tumor cells. Science 1974; 184:468–471.
81. Allison AC. On the role of mononuclear phagocytes in immunity against viruses. Prog Med Virol 1974; 18:15–31.
82. Chedid L, Audibert F, Johnson AG. Biological activities of muramyl dipeptide, a synthetic glycopeptide analogous to bacterial immunoregulating agents. Prog Allergy 1978; 25:63–80.
83. Lederer E. Synthetic immunostimulants derived from the bacterial cell wall. J Med Chem 1980; 23:819–825.
84. Fogler WE, Fidler IJ. The activation of tumoricidal properties in human blood monocytes by muramyl peptides requires a specific intracellular interaction. J Immunol 1986; 136:2311–2316.
85. Gisler RH, Dietrich FM, Baschang G, Brownbill A, Schumann G, Stabler FG, Tarcsay L, Wachsmuth ED, Dukor P. New development in drugs enhancing the immune response: activation of lymphocytes and accessory cells by muramyl-dipeptides. In: Turk JL, Parker D, eds. Drugs and Immune Responsiveness. Baltimore: University Park Press, 1979: 133–160.
86. Fogler WE, Fidler IJ. Nonselective destruction of murine macrophages by syngeneic tumoricidal macrophages. Cancer Res 1985; 45:14–18.
87. Parant M, Parant F, Chedid L, Yapo A, Petit JF, Lederer E. Fate of the synthetic immunoadjuvant, muramyl dipeptide (^{14}C-labelled) in the mouse. Int J Immunopharm 1979; 1:35–41.
88. Fidler IJ. Targeting of immunomodulators to mononuclear phagocytes for therapy of cancer. Adv Drug Deliv Rev 1988; 1:69–106:
89. Fidler IJ, Schroit AJ. Synergism between lymphokine and muramyl dipeptide encapsulated in liposomes: in situ activation of macrophages and therapy of spontaneous cancer metastasis. J Immunol 1984; 133:515–518.
90. Kleinerman ES, Fidler IJ. Production and utilization of human lymphokines containing macrophage-activating factor (MAF) activity. Lymphokine Res 1983; 2:7–12:

91. Kleinerman ES, Erickson KL, Schroit AJ, Fogler WE, Fidler IJ. Activation of tumoricidal properties in human blood monocytes by liposomes containing lipophilic muramyl tripeptide. Cancer Res 1983; 43:2010–2014.
92. Saiki I, Dunegan MA, Fann AV, Koff WC. Regulatory effects on macrophages of human recombinant interferons-α. J Interferon Res 1986; 6:603–611.
93. Pace JL. Synergistic interactions between IFN-γ and IFN-β in priming murine macrophages for tumor cell killing. J Leukoc Biol 1988; 44:514–520.
94. Hori K, Mihich E, Ehrike MJ. Role of tumor necrosis factor and interleukin 1 in γ-interferon-promoted activation of mouse tumoricidal macrophages. Cancer Res 1989; 49:2606–2614.
95. Poste G, Kirsh R, Fidler IJ. Cell surface receptors for lymphokines. Cell Immunol 44 1979; 44:71–88.
96. Poste G, Kirsh R, Fidler IJ. Rapid decay of tumoricidal activity and loss of responsiveness to lymphokines in inflammatory macrophages. Cancer Res 1979; 39:2582–2590.
97. Schroit AJ, Hart IR, Madsen J, Fidler IJ. Selective delivery of drugs encapsulated in liposomes: natural targeting to macrophages involved in various disease states. J Biol Resp Modif 1983; 2:97–100.
98. Alving CR. Delivery of liposome-encapsulated drugs to macrophages. Pharmacol Rev 1983; 22:407–424.
99. Fidler IJ. Therapy of spontaneous metastasis by intravenous injection of liposome containing lymphokines. Science 1980; 208:1469–1471.
100. Schroit AJ, Fidler IJ. Effects of liposome structure and lipid composition on the activation of the tumoricidal properties of macrophages by liposomes containing muramyl dipeptide. Cancer Res 1982; 42:161–167.
101. Xu L, Fidler IJ. The in situ activation of cytotoxic properties in murine Kupffer cells by the systemic administration of whole mycobacterium bovis organisms or muramyl tripeptide. Cancer Immunol Immunother 1984; 18:118–122.
102. Raz A, Bucana C, Fogler WE, Poste G, Fidler IJ. Biochemical, morphological and ultrastructural studies on the uptake of liposomes by murine macrophages. Cancer Res 1981; 41:487–494.
103. Sone S, Tsubura E. Human alveolar macrophages: potentiation of the tumoricidal activity by liposome-encapsulated muramyl-dipeptide analogs. J Immunol 1982; 129:1313–1317.
104. Fidler IJ, Raz A, Fogler WE, Kirsh R, Bugelski P, Poste G. The design of liposomes to improve delivery of macrophage-augmenting agents to alveolar macrophages. Cancer Res 1980; 40:4460–4466.
105. Key ME, Talmadge JE, Fogler WE, Bucana C, Fidler IJ. Isolation of tumoricidal macrophages from lung melanoma metastases of mice treated systemically with liposomes containing a lipophilic derivative of muramyl dipeptide. J Natl Cancer Inst 1982; 69:1189–1198.
106. Papahadjopoulos D. Liposomes in drug delivery: from serendipity to tumor targeting. In: Lopez-Berestein G, Fidler IJ, eds. Liposomes in the Therapy of Infectious Diseases and Cancer. New York: Liss, 1989:135–155.
107. Fidler IJ, Sone S, Fogler WE, Smith D, Braun DG, Tarcsay L, Gisler RH, Schroit AJ.

Efficacy of liposomes containing a lipophilic muramyl dipeptide derivative for activating the tumoricidal properties of alveolar macrophages in vivo. J Biol Resp Modif 1982; 1:43–55.
108. Fidler IJ, Fogler WE, Tarcsay L, Schumnann G, Graun DG, Schroit AJ. Systemic activation of macrophages and treatment of cancer metastases by liposomes containing hydrophilic or lipophilic muramyl dipeptide. Immunopharmacology 1983; 2: 235–353.
109. Fogler WE, Fidler IJ. Modulation of the immune response by muramyl dipeptide. In: Chirigos MA, Fenichel RL, eds. Immunomodulation Agents and Their Mechanisms. New York: Marcel Dekker, 1984:499–512.
110. Fidler IJ, Poste G. Macrophage-mediated destruction of malignant tumor cells and new strategies for the therapy of metastatic disease. Springer Semin Immunopathol 1982; 5:161–174.
111. Fidler IJ, Sone S, Fogler WE, Barnes ZL. Eradication of spontaneous metastasis and activation of alveolar macrophages by intravenous injection of liposomes containing muramyl dipeptide. Proc Natl Acad Sci USA 1981; 78:1680–1684.
112. Fidler IJ, Fan D, Ichinose Y. Potent in situ activation of murine lung macrophages and therapy of melanoma metastases by systemic administration of liposomes containing muramyltripeptide phosphatidylethanolamine and interferon gamma. Invasion Metastasis 1989; 9:75–88.
113. Fidler IJ. Immunomodulation of macrophages for cancer and antiviral therapy. In: Tomlinson E, Davis SS, eds. Site-Specific Drug Delivery. New York: Wiley, 1986: 111–135.
114. Fidler IJ. Optimization and limitations of systemic treatment of murine melanoma metastasis with liposomes containing muramyl tripeptide phosphatidylethanolamine. Cancer Immunol Immunother 1986; 21:169–173.
115. Deodhar SD, Barna BP, Edinger M, Chiang T. Inhibition of lung metastasis by liposomal immunotherapy in a murine fibrosarcoma model. J Biol Resp Modif 1982; 1: 27–34.
116. Eppstein DA, Van der Pas MA, Fraser-Smith EB. Liposome-encapsulated muramyl dipeptide analogue enhances non-specific host immunity. Int J Immunother 1986; 2: 115–126.
117. Lopez-Berestein G, Milas L, Hunter N, Mehta E, Eppstein D, Van der Pas MA, Mathews TR, Hersh EM. Prophylaxis and treatment of experimental lung metastases in mice after treatment with liposome-encapsulated 6-0-steroyl-N-acetylmuramyl-L-amino-butryl-D-isoglutamine. Clin Exp Metastasis 1984; 2:366–367.
118. Phillips NC, Mora ML, Chedid L, Lefrancier P, Bernard JM. Activation of tumoricidal activity and eradication of experimental metastases by freeze-dried liposomes containing a new lipophilic muramyl dipeptide derivative. Cancer Res 1985; 39: 2582–2590.
119. Phillips NC, Tsao MS. Inhibition of experimental liver tumor growth in mice by liposomes containing a lipophilic muramyl dipeptide derivative. Cancer Res 1989; 49: 936–939.
120. Brodt P, Blore J, Philips NC, J.S. M, Rioux JD. Inhibition of murine hepatic tumor growth by liposomes containing a lipophilic muramyl dipeptide. Cancer Immunol Immunother 1989; 28:54–60.

121. Thrombe P, Deodhar SD. Inhibition of liver metastasis in murine colon adenocarcinoma by liposomes containing human C-reactive protein or crude lymphokine. Cancer Immunol Immunother 1984; 16:145–150.
122. Brodey RS, Abt DA. Results of surgical treatment in 65 dogs with osteosarcoma. J Am Vet Med 1976; 168:1032–1038.
123. Gillete EL. Spontaneous canine neoplasms as models for therapeutic agents. In: Fidler IJ, White RJ, eds. Design of Models for Testing Cancer Therapeutic Agents. New York: Van Nostrand-Reinhold, 1982:185–192.
124. MacEwen EG, Kurzman ID, Rosenthal RC, Smith BW, Mainley PA, Roush JK, Howard EH. Therapy for osteosarcoma in dogs with intravenous injection of liposome encapsulated muramyl tripeptide. J Natl Cancer Inst 1989; 81:935–938.
125. Fidler IJ. Macrophages and metastasis: a biological approach to cancer therapy. Cancer Res 1985; 45:4714–4718.
126. Hisano G, Fidler IJ. Systemic activation of macrophages by liposome-entrapped muramyl tripeptide in mice pretreated with the chemotherapeutic agent adriamycin. Cancer Immunol Immunother 1990; 14:61–66.
127. Giavazzi R, Bucana CD, Hart IR. Correlation of tumor growth inhibitory activity of macrophages exposed to adriamycin and adriamycin sensitivity of the target tumor cells. J Natl Cancer Inst 1984; 73:447–455.
128. Hisano G, Fidler IJ. Systemic activation of macrophages by liposome-entrapped muramyl tripeptide in mice pretreated with the chemotherapeutic agent Adriamycin. Cancer Immunol Immunother 1982; 14:61–66.
129. Killion JJ, Kleinerman ES, Wilson MR, Tanaka M, Fidler IJ. Sequential therapy with chemotherapeutic drugs and liposome-encapsulated muramyl tripeptide: determination of potential interactions between these agents. Oncol Res 1992; 4: 413–418.
130. Fidler IJ, Nii A, Utsugi T, Brown D, Bakouche O, Kleinerman ES. Differential release of TNF-α, IL1, and PGE_2 by human blood monocytes subsequent to interaction with different bacterial derived agents. Lymphokine Res 1990; 9:449–463.
131. Utsugi T, Dinney CPN, Killion JJ, Fidler IJ. In situ activation of mouse macrophages and therapy of spontaneous renal cell cancer metastasis by liposomes containing the lipopeptide CGP 31362. Cancer Immunol Immunother 1991; 33:375–381.
132. Utsugi T, Nii A, Fan D, Pak CC, Denkins Y, van Hoogevest P, Fidler IJ. Comparative efficacy of liposomes containing synthetic bacterial cell wall analogues for tumoricidal activation of monocytes and macrophages. Cancer Immunol Immunother 1991; 33:285–292.
133. Utsugi T, Dinney CPN, Killion JJ, Brown D, Fidler IJ. In situ activation of mouse lung macrophages by coadministration of liposomes containing the lipopeptide CGP 31362 and interleukin 2 involves interaction with T lymphocytes and natural killer cells. Lymphokine Cytokine Res 1991; 10:487–493.
134. Nii A, Utsugi T, Fan D, Denkins Y, Pak C, Brown D, van Hoogevest P, Fidler IJ. Optimization of the liposomes encapsulating a new lipopeptide CGP 31362 for efficient activation of tumoricidal properties in monocytes and macrophages. J Immunother 1991; 10:236–246.
135. Dinney CPN, Bucana CD, Utsugi T, Fidler IJ, von Eschenbach AC, Killion JJ. Therapy of spontaneous lung metastasis of murine renal adenocarcinoma by sys-

temic administration of liposomes containing the macrophage activator CGP 31362. Cancer Res 1991; 51:3741–3747.

136. Dinney CPN, Utsugi T, Fidler IJ, von Eschenbach AC, Killion JJ. Immunotherapy of murine renal adenocarcinoma by systemic administration of liposomes containing the synthetic macrophage activator CGP 31362 or CGP 19835A in combination with interleukin 2 or γ-interferon. Cancer Res 1992; 52:1155–1161.
137. Fidler IJ, Fogler WE, Brownbill AF, Schumann G. Systemic activation of tumoricidal properties in mouse macrophages and inhibition of melanoma metastases by the oral administration of MTP-PE, a lipophilic muramyl dipeptide. J Immunol 1987; 138:4509–4514.
138. Tanguay S, Bucana CD, Wilson MR, Fidler IJ, von Eschenbach AC, Killion JJ. In vivo modulation of macrophage tumoricidal activity by the oral administration of the liposome-encapsulated macrophage activator CGP 19835A. Cancer Res 1994: 54: 5882–5888.
139. Saiki I, Milas L, Hunter N, Fidler IJ. Treatment of experimental lung metastasis with local thoracic irradiation followed by systemic macrophage activation with liposomes containing muramyl tripeptide. Cancer Res 1986; 46:4966–4970.
140. Suit HD. A basis for radiotherapy. In: Textbook of Radiotherapy. Philadelphia: Lea & Febiger, 1966:65–97.
141. Fidler IJ, Jessup JM, Fogler WE, Staerkel R, Mazumder A. Activation of tumoricidal properties in peripheral blood monocytes of patients with colorectal carcinoma. Cancer Res 1986; 46:994–998.
142. Hudson MM, Snyder JS, Jaffe N, Kleinerman ES. In vitro and in vivo effect of adriamycin therapy on monocyte activation by liposome-encapsulated immunomodulators. Cancer Res 1988; 48:5256–5263.
143. Creaven PJ, Cowens JW, Brenner DE, Dadey BM, Han T, Buben R, Karakousis C, Frost H, LeSher D, Hanagan J, Andrejcio K, Cushman MK. Initial clinical trial of the macrophage activator muramyl tripeptide-phosphatidylethanolamine encapsulated in liposomes in patients with advanced cancer. J Biol Resp Modif 1990; 9:492–498:
144. Murray JL, Kleninerman ES, Cunningham JE, Tatom JR, Andrejcio K, Lepe-Zungia J, Lamki LM, Rosenblum MG, Frost H, Gutterman JU, Fidler IJ, Krakoff IH. Phase I trial of liposomal muramyl-tripeptide-phosphatidylethanolamine [MTP-PE (CGP19835A)] in cancer patients. J Clin Oncol 1989; 7:1915–1925.
145. Moore MAS, Warren DJ. Synergy of interleukin-1 and granulocyte colony stimulating factor. in vivo stimulation of stem cell recovering and hematopoeitic regeneration following 5-fluorouracil treatment of mice. Proc Natl Acad Sci USA 1987; 84: 7144–7148.
146. Shah RG, Caporale LH, Moore MAS. Characterization of colony-stimulating activity produced by human monocytes and phytohemagglutinin-stimulated lymphocytes. Blood 1977; 50:811–821.
147. Kleinerman ES, Murray JL, Snyder JS, Cunningham JE, Fidler IJ. Activation of tumoricidal properties in monocytes from cancer patients following intravenous administration of liposomes containing muramyl tripeptide phosphatidylethanolamine. Cancer Res 1989; 49:4665–4670.
148. Fey GH, Fuller GM. Regulation of acute phase gene expression by inflammatory mediators. Mol Biol Med 1987; 4:323–338.

149. Goorin AM, Shuster JJ, Baker A, Horowitz ME, Meyer WH, Link MP. Changing pattern of pulmonary metastases with adjuvant chemotherapy in patients with osteosarcoma. Results from the multi-institutional osteosarcoma study. J Clin Oncol 1991; 73:447–455.
150. Kleinerman ES, Jia SF, Griffin J, Seibel NL, Benjamin RS, Jaffe N. Phase II study of liposomal muramyl tripeptide in osteosarcoma: The cytokine cascade and monocyte activation following administration. J Clin Oncol 1992; 10:1310–1316.
151. Kleinerman ES, Raymond AK, Bucana CD, Jaffe N, Harris MB, Krakoff IH, Benjamin R, Fidler IJ. Unique histological changes in lung metastases of osteosarcoma patients following therapy with liposomal muramyl tripeptide (CGP 19835A lipid). Cancer Immunol Immunother 1992; 34:211–220.
152. Dannenberg JAM, Thomashefski JJF. Pathogenesis of pulmonary tuberculosis. In Fishman AP, ed. Pulmonary Diseases and Disorders. New York: McGraw-Hill, 1988.
153. Asano T, Kleinerman ES. Liposome-encapsulated MTP-PE: A novel biologic agent for cancer therapy. J Immunother 1993; 14:286–292.
154. Kleinerman ES, Meyers PA, Raymond AK, Gano JB, Jia SF, Jaffe N. Combination therapy with ifosfamide and liposome-encapsulated muramyl tripeptide: tolerability, toxicity, and immune stimulation. J Immunother 1995; 17:181–193.
155. Asano T, McWatters A, An T, Matshushima K, Kleinerman ES. Liposomal muramyl tripeptide up-regulates interleukin-1α, interleukin-1β, tumor necrosis factor-α, interleukin-6 and interleukin-8 gene expression in human monocytes. J Pharmacol Exp Ther 1994; 268.1032 1039.
156. Nabioullin R, Sone S, Mizuno K, Yano S, Nishioka Y, Haku T, Ogura T. Interleukin-10 is a potent inhibitor of tumor cytotoxicity by human monocytes and alveolar macrophages. J Leukoc Biol 1994; 55:437–442.
157. Nishioka Y, Sone S, Orino E, Nii A, Ogura T. Down-regulation by interleukin-4 of activation of human alveolar macrophages to the tumoricidal state. Cancer Res 1991; 51:5526–5531.
158. Crawford RM, Finbloom DS, Ohara J, Paul WE, Meltzer MS. B cell stimulatory factor-1 (interleukin 4) activates macrophages for increased tumoricidal activity and expression of Ia antigens. J Immunol 1987; 135:135–141.
159. Mosmann TR, Coffman RL. Th1 and Th2 cells: different patterns of lymphokine secretion lead to different functional properties. Annu Rev Immunol 1989; 7: 145–173:
160. Talmadge JE, Schneider M, Collins M, Phillips H, Herberman RB, Wiltrout RH. Augmentation of NK cell activity in tissue specific sites by liposomes incorporating MTP-PE. J Immunol 1985; 135:1477–1483.
161. Sone S, Utsugi T, Nii A, Ogura T. Effects of human alveolar macrophages on the induction of lymphokine (IL-2)–activated killer cells. J Immunol 1987; 139:29–34.
162. Trinchieri G, Scott P. The role of interleukin 12 in the immune response, disease and therapy. Immunol Today 1994; 15:460–463.
163. Ling PD, Warren MK, Vogel SN. Antagonistic effect of interferon-β on the interferon-γ-induced expression of Ia antigen in murine macrophages. J Immunol 1985; 135:1857–1863.

26

Cells of the Dendritic Cell Lineage in Human Lung Carcinomas and Pulmonary Histiocytosis X

ABDELLATIF TAZI, ALLAN J. HANCE, and PAUL SOLER

Institut National de la Santé et de la Recherche Médicale
Paris, France

I. Introduction

Dendritic cells (DCs) and Langerhans' cells (LCs) are potent accessory cells which belong to the dendritic cell lineage. They are widely distributed and have been identified in various organs, including the lung. Functional studies have indicated that lung DCs/LCs, like their counterparts in other tissues, have strong accessory cell activity, suggesting they may play an important role in initiating lung immune responses. In this regard, they are ideally distributed to interact with antigens that have penetrated airway epithelial barriers and against which an immune response is required. Evidence is accumulating that lung DCs/LCs, like DCs/LCs in other tissues, play an essential role in capturing and transporting previously unencountered antigens to lymphoid tissues and subsequently activating "naive" T cells recognizing these antigens. Current evidence also suggests that DCs/LCs are involved in the pathogenesis of certain lung diseases of immune origin, including pulmonary histiocytosis X, and could play a role in antitumor immunity. The purpose of this chapter is to summarize briefly current information concerning lung DCs/LCs and to discuss their involvement in human lung carcinomas and pulmonary histiocytosis X.

II. Cells of the Dendritic Cell Lineage in Normal Lung

A. Distribution of Lung Dendritic and Langerhans' Cells

DCs are widely distributed in the normal lung and have been identified by several groups both in humans and rodents. They are essentially localized in subpleural and peribronchiolar connective tissues, in the alveolar interstitium, in pulmonary capillaries, and bronchus-associated lymphoid tissue (BALT) (1–6). In humans, these cells strongly express CD1c antigen, and immunoelectron microscopy has shown that they are present within alveolar interstitium, between the cytoplasmic processes of mesenchymal cells and connective tissue components that form channels thought to be pathways for fluid and migrating cells (3). LCs, which are thought to be derived from DCs, differ from the latter by a strong expression of CD1a antigen and by the presence of characteristic pentalaminar plate-like cytoplasmic organelles, Birbeck's granules, visible only by electron microscopy (7–9). $CD1a^+$ LCs are present in the human lung, but they have a more limited distribution than DCs (3). They are present almost exclusively in airways intercalated between epithelial cells (Fig. 1a). $CD1c^+$ DCs are also present in bronchial and bronchiolar tissue, but they are limited to the submucosa (3) (Fig. 1b). Consistent with the possibility that submucosal DCs give rise to intraepithelial LCs, cells with a $CD1a^+/CD1c^+$ intermediate phenotype are observed in both locations. Electron microscopic studies indicate that the majority of cells present within the airway epithelium of humans that have morphological criteria typical of DCs also contain Birbeck's granules, but the number of these granules is usually low (Fig. 2), and serial sections must often be evaluated to demonstrate their presence (10). In contrast, Birbeck's granules have not been identified in murine airway DCs, and these cells are commonly referred to as intraepithelial DCs (1,5,11–13). When airway epithelium is cut tangentially (i.e., parallel to the lumen), DCs are found to form a highly developed network resulting from the interdigitation of their long cytoplasmic processes between the epithelial cells (2,5,13) that is highly reminiscent of the network formed by LCs in the skin.

Cells of the dendritic cell lineage, which are derived from bone marrow precursors (7–9), are thought to be released into the blood. DCs are indeed present in the peripheral blood, but they represent less than 0.5% of mononuclear cells (14). As in other tissues, DCs are thought to arrive in the lung via the capillary bed, and they are present in the lung before birth (12,13). The cells are mobile, and pulmonary LCs are most likely derived from DCs which have migrated into the airway epithelium and have differentiated in the epithelial microenvironment, as it occurs in other tissues.

B. Functional Capacities of Lung Dendritic and Langerhans' Cells

Studies from several groups have indicated that pulmonary DCs/LCs, like their counterparts in other tissues, are potent accessory cells (1,2,15–18). In particular,

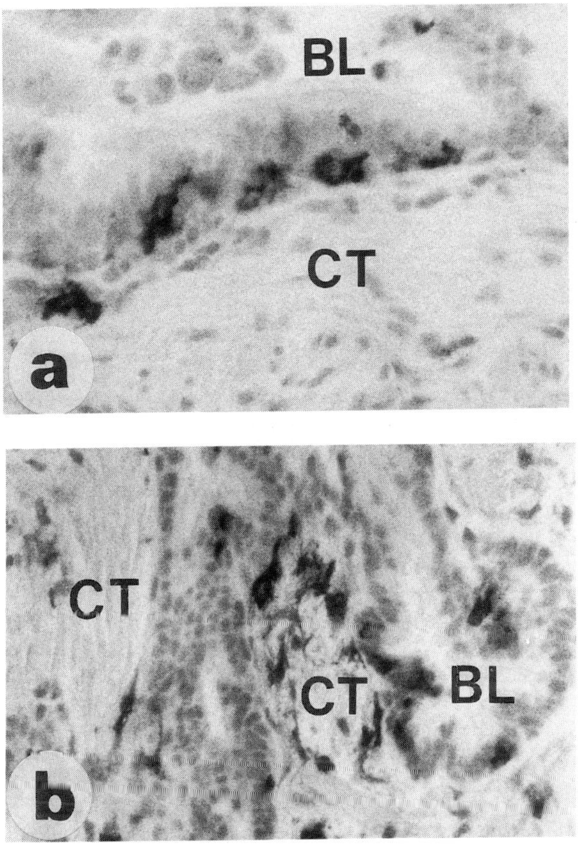

Figure 1 Distribution of DCs/LCs expressing CD1 antigens in normal human airways. (a) CD1a$^+$ cells with a dendritic shape are present almost exclusively intercalated between bronchiolar epithelial cells. (b) CD1c$^+$ cells are also present but predominantly in the submucosa. BL, bronchiolar lumen; CT, peribronchiolar connective tissue. (×125)

lung DCs/LCs are much more potent in inducing lymphocyte proliferation in response to soluble antigens or in allogeneic mixed lymphocyte reactions than pulmonary macrophages or blood monocytes, and only small numbers of DCs/LCs are required to induce significant proliferative responses (2,17,18). In contrast, alveolar macrophages are relatively weak accessory cells, and they may actually suppress the lymphocyte proliferation induced by lung DCs/LCs (2,17–20). Furthermore, some types of immune responses are apparently only possible when cells of DC lineage serve as accessory cells. In this regard, T lymphocytes recognizing certain haptens and antigens can be activated only when DCs are the accessory cells (21–23), and DCs/LCs appear to be unique in their ability to

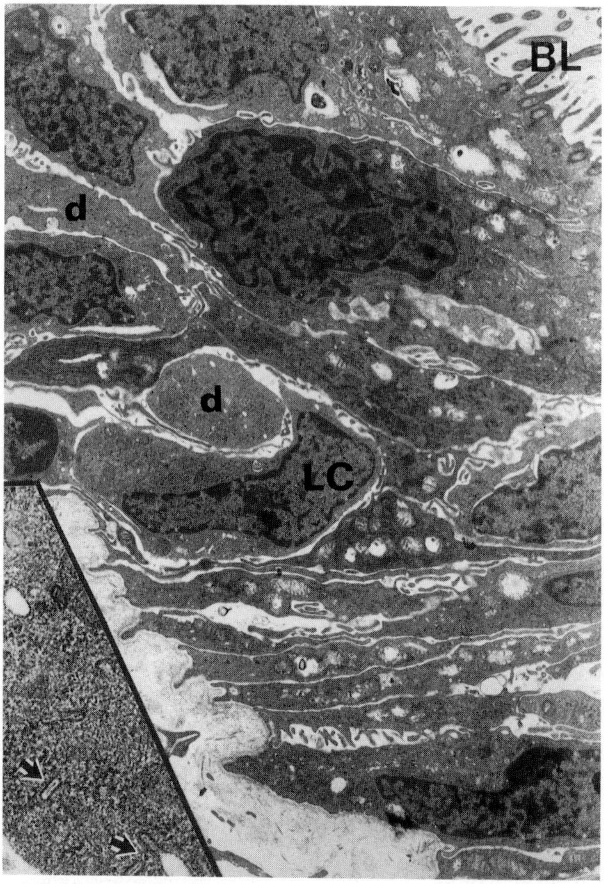

Figure 2 Electron microscopy. Bronchiolar epithelium containing an intraepithelial LC with dendritic processes (d) interdigitating between the epithelial cells. BL, bronchiolar lumen. (×4,500) The lower left part of the micrograph shows a detail of the LC with typical Birbeck's granules (arrows). (×30,000)

activate "naive" $CD4^+$ T cells and to activate $CD8^+$ T lymphocytes in the absence of help from $CD4^+$ T lymphocytes (24–26).

Results obtained in a number of studies evaluating the functional capacities of DCs/LCs (27–33) have led to the concept that intraepidermal LCs, thought to be derived from dermal DCs, serve as "sentinels" that capture, process, and transport exogenous antigens to lymphoid tissues without generating immune responses at the original site of deposition. Similar considerations apply to lung DCs/LCs. Airway intraepithelial LCs are ideally positioned to serve as sentinels that capture antigens penetrating into the lung, and subsequently process and transport them to regional lymphatic tissues (via the pulmonary lymphatics) or the spleen (via the peripheral circulation). In this regard, studies comparing intra-epithelial airway DCs and parenchymal DCs in rodents have shown that airway DCs, like epidermal LCs, may be better at antigen processing than at T-cell activation (11,34). Further studies are necessary, however, to characterize better the pathways followed by lung DCs/LCs in antigen transport and to define the factors which regulate this important process.

Because they are potent accessory cells, DCs/LCs are likely to play an essential role in pulmonary immune responses. The functional properties of DCs/LCs are strongly modulated by inflammatory cytokines and other factors (14,27, 30,35–39), suggesting that these cells are likely to play an active role in pathological situations.

Little is known, however, concerning the number, distribution, and differentiated state of DCs/LCs in human lung diseases. Although LCs have been identified in various numbers in a variety of different pathological situations, in only a few circumstances have these cells been evaluated in some detail. First, DCs/LCs have been shown to accumulate at sites of alveolar epithelial hyperplasia or metaplasia associated with cigarette smoking and certain interstitial lung diseases. Second, they have been shown to infiltrate, occasionally in very large numbers, some, but not all, bronchopulmonary carcinomas. Finally, they are known to accumulate in impressive numbers in the course of histiocytosis X, a granulomatous lung disease also called Langerhans' cell granulomatosis. Current information concerning DCs/LCs in each of these three pathological situations is discussed below.

III. Cells of the Dendritic Cell Lineage in Human Lung Disease Associated with Alveolar Epithelial Abnormalities

LCs are known to infiltrate the alveolar epithelium of smokers (3,40). They are usually observed at sites of epithelial hyperplasia (Fig. 3a), and their dendritic processes are found frequently in close apposition to the alveolar epithelial cells (Fig. 3b). LCs have been also described in the alveolar parenchyma of nonsmoking individuals with interstitial lung diseases such as idiopathic pulmonary fibrosis, pulmonary fibrosis associated with collagen-vascular diseases, and hypersensitivity pneumonitis (Fig. 4) (41,42). In all cases, they are present also at sites of alveolar epithelial hyperplasia or metaplasia in close association with epithelial cells. The accumulation of intraepithelial LCs has been also reported in animals with bronchial abnormalities (43).

In normal tissues, the presence of fully differentiated LCs (i.e., cells of

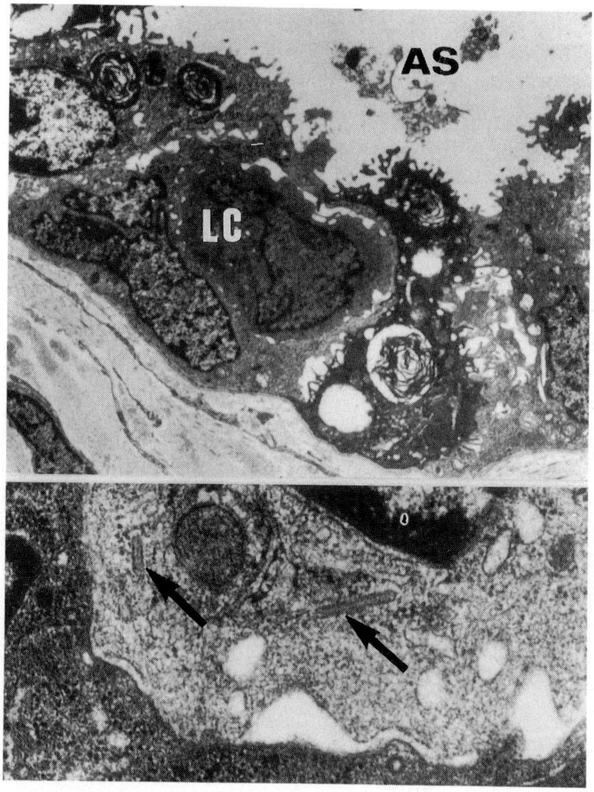

Figure 3 Electron microscopy. Hyperplastic alveolar epithelium from a cigarette smoker containing a LC intercalated between hyperplastic type II cells. AS, alveolar space. (×4,500) The lower pannel shows a detail of the LC containing typical Birbeck's granules (arrows). (×30,000)

dendritic appearance expressing CD1a antigens and containing intracytoplasmic Birbeck's granules) occurs only in the context of an epithelial microenvironment. Given this strong predilection for epithelia, it seems likely that epithelial changes are by themselves responsible for the accumulation of LCs in the lung parenchyma of smokers or patients with interstitial lung disease. The need of an epithelial environment for the differentiation of LCs suggests that contact with epithelial cells and/or soluble mediators produced by the epithelial cells may have an important influence on the recruitment and differentiation of DCs/LCs in the lung. Granulocyte-macrophage colony-stimulating factor (GM-CSF), which has been demonstrated to have important effects on the survival and function of LCs (27, 30,35,36,39), is likely to be one of these factors. LCs are known to express

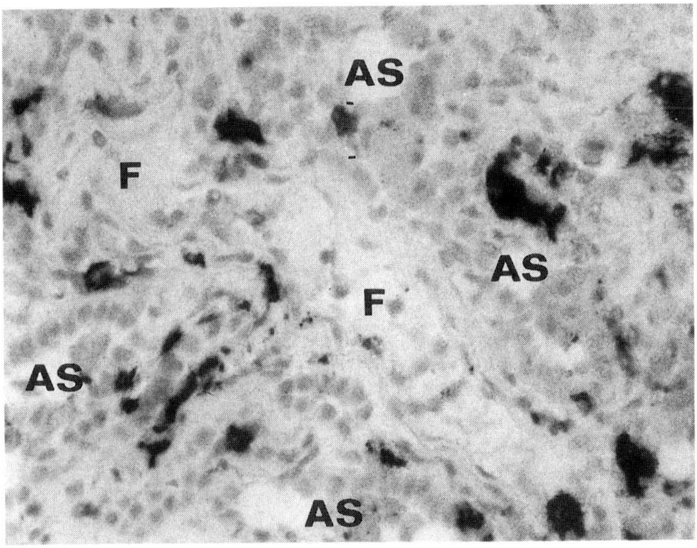

Figure 4 Parenchymal tissue from a patient with idiopathic pulmonary fibrosis. Remnants of alveolar spaces (AS) lined by cuboidal type II cells and fibrotic interstitial tissue (F) are seen. The hyperplastic alveolar epithelium is infiltrated by numerous $CD1a^+$ cells with a dendritic shape. (×250)

receptors for GM-CSF (44), and this cytokine is expressed in the lung at all sites where LCs are known to accumulate but not at other locations. Normal bronchiolar epithelium, hyperplastic alveolar epithelium, and those carcinomas that are infiltrated by LCs (see below) spontaneously produce GM-CSF (Fig. 5), whereas normal alveolar epithelium and tumors without LC infiltration do not express this factor (40). These findings strongly support the idea that GM-CSF plays an important role in the recruitment and/or differentiation of airway LCs. Intratracheal instillation of lipopolysaccharide, administration of interferon-γ, or exposure to allergens also increases the number of intraepithelial DCs, but the mechanisms of action have not been defined (4,45).

IV. Cells of the Dendritic Cell Lineage in Lung Carcinomas

A. Histopathological Aspects

The presence of DCs/LCs in lung carcinomas was first reported by Basset in 1974 in her description of LCs infiltrating a bronchioloalveolar carcinoma (46).

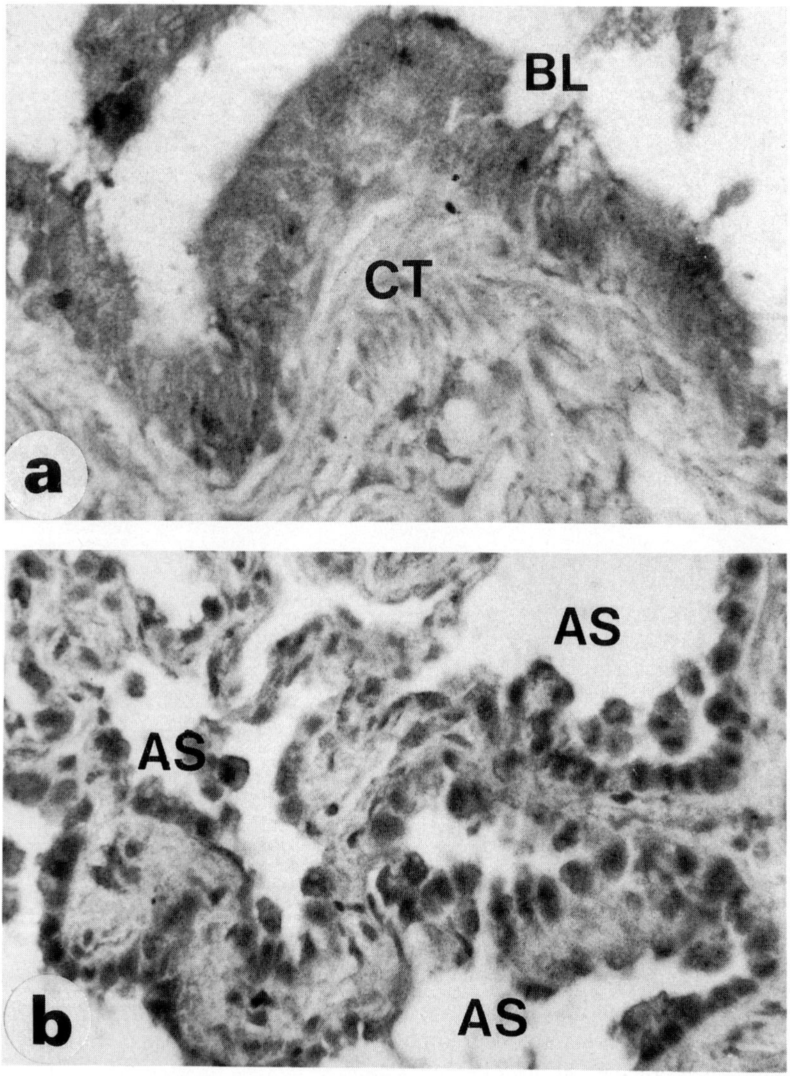

Figure 5 Production of GM-CSF in human lung tissues. GM-CSF is detected at all sites where CD1a$^+$ LCs are known to accumulate. (a) Normal bronchiolar epithelium. (b) Hyperplastic alveolar epithelium. (c) Tumor nodule (T) from a squamous cell carcinoma heavily infiltrated by LCs (not shown). BL, bronchiolar lumen; CT, peribronchiolar connective tissue; AS, alveolar space. (×250)

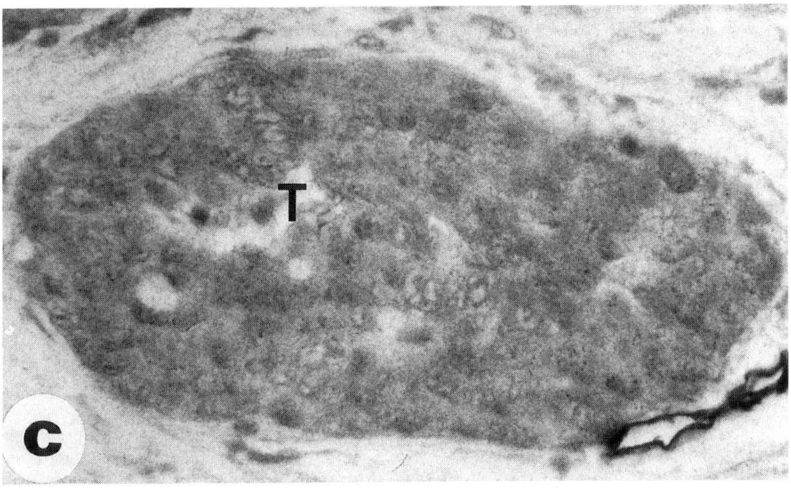

Subsequently, several groups have evaluated the density and the phenotype of these cells in lung tumors of various histological type using both immunohistochemistry and ultrastructural techniques (40,42,47–51). DCs/LCs have been identified in both adenocarcinomas and squamous cell carcinomas but are infrequently seen or absent in small cell carcinomas and carcinoid tumors (42,49,50).

Evaluation of the surface phenotype of DCs/LCs in lung carcinomas has shown that they are predominantly CD1a$^+$ LCs (40,47,49,51) (Fig. 6). These cells express S100 protein and CD1a antigens and contain characteristic intracytoplasmic Birbeck's granules. They are almost always localized within tumor nodules, whereas their number is considerably lower in the tumor stroma (40). LCs with a characteristic dendritic shape are interspersed among and intimately associated with tumor cells. Long cytoplasmic processes extending between the neoplastic cells are frequently present (40,42,50), and close contacts are observed between LCs and tumor cells by electron microscopy.

The number of LCs within lung tumors varies considerably in different specimens. Although infiltration by DCs/LCs has been found in some studies to occur more frequently in certain histological types (e.g., nonmucinous bronchioloalveolar carcinomas and papillary adenocarcinomas) than in others (e.g., small cell carcinomas), the histological type per se is not predictive of the number of CD1-positive cells present in the specimen (40). Similarly, attempts to correlate the differentiated state of the tumor with the density of LCs present have given conflicting results. In this regard, some investigators have found that differentiated lung carcinomas contained the highest numbers of LCs (48,50). In many cases, however, the degree of differentiation differs in different areas of the tumor. We

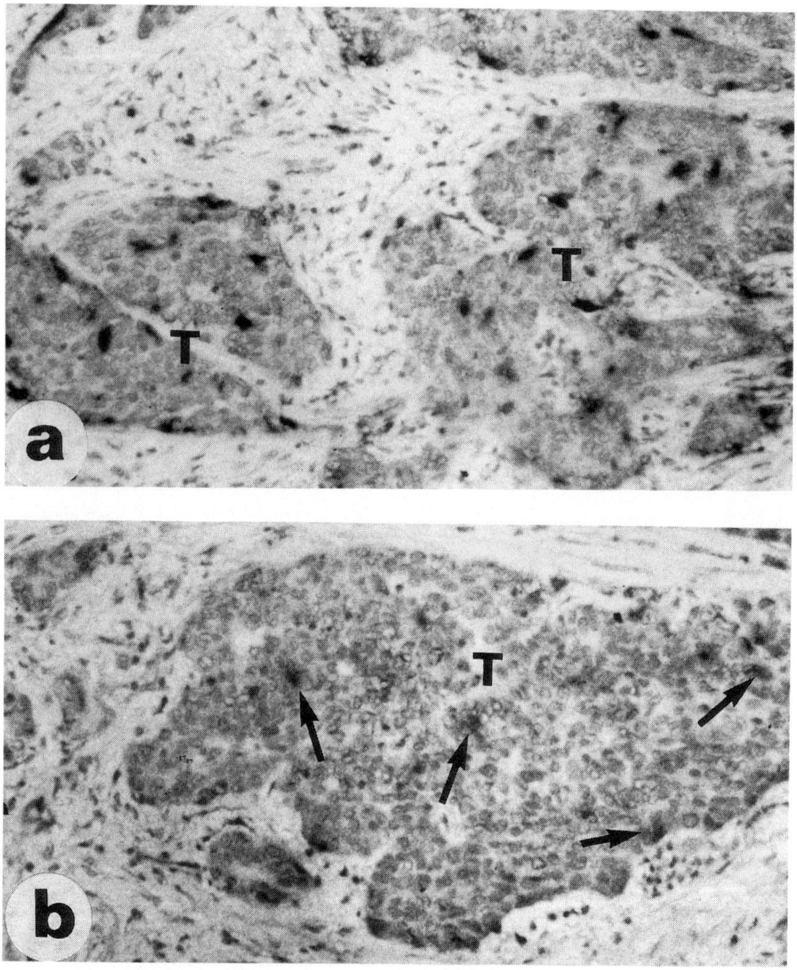

Figure 6 DCs/LCs infiltrating a squamous cell carcinoma. (a) Numerous strongly CD1a-positive cells are present within tumor nodules (T). (b) Adjacent section of the same area showing that cells expressing CD1c antigens are also present, but these cells are less numerous and only weakly positive (arrows). (×125)

found that in such cases, LCs are often more abundant in areas that are poorly differentiated (40). It is also noteworthy that the extent of infiltration by LCs can vary considerably from one tumor nodule to another in the same biopsy, and this variation is often independent of obvious histological differences in the tumor cells (40,42).

B. Factors Influencing the Accumulation of Cells of Dendritic Cell Lineage in Lung Carcinomatous Lesions

The variation of density of LCs in lung cancers suggests that the tumor microenvironment modulates the presence of these cells. We have shown that the capacity of tumor cells to produce GM-CSF and the amount of this factor produced have an important bearing on the infiltration of the tumor by DCs/LCs, thereby explaining the variable degree of infiltration in different tumors (40). Without exception, a close correlation was observed between the number of $CD1a^+$ cells present in a given tumor and the reactivity of the tumor cells with the anti–GM-CSF antibody. Thus, tumors that were strongly positive for GM-CSF were also heavily infiltrated by $CD1a^+$ cells. The distribution of DCs/LCs within a given carcinoma was also closely correlated with the distribution of GM-CSF–containing cells. Tumors which were diffusely positive for GM-CSF were also diffusely infiltrated by DCs/LCs. In a recent study evaluating the production of different cytokines by lung carcinomas, Colasante et al. confirmed that GM-CSF plays an important role in determining the number and distribution of LCs infiltrating these tumors (52). Studies in murine systems have also indicated that LCs migrate into cutaneous neoplasms, possibly as a result of chemotactic effects of tumor-derived cytokines (53). It has also been suggested that some tumors can release factors that inhibit the penetration of DCs/LCs (54).

C. Functional Aspects of Cells of the Dendritic Cell Lineage in Tumor Immunity

The presence of DCs/LCs in lung cancers raises the question as to their functional relevance. Although the functional capacities of DCs/LCs present in lung tumors has received scant attention, results obtained in other tissues give some insights into the potential role of these cells.

One hypothesis, supported by a minority, is that the presence of DCs/LCs infiltrating tumoral tissues merely reflects the production of chemotactic factors by malignant cells, and these "bystander" cells are not involved in the defense against the tumor. For example, nonimmunogenic ultraviolet (UV) light–induced murine tumor lines have been shown to attract large numbers of Ia^+ DCs, whereas an immunogenic tumor line did not (55); similar results were obtained whether the tumors were transplanted into syngenic immunocompetent or athymic immunodeficient mice. Furthermore, no relationship could be established between the density of infiltration of tumor by Ia^+ DCs and tumor immunogenicity, host immunocompetence, the extent of T-cell infiltration into the tumor, or the CD4/CD8 ratio of T cells present (55).

Currently, the majority of investigators support the idea that DCs/LCs present among malignant cells may participate in the development of an immune response against the tumor (54,56). In humans, most studies have reported a positive correlation between the density of DCs/LCs infiltrating the tumor and clinical prognosis (survival and/or development of metastasis) (48,50,54,56), although some data are partially conflicting (49). Likewise, in several studies, the infiltration of tumors by DCs/LCs has been found to correlate with the number of lymphocytes present in the carcinomas, and the two cell types have frequently been reported to form clusters within the tumor (40,47,48,50,51,54).

Studies in animal models have shown that DCs/LCs are critically involved in sensitization against tumor antigens and demonstrated that DCs/LCs from various tissues can process and present tumor antigens to T lymphocytes both in vitro and in vivo (56,57). These cells have been implicated in the generation of both class II and class I major histocompatibility complex (MHC)–restricted host responses to tumors (56–60). In various models, the injection of DCs previously pulsed with tumor antigens in vitro is able to induce regression of established tumors in vivo (56).

It remains unclear, however, why immunogenic tumors emerge at all and how they manage to grow in the vicinity of fully competent DCs/LCs. It is important to note that the effect resulting from the generation of an immune response against tumor antigens can vary from immunological rejection to tolerance induction (56,61). A number of factors have been identified that can influence the appearance of immune unresponsiveness against tumors, including the presence of suppressor T cells (56), the dose of antigen used for immunization, and the tumor burden (56,62). Substantial data suggest, however, that the functional state of the cells responsible for presenting tumor antigens may be of critical importance for the effectiveness of the ensuing antitumor immune response. For example, UVB radiation, which is known to induce skin tumors, converts LCs from immunogenic to tolerogenic cells (63). In this respect, although LCs present in lung tumors are generally HLA-DR$^+$, the expression of several surface molecules associated with LC activation (CD4, CD18, CD25, CD54) is weak or absent on LCs infiltrating tumors (51). Thus, the state of activation/differentiation of these cells may not be optimal for the generation of strong antitumor responses.

The expression of costimulatory signals by DCs/LCs is known to be essential for their capacity to stimulate resting unprimed T cells (9,10). Recently, numerous studies have emphasized the importance of interactions between B7-1 (CD80) and B7-2 (CD86) surface molecules expressed on accessory cells and CD28 on T cells in the induction of T-cell proliferation and cytokine production (64). These molecules have also been reported to play an important role in the development of tumor immunity (56,65,66). Furthermore, in a murine system, inhibition of the interaction between B7 molecules and CD28 using the fusion protein CTLA4-Ig has been reported virtually to abrogate the antitumor effects of DCs in vivo (65). Thus, one factor that may influence the ability of LCs to induce antitumor immunity is their expression of B7 molecules. The expression of these molecules is clearly susceptible to regulation. For example, epidermal LCs have been shown to be B7 negative in situ, but expression can be induced by cytokines

(67). Preliminary results obtained in our laboratory suggest that DCs/LCs in lung tumors do not express B7 molecules (A. Tazi et al., in preparation). These findings may help explain why the presence of DCs/LCs is not necessarily associated with effective antitumor immunity, and they raise the possibility that therapies designed to induce the expression of these or other costimulatory molecules may be useful in stimulating immune responses against lung tumors. The extent that such induction will be favorable may depend on additional factors. First, the functional properties of T cells expressed after activation is different for cells stimulated with B7-1 and B7-2 (68), and costimulatory signals provided by B7-1 may be more effective in inducing antitumor immunity (69). In addition, it is now recognized that B7 molecules also bind to a second receptor, CTLA-4, which is transiently expressed on T cells following activation. Because B7–CTLA-4 interactions transmit inhibitory signals to T cells, it has been suggested that inhibition of signaling through CTLA-4 might improve antitumor immunity (70). Thus, both the nature of the B7 molecules expressed on LCs and the timing of LC-lymphocyte interactions are likely to influence the efficacy of the resulting antitumor immune response.

Various cytokines, among which GM-CSF is the most active (56,57,66), are crucially involved in the regulation of the presentation of tumor antigens by DCs/LCs. In most studies, incubation of DCs/LCs in the presence of GM-CSF was necessary to induce antitumor immunity, particularly in unprimed systems (56,57,66). Transfection of the GM-CSF gene into MHC class II–negative tumors induced a specific immune response against the tumor dependent on both $CD4^+$ and $CD8^+$ T cells (71). Since the protective antitumor immunity was obtained despite the lack of expression of MHC class II antigens by tumor cells, these results strongly suggest that the reaction was mediated via an upregulation of antigen presentation function (71). It should be noted that among the various cytokines known to regulate the expression of B7 molecules at the surface of DC/LC in vitro, GM-CSF is the most potent (67,72). This cytokine is produced, however, by many progressively growing tumors, and although these tumors are infiltrated by DCs/LCs, no evidence of antitumor immunity is present. Thus, other factors must be required to permit the development of an antitumoral response, or these tumors are also producing other activities that downregulate the stimulatory effects of GM-CSF on tumor antigen presentation. In this regard, it was shown that the incubation of bone marrow–derived DCs in the presence of both GM-CSF and interleukin-4 (IL-4) was more effective in inducing antitumoral immunity than GM-CSF alone (57). Likewise, administration of tumor cells transfected with GM-CSF alone was found not to be sufficient to induce tumor rejection in vivo and had to be associated with a stimulus that produced partial tumor destruction (e.g., x-irradiation or concomittant IL-2 gene transfection) (56,71). Conversely, IL-1α and IL-10 inhibit the effects of GM-CSF on epidermal DC/LC presentation of tumor-associated antigens (73,74). Similarly, incubation of DCs/LCs in the presence of tumor necrosis factor-α (TNF-α) reversed the effects of GM-CSF on the induction of antitumor immunity in unprimed animals, although TNF-α apparently enhanced antitumor responses in primed systems (75). Interestingly, lung tumors which produce GM-CSF were recently shown constitutively to synthesize TNF-α as well (52).

Taken together, current information suports the idea that DCs/LCs are important cells for the induction of antitumor immunity in several animal models. Clearly, further studies are needed to evaluate the extent that human DCs/LCs can present tumor antigens and elicit specific T-cell immune responses and to define the factors that regulate these responses. Numerous strategies for improving the ability of DCs/LCs to initiate antitumor immune responses are being explored and may lead to the development of new therapeutic approaches. The extent that DCs/LCs will be able to induce clinically useful responses against established tumors, particularly in individuals with high tumor burden, remains unknown.

V. Pulmonary Langerhans' Cell Granulomatosis (Histiocytosis X)

Pulmonary Langerhans' cell granulomatosis (LCG), also called histiocytosis X, is a disorder of unknown etiology characterized by the presence in the lung of destructive granulomas containing large number of LCs (76–78). It is an uncommon lung disease that can occur at any age, but it usually affects young adult smokers. Multifocal involvement of several tissues (especially skin, bone, or lung) can occur in adults but is more commonly encountered in children (79). Diffuse involvement of multiple visceral organs associated with systemic symptoms and poor prognosis can also occur, but this is seen most often in infants and adults of advanced age (80,81).

Despite their extremely different clinical presentations, striking similarities in pathological findings are observed in all forms of the disease, in particular the presence of characteristic clusters of cells of DC/LC lineage organized in granulomas. Consequently, the different forms of LCG are usually considered to be a single nosological entity.

A. Histopathological Aspects

LCG Granulomas

The characteristic LCG lesions consist of poorly delimited granulomas made of activated LCs associated with inflammatory cells, especially lymphocytes, macrophages, and eosinophils (78,82,83). The histopathological findings depend on the evolution of the process. In the early stages, the lesions are granulomatous in appearance, mainly composed of large aggregates of typical LCs similar to those

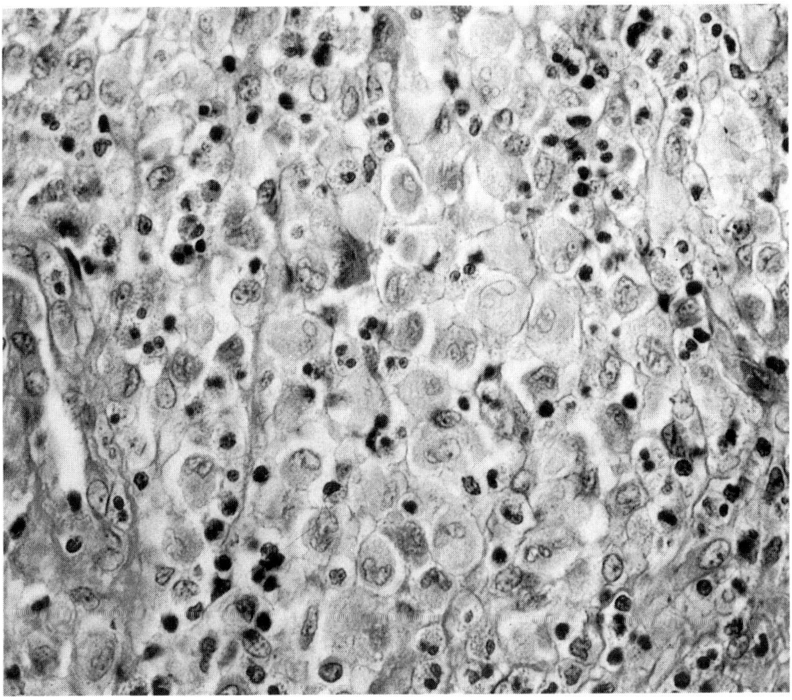

Figure 7 Light microscopy. Early pulmonary LC granuloma. Typical LCs are recognizable by their light cytoplasm and indented nucleus. Lymphocytes and eosinophils are also present interspersed between the LCs (HES ×400).

observed in normal tissues. They are recognizable by their weakly acidophilic cytoplasm containing no phagocytized inclusions and by their deeply indented or folded nucleus (Fig. 7). Lymphocytes, both interspersed between LCs and present at the periphery of the lesions, are the only other cell type consistently present in large numbers (51,83,84). These lymphocytes are predominantly $CD4^+$ T cells that express α/β T-cell receptors (84) and have a surface phenotype typical of activated cells (51). Differentiated contacts between LCs and these lymphocytes are seen by electron microscopy (Fig. 8). These contacts appear quite close, with an increased density in the cell membranes and the presence in the intercellular space of filamentous material forming bridges between the LCs and lymphocytes (84). Rosette-like clusters are sometimes observed composed of a central LCs in close association with surrounding lymphocytes. Similar rosette-like clusters may be also observed inside lymphatics. Inflammatory cells, especially macrophages and eosinophils, are also present in variable numbers, usually concentrated at the

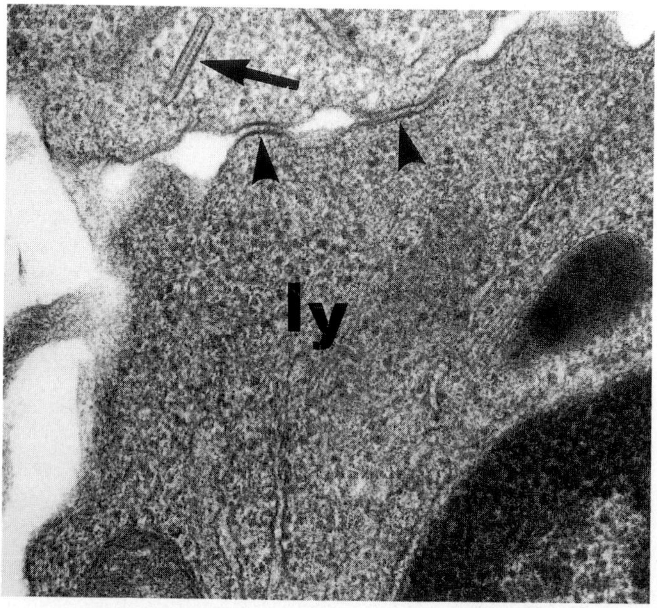

Figure 8 Electron microscopy. Early pulmonary LC granuloma. Close contacts (arrow heads) between a lymphocyte (Ly) and a LC containing Birbeck's granules (arrow). These contacts are characterized by an increased density of the adjoining cell membranes and by the presence of an intercellular filamentous material (×30,000)

periphery of the lesions. These early granulomas are sometimes compact but more often are cavitary. They are poorly demarcated and the adjacent alveolar walls are usually invaded and thickened by cellular infiltrates extending from the granulomatous mass. Adjacent alveolar spaces can contain large numbers of macrophages (desquamative interstitial pneumonia-like reaction) (77) associated with small numbers of LCs.

Middle-stage lesions of LCG are characterized by smaller proportions of LCs gathered as nests surrounded by large numbers of lymphocytes, macrophages, and eosinophils. Localized lymphoid aggregates and fibrotic changes begin to occur at the periphery of the lesion.

In late stages of LCG, LCs are rare or are entirely absent, although macrophages laden with pigments or lipidic inclusions are commonly found. Lymphoid aggregates are typically present at the periphery of the lesions. In contrast, fibrotic changes are prominent and are present as stellate scars or fibrous rings surrounding cystic spaces of variable diameter.

Distribution and Topography of LCG Lesions

Pulmonary LC granulomas are almost exclusively centered on bronchioles. Because of this strong predilection to occur around small airways, apparently normal lung parenchyma is present between adjacent lesions. The lesions are destructive and invade the bronchial wall in an eccentric fashion. As the lesion evolves, the granulomatous process progressively surrounds and destroys the bronchiole (Fig. 9). When the preexisting bronchiolar structures have been largely destroyed, it is difficult to recognize that the lesion was initially centered on a small airway. Early granulomas appear very often to be cavitary. The central cavity do not result from necrosis. Rather the central space represents the lumen of the preexisting bronchiole (77,78,83).

Reconstruction of serial sections strongly suggests that the lesions of LCG develop in terminal bronchioles or the proximal portion of respiratory bronchioles and extend without interruption up to and including alveolar ducts (78,83). Unlike other granulomatous lesions, which have a spherical shape, LC granulomas are elongated and form a continuous sheath surrounding distal airways. The granu-

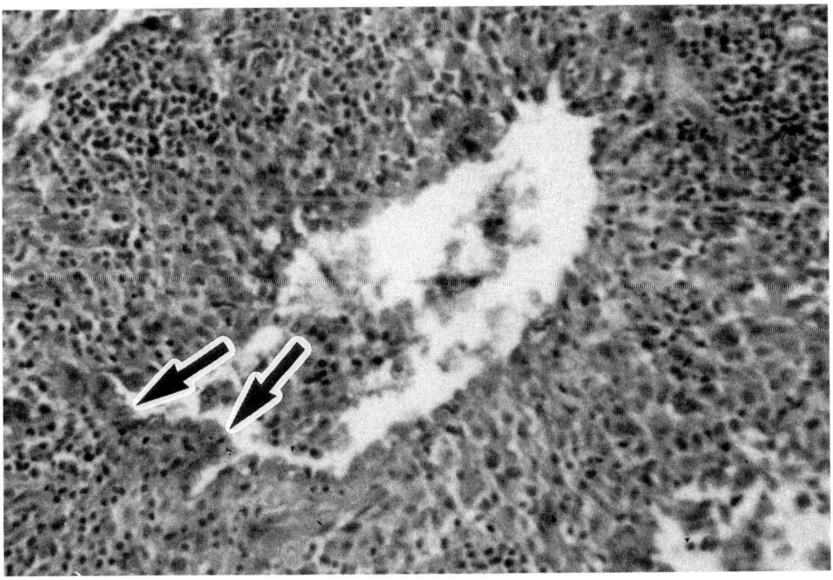

Figure 9 Light microscopy. Early pulmonary LC granuloma. The central cavity corresponds to the lumen of a preexisting bronchiole that has been completely destroyed by the granulomatous process. Remnants of bronchiolar epithelium are still visible (arrows). (HES, ×125)

lomatous reaction also progresses over time, with the shape and cellular composition of the lesions being quite variable according to their location along the bronchiolar axis. As a consequence, lesions of apparently different age can be typically found in a single biopsy.

Morphology and Surface Phenotype of LCs in LCG Granulomas

In most respects, the morphological features of LCs in LC granulomas are similar to those of LCs in normal tissues (Fig. 10). They are moderately sized cells (15–20 μm) and possess an irregular outline due to the presence of numerous dendritic processes. The nucleus is very irregular and typically indented or folded, with a thin band of heterochromatin. The cytoplasm contains a well-developed Golgi complex, abundant rough endoplasmic reticulum, and mitochondria. Although LCs are known to be capable of phagocytosis, phagolysosomes are found rarely (85,86). Birbeck's granules are similar morphologically to those present in normal LCs. The granules present in LCs within granulomas differ, however, in two respects. First, they are more numerous than in LCs present in normal lung or other tissues, and they can be found sometimes in extremely large numbers (84) (Fig. 10). Second, they are more often observed in direct continuity with the cell surface than in normal LC, but this may simply be due to their greater number.

The surface phenotype of LCs in LCG granulomas has been extensively evaluated. As for normal LCs, they express class I and class II MHC antigens, β_2-integrins (CD11a, CD11c, CD18), adhesins (CD54, CD58), leukocyte common antigen (CD45RO), and the intracellular S-100 protein (51,87). A characteristic feature of LCs in LC granulomas is the expression of both CD1a and CD1c antigens. It is noteworthy that the expression of both of these surface antigens is stronger than that observed on either DCs or LCs in normal lung (84) (Fig. 11). Similarly, a variety of other surface molecules are strongly expressed by LCs in LCG granulomas that are weakly expressed by normal LCs, including CD4 and the α-chain of the IL-2 receptor, CD25 (51,84,88). Studies performed in our laboratory demonstrate they also upregulate B7-1, B7-2, and CD40 costimulatory molecules (A. Tazi et al., in preparation).

B. Pathogenesis

The pathogenesis of pulmonary LCG remains unknown, and it is possible that different mechanisms apply to different forms of the disease. A major question that remains to be elucidated is whether some forms of LCG are due to malignant transformation of precursor cells that are committed to DC/LC differentiation. Strong evidence suggests that localized forms of the disease are reactive in origin. In this regard, it has been recently suggested that pulmonary LCG in adults results from an uncontrolled immune response initiated by LCs (78,83,84). A major argument in favor of this hypothesis is that activated LCs and $CD4^+$ T lympho-

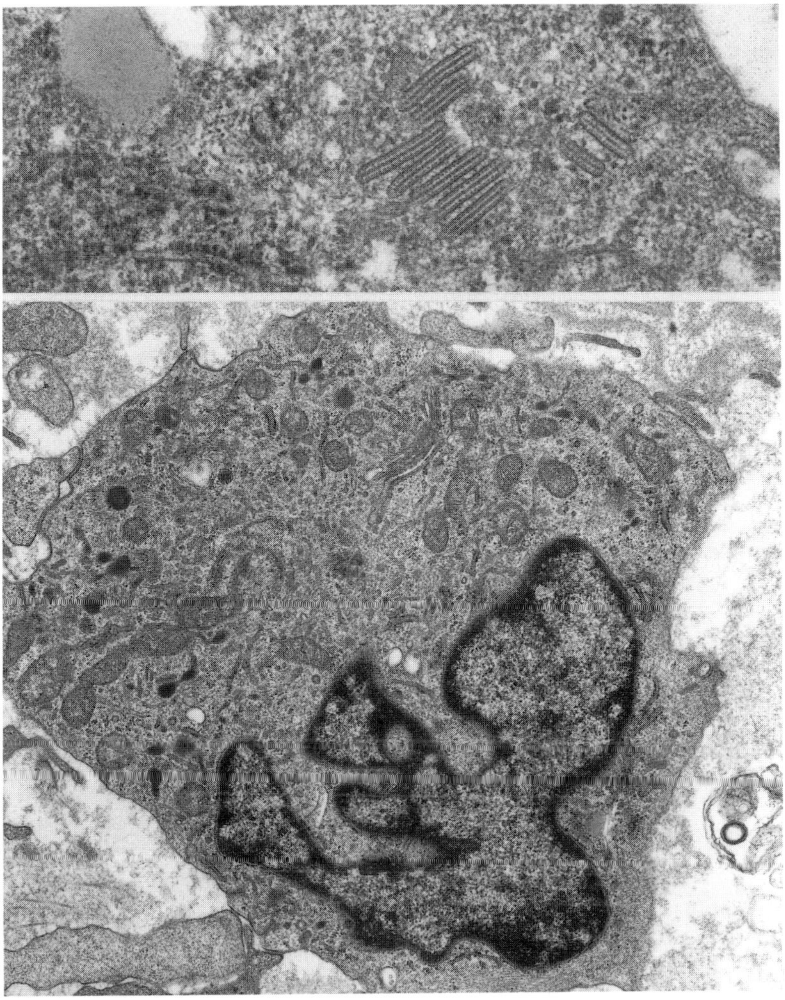

Figure 10 Electron microscopy. Early pulmonary LC granuloma. Typical LC characterized by an indented nucleus lined by a thin band of heterochromatin. The cytoplasm contains numerous usual organelles. (×4,500) The upper panel represents a high-power view of the LC showing numerous Birbeck's granules. (×35,000)

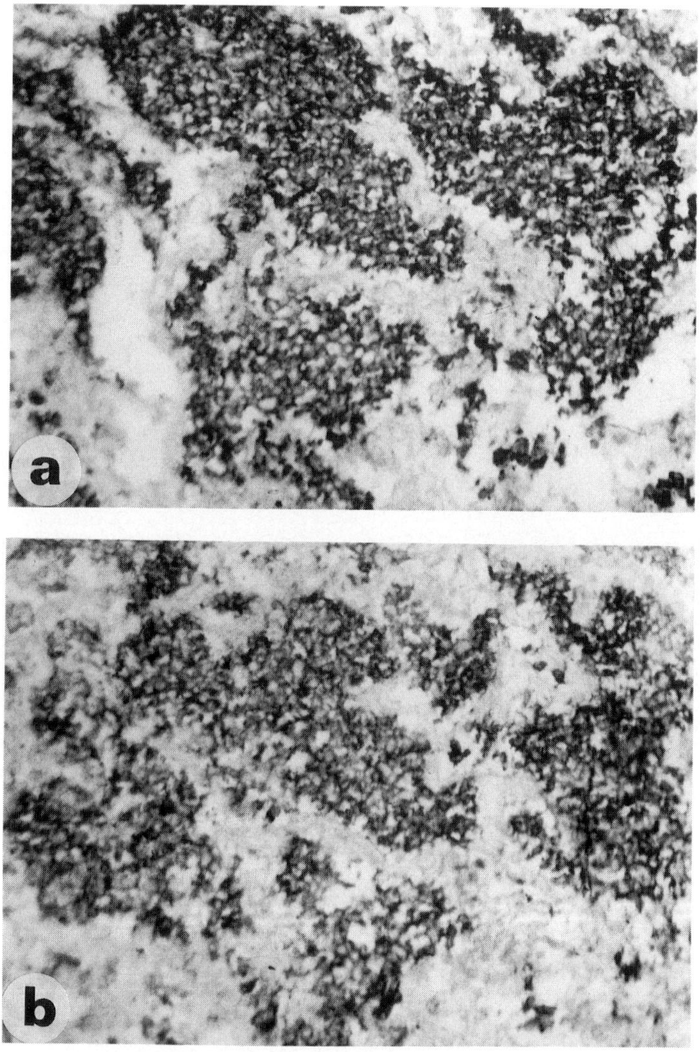

Figure 11 Surface phenotype of LC in pulmonary LCG. (a) Early granuloma mainly composed of LCs that strongly express CD1a antigens. (b) Adjacent section of the same lesion showing that the LCs also express CD1c antigens. (×125)

cytes displaying close differentiated contacts are prominent in early lesions, and the only known function of LCs is to initiate immune reactions by serving as accessory cells. Another characteristic of pulmonary LCG lesions is that they are destructive in nature. Accumulations of LCs are known to occur in other conditions, especially carcinomas, but without producing tissue destruction (40,42). Thus, one can consider that two successive events may be important in the pathogenesis of the disease: an initial event that leads to the accumulation of LCs adjacent to bronchioles and the subsequent initiation of an immune response by these cells. It should be noted, however, that these two events may be linked, with the factors inducing the attraction of LCs being able also to modify their state of activation such that these cells are predisposed to initiate an immune response locally.

Mechanisms of Accumulation of LCs in LCG

In normal lung, $CD1a^+$ LC containing Birbeck's granules are present almost exclusively intercalated between epithelial cells of airways. In general, LCs are present in an epithelial microenvironment, suggesting that their differentiation is mediated by direct contact with epithelial cells or by factors produced locally by the epithelial cells. Given the bronchocentric nature of the lesions of pulmonary LCG and the strong association with cigarette smoking, it is tempting to suggest that modifications in the bronchiolar epithelium could be a primary cause of the accumulation and activation of LCs in this disorder. In this respect, cigarette smoking has been shown to be associated with an increase in the number of DCs/LCs in the lung parenchyma of smokers and with a striking accumulation of $CD1a^+$ LCs containing Birbeck's granules within the alveolar epithelium at sites of epithelial hyperplasia (3). GM-CSF, which has important effects on LC survival and function, could be one factor possibly involved in the differentiation of pulmonary LCs (see above). This factor has been demonstrated to be produced in the human lung at all sites where LCs are known to accumulate. Thus, an increased expression of GM-CSF by bronchiolar epithelial cells associated with smoking-induced respiratory bronchiolitis could predispose to the accumulation and activation of LCs in pulmonary LCG. Consistent with this idea, remnants of bronchiolar epithelium intensely positive for GM-CSF have been recently reported in early pulmonary LCG lesions, whereas the expression of GM-CSF is not increased in adjacent bronchioles not affected by the disease (89). It is noteworthy that bronchiolar epithelial cells are rapidly destroyed by the granulomatous process, and GM-CSF production by these cells cannot sustain LCs at these sites. LCs have been found to express moderately GM-CSF in LCG lesions (89), suggesting that GM-CSF production by LCs themselves could play a part in the maintenance of granulomas. The presence of GM-CSF alone cannot explain, however, all the phenotypic features of LCs accumulating in LCG lesions. Although it is known to be a growth factor for DC/LC precursors, GM-CSF does not induce in vitro the presence of Birbeck's granules in these cells (90). Moreover, LCs accumulating at sites of alveolar hyperplasia or in carcinomas producing GM-CSF do not exhibit the large number of Birbeck's granules present in LCG lesions nor the intensely positive staining with both CD1a and CD1c antibodies (3,84).

Lung epithelial cells are able to produce several other cytokines that could contribute to this process by stimulating DCs/LCs (35–39), but the role of these cytokines in the accumulation of LCs in early LCG lesions has not been investigated. In addition, neuroendocrine epithelial cells, known to produce a variety of neuropeptides (91–93), are increased in number in tissues from patients with pulmonary LCG compared with individuals with other lung disorders, suggesting that these cells could play a role in the pathogenesis of pulmonary LCG (94). Neuropeptides are known to have strong immunomodulatory activities on immune and inflammatory cells, but their effects on pulmonary DCs/LCs have not yet been evaluated. Increased expression of enzymes involved in neuropeptide metabolism by airway epithelium from patients with LCG has also been demonstrated (95), and abnormalities in the balance of neuropeptides caused by airway abnormalities could modulate DC/LC function. Finally, LCs themselves can produce a variety of cytokines following activation (37,51,96,97) that could also contribute to the recruitment and the unusual state of activation observed for these cells in LCG.

Is Pulmonary LCG an Immune Reactive Disease?

A number of arguments strongly suggest that the granulomatous process of pulmonary LCG in adult patients results from an uncontrolled immune response initiated by LC:

The presence of activated LCs is a distinctive feature of early lesions, and the major function recognized for these cells is to serve as accessory cells. They can also express numerous molecules necessary for interactions with T cells, including B7, CD40, intracellular adhesion molecule-1 (ICAM-1), and lymphocyte function–associated antigen-3 (LFA-3) (98–103), and have remarkable lymphostimulatory activities (see above).

Lymphocytes are the other cell type prominent in LCG lesions, particularly in early stages. As expected for T cells responding to extrinsic antigens presented by LCs, these lymphocytes are predominantly $CD4^+$ T lymphocytes expressing α/β T-cell receptors (84). Contacts between LCs and T cells are frequently found that are identical to those observed in the course of hypersensitivity reactions in the skin (84,96). Evidence for a local activation of T cells infiltrating pulmonary LCG lesions has been also provided (51).

The structure of lung LCG lesions resembles that of "immune" granulomas, with a central core of accessory cells surrounded by and, to some extent, infiltrated with T cells and inflammatory cells.

The evolution of lung LCG granulomas is suggestive of an immune response. The accumulation of accessory cells and lymphocytes, the subsequent recruitment of other inflammatory cells, and the progressive replacement of these cells by macrophages and fibrotic tissue are all observed in other granulomatous processes.

Functional studies of DCs/LCs in various tissues, including lung, strongly suggest that intraepithelial LCs develop the capacity to activate efficiently T cells during their migration to regional lymphoid organs (see above). According to this model, the development of an immune response initiated by LCs in pulmonary LCG implies the presence of factors that not only increase the number of LCs at the sites of the disease but also enhance their lymphostimulatory activities and prevent their migration to regional lymph nodes (51,87). The characterization of stimuli necessary for the switch from antigen processing to lymphostimulatory phenotype by LCs is an area of intense study. A variety of cytokines or other inflammatory mediators (30,36–39,104) and interactions with surface molecules expressed on other cell types (e.g., CD40 ligand) (105) can influence the ability of cells to perform these two functions. Thus, many of the signals discussed above that may play a role in the accumulation of LCs in early LCG lesions also could promote the differentiation of theses cells toward a state in which they not only can process antigens efficiently but also can serve a lymphostimulatory role.

The idea that an uncontrolled immune response in which LCs serve as accessory cells is involved in the pathogenesis of pulmonary LCG implies the existence of an initiating antigen. Given the bronchocentric distribution of the lesions and the strong association with cigarette smoking in adult patients, it has been suggested that a component of cigarette smoke might be involved (82,83). Other initiating antigens, however, would have to be evoked to explain the presence of pulmonary LCG in nonsmokers and in children. We have previously suggested that stressed or transformed bronchiolar epithelial cells could be the target of the immune response initiated by LCs (83). As indicated above, hyperplastic epithelial cells are a potential source of mediators susceptible to induce both the accumulation and the activation of LCs. Hyperplastic cells could also express neoantigens that serve as the antigenic stimulus. This hypothesis would account for both the involvement and the destruction of bronchiolar structures in the course of the disease. It also explains why cigarette smoking, which is a common cause of bronchiolar hyperplasia, would predispose to pulmonary LCG. Likewise, it could explain why LCG may occur in nonsmoking individuals who develop bronchiolar abnormalities for other reasons (e.g., radiation therapy, chemotherapy, or occupational exposures). Neither of these hypotheses, however, easily accomodates the presence of lesions in other tissues. It remains possible that the pathogenetic mechanisms responsible for localized and extrapulmonary multifocal forms of LCG are distinct.

Is Pulmonary LCG a Malignant Disease?

The accumulation of large numbers of LCs in early lesions is often considered as the consequence of an abnormal proliferation of these cells. The mechanism leading to this accumulation is not known, however, and the relative contribution of recruitment and local proliferation of LCs remains to be established. It should be noted that LCs in mitosis are not commonly observed in the lesions of the different forms of the disease (78), suggesting that local proliferation of these cells cannot explain by itself the massive accumulation of LCs in early lesions. Recently, it has been reported that LCs in LC granulomas, at least in children, can express the proliferative cell nuclear antigen (PCNA) (106), a nuclear protein known to accumulate in proliferating cells, particularly during the S phase (107). It should be noted, however, that caution is necessary in the use of PCNA as a marker of growth fraction, since evidence is accumulating that PCNA may be expressed by nonproliferating cells (108). Studies performed in our laboratory evaluating several histopathological techniques currently used in assessment of malignancy demonstrate that LCs in lung, skin, or bone lesions in adults and children are not a rapidly dividing cell population (P. Soler et al., in preparation). Further studies are needed to determine whether the proliferating fraction of LCs is different in severe diffuse forms of the disease encountered in children.

It has been also suggested that LCG results from malignant transformation of cells of DC/LC lineage (106,109–111). It should be emphasized that studies supporting a malignant origin for LCG have generally analysed patients with diffuse forms of the disease. Several strong arguments indicate, however, that pulmonary LCG occuring in adults is not a malignant process. Cellular atypia is uncommon in LCG. The infiltration of tissues with malignant cells does not easily explain the bronchocentric distribution of the lesions. Some clinical features, such as the high rate of spontaneous remission, are not typical of malignancy. The virtual absence of LCs in late lesions and the failure to demonstrate aneuploidy in most cases (112–114) also argue against malignancy.

Recent studies using methods based on the assessment of X chromosome inactivation in females have shown that LCs in lesions from patients with both diffuse and localized forms of the disease are of clonal origin (115,116). Although the existence of clonal LCs in pulmonary LCG in adults has not been evaluated, these results suggest that LCG could be a neoplastic disorder. It is important to emphasize, however, that clonality does not necessarily mean malignancy. Clonal T cells populations have been reported in several circumstances without obvious evidence of malignancy, and it has been suggested that this clonal expansion could represent a normal host response to chronic antigenic stimulation (117). Similar considerations could apply to DCs/LCs. Evidence of clonal LCs in LCG lesions might also suggest the existence of a possible underlying genetic defect in cells committed to DC/LC differentiation. In this regard, given the arguments developed above, if clonal LCs are present in pulmonary LCG, these cells are unlikely to be autonomous malignant cells. Rather they could represent a subset of cells susceptible to factors that modify their behavior locally. For example, abnormalities leading to clonal expansion of LCs could be linked to the metabolism of GM-CSF, since this factor has been shown to have important effects on LC activation and differentiation (see above). A number of important questions remains to be studied. It remains unknown whether clonality is also present in the normal LC population in these patients or in the progenitor cell populations present in the bone marrow and peripheral blood. Similarly, the existence of the same LC clone at all sites of the disease in patients with multifocal LCG has yet to be demonstrated, and it remains unknown if the same clonal population persists over time. Studies attempting to answer to these questions, as well as studies directly comparing the pathological features of lesions from patients with different forms of the disease, are needed to determine clearly whether distinct patterns of disease of reactive and malignant origin can be identified.

The relationship between LCG and cancer is further clouded by the following observations: (1) LCG has been observed adjacent to sites of lymphomatous involvement (118–121); (2) pulmonary LCG has been reported in patients with Hodgkin's disease following radiotherapy and/or chemotherapy (118–120); and (3) an association of LCG with various malignant neoplasms has been also described (119,122–125). Whether LCG is a risk factor for the development of malignancy remains questionable.

VI. Conclusions

Cells of the dendritic cell lineage are widely distributed in normal human lung. They are potent accessory cells that likely play an essential role in the development of pulmonary immune responses. Evidence is accumulating that they may also play an important role in certain lung diseases. LCs play a central role in the abnormal immune response that characterizes pulmonary LCG, and a number of studies suggest they could play a role in tumor immunity. Further studies are needed to better understand the fundamental mechanisms and antigens involved in pulmonary LCG. Likewise, a better knowledge of the functional activities of LCs infiltrating lung carcinomas is strongly needed. Progress in both these areas of investigation might lead to the development of new specific therapeutic strategies for these disorders.

References

1. Sertl K, Takemura T, Tschachler E, Ferrans VJ, Kaliner MA, Shevach EM. Dendritic cells with antigen-presenting capability reside in airway epithelium, lung parenchyma, and visceral pleura. J Exp Med 1986; 163:436–451.
2. Holt PG, Schon-Hegrad MA, Oliver J. MHC class II antigen-bearing dendritic cells in pulmonary tissues of the rat: regulation of antigen presentation activity by endogenous macrophage populations. J Exp Med 1988; 167:262–274.
3. Soler P, Moreau A, Basset F, Hance AJ. Cigarette smoking-induced changes in the number and differentiated state of pulmonary dendritic cells/Langerhans' cells. Am Rev Respir Dis 1989; 139:1112–1117.
4. Kradin RL, McCarthy KM, Xia W, Lazarus D, Schneeberger EE. Accessory cells of the lung. I. Interferon-gamma increases Ia^+ dendritic cells in the lung without augmenting their accessory activities. Am J Respir Cell Mol Biol 1991; 4:210–218.
5. Schon-Hegrad MA, Oliver J, McMenamin PG, Holt PG. Studies on the density, distribution, and surface phenotype of intraepithelial class II major histocompatibility complex antigen (Ia)–bearing dendritic cells (DC) in the conducting airways. J Exp Med 1991; 173:1345–1356.
6. van Haarst JMW, de Wit HJ, Drexhage HA, Hoogsteden HC. Distribution and immunophenotype of mononuclear phagocytes and dendritic cells in the human lung. Am J Respir Cell Mol Biol 1994; 10:487–492.
7. Rowden G. The Langerhans' cell. CRC Crit Rev Immunol 1981; 3:95–180.
8. Austyn JM. Lymphoid dendritic cells. Immunology 1987; 62:161–170.
9. Steinman RM. The dendritic cell system and its role in immunogenicity. Annu Rev Immunol 1991; 9:271–296.
10. Hance AJ. Pulmonary immune cells in health and disease: Dendritic cells and Langerhans' cells. Eur Respir J 1993; 6:1213–1220.
11. Gong JL, Mc Carthy KM, Telford J, Tamatani T, Miyasaka M, Schneeberger EE. Intraepithelial airway dendritic cells: a distinct subset of pulmonary dendritic cells obtained by microdissection. J Exp Med 1992; 175:797–807.
12. McCarthy KM, Gong JL, Telford JR, Schneeberger EE. Ontogeny of Ia-positive accessory cells in the fetal and newborn rat lung. Am J Respir Cell Mol Biol 1992; 6:349–356.
13. Nelson DJ, McMenamin C, McWilliam AS, Brenan M, Holt PG. Development of the airway intraepithelial dendritic cell network in the rat from class II major histocompatibility (Ia)–negative precursors: differential regulation of Ia expression at different levels of the respiratory tract. J Exp Med 1994; 179:203–212.
14. Markowicz S, Engleman EG. Granulocyte-macrophage colony-stimulating factor promotes differentiation and survival of human peripheral blood dendritic cells in vitro. J Clin Invest 1990; 85:955–961.
15. Nicod LP, Lipscomb MF, Weissler JC, Lyons CR, Albertson J, Toews GB. Mononuclear cells in human lung parenchyma. Characterization of accessory cell not obtained by bronchoalveolar lavage. Am Rev Respir Dis 1987; 136:818–823.
16. Rochester CL, Goodell EM, Stoltenborg JK, Bowers WE. Dendritic cells from rat lung are potent accessory cells. Am Rev Respir Dis 1998; 138:121–128.

17. Nicod LP, Lipscomb MF, Toews GB, Weissler JC. Separation of potent and poorly functional human lung accessory cells based on autofluorescence. J Leukoc Biol 1989; 45:458–465.
18. Nicod LP, Lipscomb MF, Weissler JC, Toews GB. Mononuclear cells from human lung parenchyma support antigen-induced T-lymphocyte proliferation. J Leukoc Biol 1989; 45:336–344.
19. Holt PG, Degebrodt A, O'Leary C, Krska K, Plozza T. T-cell activation by antigen-presenting cells from lung tissue digests: suppression by endogenous macrophages. Clin Exp Immunol 1985; 62:586–593.
20. Holt PG, Oliver J, Bilyk N, McMenamin C, McMenamin PG, Kraal G, Thepen T. Downregulation of the antigen presenting cell function(s) of pulmonary dendritic cells in vivo by resident alveolar macrophages. J Exp Med 1993; 177:397–402.
21. Kapsenberg ML, Res P, Bos JD, Schootemijer A, Teunissen MBM, Van Schooten W. Nickel-specific T lymphocyte clones derived from allergic nickel-contact dermatitis lesions in man: heterogeneity based on requirement of dendritic antigen-presenting cell subsets. Eur J Immunol 1987; 17:861–865.
22. Kast WM, Boog CJP, Reep BO, Voordouw AC, Melief CJM. Failure or success in the restoration of virus-specific cytotoxic T lymphocyte response defects by dendritic cells. J Immunol 1988; 140:3186–3193.
23. Boog CJP, Kast WM, Timmers HThM, Boes J, de Waal LP, Melief CJM. Abolition of specific immune response defect by immunization with dendritic cells. Nature 1985; 318:59–62.
24. Inaba K, Young JW, Steinman RM. Direct activation of $CD8^+$ cytotoxic T lymphocytes by dendritic cells. J Exp Med 1987; 166:182–194.
25. Inaba K, Metlay JP, Crowley MT, Steinman RM. Dendritic cells pulsed with protein antigens in vitro can prime antigen-specific, MHC-restricted T cells in situ. J Exp Med 1990; 172:631–640.
26. Ronchese F, Haussmann B. B lymphocytes in vivo fail to prime naive T cells but can stimulate antigen-experienced T lymphocytes. J Exp Med 1993; 177:679–690.
27. Heufler C, Koch F, Schuler G. Granulocyte/macrophage colony-stimulating factor and interleukin-1 mediate the maturation of murine epidermal Langerhans' cells into potent immunostimulatory dendritic cells. J Exp Med 1988; 167:700–705.
28. Bujdoso R, Hopkins J, Dutia BM, Young P, McConnell I. Characterization of sheep afferent lymph dendritic cells and their role in antigen carriage. J Exp Med 1989; 170:1285–1302.
29. Romani N, Koide S, Crowley M, Witmer-Pack M, Livingstone AM, Fathman GC, Inaba K, Steinman RM. Presentation of exogenous protein antigens by dendritic cells to T-cells clones: intact protein is presented best by immature epidermal Langerhans' cells. J Exp Med 1989; 169:1168–1178.
30. Streilen JW, Grammer SF. In vitro evidence that Langerhans' cells can adopt two functionally distinct forms capable of antigen presentation to T-lymphocytes. J Immunol 1989; 143:3925–3933.
31. Kripke ML, Munn CG, Jeevan A, Tang JM, Bucana C. Evidence that cutaneous antigen-presenting cells migrate to regional lymph nodes during contact sensitization. J Immunol 1990; 145:2833–2838.

32. Demidem A, Taylor JR, Grammer SF, Steilen JW. Influence of microenvironmental factors on human Langerhans cell function in vitro. Reg Immunol 1991; 3:297–304.
33. Dai R, Grammer SF, Streilein JW. Fresh and cultured Langerhans cells display differential capacities to activate hapten-specific T cells. J Immunol 1993; 150: 59–66.
34. Pollard AM, Lipscomb MF. Characterization of murine lung dendritic cells: similarities to Langerhans' cells and thymic dendritic cells. J Exp Med 1990; 172: 159–167.
35. Koch F, Heufler C, Kämpgen E, Schneeweiss D, Böck G, Schuler G. Tumor necrosis factor-α maintains the viability of murine epidermal Langerhans' cells in culture, but in contrast to granulocyte/macrophage colony-stimulating factor, without inducing their functional maturation. J Exp Med 1990; 171:159–169.
36. Caux C, Dezutter-Dambuyant C, Schmitt D, Banchereau J. GM-CSF and TNF-α cooperate in the generation of dendritic Langerhans cells. Nature 1992; 360: 258–261.
37. Enk AH, Angeloni VL, Udey MC, Katz SI. An essential role for Langerhans cell–derived IL-1β in the initiation of primary immune responses in skin. J Immunol 1993; 150:3698–3704.
38. Macatonia SE, Doherty TM, Knight SC, O'Garra A. Differential effect of IL-10 on dendritic cell-induced T cell proliferation and IFN-γ production. J Immunol 1993; 150:3755–3765.
39. Sallusto F, Lanzavecchia A. Efficient presentation of soluble antigen by cultured human dendritic cells is maintained by granulocyte/macrophage colony-stimulating factor plus interleukin 4 and downregulated by tumor necrosis factor α. J Exp Med 1994; 179:1109–1118.
40. Tazi A, Bouchonnet F, Grandsaigne M, Boumsell L, Hance AJ, Soler P. Evidence that granulocyte-macrophage colony-stimulating factor (GM-CSF) regulates the distribution and differentiated state of cells of dendritic cells/Langerhans cells lineage in human lung and lung cancer. J Clin Invest 1993; 91:566–576.
41. Basset F, Soler P, Hance AJ. The Langerhans cells in human pathology. Ann NY Acad Sci 1986; 465:324–339.
42. Hammar S, Bockus D, Remington F, Bartha M. The widespread distribution of Langerhans cells in pathologic tissues: an ultrastructural and immunohistochemical study. Human Pathol 1986; 17:894–905.
43. Uyama T, Winter JB, Sakiyama S, Monden Y, Groen G, Prop J. Replacement of dendritic cells in the airways of rat lung allografts. Am Rev Respir Dis 1993; 148: 760–767.
44. Kämpgen E, Koch F, Heufler C, Eggert A, Gill LL, Gillis S, Dower SK, Romani N, Schuler G. Understanding the dendritic cell lineage through a study of cytokine receptors. J Exp Med 1994; 179:1767–1776.
45. McWilliam A, Nelson D, Thomas JA, Holt PG. Rapid dendritic cell recruitment is a hallmark of the acute inflammatory response at mucosal surfaces. J Exp Med 1994; 179:1331–1336.
46. Basset F, Soler P, Wyllie L, Abelanet R, Le Charpentier M, Kreis B, Breathnach AS. Langerhans cells in a bronchiolar-alveolar tumour of lung. Virchows Arch. [A] Pathol Anat and Histol 1974; 362:315–330.

47. Watanabe S, Sato Y, Kodama T, Shimosato Y. Immunohistochemical study with monoclonal antibodies on immune response in human lung cancers. Cancer Res 1983; 43:5883–5889.
48. Furukawa T, Watanabe S, Kodma T, Sato Y, Shimosato Y, Suemasu K. T-zone histiocytes in adenocarcinoma of the lung in relation to postoperative prognosis. Cancer 1985; 56:2651–2656.
49. Fox SB, Jones M, Dunnill MS, Gatter KC, Mason DY. Langerhans cells in human lung tumours: an immunohistochemical study. Histopathology 1989; 14:269–275.
50. Zeid NA, Muller HK. S100 positive dendritic cells in human lung tumors associated with cell differentiation and enhance survival. Pathology 1993; 25:338–343.
51. Colasante A, Poletti V, Rosini S, Ferracini R, Musiani P. Langerhans cells in Langerhans cell histiocytosis and peripheral adenocarcinomas of the lung. Am Rev Respir Dis 1993; 148:752–759.
52. Colasante A, Castrilli G, Aiello FB, Brunetti M, Musiani P. Role of cytokines in distribution and differentiation of dendritic cell/Langerhans's cell lineage in human primary carcinomas of the lung. Hum Pathol 1995; 26:866–872.
53. Halliday GM, Lucas AD, Barnetson RSC. Control of Langerhans' cell density by a skin tumour-derived cytokine. Immunology 1992; 77:13–18.
54. Becker Y. Dendritic cell activity against primary tumors: an overview. In vivo 1993; 7:187–192.
55. Halliday GM, Reeve VE, Barnetson R StC. Langerhans cell migration into ultraviolet light-induced squamous skin tumours is unrelated to anti-tumor immunity. J Invest Dermatol 1991; 97:830–834.
56. Grabbe S, Beissert S, Schwarz T, Granstein RD. Dendritic cells as initiators of tumor immune responses: a possible strategy for tumor immunotherapy? Immunol Today 1995; 16:117–121.
57. Mayordomo JI, Zorina T, Storkus WJ, Zitgovel L, Celluzi C, Falo LD, Melief CJ, Ildstad ST, Kast WM, DeLeo AB, Lotze MT. Bone marrow–derived dendritic cells pulsed with synthetic tumour peptides elicit protective and therapeutic antitumour immunity. Nature Med 1995; 1:1297–1302.
58. Young JW, Inaba K. Dendritic cells as adjuvants for classe I major histocompatibility complex-restricted antitumor immunity. J Exp Med 1996; 183:7–11.
59. Celluzi CM, Mayordomo JI, Storkus WJ, Lotze MT, Falo LD Jr. Peptide-pulsed dendritic cells induce antigen-specific, CTL-mediated protective tumor immunity. J Exp Med 1996; 183:283–287.
60. Paglia P, Chiodoni C, Rodolfo M, Colombo MP. Murine dendritic cells loaded in vitro with soluble protein prime CTL against tumor antigen in vivo. J Exp Med 1996; 183:317–322.
61. Tan K-C, Hosoi J, Grabbe S, Asahina A, Granstein RD. Epidermal cell presentation of tumor-associated antigens for induction of tolerance. J Immunol 1994; 153: 760–767.
62. Knight SC, Hunt R, Dore C, Medawar B. Influence of dendritic cells on tumor growth. Proc Natl Acad Sci USA 1985; 82:4495–4497.
63. Simon JC, Tigelaar RE, Bergstresser PR, Edelbaum D, Cruz PD JR. Utraviolet B radiation converts Langerhans cells from immunogenic to tolerogenic antigen-presenting cells. J Immunol 1991; 146:485–491.

64. June CH, Bluestone JA, Nadler LM, Thompson CB. The B7 and CD28 receptor families. Immunol Today 1994; 15:321–331.
65. Zitgovel L, Mayordomo JI, Tjandrawan T, DeLeo AB, Clarke MR, Lotze MT, Storkus WJ. Therapy of murine tumors with tumor peptide-pulsed dendritic cells: dependence on T cells, B7 costimulation and TH1 associated cytokines. J Exp Med 1996; 183:87–97.
66. Allison JP, Hurwitz AA, Leach DR. Manipulation of costimulatory signals to enhance antitumor T-cell responses. Curr Opin Immunol 1995; 7:682–686.
67. Kawamura T, Furue M. Comparative analysis of B7-1 and B7-2 expression in Langerhans cells: differential regulation by T helper type 1 and T helper type 2 cytokines. Eur J Immunol 1995; 25:1913–1917.
68. Thompson CB. Distinct roles for the costimulatory ligands B7-1 and B7-2 in T helper cell differentiation? Cell 1995; 81:979–982.
69. Matulonis U, Dosiou C, Freeman G, Lamont C, Mauch P, Nadler LM, Griffin JD. B7-1 is superior to B7-2 costimulation in the induction and maintenance of T cell–mediated antileukemia immunity. J Immunol 1996; 156:1126–1131.
70. Krummel MF, Allison JP. CD28 and CTLA-4 have opposing effects on the response of T cells to stimulation. J Exp Med 1995; 182:459–465.
71. Pardoll DM. Paracrine cytokine adjuvants in cancer immunotherapy. Annu Rev Immunol 1995; 13:399–415.
72. Chang C-H, Furue M, Tamaki K. B7-1 expression of Langerhans cells is up-regulated by proinflammatory cytokines, and is down-regulated by interferon-γ or by interleukin-10. Eur J Immunol 1995; 25:394–398.
73. Grabbe S, Bruvers S, Granstein RD. Interleukin 1α but not transforming growth factor β inhibits tumor antigen presentation by epidermal antigen-presenting cells. J Invest Dermatol 1994; 102:67–73.
74. Beissert S, Hosoi J, Grabbe S, Asahina A, Granstein RD. IL-10 inhibits tumor antigen presentation by epidermal antigen-presenting cells. J Immunol 195; 154: 1280–1286.
75. Grabbe S, Bruvers S, Lindgren AM, Hosoi J, Tan KC, Granstein RD. Tumor antigen presentation by epidermal antigen-presenting cells in the mouse: modulation by granulocyte-macrophage colony-stimulating factor, tumor necrosis factor α, and ultraviolet radiation. J Leukoc Biol 1992; 52:209–217.
76. Friedman PJ, Liebow AA, Sokoloff J. Eosinophilic granuloma of lung. Medicine 1981; 60:385–396.
77. Colby TV, Lombard C. Histiocytosis X in the lung. Hum Pathol 1983; 14:847–856.
78. Soler P, Tazi A, Hance AJ. Pulmonary Langerhans cell granulomatosis. Curr Opin Pulm Med 1995; 1:406–416.
79. Favara BE, McCarthy RC, Mierau GW. Histiocytosis X. Hum Pathol 1983; 14: 663–676.
80. Nezelof C, Frileux H, Cronier-Sachot J. Disseminated histiocytosis X. Analysis of prognostic factors based on a retrospective study of 50 cases. Cancer 1979; 44:1824–1838.
81. Massachusetts General Hospital: Case records of the Massachusetts General Hospital. Case 40-1993. N Engl J Med 1993; 329:1108–1115.
82. Travis WD, Borok Z, Roum JH, Zhang J, Feuerstein I, Ferrans VJ, Crystal RG.

Pulmonary Langerhans granulomatosis (histiocytosis X). A clinicopathologic study in 48 cases. Am J Surg Pathol 1993; 17:971–986.
83. Soler P, Kambouchner M, Valeyre D, Hance AJ. Pulmonary Langerhan's cell granulomatosis (histiocytosis X). Annu Rev Med 1992; 43:105–115.
84. Tazi A, Bonay M, Grandsaigne M, Battesti JP, Hance AJ, Soler P. Surface phenotype of Langerhans cells and lymphocytes in granulomatous lesions from patients with pulmonary histiocytosis X. Am Rev Respir Dis 1993; 147:1531–1536.
85. Inaba K, Inaba M, Naito M, Steinman RM. Dendritic cell progenitors phagocytose particulates, including bacillus Calmette-Guerin organisms, and sensitize mice to mycobacterial antigens in vivo. J Exp Med 1993; 178:479–488.
86. Reise Sousa C, Stahl PD, Austyn JM. Phagocytosis of antigens by Langerhans cells in vitro. J Exp Med 1993; 178:509–519.
87. de Graaf JH, Tamminga RYJ, Kamps WA, Timens W. Langerhans' cell histiocytosis: expression of leukocyte cellular adhesion molecules suggests abnormal homing and differentiation. Am J Pathol 1994; 144:466–472.
88. O'Doherty U, Steinman RM, Peng M, Cameron PU, Gezelter S, Kopeloff I, Swiggard WJ, Popoe M, Bhardwaj N. Dendritic cells freshly isolated from human blood express CD4 and mature into typical immunostimulatory dendritic cells after culture in monocyte-conditioned medium. J Exp Med 1993; 178:1067–1078.
89. Tazi A, Bonay M, Bergeron A, Grandsaigne M, Hance AJ, Soler P. Role of granulocyte-macrophage colony-stimulating factor (GM-CSF) in the pathogenesis of pulmonary histiocytosis X. Thorax 1996; 51:611–614.
90. Stössel H, Koch F, Kämpgen E, Stöger P, Lenz A, Heufler C, Romani N, Schuler G. Disappearance of certain acidic organelles (endosomes and Langerhans cell granules) accompanies loss of antigen processing capacity upon culture of epidermal Langerhans cells. J Exp Med 1990; 172:1471–1482.
91. Gould VE, Linnoila RI, Memoli VA, Warren WH. Neuroendocrine components of the bronchopulmonary tract: hyperplasias, dysplasias, and neoplasms. Lab Invest 1983; 49:519–537.
92. Johnson DE, Wobken JD, Landrum BG. Changes in bombesin, calcitonin, and serotonin immunoreactive pulmonary neuroendocrine cells in cystic fibrosis and after prolonged mechanical ventilation. Am Rev Respir Dis 1988; 137:123–131.
93. Payan DG. The role of neuropeptides in inflammation. In: Gallin JI, Goldstein IM, Snyderman R, eds. Inflammation: Basic Principles and Clinical Correlates. New York: Raven Press, 1992:177–192.
94. Aguayo SM, King TE, Waldron JA, Sherrit KM, Kane MA, Miller YE. Increased pulmonary neuroendocrine cells with bombesin-like immunoreactivity in adult patients with eosinophilic granuloma. J Clin Invest 1990; 86:838–844.
95. Cohen AJ, King TE Jr, Franklin W, Miller YE. Neutral endopeptidase expression in eosinophilic granuloma of the lung. Am J Respir Crit Care Med 1994; 149:A41.
96. Enk AH, Katz SI. Early molecular events in the induction phase of contact sensitivity. Proc Natl Acad Sci USA 1992; 89:1398–1402.
97. Heufler C, Topar G, Koch F, Trockenbacher B, Kämpgen E, Romani N, Schuler G. Cytokine gene expression in murine epidermal cell suspensions: interleukin 1 β and macrophage inflammatory protein 1α are selectively expressed in Langerhans cells but are differentially regulated in culture. J Exp Med 1992; 176:1221–1226.

98. Hanau D, Fabre M, Schmitt DA, Garaud JC, Pauly G, Tongio MM, Mayer S, Cazenave JP. Human epidermal Langerhans cells cointernalize by receptor-mediated endocytosis "nonclassical" major histocompatibility complex class I molecules (T6 antigens) and class II molecules (HLA-DR antigens). Proc Natl Acad Sci USA 1987; 84:2901–2905.
99. Hopkins J, Dutia BM, Bujdoso R, McConnell I. In vivo modulation of CD1 and MHC class II expression by sheep afferent lymph dendritic cells. J Exp Med 1989; 170:1303–1318.
100. Puré H, Inaba K, Crowley MT, Tardelli L, Witmer-Pack MD, Ruberti G, Fathman G, Steinman RM. Antigen processing by epidermal Langerhans cells correlates with the level of biosynthesis of major histocompatibility complex class II molecules, and expression of invariant chain. J Exp Med 1990; 172:1459–1469.
101. Symington FW, Brady W, Linsley PS. Expression and function of B7 on human epidermal Langerhans cells. J Immunol 1993; 150:1286–1295.
102. Teunissen MB, Rongen HA, Bos JD. Function of adhesion molecules, lymphocyte function–associated antigen-3 and intercellular adhesion molecule-1 on human epidermal Langerhans cells in antigen-specific T cell activation. J Immunol 1994; 152:3400–3409.
103. Larsen CP, Ritchie SC, Hendrix R, Linsley PS, Hathcock KS, Hodes RJ, Lowry RP, Pearson TC. Regulation of immunostimulatory function and costimulatory molecule (B7-1 and B7-2) expression on murine dendritic cells. J Immunol 1994; 152:5208–5219.
104. Hosoi J, Murphy GF, Egan CL, Lerner EA, Grabbe S, Asahina A, Granstein RD. Regulation of Langerhans cell function by nerves containing calcitonin gene–related peptide. Nature 1993; 363:159–163.
105. Caux C, Massacrier C, Vanbervliet B, Dubois B, Van Kooten C, Durand I, Banchereau J. Activation of human dendritic cells through CD40 cross-linking. J Exp Med 1994; 180:1263–1272.
106. Hage C, Willman CL, Favara BE, Isaacson PG. Langerhans' cell histiocytosis (histiocytosis X): immunophenotype and growth fraction. Hum Pathol 1993; 24: 840–845.
107. Kurki P, Vanderlaan M, Dolbeare F, Gray J, Tan EM. Expression of proliferating cell nuclear antigen (PCNA)/cyclin during the cell cycle. Exp Cell Res 1986; 166: 209–219.
108. McCormick D, Hall PA. The complexities of proliferating cell nuclear antigen. Histopathology 1992; 21:591–594.
109. Goldberg NS, Bauer K, Rosen ST, Caro W, Marder RJ, Zugerman CH, Rao S, Variakojis D. Histiocytosis X. Flow cytometric DNA-content and immunohistochemical and ultrastructural analysis. Arch Dermatol 1986; 122:446–450.
110. Wood C, Wood GS, Deneau DG, Oseroff A, Beckstead JH, Malin J. Malignant histiocytosis X: a report of a rapidly fatal case in an elderly man. Cancer 1984; 54: 347–352.
111. Ben-Ezra J, Bailey A, Azumi N, Delsol G, Stroup R, Sheibani K, Rappaport H. Malignant histiocytosis X: a distinct clinicopathologic entity. Cancer 1991; 68: 1050–1060.
112. Rabkin MS, Wittmer CT, Kjeldsberg CR, Piepkorn MW. Flow-cytometric DNA

content of histiocytosis X (Langerhans cell histiocytosis). Am J Pathol 1988; 131: 283–289.

113. McLelland J, Newton JA, Malone M, Camplejohn RS, Chu AC. A flow cytometric study of Langerhans cell histiocytosis. Br J Dermatol 1989; 120:485–491.
114. Ornvold K, Carstensen H, Larsen JK, Christensen IJ, Ralfkiaer E. Flow cytometric DNA analysis of lesions from 18 children with Langerhans cell histiocytosis (histiocytosis X). Am J Pathol 1990; 136:1301–1307.
115. Willman CL, Busque L, Griffith BB, Favara BE, McClain KL, Duncan MH, Gilliland DG. Langerhans'-cell histiocytosis (histiocytosis X): a clonal proliferative disease. N Engl J Med 1994; 331:154–160.
116. Yu RC, Chu C, Buluwela L, Chu AC. Clonal proliferation of Langerhans cells in Langerhans cell histiocytosis X. Lancet 1994; 343:767–768.
117. Beverley PCL, Abbas AK. The scientific challenge of Langerhans cell histiocytosis. Br J Cancer 1994; 70(suppl):61–63.
118. Sajjad SM, Luna MA. Primary histiocytosis X in two patients with Hodgkin's disease. Thorax 1982; 37:110–113.
119. Neumann MP, Frizzera G. The coexistence of Langerhans' cell granulomatosis (LCG) and malignant lymphoma may take different forms: report of seven cases with the review of the literature. Hum Pathol 1986; 17:1060–1065.
120. Shanley DJ, Lerud KS, Luetkans TJ. Development of pulmonary histiocytosis X after chemotherapy for Hodgkin's disease. AJR Am J Roentgenol 1990; 155:741–742.
121. Burns BF, Colby TV, Dorfman RF. Langerhans' cell granulomatosis (histiocytosis X) associated with malignant lymphomas. Am J Surg Pathol 1983; 7:529–531.
122. Lombard CM, Medeiros J, Colby TV. Pulmonary histiocytosis X and carcinoma. Arch Pathol Lab Med 1987; 111:339–341.
123. Thomashefski JF, Khiyami A, Kleinerman J. Neoplasms associated with pulmonary eosinophilic granuloma. Arch Pathol Lab Med 1991; 115:499–506.
124. Sadoun D, Vaylet F, Valeyre D, Natali D, Georges R, Allard P, Battesti JP. Bronchogenic carcinoma in patients with pulmonary histiocytosis X. Chest 1992; 101: 1610–1613.
125. Egeler RM, Neglia JP, Pucetti DM, Brennan CA, Nesbit ME. Association of Langerhans cell histiocytosis with malignant neoplasms. Cancer 1993; 71:865–873.

AUTHOR INDEX

Italic numbers give the page on which the complete reference is listed.

C

D

E

F

G

H

I

J

K

L

M

P

S

T

U

V

W

X

Y

Z

SUBJECT INDEX

A

B

C

D

I

L

M

S